I0063373

Polymers from Renewable Resources

Polymers from Renewable Resources

Special Issue Editor
George Z. Papageorgiou

MDPI • Basel • Beijing • Wuhan • Barcelona • Belgrade

MDPI

Special Issue Editor
George Z. Papageorgiou
University of Ioannina
Greece

Editorial Office
MDPI
St. Alban-Anlage 66
4052 Basel, Switzerland

This is a reprint of articles from the Special Issue published online in the open access journal *Polymers* (ISSN 2073-4360) from 2017 to 2018 (available at: https://www.mdpi.com/journal/polymers/special_issues/polymers_from_renewable_resources)

For citation purposes, cite each article independently as indicated on the article page online and as indicated below:

LastName, A.A.; LastName, B.B.; LastName, C.C. Article Title. *Journal Name* **Year**, *Article Number*, Page Range.

ISBN 978-3-03897-451-2 (Pbk)
ISBN 978-3-03897-452-9 (PDF)

Cover image courtesy of George Papageorgiou.

© 2018 by the authors. Articles in this book are Open Access and distributed under the Creative Commons Attribution (CC BY) license, which allows users to download, copy and build upon published articles, as long as the author and publisher are properly credited, which ensures maximum dissemination and a wider impact of our publications.

The book as a whole is distributed by MDPI under the terms and conditions of the Creative Commons license CC BY-NC-ND.

Contents

About the Special Issue Editor

George Z. Papageorgiou received his Diploma in Chemical Engineering from Aristotle University of Thessaloniki (AUTh). He also holds an MSc (1998) and a PhD (2002) in Polymer Science (AUTh). His current research focuses on polymers from renewable resources and biodegradable polymers. Polymer matrix nanocomposites, pharmaceutical technology, and thermal analysis are also among his research interests. He is the author and co-author of over 120 journal papers and book chapters. A significant number of his publications deal with the synthesis, phase transitions, and crystallization of polyesters of 2,5-furandicarboxylic acid and other polymers from monomers derived from biomass.

polymers

MDPI

Editorial

Thinking Green: Sustainable Polymers from Renewable Resources

George Z. Papageorgiou

Department of Chemistry, University of Ioannina, P.O. Box 1186, 45110 Ioannina, Greece; gzpap@uoi.gr

Received: 9 August 2018; Accepted: 21 August 2018; Published: 27 August 2018

The use of polymeric materials from renewable resources has a long history, with naturally occurring polymers being among the first materials used by men. In the 19th century, natural materials, such as casein, natural rubber, and cellulose, were modified to obtain useful polymeric materials. Over the past few decades, the production and application of synthetic polymers have seen an almost exponential increase. However, concerns regarding depletion of fossil resources, disposal and related issues, as well as government policies, have led to a continuously growing interest in the development of sustainable, safe, and environmentally friendly plastics from renewable resources [1,2].

Sustainable polymers from renewable resources can be obtained through chemical modification of natural polymers, such as starch, cellulose, or chitin [3,4].

Biobased polymers can also be synthesized through a two-step process from biomass (lignin, cellulose, starch, plant oils) [5–7].

Traditional (drop-in) monomers such as ethylene, 1,2-ethanediol, terephthalic acid, or novel monomers like lactide, 2,5-furandicarboxylic acid, 1,4-cyclohexane dicarboxylic acid, furfuryl alcohol, or isosorbide can be obtained through chemical or biochemical conversion [8,9]. All the above can then be polymerized to produce biobased plastics. Therefore, biopolyethylene (bio-PE), bio-poly(ethylene terephthalate) (bio-PET), new polymers such as poly(lactic acid) and poly(ethylene 2,5-furandicarboxylate) (PEF), and even thermosetting polymers can be synthesized from renewable monomers and are expected to play a key role in biobased economy in the near future [10,11].

Finally, polymer synthesis can be achieved in plants through photosynthesis using carbon dioxide (CO_2) or by microorganisms, e.g., synthesis of poly(hydroxy-alkanoate)s [2]. CO_2 is also used to synthesize nonisocyanate polyurethanes [12,13].

In general, there are four main strategies to arrive at polymers with tailored properties: (a) selection of the most appropriate monomers for homopolymer production; (b) copolymerization; (c) blending; and (d) use of fillers, fibers, and additives to obtain composites. Apart from the commercialization of the abovementioned biobased and recyclable, but nondegradable, homopolymers (PEF, bio-PE, bio-PET), the use of biodegradable polyesters, such as poly(lactic acid) (PLA), poly(β-hydroxybutyrate) (PHB), poly(butylene adipate) (PBA), poly(butylene succinate) (PBS), or poly(ε-caprolactone) is expected to expand in near future due to their favoring life cycle [14,15].

Biocomposites can be prepared by exploiting the characteristics of the above polymers as well as those of lignocellulose fibers or other biobased fillers. Other biomass-derived materials can also be elaborated for the modification of renewable or fossil-based polymers, e.g., internal plasticizers of polyvinylchloride (PVC).

Uses of polymeric materials from biomass include packaging applications, antimicrobial films, fibers, foams, or coatings production as well as applications in medicine and pharmaceutics [16]. Drug delivery systems, such as drug-loaded micro- or nanoparticles, are a topic of particular interest, considering applications of biodegradable and biocompatible polymers from biomass.

Contents of This Issue

In this issue, the recent developments in biobased polymers toward general and engineering applications are reviewed [17]. The development of antimicrobial films using plant secondary metabolite-derived polymers is also discussed [18]. Trends in PLA and poly(hydroxy-alkanoate)s (PHA) nanocomposites are presented [19], and important features such as crystallization and stereocomplexation of PLA in multiblock copolymers, as well as biodegradation and mechanical properties of PLA in biocomposites, are studied [20–23].

Details of PEF synthesis applying solid state polymerization are given by Bikiaris and co-workers [24], and the results of structural investigation of the polymorphic forms of this most promising biobased polyester are reported by Maini et al. [25].

Studies on synthesis of poly(butylene 2,5-furandicarboxylate) (PBF)-related copolyesters containing isophthalate units via ring-opening polymerization and also synthesis and characterization of nanocomposites containing bacterial cellulose based on PBF and related copolymers with butylene diglycolate are included in this issue [26,27].

Sanchez-Lopez et al. have prepared renewable polyesters from isosorbide, 2,5-furandicarboxylic acid (FDCA) succinic acid, 1,3-propanediol, and 1,5-pentanediol for coating applications, while copolyesters based on cyclohexanedicarboxylic acid are synthesized and the gas barrier properties of them are evaluated [28,29].

Polymerization of furfuryl alcohol (FA), which is also a biobased monomer derived from lignocellulosic biomass, is also investigated [30].

Starch-based materials play their own role in biobased materials. Thermoplastic potato starch/halloysite nano-biocomposites are prepared and characterized in a paper in this issue [31].

Polyethylene biocomposites for 3D printing, as well as biocomposites of lignocellulosic biomass and recycled PET, are prepared and characterized [32,33]. Biodegradable poly(ε-caprolactone) blends with ionic liquid are studied in regard to their crystallization characteristics [34]. Furthermore, the use of a green binder based on enzymatically polymerized eucalypt kraft lignin for fiberboard applications is tested [35].

Foamed polymeric composite materials based on polyurethane or PHB with lignin or cellulose are prepared and studied. Lipase-catalyzed synthesis, properties, and application of biobased dimer acid cyclocarbonate with potential applications in nonisocyanate polyurethanes are also studied [36–39].

Referring to improvements in thermoplastics with internal or external plasticization using renewable materials, poly(vinyl chloride) is modified and plasticized by grafting cardanol groups. Polystyrene is also modified with eugenol for liquid crystal orientation [40,41].

Membranes made of porous regenerated cellulose—suitable bioadsorbents for wastewater treatment—are prepared in two modification stages involving oxidation on both sides and then functionalization with polyethylenimine [42].

Pyrolysis is a technique that can be applied to arrive at monomers and starting materials from renewable resources. In this context, Jiang et al. evaluate the effect of glycerol pretreatment on levoglucosan production by fast pyrolysis [43].

The applications of biopolymers in medicine and pharmaceutics are also included in the scope of this issue. The role of hyaluronic acid in promoting the osteogenesis of the human bone morphogenetic protein-2 in an absorbable collagen sponge is investigated by Huang et al. [44].

Moreover, some cases of applications of polymers from renewable resources in drug delivery systems are examined. An example of such a polymer with interest in drug delivery is chitosan. The potential for tailoring drug release rates by changes in the particle engineering of chitosan-based powders is examined, while thiolated chitosan masked microspheres with mesocellular silica foam are proposed for intranasal delivery of paliperidone [45]. Starch-chitosan polyplexes are also tested as carrier for anti-infectives and gene delivery by Yasar et al. [46,47].

Fucoidan is a polysaccharide composed of chemical units that can be specifically recognized by alveolar macrophages. Inhalable fucoidan microparticles combining two antitubercular drugs—isoniazid and rifabutin—are prepared and evaluated [48]. Furthermore, poly(lactic acid) and poly(lactic acid-co-glycolic acid) (PLGA) nanoparticles are extensively used in drug delivery. In this issue,

Polymers **2018**, *10*, 952

the dual drug delivery of sorafenib and doxorubicin from PLGA and poly(ethylene glycol)-poly(lactic acid-co-glycolic acid) (PEG-PLGA) polymeric nanoparticles is investigated by Babos et al. [49].

Finally, novel isocyanate-modified carrageenans are prepared and characterized as sorbent materials for preconcentration and removal of diclofenac (DCF) and carbamazepine (CBZ) in different aqueous matrices (surface waters and wastewaters) [50].

Conflicts of Interest: The authors declare no conflict of interest.

References

1. Schneiderman, D.K.; Hillmyer, M.A. 50th anniversary perspective: There is a great future in sustainable polymers. *Macromolecules* **2017**, *50*, 3733–3749. [CrossRef]
2. Zhu, Y.; Romain, C.; Williams, C.K. Sustainable polymers from renewable resources. *Nature* **2016**, *540*, 354–362. [CrossRef] [PubMed]
3. HStorz, H.; Vorlop, K.-D. Bio-based plastics: Status, challenges and trends. *Landbauforsch.-Ger.* **2013**, *63*, 321–332.
4. Upton, B.M.; Kasko, A.M. Strategies for the conversion of lignin to high-value polymeric materials: Review and perspective. *Chem. Rev.* **2015**, *116*, 2275–2306. [CrossRef] [PubMed]
5. Gandini, A.; Lacerda, T.M.; Carvalho, A.J.; Trovatti, E. Progress of polymers from renewable resources: Furans, vegetable oils, and polysaccharides. *Chem. Rev.* **2015**, *116*, 1637–1669. [CrossRef] [PubMed]
6. Isikgor, F.H.; Becer, C.R. Lignocellulosic biomass: A sustainable platform for the production of bio-based chemicals and polymers. *Polym. Chem.* **2015**, *6*, 4497–4559. [CrossRef]
7. Zhang, C.; Garrison, T.F.; Madbouly, S.A.; Kessler, M.R. Recent advances in vegetable oil-based polymers and their composites. *Prog. Polym. Sci.* **2017**, *71*, 91–143. [CrossRef]
8. Mülhaupt, R. Green polymer chemistry and bio-based plastics: Dreams and reality. *Macromol. Chem. Phys.* **2013**, *214*, 159–174. [CrossRef]
9. Hernández, N.; Williams, R.C.; Cochran, E.W. The battle for the "green" polymer. Different approaches for biopolymer synthesis: Bioadvantaged vs. bioreplacement. *Organ. Biomol. Chem.* **2014**, *12*, 2834–2849. [CrossRef] [PubMed]
10. Nguyen, H.T.H.; Qi, P.; Rostagno, M.; Feteha, A.; Miller, S.A. The quest for high glass transition temperature bioplastics. *J. Mater. Chem. A* **2018**, *6*, 9298–9331. [CrossRef]
11. Raquez, J.-M.; Deléglise, M.; Lacrampe, M.-F.; Krawczak, P. Thermosetting (bio) materials derived from renewable resources: A critical review. *Prog. Polym. Sci.* **2010**, *35*, 487–509. [CrossRef]
12. Darensbourg, D.J. Making plastics from carbon dioxide: Salen metal complexes as catalysts for the production of polycarbonates from epoxides and CO$_2$. *Chem. Rev.* **2007**, *107*, 2388–2410. [CrossRef] [PubMed]
13. Cokoja, M.; Bruckmeier, C.; Rieger, B.; Herrmann, W.A.; Kuehn, F.E. Transformation of carbon dioxide with homogeneous transition-metal catalysts: A molecular solution to a global challenge? *Angew. Chem. Int. Ed.* **2011**, *50*, 8510–8537. [CrossRef] [PubMed]
14. Miller, S.A. Sustainable Polymers: Opportunities for the Next Decade. *ACS Macro Lett.* **2013**, *2*, 550–554. [CrossRef]
15. Tschan, M.J.-L.; Brulé, E.; Haquette, P.; Thomas, C.M. Synthesis of biodegradable polymers from renewable resources. *Polym. Chem.* **2012**, *3*, 836–851. [CrossRef]
16. Vilela, C.; Sousa, A.F.; Fonseca, A.C.; Serra, A.C.; Coelho, J.F.; Freire, C.S.; Silvestre, A.J. The quest for sustainable polyesters–insights into the future. *Polym. Chem.* **2014**, *5*, 3119–3141. [CrossRef]
17. Nakajima, H.; Dijkstra, P.; Loos, K. The recent developments in biobased polymers toward general and engineering applications: Polymers that are upgraded from biodegradable polymers, analogous to petroleum-derived polymers, and newly developed. *Polymers* **2017**, *9*, 523. [CrossRef]
18. Al-Jumaili, A.; Kumar, A.; Bazaka, K.; Jacob, M.V. Plant Secondary Metabolite-Derived Polymers: A Potential Approach to Develop Antimicrobial Films. *Polymers* **2018**, *10*, 515. [CrossRef]
19. Sun, J.; Shen, J.; Chen, S.; Cooper, M.A.; Fu, H.; Wu, D.; Yang, Z. Nanofiller Reinforced Biodegradable PLA/PHA Composites: Current Status and Future Trends. *Polymers* **2018**, *10*, 505. [CrossRef]
20. D'Ambrosio, R.M.; Michell, R.M.; Mincheva, R.; Hernández, R.; Mijangos, C.; Dubois, P.; Müller, A.J. Crystallization and Stereocomplexation of PLA-mb-PBS Multi-Block Copolymers. *Polymers* **2017**, *10*, 8. [CrossRef]

21. Castro-Aguirre, E.; Auras, R.; Selke, S.; Rubino, M.; Marsh, T. Impact of nanoclays on the biodegradation of poly (lactic acid) nanocomposites. *Polymers* **2018**, *10*, 202. [CrossRef]

22. Aguilar, M.D.; Corominas, R.R.; Farrés, Q.T.; Orús, X.E.; Pujol, P.M.; González, J.A.M. Bleached Kraft Eucalyptus Fibers as Reinforcement of Poly (Lactic Acid) for the Development of High-Performance Biocomposites. *Polymers* **2018**, *10*, 699. [CrossRef]

23. Aranberri, I.; Montes, S.; Azcune, I.; Rekondo, A.; Grande, H.-J. Fully Biodegradable Biocomposites with High Chicken Feather Content. *Polymers* **2017**, *9*, 593. [CrossRef]

24. Kasmi, N.; Papageorgiou, G.Z.; Achilias, D.S.; Bikiaris, D.N. Solid-State Polymerization of Poly (Ethylene Furanoate) Biobased Polyester, II: An Efficient and Facile Method to Synthesize High Molecular Weight Polyester Appropriate for Food Packaging Applications. *Polymers* **2018**, *10*, 471. [CrossRef]

25. Maini, L.; Gigli, M.; Gazzano, M.; Lotti, N.; Bikiaris, D.N.; Papageorgiou, G.Z. Structural investigation of poly (ethylene furanoate) polymorphs. *Polymers* **2018**, *10*, 296. [CrossRef]

26. Morales-Huerta, J.C.; de Ilarduya, A.M.; Muñoz-Guerra, S. Partially Renewable Poly (butylene 2,5-furandicarboxylate-co-isophthalate) Copolyesters Obtained by ROP. *Polymers* **2018**, *10*, 483. [CrossRef]

27. Matos, M.; Sousa, A.F.; Silva, N.H.C.S.; Freire, C.S.R.; Andrade, M.; Mendes, A.; Silvestre, A.J.D. Furanoate-Based Nanocomposites: A Case Study Using Poly(Butylene 2,5-Furanoate) and Poly(Butylene 2,5-Furanoate)-co-(Butylene Diglycolate) and Bacterial Cellulose. *Polymers* **2018**, *10*, 810. [CrossRef]

28. Lomelí-Rodríguez, M.; Corpas-Martínez, J.R.; Willis, S.; Mulholland, R.; Lopez-Sanchez, J.A. Synthesis and Characterization of Renewable Polyester Coil Coatings from Biomass-Derived Isosorbide, FDCA, 1,5-Pentanediol, Succinic Acid, and 1,3-Propanediol. *Polymers* **2018**, *10*, 600. [CrossRef]

29. Siracusa, V.; Genovese, L.; Ingrao, C.; Munari, A.; Lotti, N. Barrier Properties of Poly (Propylene Cyclohexanedicarboxylate) Random Eco-Friendly Copolyesters. *Polymers* **2018**, *10*, 502. [CrossRef]

30. Falco, G.; Guigo, N.; Vincent, L.; Sbirrazzuoli, N. FA Polymerization Disruption by Protic Polar Solvents. *Polymers* **2018**, *10*, 529. [CrossRef]

31. Ren, J.; Dang, K.M.; Pollet, E.; Avérous, L. Preparation and Characterization of Thermoplastic Potato Starch/Halloysite Nano-Biocomposites: Effect of Plasticizer Nature and Nanoclay Content. *Polymers* **2018**, *10*, 808. [CrossRef]

32. Filgueira, D.; Holmen, S.; Melbø, J.K.; Moldes, D.; Echtermeyer, A.T.; Chinga-Carrasco, G. 3D Printable Filaments Made of Biobased Polyethylene Biocomposites. *Polymers* **2018**, *10*, 314. [CrossRef]

33. Santos, R.P.d.; Rossi, P.F.; Ramos, L.A.; Frollini, E. Renewable Resources and a Recycled Polymer as Raw Materials: Mats from Electrospinning of Lignocellulosic Biomass and PET Solutions. *Polymers* **2018**, *10*, 538. [CrossRef]

34. Yang, C.-T.; Lee, L.-T.; Wu, T.-Y. Isothermal and Nonisothermal Crystallization Kinetics of Poly (ε-caprolactone) Blended with a Novel Ionic Liquid, 1-Ethyl-3-propylimidazolium Bis (trifluoromethanesulfonyl) imide. *Polymers* **2018**, *10*, 543. [CrossRef]

35. Gouveia, S.; Otero, L.A.; Fernández-Costas, C.; Filgueira, D.; Sanromán, Á.; Moldes, D. Green Binder Based on Enzymatically Polymerized Eucalypt Kraft Lignin for Fiberboard Manufacturing: A Preliminary Study. *Polymers* **2018**, *10*, 642. [CrossRef]

36. Zhang, X.; Jeremic, D.; Kim, Y.; Street, J.; Shmulsky, R. Effects of Surface Functionalization of Lignin on Synthesis and Properties of Rigid Bio-Based Polyurethanes Foams. *Polymers* **2018**, *10*, 706. [CrossRef]

37. Leng, W.; Li, J.; Cai, Z. Synthesis and Characterization of Cellulose Nanofibril-Reinforced Polyurethane Foam. *Polymers* **2017**, *9*, 597. [CrossRef]

38. Ventura, H.; Sorrentino, L.; Laguna-Gutierrez, E.; Rodriguez-Perez, M.; Ardanuy, M. Gas Dissolution Foaming as a Novel Approach for the Production of Lightweight Biocomposites of PHB/Natural Fibre Fabrics. *Polymers* **2018**, *10*, 249. [CrossRef]

39. He, X.; Wu, G.; Xu, L.; Yan, J.; Yan, Y. Lipase-Catalyzed Synthesis, Properties Characterization, and Application of Bio-Based Dimer Acid Cyclocarbonate. *Polymers* **2018**, *10*, 262.

40. Jia, P.; Zhang, M.; Hu, L.; Wang, R.; Sun, C.; Zhou, Y. Cardanol Groups Grafted on Poly(vinyl chloride)—Synthesis, Performance and Plasticization Mechanism. *Polymers* **2017**, *9*, 621. [CrossRef]

41. Ju, C.; Kim, T.; Kang, H. Renewable, Eugenol—Modified Polystyrene Layer for Liquid Crystal Orientation. *Polymers* **2018**, *10*, 201. [CrossRef]

42. Wang, W.; Bai, Q.; Liang, T.; Bai, H.; Liu, X. Two-Sided Surface Oxidized Cellulose Membranes Modified with PEI: Preparation, Characterization and Application for Dyes Removal. *Polymers* **2017**, *9*, 455. [CrossRef]

43. Jiang, L.; Wu, N.; Zheng, A.; Wang, X.; Liu, M.; Zhao, Z.; He, F.; Li, H.; Feng, X. Effect of Glycerol Pretreatment on Levoglucosan Production from Corncobs by Fast Pyrolysis. *Polymers* **2017**, *9*, 599. [CrossRef]

44. Huang, H.; Feng, J.; Wismeijer, D.; Wu, G.; Hunziker, E. Hyaluronic Acid Promotes the Osteogenesis of BMP-2 in an Absorbable Collagen Sponge. *Polymers* **2017**, *9*, 339. [CrossRef]

45. Nanaki, S.; Tseklima, M.; Christodoulou, E.; Triantafyllidis, K.; Kostoglou, M.; Bikiaris, D. Thiolated Chitosan Masked Polymeric Microspheres with Incorporated Mesocellular Silica Foam (MCF) for Intranasal Delivery of Paliperidone. *Polymers* **2017**, *9*, 617. [CrossRef]

46. Yasar, H.; Ho, D.-K.; de Rossi, C.; Herrmann, J.; Gordon, S.; Loretz, B.; Lehr, C.-M. Starch-Chitosan Polyplexes: A Versatile Carrier System for Anti-Infectives and Gene Delivery. *Polymers* **2018**, *10*, 252. [CrossRef]

47. Do Nascimento, E.G.; de Caland, L.B.; de Medeiros, A.S.; Fernandes-Pedrosa, M.F.; Soares-Sobrinho, J.L.; Santos, K.S.D.; da Silva-Júnior, A.A. Tailoring Drug Release Properties by Gradual Changes in the Particle Engineering of Polysaccharide Chitosan Based Powders. *Polymers* **2017**, *9*, 253. [CrossRef]

48. Cunha, L.; Rodrigues, S.; da Costa, A.R.; Faleiro, M.; Buttini, F.; Grenha, A. Inhalable Fucoidan Microparticles Combining Two Antitubercular Drugs with Potential Application in Pulmonary Tuberculosis Therapy. *Polymers* **2018**, *10*, 636. [CrossRef]

49. Babos, G.; Biró, E.; Meiczinger, M.; Feczkó, T. Dual drug delivery of sorafenib and doxorubicin from PLGA and PEG-PLGA polymeric nanoparticles. *Polymers* **2018**, *10*, 895. [CrossRef]

50. Papageorgiou, M.; Nanaki, S.; Kyzas, G.; Koulouktsi, C.; Bikiaris, D.; Lambropoulou, D. Novel Isocyanate-Modified Carrageenan Polymer Materials: Preparation, Characterization and Application Adsorbent Materials of Pharmaceuticals. *Polymers* **2017**, *9*, 595. [CrossRef]

© 2018 by the author. Licensee MDPI, Basel, Switzerland. This article is an open access article distributed under the terms and conditions of the Creative Commons Attribution (CC BY) license (http://creativecommons.org/licenses/by/4.0/).

polymers

MDPI

Review

The Recent Developments in Biobased Polymers toward General and Engineering Applications: Polymers that Are Upgraded from Biodegradable Polymers, Analogous to Petroleum-Derived Polymers, and Newly Developed

Hajime Nakajima, Peter Dijkstra and Katja Loos *

Macromolecular Chemistry and New Polymeric Materials, Zernike Institute for Advanced Materials, University of Groningen, Nijenborgh 4, 9747 AG Groningen, The Netherlands; hnkajima@gmail.com (H.N.); peter.dijkstra@rug.nl (P.D.)
* Correspondence: k.u.loos@rug.nl; Tel.: +31-50-363-6867

Received: 31 August 2017; Accepted: 18 September 2017; Published: 18 October 2017

Abstract: The main motivation for development of biobased polymers was their biodegradability, which is becoming important due to strong public concern about waste. Reflecting recent changes in the polymer industry, the sustainability of biobased polymers allows them to be used for general and engineering applications. This expansion is driven by the remarkable progress in the processes for refining biomass feedstocks to produce biobased building blocks that allow biobased polymers to have more versatile and adaptable polymer chemical structures and to achieve target properties and functionalities. In this review, biobased polymers are categorized as those that are: (1) upgrades from biodegradable polylactides (PLA), polyhydroxyalkanoates (PHAs), and others; (2) analogous to petroleum-derived polymers such as bio-poly(ethylene terephthalate) (bio-PET); and (3) new biobased polymers such as poly(ethylene 2,5-furandicarboxylate) (PEF). The recent developments and progresses concerning biobased polymers are described, and important technical aspects of those polymers are introduced. Additionally, the recent scientific achievements regarding high-spec engineering-grade biobased polymers are presented.

Keywords: biobased polymers; biodegradable polymers; polylactides (PLA); poly(hydroxy alkanoates) (PHAs); bio-poly(ethylene terephthalate) (bio-PET); poly(ethylene 2,5-furandicarboxylate) (PEF); biobased polyamides; succinate polymers; polyterpenes; modified lactide

---

1. Introduction

In the mid-20th century, the polymer industry completely relied on petroleum-derived chemistry, refinery, and engineering processes. The negative impacts of these processes on the environment was scientifically discussed in this period, but the processes were not changed in industrial settings until their negative effects reached a critical level around the 1980s. At this point, biodegradable polymers such as polylactides (PLA), poly(hydroxy alkanoates) (PHAs) succinate derived polymers, and others began to develop, and practical biodegradable polymers were commercialized and launched, solving many waste problems in the agricultural, marine fishery, and construction industries, among others [1,2]. The development of biodegradable polymers is recognized as one of the most successful innovations in the polymer industry to address environmental issues.

Since the late 1990s, the polymer industry has faced two serious problems: global warming and depletion of fossil resources. One solution in combating these problems is to use sustainable resources instead of fossil-based resources. Biomass feedstocks are a promising resource because of their sustainability. Although biomass is the oldest energy source, having been used for direct combustion since the Stone Age, it is still uncommon to utilize biomass as chemical building blocks and fuel during refinery processes [3]. The development of refinery processes has been dramatically accelerated due to improvements in the combinations of chemical and biological pathways for production of, for example, bio-ethanol, bio-diesel, and bio-olefins [4–6]. Biomass feedstock can be converted into raw materials for polymer production, and the resulting polymers are called "biobased polymers" [7–10]. As the term "biobased polymers" is still relatively new in polymer science and industry, it is sometimes confused with other terms such as biopolymers, biodegradable polymers, and bioabsorbable polymers. More specifically a biopolymer is classified as a natural polymer formed by plants, microorganisms, and animals. Naturally derived biomass polymers are termed "1st class biobased polymers" for and bio-engineered polymers (vide infra) as "2nd class biobased polymers". Biopolymers show biodegradability, but this class of polymers does not include artificially synthesized biodegradable polymers. Biodegradable polymers include both naturally derived ones and artificially synthesized ones. They are sometimes defined as biocompostable polymers, especially in waste, agricultural, fishery and construction industries. The term biodegradable polymer is also used for medical, pharmaceutical, and bioengineering applications. Biodegradable polymers consisting of naturally derived building blocks are also called bioabsorbable polymers, when they are specifically applied for medical, pharmaceutical, or other bioengineering applications.

The importance of biobased polymers is well known, and much research and development activities concerns the use of biobased polymers in science, engineering, and industry. Generally, biobased polymers are classified into three classes:

- 1st class; naturally derived biomass polymers: direct use of biomass as polymeric material including chemically modified ones such as cellulose, cellulose acetate, starches, chitin, modified starch, etc.;
- 2nd class; bio-engineered polymers: bio-synthesized by using microorganisms and plants such as poly(hydroxy alkanoates (PHAs), poly(glutamic acid), etc.;
- 3rd class; synthetic polymers such as polylactide (PLA), poly(butylene succinate) (PBS), bio-polyolefins, bio-poly(ethylene terephtalic acid) (bio-PET) [8,9].

Usually, 1st class is directly used without any purification and 2nd class polymers are directly produced from naturally derived polymers without any breakdown, and they play an important role in situations that require biodegradability. Direct usage of 1st and 2nd class polymers allows for more efficient production, which can produce desired functionalities and physical properties, but chemical structure designs have limited flexibility. Monomers used in 3rd class polymers are produced from naturally derived molecules or by the breakdown of naturally derived macromolecules through the combination of chemical and biochemical processes. As breakdown processes allow monomers to have versatile chemical structures, polymers comprised of these monomers also have extremely versatile chemical structures. It is practically possible to introduce monomers in 3rd class polymers into the existing production system of petroleum-derived polymers. For the above reasons, the 3rd class of biobased polymers is the most promising. Some of these 3rd class polymers such as bio-polyolefines and bio-PET are not supposed to enter natural biological cycles after use. Thus, the contribution for reducing environmental impact from these polymer classes is mainly derived from reducing the carbon footprint. In Table 1, the chronological development and categorization of biobased polymers that are based on application fields are displayed and compared with those of petroleum-derived polymers. From 1970 to 1990, PLA (low L-content) and poly(hydroxy alkanoates) (PHAs) are the most important and representative development of biobased polymers [11]. During that period, scientists developed a fundamental understanding of biobased polymers for future applications. Since the 1990s, biobased polymers have gradually shifted from biodegradable applications to general and engineering applications. High L-content PLLA, high molecular weight PHAs, and stereocomplexed-PLA (sc-PLA) (low T_m grade) are important examples of this development. The deliverables of these were effectively applied to industrialize general applications of PLLA, PHAs, and succinate polymers. In this period, high-performance new-generation PLAs, such as sc-PLA (high T_m grade) [12] and stereoblock-PLA (sb-PLA) [13], were proactively created. After the successful upgrading of these biodegradable polymers, more promising building blocks have been identified to create more attractive chemical structures for biobased polymers [6], which are known as the US Department of Energy's (DOE's) 12 top biobased molecules [14]. Around 2010, engineering-grade biobased polymers that were analogous to petroleum-derived polymers such as poly(ethylene terephthalate) (PET) and polyamides began to be applied to industry. Completely new biobased polymers with the potential for super-engineering applications also started to appear around 2010. It is expected these will be applied to new applications and conventional petroleum-derived polymers will rarely be used in the distant future.

Table 1. Development of biobased polymers and comparison with petroleum-derived polymers.

	Petroleum-derived polymers	Biobased polymers	
	Industry	Industrial approach	Scientific approach
Super-engineering applications	since 1960 PEEK, PSU, PES, PPS, PEI, PAI, LCP	not yet	since 2010 bio-LCP, bio-PEEK (new generation)
Engineering/semi-engineering applications	since 1950 Polyamide, POM, PC, PPO, PET, PTT, PBT, ultra-high MW PE, HIPS	since 2010 bio-PET, bio-PTT, bio-PBT, bio-polyamide (analogous to petroleum-derived ones)	since 2000 polyterpenes, PEF, bio-polyamide, sc-PLA (high T_m), sb-PLA (high T_m) (new generation)
General applications	since 1930 PE, PP, PS, PMMA, PVC, ABS	since 2000 PLLA (high-L content) reinforced PHAs, PHAs blends, succinate polymers, bio-PE/PP	since 1990 sc-PLA (low T_m), PHAs (super high MW), succinate polymers (upgrading from biodegradable polymers)
Biodegradable/biocompatible applications	since 1970 PCL, PEG	since 1990 PLLA (low-L content) PBS, PHAs, PGA, polysaccharides	since 1970 PLA, PHAs, succinate polymers

Recent economic studies have revealed that biobased polymers can create new business opportunities and stable growth in new, plastic markets [15,16]. Actual growth is influenced by the current events and issues concerning economics, politics, and international affairs, but stable growth of the biobased polymer industry was observed in all proposed scenarios [15]. The strong social interest in a sustainable society is still the most important factor in the development of these polymers, but recent improvements in the quality and functionality of biobased polymers have led the growth of these plastic markets. There are several successful examples of industrialization of these polymers, including pilot-scale production of polylactide (PLA) at NatureWorks and Corbion/Total; poly(trimethylene terephthalate) (PTT) at DuPont; poly(isosorbide carbonate) at Mitsubishi Chemicals; biobased polyamides at Arkema, Toray, BASF, DSM, and others; and poly(ethylene 2,5-furandicarboxylate) (PEF) at Synvina.

Biobased polymers are being applied to general and engineering situations. For example, because of improvements in the physical durability and processability of PLA, it has been used in the packaging industry [17–19]. In addition, due to the superior gas barrier properties of PEF, it is being used for bottles, films, and other packaging materials in the food and beverage industry [20,21]. Further, biobased PTT is analogous to petroleum-derived PTT, and its biobased, sustainable nature and intrinsic flexible chain properties allow comfortable stretching and shape recovery properties are attractive promising [22]. However, the stability of the production and processability of these biobased polymers can still be improved. The current general approach to improving processability is physical modification and optimization of polymer processing, including optimization of processing parameters, extruder screw design, selection of appropriate additives, and post-orientation for strain-induced crystallization. These developments in processing conditions have made biobased polymers analogous to certain petroleum-derived polymers. In addition to physical modification and optimization, the importance of chemical modification and optimization has been emphasized as they allow for further improvement and new functionalities of biobased polymers. This review introduces recent important developments in chemical modifications of biobased polymers and development of new biobased building blocks for new generation biobased polymers.

2. Biobased Polymers: Upgraded from Biodegradable-Grade Polymers

PLA, PHAs, and succinate polymers are the most common biobased polymers since they have been successfully applied in the biodegradable plastic industry. The biodegradability of these polymers has been utilized to solve environmental issues, such as waste and public pollution. Due to changes in the social requirements for biodegradable polymers, it is necessary to improve the performance of biodegradable polymers so they can be used for general and engineering applications. The recent examples of development and applications of PLA, PHAs, and succinate polymers are described. Their fundamental properties and chemistries are also introduced.

2.1. Polylactide (PLA)

2.1.1. High L-Content PLA (PLLA)

PLA is generally prepared via ring-opening polymerization (ROP) of lactide, which is a cyclic dimer from lactic acid. Direct polycondensation from lactic acid is also performed, but ROP is the standard process in most industries. PLA has a chiral active chain structure, and controlling it allows one to determine the physical properties of PLA. The relationship between the physical properties and L-unit content of PLA has been comprehensively studied [23]. Regarding the effectiveness of biological production, L-lactic acid has superior productivity compared to D-lactic acid. Therefore, poly(L-lactide) (PLLA) is more commonly commercialized. The parameters listed in Table 2 are related to crystallinity. The table shows a clear trend in which the physical properties of PLA are improved by increasing the purity of the L-unit content. The growth rate of spherulite and increase in L-unit content are almost proportional; when the L-unit content is increased 1.0%, the growth rate of spherulite is increased about 2.0 times. Other parameters concerning the crystallization properties of PLA show a similar trend; crystallinity depends on crystallization conditions, but in this report, it is described that crystallinity increases 1.3 times when L-unit content is increased 1.0%. The common crystal structure of highly pure homo-chiral PLA is pseudo-orthorhombic and consists of left-handed 10_3-helical chains, which are generally called α-forms [24–26]. A slightly disordered pseudo-orthorhombic PLA is called an α'-form [27,28]. Because of the slightly disordered structure of the α'-form, an α'-form based PLA crystal has lower thermal and physical properties than those of α-form. Table 3 summarizes the infrared spectroscopy (IR) frequencies of α-forms and α'-forms. Although both α-forms and α'-forms have the same helical conformation, IR analysis of these forms reveals different results, which can be utilized for detection of crystallization form of PLA [29]. The chemical structure and conformation of homo-chiral PLLA are shown in Figure 1.

Table 2. Poly(L-lactide) (PLLA) crystallization parameters [23].

L-Purity (%)	M_w ($\times 10^5$)	Approximate value of growth rate of spherulite (μm/min) [1]	t_s (min) [2]	$t_{1/2}$ (min) [3]	t_e (min) [4]	Crystallinity (%) [5]
99.75	1.39	5.2	0.97	3.02	8.12	37.8
98.82	1.55	4.2	2.47	8.04	16.48	31.9
97.79	1.42	2.4	5.19	14.2	28.69	23.7

[1] From analysis performed using polarized optical microscopy for isothermal crystallization at 130 °C (Approximate values from plot of Figure 2 in Reference [23]); [2] starting time of crystallization at 110 °C; [3] half-crystallization time at 110 °C; [4] ending time of crystallization at 110 °C; [5] Crystallinity after completion of crystallization at 110 °C.

Table 3. IR frequencies of amorphous, α'-form, and α-form PLLA [29].

	Amorphous (cm^{-1})	α'-Form (cm^{-1})	α-Form (cm^{-1})
ν_{as} (CH$_3$)	2995	2997	2997 3006
ν_s (CH$_3$)	2945	2946	2946 2964
ν (C=O)	1757	1761	1759 1749
δ_{as} (CH$_3$)	1454	1457	1457 1444
δ_s (CH$_3$)	1387	1386	1386 1382

2.1.2. Stereocomplexed PLA

Sc-PLA is a complex form of PLLA and poly(D-lactide) (PDLA) that was initially reported as an insoluble precipitant for solutions [30]. As the chemical properties of PLA change during the formation of sc-PLA, the original solubilities of homo-chiral PLAs are lost. As a result, sc-PLA is selectively precipitated as granules made from sc-PLA crystallites. A sc-PLA film is created from the Langmuir–Blodgett membrane when PLLA and PDLA are combined [31]. In addition, PLLA and PDLA with molecular weights as high as 1000 kDa have preferable stereocomplexation on the water surface. Further, sc-PLA is assembled on a quartz crystal microbalance (QCM) substrate by stepwise immersion of the QCM in acetonitrile solutions of PLLA and PDLA [32]. The Langmuir–Blodgett membrane and assembled methods are interesting new approaches to achieve nano-ordered structural control of sc-PLA layers.

A striking property of sc-PLA is its high T_m (around 230 °C). This is 50 °C higher than the conventional T_m of homo-chiral high L-content PLLA. In contrast to the stereocomplexation of high molecular weight PLA in a solution state, a simple polymer melt-blend of PLLA and PDLA is usually accompanied by homo-chiral crystallization of PLLA and PDLA, particularly when their molecular weight is sufficient for general industrial applications.

The homo-chiral crystals deteriorate the intrinsic properties of sc-PLA, but this drawback can be overcome using sb-PLA. A sb-PLA with an equimolar or moderate non-equimolar PLLA to PDLA ratio features 100% selective stereocomplexation [13]. Therefore, formation of homo-chiral PLA-derived crystallization, which is known to cause poor physical performance of sc-PLA produced from direct combination of PLLA and PDLA, is prevented. An important issue with sb-PLA is that its T_g is identical to that of homo-chiral PLA, and thus the final thermal durability of sb-PLA is controlled by T_g due to its relatively low crystallinity. The chemical structure of sb-PLA and conformation of sc-PLA from a combination of PLLA/PDLA are shown in Figure 1.

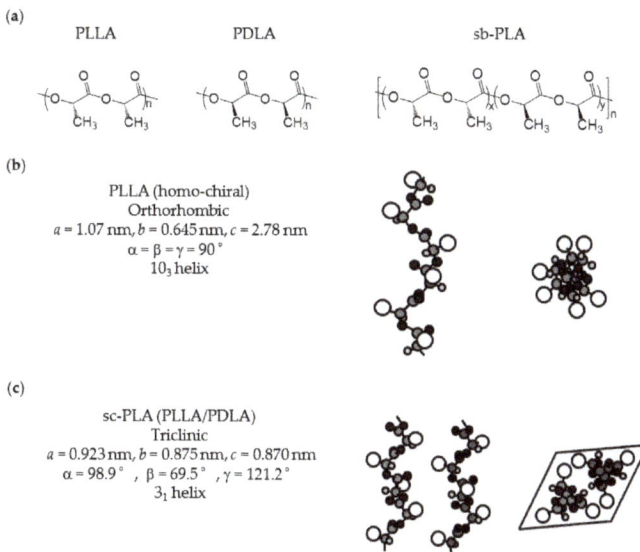

Figure 1. Chemical structures and conformation of PLA: (**a**) chemical structures and chirality; (**b**) conformation of PLLA (homo-chiral) [33]; and (**c**) conformation of sc-PLA from a combination of PLLA and PDLA [34].

2.1.3. Examples of PLA Applications

Although the application of PLA was limited to biodegradable plastics in the early stage of its development, it has been successfully applied to general and semi-engineering situations and achieved successful commercialization. The most common commercialized PLA in the world is made by NatureWorks, which trademarked their PLA "Ingeo" [35]. Currently, there are more than 20 commercialized types of Ingeo with both amorphous and semi-crystalline structures, allowing customers to choose the PLA that is appropriate for their specific situations (Table 4). Another important player affecting the industrialization of PLA is Corbion/Total. Now, there are many commercial-grade PLAs on the market, such as Biofoam, made by Synbra; Revode, made by Zhejiang Hisun Biomaterials Biological Engineering; Futerro, made by Futerro; Lacea, made by Mitsu Chemicals; and Terramac, made by Unitika. sc-PLA will play a key role in future engineering applications of PLA. Biofront, made by Teijin, is a good example of the industrial development of sc-PLA [36]. This product features high physical properties, including a melting point of 215 °C, HDT of 130 °C to 0.45 MPa, and a modulus of 115 MPa at 23 °C. These properties are considered suitable enough for sc-PLA to replace petroleum-derived engineering plastics.

Table 4. Properties of commercial-grade Ingeo PLA [9].

Ingeo type	Application	MFR (g/10 min, 210 °C/2.16 kg)	T_m (°C)	T_g (°C)
2003D		6	145–160	55–60
3001D	extrusion, injection	22	155–170	55–60
3251D		80	155–170	55–60
3801X			155–170	45
4032D	film, sheet	7	155–170	55–60
4060D		10	-	55–60
6060D		8	122–135	55–60
6252D	fiber, non-woven	80	155–170	55–60
6752D		14	145–160	55–60

2.2. Poly(hydroxyalkanoates) (PHAs)

PHAs are members of a family of polyesters that consist of hydroxyalkanoate monomers. In nature, they exist as homopolymers such as poly(3-hydroxybutyrate) (P3HB) or copolymer poly(3-hydroxybutyrate-*co*-3-hydroxyvalerate) (P(3HB-*co*-3HV)) [37]. PHAs exist as granules of pure polymer in bacteria, which are used as an energy storage medium (akin to fat for animals and starch for plants). PHAs are commercially produced using energy-rich feedstock, which is transformed into fatty acids on which the bacteria feed. During industrial production of PHAs, after a few "feast–famine" cycles, cells are isolated and lysed. The polymer is extracted from the remains of the cells, purified, and processed into pellets or powder [37]. In addition to using pure feedstock as a source of energy for PHAs production, there are on-going efforts to use energy-rich waste water as feedstock and thus as PHAs [38]. Production of PHAs can be improved using genetic modification, either by increasing the amount of PHAs-producing bacteria or by modifying plants to start making PHAs [39,40]. As chemical synthesis of PHAs via the ROP of a corresponding lactone is feasible, ROP of lactones for PHAs can be done via metal-based or enzymatic catalysts [41]. However, the chain of chemically synthesized PHAs is shorter in length than that of biologically synthesized PHAs. The latter also ensures great stereo control and enantiomeric pure (*R*) configuration in almost all PHAs. Through depolymerization, enantiomeric purity allows for the creation of an enantiomeric monomer that can be used as a building block [42]. On the other hand, when pure (*S*)-methyl 3-hydroxybutyrate is used as feedstock for the production of PHAs, the corresponding (*S*)-configuration polymer is produced [43].

The biological synthesis of P3HB is displayed in Figure 2. Sugars in the feedstock are converted to acetates, which are complexed to coenzyme A and form acetyl CoA. This product is dimerized to acetoacetyl A. Additionally, through reduction, hydroxy butyryl CoA is polymerized.

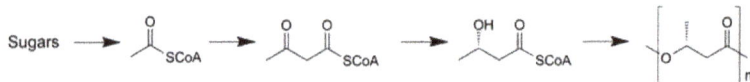

Figure 2. Biological synthesis scheme of P3HB.

PHAs consisting of 4–14 carbon atoms in the repeating unit are called "short chain length PHAs" (sCL-PHAs) or "medium chain length PHAs" (mCL-PHAs) [44]. Some of these PHAs are commercialized. The average molecular weight (M_w) of PHAs corresponds to their chain length. Typically, the M_w of sCL-PHAs is around 500,000, while that of mCL-PHAs is lower than 100,000. In large part, the chain length of PHAs determines the flexibility of the polymer, with short chain butyrate providing the most rigidity and longer side chains disturbing crystal packing, resulting in more flexibility. Long chain length PHAs, which consist of repeating units of more than 14 carbon atoms, and PHAs that consist of either aromatic or unsaturated side-chains are rarely commercialized. The most commonly commercialized PHAs are P3HB, P(3HB-*co*-3HV) and P(3HB-*co*-3-hydroxyhexanoate) (P(3HB-*co*-3HH)), the thermal and physical properties of which are displayed in Table 5. P3HB has a T_g of 4 °C, which becomes lower when the PHAs has a longer chain length. The T_m of PHAs decreases with increasing chain length; P3HB has a melt temperature of 160 °C, while the melt temperature of P3HB-*co*-3HV is only 145 °C. Both T_g and T_m can be altered by changing the ratio of repeating units. The chemical structures of PHAs are shown in Figure 3.

Table 5. Thermal and mechanical properties of representative PHAs [45].

	P3HB	P(3HB-*co*-20% 3HV)	P(3HB-*co*-12% 3HH)	Poly(4-hydroxybutyrate) (P4HB)	P(3HB-*co*-16% 4HB)
T_m (°C)	177	145	61	60	152
T_g (°C)	4	−1	−35	−50	−8
Tensile (MPa)	40	32	9	104	26
Elongation at break (%)	6	50	380	1000	444

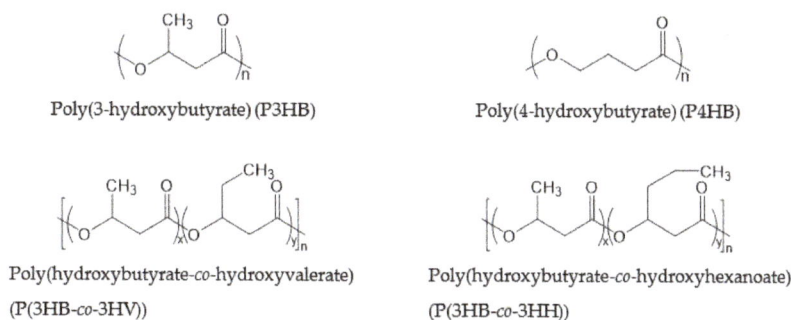

Poly(3-hydroxybutyrate) (P3HB)

Poly(4-hydroxybutyrate) (P4HB)

Poly(hydroxybutyrate-*co*-hydroxyvalerate) (P(3HB-*co*-3HV))

Poly(hydroxybutyrate-*co*-hydroxyhexanoate) (P(3HB-*co*-3HH))

Figure 3. Chemical structures of PHAs.

P3HB crystalizes in an orthorhombic structure (P3HB: a = 5.76 Å, b = 13.20 Å, and c = 5.96 Å), and its crystallinity can reach 80% [46]. Pure P3HB has poor nucleation density, leading to slow crystallization, due to the formation of large crystallites induced by poorly dispersed nucleation points. A promising way to improve crystallization speed is quiescent crystallization in isothermal conditions, which are 10–20 °C lower in temperature than crystallization conditions. This allows crystallization with the most possibility for arranging chains. It should be applied with appropriate nucleation agents for optimum processing in industry. Processing PHAs is challenging compared to conventional petroleum-derived polymers because of their sensitivity to thermal degradation and slow solidification due to slow crystallization. The degradation temperature of PHAs is around 180 °C, which is near the optimum processing temperature for polyester. A rapid increase of shear-induced internal heat can cause severe degradation, leading to a drop in molecular weight and discoloration. For these reasons, it is important to precisely monitor and control the practical temperature in extruders during processing of PHAs. Processing PHAs is challenging also due to their low durability and tackiness in the final product due to insufficient crystallinity. Cooling below the T_g can easily decrease tackiness, but the T_g of PHAs is 0 °C or lower, which is not an easily controllable temperature for the conventional extruders and molders used in the plastic industry.

2.3. Polysaccharides

Carbohydrates are probably the most prevalent group of organic chemicals on earth. Encompassing monosaccharides, disaccharides (commonly known as sugars), oligosaccharides, and polysaccharides, they are present in all lifeforms. Polysaccharides include well known polymers, such as cellulose and starch and their derivatives, as well as more exotic polymers, such as chitosan and pectin. In this review, we will briefly focus on cellulose and starch. The chemical structures of cellulose and starch are shown in Figure 4.

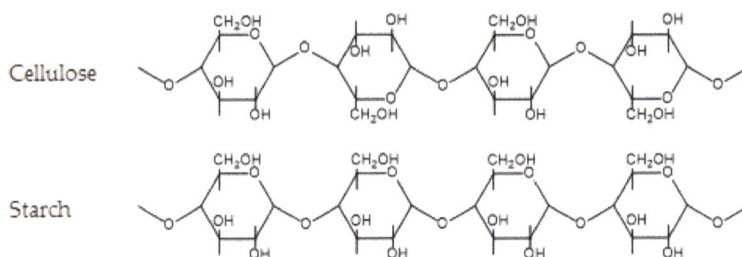

Figure 4. Chemical structure of cellulose and starch.

Cellulose, or more specifically, cellulose nitrate, has a special place in the history of polymer chemistry: it is the first polymer to be deliberately synthesized by human beings during the quest for synthetic ivory. Cellulose nitrate resulted in further derivatives of cellulose, such as cellulose acetate because of the safety aspects in handling and processing. Cellulose derivatives are still used on a wide scale in film, cigarette filters, and biomedical applications [47]. Other cellulose products, such as paper and cotton clothes, can be viewed as polymeric products. Cellulose is important to the polymer industry due to its abundance in plant fibers. It is not used as a polymer matrix but as an additive; the incorporation of natural fibers (e.g., wood, hemp, and flax) into a polymer compound improves the mechanical properties of the final product. The current focus of cellulose research is nano-cellulose, including cellulose nano-fibers and nano-crystalline cellulose [48–50]. Cellulose nanofibers are delaminated fibrils with a small diameter (5–25 nm) and long length (micrometer scale). Cellulose nano-fibers have interesting properties, such as high tensile strength and absorbance ratio. Nano-crystalline cellulose—tiny crystals of cellulose—is of interest due to its high mechanical load and shear thinning properties. Both materials are produced from wood fibers after intensive physical, chemical, and separation procedures.

As previously mentioned, starch is a means for obtaining and storing energy in plants. Starch-rich plants have been used for ages as sources of food, and starch is very commonly extracted for use in industry. Starch is stored in granules containing linear amylose and branched amylopectin. Both feature repeating units of D-glucose linked in α 1,4 fashion, with amylopectin containing about 6% of 1,6 linkages. A natural starch is not directly applicable for a processing, rather starch and water are passed through an extruder which produces thermoplastic starch (TPS) [51,52]. TPS is however not stable and retrodegradation is an issue; i.e., TPS tries to revert to its natural starch form. The main process hereby is the gelatinization of the starch granules which causes swelling of the amorphous parts of the granules. To stabilize the thermoplastic starch plasticizers (e.g., glycol and sugars) are added. An interesting approach to improve starch processability is a formation of amylose–lysophosphatidylcholine complexation to control rheological behaviors [53]. The induced lower modulus proved the formation of particle gel, resulting in less retrogradation. The complexation is also able to decrease the susceptibility of starch granules against amylase digestion [54,55].

2.4. Succinate Polymers

As biobased succinic acid (SA) becomes more commercially available, more biobased succinate polymers are being developed [56,57]. Poly(butylene succinate) (PBS), which is produced by direct polycondensation of SA and butanediol (BD), is one of the most well-known succinate polymers [58]. Both SA and BD for commercialized PBS were only produced from fossil fuel resources until recently, but the high interest in green sources led to the discovery that the two monomers can be obtained from refined biomass feedstock.

The widely commercialized PBS named "Bionolle" was launched by Showa Denko in 1993, and 3000–10,000 tons are produced per year. Since 2013, Succinity, a joint venture of BASF and Corbion, has been able to produce 10,000 tons of 100% biobased PBS. Poly(ethylene succinate) (PES) produced via polymerization of succinic acid and ethylene glycol is biodegradable and could also be sourced from biobased building blocks [59]. PES was commercialized by Nippon Shokubai from fossil resources. It has been claimed that PES is suited for film applications due to its good oxygen barrier properties and elongation. Copolymers of succinic acid and other dicarboxylic acids, such as adipic acid for poly(butylene succinate-*co*-butylene adipate) [60], poly(butylene succinate-*co*-butylene terephthalic acid) [61], and poly(butylene succinate-*co*-butylene furandicarboxylate) [62], have been reported. Figure 5 shows the chemical structures of the presented succinate polymers. Because of these polymers' relatively long alkyl chains, they usually have soft properties; for instance, PBS has a T_m of 115 °C and tensile strength of 30–35 MPa. Thus, succinate polymers are usually considered an alternative to polyolefins in the packaging industry.

Poly(butylene succinate) (PBS)

Poly(ethylene succinate) (PES)

Poly(butylene succinate-*co*-butylene adipate) (PBSA)

Poly(butylene succinate-*co*-butylene terephthalic acid) (PBST)

Poly(butylene succinate-*co*-butylene furandicarboxylate) (PBSF)

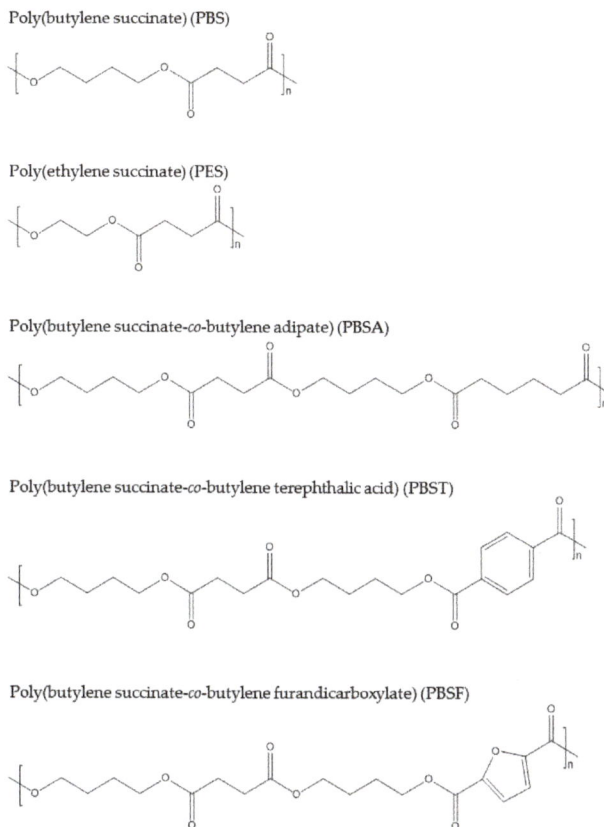

Figure 5. Chemical structures of succinate polymers.

3. Biobased Polymers Analogous to Conventional Petroleum-Derived Polymers

3.1. Biobased Polyethylene (Bio-PE)

Due to soaring oil prices, bio-ethanol produced by fermentation of sugar streams attracted the fuel industry in the 1970s. Bio-ethanol could also be chemically converted to bio-ethylene for production of biobased polyethylene (bio-PE) [63]. A drop in the price of oil diminished the bio-PE market, but the polymer continues to be exploited by important players such as Braskem due to increasing oil prices and environmental awareness [64]. The big advantage of bio-PE is the fact that its properties are identical to fossil-based PE, which has a complete infrastructure for processing and recycling. However, it faces direct competition with fossil-based feedstock, the price of which heavily fluctuates (e.g., shale gas is cheap) [65]. The downside of biobased PE is that it is not biodegradable. However, as will be shown next, some plastics produced from fossil fuel feedstock are biodegradable.

3.2. Biobased Poly(Ethylene Terephthalate) (PET) and Poly(Trimethylene Terephthalate) (PTT)

PET is a high-performance engineering plastic with physical properties that are suitable enough to be applied to bottles, fibers, films, and engineering applications. While PET plays an important role in the plastic market, huge consumption of this polymer results in serious environmental issues, especially regarding waste, because of its poor sustainability and degradability. Polymer-to-polymer material recycling of PET has been launched in some fields, but it is always accompanied by non-negligible deterioration of the polymer's physical properties in the final recycled products due to side-reactions and thermal degradation, hydrolysis, and thermo-oxidative degradation during recycling. Ways to chemically recycle PET are under development, but many technical difficulties, such as the high stability of PET under normal hydrolysis, alcoholysis, or breakdown processes, must be overcome. To realize a truly sustainable PET industry, it is important to establish sustainable production of monomers for biobased PET (bio-PET) from sustainable resources, such as biomass. First, ethylene glycol (EG) from petroleum-derived sources must be replaced by EG from biobased sources. The Coca-Cola Company (TCCC), a beverage giant, has accelerated the production of bio-PET known as "PlantBottle" [66]. PlantBottle, which was launched in 2009, consists of 30% biobased materials, 100% biobased EG (bio-EG) and petroleum-derived terephthalic acid (TPA).

Following this, biobased terephthalic acid (bio-TPA) is being developed to further improve the sustainability of PET, as bio-TPA is produced from naturally derived sustainable biomass feedstock. Theoretically, combining bio-EG and bio-TPA could achieve 100% natural biomass feedstock derived bio-PET. Figure 6 shows the proposed development of bio-TPA from biomass feedstock. One of the most important players in the development of bio-TPA is Gevo [67]. Based on technology announced by Gevo, biobased isobutylene obtained from *iso*-butanol, which is produced by dehydration of sugar, is a key building block in various chemicals, such as ethyl *tert*-butyl ether, methyl methacrylate, isooctane, and other alkanes. For bio-TPA production, *p*-xylene is first produced by cyclization of two isooctane molecules via dehydrogenation. Second, the *p*-xylene is converted to bio-TPA via oxidation. However, this is not the only way to obtain bio-TPA; it has been proposed that bio-TPA could be obtained from other biomass-derived products. Muconic acid, which is produced from sugar through a combination of chemical processes and biorefinery, is one interesting building block for bio-TPA [68]. After a series of stepwise *cis–cis* to *trans–trans* transitions of muconic acid, a tetrahydro terephthalic acid (THTA) can be produced by an ethylene addition reaction, dehydrogenation of which produces bio-TPA. Bio-TPA produced from limonene-derived building blocks is also under development. *p*-cymene is a limonene-derived precursor that can be produced from chemical refinery of limonene [69]. Oxidation uses concentrated nitric acid for the *iso*-propyl group, which reacts with potassium permanganate. This oxidation results in 85% overall conversion from limonene, which is the target in industrial applications. Bio-TPA from furan derivatives should be also featured, as biobased furan derivatives such as 2,5-furan dicarboxylic acid (FDCA) are already in large scale pilot production state [70–73]. Diels-Alder (DA) reaction is the key chemical reaction in the bio-TPA production from furan derivatives. First, furfural is oxidized and dehydrated to produce maleic anhydride, which is then reacted with furan to produce a DA adduct. Dehydration of the DA adduct results in phthalic anhydride, which is converted to bio-TPA via phthalic acid and dipotassium phthalate. Another interesting bio-TPA synthesis pathway involving DA was reported by Avantium, the leading developer of biobased FDCA. Hydroxymethylfurfural (HMF) from fructose is an important precursor of FDCA and is produced by hydrogenation to convert HMF to dimethyl furan (DMF). DMF is converted to *p*-xylene through several steps, such as cyclization with ethylene by DA and dehydration, and then the *p*-xylene is converted to bio-TPA. Another interesting approach is reported by BioBTX, which is building a pilot plant to produce aromatics (benzene, toluene, and terephthalate or BTX) by means of catalytic pyrolysis of biomass (e.g., wood and other lignin-rich biomass resources) [74]. Figure 6 shows the four methods of bio-TPA production discussed above. Similar to bio-EG, bio-based 1,3-propoane diol (bio-PDO) is used for development of biobased poly(trimethylene terephthalate) (PTT). Because of the good shape recovery properties of PTT due to its unique chain conformation, PTT fibers are used in the carpet and textile industries. Partnering with Tate & Lyle and Genencor, DuPont produces bio-PDO named "Susterra" by fermenting sugars from starches [75]. Susterra is used as building block for biobased PTT named "Sorona", which consists of 37 wt % sustainable components. Biobased poly(butylene terephthalate) (PBT) will become an available biobased polyester, since biobased production of monomer component BD is under steady development [76]. PBT is used for special applications that require high dimensional stability and excellent slidability. It explores new applications of biobased polymers when PET and PTT are rarely used.

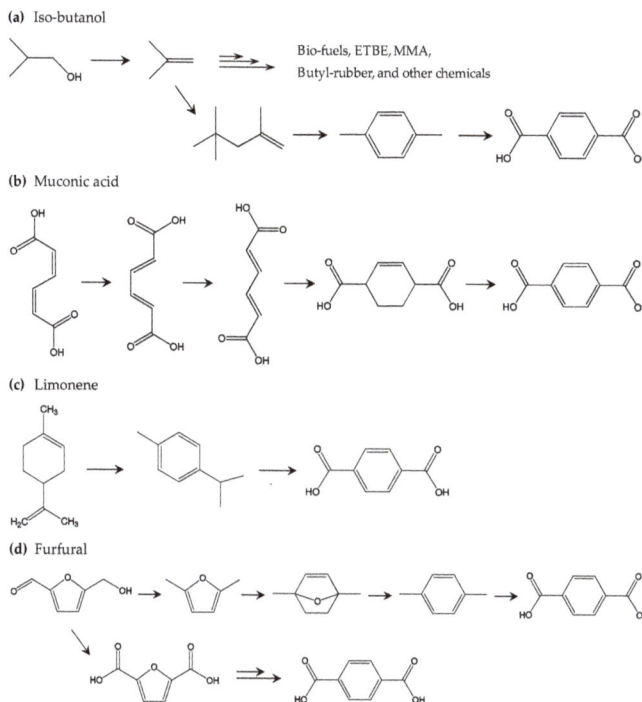

Figure 6. Proposed methods to achieve biobased TPA: (**a**) the *iso*-butanol method [67]; (**b**) the muconic acid method [68]; (**c**) the limonene method [69]; and (**d**) the furfural method [70–73].

3.3. Biobased Polyamides

Development of biobased polyamides is accelerated by the recent progress in the refinery of biobased building blocks. Table 6 lists the chemical structures, typical T_m, moduli, and suppliers of commercialized biobased polyamides. Polyamides 6 and 6.6 are representative petroleum-derived polyamides that are widely used for many general and engineering applications. Thus, making the properties of biobased polyamides similar to those of polyamides 6 and 6.6 is a reasonable milestone to set when creating realistic development strategies. Figure 7 shows the typical methods producing building blocks and polymerizing biobased polyamides. As shown in Table 6, the thermal properties of biobased polyamides containing four carbons (4C) are comparable or higher than those of polyamides 6 and 6.6. The high T_m of 4C polyamides is accompanied by high thermal durability and mechanical strength, but the rigidity of these polyamides should be moderated to ensure stable processing in practical extrusion and injection molding.

Table 6. Chemical structures, suppliers, T_m, and moduli of biobased polyamides [77–79].

Source	Chemical structure	Examples of commercial suppliers	T_m (°C)	Modulus (GPa)
Biobased	Polyamide 4	N.A.	265	
	Polyamide 4.6	DSM	295	
	Polyamide 4.10	DSM	250	1.3
	Polyamide 6.10	Evonik	206	2.1
	Polyamide 10.10	Arkema, Evonik	191	1.8
	Polyamide 11	Arkema	185	1.0
	Polyamide 12	Evonik	178	1.6
Petroleum derived	Polyamide 6	Chemical companies	218	3.0
	Polyamide 6.6	Chemical companies	258	2.5

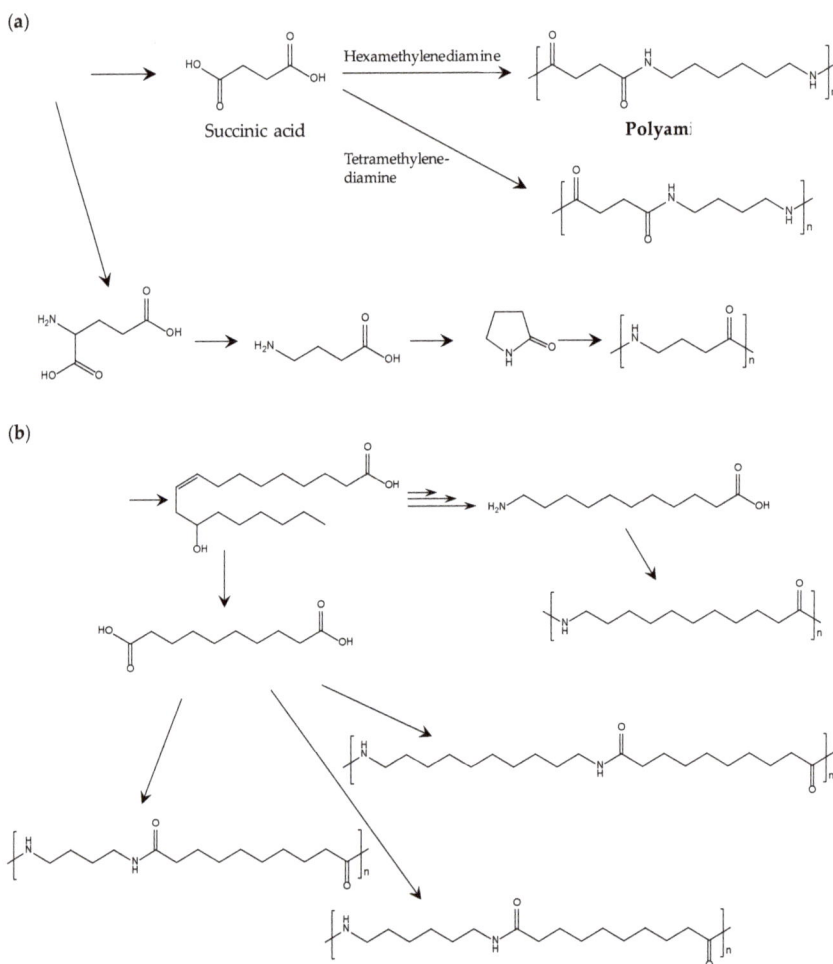

Figure 7. Method of building block production and biobased polyamide polymerization: (**a**) biobased polyamides from sugar; (**b**) from castor oil [78].

Among the general techniques for moderation of rigidness, branching in the main chain of a polymer may be the most promising for polyamide 4 [77]. In this report, 3- and 4-arm branched polyamide 4 were prepared with high molecular weight, comparable to linear polyamides. The branched structure improved mechanical properties without decreasing T_m. Although promising improvements were made in physical properties, there are some technical issues regarding polyamide 4 that must be overcome, including the level of gel formation during polymerization. When the amount of initiator for branched polyamide 4 is higher than 3.0 mol %, some gelation occurs, which might negatively affect the physical properties of the final product. For stable industrial production of polyamide 4, the optimum polymerization process must be determined. Table 6 displays polyamides that consist of 4C, 10-carbon (10C), 11-carbon (11C), and 12-carbon (12C) biobased building blocks [77–79]. The 10C, 11C, and 12C comprising biobased polyamides have milder and softer physical properties. However, these properties are prized for applications such as automotive fuel lines, bike tubing, and cable coating, which require flexibility. Sebacic acid for C10 and 11-aminoundecanoic acid for C11 are the building blocks for polyamides 4.10, 6.10, 10.10 and 11. The long alkyl chains of these result in low water uptake and low density, which are advantages over conventional polyamides. Besides the relatively low T_m of polyamides 6.10, 10.10 and 11 compared to polyamides 6 and 6.6, the flexibility of long alkyl chains is attractive for engineering applications that require high impact resistance and resilience.

Another interesting approach to creating biobased polyamides is the development of biobased lactams [80–82]. The proposed steps of lactam synthesis are rather complicated, but it is expected that the tunable aspects of biobased lactams will lead to new functionalized polyamides in the near future.

4. Newly Developed Biobased Polymers

4.1. Poly(Ethylene 2,5-Furandicarboxylate) (PEF)

4.1.1. PEF from Condensation of Diol and FDCA

In the past, PEF was not considered special; it was a downgraded PET because of its slow crystallization and low T_m. Although the general flow of polymerization processes, physical properties, and fundamental crystallography of PEF had been reported in the 1940s, the available information was not sufficient for practical application. However, in 2008, some reliable information about PEF, including its currently known polymerizations, was reported [83]. Around the same time, the widely known thermal properties of PEF—T_m around 210 °C and T_g around 80 °C—were reported [84]. Other studies followed, increasing the scientific understanding of the physical properties of PEF. The thermal decomposition temperature of PEF is approximately 300 °C, which also results in β-hydrogen bonds [85]. The brittleness and rigidity of PEF result in about 4% elongation at break [86]. PEF is generally produced by polycondensation and polytransesterification of EG and FDCA, derivatives of dichloride-FDCA, dimethyl-FDCA, diethyl-FDCA, or bis-(hydroxyethyl)-FDCA [85]. Solid-state polymerization (SSP) is the key to obtaining high molecular weight, which enables PEF to be suitable for engineering applications. These steps are analogous to industrial processes for producing PET. The results of scientific studies have been successfully applied to pilot and upcoming industrial production of PEF. The most widely known example of industrial PEF production is that of Synvina, a joint venture of Avantium and BASF. Figure 8 shows Avantium's plan for production of PEF from FDCA derived from fructose [87]. First, fructose is converted to 5-methoxy methyl furfural (MMF) by dehydration. MMF is then treated by oxidation to produce crude FDCA, which can be highly purified to achieve high-grade FDCA that can be used for production of PEF. With optimal modification, MMF and hydromethyl furfural can serve as important intermediate biobased building blocks for fine chemicals. Therefore, the side products of FDCA may create a new biobased industry. In Avantium's plan, the side product methyl levulinate is also considered an interesting chemical for development of biobased building blocks. The important properties and functionalities of PEF are compared with those of PET in Table 7 [88]. The remarkably high gas barrier properties of PEF should be emphasized; the high O_2 barrier is advantageous for packaging, leading to PEF's practical application in the food and beverage industry. Thermal properties of other poly(alkylene furanoates) (PAF) from FDCA and other biobased aliphatic diols such as C3–C18 long alkyl chain liner alkyls are also reported [85,86,89–94]. The T_m and T_g of them constantly drop, as length of alkylene chain becomes longer, as it is represented by T_m and T_g of PEF are 211 °C and 86 °C, poly(trimethylene furanoate) are 172 °C and 57 °C, poly(butylene furanoate) (PBF) are 172 °C and 44 °C, and poly(1,6-hexanediol furanoate) are 144 °C and 13 °C, respectively [92]. PAF consisting of isosorbide which has rigid and bulky structure is also an interesting polymer because of its outstanding T_g [93]. The reported isosorbide containing PAF shows T_g 196 °C with excellent amorphous properties. Long alkylene chain containing PAF are also prepared by environmentally benign process i.e., enzymatic polymerization [95]. In this report, the structure–property relationships for example, alkylene chain length and thermal properties, crystallinity, and alkylene component were scientifically discussed and summarized. The enzymatic polymerization was applied for FDCA based polyamide [96] and furan containing polyester from 2,5-bis(hydroxymethyl)furan and aliphatic dicarboxylic acid [97].

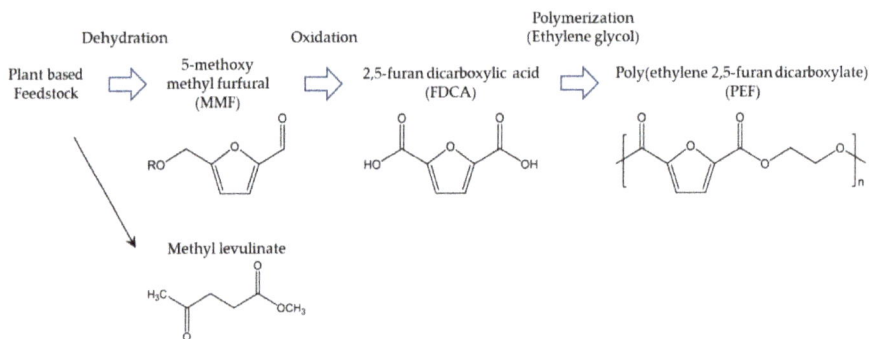

Figure 8. Avantium's PEF production process [87].

Table 7. Comparison of the physical properties of PEF and PET [88].

	PEF	PET
Density (g/cm^3)	1.43	1.36
O_2 permeability	0.0107	0.114
CO_2 permeability	0.026	0.46
T_g (°C)	88	76
T_m (°C)	210–230	250–270
E-modulus (GPa)	3.1–3.3	2.1–2.2
Yield stress (MPa)	90–100	50–60
Quiescent crystallization time (min)	20–30	2–3

4.1.2. PEF from ROP

Although the improvements in polycondensation, polytransesterification, and SSP have enabled PEF with a high molecular weight to be consistently and stably produced, it is still important to develop an alternative method of polymerization of PEF for further functionalization and minimization of side reactions. One interesting alternative is ROP from cyclic compounds consisting of FDCA and liner alkyl diols (Figure 9). ROP is advantageous as it can precisely control molecular weight by adjusting initiator and monomer ratio. Precise control of polydispersity can also be attained by reducing trans-esterification. In addition, various sequence structures can be prepared by copolymerization with other lactones, and end group functionalization. One study reported ROP of poly(butylene 2,5-furandicarboxylate) (PBF) [98]. In this report, cyclic oligomers of PBF are synthesized by reaction of furandicarbonyl dichloride (FDCC) and 1,4-butanediol in solution, and the obtained cyclic oligomers were used for ROP at 270 °C in a bulk state. The molecular weight of the obtained PBF was too low for practical applications, but the thermal properties were comparable to those achieved using conventional polymerization.

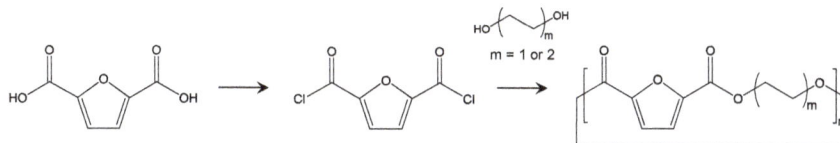

Figure 9. Synthetic scheme of cyclic oligomers for PEF and PBF [98,99].

High molecular weight PBF and PEF were also obtained in another study [99]. The starting cyclic oligomers were prepared by reaction of FDCC and corresponding diols, and the remaining liner oligomer was carefully removed. The obtained cyclic oligomers were used for ROP, which was catalyzed by stannous octoate. The final molecular weight was 50,000 with M_w/M_n of 1.4 for PEF and 65,000 with M_w/M_n of 1.9 for PBF. Differences in the physical properties of PEF obtained using polycondensation/SSP and ROP have not been deeply studied yet, but these differences will lead to new applications of PEF.

4.2. High-Performance PLA from Modified Lactides

The functional groups in the main chains of polylactones such as methylene, ester, and ether control the properties and functionalities of these polylactones, but the functional groups in side chains also play an important role. A simple example of an effect from difference in side group structure is the difference in the thermal properties of polyglycolide (PGL) and PLA. The T_g and T_m of PGL are around 40 °C and 230 °C, respectively, and those of PLA are 55 °C and 170 °C, respectively. In addition, methyl substitution results in higher hydrophobicity, so the hydrolytic stability of PLA is higher than that of PGL. This indicates that a desired functionalization of PLA can be managed by optimum substitution of the methyl group of lactide for other functional groups to overcome the drawbacks of PLA, such as low T_g and transparency.

It has been proposed that T_g can be improved by substituting methyl for a bulky functional group, such as a phenyl group, and a phenyl-substituted lactide is practically reported by using naturally derived phenyl containing mandelic acid [100]. Mandelic acid is a biobased α-hydroxy acid that is widely used as a precursor for cosmetics, food additives, and other chemicals. Phenyl-substituted lactide can be synthesized by cyclic dimerization of mandelic acid, also called mandelide, and the reported T_g of polymandelide (PMA) is higher than 100 °C, which is high enough to be an alternative to high-T_g petroleum-derived amorphous polymers such as polystyrene. The reported ROP of mandelide is only applicable for *meso*-mandelide, which produces completely amorphous PMA, because the high T_m and poor solubility of *racemic* mandelide are not suitable for ordinal ROP in bulk or solution state.

Another interesting polymerization of PMA is ROP of phenyl containing 1,3-dioxolan-4-one (Ph-DXO) [101]. This method allows for control of the chiral structure of the final PMA by preparing homo-chiral Ph-DXO, as Ph-DXO with any chirality has moderate solubility.

In addition, ROP of cyclic *o*-carboxyanhydride can be used for synthesis of *isotactic* PMA. This report is the first about *isotactic* PMA for crystalline PMA, and its T_m is reportedly higher than 310 °C [102]. These thermal properties of PMA allow for new applications of biobased polymers.

Norbornene-substituted lactide also yields high-T_g PLA [103]. For norbornene substitution, L-lactide is brominated, and then an elimination reaction produces (6S)-3-methylene-6-methyl-1,4-dioxane, which is modified by DA reaction, producing norbornene modified lactide is produced. When the norbornene-modified lactide is involved in a ring-opening metathesis reaction, a polymer with narrow polydispersity and T_g of 192 °C is obtained.

Substitution of methyl with a longer alkyl chain can be used to soften PLA by decreasing its T_g. For example, ethylene-modified lactide results in T_g of 66 °C, *n*-hexyl-modified lactide results in T_g of −37 °C, and *iso*-butyl modified lactide results in T_g of 22 °C [104]. The ROP scheme of high-T_g PLA is shown in Figure 10, and the T_g values of the abovementioned substituted PLAs are listed in Table 8.

(a) **(b)**

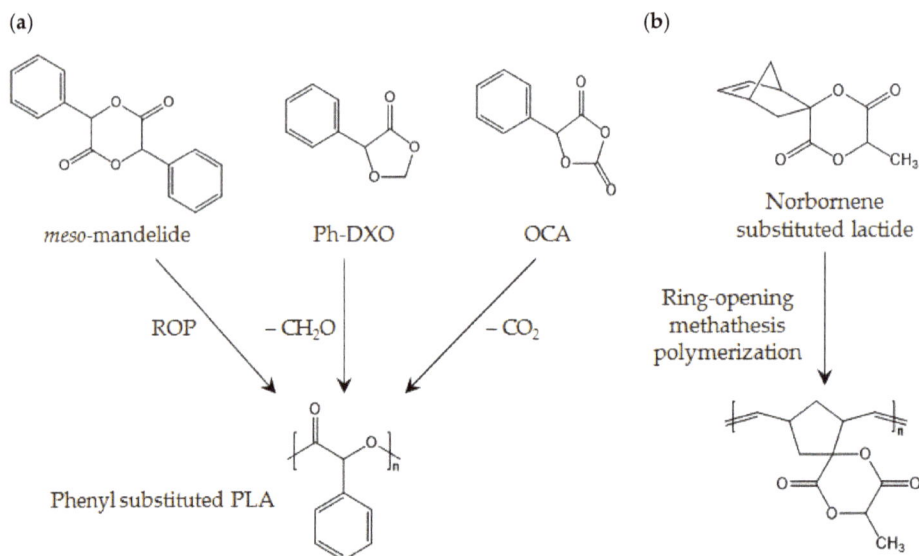

Figure 10. (a) Phenyl-substituted PLA; and (b) high-T_g polymer produced from norbornene-substituted lactide.

Table 8. Chemical structure of modified lactides and their T_g values [100–104].

Modified lactide	T_g of Polymers (°C)
Glycolide	40
methyl glycolide(lactide)	66
ethylglycolide	15
hexyl glycolide	−37
isobutyl glycolide	22
cyclohexyl glycolide (meso)	96
cyclohexyl glycolide (iso)	104
meso-mandelide	100
Norbornene	192

4.3. Terpen-Derived Biobased Polymers

Terpens are a class of naturally abundant organic compounds that are the main component of resins from a variety of plants, especially conifers. Terpens are used by plants for defense, deterring herbivores and attracting predators or parasites to those herbivores. In addition, some insects emit terpens from their osmeteria, such as termites and the caterpillars of swallowtail butterflies, also for defensive reasons. A variety of chemical modifications and functionalizations, such as oxidation, hydrogenation, and rearrangement of the carbon skeleton, can be applied to terpens, resulting in compounds called terpenoids. Both terpens and terpenoids are used in essential oils and fragrances for perfumes, cosmetics, and pharmaceuticals. The polymerizability of economically reasonable terpens and terpenoids is being studied, and there are interesting reports of biobased polyterpenes, especially in high-T_g polymers with excellent amorphous properties.

Pinenes are an important and widely known class of terpen. Polypinenes consisting of alicyclic hydrocarbon polymers comprised of β-pinene or α-phellandrene have high T_g (>130 °C), excellent transparency, and amorphous character [105–107]. In the early stage of development, polypinenes with high molecular weight are prepared by cationic polymerization using an optimum Lewis acid under polymerization conditions. However, the temperature required for polymerization (lower than −70 °C) is too low for industrial production.

Radical polymerization of modified pinenes has been reported to be an alternative to cationic polymerization of pinenes [108]. In this report, α-pinene is transformed into pinocarvone, which contains a reactive exo-methylene group involved in radical polymerization by chemical photo-oxidation and visible-light irradiation. Radical polymerization of pinocarvone is performed in relatively uncommon solvent to achieve high molecular weight and conversion as well as excellent thermal properties (T_g higher than 160 °C). The above mentioned cationic and radical polymerization methods are shown in Figure 11.

Limonene is classified as a cyclic terpen and the reason for the attractive smell of citrus fruits. Limonene is an optically active molecule, and its D-isomer is common in nature. D-limonene is widely used in the cosmetics and food industries. In addition to the economic value of limonene, it features high reactivity during radical polymerization in biobased polymer applications. High-T_g limonene homo-polymers can achieve excellent glass morphology [109]. The kinetics study in that report also indicates the possibility of high molecular weight and high polylimonene yield by optimizing the polymerization conditions.

Copolymerization of limonene and other vinyl groups containing monomers is an approach to synthesis of limonene copolymers [110]. One report presented a striking example of copolymerization of limonene and carbon dioxide to yield a high molecular weight polycarbonate [111]. In this report, limonene is converted to limonene-oxide, and polycarbonate obtained from copolymerization and thiol-ene coupling achieved excellent T_g (>150 °C). There are also interesting reports of chiral active polylimonens, as these show unique properties and stereocomplexability due to the interaction of chiral counterparts [112–114]. A interaction of L-configured and D-configured polylimonene carbonate forms a stereocomplex with T_g of >120 °C. Interestingly, the preferred crystallization of poly(limonene carbonate) occurs only in a stereocomplexed formation.

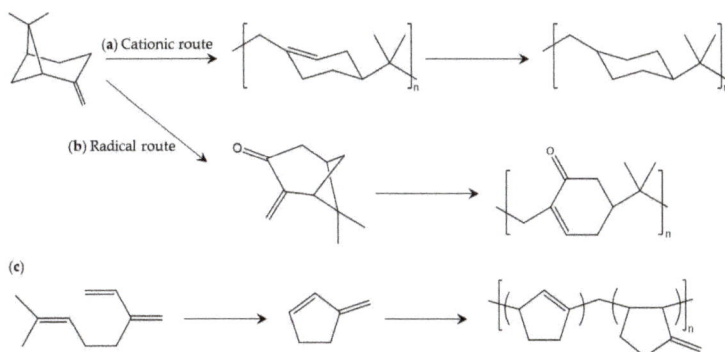

Figure 11. Production of polyterpenes from β-pinene using: (**a**) cationic polymerization [105]; and (**b**) radical polymerization [108]; and (**c**) production from myrcene [115].

Myrcene is an organic olefinic hydrocarbon consisting of optically active carbon. β-myrcene is one of the main components of essential oils, but α-myrcene has not been found in nature and is used in extremely few situations. In industry, β-myrcene can be cheaply produced by pyrolysis of pinene. There are several interesting reports concerning polymers comprised of myrcenes and their derivatives. For example, myrcene can be converted to cyclic diene monomer, which produces an amorphous polymer (Figure 11) [115]. In addition, a copolymer of myrcene and dibutyl itaconate can be used for functionalized applications [116], and the low T_g of the copolymer is promising for biobased elastomer applications. The polymerization process and T_g of the featured polyterpenes are listed in Table 9.

Table 9. Polymerization process and thermal properties of polyterpenes [105–115].

	Polymerization	T_g (°C)
α-pinene	free radical	162
β-pinene	cationic	132
β-pinene	cationic	90
	cationic	130
limonene oxide	trans-carbonation	95
	trans-carbonation	114
limonene oxide/phthalic anhydride	ROP, ester	82
Myrcene/styrene	emulsion	−61
myrcene(3-methylenecyclopentene)	cationic polymerization	11

4.4. Other Noteworthy Biobased Polymers

This section presents recent studies about new biobased polymers with physical properties that are superior to conventional petroleum-derived polymers. By utilizing naturally derived phenols, which contain aromatic rings, biobased liquid crystalline polymers (bio-LCP) can be developed. As the main chemical bonds of bio-LCP are ester bonds, hydroxy and carboxylic acid, biobased building blocks from natural phenols, can be used to produce bio-LCP. For example, 4-hydroxycinnamic acid (4HCA) is one important phenol that can be used to introduce liquid crystalline properties into polyesters via its aromatic function. 4HCA exists in plant cells that are intermediates of metabolites of the biosynthetic pathway of lignin. The mechanical properties of 4HCA-derived bio-LCP are superior to those of other commercialized biobased plastics, with a mechanical strength (σ) of 63 MPa, a Young's modulus (E) of 16 GPa, and a maximum softening temperature of 169 °C [117,118].

One study reported a high-performance biobased polyamide with T_g of >250 °C [119]. This polyamide consists of repeating units from {(4,4′-diyl-α-truxillic acid dimethyl ester) 4,4′-diacetamido-α-truxillamide}. Monomers are prepared through conversion from naturally derived 4-aminophenylalanine, which involves UV coupling of cinnamic acid-derived vinyl groups from each monomer. Scientific investigation into monomer production and polymerization processes is still being performed, but this innovation indicates the possibility of development of super-engineering-grade polymers from naturally derived feed stock.

High-T_m biobased esterified poly(α-glucan) can undergo in vitro enzymatic synthesis [120]. Naturally available sucrose is used as the starting material for polymerization of linear poly(α-glucan), which is then enzymatically catalyzed and esterified on acetic or propionic anhydride. The T_g and T_m of acetated poly(α-glucan) are 177 °C and 339 °C, respectively, and the T_g and T_m of propionated poly(α-glucan) are 119 °C and 299 °C, respectively. The molecular weight of these polymers is higher than 150,000; therefore, it is expected that they have reliable mechanical strength when processed using the right procedures. The in vitro process is technically challenging in terms of scaling up and stabilizing production, but the promising thermal properties should be featured for future applications.

Biobased poly(ether-ether ketone) (bio-PEEK) consisting of FDCA derivatives is a representative super-engineering biobased polymer. Bio-PEEK has a T_m of >300 °C, which is comparable to that of PEEK created from petroleum-derived resources [121]. Synthesis with TPA-derived biobased monomers is a way to replicate conventional PEEK. Thus, it is easily applicable to industrial processes as long as a supply chain of biobased furan derivatives are created. Figure 12 displays the chemical structures of the aforementioned biobased polymers.

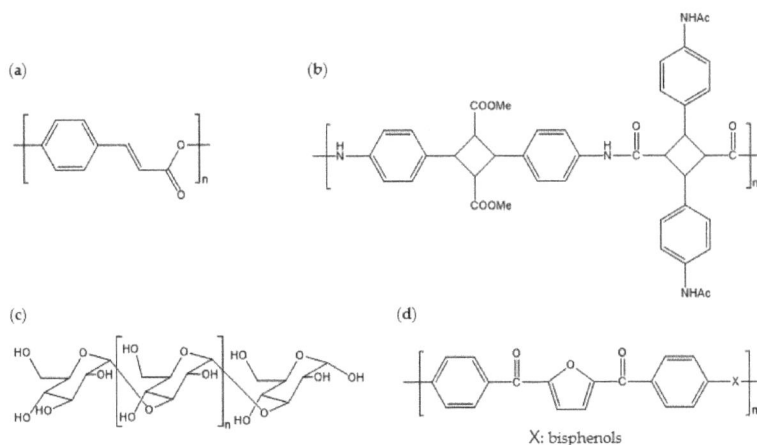

Figure 12. Chemical structures of: (**a**) poly(4-hydroxycinnamic acid); (**b**) poly((4,4′-diyl-α-truxillic acid dimethyl ester) 4,4′-diacetamido-α-truxillamide); (**c**) poly(α-glucan); and (**d**) poly(ether-ether ketone) consisting of FDCA derivatives.

5. Discussion

There have been constant and stable improvements in the production of biobased polymers (i.e., in the polymerization and refinery processes that yield biobased building blocks) in the past few decades. As a result, the application of biobased polymers has been expanded. In the early stage of development of biobased polymers, they were recognized as biodegradable polymers for temporary applications, which is still an important part of their applications. However, upgraded biodegradable polymers can now be used for general and engineering applications. These polymers as well as those that are analogous to conventional petroleum-derived polymers play an important role in further growth of biobased polymer applications. As the scaling-up of production of monomers for polymers that are analogous to conventional polymers has been successful, production of biobased polymers will also be scaled up. This will result in prices that are competitive with those of petroleum-derived polymers. Moreover, new biobased polymers comprised only of biobased building blocks, such as PEF and biobased polyamides, have unique and promising functionalities and applications. Thus, the goal of biobased polymer production is no longer to simply replace petroleum-derived polymers. Explorations into the topic will be accelerated by the development of high-spec engineering-grade biobased polymers, which have already been reported at the scientific level. We are confident that industrialization of these polymers will be discussed in the near future.

Acknowledgments: Financial support by the Netherlands Organization for Scientific Research (NWO) via a VICI innovational research grant is greatly acknowledged.

Author Contributions: Hajime Nakajima was in charge of designing contents and structures of this paper. Hajime Nakajima was also the main author and writer of this paper. Peter Dijkstra was responsible for the scientific contents of 2.2 Poly(hydroxyalkanoates) (PHAs), 2.3. Polysaccharides, and 2.4 Succinate Polymers. Katja Loos was responsible for supervision of scientific contents and discussions of the entire part of this paper.

Conflicts of Interest: The authors declare no conflict of interest.

References

1. Steinbuechel, A. *Biopolymers*; Wiley-VCH: Weinheim, Germany, 2001.
2. Domb, A.J.; Kost, J.; Wiseman, D.M. *Handbook of Biodegradable Polymers*; Harwood Academic Publishers: London, UK, 1997; ISBN 90-5702-153-6.

3. Klass, D.L. *Biomass for Renewable Energy, Fuels, and Chemicals*; Academic Press: San Diego, CA, USA, 1998; ISBN 0-12-410950-0.

4. Mohsenzadeh, A.; Zamani, A.; Taherzadeh, M.J. Bioethylene Production from Ethanol: A review and Techno-economical Evaluation. *ChemBioEng Rev.* **2017**, *4*, 75–91. [CrossRef]

5. Atabani, A.E.; Silitonga, A.S.; Badruddin, I.A.; Mahlia, T.M.I.; Masjuki, H.H.; Mekhilef, S. A comprehensive review on biodiesel as an alternative energy resource and its characteristics. *Renew. Sust. Energy Rev.* **2012**, *16*, 2070–2093. [CrossRef]

6. Corma, A.; Iborra, S.; Velty, A. Chemical Routes for the Transformation of Biomass into Chemicals. *Chem. Rev.* **2007**, *107*, 2411–2502. [CrossRef] [PubMed]

7. Im, S.S.; Kim, Y.H.; Yoon, J.S.; Chin, I.-J. *Biobased-Polymers: Recent Progress*; Wiley-VCH: New York, NY, USA, 2005; ISBN 3527313273.

8. Kimura, Y. Molecular, Structural, and Material Design of Bio-Based Polymers. *Polym. J.* **2009**, *41*, 797–807. [CrossRef]

9. Nakajima, H.; Kimura, Y. Chapter 1, General introduction: Overview of the current development of biobased polymers. In *Bio-Based Polymers*, 1st ed.; Kimura, Y., Ed.; CMC Publishing Co., Ltd.: Tokyo, Japan, 2013; pp. 1–23. ISBN 978-4-7813-0271-3.

10. Babu, R.P.; O'Connor, K.; Seeram, R. Current progress on bio-based polymers and their future trends. *Prog. Biomater.* **2013**, *2*, 1–16. [CrossRef]

11. Tsuji, H. Polylactide. In *Biopolymers, Vol.4, Polyesters III*; Steinbuchel, A., Doi, Y., Eds.; Wiley-VCH Verlag GmBH: Weinheim, Germany, 2002; pp. 129–177. ISBN 978-3-527-30225-3.

12. Tsuji, H. Poly(lactide) Stereocomplexes: Formation, Structure, Properties, Degradation, and Applications. *Macromol. Biosci.* **2005**, *5*, 569–597. [CrossRef] [PubMed]

13. Fukushima, K.; Yoshiharu, K. Stereocomplexed polylactides (Neo-PLA) as high-performance bio-based polymers: Their formation, properties, and application. *Polym. Int.* **2006**, *55*, 626–642. [CrossRef]

14. National Renewable Energy Laboratory Report. Available online: https://www.nrel.gov/docs/fy04osti/35523.pdf (accessed on 16 August 2017).

15. PRO-BIP2009. Available online: https://www.uu.nl/sites/default/files/copernicus_probip2009_final_june_2009_revised_in_november_09.pdf (accessed on 15 July 2017).

16. Bio-Based Chemicals. Available online: http://www.iea-bioenergy.task42-biorefineries.com/upload_mm/b/a/8/6d099772-d69d-46a3-bbf7-62378e37e1df_Biobased_Chemicals_Report_Total_IEABioenergyTask42.pdf (accessed on 16 August 2017).

17. Corbion/Total Announcement. Available online: https://www.total-corbion.com/products/pla-polymers/ (accessed on 16 August 2017).

18. Mochizuki, M. Application of Bio-based Polymers. In *Bio-Based Polymers*, 1st ed.; Kimura, Y., Ed.; CMC Publishing Co., Ltd.: Tokyo, Japan, 2013; Chapter 5; pp. 165–174, ISBN 978-4-7813-0271-3.

19. Mochizuki, M. *Biopolymers, Vol.4, Polyesters III*; Wiley-VCH Verlag GmBH: Weinheim, Germany, 2002; pp. 1–23. ISBN 978-3-527-30225-3.

20. Avantium Report. Renewable Chemicals into Bio-Based Materials: From Lignocellulose to PEF. Available online: http://biobasedperformancematerials.nl/upload_mm/3/5/7/651bed82-390b-4435-a006-7909570de736_BPM%202017%20-%20Speaker%2006%20-%20Ed%20de%20Jong%20-%20Renewable%20chemicals%20into%20bio-based%20materials%20-%20from%20lignocellulose%20to%20PEF.pdf (accessed on 16 August 2017).

21. Avantium Report. PEF, a 100% Bio-Based Polyester: Synthesis, Properties & Sustainability. Available online: http://euronanoforum2015.eu/wp-content/uploads/2015/06/PlenaryII_PEF_a_100_bio-based_polyester_Gert-JanGruter_11062015_final.pdf (accessed on 16 August 2017).

22. Kaku, M. Poly(trimethylene terephthalate, PTT). In *Bio-Based Polymers*, 1st ed.; Kimura, Y., Ed.; CMC Publishing Co., Ltd.: Tokyo, Japan, 2013; Chapter 3.4; pp. 86–94, ISBN 978-4-7813-0271-3.

23. Mochizuki, M. Crystallization Behaviors of highly LLA-rich PLA Effects of D-isomer ratio of PLA on the rate of crystallization, crystallinity, and melting point. *Sen'I Gakkaishi* **2010**, *66*, 70–77. [CrossRef]

24. Marega, C.; Marigo, A.; Noto, V.D.; Zannetti, R.; Martorana, A.; Paganetto, G. Structure and crystallization kinetics of poly(L-lactic acid). *Macromol. Chem. Phys.* **1992**, *193*, 1599–1606. [CrossRef]

25. Sasaki, S.; Asakura, T. Helix Distortion and Crystal Structure of the α-Form of Poly(L-lactide). *Macromolecules* **2003**, *36*, 8385–8390. [CrossRef]

26. Alemán, C.; Lotz, B.; Puiggali, J. Crystal Structure of the α-Form of Poly(L-lactide). *Macromolecules* **2001**, *34*, 4795–4801. [CrossRef]

27. Wasanasuk, K.; Tashiro, K. Crystal structure and disorder in poly(L-lactic acid) δ form (α' form) and the phase transition mechanism to the ordered α form. *Polymer* **2011**, *52*, 6097–6109. [CrossRef]

28. Zhang, J.; Tashiro, K.; Tsuji, H.; Domb, A.J. Disorder-to-Order Phase Transition and Multiple Melting Behavior of Poly(L-lactide) Investigated by Simultaneous Measurements of WAXD and DSC. *Macromolecules* **2008**, *41*, 1352–1357. [CrossRef]

29. Zhang, J.; Duan, Y.; Sato, H.; Tsuji, H.; Noda, I.; Yan, S.; Ozaki, Y. Crystal Modifications and Thermal Behavior of Poly(L-lactic acid) Revealed by Infrared Spectroscopy. *Macromolecules* **2005**, *38*, 8012–8021. [CrossRef]

30. Ikada, Y.; Jamshidi, K.; Tsuji, H.; Hyon, S.H. Stereocomplex formation between enantiomeric poly(lactides). *Macromolecules* **1987**, *20*, 904–906. [CrossRef]

31. Duan, Y.; Liu, J.; Sato, H.; Zhang, J.; Tsuji, H.; Ozaki, Y.; Yan, S. Molecular Weight Dependence of the Poly(L-lactide)/Poly(D-lactide) Stereocomplex at the Air−Water Interface. *Biomacromolecules* **2006**, *7*, 2728–2735. [CrossRef] [PubMed]

32. Serizawa, T.; Yamashita, H.; Fujiwara, T.; Kimura, Y.; Akashi, M. Stepwise Assembly of Enantiomeric Poly(lactide)s on Surfaces. *Macromolecules* **2001**, *34*, 1996–2001. [CrossRef]

33. Hoogsteen, W.; Postema, A.R.; Pennings, A.J.; Brinke, G.T.; Zugenmaier, P. Crystal structure, conformation and morphology of solution-spun poly(L-lactide) fibers. *Macromolecules* **1990**, *23*, 634–642. [CrossRef]

34. Okihara, T.; Tsuji, M.; Kawaguchi, A.; Katayama, K.; Tsuji, H.; Hyon, S.-H.; Ikada, Y. Crystal structure of stereocomplex of poly(L-lactide) and poly(D-lactide). *J. Macromol. Sci. Phys.* **1991**, *30*, 119–140. [CrossRef]

35. NatureWorks Website. Available online: http://www.natureworksllc.com/What-is-Ingeo (accessed on 16 August 2017).

36. Niaounakis, M. Chapter 1, Definition of Terms and Types of Biopolymers. In *Biopolymers: Applications and Trends*, 1st ed.; Niaounakis, M., Ed.; Elsevier: Amsterdam, The Netherlands, 2015; pp. 1–90. ISBN 9780323353991.

37. Madison, L.L.; Huisman, G.W. Metabolic Engineering of Poly(3-Hydroxyalkanoates): From DNA to Plastic. *Microbiol. Mol. Biol. Rev.* **1999**, *63*, 21–53. [PubMed]

38. Morgan-Sagastume, F.; Valentino, F.; Hjort, M.; Cirne, D.; Karabegovic, L.; Gerardin, F.; Johansson, P.; Karlsson, A.; Magnusson, P.; Alexandersson, T.; et al. Polyhydroxyalkanoate (PHA) production from sludge and municipal wastewater treatment. *Water Sci. Technol.* **2014**, *69*, 177–184. [CrossRef] [PubMed]

39. Chatterjee, R.; Yuan, L. Directed evolution of metabolic pathways. *Trends Biotechnol.* **2006**, *24*, 28–38. [CrossRef] [PubMed]

40. Witholt, B.; Kessler, B. Perspectives of medium chain length poly(hydroxyalkanoates), a versatile set of bacterial bioplastics. *Curr. Opin. Biotechnol.* **1999**, *10*, 279–285. [CrossRef]

41. Gerngross, T.U.; Martin, D.P. Enzyme-catalyzed synthesis of poly[(R)-(-)-3-hydroxybutyrate]: Formation of macroscopic granules in vitro. *Proc. Natl. Acad. Sci. USA* **1995**, *92*, 6279–6283. [CrossRef] [PubMed]

42. Ren, Q.; Grubelnik, A.; Hoerler, M.; Ruth, K.; Hartmann, R.; Felber, H.; Zinn, M. Bacterial Poly(hydroxyalkanoates) as a Source of Chiral Hydroxyalkanoic Acids. *Biomacromolecules* **2005**, *6*, 2290–2298. [CrossRef] [PubMed]

43. Haywood, G.W.; Anderson, A.J.; Williams, D.R.; Dawes, E.A.; Ewing, D.F. Accumulation of a poly(hydroxyalkanoate) copolymer containing primarily 3-hydroxyvalerate from simple carbohydrate substrates by *Rhodococcus* sp. NCIMB 40126. *Int. J. Biol. Macromol.* **1991**, *13*, 83–88. [CrossRef]

44. Matsumoto, K.; Murata, T.; Nagao, R.; Nomura, C.T.; Arai, S.; Arai, Y.; Takase, K.; Nakashita, H.; Taguchi, S.; Shimada, H. Production of short-chain-length/medium-chain-length polyhydroxyalkanoate (PHA) copolymer in the plastid of Arabidopsis thaliana using an engineered 3-ketoacyl-acyl carrier protein synthase III. *Biomacromolecules* **2009**, *10*, 686–690. [CrossRef] [PubMed]

45. Pollet, E.; Averous, L.; Plackett, D. *Biopolymers: New Materials for Sustainable Films and Coatings*; Wiley-VCH: New York, NY, USA, 2011; ISBN 9780470683415.

46. Yokouchi, M.; Chatani, Y.; Tadokoro, H.; Teranishi, K.; Tani, H. Structural studies of polyesters: 5. Molecular and crystal structures of optically active and racemic poly (β-hydroxybutyrate). *Polymer* **1973**, *14*, 267–272. [CrossRef]

47. Hoenich, N.A. Cellulose for medical applications: Past, present, and future. *BioResources* **2006**, *1*, 270–280.

48. Dufresne, A. Nanocellulose: A new ageless bionanomaterial. *Mater. Today* **2013**, *16*, 220–227. [CrossRef]

49. Vshivkov, S.A.; Rusinova, E.V.; Galyas, A.G. Phase diagrams and rheological properties of cellulose ether solutions in magnetic field. *Eur. Polym. J.* **2014**, *59*, 326–332. [CrossRef]

50. Morán, J.I.; Alvarez, V.A.; Cyras, V.P.; Vázquez, A. Extraction of cellulose and preparation of nanocellulose from sisal fibers. *Cellulose* **2008**, *15*, 149–159. [CrossRef]

51. Khan, B.; Niazi, M.B.K.; Samin, G.; Jahan, Z. Thermoplastic Starch: A Possible Biodegradable Food Packaging Material—A Review. *J. Food Proc. Eng.* **2017**, *40*, e12447. [CrossRef]

52. Halley, P.J.; Truss, R.W.; Markotsis, M.G.; Chaleat, C.; Russo, M.; Sargent, A.L.; Tan, I.; Sopade, P.A. A Review of Biodegradable Thermoplastic Starch Polymers. *ACS Symp. Ser.* **2007**, *978*, 287–300.

53. Ahmadi-Abhari, S.; Woortman, A.J.; Hamer, R.J.; Loos, K. Rheological properties of wheat starch influenced by amylose–lysophosphatidylcholine complexation at different gelation phases. *Carbohydr. Polym.* **2015**, *122*, 197–201. [CrossRef] [PubMed]

54. Ahmadi-Abhari, S.; Woortman, A.J.J.; Oudhuis, A.A.C.M.; Hamer, R.J.; Loos, K. The effect of temperature and time on the formation of amylose–lysophosphatidylcholine inclusion complexes. *Starch* **2014**, *66*, 251–259. [CrossRef]

55. Ahmadi-Abhari, S.; Woortman, A.J.J.; Hamer, R.J.; Loos, K. Assessment of the influence of amylose-LPC complexation on the extent of wheat starch digestibility by size-exclusion chromatography. *Food Chem.* **2013**, *14*, 4318–4323. [CrossRef] [PubMed]

56. Thakker, C.; Martínez, I.; San, K.-Y.; Bennett, G.N. Succinate production in *Escherichia coli*. *Biotechnol. J.* **2012**, *7*, 213–224. [CrossRef] [PubMed]

57. Zeikus, J.G.; Jain, M.K.; Elankovan, P. Biotechnology of succinic acid production and markets for derived industrial products. *Appl. Microbiol. Biotechnol.* **1999**, *51*, 545–552. [CrossRef]

58. Xu, J.; Guo, B.-H. Poly(butylene succinate) and its copolymers: Research, development and industrialization. *Biotechnol. J.* **2010**, *5*, 1149–1163. [CrossRef] [PubMed]

59. Niaounakis, M. *Biopolymers: Applications and Trends*, 1st ed.; William Andrew: New York, NY, USA, 2015; ISBN 9780323353991.

60. Siracusa, V.; Lotti, N.; Munari, A.; Rosa, M.D. Poly(butylene succinate) and poly(butylene succinate-*co*-adipate) for food packaging applications: Gas barrier properties after stressed treatments. *Polym. Degrad. Stab.* **2015**, *119*, 35–45. [CrossRef]

61. Luo, S.; Li, F.; Yu, J.; Cao, A. Synthesis of poly(butylene succinate-*co*-butylene terephthalate) (PBST) copolyesters with high molecular weights via direct esterification and polycondensation. *J. Appl. Polym. Sci.* **2010**, *115*, 2203–2211. [CrossRef]

62. Wu, L.; Mincheva, R.; Xu, Y.; Raquez, J.-M.; Dubois, P. High Molecular Weight Poly(butylene succinate-*co*-butylene furandicarboxylate) Copolyesters: From Catalyzed Polycondensation Reaction to Thermomechanical Properties. *Biomacromolecules* **2012**, *13*, 2973–2981. [CrossRef] [PubMed]

63. Morschbacker, A. Bio-Ethanol Based Ethylene. *Polym. Rev.* **2009**, *49*, 79–84. [CrossRef]

64. Braskem report. Development of Bio-Based Olefins. Available online: http://www.inda.org/BIO/vision2014_659_PPT.pdf (accessed on 22 August 2017).

65. Hess, G.; Johnson, J. Deconstructing Inherently Safer Technology. *Chem. Eng. News* **2014**, *92*, 11–16.

66. The Coca Cola Company Website. Available online: http://www.coca-colacompany.com/plantbottle-technology (accessed on 19 August 2017).

67. Gevo Report. Available online: http://www.gevo.com/wp-content/uploads/PDF/gevo-roadshow-2011-web.pdf (accessed on 19 August 2017).

68. Carraher, J.M.; Pfennig, T.; Rao, R.G.; Shanks, B.H.; Tessonnier, J.-P. *Cis,cis*-Muconic acid isomerization and catalytic conversion to biobased cyclic-C_6-1,4-diacid monomers. *Green Chem.* **2017**, *19*, 3042–3050. [CrossRef]

69. Colonna, M.; Berti, C.; Fiorini, M.; Binassi, E.; Mazzacurati, M.; Vannini, M.; Karanam, S. Synthesis and radiocarbon evidence of terephthalate polyesters completely prepared from renewable resources. *Green Chem.* **2011**, *13*, 2543–2548. [CrossRef]

70. Shiramizu, M.; Toste, F.D. On the Diels-alder Approach to Solely Biomass-derived Polyethylene terephthalate (PET): Conversion of 2,5-Dimethylfuran and Acrolein into *p*-Xylene. *Chem. Eur. J.* **2011**, *17*, 12452–12457. [CrossRef] [PubMed]

71. Agirrezabal-Telleria, I.; Gandarias, I.; Arias, P.L. Heterogeneous acid-catalysts for the production of furan-derived compounds (furfural and hydroxymethylfurfural) from renewable carbohydrates. *Rev. Catal. Today* **2014**, *234*, 42–58. [CrossRef]

72. Tachibana, Y.; Kimura, S.; Kasuya, K. Synthesis and Verification of Biobased Terephthalic Acid from Furfural. *Sci. Rep.* **2015**, *5*, 8249. [CrossRef] [PubMed]
73. Collias, D.I.; Harris, A.M.; Nagpal, V.; Cottrell, I.W.; Schultheis, M.W. Biobased Terephthalic Acid Technologies: A Literature Review. *Ind. Biotech.* **2014**, *10*, 91–105. [CrossRef]
74. Schenk, N.J.; Biesbroek, A.; Heeres, A.; Heeres, H.J. Process for the Preparation of Aromatic Compounds. Patent WO 2,015,047,085 A1, 2 April 2015.
75. DuPont Tate & Lyle BioProducts Report. Available online: http://www.cosmoschemicals.com/uploads/products/pdf/technical/susterra-propanediol-89.pdf (accessed on 01 October 2017).
76. Bio-Based World News Report. Available online: https://www.biobasedworldnews.com/novamont-opens-worlds-first-plant-for-the-production-of-bio-based-butanediol-on-industrial-scale (accessed on 19 August 2017).
77. Kawasaki, N.; Nakayama, A.; Yamano, N.; Takeda, S.; Kawata, Y.; Yamamoto, N.; Aiba, S. Synthesis, thermal and mechanical properties and biodegradation of branched polyamide 4. *Polymer* **2005**, *46*, 9987–9993. [CrossRef]
78. Winnacker, M.; Rieger, B. Biobased Polyamides: Recent Advances in Basic and Applied Research. *Macromol. Rapid Commun.* **2016**, *37*, 1391–1413. [CrossRef] [PubMed]
79. Moran, C.S.; Barthelon, A.B.; Pearsall, A.; Mittal, V.; Dorgan, J.R. Biorenewable blends of polyamide-4,10 and polyamide-6,10. *J. Appl. Polym. Sci.* **2016**, *133*, 43626. [CrossRef]
80. Schouwer, F.D.; Claes, L.; Claes, N.; Bals, S.; Degrèvec, J.; Vos, D.E.D. Pd-catalyzed decarboxylation of glutamic acid and pyroglutamic acid to bio-based 2-pyrrolidone. *Green Chem.* **2015**, *17*, 2263–2270. [CrossRef]
81. Winnacker, M.; Tischner, A.; Neumeier, M.; Rieger, B. New insights into synthesis and oligomerization of ε-lactams derived from the terpenoid ketone (−)-menthone. *RSC. Adv.* **2015**, *5*, 77699–77705. [CrossRef]
82. Winnacker, M.; Vagin, S.; Auer, V.; Rieger, B. Synthesis of Novel Sustainable Oligoamides Via Ring-Opening Polymerization of Lactams Based on (−)-Menthone. *Macromol. Chem. Phys.* **2014**, *215*, 1654–1660. [CrossRef]
83. Gandini, A. Polymers from Renewable Resources: A Challenge for the Future of Macromolecular Materials. *Macromolecules* **2008**, *41*, 9491–9504. [CrossRef]
84. Gandini, A.; Silvestre, A.J.D.; Neto, C.P.; Sousa, A.F.; Gomes, M. The furan counterpart of poly (ethylene terephthalate): An alternative material based on renewable resources. *J. Polym. Sci. Part A Polym. Chem.* **2009**, *47*, 295–298. [CrossRef]
85. Sousa, A.F.; Vilela, C.; Fonseca, A.C.; Matos, M.; Freire, C.S.R.; Gruter, G.-J.M.; Coelho, J.F.J.; Silvestre, A.J.D. Biobased polyesters and other polymers from 2,5-furandicarboxylic acid: A tribute to furan excellency. *Polym. Chem.* **2015**, *6*, 5961–5983. [CrossRef]
86. Knoop, R.J.; Vogelzang, W.; Haveren, J.V.; Es, D.S.V. High molecular weight poly(ethylene-2,5-furanoate); critical aspects in synthesis and mechanical property determination. *J. Polym. Sci. Part A Polym. Chem.* **2013**, *51*, 4191–4199. [CrossRef]
87. Avantium YXY Technology Website. Available online: https://www.avantium.com/yxy/yxy-technology/ (accessed on 19 August 2017).
88. Avantium Report. Available online: https://www.coebbe.nl/sites/default/files/documenten/nieuwsbericht/491/PEF%20Polyester%20-%20Ed%20de%20Jong.pdf (accessed on 7 August 2017).
89. Gomes, M.; Gandini, A.; Silvestre, A.J.D.; Reis, B. Synthesis and characterization of poly(2,5-furan dicarboxylate)s based on a variety of diols. *J. Polym. Sci. Part A Polym. Chem.* **2011**, *49*, 3759–3768. [CrossRef]
90. Tsanaktsis, V.; Vouvoudi, E.; Papageorgiou, G.Z.; Papageorgiou, D.G.; Chrissafis, K.; Bikiaris, D.N. Thermal degradation kinetics and decomposition mechanism of polyesters based on 2,5-furandicarboxylic acid and low molecular weight aliphatic diols. *J. Anal. Appl. Pyrolysis* **2014**, *112*, 369–378. [CrossRef]
91. Jiang, M.; Liu, Q.; Zhang, Q.; Ye, C.; Zhou, G. A series of furan-aromatic polyesters synthesized via direct esterification method based on renewable resources. *J. Polym. Sci. Part A Polym. Chem.* **2012**, *50*, 1026–1036. [CrossRef]
92. Papageorgiou, G.Z.; Papageorgiou, D.G.; Terzopoulou, Z.; Bikiaris, D.N. Production of bio-based 2,5-furan dicarboxylate polyesters: Recent progress and critical aspects in their synthesis and thermal properties. *Eur. Polym. J.* **2016**, *83*, 202–229. [CrossRef]
93. Avantium Report. Furanics: Versatile Molecules Applicable for Biopolymers Applications. Available online: http://www.soci.org/-/media/Files/Conference-Downloads/2009/Bioplastic-Processing-Apr-09/Jong.ashx?la=en (accessed on 7 August 2017).

94. Storbeck, R.; Ballauff, M. Synthesis and properties of polyesters based on 2,5-furandicarboxylic acid and 1,4:3,6-dianhydrohexitols. *Polymer* **1993**, *34*, 5003–5006. [CrossRef]

95. Jiang, Y.; Woortman, A.J.J.; Ekensteina, G.O.R.A.V.; Loos, K. A biocatalytic approach towards sustainable furanic–aliphatic polyesters. *Polym. Chem.* **2015**, *6*, 5198–5211. [CrossRef]

96. Jiang, Y.; Maniar, D.; Woortman, A.J.J.; Loos, K. Enzymatic synthesis of 2,5-furandicarboxylic acidbased semi-aromatic polyamides: Enzymatic polymerization kinetics, effect of diamine chain length and thermal properties. *RSC Adv.* **2016**, *6*, 67941–67953. [CrossRef]

97. Jiang, Y.; Woortman, A.J.J.; Ekensteina, G.O.R.A.V.; Petrović, D.M.; Loos, K. Enzymatic Synthesis of Biobased Polyesters Using 2,5-Bis(hydroxymethyl)furan as the Building Block. *Biomacromolecules* **2014**, *15*, 2482–2493. [CrossRef] [PubMed]

98. Pfister, D.; Storti, G.; Tancini, F.; Costa, L.I.; Morbidelli, M. Synthesis and Ring-Opening Polymerization of Cyclic Butylene 2,5-Furandicarboxylate. *Macromol. Chem. Phys.* **2015**, *216*, 2141–2146. [CrossRef]

99. Morales-Huerta, J.C.; Ilarduya, A.M.D.; Munoz-Guerra, S. Poly(alkylene 2,5-furandicarboxylate)s (PEF and PBF) by ring opening polymerization. *Polymer* **2016**, *87*, 148–158. [CrossRef]

100. Liu, T.; Simmons, T.L.; Bohnsack, D.A.; Mackay, M.E.; Smith, M.R.; Baker, G.L. Synthesis of Polymandelide: A Degradable Polylactide Derivative with Polystyrene-like Properties. *Macromolecules* **2007**, *40*, 6040–6047. [CrossRef]

101. Cairns, S.A.; Schultheiss, A.; Shaver, M.P. A broad scope of aliphatic polyesters prepared by elimination of small molecules from sustainable 1,3-dioxolan-4-ones. *Polym. Chem.* **2017**, *8*, 2990–2996. [CrossRef]

102. Buchard, A.; Carbery, D.R.; Davidson, M.G.; Ivanova, P.K.; Jeffery, B.J.; Kociok-Kohn, G.I.; Lowe, J.P. Preparation of Stereoregular Isotactic Poly(mandelic acid) through Organocatalytic Ring-Opening Polymerization of a Cyclic O-Carboxyanhydride. *Angew. Chem. Int. Ed.* **2014**, *53*, 13858–13861. [CrossRef] [PubMed]

103. Jing, F.; Hillmyer, M.A. Bifunctional Monomer Derived from Lactide for Toughening Polylactide. *J. Am. Chem. Soc.* **2008**, *130*, 13826–13827. [CrossRef] [PubMed]

104. Yin, M.; Baker, G.L. Preparation and Characterization of Substituted Polylactides. *Macromolecules* **1999**, *32*, 7711–7718. [CrossRef]

105. Satoh, K.; Sugiyama, H.; Kamigaito, M. Biomass-derived heat-resistant alicyclic hydrocarbon polymers: Poly(terpenes) and their hydrogenated derivatives. *Green Chem.* **2006**, *8*, 878–882. [CrossRef]

106. Satoh, K.; Nakahara, A.; Mukunoki, K.; Sugiyama, H.; Saito, H.; Kamigaito, M. Sustainable cycloolefin polymer from pine tree oil for optoelectronics material: Living cationic polymerization of β-pinene and catalytic hydrogenation of high-molecular-weight hydrogenated poly(β-pinene). *Polym. Chem.* **2014**, *5*, 3222–3230. [CrossRef]

107. Li, A.-Y.; Sun, X.-D.; Zhang, H.-B.; Zhang, Y.-C.; Wang, B.; Shi, L.-Q. Cationic copolymerization of 1,3-pentadiene with α-pinene. *J. Polym. Eng.* **2014**, *34*, 583–589. [CrossRef]

108. Miyaji, H.; Satoh, K.; Kamigaito, M. Bio-Based Polyketones by Selective Ring-Opening Radical Polymerization of a-Pinene-Derived Pinocarvone. *Angew. Chem. Int. Ed.* **2016**, *55*, 1372–1376. [CrossRef] [PubMed]

109. Singh, A.; Kamal, M. Synthesis and Characterization of Polylimonene: Polymer of an Optically Active Terpene. *J. Appl. Polym. Sci.* **2012**, *125*, 1456–1459. [CrossRef]

110. Sharma, S.; Srivastava, A. Radical co-polymerization of limonene with N-vinyl pyrrolidone: Synthesis and characterization. *Des. Monomer Polym.* **2006**, *9*, 503–516. [CrossRef]

111. Martín, C.; Kleij, A.W. Terpolymers Derived from Limonene Oxide and Carbon Dioxide: Access to Cross-Linked Polycarbonates with Improved Thermal Properties. *Macromolecules* **2016**, *49*, 6285–6295. [CrossRef]

112. Byrne, C.; Allen, S.; Lobkovsky, E.; Coates, W.G. Alternating Copolymerization of Limonene Oxide and Carbon Dioxide. *J. Am. Chem. Soc.* **2004**, *126*, 11404–11405. [CrossRef] [PubMed]

113. Auriemma, F.; Rosa, C.D.; Caprio, M.R.D.; Girolamo, R.D.; Ellis, W.C.; Coates, G.W. Stereocomplexed Poly(Limonene Carbonate): A Unique Example of the Cocrystallization of Amorphous Enantiomeric Polymers. *Angew. Chem. Int. Ed.* **2015**, *54*, 1215–1218. [CrossRef] [PubMed]

114. Auriemma, F.; Rosa, C.D.; Caprio, M.R.D.; Girolamo, R.D.; Coates, G.W. Crystallization of Alternating Limonene Oxide/Carbon Dioxide Copolymers: Determination of the Crystal Structure of Stereocomplex Poly(limonene carbonate). *Macromolecules* **2015**, *48*, 2534–2550. [CrossRef]

Polymers **2017**, *9*, 523

115. Kobayashi, S.; Cheng, L.; Hoye, T.; Hillmyer, M.A. Controlled Polymerization of a Cyclic Diene Prepared from the Ring-Closing Metathesis of a Naturally Occurring Monoterpene. *J. Am. Chem. Soc.* **2009**, *131*, 7960–7961. [CrossRef] [PubMed]

116. Sarkar, P.; Bhowmick, A.K. Green Approach toward Sustainable Polymer: Synthesis and Characterization of Poly(myrcene-*co*-dibutyl itaconate). *ACS Sustain. Chem. Eng.* **2016**, *4*, 2129–2141. [CrossRef]

117. Kaneko, T.; Matsusaki, M.; Hang, T.T.; Akashi, M. Thermotropic Liquid-Crystalline Polymer Derived from Natural Cinnamoyl Biomonomers. *Macromol. Rapid. Commun.* **2004**, *25*, 673–677. [CrossRef]

118. Kaneko, T.; Thi, T.H.; Dong, J.-S.; Akashi, M. Environmentally degradable, high-performance thermoplastics from phenolic phytomonomers. *Nat. Mater.* **2006**, *5*, 966–970. [CrossRef] [PubMed]

119. Tateyama, S.; Masuo, S.; Suvannasara, P.; Oka, Y.; Miyazato, A.; Yasaki, K.; Teerawatananond, T.; Muangsin, N.; Zhou, S.; Kawasaki, Y.; et al. Ultrastrong, Transparent Polytruxillamides Derived from Microbial Photodimers. *Macromolecules* **2016**, *49*, 3336–3342. [CrossRef]

120. Puanglek, S.; Kimura, S.; Enomoto-Rogers, Y.; Kabe, T.; Yoshida, M.; Wada, M.; Iwata, T. In vitro synthesis of linear α-1,3-glucan and chemical modification to ester derivatives exhibiting outstanding thermal properties. *Sci. Rep.* **2016**, *6*, 1–8. [CrossRef] [PubMed]

121. Kanetaka, Y.; Yamazaki, S.; Kimura, K. Preparation of Poly(ether ketone)s Derived from 2,5-Furandicarboxylic Acid via Nucleophilic Aromatic Substitution Polymerization. *J. Polym. Sci. Part A Polym. Chem.* **2016**, *54*, 3094–3101. [CrossRef]

© 2017 by the authors. Licensee MDPI, Basel, Switzerland. This article is an open access article distributed under the terms and conditions of the Creative Commons Attribution (CC BY) license (http://creativecommons.org/licenses/by/4.0/).

polymers

MDPI

Review

Plant Secondary Metabolite-Derived Polymers: A Potential Approach to Develop Antimicrobial Films

Ahmed Al-Jumaili [1,2], Avishek Kumar [1], Kateryna Bazaka [1,3] and Mohan V. Jacob [1,*]

[1] Electronics Materials Lab, College of Science and Engineering, James Cook University, Townsville, QLD 4811, Australia; Ahmed.Aljumaili@my.jcu.edu.au (A.A.-J.); Avishek.kumar@my.jcu.edu.au (A.K.); kateryna.bazaka@qut.edu.au (K.B.)
[2] Physics Department, College of Science, Ramadi, Anbar University, Ramadi 11, Iraq
[3] School of Chemistry, Physics, Mechanical Engineering, Queensland University of Technology, Brisbane, QLD 4000, Australia
*   Correspondence: mohan.jacob@jcu.edu.au; Tel.: +61-7-4781-4379

Received: 4 April 2018; Accepted: 2 May 2018; Published: 10 May 2018

Abstract: The persistent issue of bacterial and fungal colonization of artificial implantable materials and the decreasing efficacy of conventional systemic antibiotics used to treat implant-associated infections has led to the development of a wide range of antifouling and antibacterial strategies. This article reviews one such strategy where inherently biologically active renewable resources, i.e., plant secondary metabolites (PSMs) and their naturally occurring combinations (i.e., essential oils) are used for surface functionalization and synthesis of polymer thin films. With a distinct mode of antibacterial activity, broad spectrum of action, and diversity of available chemistries, plant secondary metabolites present an attractive alternative to conventional antibiotics. However, their conversion from liquid to solid phase without a significant loss of activity is not trivial. Using selected examples, this article shows how plasma techniques provide a sufficiently flexible and chemically reactive environment to enable the synthesis of biologically-active polymer coatings from volatile renewable resources.

Keywords: volatile renewable resources; microbial infection; plant secondary metabolites; antimicrobial essential oils; biologically-active polymers; plasma-assisted technique

1. Introduction

In 1963, Lieutenant W. Sanborn was the first to systematically relate surface contamination to the transmission of microorganisms [1]. Later, numerous studies have confirmed the attachment and proliferation of microbial cells on artificial surfaces, such as that of medical devices [2,3]. In spite of significant progress in the development of antibacterial and antifouling surfaces, microbial adhesion and the resulting development of a thick sessile layer, i.e., the biofilm, on the surfaces of synthetic implants remains a major issue with their clinical use [4]. Therapeutic statistics have demonstrated that approximately 80% of worldwide surgical site associated-infections may relate to microscopic biofilm formation [5]. Further, owing to microbial infection, and the subsequent failure of medical devices, there has been a significant increase in the number of revision surgeries [6,7]. In the United States alone, approximately 17 million new biofilm-related infections are reported annually, leading to approximately 550,000 fatalities each year [8].

The emergence of bacteria that are resistant to typically used antibiotics is now well recognized [9,10]. The most serious problem caused by antibiotic resistance is that some pathogenic bacteria have now become resistant to virtually all standard antibiotics [11,12]. Significant examples are methicillin-resistant *Staphylococcus aureus* (MRSA), vancomycin-resistant *Enterococcus* (VRE), multi-drug-resistant *Mycobacterium tuberculosis* (MDR-TB), and *Klebsiella pneumoniae* carbapenemase-producing bacteria [13]. Moreover, today, MRSA, a leading cause of most common hospital infections, and *Neisseria gonorrhoeae*, the pathogen responsible for gonorrhea, are almost resistant to benzyl penicillin, while in the past, these pathogens were highly susceptible to the drug [14]. The impact of microbial resistance can be diminished considerably through reduced antibiotic consumption.

Renewable resources have attracted some research attention as precursors for developing tailored bioactive polymers that are capable of minimizing the rate of bacterial adhesion and biofilm growth in healthcare facilities. Within the therapeutic arsenal of naturally-available alternatives that have been explored, plant secondary metabolites (PSMs), such as essential oils and herb extracts, have revealed relatively powerful broad-spectrum antibacterial activities [15,16]. Good examples of currently used PSMs are tea tree (*Melaleuca alternifolia*), geranium, zataria, and cinnamon oils that have shown inherent bactericidal performance in their liquid and/or vapor form toward important pathogenic microbes. Due to the presence of a large number of active molecules within a single essential oil or plant extract, their antimicrobial pathway is not fully understood and cannot be attributed to a particular mechanism [17]. However, the pharmaceutical, cosmetic, and food industries have recently paid great attention to bioactive PSMs, by way of the usage of natural additives as a substitute for synthetic preservatives [18]. Indeed, PSMs are a relatively low-cost renewable resource available in commercial quantities, with limited toxicity, and potentially, different biocidal mechanisms to synthetic antibiotics, which make them an appropriate precursor for "green" functional polymeric materials. On the other hand, using PSMs for surface functionalization through immobilization or synthesis of coatings without loss of functionality is challenging, in part due to the issues with solubility and volatility of these precursors. The plasma-assisted technique overcomes these challenges, allowing the fabrication of a polymerized 3D matrix from renewable precursors with control over its surface properties and chemical functionality. Under appropriate fabrication conditions, plasma-enabled synthesis may help preserve/retain the inherent antimicrobial functionality of PSMs within the solid polymer-like thin films. Plasma polymers of PSMs (PP-PSMs) have several advantages including low cytotoxicity, long-term stability, and a reduced risk of developing microbial resistance. These advantageous properties render PP-PSMs a suitable candidate for bioactive coating applications.

Thus, the focus of this article is on:

- The challenge of bacterial adhesion, biofilms formation, and medical device-associated infections.
- The retention of inherent antimicrobial activity of sustainable monomers, e.g., plant secondary metabolites within solid polymers with the aim of applying them as bioactive coatings.

2. Microbial Contamination

Global production of medical devices and associated materials is an industry worth over $180 billion, and is expanding swiftly [19]. Microbial contamination of these biomaterials is a serious and widespread problem facing current health systems, because it often leads to devastating infections and the failure of the affected device. Adhesion of planktonic microorganisms (e.g., bacteria and fungi) to surfaces is the first stage during surface colonization, followed by the subsequent formation of biofilms which provide an ideal environment for the microbial community to flourish and effectively evade treatment. An active biofilm can be up to 1000 times more resistant to an antimicrobial treatment than planktonic bacteria of the same species [20,21]. Biofilms act as a nidus for systemic pathogenic infections, including dental cavities, periodontal disease, pneumonia associated with cystic fibrosis, otitis media, osteomyelitis, bacterial prostatitis, native valve endocarditis, meloidosis, and musculoskeletal infections [22,23]. Thus, a thorough understanding of the mechanisms by which microorganisms attach to the substrate, and the structure and dynamics of biofilm formation is necessary to develop bio-active coatings that reduce or prevent medical device-associated infections.

2.1. Bacterial Adhesion

Bacterial cells are essentially capable of attaching to all natural and artificial surfaces [24]. Yet it has been assumed that bacteria favorably stick to rougher surfaces for three reasons: (i) A higher surface area available for attachment; (ii) protection from shear forces; and (iii) chemical changes that cause preferential physico-chemical interactions [25]. Also, there is consensus among scientists that the solid–liquid interface between a surface and an aqueous medium (e.g., water and blood) provides a suitable environment for the adhesion and propagation of bacteria [26].

Before the first microorganism reaches the surface, water, salt ions, or proteins that exist in the environment will adhere to the substrate because of the nature of the attachment, which is dependent on the properties of the material [27] and the chemistry of the environment. Consequently, a single layer of organic macromolecules called a 'conditioning film' is formed [28]. The characteristics of conditioning films in turn significantly influence the surface colonization. As the bacterial cell approaches the surface (a few nanometers), the initial stage of adhesion is governed by a number of physico-chemical effects, which include long-range and short-range forces. The long range forces include gravitational, van der Waals, and electrostatic interactions, while the short range forces include hydrogen bonding, dipole–dipole, ionic, and hydrophobic interactions [21,29]. The initial microbial attachment is considered reversible, as the cell will attach to the conditioning film not the surface itself. During adhesion to the surface, various bacteria can transiently produce flagella that render them very motile. Depending on the species, microorganisms may have appendages such as fimbriae, or polymeric fibers, also called pili or curli, which enhance attachment to surfaces [30]. For example, the curli fibres of *E. coli* are 4–6 nm wide unbranched filaments, having a distinctive morphology that can be easily detected by electron microscopy [31]. If the microorganisms are not immediately removed from the surface, they can anchor themselves more permanently by producing a large amount of fibrous glycocalyx that performs the role of 'cement' to attach cells to the targeted surface [32].

2.2. Biofilm Formation

After adhering to solid surfaces, the next step of permanent attachment is growing a bacterial "sanctuary", which is the biofilm. Biofilm formation is a four stage process which includes: (i) irreversible attachment; (ii) early development; (iii) maturation; and (iv) detachment or dispersal of cells, as seen Figure 1 [29]. In the case of irreversible adhesion, major changes occur in gene/protein expression of microbial cells. It has been shown conclusively that bacteria secrete a highly hydrated layer (biofilm) that provides a shield against host defense system and antibiotics, and strengthens the attachment of the microorganisms to the surface. Early steps of biofilm formation are controlled by physical adsorption processes and evolution dynamics of planktonic pathogens [33].

A biofilm cluster consists of accumulations of extracellular polymeric substances (EPS), primarily polysaccharides, proteins, nucleic acids, and lipids [34,35]. Typically, a viable biofilm involves three organic layers. The first layer is attached to the surface of the tissue or biomaterial, the second layer is called the "biofilm base", which holds the bacterial aggregation, and the third layer, known as the "surface film", performs as an outer layer where planktonic organisms are released [6]. Biofilm architecture is heterogeneous both in space and time. The thickness of a biofilm varies depending on the microbial species. For example, the mean thickness of a *P. aeruginosa* biofilm is about 24 µm, while *S. epidermidis* has a mean biofilm thickness of 32.3 µm; thickness can reach more than 400 µm in some species [36]. Active biofilms are highly hydrated, with 50%–90% of the overall area at each sectioning depth comprising EPS and liquid [37]. Direct microscopic observation has shown that biofilm clusters accumulate a large quantity of pathogens within a small area, with microorganism cell densities on an infected surface reaching 10^6 cells/cm^2 [38]. Microorganisms communicate with each other inside a biofilm by producing chemotactic particles or pheromones, in a process called "quorum sensing" [39]. Biofilm sanctuaries can include a single infectious species or multiple infectious species, as well as non-pathogenic microorganisms which nevertheless can produce substances that would benefit the survival and proliferation of the pathogenic species. In the case of the infection of medical devices and implants, a single bacterial species is usually responsible for biofilm formation. While in environmental surfaces, groupings of various species will usually dominate the biofilm [40].

Hydrodynamic, physiological, and ecological conditions, along with presence of other colonizers and harmful agents (e.g., antibiotics and antimicrobial nanoparticles), considerably influence the biofilm structure. For example, biofilm structures of *P. fluorescens* and *P. aeruginosa* are significantly affected by nutritional cues, e.g., carbon and iron availability in their surroundings, respectively [41]. It has been reported that shear forces affect the distribution of microcolonies due to the passage of fluid over the biofilm. At low shear forces, the colonies are formed like a channel, while at high shear forces, the colonies are extended and susceptible to rapid vibrations [42]. These channels are essential for bacteria to transport the necessary water, nutrients, and oxygen to the bacterial community within the biofilm [43]. It has been shown that an increasing loading rate applied under a stable shear stress induced the formation of thicker and rougher biofilms [44]. Detachment is a fundamental process in biofilm development that benefits the bacterial life cycle by allowing planktonic cells to return to the environment and settle new territories [45]. Three different biofilm strategies have been suggested to elucidate biofilm detachment: (i) swarming dispersal, where planktonic cells are freed from a bacterial cluster; (ii) clumping dispersal, where aggregates of microbial cells are separated as clumps; and (iii) surface dispersal, where biofilm matrices move across infected surfaces through shear-mediated transport [46]. Detachment initiation has been proposed to initiate in response to specific endogenous or/and exogenous cues (e.g., a lack of nutrients that causes starvation, or high cell densities) [47]. The event of detachment is complex and random. In some cases, separation of large masses (cell clusters larger than 1000 μm^2) from mature biofilms represents only 10% of the detachment process, yet accounts for more than 60% of the microorganisms detached [48]. A considerable amount of literature has been published on biofilms, yet the mechanisms of biofilm detachment are poorly understood [49,50]. Better understanding of the detachment mechanisms is necessary to accurately evaluate the spatial distribution of the bacterial cells in their environment, their ability to survive, as well as their resistance to biocides.

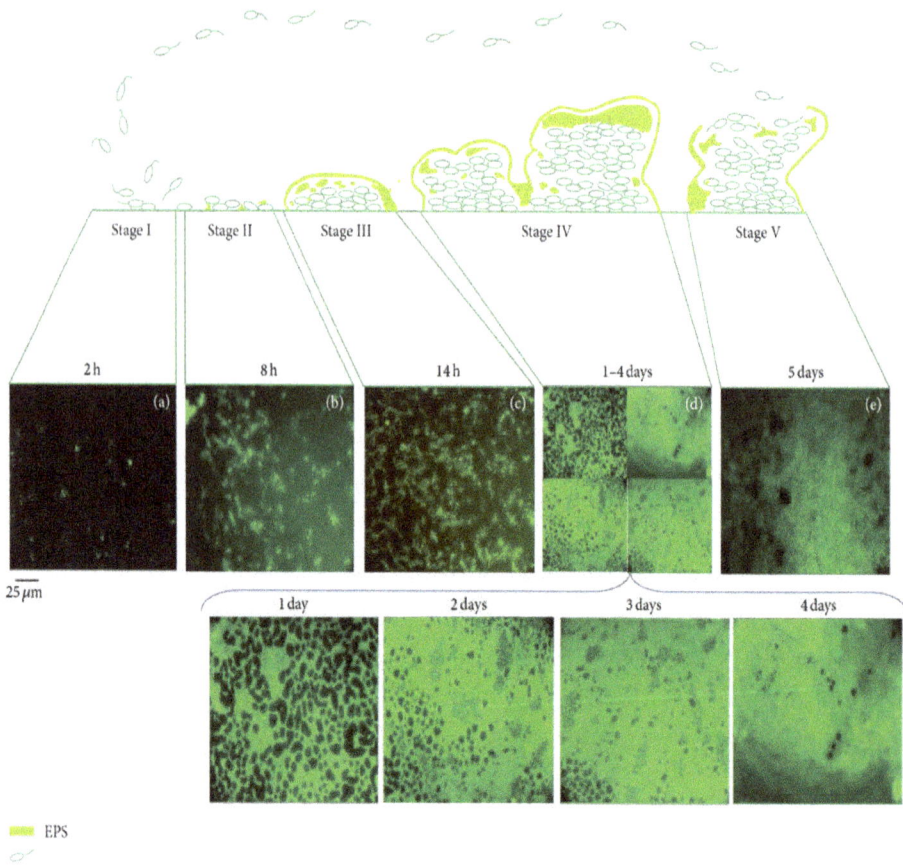

Figure 1. Schematic of the lifecycle of *P. aeruginosa* grown in glucose media. Images of inverted fluorescence microscopy with 400× magnification present stages of biofilm development. In stage I, planktonic bacteria attach to a solid surface. In stage II, the attachment becomes irreversible. Stage III elucidates the microcolony foundation. Stage IV illustrates the biofilm maturation and growth of the three-dimensional bacterial sanctuaries. In stage V, dispersion occurs and free planktonic cells are released from the cluster biofilm to colonize new locations. Images characterize a 250 × 250 μm^2 field. Reproduced from [51].

2.3. The Impact of Biofilm Formation in the Healthcare Environment

Microbial infections related to bacterial attachment and biofilm formation have been detected on various medical devices including prosthetic heart valves, orthopedic implants, intravascular catheters, artificial hearts, left ventricular assist devices, cardiac pacemakers, vascular prostheses, cerebrospinal fluid shunts, urinary catheters, ocular prostheses, contact lenses, and intrauterine contraceptive devices [52]. The three most common device-related infections are central line-associated bloodstream infection, ventilator-associated pneumonia (VAP), and Foley catheter-associated urinary tract infection (UTI) [53]. Studies have shown that 60–70% of nosocomial infections are associated with some type of an implanted medical device [54]. More specifically, the Centre for Disease Control and Prevention in the USA reported that of the infections in medical devices, 32% are urinary tract infections, 22% are surgical site infections, 15% can be attributed to pneumonia and lung infections, and 14% constitute bloodstream infections [55]. Microorganisms also form biofilms on the damaged vascular endothelium of native heart valves in patients with pre-existing cardiac disease, causing Candida infectious endocarditis [35]. It is known that biofilms of Candida species cause malfunctioning of the valve in tracheo-esophageal voice prostheses, leading to an increase in air flow resistance and potential fluid leakage [56]. Furthermore, scanning electron microscopy confirmed biofilm development at the tip of urinary catheters even after a short period of exposure [29].

While a large number of microorganisms are capable of causing infections, those that are able to survive and thrive in clean sites, such as that of clinics and hospitals, present a considerable threat [57]. These organisms include Gram-positive *Enterococcus faecalis, Candida albicans, Staphylococcus aureus, Staphylococcus epidermidis*, and *Streptococcus viridans*; and Gram-negative *Escherichia coli, Klebsiella pneumoniae, Proteus mirabilis*, and *Pseudomonas aeruginosa*. Prevalence of these pathogens is a serious problem in modern societies. For example, *C. albicans* causes superficial and serious systemic diseases, and is known as one of the major agents of contamination in indwelling medical devices [58–60]. *P. aeruginosa* is an opportunistic pathogen of immunocompromised hosts and can cause native acute and chronic lung infections that result in significant morbidity and mortality, especially in cystic fibrosis patients [61,62]. *S. aureus* and *S. epidermidis* have been shown to strongly adhere and form biofilms on metallic implants, e.g., orthopaedic screws, leading to potential device failure [6].

Overview on plant extracts:

In ancient times, plant extracts and natural oils were used in various treatment procedures as antiviral, antimitotic, and antitoxigenic agents due to their strong and broad-spectrum antimicrobial activity. These products can be extracted from all plant organs such as leaves, buds, flowers, roots, stems, seeds, fruits, bark, twigs, or wood. The earliest recorded reference to the techniques and methods used to yield essential oils is believed to be that of Ibn al-Baitar (1188–1248) [63]. Nowadays, natural oils are used in numerous pharmaceutical and therapeutic applications, including ethical medicines for colds, perfumes, make-up products, in dentistry, as food preservatives, and recently, also in the field of sustainable conservation of cultural heritage [64–68]. More than 250 types of these naturally generated oils are traded annually on the global market, at a value of 1.2 billion USD [69].

PSMs are extracted as part of highly complex mixtures of various individual constituents (often hundreds of components) [70]. PSMs were reported to contain a variety of chemical groups in their structure, such as alcohols (terpineol, menthol, geraniol, linalool, citronellol, and borneol), aldehydes (benzaldehyde, citral, cinnamaldehyde, citronellal, and vanillin), acids (benzoic, cinnamic, isovaleric, and myristic), esters (acetates, cinnamates, benzoates, and salicylates), hydrocarbons (cymene, sabinene, myrcene, and storene), ketones (carvone, camphor, pulegone, menthone, and thujone), phenol ethers (safrol and anethol), phenols (carvacrol, eugenol, and thymol), terpenes (camphene, cedrene, limonene, pinene, and phellandrene), and oxides (cineol) [71].

3. The Antibacterial Activities of PSMs

Even though synthetic antibiotics have been the best weapon for eradicating microbial infections since the arrival of penicillin, the overuse of these medications is gradually rendering them ineffective. It is anticipated that if new strategies are not developed soon, medical treatments could retreat to the era where slight injuries and common infections develop into a serious medical problems. One promising strategy has been inspired by the inherent bioactivity of plant secondary metabolites [72]. It is known that most plants produce these organic molecules as antimicrobial agents to combat harmful microorganisms [73,74]. In the past few decades, the progress in the synthesis of nanoscale materials, in particular plasma-assisted fabrication provides the means to retain the antimicrobial activities of PSMs within bioactive coatings. This family of techniques is compatible with PSMs, and offers several advantages such as being an environmentally friendly, versatile, and low-cost technology (discussed further in this article).

In their liquid form, lavender, garlic, oregano, lemongrass, and cinnamon oils are good examples of naturally-occurring substances with strong antibacterial activity [75,76]. Their individual constituents, e.g., citronellol and geraniol are aromatic acyclic monoterpene alcohols that are very powerful bactericides [77–80]. Terpinene-4-ol, a major component of tea tree oil, is a broad-spectrum nonspecific biocide well-known as a natural agent against microbial species such as *E. coli*, *P. aeruginosa*, *Acinetobacter baumannii*, and several drug-resistant bacteria (e.g., MRSA) [81]. A number of PSMs have been used against cancer cells, whereas others are currently used in food preservation [82,83]. In their vapor phase, a number of PSMs have demonstrated strong antibacterial activities [84,85]. So far, there are thousands of natural oils currently known. Among them, 300 oils are important and commonly used in the pharmaceutical, food, sanitary, agronomic, perfume, and cosmetic productions [86].

3.1. The Antibacterial Mechanisms of PSMs

The antibacterial action of PSMs (in their liquid form) is complex and not yet fully understood; it potentially involves several mechanisms, as summarized in Figure 2. A number of researchers have proposed that the hydrophobic nature of PSMs allows them to accumulate and perturb the structure and function of lipids of the microbial cell membrane, disturbing biological function, and causing the failure of chemiosmotic control, thus rendering the membrane more permeable [75,87]. An increase in membrane fluidity and permeability results in membrane expansion and the damage of membrane-embedded proteins which triggers inhibition of the respiration system and alteration of ion transport activities of bacterial cells [88]. For example, carvacrol oil was reported to make the cell membrane permeable to K^+ and H^+, consequently dissipating the proton motive force and inhibiting ATP production [89]. Similarly, menthol and citronellol cause an expansion of the cell membrane, leading to the passive diffusion of ions between the stretched phospholipids [69]. Ultee and Smid (2001) hypothesized that during exposure to PSMs, the driving force for the optimal secretion of the toxin (ATP or the proton motive force) is not sufficient, causing accumulation of the toxin inside the cell, which in turn inhibits normal microbial metabolism [90]. Some active PSMs are capable to coagulate the microbial cytoplasm, leading to cell inactivation [91]. For example, it has been observed that coagulated materials (related to denatured membranes, cytoplasmic constituents, and proteins) were formed outside of the bacterial body when cells (*E. coli*) were grown in the presence of tea tree oil. These coagulates were released through microscopic holes produced in the cell wall as a result of the interaction with the oil [92].

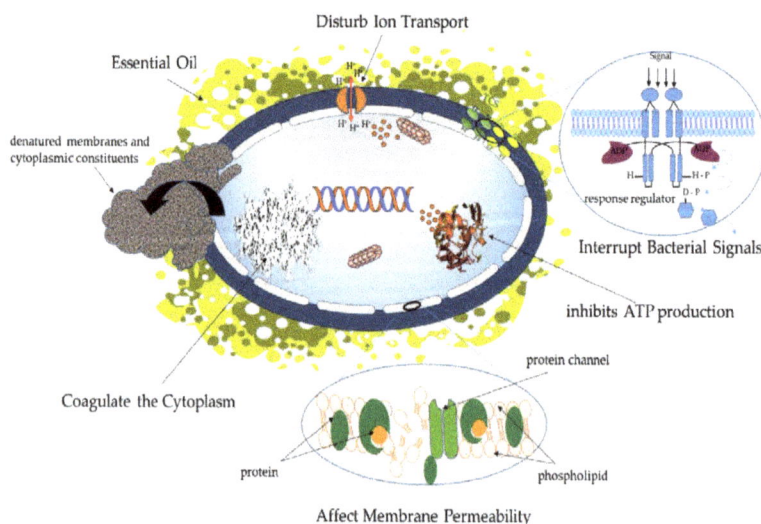

Figure 2. Scheme represents the proposed antibacterial mechanisms of secondary planet metabolizes in their liquid form.

Exposure to PSMs can lead to the reduction in enzymatic activities, loss of turgor pressure, changes in DNA synthesis and inhibition of different metabolic functions [89]. Moreover, some oils, such as rose, geranium, lavender, and rosemary have been shown to inhibit cell–cell communication, affecting the quorum sensing (QS) network in the bacterial community [93]. The QS system is vital for bacterial growth, and hence, any interference with the sensing network may reduce pathogenicity, biofilm formation, and antibiotic resistance during infection events.

The antimicrobial performance of PSMs is linked to their chemical structure, particularly the presence of –OH functional groups [94]. Each compound may reveal a different biocidal mechanism toward microorganisms [69]. The bioactivity of several active oils is associated with the presence of phenolic groups. For example, the antimicrobial efficacy of clove, thyme, and oregano oils is related to the presence of phenol-containing eugenol, thymol, and carvacrol, respectively [95]. However, other findings indicate that the components present in high quantities in the oil are not necessarily responsible for the entire biological activity of a PSM. The antibacterial performance of these complex mixtures relies on a variety of synergistic effects of different sub-components in the oil. Furthermore, it can also be attributed to the presence of other components that may be effective even in small quantities [96,97]. In the case of essential oils containing a high percentage of phenolic compounds (e.g., carvacrol, thymol), it can be assumed that their bactericidal action would be similar to other phenolic groups, e.g., by way of the disturbance of the membrane, disorder the proton motive force, electron flow, and coagulation of cell contents [87]. In the case of complex mixtures, where numerous active molecules are present, potential synergistic and antagonistic influences, as well as minor compounds that can have an important contribution to the oil's activity, need to be considered [64,98]. It is important to state that the biocidal mechanisms of PSMs are dissimilar from currently used synthetic antibiotics, which should minimize the likelihood of the development of microbial cross-drug resistance [99].

3.2. Sustainable Polymers from Bioactive Essential Oils

The ecological concerns of current petroleum processing, along with the economic recession, depleting reserves, and political aspects, have led to increased interest in the production of sustainable polymers derived from renewable resources [100,101]. These eco-friendly polymers can be derived from a wide range of possible precursor materials, including oxygen-rich monomers (e.g., carboxylic acids), hydrocarbon-rich monomers (fatty acids, terpenes, and vegetable oils), and non-hydrocarbon monomers (carbon dioxide) [102]. So far, polymers derived from essential oils, vegetable oils, bio-ethanol, cellulose, fats, resins, naturally occurring polysaccharides, microbial syntheses, and other natural ingredients have been widely used for a variety of applications [103–108]. Essential oils, in particular, are renewable in nature, relatively inexpensive, available in commercial quantities, and display minimal toxicity compared to many conventionally-used precursors, which make them an appropriate precursor for "green" functional materials. Among them, terpenes (major components in a large number of essential oils) have received considerable attention. Their structure contains one or more carbon–carbon double bonds, showing a carbon skeleton of isoprene. The abundance of double bonds allows for cationic and radical polymerization of terpenes, along with epoxidation as a path to biodegradable oxygenated polymers [109]. Cationic polymerization has been generally accepted to be the most appropriate kind of chain reaction for these monomers [110]. However, essential oils have not been widely applied to the production of bioactive polymers due to limitations associated with fabrication systems and oil properties [111,112]. These limitations include challenges in controlling the surface chemistry and morphology of the synthesized materials, and solubility and/or volatility of the natural monomers.

Recent technological advances in the field of controlled polymerization, catalysis, nanoencapsulation, and effective organic functionalization, give great potential for the application of essential oils in manufacturing of sustainable polymers with innovative designs and characteristics. This allows the fabrication of organic films with good control over film thickness, physico-chemical properties, and, importantly, biological functionality. For instance, it was possible to successfully engineer antibacterial UV-cured networks by using a thiol-ene route with covalent immobilization of natural terpenes (linalool and a trithiol) as antibacterial agents, without employing any organic solvent. These bio-based materials exhibited attractive thermal properties, were not affected by water penetration under high moisture conditions, and displayed strong inhibition of microorganisms [113]. Chen et al., (2012) developed the reversible transfer polymerization approach to design a series of cationic rosin-containing methacrylate bioactive-copolymers. The antibacterial activities of these rosin-containing copolymers were found to be dependent on both the degree of quaternization of the rosin group, the molecular weight of copolymers, and the conformation of hydrophobic group [114]. Furthermore, a cinnamon essential oil/cyclodextrin integrated into a polylactic acid nanofilm made by electrospinning and co-precipitation showed strong antimicrobial activity [115].

Several studies have been carried out in order to incorporate active essential oils into selected polymers through applying emulsification or homogenization methods, where ultra-fine emulsions of oils are formed containing polymer at the continuous aqueous phase. Upon drying, lipid droplets remain incorporated into the polymer structure. The releasing rate of the embedded-oils from films is subject to multiple factors, such as electrostatic interactions between the oil and the polymer chains, osmosis, structural variations induced by the presence of the oil, as well as environmental circumstances [116]. Remarkably, a small fraction of an incorporated essential oil within a polymer structure is sufficient to achieve the desired antimicrobial properties. For example, quince seed mucilage films containing a low percentage (1.5–2%) of oregano essential oil were reported to be very effective against several microorganisms, including *S. aureus*, *E. coli*, and *S. putrefaciens* [117]. Other findings showed that inactive chitosan films were transformed into bioactive materials when a small quantity (~1–2%) of extract from two endemic herbs (*Thymus moroderi* or *Thymus piperella*) were integrated within the films [118].

Encapsulation of oils has been developed as one such technology that has great potential to improve the physical stability of the active components, protecting them from degradation due to environmental aspects (e.g., oxygen, light, moisture, and pH) [119]. Among the nanometric encapsulation structures currently being used, nanoemulsions are mainly utilized due to the possibility of formulation with natural components and the compatibility with industrially scalable manufacturing processes by high pressure homogenization [120]. Nanoemulsions are defined as emulsions with ultra-small droplet sizes of approximately 100 nm. At this scale, there is a potential of enhancing physico-chemical properties and stability of the active compound. In addition, the oil bioactivity can be considerably increased, since significant increases in the surface area per unit of mass can be achieved, improving the passive mechanisms of cell absorption, which again allows for the reduction of the oil quantity required to ensure antimicrobial action [121]. The encapsulated essential oils are promising antimicrobial agents for biodegradable/edible coatings in food packaging industries to inhibit pathogenic microorganisms [122]. It has been reported that the encapsulation in nanoemulsion formulation of a terpenes mixture and limonene increased the antimicrobial performance of the pure compounds against various microorganisms such as *E. coli* and *Saccharomyces cerevisiae*, through increases of transport mechanisms in the membrane of the target cell [120]. Mohammadi et al., (2015) also encapsulated *Zataria multiflora* essential oil in chitosan nanoparticles (average size of 125–175 nm) and reported a controlled and sustained release of essential oil for 40 days, along with a superior antifungal performance in comparison with the un-encapsulated oil [123]. Moreover, films with 1.5% nanocomposite marjoram oil diminished the numbers of *E. coli*, *S. aureus*, and *Listeria monocytogenes* populations with respect to the control of up to 4.52, 5.80, and 6.33 log, respectively [124]. Similarly, introduction of carvacrol nanoemulsions into modified chitosan have led to the development of a bioactive film, which was active against Gram-negative pathogenic bacteria [125].

It is worth mentioning that, in many cases, the vapor phase of essential oils exhibits strong inhibitive performance against pathogens, even more effective than direct application [126,127]. For instance, Avila-Sosa et al. (2012) found that chitosan films incorporating cinnamon or Mexican oregano essential oils can inhibit fungi by vapor contact at lower oil concentrations than those required for amaranth and starch polymeric coatings [116].

3.3. Plasma-Assisted Fabrication of PSMs

Among fabrication techniques, cold plasma polymerization is a multipurpose approach that is a relatively fast and low-cost method for fabricating coatings from a wide array of natural precursors, including those that do not usually polymerize by conventional methods, and do not require further chemical or physical processing (e.g., annealing and catalysts) [128]. From a processing point of view, many PSMs are compatible with plasma polymerization, which is in essence a chemical vapor deposition process enhanced by the catalytic activity of plasma, because they are highly volatile at room temperature no external heat or carrier gas are required to deliver the precursor macromolecules to the fabrication zone.

Introduction of PSMs molecules, in vapor phase, into a highly reactive plasma field triggers a wide range of reactions including fragmentation, oligomerization, rearrangement, and polymerization. The degree of dissociation is highly dependent on the amount of energy provided into the plasma system and the pressure in the chamber. Fragmentation is initiated by active electrons rather than thermal excitation or chemical reactions, creating a unique mixture of chemically diverse species (e.g., unsaturated bonds, ions, neutrals, and free radicals), which may not be reachable under other conditions [129]. It is believed that weakly ionized plasma and relatively low substrate temperature during deposition promotes condensation and adsorption of non-excited species, which help to increase the proportion of non/partially-fragmented precursor molecules on the substrate [130]. The recombination of the reactive species and precursor molecules may lead to the formation of the organic thin layer (polymer) on the surface of a given substrate. Due to the diversity of functional groups and reactive species, the polymer can be formed in several ways, involving free-radicals and induced-polymerization of fragments containing unsaturated carbon–carbon bonds; recombination fragment/recombination is initiated by the plasma-generated and surface-attached reactive ions [131]. The formed polymer is often highly branched and highly cross-linked (amorphous), comprising large quantities of trapped free radicals in its structure [132].

A large number of species that exists in the discharge (e.g., ions, electrons, stable molecules, radicals, and photons) can react with each other and the forming chains through a range of interaction mechanisms, as seen in Figure 3. The complexity of the process of PSM plasma makes the evaluation of each specific reaction, along with the prediction of material properties, very challenging. In some cases, few specific reactions can dominate the formation of the film, especially at low input power. Thus, it is rational to propose that films fabricated from PSMs, using plasma under specific deposition conditions (e.g., specific input power, frequency, flow rate, and temperature), could retain some/most of the functional groups of the original PSMs within the bioactive three-dimensional solid film. In addition, the unfragmented precursor molecules trapped within the polymer during the fabrication may elute over time, acting as a drug release coating, with the capacity to retard microbial attachment and biofilm development on the surface [133].

A number of attempts have been made to manufacture antibacterial surfaces, based on plasma polymerization of essential oils, where antibacterial performance is based only on the natural bioactivity of the polymerized surfaces, in the absence of synthetic additives, inorganic nanoparticles, or conventional antibiotics. Using this information, we strongly encourage the reader to further research this rapidly growing and highly-promising arena. Here, we highlight the successful manufacturing of antimicrobial coatings from different PSMs using the cold plasma polymerization technique.

3.3.1. Terpinen-4-ol

Terpinen-4-ol is a monocyclic terpene alcohol that is an active component of tea tree oil. Terpinen-4-ol has demonstrated powerful antimicrobial and anti-inflammatory properties [134,135]. Upon interaction with microorganisms, cyclic terpene hydrocarbons have been shown to accumulate in the cell membrane. This disturbs membrane integrity, triggering an increased passive flux of protons through the membrane and dissipation of the proton motive force [136]. Bazaka et al. (2011) prepared plasma polymerized coatings derived from terpinen-4-ol at various input power levels, showing a considerable potential in minimizing bacterial attachment and metabolic activity of *S. aureus* and *P. aeruginosa*. Fabrication at a low input power level, 10 W, resulted in a partial retention of biologically-active groups of the original precursor, which led to significant antimicrobial and antibiofouling activities of the terpenol-derived coatings [137]. Confocal laser scanning microscopy evidently showed that around 90% of *S. aureus* cells retained on the films of the 10 W substrata were non-viable, in comparison to that retained on the surface of 25 W films [138,139]. However, when fabricated at higher input power (25 W), these films lost their biocidal activity, and promoted adhesion and proliferation of tested bacterial cells and biofilm development. In a recent report, the decrease in antibacterial activity with increasing radio frequency (RF) energy was also observed in the plasma polymerization of polyterpenol films [140].

3.3.2. Carvone

Carvone is found in various essential oils, such as caraway, spearmint, and dill. This PSM has shown a variety of antiproliferative effects with regards to microbial cells; likely due to the presence of a monoterpene group in its structure [141,142]. In addition, carvone and its related compounds were shown to be potential chemopreventive agents, due to their ability to induce increased activity of detoxifying enzymes. The α,β-unsaturated ketone system in carvone is generally expected to be responsible for the high enzyme-inducing action [143]. Recently, Chan et al., (2016) fabricated polymer coatings resultant from plasma polymerization of carvone [144]. At an input power of 10 W, carvone polymerized coatings demonstrated almost equal antimicrobial performance against both Gram-negative and Gram-positive bacteria (86% decrease in *E. coli* and 84% reduction in *S. aureus*), with no cytotoxic effect towards primary human endothelial cells. In addition, these coatings were smooth, highly cross-linked hydrocarbons, with low fractions of carboxyl, hydroxyl, and amine-amide functionalities. Although the carvone surfaces reduce bacterial adhesion, it was observed that some cells were damaged and died after attaching to the surface. The scanning electron microscope (SEM) images clearly exhibited membrane distortion, pore creation, and membrane rupture of microorganisms attached on the surface of plasma polymers of carvone.

3.3.3. Eucalyptol

Eucalyptol, a major component of eucalyptus oil and a minor component of tea tree oil, is a saturated monoterpene known by a variety of synonyms, such as 1,8-cineole, 1,8-epoxy-p-menthane, and cajeputol. This PSM has been demonstrated to retain strong biological activities, including anti-inflammatory, antifungal, antibiofilm, and antiseptic properties toward a range of bacteria [145–148]. The retention of the natural bio-active groups of the 1,8-cineole oil was also achieved using plasma polymerization. Fabricated at 20 W, moderate hydrophobic coatings were achieved, with the ability to reduce the attachment of *E. coli* and *S. aureus* cells by 98% and 64%, respectively, compared to unmodified glass. In addition, the 1,8-cineole plasma films resisted biofilm formation after 5 days of incubation in the presence of bacterial cells. The polymer surface and any products that may be released from the film were also found not to be cytotoxic to mammalian cells [149]. In the same way, Mann and Fisher (2017) used a range of applied RF powers (P = 50–150 W) and $H_2O_{(v)}$ plasma-treatment during the plasma fabrication of 1,8-cineole polymers. The fabricated films retained some antimicrobial behaviors characteristic of the precursor, in addition to the desired properties, such as being highly adherent to the substrate, conformal, and with smooth surfaces. The in vitro studies showed that *E. coli* were largely nonviable and unable to colonize the plasma–cineole surface over the 5 day biofilm development assay period. The biofilm coverage on these surfaces was significantly lower (<10%) than the glass control [150].

3.3.4. Geranium

Geranium (*Pelargonium graveolens*) oil produces a mixture of various components (more than 80), such as linalool, citronellol, and geraniol [151]. Studies have revealed that geranium oil is able to combat pathogens, including both Gram-negative and Gram-positive bacterial strains [82,152]. More recently, geranium oil-derived coatings were also found to have the potential to reduce the microbial adhesion and biofilm formation of select human pathogens, such as *S. aureus*, *P. aeruginosa*, and *E. coli*. The input RF power, in particular, played a substantial role in controlling the surface biochemistry and extensively enhanced the biocidal activity of the fabricated coatings. Films deposited at 10 W caused a significant decrease in the number of cells, biovolume, and biofilm thickness. In contrast, there was no significant change in the bacterial colonization between films fabricated at 50 W and an unmodified glass control. In addition to their biological activities, geranium polymer films showed several advantages, including low density, uniform coverage, good adhesion, and considerable physical stability [153,154].

Despite the fact that the mechanism by which the deactivation process takes place is not fully understood, the attractive antibacterial performance of PP-PSMs surfaces indicate that the original active chemistry of the oils are partially retained within the structure of the fabricated films. Undeniably, plasma parameters are the key factors that determine the extent of retention of biological functionality. The degree of precursor fragmentation is directly related to the amount of applied energy (RF power). For example, during the polymerization of geranium oil and terpinen-4-ol, a slight increase in the input power resulted in the failure to preserve the desired functional groups within the polymer. One reason for this loss of bactericidal activity could be the complete dissociation of the precursor functionalities upon plasma exposure. Furthermore, these polymerized films demonstrated a wide range of functional groups in their structure, such as primarily methyl/methylene functionalities, as well as hydroxyl, alkene, and carbonyl groups. The hydroxyl group particularly is broadly accepted to be an antimicrobial agent of polymer surfaces. It was previously reported that *S. aureus* cells do not preferentially attach to polymers comprising –OH functionality than those bearing carboxylic and methyl groups [155]. However, other surface parameters should be carefully considered during plasma fabrication. It is well known that surface chemistry, hydrophobicity, free energy, and architecture of polymer films have the potential to significantly influence the final antibacterial outcome. The synergistic effects of these parameters may determine the inhibition of bacterial attachment and proliferation.

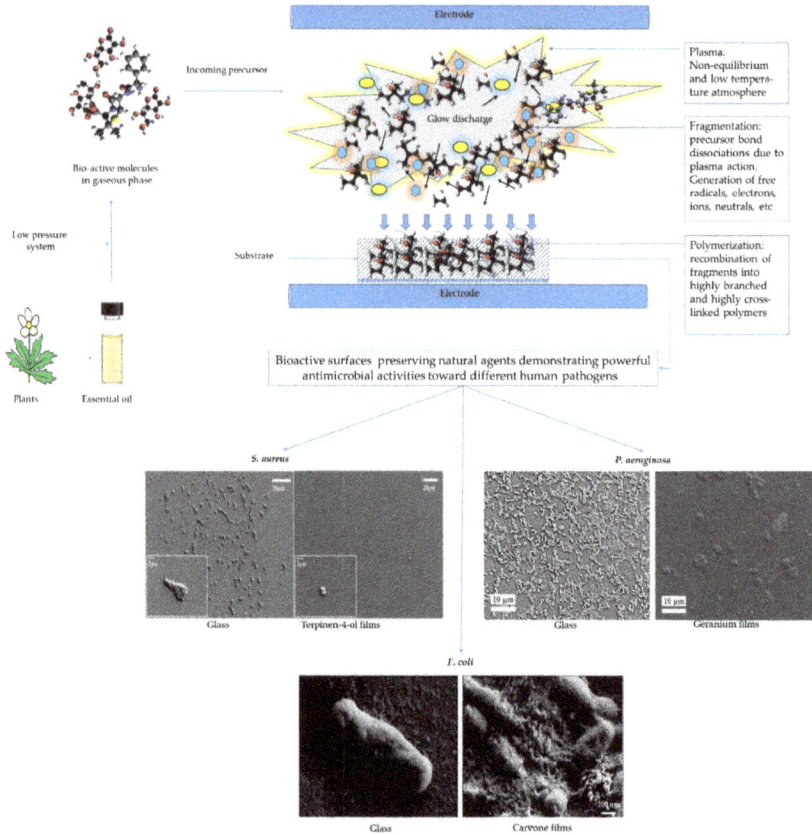

Figure 3. Representative examples of plasma polymerization of plant secondary metabolites, where retention of the antimicrobial activity was achieved. As soon as a bioactive secondary plant metabolite (or an essential oil) is placed under low pressure, the molecules gain sufficient kinetic energy to separate and begin independently moving towards the glow region within the deposition chamber. Exposure of the molecules to the highly reactive plasma initiates various chemical reactions, such as bonds fragmentation, oligomerization, and polymerization. At the chosen plasma parameters, the process allows for the preservation of active functional groups of PSMs within the cross-linked solid polymeric films. Direct observations of SEM demonstrated the powerful antimicrobial performance of geranium, terpenen-4-ol, and carvone films in contact with different pathogens. The antimicrobial activities of these films included antibiofouling effects and/or bactericidal actions (e.g., membrane distortion, pores creation, and membrane damage). The SEM images are reproduced with permission from [137,144,153].

3.4. Properties of PSM-Derived Polymers

For a successful polymeric antibacterial coating to satisfy the requirements of biomedical applications, the material should possess a range of specific biological, physical, and chemical properties. Films fabricated from PSMs display a wide range of desired properties, including optical transparency, moderate hydrophilicity, relatively high degradation temperature, low post-annealing retention, and good biocompatibility, forming simple, useful, and versatile bioactive coatings. Hence, a brief description of some important physico-chemical characteristics of PP-PSMs fabricated at a low input power (below 100 W) is provided below.

As a general trend observed in PP-PSMs, polymers deposited at a higher input power are typically less susceptible to mechanical deformation. This trend is owing to an increase in the degree of cross-linking correlated with higher input power, and hence, films are likely to be more stable and less susceptible to wear [156]. Highly cross-linked polymers are expressively more stiff and dense compared to conventional polymers (amorphous or crystalline arrangements). This is related to the vibrational movement of the carbon backbone of the polymeric structure that is constrained by the presence of a multiple covalent bonds between polymer chains [157].

The topographical features of PP-PSMs fabricated at suitable parameters have been shown to be uniform, pinhole free, with films being highly-adherent to the substrate [154,158–163]. The uniformity indicates that polymerization reactions occurred on the surface of the substrate in preference to the gas phase. Moreover, ultra-smooth surfaces (with an average roughness of less than 1 nm) were attained for plasma polymerization of various PSMs, which is a particularly significant factor that may influence the initial microbial adhesion [164,165]. It is worth to mention that the properties of the surface of plasma polymerized films make them highly susceptible to growth conditions, especially the input power, where more energetic ions can cause more surface bombardment and etching. Furthermore, the chosen precursor plays an important role in determining the overall surface properties, since to a degree, it defines the chemical functionalities and determines the quantities of free radicals in the plasma system [166,167].

A large number of plasma polymers developed from PSMs were reported to have favorable optical properties. Although optical properties were affected by processing parameters during film deposition, PP-PSMs were found to be optically transparent in the visible region and have high absorption in the infrared region. The refractive index and extinction coefficient were in the range of 1.5 and 0.001 (at 500 nm), respectively [168,169]. In addition, PSMs-derived polymer materials had optical energy gap (E_g) values in the insulating and semiconducting region. For example, films fabricated from terpinen-4-ol, linalool, γ-terpinene, and geranium have E_g = 2.5, 2.9, 3.0, and 3.6 eV, respectively [153,165,170]. It is important to note that the optical properties of plasma films are characteristically dependent on the structure of the *p*-conjugated chains in both the ground and excited stats, as well as on the inter-chain orientation [171].

In general, PP-PSMs were moderately hydrophilic, with values of contact angles ranging from ~50° to 80°. The wetting characteristics were defined largely by plasma conditions and the chemistry of the chosen precursor. For example, improved hydrophobicity of the surface was observed for films fabricated from γ-terpinene when increasing the deposition RF power from 61.0° (10 W) to 80.7° (75 W). This polymer showed strong electron donor and negligible electron acceptor behavior [172]. The range of contact angle values of the plasma polymer are well-suited for biological uses, since they enable and promote the adhesion of various cell types [153].

Given their potential application as an antibacterial coating for implants, the cytocompatibility of PP-PSMs was examined for several types of mammalian cells. A study that tested the biocompatibility of coatings fabricated from various oils (e.g., limonene, tea tree, lavender, and eucalyptus) at different deposition powers showed minimum toxic effects. After being implanted in mice for 3 days, 14 days, and 28 days, all PP-PSM films were demonstrated to be biocompatible. While in most cases, these coatings did not produce an unwanted host or material response, in a number of cases mice sinus formation was observed, however, it was deemed not significant [173]. The biocompatibility of polymer films is a property that should be addressed carefully for protective coatings in medical applications, in particular for implantable devices, since the film surface directly interfaces with various bio-components including blood, proteins, cells, and tissue growth. Hence, non-biocompatible coatings may lead to failure, toxic responses, abnormal cell/tissue responses, and device degradation.

It should be noted that PP-PSMs have shown some limitations. For example, they are generally insoluble in organic solvents owing to their high degree of cross-linking. This feature, in particular, greatly complicates the characterization of the polymers [131]. It was also observed that PP-PSMs are highly susceptible to changes brought by the chemical composition of the medium (e.g., the aqueous solution and body fluid) that may affect their operation in some applications [174].

Essential oils variations:

Generally, it is accepted that the chemical composition of essential oils varies by plant health, growth stage, climate, edaphic factors, and harvest time. On the other hand, the degradation kinetics of these oils, due to external factors (e.g., temperature, light, and atmospheric oxygen exposure, presence of impurities), should be thoroughly taken into account [175]. For example, pure cinnamaldehyde was reported to decompose to benzaldehyde at temperatures approaching 60 °C. But, once it combined with eugenol or cinnamon leaf oil, cinnamaldehyde remained stable at 200 °C [176]. The molecular structures of natural oils have a substantial effect on the degree of degradation. Compounds rich in allylic hydrogen atoms could be potential targets for autoxidation, where hydrogen atom abstraction is giving rise to resonance-stabilized radicals, which are highly preferable due to their lower activation energy [177]. Furthermore, essential oil components are generally known to easily convert into each other (through processes such as isomerization, oxidation, cyclization, or dehydrogenation reactions), because of their structural relationship within the same chemical group [175,177].

It is important to mention that several essential oils (e.g., tea tree, lavender, and terpenene-4-ol) have shown some irritation and allergies in users (via inhalation or direct contact) [178–180]. The allergic reactions typically arise from certain components such as benzyl alcohol, cinnamyl alcohol, iso-eugenol, eugenol, hydroxycitronellal, geraniol, and various others constituents [181–183]. However, sensitive symptoms due to essential oils can range from relatively minor incidences of irritation and sensitization, to contact dermatitis and the most serious anaphylactic reaction, thus, should be well considered [184,185].

4. Challenges

In the scientific and manufacturing field, replication or reproduction of consistent systematic results is the key to success [186]. A major issue of plasma techniques is the constancy of the result, particularly across different plasma systems, due to differences in processing parameters (e.g., power, pressure, temperature, flow rate, and tube geometry). For example, changes in the design of plasma equipment can affect the flow dynamics of vapors through the system, and the profile of the plasma discharge zone, which could potentially alter the nature, homogeneity, and density of the gas phase species inside the reactor. Indeed, this problem becomes more obvious during the fabrication of functional coatings from PSMs, where retention of certain chemical moieties is essential [187]. To minimize the variation of films produced across different plasma systems, a scaling factor route can be applied that takes into account both the actual energy consumed in the active plasma field, and the differences in the geometry of the utilized reactors [188].

Another concern comes from the varying properties of the renewable precursor. It is well documented that essential oil composition is very complex and depends on multiple interacting factors. In addition, De Masi et al. (2006) reported that the chemical compounds of essential oils were found to be extremely variable in various cultivars/genotypes of the same plant species and they these differences were not necessarily correlated with genetic relationships [189]. Additionally, the oil quality and biological activity can be affected storage conditions, e.g., temperature [190]. The potential to obtain biopolymer films with consistent properties regardless of the base material source, method, or time of harvest is important for successful integration into industrialized processes.

As mentioned previously, typical plasma polymerization of PSMs (continuous mode) yields the fragmentation of large quantities of precursor molecules. The random recombination of fragments, radicals, and atoms renders the chemical structure and configuration completely irregular. In fact, the density of desired functional groups remain relatively insufficient even if the used fabrication power is low. Pulsed-plasma polymerization can address this issue. This technique offers a sequence of on-periods (a few μs-long periods during which fragmentation takes place) and off-periods (μs to

Polymers **2018**, *10*, 515

ms-long periods during which recombination and polymerization occurs), where the resultant polymer should consist of more chemically regular structures than those of the continuous mode [191]. The idea is to further reduce the degree of dissociation/fragmentation of the precursor molecules, and hence the off-period reactions contributing more non-fragmented functionalities into the formed polymer. To date, the pulsed-plasma polymerization has not been used for the synthesis of antibacterial surfaces from PSMs. We highly encourage researchers to explore and expand the usage of the pulsed-plasma method, where the optimization of the desired functionality will essentially include the increasing/decreasing the off-period in pulsed polymerization.

5. Conclusions

A better understanding of the way to preserve/retain the bioactivity of essential oils within a thin film is critical for the development of a wide range of bactericidal coatings suitable for medical devices. The aforementioned polymer materials that were derived from renewable resources present a promising approach toward producing antimicrobial and biocompatible materials and tissue contact coatings. However, information on the long-term performance of plasma polymerized PSMs thin films requires further exploration. Also, although a small number of systematic studies showed promising antimicrobial activity using encapsulating essential oils, further research in this direction is warranted.

Funding: This research received no external funding.

Acknowledgments: Ahmed Al-Jumaili acknowledges the post graduate scholarship offered by the Ministry of Higher Education and Scientific Research, Iraq, and is grateful to JCUPRS for the financial support.

Conflicts of Interest: The authors declare no conflicts of interest.

References

1. Sanborn, L.W.R. The relation of surface contamination to the transmission of disease. *Am. J. Public Health Nations Health* **1963**, *53*, 1278–1283. [CrossRef] [PubMed]
2. Dancer, S.J. Importance of the environment in meticillin-resistant *Staphylococcus aureus* acquisition: The case for hospital cleaning. *Lancet Infect. Dis.* **2008**, *8*, 101–113. [CrossRef]
3. Knetsch, M.L.W.; Koole, L.H. New strategies in the development of antimicrobial coatings: The example of increasing usage of silver and silver nanoparticles. *Polymers* **2011**, *3*, 340–366. [CrossRef]
4. Wu, S.; Liu, X.; Yeung, A.; Yeung, K.W.; Kao, R.Y.; Wu, G.; Hu, T.; Xu, Z.; Chu, P.K. Plasma-modified biomaterials for self-antimicrobial applications. *ACS Appl. Mater. Interfaces* **2011**, *3*, 2851–2860. [CrossRef] [PubMed]
5. Edmiston, C.E., Jr.; McBain, A.J.; Roberts, C.; Leaper, D. Clinical and microbiological aspects of biofilm-associated surgical site infections. In *Biofilm-Based Healthcare-Associated Infections*; Springer: International Publishing: Cham, Switzerland, 2015; pp. 47–67.
6. Veerachamy, S.; Yarlagadda, T.; Manivasagam, G.; Yarlagadda, P.K. Bacterial Adherence and Biofilm Formation on Medical Implants: A Review. *Proc. Inst. Mech. Eng. Part H J. Eng. Med.* **2014**, *228*, 1083–1099. [CrossRef] [PubMed]
7. Batoni, G.; Maisetta, G.; Esin, S. Antimicrobial peptides and their interaction with biofilms of medically relevant bacteria. *Biochim. Biophys. Acta Biomembr.* **2016**, *1858*, 1044–1060. [CrossRef] [PubMed]
8. Joseph, R.; Naugolny, A.; Feldman, M.; Herzog, I.M.; Fridman, M.; Cohen, Y. Cationic pillararenes potently inhibit biofilm formation without affecting bacterial growth and viability. *J. Am. Chem. Soc.* **2016**, *138*, 754–757. [CrossRef] [PubMed]
9. Mingeot-Leclercq, M.-P.; Décout, J.-L. Bacterial lipid membranes as promising targets to fight antimicrobial resistance, molecular foundations and illustration through the renewal of aminoglycoside antibiotics and emergence of amphiphilic aminoglycosides. *MedChemComm* **2016**, *7*, 586–611. [CrossRef]
10. Chopra, I.; Roberts, M. Tetracycline antibiotics: Mode of action, applications, molecular biology, and epidemiology of bacterial resistance. *Microbiol. Mol. Biol. Rev.* **2001**, *65*, 232–260. [CrossRef] [PubMed]
11. Woon, S.-A.; Fisher, D. Antimicrobial agents–optimising the ecological balance. *BMC Med.* **2016**, *14*, 114. [CrossRef] [PubMed]

12. Mathur, S.; Singh, R. Antibiotic resistance in food lactic acid bacteria—A review. *Int. J. Food Microbiol.* **2005**, *105*, 281–295. [CrossRef] [PubMed]

13. Willers, C.; Wentzel, J.F.; Plessis, L.H.d.; Gouws, C.; Hamman, J.H. Efflux as a mechanism of antimicrobial drug resistance in clinical relevant microorganisms: The role of efflux inhibitors. *Expert Opin. Ther. Targets* **2017**, *21*, 23–36. [CrossRef] [PubMed]

14. Russo, A.; Concia, E.; Cristini, F.; de Rosa, F.G.; Esposito, S.; Menichetti, F.; Petrosillo, N.; Tumbarello, M.; Venditti, M.; Viale, P.; et al. Current and future trends in antibiotic therapy of acute bacterial skin and skin-structure infections. *Clin. Microbiol. Infect.* **2016**, *22*, S27–S36. [CrossRef]

15. Hasan, J.; Crawford, R.J.; Ivanova, E.P. Antibacterial surfaces: The quest for a new generation of biomaterials. *Trends Biotechnol.* **2013**, *31*, 295–304. [CrossRef] [PubMed]

16. O'Bryan, C.A.; Pendleton, S.J.; Crandall, P.G.; Ricke, S.C. Potential of plant essential oils and their components in animal agriculture–in vitro studies on antibacterial mode of action. *Front. Vet. Sci.* **2015**, *2*, 35. [CrossRef] [PubMed]

17. Nazzaro, F.; Fratianni, F.; De Martino, L.; Coppola, R.; De Feo, V. Effect of essential oils on pathogenic bacteria. *Pharmaceuticals* **2013**, *6*, 1451–1474. [CrossRef] [PubMed]

18. Murbach Teles Andrade, B.F.; Nunes Barbosa, L.; da Silva Probst, I.; Fernandes Júnior, A. Antimicrobial activity of essential oils. *J. Essent. Oil Res.* **2014**, *26*, 34–40. [CrossRef]

19. Altenstetter, C. Global and local dynamics: The regulation of medical technologies in the European Union, Japan and the United States. Presented to panel 6E Context and Regulatory Design, Third Biennial Conference 'Regulation in the Age of Crisis', Dublin, Ireland, 17–19 June 2010.

20. Méndez-Vilas, A.; Díaz, J. *Microscopy: Science, Technology, Applications and Education*; Formatex Research Center: Badajoz, Spain, 2010.

21. Hetrick, E.M.; Schoenfisch, M.H. Reducing implant-related infections: Active release strategies. *Chem. Soc. Rev.* **2006**, *35*, 780–789. [CrossRef] [PubMed]

22. Chandki, R.; Banthia, P.; Banthia, R. Biofilms: A microbial home. *J. Indian Soc. Periodontol.* **2011**, *15*, 111–114. [PubMed]

23. De la Fuente-Núñez, C.; Reffuveille, F.; Fernández, L.; Hancock, R.E.W. Bacterial biofilm development as a multicellular adaptation: Antibiotic resistance and new therapeutic strategies. *Curr. Opin. Microbiol.* **2013**, *16*, 580–589. [CrossRef] [PubMed]

24. Chen, Y.; Harapanahalli, A.K.; Busscher, H.J.; Norde, W.; van der Mei, H.C. Nanoscale cell wall deformation impacts long-range bacterial adhesion forces on surfaces. *Appl. Environ. Microbiol.* **2014**, *80*, 637–643. [CrossRef] [PubMed]

25. Scheuerman, T.R.; Camper, A.K.; Hamilton, M.A. Effects of substratum topography on bacterial adhesion. *J. Colloid Interface Sci.* **1998**, *208*, 23–33. [CrossRef] [PubMed]

26. Donlan, R.M. Biofilms: Microbial life on surfaces. *Emerg. Infect. Dis.* **2002**, *8*, 881–890. [CrossRef] [PubMed]

27. Crawford, R.J.; Ivanova, E.P. *Superhydrophobic Surfaces*; Elsevier: New York, NY, USA, 2015.

28. Boland, T.; Latour, R.A.; Stutzenberger, F.J. Molecular basis of bacterial adhesion. In *Handbook of Bacterial Adhesion*; Springer: Berlin/Heidelberg, Germany, 2000; pp. 29–41.

29. Basak, S.; Rajurkar, M.N.; Attal, R.O.; Mallick, S.K. Biofilms: A challenge to medical fraternity in infection control. *Infect. Control* **2013**, *57*. [CrossRef]

30. Anselme, K.; Davidson, P.; Popa, A.M.; Giazzon, M.; Liley, M.; Ploux, L. The interaction of cells and bacteria with surfaces structured at the nanometre scale. *Acta Biomater.* **2010**, *6*, 3824–3846. [CrossRef] [PubMed]

31. Epstein, E.A.; Reizian, M.A.; Chapman, M.R. Spatial clustering of the curlin secretion lipoprotein requires curli fiber assembly. *J. Bacteriol.* **2009**, *191*, 608–615. [CrossRef] [PubMed]

32. Fletcher, M.; Savage, D.C. *Bacterial Adhesion: Mechanisms and Physiological Significance*; Springer Science & Business Media: Berlin/Heidelberg, Germany, 2013.

33. Moss, J.A.; Nocker, A.; Lepo, J.E.; Snyder, R.A. Stability and change in estuarine biofilm bacterial community diversity. *Appl. Environ. Microbiol.* **2006**, *72*, 5679–5688. [CrossRef] [PubMed]

34. Flemming, H.-C.; Wingender, J. The biofilm matrix. *Nat. Rev. Microbiol.* **2010**, *8*, 623–633. [CrossRef] [PubMed]

35. Douglas, L.J. *Candida* biofilms and their role in infection. *Trends Microbiol.* **2003**, *11*, 30–36. [CrossRef]

36. Ma, L.; Conover, M.; Lu, H.; Parsek, M.R.; Bayles, K.; Wozniak, D.J. Assembly and development of the *Pseudomonas aeruginosa* biofilm matrix. *PLoS Pathog.* **2009**, *5*, e1000354. [CrossRef] [PubMed]

37. Karimi, A.; Karig, D.; Kumar, A.; Ardekani, A. Interplay of physical mechanisms and biofilm processes: Review of microfluidic methods. *Lab Chip* **2015**, *15*, 23–42. [CrossRef] [PubMed]
38. Sharafat, I.; Saeed, D.K.; Yasmin, S.; Imran, A.; Zafar, Z.; Hameed, A.; Ali, N. Interactive effect of trivalent iron on activated sludge digestion and biofilm structure in attached growth reactor of waste tire rubber. *Environ. Technol.* **2017**, 1–37. [CrossRef] [PubMed]
39. Hassan, A.; Usman, J.; Kaleem, F.; Omair, M.; Khalid, A.; Iqbal, M. Evaluation of different detection methods of biofilm formation in the clinical isolates. *Braz. J. Infect. Dis.* **2011**, *15*, 305–311. [CrossRef]
40. Nguyen, S.H.; Webb, H.K.; Crawford, R.J.; Ivanova, E.P. Natural antibacterial surfaces. In *Antibacterial Surfaces*; Springer: Berlin/Heidelberg, Germany, 2015; pp. 9–26.
41. Jackson, D.W.; Suzuki, K.; Oakford, L.; Simecka, J.W.; Hart, M.E.; Romeo, T. Biofilm formation and dispersal under the influence of the global regulator csra of *Escherichia coli*. *J. Bacteriol.* **2002**, *184*, 290–301. [CrossRef] [PubMed]
42. Socransky, S.S.; Haffajee, A.D. Dental biofilms: Difficult therapeutic targets. *Periodontology 2000* **2002**, *28*, 12–55. [CrossRef] [PubMed]
43. Donlan, R.M. Biofilm formation: A clinically relevant microbiological process. *Clin. Infect. Dis.* **2001**, *33*, 1387–1392. [CrossRef] [PubMed]
44. Derlon, N.; Coufort-Saudejaud, C.; Queinnec, I.; Paul, E. Growth limiting conditions and denitrification govern extent and frequency of volume detachment of biofilms. *Chem. Eng. J.* **2013**, *218*, 368–375. [CrossRef]
45. Barraud, N.; Hassett, D.J.; Hwang, S.-H.; Rice, S.A.; Kjelleberg, S.; Webb, J.S. Involvement of nitric oxide in biofilm dispersal of *Pseudomonas aeruginosa*. *J. Bacteriol.* **2006**, *188*, 7344–7353. [CrossRef] [PubMed]
46. Hall-Stoodley, L.; Costerton, J.W.; Stoodley, P. Bacterial biofilms: From the natural environment to infectious diseases. *Nat. Rev. Microbiol.* **2004**, *2*, 95–108. [CrossRef] [PubMed]
47. Alexander, S.-A.; Schiesser, C.H. Heteroorganic molecules and bacterial biofilms: Controlling biodeterioration of cultural heritage. *Org. Chem.* **2017**, 180–222.
48. Fish, K.E.; Osborn, A.M.; Boxall, J. Characterising and understanding the impact of microbial biofilms and the extracellular polymeric substance (EPS) matrix in drinking water distribution systems. *Environ. Sci. Water Res. Technol.* **2016**, *2*, 614–630. [CrossRef]
49. Kaplan, J.Á. Biofilm dispersal: Mechanisms, clinical implications, and potential therapeutic uses. *J. Dent. Res.* **2010**, *89*, 205–218. [CrossRef] [PubMed]
50. Parsek, M.R. Controlling the connections of cells to the biofilm matrix. *J. Bacteriol.* **2016**, *198*, 12–14. [CrossRef] [PubMed]
51. Rasamiravaka, T.; Labtani, Q.; Duez, P.; El Jaziri, M. The formation of biofilms by *Pseudomonas aeruginosa*: A review of the natural and synthetic compounds interfering with control mechanisms. *BioMed Res. Int.* **2015**, *2015*, 1–17. [CrossRef] [PubMed]
52. Bryers, J.D. Medical biofilms. *Biotechnol. Bioeng.* **2008**, *100*, 1–18. [CrossRef] [PubMed]
53. Device-Related Infections. Available online: http://www.infectioncontroltoday.com/articles/2006/11/device-related-infections.aspx (accessed on 3 May 2018).
54. Pradeep, K.S.; Easwer, H.; Maya, N.A. Multiple drug resistant bacterial biofilms on implanted catheters—A reservoir of infection. *J. Assoc. Phys. India* **2013**, *61*, 702–707.
55. Klevens, R.M.; Edwards, J.R.; Gaynes, R.; System, N.N.I.S. The impact of antimicrobial-resistant, health care–associated infections on mortality in the United States. *Clin. Infect. Dis.* **2008**, *47*, 927–930. [CrossRef] [PubMed]
56. Coenye, T.; De Prijck, K.; Nailis, H.; Nelis, H.J. Prevention of *Candida albicans* biofilm formation. *Open Mycol. J.* **2011**, *5*, 9–20. [CrossRef]
57. Deorukhkar, S.C.; Saini, S. Why *Candida* species have emerged as important nosocomial pathogens? *Int. J. Curr. Microbiol. Appl. Sci.* **2016**, *5*, 533–545. [CrossRef]
58. Sanclement, J.A.; Webster, P.; Thomas, J.; Ramadan, H.H. Bacterial biofilms in surgical specimens of patients with chronic rhinosinusitis. *Laryngoscope* **2005**, *115*, 578–582. [CrossRef] [PubMed]
59. Ramage, G.; Martínez, J.P.; López-Ribot, J.L. *Candida* biofilms on implanted biomaterials: A clinically significant problem. *FEMS Yeast Res.* **2006**, *6*, 979–986. [CrossRef] [PubMed]
60. Jabra-Rizk, M.A.; Kong, E.F.; Tsui, C.; Nguyen, M.H.; Clancy, C.J.; Fidel, P.L.; Noverr, M. *Candida albicans* pathogenesis: Fitting within the host-microbe damage response framework. *Infect. Immun.* **2016**, *84*, 2724–2739. [CrossRef] [PubMed]

61. Alhede, M.; Kragh, K.N.; Qvortrup, K.; Allesen-Holm, M.; van Gennip, M.; Christensen, L.D.; Jensen, P.Ø.; Nielsen, A.K.; Parsek, M.; Wozniak, D. Phenotypes of non-attached *Pseudomonas aeruginosa* aggregates resemble surface attached biofilm. *PLoS ONE* **2011**, *6*, e27943. [CrossRef] [PubMed]

62. Wagner, V.E.; Iglewski, B.H. *P. aeruginosa* biofilms in CF infection. *Clin. Rev. Allergy Immunol.* **2008**, *35*, 124–134. [CrossRef] [PubMed]

63. Firenzuoli, F.; Jaitak, V.; Horvath, G.; Bassolé, I.H.N.; Setzer, W.N.; Gori, L. Essential oils: New perspectives in human health and wellness. *Evid.-Based Complement. Altern. Med.* **2014**, *2014*, 1–2. [CrossRef] [PubMed]

64. Wang, W.; Li, N.; Luo, M.; Zu, Y.; Efferth, T. Antibacterial Activity and Anticancer Activity of *Rosmarinus officinalis* L. Essential Oil Compared to That of Its Main Components. *Molecules* **2012**, *17*, 2704–2713. [CrossRef] [PubMed]

65. Inouye, S.; Takizawa, T.; Yamaguchi, H. Antibacterial activity of essential oils and their major constituents against respiratory tract pathogens by gaseous contact. *J. Antimicrob. Chemother.* **2001**, *47*, 565–573. [CrossRef] [PubMed]

66. Stupar, M.; Grbić, M.L.; Džamić, A.; Unković, N.; Ristić, M.; Jelikić, A.; Vukojević, J. Antifungal activity of selected essential oils and biocide benzalkonium chloride against the fungi isolated from cultural heritage objects. *S. Afr. J. Bot.* **2014**, *93*, 118–124. [CrossRef]

67. Rotolo, V.; Barresi, G.; Di Carlo, E.; Giordano, A.; Lombardo, G.; Crimi, E.; Costa, E.; Bruno, M.; Palla, F. Plant extracts as green potential strategies to control the biodeterioration of cultural heritage. *Int. J. Conserv. Sci.* **2016**, *2*, 839–846.

68. Borrego, S.; Valdés, O.; Vivar, I.; Lavin, P.; Guiamet, P.; Battistoni, P.; Gómez de Saravia, S.; Borges, P. Essential oils of plants as biocides against microorganisms isolated from cuban and argentine documentary heritage. *ISRN Microbiol.* **2012**, *2012*, 1–7. [CrossRef] [PubMed]

69. Swamy, M.K.; Akhtar, M.S.; Sinniah, U.R. Antimicrobial properties of plant essential oils against human pathogens and their mode of action: An updated review. *Evid.-Based Complement. Altern. Med.* **2016**, *2016*, 1–21. [CrossRef] [PubMed]

70. Calo, J.R.; Crandall, P.G.; O'Bryan, C.A.; Ricke, S.C. Essential oils as antimicrobials in food systems—A review. *Food Control* **2015**, *54*, 111–119. [CrossRef]

71. Eze, U.A. In vitro antimicrobial activity of essential oils from the lamiaceae and rutaceae plant families against β lactamse-producing clinical isolates of *Moraxella Catarrhalis*. *EC Pharm. Sci.* **2016**, *2*, 325–337.

72. Kateryna, B.; Olha, B.; Igor, L.; Shuyan, X.; Elena, P.I.; Michael, K.; Kostya, O. Plasma-potentiated small molecules—Possible alternative to antibiotics? *Nano Futur.* **2017**, *1*, 025002.

73. Silva, N.; Fernandes Júnior, A. Biological properties of medicinal plants: A review of their antimicrobial activity. *J. Venom. Anim. Toxins Incl. Trop. Dis.* **2010**, *16*, 402–413. [CrossRef]

74. Hammami, I.; Triki, M.A.; Rebai, A. Chemical compositions, antibacterial and antioxidant activities of essential oil and various extracts of *Geranium sanguineum* L. Flowers. *Sch. Res. Libr.* **2011**, *3*, 135–144.

75. Hui, L.; He, L.; Huan, L.; XiaoLan, L.; AiGuo, Z. Chemical composition of *Lavender* essential oil and its antioxidant activity and inhibition against rhinitis-related bacteria. *Afr. J. Microbiol. Res.* **2010**, *4*, 309–313.

76. Glinel, K.; Thebault, P.; Humblot, V.; Pradier, C.-M.; Jouenne, T. Antibacterial surfaces developed from bio-inspired approaches. *Acta Biomater.* **2012**, *8*, 1670–1684. [CrossRef] [PubMed]

77. Hierro, I.; Valero, A.; Perez, P.; Gonzalez, P.; Cabo, M.; Montilla, M.; Navarro, M. Action of different monoterpenic compounds against *Anisakis simplex* s.l. L3 larvae. *Phytomedicine* **2004**, *11*, 77–82. [CrossRef] [PubMed]

78. Bigos, M.; Wasiela, M.; Kalemba, D.; Sienkiewicz, M. Antimicrobial activity of geranium oil against clinical strains of *Staphylococcus aureus*. *Molecules* **2012**, *17*, 10276–10291. [CrossRef] [PubMed]

79. Friedman, M.; Henika, P.R.; Levin, C.E.; Mandrell, R.E. Antibacterial activities of plant essential oils and their components against *Escherichia coli* O157:H7 and *Salmonella enterica* in apple juice. *J. Agric. Food Chem.* **2004**, *52*, 6042–6048. [CrossRef] [PubMed]

80. Tisserand, R.; Young, R. *Essential Oil Safety: A Guide for Health Care Professionals*; Elsevier Health Sciences: New York, NY, USA, 2013.

81. Sun, L.-M.; Zhang, C.-L.; Li, P. Characterization, antibiofilm, and mechanism of action of novel peg-stabilized lipid nanoparticles loaded with terpinen-4-ol. *J. Agric. Food Chem.* **2012**, *60*, 6150–6156. [CrossRef] [PubMed]

82. Prabuseenivasan, S.; Jayakumar, M.; Ignacimuthu, S. In vitro antibacterial activity of some plant essential oils. *BMC Complement. Altern. Med.* **2006**, *6*, 39. [CrossRef] [PubMed]

83. Faid, M.; Bakhy, K.; Anchad, M.; Tantaoui-Elaraki, A. Almond paste: Physicochemical and microbiological characterization and preservation with sorbic acid and cinnamon. *J. Food Prot.* **1995**, *58*, 547–550. [CrossRef]

84. Maruzzella, J.C.; Sicurella, N.A. Antibacterial activity of essential oil vapors. *J. Pharm. Sci.* **1960**, *49*, 692–694. [CrossRef]

85. Maruzzella, J.C.; Chiaramonte, J.S.; Garofalo, M.M. Effects of vapors of aromatic chemicals on fungi. *J. Pharm. Sci.* **1961**, *50*, 665–668. [CrossRef]

86. Akthar, M.S.; Degaga, B.; Azam, T. Antimicrobial activity of essential oils extracted from medicinal plants against the pathogenic microorganisms: A review. *Issues Biol. Sci. Pharm. Res.* **2014**, *2*, 1–7.

87. Burt, S. Essential oils: Their antibacterial properties and potential applications in foods—A review. *Int. J. Food Microbiol.* **2004**, *94*, 223–253. [CrossRef] [PubMed]

88. Devi, K.P.; Nisha, S.A.; Sakthivel, R.; Pandian, S.K. Eugenol (an essential oil of clove) acts as an antibacterial agent against salmonella typhi by disrupting the cellular membrane. *J. Ethnopharmacol.* **2010**, *130*, 107–115. [CrossRef] [PubMed]

89. Ultee, A.; Kets, E.; Smid, E. Mechanisms of action of carvacrol on the food-borne pathogen *Bacillus cereus*. *Appl. Environ. Microbiol.* **1999**, *65*, 4606–4610. [PubMed]

90. Ultee, A.; Smid, E.J. Influence of Carvacrol on growth and toxin production by *Bacillus cereus*. *Int. J. Food Microbiol.* **2001**, *64*, 373–378. [CrossRef]

91. Raut, J.S.; Karuppayil, S.M. A status review on the medicinal properties of essential oils. *Ind. Crops Prod.* **2014**, *62*, 250–264. [CrossRef]

92. Gustafson, J.; Liew, Y.; Chew, S.; Markham, J.; Bell, H.; Wyllie, S.; Warmington, J. Effects of tea tree oil on *Escherichia coli*. *Lett. Appl. Microbiol.* **1998**, *26*, 194–198. [CrossRef] [PubMed]

93. Szabó, M.Á.; Varga, G.Z.; Hohmann, J.; Schelz, Z.; Szegedi, E.; Amaral, L.; Molnár, J. Inhibition of quorum-sensing signals by essential oils. *Phytother. Res.* **2010**, *24*, 782–786. [CrossRef] [PubMed]

94. Christaki, E.; Bonos, E.; Giannenas, I.; Florou-Paneri, P. Aromatic plants as a source of bioactive compounds. *Agriculture* **2012**, *2*, 228–243. [CrossRef]

95. Ben Arfa, A.; Combes, S.; Preziosi-Belloy, L.; Gontard, N.; Chalier, P. Antimicrobial activity of *Carvacrol* related to its chemical structure. *Lett. Appl. Microbiol.* **2006**, *43*, 149–154. [CrossRef] [PubMed]

96. Bassolé, I.H.N.; Juliani, H.R. Essential oils in combination and their antimicrobial properties. *Molecules* **2012**, *17*, 3989–4006. [CrossRef] [PubMed]

97. Trombetta, D.; Castelli, F.; Sarpietro, M.G.; Venuti, V.; Cristani, M.; Daniele, C.; Saija, A.; Mazzanti, G.; Bisignano, G. Mechanisms of antibacterial action of three monoterpenes. *Antimicrob. Agents Chemother.* **2005**, *49*, 2474–2478. [CrossRef] [PubMed]

98. Ćavar, S.; Maksimović, M.; Vidic, D.; Parić, A. Chemical composition and antioxidant and antimicrobial activity of essential oil of *Artemisia annua* L. from Bosnia. *Ind. Crops Prod.* **2012**, *37*, 479–485. [CrossRef]

99. Stoica, P.; Chifiriuc, M.C.; Râpă, M.; Bleotu, C.; Lungu, L.; Vlad, G.; Grigore, R.; Berteşteanu, Ş.; Stavropoulou, E.; Lazăr, V. Fabrication, characterization and bioevaluation of novel antimicrobial composites based on polycaprolactone, chitosan and essential oils. *Rom. Biotechnol. Lett.* **2015**, *20*, 10521–10535.

100. Balat, M.; Balat, M. Political, economic and environmental impacts of biomass-based hydrogen. *Int. J. Hydrog. Energy* **2009**, *34*, 3589–3603. [CrossRef]

101. Balzani, V.; Armaroli, N. *Energy for a Sustainable World: From the Oil Age to a Sun-Powered Future*; John Wiley & Sons: Hoboken, NJ, USA, 2010.

102. Winnacker, M.; Rieger, B. Recent progress in sustainable polymers obtained from cyclic terpenes: Synthesis, properties, and application potential. *ChemSusChem* **2015**, *8*, 2455–2471. [CrossRef] [PubMed]

103. Saggiorato, A.G.; Gaio, I.; Treichel, H.; de Oliveira, D.; Cichoski, A.J.; Cansian, R.L. Antifungal activity of basil essential oil (*Ocimum basilicum* L.): Evaluation in vitro and on an italian-type sausage surface. *Food Bioprocess Technol.* **2012**, *5*, 378–384. [CrossRef]

104. Teramoto, Y.; Nishio, Y. Cellulose diacetate-*graft*-poly(lactic acid)s: Synthesis of wide-ranging compositions and their thermal and mechanical properties. *Polymer* **2003**, *44*, 2701–2709. [CrossRef]

105. Arrieta, M.P.; López, J.; Hernández, A.; Rayón, E. Ternary PLA–PHB–Limonene blends intended for biodegradable food packaging applications. *Eur. Polym. J.* **2014**, *50*, 255–270. [CrossRef]

106. Robertson, M.L.; Paxton, J.M.; Hillmyer, M.A. Tough blends of polylactide and castor oil. *ACS Appl. Mater. Interfaces* **2011**, *3*, 3402–3410. [CrossRef] [PubMed]

107. Morschbacker, A. Bio-ethanol based ethylene. *Polym. Rev.* **2009**, *49*, 79–84. [CrossRef]

108. Mathers, R.T.; Damodaran, K.; Rendos, M.G.; Lavrich, M.S. Functional hyperbranched polymers using ring-opening metathesis polymerization of dicyclopentadiene with monoterpenes. *Macromolecules* **2009**, *42*, 1512–1518. [CrossRef]

109. Quilter, H.C.; Hutchby, M.; Davidson, M.G.; Jones, M.D. Polymerisation of a terpene-derived lactone: A bio-based alternative to ε-caprolactone. *Polym. Chem.* **2017**, *8*, 833–837. [CrossRef]

110. Gandini, A. The irruption of polymers from renewable resources on the scene of macromolecular science and technology. *Green Chem.* **2011**, *13*, 1061–1083. [CrossRef]

111. Wilbon, P.A.; Chu, F.; Tang, C. Progress in renewable polymers from natural terpenes, terpenoids, and rosin. *Macromol. Rapid Commun.* **2013**, *34*, 8–37. [CrossRef] [PubMed]

112. Shit, S.C.; Shah, P.M. Edible polymers: Challenges and opportunities. *J. Polym.* **2014**, *2014*, 1–13. [CrossRef]

113. Modjinou, T.; Versace, D.-L.; Abbad-Andallousi, S.; Bousserrhine, N.; Babinot, J.; Langlois, V.; Renard, E. Antibacterial networks based on isosorbide and linalool by photoinitiated process. *ACS Sustain. Chem. Eng.* **2015**, *3*, 1094–1100. [CrossRef]

114. Chen, Y.; Wilbon, P.A.; Chen, Y.P.; Zhou, J.; Nagarkatti, M.; Wang, C.; Chu, F.; Decho, A.W.; Tang, C. Amphipathic antibacterial agents using cationic methacrylic polymers with natural rosin as pendant group. *RSC Adv.* **2012**, *2*, 10275–10282. [CrossRef]

115. Wen, P.; Zhu, D.-H.; Feng, K.; Liu, F.-J.; Lou, W.-Y.; Li, N.; Zong, M.-H.; Wu, H. Fabrication of electrospun polylactic acid nanofilm incorporating cinnamon essential oil/β-cyclodextrin inclusion complex for antimicrobial packaging. *Food Chem.* **2016**, *196*, 996–1004. [CrossRef] [PubMed]

116. Avila-Sosa, R.; Palou, E.; Jiménez Munguía, M.T.; Nevárez-Moorillón, G.V.; Navarro Cruz, A.R.; López-Malo, A. Antifungal activity by vapor contact of essential oils added to amaranth, chitosan, or starch edible films. *Int. J. Food Microbiol.* **2012**, *153*, 66–72. [CrossRef] [PubMed]

117. Jouki, M.; Yazdi, F.T.; Mortazavi, S.A.; Koocheki, A. Quince seed mucilage films incorporated with oregano essential oil: Physical, thermal, barrier, antioxidant and antibacterial properties. *Food Hydrocoll.* **2014**, *36*, 9–19. [CrossRef]

118. Ruiz-Navajas, Y.; Viuda-Martos, M.; Sendra, E.; Perez-Alvarez, J.A.; Fernández-López, J. In vitro antibacterial and antioxidant properties of chitosan edible films incorporated with thymus moroderi or thymus piperella essential oils. *Food Control* **2013**, *30*, 386–392. [CrossRef]

119. Quirós-Sauceda, A.E.; Ayala-Zavala, J.F.; Olivas, G.I.; González-Aguilar, G.A. Edible coatings as encapsulating matrices for bioactive compounds: A review. *J. Food Sci. Technol.* **2014**, *51*, 1674–1685. [CrossRef] [PubMed]

120. Donsì, F.; Annunziata, M.; Sessa, M.; Ferrari, G. Nanoencapsulation of essential oils to enhance their antimicrobial activity in foods. *LWT Food Sci. Technol.* **2011**, *44*, 1908–1914. [CrossRef]

121. Sessa, M.; Ferrari, G.; Donsì, F. Novel edible coating containing essential oil nanoemulsions to prolong the shelf life of vegetable products. *Chem. Eng. Trans.* **2015**, *43*, 55–60.

122. Peng, Y.; Li, Y. Combined effects of two kinds of essential oils on physical, mechanical and structural properties of chitosan films. *Food Hydrocoll.* **2014**, *36*, 287–293. [CrossRef]

123. Mohammadi, A.; Hashemi, M.; Hosseini, S.M. Nanoencapsulation of zataria multiflora essential oil preparation and characterization with enhanced antifungal activity for controlling *Botrytis cinerea*, the causal agent of gray mould disease. *Innov. Food Sci. Emerg. Technol.* **2015**, *28*, 73–80. [CrossRef]

124. Alboofetileh, M.; Rezaei, M.; Hosseini, H.; Abdollahi, M. Antimicrobial activity of alginate/clay nanocomposite films enriched with essential oils against three common foodborne pathogens. *Food Control* **2014**, *36*, 1–7. [CrossRef]

125. Severino, R.; Ferrari, G.; Vu, K.D.; Donsì, F.; Salmieri, S.; Lacroix, M. Antimicrobial effects of modified chitosan based coating containing nanoemulsion of essential oils, modified atmosphere packaging and gamma irradiation against *Escherichia coli* O157:H7 and *Salmonella* typhimurium on green beans. *Food Control* **2015**, *50*, 215–222. [CrossRef]

126. Goñi, P.; López, P.; Sánchez, C.; Gómez-Lus, R.; Becerril, R.; Nerín, C. Antimicrobial activity in the vapour phase of a combination of cinnamon and clove essential oils. *Food Chem.* **2009**, *116*, 982–989. [CrossRef]

127. López, P.; Sánchez, C.; Batlle, R.; Nerín, C. Solid- and vapor-phase antimicrobial activities of six essential oils: Susceptibility of selected foodborne bacterial and fungal strains. *J. Agric. Food Chem.* **2005**, *53*, 6939–6946. [CrossRef] [PubMed]

128. Jacob, M.V.; Easton, C.D.; Anderson, L.J.; Bazaka, K. RF plasma polymerised thin films from natural resources. *Int. J. Mod. Phys. Conf. Ser.* **2014**, *32*, 1–10. [CrossRef]

129. Kumar, A.; Grant, D.S.; Bazaka, K.; Jacob, M.V. Tailoring terpenoid plasma polymer properties by controlling the substrate temperature during pecvd. *J. Appl. Polym. Sci.* **2018**, *135*, 45771. [CrossRef]

130. Lopez, G.P.; Ratner, B.D. Substrate temperature effects on film chemistry in plasma deposition of organics. 1. Nonpolymerizable precursors. *Langmuir* **1991**, *7*, 766–773. [CrossRef]

131. Shi, F.F. Recent advances in polymer thin films prepared by plasma polymerization synthesis, structural characterization, properties and applications. *Surf. Coat. Technol.* **1996**, *82*, 1–15. [CrossRef]

132. Bazaka, K.; Jacob, M.V.; Crawford, R.J.; Ivanova, E.P. Efficient surface modification of biomaterial to prevent biofilm formation and the attachment of microorganisms. *Appl. Microbiol. Biotechnol.* **2012**, *95*, 299–311. [CrossRef] [PubMed]

133. Bazaka, K.; Jacob, M.; Chrzanowski, W.; Ostrikov, K. Anti-bacterial surfaces: Natural agents, mechanisms of action, and plasma surface modification. *RSC Adv.* **2015**, *5*, 48739–48759. [CrossRef]

134. Brady, A.; Loughlin, R.; Gilpin, D.; Kearney, P.; Tunney, M. In vitro activity of tea-tree oil against clinical skin isolates of meticillin-resistant and-sensitive *Staphylococcus aureus* and coagulase-negative staphylococci growing planktonically and as biofilms. *J. Med. Microbiol.* **2006**, *55*, 1375–1380. [CrossRef] [PubMed]

135. Mondello, F.; De Bernardis, F.; Girolamo, A.; Salvatore, G.; Cassone, A. In vitro and in vivo activity of tea tree oil against azole-susceptible and-resistant human pathogenic yeasts. *J. Antimicrob. Chemother.* **2003**, *51*, 1223–1229. [CrossRef] [PubMed]

136. Carson, C.F.; Mee, B.J.; Riley, T.V. Mechanism of action of *Melaleuca alternifolia* (tea tree) oil on *Staphylococcus aureus* determined by time-kill, lysis, leakage, and salt tolerance assays and electron microscopy. *Antimicrob. Agents Chemother.* **2002**, *46*, 1914–1920. [CrossRef] [PubMed]

137. Bazaka, K.; Jacob, M.V.; Truong, V.K.; Crawford, R.J.; Ivanova, E.P. The effect of polyterpenol thin film surfaces on bacterial viability and adhesion. *Polymers* **2011**, *3*, 388–404. [CrossRef]

138. Bazaka, K.; Jacob, M.; Truong, V.; Wang, F.; Pushpamali, W.; Wang, J.; Ellis, A.; Berndt, C.; Crawford, R.; Ivanova, E. Effect of plasma-enhanced chemical vapour deposition on the retention of antibacterial activity of terpinen-4-ol. *Biomacromolecules* **2010**, *11*, 2016. [CrossRef] [PubMed]

139. Bazaka, K.; Jacob, M.V.; Ivanova, E.P. A study of a retention of antimicrobial activity by plasma polymerized terpinen-4-ol thin films. In *Materials Science Forum*; Trans Tech Publ: Zürich, Switzerland, 2010; pp. 2261–2264.

140. Bayram, O.; Simsek, O. Investigation of the effect of RF energy on optical, morphological, chemical and antibacterial properties of polyterpenol thin films obtained by RF-PECVD technique. *J. Mater. Sci. Mater. Electron.* **2018**, *29*, 6586–6593. [CrossRef]

141. Aggarwal, K.K.; Khanuja, S.P.S.; Ahmad, A.; Santha Kumar, T.R.; Gupta, V.K.; Kumar, S. Antimicrobial activity profiles of the two enantiomers of limonene and carvone isolated from the oils of *Mentha spicata* and *Anethum sowa*. *Flavour Fragr. J.* **2002**, *17*, 59–63. [CrossRef]

142. McGeady, P.; Wansley, D.L.; Logan, D.A. Carvone and perillaldehyde interfere with the serum-induced formation of filamentous structures in *Candida albicans* at substantially lower concentrations than those causing significant inhibition of growth. *J. Nat. Prod.* **2002**, *65*, 953–955. [CrossRef] [PubMed]

143. De Carvalho, C.C.C.R.; da Fonseca, M.M.R. Carvone: Why and how should one bother to produce this terpene. *Food Chem.* **2006**, *95*, 413–422. [CrossRef]

144. Chan, Y.W.; Siow, K.S.; Ng, P.Y.; Gires, U.; Majlis, B.Y. Plasma polymerized carvone as an antibacterial and biocompatible coating. *Mater. Sci. Eng. C* **2016**, *68*, 861–871. [CrossRef] [PubMed]

145. Juergens, U.; Dethlefsen, U.; Steinkamp, G.; Gillissen, A.; Repges, R.; Vetter, H. Anti-inflammatory activity of 1,8-cineol (Eucalyptol) in bronchial asthma: A double-blind placebo-controlled trial. *Respir. Med.* **2003**, *97*, 250–256. [CrossRef] [PubMed]

146. Dalleau, S.; Cateau, E.; Bergès, T.; Berjeaud, J.-M.; Imbert, C. In vitro activity of terpenes against *Candida* biofilms. *Int. J. Antimicrob. Agents* **2008**, *31*, 572–576. [CrossRef] [PubMed]

147. Serafino, A.; Vallebona, P.S.; Andreola, F.; Zonfrillo, M.; Mercuri, L.; Federici, M.; Rasi, G.; Garaci, E.; Pierimarchi, P. Stimulatory effect of eucalyptus essential oil on innate cell-mediated immune response. *BMC Immunol.* **2008**, *9*, 17. [CrossRef] [PubMed]

148. Juergens, U.R.; Engelen, T.; Racké, K.; Stöber, M.; Gillissen, A.; Vetter, H. Inhibitory activity of 1,8-cineol (eucalyptol) on cytokine production in cultured human lymphocytes and monocytes. *Pulm. Pharmacol. Ther.* **2004**, *17*, 281–287. [CrossRef] [PubMed]

149. Pegalajar-Jurado, A.; Easton, C.D.; Styan, K.E.; McArthur, S.L. Antibacterial activity studies of plasma polymerised cineole films. *J. Mater. Chem. B* **2014**, *2*, 4993–5002. [CrossRef]
150. Mann, M.N.; Fisher, E.R. Investigation of antibacterial 1,8-cineole-derived thin films formed via plasma-enhanced chemical vapor deposition. *ACS Appl. Mater. Interfaces* **2017**, *9*, 36548–36560. [CrossRef] [PubMed]
151. Jalali-Heravi, M.; Zekavat, B.; Sereshti, H. Characterization of essential oil components of iranian geranium oil using gas chromatography–mass spectrometry combined with chemometric resolution techniques. *J. Chromatogr. A* **2006**, *1114*, 154–163. [CrossRef] [PubMed]
152. Edwards-Jones, V.; Buck, R.; Shawcross, S.G.; Dawson, M.M.; Dunn, K. The effect of essential oils on methicillin-resistant *Staphylococcus aureus* using a dressing model. *Burns* **2004**, *30*, 772–777. [CrossRef] [PubMed]
153. Al-Jumaili, A.; Bazaka, K.; Jacob, M.V. Retention of antibacterial activity in geranium plasma polymer thin films. *Nanomaterials* **2017**, *7*, 270. [CrossRef] [PubMed]
154. Al-Jumaili, A.; Alancherry, S.; Bazaka, K.; Jacob, M.V. The electrical properties of plasma-deposited thin films derived from *Pelargonium graveolens*. *Electronics* **2017**, *6*, 86. [CrossRef]
155. Tegoulia, V.A.; Cooper, S.L. *Staphylococcus aureus* adhesion to self-assembled monolayers: Effect of surface chemistry and fibrinogen presence. *Colloids Surf. B Biointerfaces* **2002**, *24*, 217–228. [CrossRef]
156. Bazaka, K.; Jacob, M.V. Nanotribological and nanomechanical properties of plasma-polymerized polyterpenol thin films. *J. Mater. Res.* **2011**, *26*, 2952–2961. [CrossRef]
157. Bae, I.; Jung, C.; Cho, S.; Song, Y.; Boo, J. Characterization of organic polymer thin films deposited using the PECVD method. *J. Korean Phys. Soc.* **2007**, *50*, 1854–1857. [CrossRef]
158. Bazaka, K.; Jacob, M.V. Effects of iodine doping on optoelectronic and chemical properties of polyterpenol thin films. *Nanomaterials* **2017**, *7*, 11. [CrossRef] [PubMed]
159. Vasilev, K.; Griesser, S.S.; Griesser, H.J. Antibacterial surfaces and coatings produced by plasma techniques. *Plasma Process. Polym.* **2011**, *8*, 1010–1023. [CrossRef]
160. Jampala, S.N.; Sarmadi, M.; Somers, E.; Wong, A.; Denes, F. Plasma-enhanced synthesis of bactericidal quaternary ammonium thin layers on stainless steel and cellulose surfaces. *Langmuir* **2008**, *24*, 8583–8591. [CrossRef] [PubMed]
161. Barnes, L.; Cooper, I. *Biomaterials and Medical Device-Associated Infections*; Elsevier: New York, NY, USA, 2014.
162. Lischer, S.; Körner, E.; Balazs, D.J.; Shen, D.; Wick, P.; Grieder, K.; Haas, D.; Heuberger, M.; Hegemann, D. Antibacterial burst-release from minimal Ag-containing plasma polymer coatings. *J. R. Soc. Interface* **2011**, *8*, 1019–1030. [CrossRef] [PubMed]
163. Bayram, O. Determination of the optical and chemical properties of aniline doped plasma polymerized cineole thin films synthesized at various RF powers. *J. Mater. Sci. Mater. Electron.* **2018**, *29*, 8564–8570. [CrossRef]
164. Bazaka, K.; Ahmad, J.; Oelgemöller, M.; Uddin, A.; Jacob, M.V. Photostability of plasma polymerized γ-terpinene thin films for encapsulation of OPV. *Sci. Rep.* **2017**, *7*, 45599. [CrossRef] [PubMed]
165. Bazaka, K.; Jacob, M. Synthesis of radio frequency plasma polymerized non-synthetic terpinen-4-ol thin films. *Mater. Lett.* **2009**, *63*, 1594–1597. [CrossRef]
166. Yasuda, H. *Plasma Polymerization*; Academic Press: Cambridge, MA, USA, 2012.
167. Ahmad, J.; Bazaka, K.; Whittle, J.D.; Michelmore, A.; Jacob, M.V. Structural characterization of γ-terpinene thin films using mass spectroscopy and X-ray photoelectron spectroscopy. *Plasma Process. Polym.* **2015**, *12*, 1085–1094. [CrossRef]
168. Easton, C.; Jacob, M. Optical characterisation of radio frequency plasma polymerised *Lavandula angustifolia* essential oil thin films. *Thin Solid Films* **2009**, *517*, 4402–4407. [CrossRef]
169. Taguchi, D.; Manaka, T.; Iwamoto, M.; Bazaka, K.; Jacob, M.V. Analyzing hysteresis behavior of capacitance–voltage characteristics of Izo/C60/Pentacene/Au diodes with a hole-transport electron-blocking polyterpenol layer by electric-field-induced optical second-harmonic generation measurement. *Chem. Phys. Lett.* **2013**, *572*, 150–153. [CrossRef]
170. Ahmad, J.; Bazaka, K.; Jacob, M. Optical and surface characterization of radio frequency plasma polymerized 1-isopropyl-4-methyl-1,4-cyclohexadiene thin films. *Electronics* **2014**, *3*, 266–281. [CrossRef]
171. Kim, M.; Cho, S.; Han, J.; Hong, B.; Kim, Y.; Yang, S.; Boo, J.-H. High-rate deposition of plasma polymerized thin films using PECVD method and characterization of their optical properties. *Surf. Coat. Technol.* **2003**, *169*, 595–599. [CrossRef]

172. Ahmad, J.; Bazaka, K.; Oelgemöller, M.; Jacob, M.V. Wetting, solubility and chemical characteristics of plasma-polymerized 1-isopropyl-4-methyl-1, 4-cyclohexadiene thin films. *Coatings* **2014**, *4*, 527–552. [CrossRef]

173. Jacob, M.; Bazaka, K. *Fabrication of Electronic Materials from Australian Essential Oils*; Australian Government: Australia; Rural Industries Research and Development Corporation: Research and Development Corporation, Canberra, 2010.

174. Bazaka, K.; Jacob, M.V.; Crawford, R.J.; Ivanova, E.P. Plasma-assisted surface modification of organic biopolymers to prevent bacterial attachment. *Acta Biomater.* **2011**, *7*, 2015–2028. [CrossRef] [PubMed]

175. Turek, C.; Stintzing, F.C. Stability of essential oils: A review. *Compr. Rev. Food Sci. Food Saf.* **2013**, *12*, 40–53. [CrossRef]

176. Friedman, M.; Kozukue, N.; Harden, L.A. Cinnamaldehyde content in foods determined by gas chromatography−mass spectrometry. *J. Agric. Food Chem.* **2000**, *48*, 5702–5709. [CrossRef] [PubMed]

177. Bäcktorp, C.; Wass, J.T.J.; Panas, I.; Sköld, M.; Börje, A.; Nyman, G. Theoretical investigation of linalool oxidation. *J. Phys. Chem. A* **2006**, *110*, 12204–12212. [CrossRef] [PubMed]

178. Selvaag, E.; Holm, J.Ø.; Thune, P. Allergic contact dermatitis in an aroma therapist with multiple sensitizations to essential oils. *Contact Dermat.* **1995**, *33*, 354–355. [CrossRef]

179. Bleasel, N.; Tate, B.; Rademaker, M. Allergic contact dermatitis following exposure to essential oils. *Aust. J. Dermatol.* **2002**, *43*, 211–213. [CrossRef]

180. Edris, A.E. Pharmaceutical and therapeutic potentials of essential oils and their individual volatile constituents: A review. *Phytother. Res.* **2007**, *21*, 308–323. [CrossRef] [PubMed]

181. Uter, W.; Schmidt, E.; Geier, J.; Lessmann, H.; Schnuch, A.; Frosch, P. Contact allergy to essential oils: Current patch test results (2000–2008) from the information network of departments of dermatology (IVDK). *Contact Dermat.* **2010**, *63*, 277–283. [CrossRef] [PubMed]

182. Brand, C.; Grimbaldeston, M.; Gamble, J.; Drew, J.; Finlay-Jones, J.; Hart, P. Tea tree oil reduces the swelling associated with the efferent phase of a contact hypersensitivity response. *Inflamm. Res.* **2002**, *51*, 236–244. [CrossRef] [PubMed]

183. Herman, A.; Tambor, K.; Herman, A. Linalool affects the antimicrobial efficacy of essential oils. *Curr. Microbiol.* **2016**, *72*, 165–172. [CrossRef] [PubMed]

184. Karbach, J.; Ebenezer, S.; Warnke, P.; Behrens, E.; Al-Nawas, B. Antimicrobial effect of Australian antibacterial essential oils as alternative to common antiseptic solutions against clinically relevant oral pathogens. *Clin. Lab.* **2015**, *61*, 61–68. [CrossRef] [PubMed]

185. Maddocks-Jennings, W. Critical incident: Idiosyncratic allergic reactions to essential oils. *Complement. Ther. Nurs. Midwifery* **2004**, *10*, 58–60. [CrossRef]

186. Al-Jumaili, A.; Alancherry, S.; Bazaka, K.; Jacob, M. Review on the antimicrobial properties of carbon nanostructures. *Materials* **2017**, *10*, 1066. [CrossRef] [PubMed]

187. Kumar, A.; Grant, D.; Alancherry, S.; Al-Jumaili, A.; Bazaka, K.; Jacob, M.V. Plasma polymerization: Electronics and biomedical application. In *Plasma Science and Technology for Emerging Economies*; Springer: Berlin/Heidelberg, Germany, 2017; pp. 593–657.

188. Hegemann, D.; Hossain, M.M.; Körner, E.; Balazs, D.J. Macroscopic description of plasma polymerization. *Plasma Process. Polym.* **2007**, *4*, 229–238. [CrossRef]

189. De Masi, L.; Siviero, P.; Esposito, C.; Castaldo, D.; Siano, F.; Laratta, B. Assessment of agronomic, chemical and genetic variability in common basil (*Ocimum basilicum* L.). *Eur. Food Res. Technol.* **2006**, *223*, 273–281. [CrossRef]

190. Mastromatteo, M.; Lucera, A.; Sinigaglia, M.; Corbo, M.R. Combined effects of thymol, carvacrol and temperature on the quality of non conventional poultry patties. *Meat Sci.* **2009**, *83*, 246–254. [CrossRef] [PubMed]

191. Friedrich, J. Pulsed-plasma polymerization. In *The Plasma Chemistry of Polymer Surfaces*; Wiley-VCH Verlag GmbH & Co. KGaA: Weinheim, Germany, 2012; pp. 377–456.

© 2018 by the authors. Licensee MDPI, Basel, Switzerland. This article is an open access article distributed under the terms and conditions of the Creative Commons Attribution (CC BY) license (http://creativecommons.org/licenses/by/4.0/).

polymers

MDPI

Review

Nanofiller Reinforced Biodegradable PLA/PHA Composites: Current Status and Future Trends

Jingyao Sun [1], Jingjing Shen [2], Shoukai Chen [1], Merideth A. Cooper [3], Hongbo Fu [1], Daming Wu [1,4,*] and Zhaogang Yang [3,*]

1   College of Mechanical and Electrical Engineering, Beijing University of Chemical Technology, Beijing 100029, China; sunjingyao5566@sina.com (J.S.); yutoujuzi@gmail.com (S.C.); andrewlinxy@gmail.com (H.F.)
2   School of Civil Engineering & Architecture, Taizhou University, Taizhou 318000, Zhejiang, China; syusejing@gmail.com
3   Department of Chemical and Biomolecular Engineering, The Ohio State University, Columbus, OH 43210, USA; cooper.1774@buckeyemail.osu.edu
4   State Key Laboratory of Organic-Inorganic Composites, Beijing 100029, China
*   Correspondence: wudaming@vip.163.com (D.W.); yang.1140@osu.edu (Z.Y.); Tel.: +86-10-64435015 (D.W.); +1-6146254075 (Z.Y.); Fax: +86-10-64435015 (D.W.); +1-6142923769 (Z.Y.)

Received: 5 April 2018; Accepted: 4 May 2018; Published: 7 May 2018

Abstract: The increasing demand for environmental protection has led to the rapid development of greener and biodegradable polymers, whose creation provided new challenges and opportunities for the advancement of nanomaterial science. Biodegradable polymer materials and even nanofillers (e.g., natural fibers) are important because of their application in greener industries. Polymers that can be degraded naturally play an important role in solving public hazards of polymer materials and maintaining ecological balance. The inherent shortcomings of some biodegradable polymers such as weak mechanical properties, narrow processing windows, and low electrical and thermal properties can be overcome by composites reinforced with various nanofillers. These biodegradable polymer composites have wide-ranging applications in different areas based on their large surface area and greater aspect ratio. Moreover, the polymer composites that exploit the synergistic effect between the nanofiller and the biodegradable polymer matrix can lead to enhanced properties while still meeting the environmental requirement. In this paper, a broad review on recent advances in the research and development of nanofiller reinforced biodegradable polymer composites that are used in various applications, including electronics, packing materials, and biomedical uses, is presented. We further present information about different kinds of nanofillers, biodegradable polymer matrixes, and their composites with specific concern to our daily applications.

Keywords: biodegradable; nanocellulose; carbon nanotubes; nanoclay; polymer composites

1. Introduction

With the depletion of energy and the worsening environmental problems associated with plastic scrap disposal from petroleum production, there has been a growing interest in biodegradable and renewable resources. With the increasing demand for biodegradable materials, the advancements in the field of degradable composites are getting more and more attention [1–6]. However, most commercial composites are carbon fiber or fiberglass-reinforced epoxy composites that are the subject of fossil-fuel-based composites and are not in line with sustainable green development because of the difficulties in recovering and disposing of these materials at their life-cycle end [7–11]. At the same time, traditional biodegradable polymer materials exhibit poor mechanical properties, a narrow processing window, low electrical and thermal properties and cannot meet the actual needs [12–14]. However, using biodegradable polymer materials as substrates, biodegradable polymer composites that can both degrade and meet practical requirements can be developed by adding appropriate reinforcing fillers using advanced technologies and methods. Biodegradable composite materials are used in many fields such as artificial joints, wound treatment, delivery of corresponding drugs and body orthopedic devices, and are widely used in food packaging and agricultural films [15–17]. To expand the range of applications of biodegradable polymers in various fields, the performance of biodegradable polymers needs to be enhanced. An example of this is the use of bio-based hybrid nanocomposites that would enhance the synergy of natural fibers in bio-based polymers while improving performance and maintaining environmental attractiveness [1,7,18]. With the changing environment, people are becoming increasingly integral to waste treatment. With the biodegradability of bio-based polymers, sustainability is also enhanced so that they can be applied in practical engineering. The rapid development of bio-based polymers is due to the significant progress in the production of bio-based components from biomass feedstocks, enabling bio-based polymers with more functional and chemical structures to achieve target performance and functionality. However, the mechanical properties and barrier properties of biopolymers limit their popularization and application [19]. There are three aspects related to the promotion of biopolymers: performance, processing and cost. All biodegradable polymers have "performance and processing" issues. Nano-enhanced biodegradable biomaterials can be used to improve the performance of biomaterials. Bio-nanocomposites are comprised of biopolymer matrices and nanoparticles (sizes less than several hundred nanometers) which are used for the reinforcement or functionalization [20–22]. With the high depth to width ratio and high superficial area of nanoparticles, bio-nanocomposites are a new type of material that has significantly improved performance compared to the basic bio-polymers [23–25]. In the packaging industry, the continuous impact of plastomer packaging castoff on the environment has also attracted worldwide attention due to limited processing methods [26–29].

Thus, the biodegradable plastic packaging receives more attentions attribute to its sustainable consumption of natural resources while the environmental burden is becoming more and more serious [30,31]. At the same time, the increase of food security led to an increased need for biodegradable wrappers made from renewable resources (biomacromolecules), as a substitute for synthetic plastic wrappers, particularly for short period wrapping and single-use (i.e., single-use tableware, single-use dishes, glass and tableware, garbage bags, drink vessels, farm mulching films, fast food boxes, medical apparatus and instruments, etc.) [25]. From the current situations, the application of nanotechnology in packaging is much faster than the food itself, because people are suspicious of the safety of nanoparticles that are directly added to the food. For food packaging materials, it is believed that nanoparticles will not affect the ingredients of the food itself. It will only migrate from packaging to the food under high temperature level and longer heating time. Therefore, the biodegradable nanocomposites used as packaging materials have minimum effect on the food security [32–34]. Prior to mercantile use of bio-based wrappers, some elementary problems must be solved including the rate of degradation, changes in mechanical behavior during conservation, possibility of microbial proliferation, and contamination of packaged foods by harmful compounds. In fact, these biopolymer wrappers have a higher hydrophilicity and are weaker during processing, resulting in industrial restrictions [35]. Although researchers have done much research on improving the packaging performance of biopolymer membranes, the issues of physics, thermology, and mechanical behavior still cannot meet the requirements in industrial activities [31,36]. Consequently, we have been working on developing biocomposites with improved mechanical behavior, separation properties, rheology and hot properties of food packaging pellicles. There are many bio-based polymer fillers available. The nature of the nanofiller can be organic or inorganic. For example, inorganic fillers include silicon dioxide (SiO_2), titanium dioxide (TiO_2), calciumcarbonate ($CaCO_3$), polyhedral oligomeric silsesquioxane (POSS) and other particles. Coconut shell nanofillers, cellulose nanofillers, and other organic and natural derivatives belong to the class of organic fillers [37–41]. Different nanofillers can improve the mechanical properties, heat resistance, barrier properties and can promote the development of biodegradable materials [6,42–44].

This article summarizes the latest research progress of biologic macromolecule materials, nanofillers and bio-nanocomposites in hardgoods, packing, electron and biomedical applications. Bio-based composites can overcome the inherent disadvantages of some bio-based materials such as narrow processing windows, poor barrier properties, poor biocompatibility and conductivity. This article also looks forward to the future processing technology, product development and application of nanofillers to enhance biocomposite materials [45].

2. Biodegradable Polymers

Biobased polymers [46–49] can be categorized into: (1) upgrades of biodegradable polylactic acid (PLA), polyhydroxyalkanoates (PHA), and the like; (2) polymerization similar to fossil oil derived materials, for instance bio aggregation (Bio-PET); and (3) new biologically based polymers, for instance 2,5- furan two formate (PEF) [50]. Polymer materials from natural crops, including corn-based isosorbide polycarbonate, can also be regarded as biologically based polymers. This article generally describes two biopolymers of polylactic acid (PLA) and polyhydroxyalkanoates (PHA). PLA, PHA and butanedioic acid polymers are the general biologically based polymers, which are often used in biodegradable plastic operations for the addressing of environmental problems. With the changing social demand for biodegradable polymers, the performance needs to be improved to make it more applicable [51,52]. Recently, examples of the development and application of PLA, PHA and succinate polymers have been described. The basic characteristics and chemical properties of the nanocomposites made from PLA, PHA, and some other polymers are also introduced in the following section.

2.1. Biodegradable Polylactic Acid (PLA)

PLA is made from the ring opening polymerization (ROP) of the ring two polymer lactide from lactic acid [53]. The nexus among the nature of PLA and the content of the L unit was studied. Polylactic acid is receiving increasing attention due to its biodegradability and potential role for replacing traditional polymers. PLA belongs to a highly crystalline polymerization because it is stereotactic. These stereoisomers can be controlled by using diverse activators. As the first large-scale produced bio-based plastic, PLA has good mechanical properties, mainly by hydrolytic degradation [54–56].

Since PLA's raw materials are based on agricultural raw materials, the continuous supply of PLA resins is of great significance to the development of the global agricultural economy. The increase in the high molecular weight of polylactic acid is the driving force for the extended application of PLA. Various techniques can be used to prepare these polymerizate, mainly through the formation of azeotropic dewatering polycondensation, direct polycondensation, or condensation of the lactide (as shown in Figure 1) [57]. Although the production technology of PLA has been greatly improved, there are still many areas for improvement in PLA applications. PLA has mainly been used to replace thermoplastics. For example, for important foods that require high separate safeguard, polyethylene terephthalate (PET) packaging may not be replaced by PLA, because the barrier properties of PLA are different from those of PET. The brittleness of PLA will also be a limiting condition for its tenacity and shock resistance. Finally, when PLA is exposed to atrocious weather conditions, it may have unpredictable characteristics. Compared to conventional thermoplastic polymers, polylactic acid has poor heat endurance and shock resistance. As a result, there is a gap between PLA and conventional polymers. However, the use of polylactic acid could be extensive if its performance was improved. In recent years, researchers have used a variety of nanofillers to improve the performance of PLA, mainly phyllosilicates, carbon nanotubes, hydroxyapatite, layered titanates, etc. [58]. The good compatibility, material characteristics and low cost of PLA will make it widely used in medicine. Furthermore, the shear-thinning characteristics of PLA can be used in conventional polymer processing technology processes, and some new technologies (such as the use of supercritical foams and electrospinning to produce nanofibers) which has positive implications for spreading the application of such polymers [59].

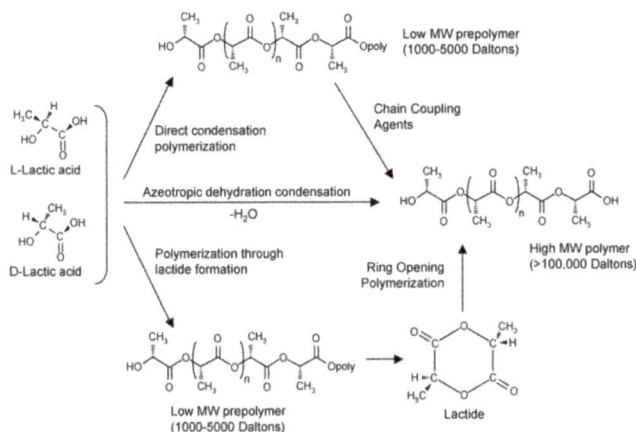

Figure 1. Synthesis of PLA from l- and d-lactic acids. Reproduced with permission from [57].

2.2. Polyhydroxyalkanoate (PHA)

Polyhydroxy chain alkanoate (PHA) is a kind of biological polyester, made naturally by various microorganisms. Research work is being done to create PHA in transgenic plants [60]. These polymeric compounds have different structures and corresponding properties. More than 150 different genres of PHA are homopolymers, which are created by diverse kinds of bacteria and growing conditions. Polyhydroxybutyrate (PHB) and poly (hydroxybutyrate Hydroxyvalerate) (PHBV) are the most famous polymerizates in polyhydroxyalkanes systems [54]. PHA is a member of the polyester family composed of hydroxyalkanate monomers. In nature, they exist in the form of homopolymers. PHA is present in pure particulate polymer form and is used as an energy storage medium in bacteria (similar to animal fats and starch-like starch plants). PHA is commercially produced using energy-rich feedstocks that are converted to fatty acid feedstocks. During PHA's industrial production process, the cells are separated and dissolved after several production cycles. The extraction of the purified cell residue from a polymer is processed into particles or powders [61].

In recent years, researchers found that a large number of bacteria can produce different polyhydroxyalkanoate bio-polyesters. In general, it does not seem too easy to manipulate PHA structure and the proportion of monomers in the copolymer. However, the weakening of the beta oxidation cycle by pseudomonad putida and pseudomonad thermophila makes various PHA structures controllable. It has become a reality to use functionalized PHA by introducing functional groups, which contains fatty acids, into polyhydroxy polymer chains in a predetermined ratio [62]. However, commercialization of PHA is costly, has molecular weight (MW) and structural instability, and, thus, the thermomechanical properties are not stable [36,63,64]. Furthermore, high costs are associated with complex biological processes such as bactericidal effects, low conversion of carbon substrates to PHA products, slow growth of microorganisms, and downstream separation [65]. Therefore, researchers are not aware of the limitations of PHA production. A growing body of research has been conducted on methods to enhance production, with the objective being sustainably producing large-scale microbial polyhydroxyalkanoates (PHAs), promoting the commercialization of PHAs and expanding their range of applications [66,67]. For example, Burniol-Figols summarized the method for the production of PHA by fermenting crude glycerol (as shown in Figure 2) and studied the effects of nitrogen under different conditions [68]. Koller focuses on the investigation of the thermal and rheological properties of PHA polymers accumulated by *Synechocystis salina*. The determined thermal and rheological properties show that PHA polymers accumulated by *S. salina* on digestate supernatant or mineral medium are comparable with the commercial available poly(3-hydroxybutyrate). However, the results demonstrated that PHA polymers generally need to be modified before the melting process to increase their stability in the molten state [69].

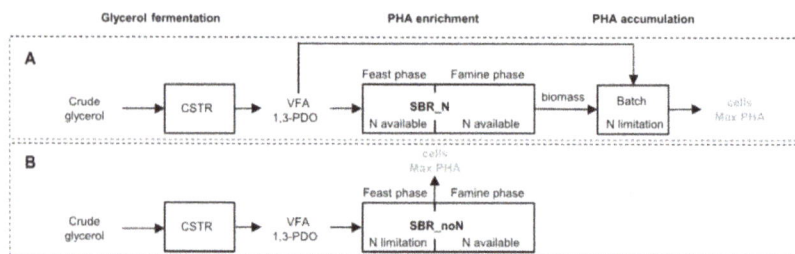

Figure 2. Summary of strategies for production of PHA from fermented crude glycerol Reproduced with permission from [68].

PHA is not only a kind of environmentally friendly biopolymer, but also has many adjustable material properties. With the further reduction in cost and the development of high value-added applications, it will become a kind of multi-application field material which can be accepted by the market. Because it is a family with a wide range of components, its performance from hard to high flexibility enables it to be applied to different applications. The structural diversity of PHA and the variability of its properties make it an important member of biomaterials. Compared with PLA, the developing history of PHA is short, but the development potential and range of applications are bigger.

3. Nanofillers for Biodegradable Nanocomposites

Nanofillers can improve or adjust the properties of the materials into which they are incorporated, such as flame retardant properties, optical or electrical properties, mechanical properties and thermal properties [57]. Nanofillers need to be incorporated into the polymer matrix in a certain proportion. There are many nanofillers used in nanocomposites, which mainly include nanoclays, carbon nanotubes and some organic nanofillers [54]. This paper mainly introduces nanofiller reinforced biodegradable polymer composites, in which the matrix need to come from renewable resources.

3.1. Nanocellulose

With the advancement of technology, the natural polymer cellulose attracts more people's attention worldwide. A new kind of "cellulose" has been used as an advanced material [70]. Cellulose is considered a product or extract of natural cellulose consisting of nanoscale structural materials. In general, the cellulose family can be divided into three kinds: (1) cellulose nanocrystals which has other names such as nanocrystalline cellulose; (2) cellulose nanofibers which is also known as nanofibrillated cellulose (NFC); and (3) bacterial cellulose (BC), also called microbial cellulose. It can be obtained from wood, flour, beets, potato tubers, ramie, algae and other plants. The BC can be reproduced quickly by converting large unit (cm) into small units (nm) and let them grow back into large units under adapt circumstance. Bacterial nanocellulose is a nanocellulose that is secreted by microorganisms and has been demonstrated to be useful for artificial blood vessels in tissue engineering. Bacterial nanocellulose as nanofiller has good mechanical properties and biocompatibility, ultrafine fiber network and high porosity [71]. Chemically induced deconstruction strategies, such as acid hydrolysis, are commonly applied to pick up CNC from natural cellulose while retaining highly crystalline structures. The process of BC construction is shown in Figure 3, which is typically synthesized by bacteria in a pure form [72]. Different types of cellulose exhibit different properties, which determine their applicability and functionality, that is, some types of cellulose are better suited for specific applications. A high Young's modulus/high tensile strength are the typical properties of cellulose that are important. Some aspect ratios that can be manipulated depend on particle type, and potential compatibility with other materials. In addition, the choice of chemistry and material affinity give rise to very versatile cleavage [73]. There are many studies on nanofibers. Jafari et al. [74] conducted a study of nanofiber coatings and found that the low DE of polymer–polymer complexes decreases the water adsorption and the solubility of maltodextrin. Second, the crystalline nanocellulose fibers increase the path of curvature and curvature in the material and reduce the water possibility of molecular penetration. Kuo et al. developed an enhanced bio-nanocomposite fiber/resin interface with a blend of toughened epoxy resins to improve resin penetration and fiber distribution [75]. Nanocellulose can control the rheology stability of dispersion and give the composite stronger mechanical properties. Synthetic modification of fibrous cellulose is a way to get chemical compatibility of the systems. However, this also limits the environmental benefits of using cellulose components. Therefore, an attractive step forward in compatibility and further expansion of nanocellulose applications is through the use of surfactants [76]. The tensile strengths of some nanocomposites are linearly correlated with the strengths of cellulose nanofibers. Better dispersion of individual cellulose nanofibers can improve the performance of the composites [77].

Linear biopolymer cellulose is naturally found in all plants. In addition to be the major natural polymer on Earth, it has a variety of functions, including excellent biocompatibility, lower density, greater strength, and has the most favorable mechanical properties at a fraction of the cost. With the recyclability, anisotropic shape, good mechanical properties, fine biocompatibility, and adjustable surface chemistry of nanocellulose, a growing number of applications of nanocellulose materials in science and biomedical engineering related fields has attracted peoples' interest. Although the topics of nanocellulose has been extensively studied over the past years, there is clear room for growth, especially regarding new developments in coatings and medical devices [73,78].

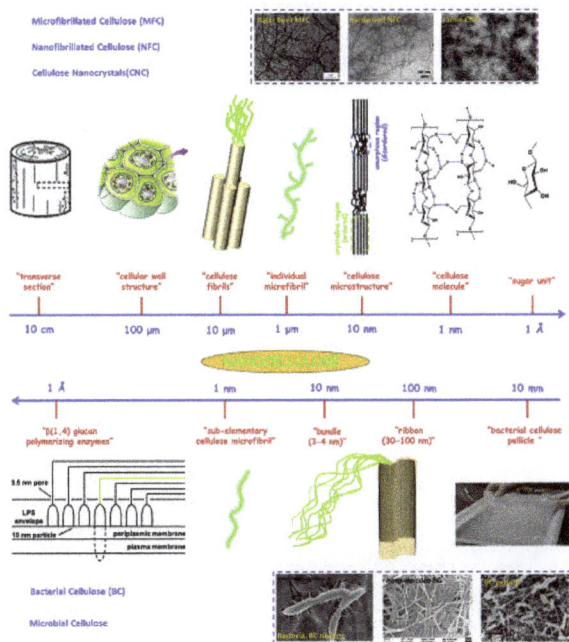

Figure 3. Hierarchical structure of cellulose; top image (from large unit to small unit): cellulose nanocrystals (CNC), micro/nanofibrillated cellulose (MFC and NFC); bottom image (from tiny unit to small unit): bacterial cellulose (BC). Reproduced with permission from [72].

3.2. Nanoclays

Layered silicate, also known as clay, is the most commonly used nanolayered silicate nanocomposite in polymer composites, which have a widespread application in the preparation of composites based on clay. The changes in size of the silicate layer depends on the source of the clay, the silicate particles and the production technology. The structure in the silicate layer will be changed, and the substitute induces a negative charge in the slime sheet that is spontaneously offset by the cations in the interlayer spacing. All electricity variation relies on layered silicate [79]. Biodegradable plastic clay nanocomposites have attracted wide attention because of their ameliorated mechanical and obstruction properties and their lower burnable points on each native polymer. Majeed researched the natural fiber filled hybrid composite in the field of food applications and obtained a wrapper with enhanced barrier and obstruction properties [80]. Malwela studied the enzymatic degradation behavior of nanometer clay reinforced degradation material. The effect of nanoclay on the degradation rate of blends at various temperatures is diverse from that of PLA/PBSA mixture composites, which provide a reference for subsequent studies. Of course, nanoclays are also used for several new and improved biodegradable polymer materials, with a growing range of applications, not only because of its strong practicality and low price point, but also because it has machinability and thermosetting properties [81–85].

3.3. Carbon Nanotubes and Graphene

3.3.1. Carbon Nanotubes

Nanotubes have been synthesized by a number of methods including arc discharge, laser ablation, chemical vapor deposition, etc. [86]. Carbon nanotubes exhibit excellent mechanical, electrical and magnetic properties which make them an ideal material for high-strength polymer composites. However, because of van der Waals interactions, carbon nanotubes usually form stable bundles that are extremely hard to disperse and align in the polymer. Due to this, the biggest issue with the manufacturing of carbon nanotube-reinforced composites is the ability to disperse and assess the dispersibility as well as arranging and controlling the carbon nanotubes in the matrix. There are some methods for dispersing nanotubes in a polymer matrix, such as solution mixing and melt mixing. The chemical vapor deposition process has progressed the construction of carbon nanotubes, which has promoted the widespread industrial application of this process.

Carbon nanotubes can be used in many situations, including polymer composites, electrochemical energy storage/conversion, hydrogen storage and others. Polymer composites are used as functional fillers that not only increase the thermal/mechanical properties, but also provide additional functions such as increase in flame resistance, and barrier properties. Its polymer composites have been extensively studied in many fields. Carbon nanotubes have been effectively exploited by researchers to develop techniques based on renewable resources, polymer materials, etc. Ma et al. [87] proposed the design of a bio-based conducting and rapidly-crystallizing nanocomposite with controlled distribution of multi-walled carbon nanotubes through an interface stereocomplex. Thin films and (semi) conductive materials show significant potential.

Hapuarachchi and co-worker [88] found that the multi-walled carbon nanotubes can be applied as flame retardants for PLA and its natural fiber reinforced composites. They found that the heat release rate (HRR) was reduced by 58% compared to native PLA because of the addition of multi-walled carbon nanotubes. The advancement of carbon nanotubes with better conductivity has created new applications for polymer composites in electromagnetic interference shielding fields.

3.3.2. Graphene

Graphene is a single layer of hybridized carbon atoms arranged in a two-dimensional lattice which can be manufactured by the peeling off of graphite nanosheets. The theoretical specific surface area of graphene sheets is 2630–2956 square meters and the aspect ratio is more than 2000. With its special structure, graphene has outstanding thermal properties and mechanical properties. One of the most advanced applications of graphene is a filler for nanocomposite polymers. However, it is hindered by poor solubility in most cases. Moreover, the large surface area of graphene leads to significant aggregation in the polymer because of the van der Waals. Thus, graphene has significant optical activity that can be observed on some substrates by simple optical methods. Actually, different numbers of atomic layers of graphene can be distinguished relatively easily using transmission optical microscopy [89].

Due to the excellent electrical, mechanical, optical and transport properties of graphene, it has been used in many different types of applications. Graphene reinforced nanocomposites have a high level of hardness and strength. These nanocomposites should have excellent mechanical properties and graphene is expected to be a reinforcing material for high performance nanocomposites. However, there is a problem in obtaining good dispersion, and there are challenges in getting graphene to completely peel into a single layer or a few layers of material having a reasonable lateral size or generating graphene oxide without significantly damages [90]. Researchers have done a lot of work on graphene reinforced nanocomposites. Graphene-based adsorbents have attracted widespread interest as effective adsorbents for the removal of heavy metals from the environment [91]. Sima Kashi et al. conducted a study on the dielectric and EMI shielding performance of graphene-based biodegradable nanocomposites. The study found that the addition of graphene nanosheets significantly enhanced the dielectric constant of both polymers [92]. Purnima Baruah et al. studied bio-based, tough, hyperbranched epoxy/graphene (HBE) nanocomposites with enhanced biodegradability. Performance studies showed that the addition of graphene oxide (GO) to HBE increased the bond strength by 189%, the toughness by 263%, the tensile strength by 161%, and the elongation at break by 159% [93]. After extensive research, it has also been necessary to guarantee that a strong interface exists between the reinforcing material and the polymer matrix to give the best properties of the graphene reinforced nano-polymer. It should be noted that apart from providing good prospects for mechanical enhancement, it can also be used to control functional properties such as conductivity, swelling, gas barrier properties and stability.

3.4. Other Functional Nanofillers

Bio-nanoparticles and a variety of functional fillers are attracting a great deal of attention because of their diverse biomedical and biotechnological applications. Nano-scale fillers play a significant role in the manufacture of biological composites because they bring a variety of desirable functions to the composite. There are a variety of functional nanofillers such as silica nanoparticles, hydroxyapatite, layered double hydroxide (LDH), polyhedral oligomeric silsesquioxanes (POSS), cellulose nanofibers, etc. [94,95]. Recently, some nanofillers have drawn more attention due to their versatility in the manufacture of biomedical applications. Hydroxyapatite is famous for its bioactive and biocompatible ceramic which is found in bones and teeth. Bio-based polymers of LDH find wide applications in tissue engineering, drug delivery and gene therapy because of their compatibility and their non-cytotoxic and nonirritating biological systems [96]. The Hap/GO nanocomposites prepared by M. Ramadas et al. provide excellent biocompatibility for use in orthopedic, drug delivery and dental applications [97].

4. Processing Methods and Applications

4.1. Processing Methods

Bio-based materials have disadvantages such as poor hydrophilicity, poor electrical conductivity, and poor mechanical properties during processing [21]. However, nanofillers can overcome the above defects and achieve the purpose of enhancing the properties of composite materials. Therefore, nanofillers are used to enhance biodegradable materials. There are many ways to prepare nanofiller biodegradable composite materials. Different nanofillers have different treatment methods. This article describes the processing of nanocellulose based biocomposites, the preparation of nanoclay based composites, the processing of polymer–carbon nanotube based biocomposites, and the processing of functional nanocomposites.

4.1.1. Processing of Nanocellulose Based Bio-Nanocomposites

At present, the main treatment methods for using nanocellulose fillers are solvent casting and melt processing [98]. These are mainly used to solve the problem that the nanocellulose cannot be evenly dispersed in a non-polar medium. Due to the polarity of nanocellulose whiskers, it is difficult to disperse homogeneously in non-polar media and therefore needs to be uniformly dispersed in polar media or in aqueous media [99]. Solvent casting and melt processing methods can help nanocellulose evenly disperse in the polymer. For the above two methods, the polymer used is different. For solvent casting, mainly three types of polymers are used: (1) water-soluble polymer; (2) polymer emulsion; and (3) water-insoluble polymer. There are two effective ways to achieve solvent casting. For polymer emulsions and water-insoluble polymers, it is possible to utilize polar head and long hydrophobic tail surfactants of polymer emulsions or water-insoluble polymers, Surfactants are coated on the surface of nanocellulose crystals. Another method is to graft hydrophobic chains onto the surface of nanocellulose crystals. Both methods allow the nanocellulose filler to be uniformly dispersed in the polymer.

Melt extrusion is the most commonly used in industry [100]. Melt extrusion refers to the process of adding the plasticized material to the extruder for forming. However, this method has several tough problems. The biggest problem currently is the use of dry nanocellulose. During the extrusion process, the nanocellulose particles easily form hydrogen bonds in the amorphous state. These hydrogen bonds have a strong adsorption force, which makes the material prone to aggregate when it is dry, so it is difficult to evenly disperse the polymer. Currently, researchers are working to overcome this by studying the feed process. Using the method of pumping suspension [99], nanocomposite-enhanced PLA biodegradable composites were obtained, which has better dispersibility. Another technique is wet extrusion [101]. Compared to melt extrusion, the wet extrusion has a lower temperature and is suitable for applying in biomedical applications. Because the temperature of the melt extrusion is too high to degrade the protein. At present, Danya M. Lavin's team [102] used the wet extrusion method to prepare a self-assembled microfiber scaffold for drug delivery. The polylactic acid solution was added dropwise into a uniformly stirred, water-insoluble solvent to make it a liquid non-woven polymer. Fibrous scaffolds, then adjust the concentration of polymer spinning dope and increase the ratio of silicone oil to petroleum ether to achieve fiber diameter control.

4.1.2. Processing of Nanoclay Based Bio-Nanocomposites

At present, nanoclay biodegradable composites are mainly mixed by means of intercalation layered silicate. There are three main methods: (1) polymer solution embedding; (2) in-situ polymerization; and (3) melt embedding method.

The so-called polymer solution embedding method mainly works by macromolecule clay intercalation solvent in the polymer [103]. By this way, the nanoclay can be embedded in the polymer and will not damage the internal structure of the polymer, forming nanoclay–polymer composites. However, this method requires a large amount of solvent, thereby displacing a large amount of waste liquid, which has an adverse effect on the environment.

In-situ polymerization utilizes the polymerization of monomers in phyllosilicates. In this method, the nanoclays expand in liquid monomer. Under the action of polymerization, the nanoclays are effectively embedded in the polymer, a process requiring catalyst initiation of polymerization. The catalyst can effectively make the nanoclay fixed in the inner layer of silicate without falling off [104].

Melt embedding is currently the most widely used method in industrial production. When the temperature reaches the melting temperature, the nanoclay is annealed [105], in which case the polymer chains can enter the silicate interlayer, forming a sandwich structure. This method does not produce waste liquid as in the in-situ polymerization method and is a more economical and green method. This method can prepare various nanocomposites of different morphologies, depending on the manner in which the polymer chains are embedded and the types of functional groups. Figure 4 shows the formation of nanoclay composites.

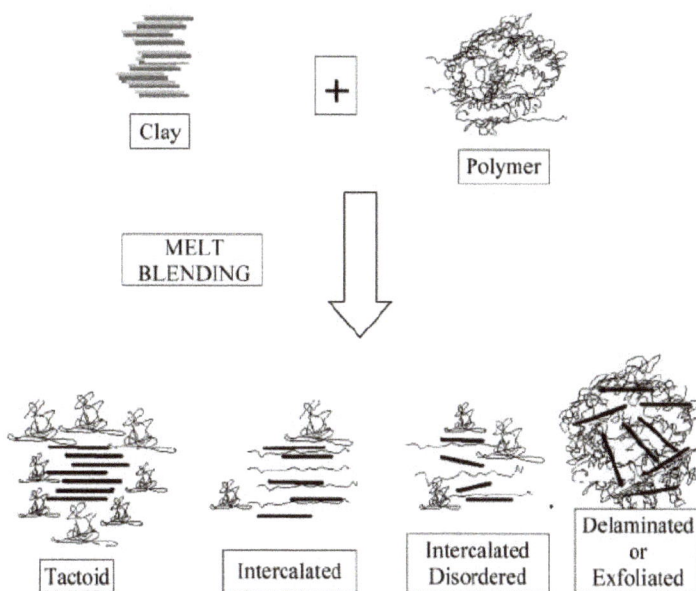

Figure 4. Schematic illustration of terminology used to describe nanocomposites formed from organoclays. Reproduced with permission from [106].

4.1.3. Processing of Polymer–Carbon Nanotubes Based Bio-Nanocomposites

Carbon nanotubes act as nanofillers to enhance biodegradable composites; the extent of their enhancement depends mainly on the molecular orientation and degree of dispersion. At present, single-walled carbon nanotubes (SWCNTs) and multi-walled carbon nanotubes (MWCNTs) are mainly dispersed in polymer composites, and van der Waals forces are used to enable the polymer matrix to condense carbon nanotubes. The degree of dispersion depends on the network structure of the structure. This uniformity depends on the molecular orientation of the matrix and its compatibility [107].

In addition, the differences in the size of the SWNTs and the larger surface energy lead to the polymerization of the MWCNTs. Therefore, it is difficult to uniformly distribute the surface of polymer. The current methods of uniformly dispersing carbon nanotubes include chemical modification [108], coating of carbon nanotubes [109], in-situ polymerization [104], ultrasonic dispersion [110], melt processing [111], addition of surfactants [112], electrospinning [113], electrochemistry and crystallization [114]. Uniform dispersion of carbon nanotubes can significantly increase the strength and toughness of composites, and improve the electrical conductivity of composites. It is an important measure for degradable biocomposites.

4.1.4. Processing of Functional Nanocomposites

Hap-based nano-natural degradation of composite materials are used in medical applications. Hap composite materials can be prepared by traditional physicochemical methods [115]. The preparation method mainly adopts solvent casting. The solution concentration determines the degree of polymer dispersion while the mixing time and the mixing method determine polymer uniformity. Hap particles are not easily reacted due to the lack of reactive functional groups (hydroxyl groups). Currently, this problem can be solved by changing the Hap particle size and increasing the surface energy [116]. It should be noted that the catalyst added in the mix must be harmless and biocompatible without changing its properties.

The other is LDH nanocomposite. The polymer is mainly prepared by three methods: (a) monomer exchange and in situ polymerization; (b) coprecipitation or polymer displacement; and (c) polymer recombination [117]. Figure 5 shows three specific preparation methods. In-situ polymerization is to fill the interlayer of nanolayers with the reaction monomers, allowing them to polymerize between layers. The co-precipitation method refers to the fact that two or more cations are contained in a solution. They are present in a homogeneous solution and a precipitant is added. After the precipitation reaction, uniform precipitation of various components can be obtained. Polymer recombination refers to the process of reacting and orderly arranging different polymer monomers and reacting to form new polymers.

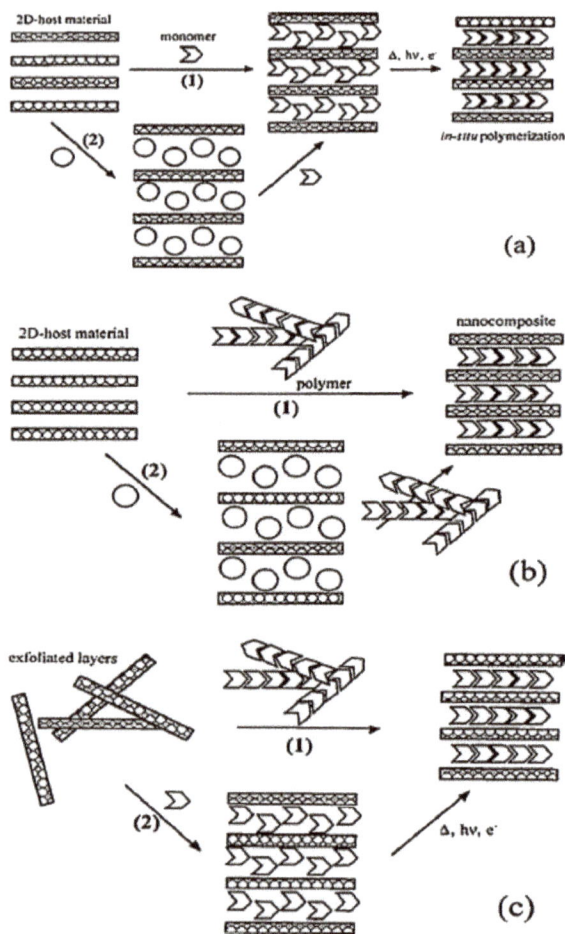

Figure 5. Preparation of LDH nanocomposites of various methods, (**a**) monomer exchange and in situ polymerization, (**b**) direct exchange, (**c**) exfoliated layers restacking. Reproduced with permission from [117].

Method (a) is used in the preparation of polymer–LDH nanocomposites. The coprecipitation method is useful for layered hydroxide. The process consists of "co-assembly" synthesis of LDH in the presence of a polymer formed between LDH sheets [118].

4.2. Application

4.2.1. Electronics

Nanofillers Enhanced Biodegradation Composites have been widely used in many fields and have many advanced developments in electronics, including diodes, solar cells, and electromagnetic applications [119]. With the continuous increase of electronic devices, abandoned electronic devices cause serious environmental pollution, and the appearance of biodegradable bio-composites greatly relieves the environmental pressure.

Nanocellulose bio-composites are widely used in medical, electronics, packaging and other fields. They are currently used in the development of flexible electronic equipment using roll-to-roll manufacturing process [120–122]. The technology relies on the substrate material and nanocellulose composite material as the base material to achieve the purpose of preparation. Masaya Nogi and co-workers [120,122] have experimentally proved the advantage of nanoscale reinforcing using cellulose nanofibers. They obtained transparent composites by enhancing various types of resin using BC nanofibers, even in the fiber content of up to 70 wt %. As BC nanofibers are bundles of semi-crystalline and extended cellulose chains, the obtained nanocomposites are not only highly transparent and flexible, but also present high mechanical strength comparable to low carbon steel and low coefficient of thermal expansion comparable to silicon, which make the composite suitable for applications. Moreover, they have succeeded in depositing an electroluminescent layer (comprised of organic light-emitting diodes) on these transparent BC nanocomposites. Petersson et al. [123] successfully prepared poly(lactic acid) cellulose-based biodegradable nanocomposites. They found that the poly(lactic acid) cellulose base can enhance the mechanical properties and thermal stability of the materials. It was concluded through experiments that the poly(lactic acid) cellulose-based biodegradable nanomaterials. The composite material is stable at 220 °C, so the material can be suitable for high temperature environments.

Carbon nanotube biodegradable composites have been applied in the development of flexible sensors. Sensors of this kind of materials can be applied in various temperature, humidity and complex chemical environments with better electromagnetic and mechanical properties [124,125]. Han and co-workers [124] reported a humidity sensor on cellulose paper using functionalized single-walled carbon nanotubes. The conductance displacement of the nanotube network wrapped on microfibril cellulose was used for humidity sensing. Compared with control sensors made on glass substrates, cellulose mediated charge transfer on paper enhances sensitivity. They furtherly prepared a similar CNT-based sensor device on cellulose paper for ammonia sensing [125]. At present, Yun et al. [105] manufactured carbon nanotube–cellulose biodegradable composites, which served as a base material for chemical vapor sensors. Cellulose solution was prepared by dissolving cotton pulp in LiCl/N, N-dimethylacetamide solution. The multi-walled carbon nanotubes (MWCNTs) were covalently grafted to cellulose by reacting imidazolides–MWCNTs with cellulose solution. Figure 6 below is a schematic diagram of a M/C paper and cross finger electrode chemical gas sensor. Its good economic benefits, biocompatibility and eco-friendly advantages led the sensor to receive extensive attention.

Carbon nanotube biodegradable composite materials are used in solar cells. Carbon nanotubes added to the flexible photovoltaic cells can directly improve the conductivity of the polymer and enhanced solar cell photon absorption capacity. Valentin et al. [126] developed a novel method of carbon nanotube and researched their electrical properties. It is concluded that the SWNTs composites can be used in organic conductive materials.

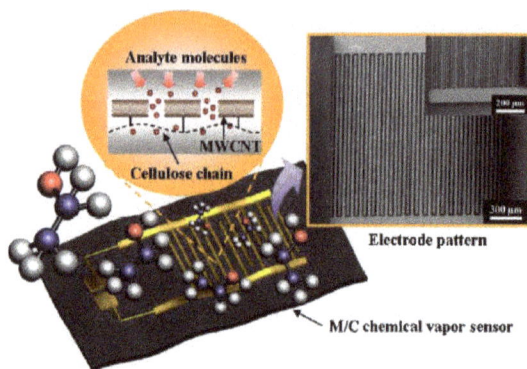

Figure 6. Schematic of chemical vapor sensor made with M/C paper and IDT shaped electrode (reproduced with permission [105]).

4.2.2. Packaging Industries

There are three major drawbacks with biodegradable plastics currently used for packaging: performance, processing and cost. Emerging nanofillers enhance biodegradable composite materials to help overcome the above problems. Nanofilling can effectively improve these problems, and nanocomposites have significant advantages over traditional composites.

The materials currently used in food packaging mainly value their stretchability and permeability [127]. Nanofiller biodegradable composite materials not only have the key features of metal-based packaging, but also other good properties including mechanical properties, thermodynamic properties, environmental harm and so on. Furthermore, nanofiller composites may significantly improve the high barrier properties of polymers [104]. Due to the presence of nanofillers, the molecular pathways increase as it passes through the substrate. The presence of the nanoclay layer allows a significant increase in the diffusion path of gas or other water vapor molecules through the polymer, resulting in a substantial reduction in the rate of permeation through the polymer, thereby effectively enhancing its barrier properties. Figure 7 shows the nanoclay filler composite diffusion path schematic [128]. The figure shows that due to the presence of the nanoclay layer, the movement path of the water vapor molecules becomes longer, and the straight line from the beginning becomes a curve, which is why the nanoclay composite material improves the barrier properties.

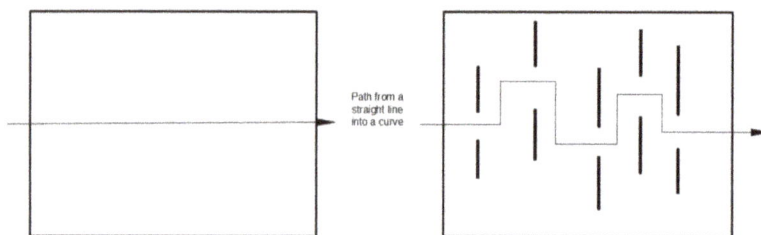

Figure 7. Nanoclay filler composite diffusion path schematic.

Barrier properties of nanofiller biodegradable composites mainly depend on the molecular orientation in the polymer matrix material and the dispersion uniformity of the nanofiller. Comparing the various nanofillers, nanoclay has the best effect. Due to cost and ease of processing, PLA, PHB and starch-based nanocomposites are the most popular in the packaging industry [129].

As the most popular biodegradable material, PLA is involved in the preparation of many products such as cups, cutlery and packaging boxes [130]. PLA is mainly pressed into the cardboard by squeezing the coating layer, which is further developed and applied as a packaging material. Using PLA as a substrate material and nanoclay as a filler to enhance the performance of PLA, not only has the characteristics of PLA, but also enhances the barrier property of composites. Chang et al. [131] studied the effect of the nanoclay modification of PLA. The researchers prepared a melt-intercalated nanocomposite. The results show that the permeability of all nanocomposites decreases. In addition, the researchers also studied the effects of shear and feed rates on the permeability of PLA nanocomposite films. The oxygen barrier properties of PLA nanocomposites have been improved by 15–48% compared to that of pure PLA materials [132]. Moreover, the shear rate and feed rate have little effect on the forming, so the barrier properties of PLA nanocomposites depend mainly on the molecular orientation of the substrate material and the dispersion uniformity of the nanofiller.

For PHA to be able to enter the researcher's field of vision depends mainly on its high hydrophobicity, as it is widely used in coatings and films [133]. In the food packaging, PHA barrier properties and polyethylene are very similar, which makes PHA useful in packaging materials. However, the workability and gas barrier properties of PHA have limited their development in the packaging field. Currently, Sanchez-Garcia et al. [134] studied the relationship between morphology and mechanical properties, including tensile strength and Young's modulus. The high dispersion improves the barrier properties of PHB/clay nanocomposites.

Degradable bags made from starch-based substrates [135–137] result in many defects due to the high hydrophilicity of the starch, which can now be overcome using nanocomposites. Park et al. [138]. studied the permeability of different nanomaterials in water demonstrating that nanocomposites have an impediment to the penetration of water, which is in favor of the development of packaging technology.

Nanocomposites would lead to a variation in the permeability of water vapor, which is related to the saturation of the polymer base material and the nanofiller [139,140]. In most cases, different nanocomposites will have different degrees of saturation, thus resulting in different permeabilities. Moreover, the semi-crystalline polymer itself has crystalline regions and non-crystalline regions that are impermeable to molecules and therefore result in different permeabilities. In addition, the decrease in permeability is mainly due to the increase in the molecular path of the molecules through the polymer by the nanofillers, which is an effective way to increase the movement of gas molecules [141]. The effect of nanofillers on the permeability of the composite depends not only on the crystallinity of the substrate itself, but also on the different types of nanofillers.

4.2.3. Medical Applications

Biodegradable composites rely on their biocompatibility and versatility and these materials have been widely used in the medical field. Biocompatible composite materials play a role in the human body without adversely affecting drugs. Therefore, bio-based composite materials have drawn more and more attention and biocompatible materials have been more clinically used [142–147]. In addition, soy-derived polymers have proven useful as bone fillers [148]. Bacterial nanocellulose has also been shown to be useful for artificial blood vessels [149]. Nanofiller biodegradable composite materials can be applied to clinical medicine.

Biodegradable nanocomposites are very useful in tissue engineering for the regeneration of primary tissue structures. Nanocomposites with three major characteristics can be applied to tissue engineering: (1) the extracellular matrix consisting of macromolecules; (2) the extracellular matrix where fiber forms exist; and (3) macromolecules in the extracellular matrix have a diameter of less than 500 nm [150].

The researchers found that the toxins in BNC can be easily eliminated by the use of sodium hydroxide treatment and water purification 13 purification methods [151]. Highly inhibited water in the fiber network prevents the adsorption of proteins in the blood, which is beneficial to blood compatibility. Bacterial nanofiber biodegradable composites with good fiber networks and pores

Polymers **2018**, *10*, 505

provide a good environment for cell growth. This is a very important aspect of artificial blood vessels, proving that artificial blood vessels can form new blood vessel tissue in animal experiments.

Tang's group [149] used two-photon reactor to prepare BNC composite artificial blood vessel. Researchers successfully prepared double-silicon (D-BNC) and single-silicon (S-BNC) artificial blood vessels, relying mainly on the unique properties of bacterial nanofillers. The researchers explained the process and nature of the biosynthesis of the artificial blood vessels. The results showed that the BNC artificial blood vessels rely on an ultra-fine, uniform fiber network and high porosity to be able to act as biomedical prostheses and to function in living organisms.

In addition, electrospun biocompatible polymer nanofiber composites are currently capable of performing hard tissue repairs in the form of porous membranes for implantation in humans [152]. Nanofiber PLA-PEG mosaic copolymer composites prepared by electrospinning can be used for bone tissue engineering scaffolds. Adding nanofiber fillers can not only increase the tensile strength and flexibility of the bones, but also increase the stiffness and strength of the bones. At present, the skeleton uses porous scaffold structure. This structure has high elasticity and plays a role in protecting the skeleton and has been successfully implanted in animal models to achieve bone regeneration [153].

5. Summary and Outlook

This article mainly introduces the research status of nanofillers to enhance biodegradable composite materials. This article describes different kinds of nanofillers and biodegradable polymers for biodegradable nanocomposites. Different nanofillers have different treatment methods and apply to different fields. Biodegradable composite materials use nanofillers to enhance performance and these biodegradable nanocomposites have broader application prospects. This paper introduces the enhancement of bio-based polymers by nanoclay, nanocellulose, carbon nanotubes and functional nanofillers, which improves the deficiencies of bio-based polymers. The process of adding fillers involves a series of different treatment methods, mainly by solvent casting and melt processing methods, which can be applied to nanocellulose. The method of polymer intercalation is mainly suitable for the addition of nanoclay filler. For carbon nanotubes, the current industry mainly uses the melt extrusion method.

Biodegradable nanocomposites have mainly been prepared for use in electronics, packaging and biomedical applications. Currently, in electronics, nanocellulose composite materials can be applied to the preparation of electronic displays, flexible sensors, light-emitting diodes, etc., to promote the development of flexible electronic devices. In the field of packaging, biodegradable polymers exist to fix three major defects: performance, processing and cost. The nanofiller enhanced biodegradable composite materials can overcome the above problems, and help to improve the performance of biodegradable polymers. In biomedicine, nano-enhanced biodegradable composite materials are mainly used in the field of tissue engineering, drug delivery and gene therapy. Bio-composites rely on their biocompatibility and has been the concern of researchers. The use of nanofillers has been explored to enhance bio-based composite materials to obtain more functions and achieve more medical breakthroughs.

Nanofiller reinforced biodegradable composite materials are more widely used, and the area has attracted many researchers. In future research, researchers are more likely to focus on nanofillers to enhance composite processes, and to study more highly industrialized and efficient processes that are very difficult for this nanotechnology. In the future, biodegradable composite materials can replace most of the current materials, which is very important for sustaining our life; therefore, it is an urgent task to study nanofillers to enhance biodegradable composite materials.

Acknowledgments: We gratefully acknowledge financial support from the National Natural Science Foundation of China (Grant Nos. 51673020 and 51173015).

Author Contributions: J.S. (Jingyao Sun), J.S. (Jingjing Shen), D.W., and Z.Y. conceived and designed the experiments; J.S. (Jingyao Sun), S.C. and H.F. performed the experiments; J.S. (Jingyao Sun) and M.C. analyzed the data; J.S. (Jingyao Sun), M.C. and Z.Y. wrote the paper.

Conflicts of Interest: The authors declare no conflict of interest.

References

1. Sridhar, V.; Lee, I.; Chun, H.H.; Park, H. Graphene reinforced biodegradable poly(3-hydroxybutyrate-co-4-hydroxybutyrate) nano-composites. *Express Polym. Lett.* **2013**, *7*, 320–328. [CrossRef]
2. Fan, H.; Wang, L.; Zhao, K.; Li, N.; Shi, Z.; Ge, Z.; Jin, Z. Fabrication, Mechanical Properties, and Biocompatibility of Graphene-Reinforced Chitosan Composites. *Biomacromolecules* **2010**, *11*, 2345–2351. [CrossRef] [PubMed]
3. Mathew, A.P.; Oksman, K.; Sain, M. Mechanical properties of biodegradable composites from poly lactic acid (PLA) and microcrystalline cellulose (MCC). *J. Appl. Polym. Sci.* **2005**, *97*, 2014–2025. [CrossRef]
4. Wang, X.; Yang, H.; Song, L.; Hu, Y.; Xing, W.; Lu, H. Morphology, mechanical and thermal properties of graphene-reinforced poly(butylene succinate) nanocomposites. *Compos. Sci. Technol.* **2011**, *72*, 1–6. [CrossRef]
5. Liu, H.; Li, J.; Ren, N.; Qiu, J.; Mou, X. Graphene oxide-reinforced biodegradable genipin-cross-linked chitosan fluorescent biocomposite film and its cytocompatibility. *Int. J. Nanomed.* **2013**. [CrossRef] [PubMed]
6. Lalwani, G.; Henslee, A.M.; Farshid, B.; Lin, L.; Kasper, F.K.; Qin, Y.-X.; Mikos, A.G.; Sitharaman, B. Two-Dimensional Nanostructure-Reinforced Biodegradable Polymeric Nanocomposites for Bone Tissue Engineering. *Biomacromolecules* **2013**, *14*, 900–909. [CrossRef] [PubMed]
7. Jia, W.; Gong, R.H.; Hogg, P.J. Poly (lactic acid) fibre reinforced biodegradable composites. *Compos. Part B Eng.* **2014**, *62*, 104–112. [CrossRef]
8. Nazhat, S.N.; Kellomäki, M.; Törmälä, P.; Tanner, K.E.; Bonfield, W. Dynamic mechanical characterization of biodegradable composites of hydroxyapatite and polylactides. *J. Biomed. Mater. Res.* **2001**, *58*, 335–343. [CrossRef] [PubMed]
9. Nam, T.H.; Ogihara, S.; Tung, N.H.; Kobayashi, S. Effect of alkali treatment on interfacial and mechanical properties of coir fiber reinforced poly(butylene succinate) biodegradable composites. *Compos. Part B Eng.* **2011**, *42*, 1648–1656. [CrossRef]
10. Barkoula, N.M.; Garkhail, S.K.; Peijs, T. Biodegradable composites based on flax/polyhydroxybutyrate and its copolymer with hydroxyvalerate. *Ind. Crops Prod.* **2010**, *31*, 34–42. [CrossRef]
11. Akil, H.M.; Omar, M.F.; Mazuki, A.A.M.; Safiee, S.; Ishak, Z.A.M.; Abu Bakar, A. Kenaf fiber reinforced composites: A review. *Mater. Des.* **2011**, *32*, 4107–4121. [CrossRef]
12. La Mantia, F.P.; Morreale, M. Green composites: A brief review. *Compos. Part A Appl. Sci. Manuf.* **2011**, *42*, 579–588. [CrossRef]
13. Abdul Khalil, H.P.S.; Bhat, A.H.; Ireana Yusra, A.F. Green composites from sustainable cellulose nanofibrils: A review. *Carbohydr. Polym.* **2012**, *87*, 963–979. [CrossRef]
14. Koronis, G.; Silva, A.; Fontul, M. Green composites: A review of adequate materials for automotive applications. *Compos. Part B Eng.* **2013**, *44*, 120–127. [CrossRef]
15. Saba, N.; Tahir, P.; Jawaid, M. A Review on Potentiality of Nano Filler/Natural Fiber Filled Polymer Hybrid Composites. *Polymers* **2014**, *6*, 2247–2273. [CrossRef]
16. Chieng, B.; Ibrahim, N.; Yunus, W.; Hussein, M.; Then, Y.; Loo, Y. Effects of Graphene Nanoplatelets and Reduced Graphene Oxide on Poly(lactic acid) and Plasticized Poly(lactic acid): A Comparative Study. *Polymers* **2014**, *6*, 2232–2246. [CrossRef]
17. Yusoff, R.B.; Takagi, H.; Nakagaito, A.N. Tensile and flexural properties of polylactic acid-based hybrid green composites reinforced by kenaf, bamboo and coir fibers. *Ind. Crops Prod.* **2016**, *94*, 562–573. [CrossRef]
18. Hamad, K.; Kaseem, M.; Ko, Y.G.; Deri, F. Biodegradable polymer blends and composites: An overview. *Polym. Sci. Ser. A* **2014**, *56*, 812–829. [CrossRef]
19. Chen, G.-Q.; Patel, M.K. Plastics Derived from Biological Sources: Present and Future: A Technical and Environmental Review. *Chem. Rev.* **2011**, *112*, 2082–2099. [CrossRef] [PubMed]
20. Carvalho, R.A.; Santos, T.A.; de Azevedo, V.M.; Felix, P.H.C.; Dias, M.V.; Borges, S.V. Bio-nanocomposites for food packaging applications: Effect of cellulose nanofibers on morphological, mechanical, optical and barrier properties. *Polym. Int.* **2018**, *67*, 386–392. [CrossRef]
21. Siqueira, G.; Bras, J.; Dufresne, A. Cellulosic Bionanocomposites: A Review of Preparation, Properties and Applications. *Polymers* **2010**, *2*, 728–765. [CrossRef]
22. Pracella, M.; Haque, M.M.-U.; Puglia, D. Morphology and properties tuning of PLA/cellulose nanocrystals bio-nanocomposites by means of reactive functionalization and blending with PVAc. *Polymer* **2014**, *55*, 3720–3728. [CrossRef]

23. Chen, Z.; Zhang, A.; Wang, X.; Zhu, J.; Fan, Y.; Yu, H.; Yang, Z. The Advances of Carbon Nanotubes in Cancer Diagnostics and Therapeutics. *J. Nanomater.* **2017**, *2017*, 1–13. [CrossRef]
24. Chen, Z.; Zhang, A.; Yang, Z.; Wang, X.; Chang, L.; Chen, Z.; James Lee, L. Application of DODMA and Derivatives in Cationic Nanocarriers for Gene Delivery. *Curr. Org. Chem.* **2016**, *20*, 1813–1819. [CrossRef]
25. Xie, J.; Yang, Z.; Zhou, C.; Zhu, J.; Lee, R.J.; Teng, L. Nanotechnology for the delivery of phytochemicals in cancer therapy. *Biotechnol. Adv.* **2016**, *34*, 343–353. [CrossRef] [PubMed]
26. Mahalik, N.P.; Nambiar, A.N. Trends in food packaging and manufacturing systems and technology. *Trends Food Sci. Technol.* **2010**, *21*, 117–128. [CrossRef]
27. Mihindukulasuriya, S.D.F.; Lim, L.T. Nanotechnology development in food packaging: A review. *Trends Food Sci. Technol.* **2014**, *40*, 149–167. [CrossRef]
28. Bugnicourt, E.; Cinelli, P.; Lazzeri, A.; Alvarez, V. Polyhydroxyalkanoate (PHA): Review of synthesis, characteristics, processing and potential applications in packaging. *Express Polym. Lett.* **2014**, *8*, 791–808. [CrossRef]
29. Ahmed, J.; Varshney, S.K. Polylactides—Chemistry, Properties and Green Packaging Technology: A Review. *Int. J. Food Prop.* **2011**, *14*, 37–58. [CrossRef]
30. Barlow, C.Y.; Morgan, D.C. Polymer film packaging for food: An environmental assessment. *Resour. Conserv. Recycl.* **2013**, *78*, 74–80. [CrossRef]
31. Rhim, J.-W.; Park, H.-M.; Ha, C.-S. Bio-nanocomposites for food packaging applications. *Prog. Polym. Sci.* **2013**, *38*, 1629–1652. [CrossRef]
32. Othman, S.H. Bio-nanocomposite Materials for Food Packaging Applications: Types of Biopolymer and Nano-sized Filler. *Agric. Agric. Sci. Procedia* **2014**, *2*, 296–303. [CrossRef]
33. Busolo, M.A.; Fernandez, P.; Ocio, M.J.; Lagaron, J.M. Novel silver-based nanoclay as an antimicrobial in polylactic acid food packaging coatings. *Food Addit. Contam. Part A* **2010**, *27*, 1617–1626. [CrossRef] [PubMed]
34. Huang, J.-Y.; Li, X.; Zhou, W. Safety assessment of nanocomposite for food packaging application. *Trends Food Sci. Technol.* **2015**, *45*, 187–199. [CrossRef]
35. Cherpinski, A.; Torres-Giner, S.; Cabedo, L.; Méndez, J.A.; Lagaron, J.M. Multilayer structures based on annealed electrospun biopolymer coatings of interest in water and aroma barrier fiber-based food packaging applications. *J. Appl. Polym. Sci.* **2018**, *135*, 45501. [CrossRef]
36. Peelman, N.; Ragaert, P.; De Meulenaer, B.; Adons, D.; Peeters, R.; Cardon, L.; Van Impe, F.; Devlieghere, F. Application of bioplastics for food packaging. *Trends Food Sci. Technol.* **2013**, *32*, 128–141. [CrossRef]
37. Chen, Z.; Cong, M.; Hu, J.; Yang, Z.; Chen, Z. Preparation of Functionalized TiO$_2$ Nanotube Arrays and Their Applications. *Sci. Adv. Mater.* **2016**, *8*, 1231–1241. [CrossRef]
38. Kuang, T.; Fu, D.; Chang, L.; Yang, Z.; Yang, J.; Fan, P.; Zhong, M.; Chen, F.; Peng, X. Enhanced Photocatalysis of Yittium-Doped TiO$_2$/D-PVA Composites: Degradation of Methyl Orange (MO) and PVC Film. *Sci. Adv. Mater.* **2016**, *8*, 1286–1292. [CrossRef]
39. Dong, Y.; Zhou, G.; Chen, J.; Shen, L.; Jianxin, Z.; Xu, Q.; Zhu, Y. A new LED device used for photodynamic therapy in treatment of moderate to severe acne vulgaris. *Photodiagn. Photodyn. Ther.* **2016**, *13*, 188–195. [CrossRef] [PubMed]
40. Sha, L.; Chen, Z.; Chen, Z.; Zhang, A.; Yang, Z. Polylactic Acid Based Nanocomposites: Promising Safe and Biodegradable Materials in Biomedical Field. *Int. J. Polym. Sci.* **2016**, *2016*, 1–11. [CrossRef]
41. Chen, Z.; Chen, Z.; Yang, Z.; Hu, J.; Yang, Y.; Chang, L.; Lee, L.J.; Xu, T. Preparation and characterization of vacuum insulation panels with super-stratified glass fiber core material. *Energy* **2015**, *93*, 945–954. [CrossRef]
42. Thakur, V.K.; Thakur, M.K.; Raghavan, P.; Kessler, M.R. Progress in Green Polymer Composites from Lignin for Multifunctional Applications: A Review. *ACS Sustain. Chem. Eng.* **2014**, *2*, 1072–1092. [CrossRef]
43. Frone, A.N.; Berlioz, S.; Chailan, J.-F.; Panaitescu, D.M. Morphology and thermal properties of PLA–cellulose nanofibers composites. *Carbohydr. Polym.* **2013**, *91*, 377–384. [CrossRef] [PubMed]
44. Georgiopoulos, P.; Kontou, E.; Niaounakis, M. Thermomechanical properties and rheological behavior of biodegradable composites. *Polym. Compos.* **2013**. [CrossRef]
45. Wu, G.; Deng, X.; Song, J.; Chen, F. Enhanced biological properties of biomimetic apatite fabricated polycaprolactone/chitosan nanofibrous bio-composite for tendon and ligament regeneration. *J. Photochem. Photobiol. B Biol.* **2018**, *178*, 27–32. [CrossRef] [PubMed]
46. Wang, P.; Zhang, D.; Zhou, Y.; Li, Y.; Fang, H.; Wei, H.; Ding, Y. A well-defined biodegradable 1,2,3-triazolium-functionalized PEG-b-PCL block copolymer: Facile synthesis and its compatibilization for PLA/PCL blends. *Ionics* **2017**, *24*, 787–795. [CrossRef]

47. Iwata, T. Biodegradable and Bio-Based Polymers: Future Prospects of Eco-Friendly Plastics. *Angew. Chem. Int. Ed.* **2015**, *54*, 3210–3215. [CrossRef] [PubMed]
48. Satoh, K. Controlled/living polymerization of renewable vinyl monomers into bio-based polymers. *Polym. J.* **2015**, *47*, 527–536. [CrossRef]
49. Okuda, T.; Ishimoto, K.; Ohara, H.; Kobayashi, S. Renewable Biobased Polymeric Materials: Facile Synthesis of Itaconic Anhydride-Based Copolymers with Poly(l-lactic acid) Grafts. *Macromolecules* **2012**, *45*, 4166–4174. [CrossRef]
50. Nakajima, H.; Dijkstra, P.; Loos, K. The Recent Developments in Biobased Polymers toward General and Engineering Applications: Polymers that are Upgraded from Biodegradable Polymers, Analogous to Petroleum-Derived Polymers, and Newly Developed. *Polymers* **2017**, *9*, 523. [CrossRef]
51. Lotz, B.; Li, G.; Chen, X.; Puiggali, J. Crystal polymorphism of polylactides and poly(Pro-alt-CO): The metastable beta and gamma phases. Formation of homochiral PLLA phases in the PLLA/PDLA blends. *Polymer* **2017**, *115*, 204–210. [CrossRef]
52. Kanetaka, Y.; Yamazaki, S.; Kimura, K. Preparation of poly(ether ketone)s derived from 2,5-furandicarboxylic acid via nucleophilic aromatic substitution polymerization. *J. Polym. Sci. Part A Polym. Chem.* **2016**, *54*, 3094–3101. [CrossRef]
53. Mochizuki, M. Crystallization Behaviors of highly LLA-rich PLA Effects of D-isomer ratio of PLA on the rate of crystallization, crystallinity, and melting point. *Sen-I Gakkaishi* **2010**, *66*, 70–77.
54. Reddy, M.M.; Vivekanandhan, S.; Misra, M.; Bhatia, S.K.; Mohanty, A.K. Biobased plastics and bionanocomposites: Current status and future opportunities. *Prog. Polym. Sci.* **2013**, *38*, 1653–1689. [CrossRef]
55. Guo, Y.; Chang, C.-C.; Halada, G.; Cuiffo, M.A.; Xue, Y.; Zuo, X.; Pack, S.; Zhang, L.; He, S.; Weil, E.; et al. Engineering flame retardant biodegradable polymer nanocomposites and their application in 3D printing. *Polym. Degrad. Stab.* **2017**, *137*, 205–215. [CrossRef]
56. Du, Y.; Wu, T.; Yan, N.; Kortschot, M.T.; Farnood, R. Fabrication and characterization of fully biodegradable natural fiber-reinforced poly(lactic acid) composites. *Compos. Part B Eng.* **2014**, *56*, 717–723. [CrossRef]
57. Lim, L.T.; Auras, R.; Rubino, M. Processing technologies for poly(lactic acid). *Prog. Polym. Sci.* **2008**, *33*, 820–852. [CrossRef]
58. Thummarungsan, N.; Paradee, N.; Pattavarakorn, D.; Sirivat, A. Influence of graphene on electromechanical responses of plasticized poly(lactic acid). *Polymer* **2018**, *138*, 169–179. [CrossRef]
59. Hamad, K.; Kaseem, M.; Yang, H.W.; Deri, F.; Ko, Y.G. Properties and medical applications of polylactic acid: A review. *Express Polym. Lett.* **2015**, *9*, 435–455. [CrossRef]
60. Insomphun, C.; Chuah, J.-A.; Kobayashi, S.; Fujiki, T.; Numata, K. Influence of Hydroxyl Groups on the Cell Viability of Polyhydroxyalkanoate (PHA) Scaffolds for Tissue Engineering. *ACS Biomater. Sci. Eng.* **2016**, *3*, 3064–3075. [CrossRef]
61. Madison, L.L.; Huisman, G.W. Metabolic engineering of poly(3-hydroxyalkanoates): From DNA to plastic. *Microbiol. Mol. Biol. Rev. MMBR* **1999**, *63*, 21–53. [PubMed]
62. Chen, G.-Q.; Jiang, X.-R.; Guo, Y. Synthetic biology of microbes synthesizing polyhydroxyalkanoates (PHA). *Synth. Syst. Biotechnol.* **2016**, *1*, 236–242. [CrossRef] [PubMed]
63. Lim, J.; You, M.; Li, J.; Li, Z. Emerging bone tissue engineering via Polyhydroxyalkanoate (PHA)-based scaffolds. *Mater. Sci. Eng. C* **2017**, *79*, 917–929. [CrossRef] [PubMed]
64. Zhang, J.; Shishatskaya, E.I.; Volova, T.G.; da Silva, L.F.; Chen, G.-Q. Polyhydroxyalkanoates (PHA) for therapeutic applications. *Mater. Sci. Eng. C* **2018**, *86*, 144–150. [CrossRef] [PubMed]
65. Chen, G.-Q.; Jiang, X.-R. Engineering bacteria for enhanced polyhydroxyalkanoates (PHA) biosynthesis. *Synth. Syst. Biotechnol.* **2017**, *2*, 192–197. [CrossRef] [PubMed]
66. Korkakaki, E.; van Loosdrecht, M.C.M.; Kleerebezem, R. Impact of phosphate limitation on PHA production in a feast-famine process. *Water Res.* **2017**, *126*, 472–480. [CrossRef] [PubMed]
67. Koller, M.; Maršálek, L.; de Sousa Dias, M.M.; Braunegg, G. Producing microbial polyhydroxyalkanoate (PHA) biopolyesters in a sustainable manner. *New Biotechnol.* **2017**, *37*, 24–38. [CrossRef] [PubMed]
68. Burniol-Figols, A.; Varrone, C.; Daugaard, A.E.; Le, S.B.; Skiadas, I.V.; Gavala, H.N. Polyhydroxyalkanoates (PHA) production from fermented crude glycerol: Study on the conversion of 1,3-propanediol to PHA in mixed microbial consortia. *Water Res.* **2018**, *128*, 255–266. [CrossRef] [PubMed]

69. Kovalcik, A.; Meixner, K.; Mihalic, M.; Zeilinger, W.; Fritz, I.; Fuchs, W.; Kucharczyk, P.; Stelzer, F.; Drosg, B. Characterization of polyhydroxyalkanoates produced by Synechocystis salina from digestate supernatant. *Int. J. Biol. Macromol.* **2017**, *102*, 497–504. [CrossRef] [PubMed]

70. Chin, K.-M.; Sung Ting, S.; Ong, H.L.; Omar, M. Surface functionalized nanocellulose as a veritable inclusionary material in contemporary bioinspired applications: A review. *J. Appl. Polym. Sci.* **2018**, *135*, 46065. [CrossRef]

71. Scherner, M.; Reutter, S.; Klemm, D.; Sterner-Kock, A.; Guschlbauer, M.; Richter, T.; Langebartels, G.; Madershahian, N.; Wahlers, T.; Wippermann, J. In vivo application of tissue-engineered blood vessels of bacterial cellulose as small arterial substitutes: Proof of concept? *J. Surg. Res.* **2014**, *189*, 340–347. [CrossRef] [PubMed]

72. Lin, N.; Dufresne, A. Nanocellulose in biomedicine: Current status and future prospect. *Eur. Polym. J.* **2014**, *59*, 302–325. [CrossRef]

73. Abitbol, T.; Rivkin, A.; Cao, Y.; Nevo, Y.; Abraham, E.; Ben-Shalom, T.; Lapidot, S.; Shoseyov, O. Nanocellulose, a tiny fiber with huge applications. *Curr. Opin. Biotechnol.* **2016**, *39*, 76–88. [CrossRef] [PubMed]

74. Akhavan Mahdavi, S.; Mahdi Jafari, S.; Assadpoor, E.; Dehnad, D. Microencapsulation optimization of natural anthocyanins with maltodextrin, gum Arabic and gelatin. *Int. J. Biol. Macromol.* **2016**, *85*, 379–385. [CrossRef] [PubMed]

75. Kuo, P.-Y.; Barros, L.d.A.; Yan, N.; Sain, M.; Qing, Y.; Wu, Y. Nanocellulose composites with enhanced interfacial compatibility and mechanical properties using a hybrid-toughened epoxy matrix. *Carbohydr. Polym.* **2017**, *177*, 249–257. [CrossRef] [PubMed]

76. Tardy, B.L.; Yokota, S.; Ago, M.; Xiang, W.; Kondo, T.; Bordes, R.; Rojas, O.J. Nanocellulose–surfactant interactions. *Curr. Opin. Colloid Interface Sci.* **2017**, *29*, 57–67. [CrossRef]

77. Lee, K.-Y.; Aitomäki, Y.; Berglund, L.A.; Oksman, K.; Bismarck, A. On the use of nanocellulose as reinforcement in polymer matrix composites. *Compos. Sci. Technol.* **2014**, *105*, 15–27. [CrossRef]

78. Mishra, R.K.; Sabu, A.; Tiwari, S.K. Materials chemistry and the futurist eco-friendly applications of nanocellulose: Status and prospect. *J. Saudi Chem. Soc.* **2018**. [CrossRef]

79. Chivrac, F.; Pollet, E.; Avérous, L. Progress in nano-biocomposites based on polysaccharides and nanoclays. *Mater. Sci. Eng. R Rep.* **2009**, *67*, 1–17. [CrossRef]

80. Majeed, K.; Jawaid, M.; Hassan, A.; Abu Bakar, A.; Abdul Khalil, H.P.S.; Salema, A.A.; Inuwa, I. Potential materials for food packaging from nanoclay/natural fibres filled hybrid composites. *Mater. Des.* **2013**, *46*, 391–410. [CrossRef]

81. Malin, F.; Znoj, B.; Šegedin, U.; Skale, S.; Golob, J.; Venturini, P. Polyacryl–nanoclay composite for anticorrosion application. *Prog. Org. Coat.* **2013**, *76*, 1471–1476. [CrossRef]

82. Hakamy, A.; Shaikh, F.U.A.; Low, I.M. Characteristics of hemp fabric reinforced nanoclay–cement nanocomposites. *Cem. Concr. Compos.* **2014**, *50*, 27–35. [CrossRef]

83. Felbeck, T.; Bonk, A.; Kaup, G.; Mundinger, S.; Grethe, T.; Rabe, M.; Vogt, U.; Kynast, U. Porous nanoclay polysulfone composites: A backbone with high pore accessibility for functional modifications. *Microporous Mesoporous Mater.* **2016**, *234*, 107–112. [CrossRef]

84. Shettar, M.; Achutha Kini, U.; Sharma, S.S.; Hiremath, P. Study on Mechanical Characteristics of Nanoclay Reinforced Polymer Composites. *Mater. Today Proc.* **2017**, *4*, 11158–11162. [CrossRef]

85. Memiş, S.; Tornuk, F.; Bozkurt, F.; Durak, M.Z. Production and characterization of a new biodegradable fenugreek seed gum based active nanocomposite film reinforced with nanoclays. *Int. J. Biol. Macromol.* **2017**, *103*, 669–675. [CrossRef] [PubMed]

86. Daenen, M.; Zhang, L.; Erni, R.; Williams, O.A.; Hardy, A.; Van Bael, M.K.; Wagner, P.; Haenen, K.; Nesladek, M.; Van Tendeloo, G. Diamond Nucleation by Carbon Transport from Buried Nanodiamond TiO$_2$ Sol-Gel Composites. *Adv. Mater.* **2009**, *21*, 670–673. [CrossRef]

87. Ma, P.; Jiang, L.; Ye, T.; Dong, W.; Chen, M. Melt Free-Radical Grafting of Maleic Anhydride onto Biodegradable Poly(lactic acid) by Using Styrene as A Comonomer. *Polymers* **2014**, *6*, 1528–1543. [CrossRef]

88. Hapuarachchi, T.D.; Peijs, T. Multiwalled carbon nanotubes and sepiolite nanoclays as flame retardants for polylactide and its natural fibre reinforced composites. *Compos. Part A Appl. Sci. Manuf.* **2010**, *41*, 954–963. [CrossRef]

89. Tan, B.; Thomas, N.L. A review of the water barrier properties of polymer/clay and polymer/graphene nanocomposites. *J. Membr. Sci.* **2016**, *514*, 595–612. [CrossRef]

90. Young, R.J.; Kinloch, I.A.; Gong, L.; Novoselov, K.S. The mechanics of graphene nanocomposites: A review. *Compos. Sci. Technol.* **2012**, *72*, 1459–1476. [CrossRef]

91. Sherlala, A.I.A.; Raman, A.A.A.; Bello, M.M.; Asghar, A. A review of the applications of organo-functionalized magnetic graphene oxide nanocomposites for heavy metal adsorption. *Chemosphere* **2018**, *193*, 1004–1017. [CrossRef]

92. Kashi, S.; Gupta, R.K.; Baum, T.; Kao, N.; Bhattacharya, S.N. Dielectric properties and electromagnetic interference shielding effectiveness of graphene-based biodegradable nanocomposites. *Mater. Des.* **2016**, *109*, 68–78. [CrossRef]

93. Baruah, P.; Karak, N. Bio-based tough hyperbranched epoxy/graphene oxide nanocomposite with enhanced biodegradability attribute. *Polym. Degrad. Stab.* **2016**, *129*, 26–33. [CrossRef]

94. Hule, R.A.; Pochan, D.J. Polymer Nanocomposites for Biomedical Applications. *MRS Bulletin* **2011**, *32*, 354–358. [CrossRef]

95. Millon, L.E.; Wan, W.K. The polyvinyl alcohol–bacterial cellulose system as a new nanocomposite for biomedical applications. *J. Biomed. Mater. Res. Part B Appl. Biomater.* **2006**, *79B*, 245–253. [CrossRef] [PubMed]

96. Bhatia, S.K.; Kurian, J.V. Biological characterization of Sorona polymer from corn-derived 1,3-propanediol. *Biotechnol. Lett.* **2007**, *30*, 619–623. [CrossRef] [PubMed]

97. Ramadas, M.; Bharath, G.; Ponpandian, N.; Ballamurugan, A.M. Investigation on biophysical properties of Hydroxyapatite/Graphene oxide (HAp/GO) based binary nanocomposite for biomedical applications. *Mater. Chem. Phys.* **2017**, *199*, 179–184. [CrossRef]

98. Dufresne, A. Processing of Polymer Nanocomposites Reinforced with Polysaccharide Nanocrystals. *Molecules* **2010**, *15*, 4111–4128. [CrossRef] [PubMed]

99. Oksman, K.; Mathew, A.P.; Bondeson, D.; Kvien, I. Manufacturing process of cellulose whiskers/polylactic acid nanocomposites. *Compos. Sci. Technol.* **2006**, *66*, 2776–2784. [CrossRef]

100. Bondeson, D.; Oksman, K. Polylactic acid/cellulose whisker nanocomposites modified by polyvinyl alcohol. *Compos. Part A Appl. Sci. Manuf.* **2007**, *38*, 2486–2492. [CrossRef]

101. Xia, Y.; Shi, C.-Y.; Fang, J.-G.; Wang, W.-Q. Approaches to developing fast release pellets via wet extrusion-spheronization. *Pharm. Dev. Technol.* **2016**. [CrossRef] [PubMed]

102. Lavin, D.M.; Harrison, M.W.; Tee, L.Y.; Wei, K.A.; Mathiowitz, E. A novel wet extrusion technique to fabricate self-assembled microfiber scaffolds for controlled drug delivery. *J. Biomed. Mater. Res. Part A* **2012**, *100A*, 2793–2802. [CrossRef] [PubMed]

103. Vaia, R.A.; Giannelis, E.P. Polymer Melt Intercalation in Organically-Modified Layered Silicates: Model Predictions and Experiment. *Macromolecules* **1997**, *30*, 8000–8009. [CrossRef]

104. Sinha Ray, S.; Okamoto, M. Polymer/layered silicate nanocomposites: A review from preparation to processing. *Prog. Polym. Sci.* **2003**, *28*, 1539–1641. [CrossRef]

105. Yun, S.; Kim, J. Multi-walled carbon nanotubes–cellulose paper for a chemical vapor sensor. *Sens. Actuators B Chem.* **2010**, *150*, 308–313. [CrossRef]

106. Dennis, H.R.; Hunter, D.L.; Chang, D.; Kim, S.; White, J.L.; Cho, J.W.; Paul, D.R. Effect of melt processing conditions on the extent of exfoliation in organoclay-based nanocomposites. *Polymer* **2001**, *42*, 9513–9522. [CrossRef]

107. Shaffer, M.S.P.; Windle, A.H. Analogies between Polymer Solutions and Carbon Nanotube Dispersions. *Macromolecules* **1999**, *32*, 6864–6866. [CrossRef]

108. Eitan, A.; Jiang, K.; Dukes, D.; Andrews, R.; Schadler, L.S. Surface Modification of Multiwalled Carbon Nanotubes: Toward the Tailoring of the Interface in Polymer Composites. *Chem. Mater.* **2003**, *15*, 3198–3201. [CrossRef]

109. Star, A.; Stoddart, J.F.; Steuerman, D.; Diehl, M.; Boukai, A.; Wong, E.W.; Yang, X.; Chung, S.-W.; Choi, H.; Heath, J.R. Preparation and Properties of Polymer-Wrapped Single-Walled Carbon Nanotubes. *Angew. Chem. Int. Ed.* **2001**, *40*, 1721–1725. [CrossRef]

110. Qian, D.; Dickey, E.C.; Andrews, R.; Rantell, T. Load transfer and deformation mechanisms in carbon nanotube-polystyrene composites. *Appl. Phys. Lett.* **2000**, *76*, 2868–2870. [CrossRef]

111. Siochi, E.J.; Working, D.C.; Park, C.; Lillehei, P.T.; Rouse, J.H.; Topping, C.C.; Bhattacharyya, A.R.; Kumar, S. Melt processing of SWCNT-polyimide nanocomposite fibers. *Compos. Part B Eng.* **2004**, *35*, 439–446. [CrossRef]

112. Schadler, L.S.; Giannaris, S.C.; Ajayan, P.M. Load transfer in carbon nanotube epoxy composites. *Appl. Phys. Lett.* **1998**, *73*, 3842–3844. [CrossRef]

113. Dror, Y.; Salalha, W.; Khalfin, R.L.; Cohen, Y.; Yarin, A.L.; Zussman, E. Carbon Nanotubes Embedded in Oriented Polymer Nanofibers by Electrospinning. *Langmuir* **2003**, *19*, 7012–7020. [CrossRef]

114. Chen, G.Z.; Shaffer, M.S.P.; Coleby, D.; Dixon, G.; Zhou, W.; Fray, D.J.; Windle, A.H. Carbon Nanotube and Polypyrrole Composites Coating and Doping. *Adv. Mater.* **2000**, *12*, 522–526. [CrossRef]

115. Šupová, M. Problem of hydroxyapatite dispersion in polymer matrices: A review. *J. Mater. Sci. Mater. Med.* **2009**, *20*, 1201–1213. [CrossRef] [PubMed]

116. Ahn, E.S.; Gleason, N.J.; Nakahira, A.; Ying, J.Y. Nanostructure Processing of Hydroxyapatite-based Bioceramics. *Nano Lett.* **2001**, *1*, 149–153. [CrossRef]

117. Leroux, F.; Besse, J.-P. Polymer Interleaved Layered Double Hydroxide: A New Emerging Class of Nanocomposites. *Chem. Mater.* **2001**, *13*, 3507–3515. [CrossRef]

118. Darder, M.; López-Blanco, M.; Aranda, P.; Leroux, F.; Ruiz-Hitzky, E. Bio-Nanocomposites Based on Layered Double Hydroxides. *Chem. Mater.* **2005**, *17*, 1969–1977. [CrossRef]

119. Jung, Y.J.; Kar, S.; Talapatra, S.; Soldano, C.; Viswanathan, G.; Li, X.; Yao, Z.; Ou, F.S.; Avadhanula, A.; Vajtai, R.; et al. Aligned Carbon Nanotube−Polymer Hybrid Architectures for Diverse Flexible Electronic Applications. *Nano Lett.* **2006**, *6*, 413–418. [CrossRef] [PubMed]

120. Nogi, M.; Yano, H. Transparent Nanocomposites Based on Cellulose Produced by Bacteria Offer Potential Innovation in the Electronics Device Industry. *Adv. Mater.* **2008**, *20*, 1849–1852. [CrossRef]

121. Eichhorn, S.J.; Dufresne, A.; Aranguren, M.; Marcovich, N.E.; Capadona, J.R.; Rowan, S.J.; Weder, C.; Thielemans, W.; Roman, M.; Renneckar, S.; et al. Review: Current international research into cellulose nanofibres and nanocomposites. *J. Mater. Sci.* **2009**, *45*, 1–33. [CrossRef]

122. Yano, H.; Sugiyama, J.; Nakagaito, A.N.; Nogi, M.; Matsuura, T.; Hikita, M.; Handa, K. Optically Transparent Composites Reinforced with Networks of Bacterial Nanofibers. *Adv. Mater.* **2005**, *17*, 153–155. [CrossRef]

123. Petersson, L.; Kvien, I.; Oksman, K. Structure and thermal properties of poly(lactic acid)/cellulose whiskers nanocomposite materials. *Compos. Sci. Technol.* **2007**, *67*, 2535–2544. [CrossRef]

124. Han, J.-W.; Kim, B.; Li, J.; Meyyappan, M. Carbon Nanotube Based Humidity Sensor on Cellulose Paper. *J. Phys. Chem. C* **2012**, *116*, 22094–22097. [CrossRef]

125. Han, J.-W.; Kim, B.; Li, J.; Meyyappan, M. A carbon nanotube based ammonia sensor on cellulose paper. *RSC Adv.* **2014**, *4*, 549–553. [CrossRef]

126. Valentini, L.; Kenny, J.M. Novel approaches to developing carbon nanotube based polymer composites: Fundamental studies and nanotech applications. *Polymer* **2005**, *46*, 6715–6718. [CrossRef]

127. Johannson, C. Bio-nanocomposites for food packaging applications. In *Nanocomposites with Biodegradable Polymers: Synthesis, Properties and Future Perspectives*; Oxford University Press: Oxford, UK, 2011; pp. 348–367. [CrossRef]

128. Bharadwaj, R.K. Modeling the Barrier Properties of Polymer-Layered Silicate Nanocomposites. *Macromolecules* **2001**, *34*, 9189–9192. [CrossRef]

129. Azeredo, H.M.C.; Mattoso, L.H.C.; Wood, D.; Williams, T.G.; Avena-Bustillos, R.J.; McHugh, T.H. Nanocomposite Edible Films from Mango Puree Reinforced with Cellulose Nanofibers. *J. Food Sci.* **2009**, *74*, N31–N35. [CrossRef] [PubMed]

130. Auras, R.A.; Singh, S.P.; Singh, J.J. Evaluation of oriented poly(lactide) polymers vs. existing PET and oriented PS for fresh food service containers. *Packag. Technol. Sci.* **2005**, *18*, 207–216. [CrossRef]

131. Chang, J.-H.; An, Y.U.; Sur, G.S. Poly(lactic acid) nanocomposites with various organoclays. I. Thermomechanical properties, morphology, and gas permeability. *J. Polym. Sci. Part B Polym. Phys.* **2003**, *41*, 94–103. [CrossRef]

132. Thellen, C.; Orroth, C.; Froio, D.; Ziegler, D.; Lucciarini, J.; Farrell, R.; D'Souza, N.A.; Ratto, J.A. Influence of montmorillonite layered silicate on plasticized poly(l-lactide) blown films. *Polymer* **2005**, *46*, 11716–11727. [CrossRef]

133. Petersen, K.; Væggemose Nielsen, P.; Bertelsen, G.; Lawther, M.; Olsen, M.B.; Nilsson, N.H.; Mortensen, G. Potential of biobased materials for food packaging. *Trends Food Sci. Technol.* **1999**, *10*, 52–68. [CrossRef]

134. Sanchez-Garcia, M.D.; Lagaron, J.M. Novel clay-based nanobiocomposites of biopolyesters with synergistic barrier to UV light, gas, and vapour. *J. Appl. Polym. Sci.* **2010**, *118*, 188–199. [CrossRef]

135. Versino, F.; Lopez, O.V.; Garcia, M.A.; Zaritzky, N.E. Starch-based films and food coatings: An overview. *Starch Stärke* **2016**, *68*, 1026–1037. [CrossRef]

136. López, O.V.; Castillo, L.A.; García, M.A.; Villar, M.A.; Barbosa, S.E. Food packaging bags based on thermoplastic corn starch reinforced with talc nanoparticles. *Food Hydrocoll.* **2015**, *43*, 18–24. [CrossRef]

137. Ibrahim, H.; Farag, M.; Megahed, H.; Mehanny, S. Characteristics of starch-based biodegradable composites reinforced with date palm and flax fibers. *Carbohydr. Polym.* **2014**, *101*, 11–19. [CrossRef] [PubMed]

138. Park, H.-M.; Lee, W.-K.; Park, C.-Y.; Cho, W.-J.; Ha, C.-S. Environmentally friendly polymer hybrids Part I Mechanical, thermal, and barrier properties of thermoplastic starch/clay nanocomposites. *J. Mater. Sci.* **2003**, *38*, 909–915. [CrossRef]

139. Gorrasi, G.; Pantani, R.; Murariu, M.; Dubois, P. PLA/Halloysite Nanocomposite Films: Water Vapor Barrier Properties and Specific Key Characteristics. *Macromol. Mater. Eng.* **2014**, *299*, 104–115. [CrossRef]

140. Liu, S.; Cai, P.; Li, X.; Chen, L.; Li, L.; Li, B. Effect of film multi-scale structure on the water vapor permeability in hydroxypropyl starch (HPS)/Na-MMT nanocomposites. *Carbohydr. Polym.* **2016**, *154*, 186–193. [CrossRef] [PubMed]

141. Chang, P.R.; Jian, R.; Yu, J.; Ma, X. Fabrication and characterisation of chitosan nanoparticles/plasticised-starch composites. *Food Chem.* **2010**, *120*, 736–740. [CrossRef]

142. Yang, Z.; Xie, J.; Zhu, J.; Kang, C.; Chiang, C.; Wang, X.; Wang, X.; Kuang, T.; Chen, F.; Chen, Z.; et al. Functional exosome-mimic for delivery of siRNA to cancer: In vitro and in vivo evaluation. *J. Controll. Release* **2016**, *243*, 160–171. [CrossRef] [PubMed]

143. Yang, Z.; Chang, L.; Chiang, C.-l.; James Lee, L. Micro-/nano-electroporation for active gene delivery. *Curr. Pharm. Des.* **2015**, *21*, 6081–6088. [CrossRef] [PubMed]

144. Yang, Z.; Chang, L.; Li, W.; Xie, J. Novel biomaterials and biotechnology for nanomedicine. *Eur. J. BioMed. Res.* **2015**, *1*, 1–2. [CrossRef]

145. Zhou, C.; Yang, Z.; Teng, L. Nanomedicine based on Nucleic Acids: Pharmacokinetic and Pharmacodynamic Perspectives. *Curr. Pharm. Biotechnol.* **2014**, *15*, 829–838. [CrossRef] [PubMed]

146. Yang, Z.; Yu, B.; Zhu, J.; Huang, X.; Xie, J.; Xu, S.; Yang, X.; Wang, X.; Yung, B.C.; Lee, L.J.; et al. A microfluidic method to synthesize transferrin-lipid nanoparticles loaded with siRNA LOR-1284 for therapy of acute myeloid leukemia. *Nanoscale* **2014**, *6*, 9742. [CrossRef] [PubMed]

147. Xie, J.; Teng, L.; Yang, Z.; Zhou, C.; Liu, Y.; Yung, B.C.; Lee, R.J. A Polyethylenimine-Linoleic Acid Conjugate for Antisense Oligonucleotide Delivery. *BioMed. Res. Int.* **2013**, *2013*, 1–7. [CrossRef] [PubMed]

148. Giavaresi, G.; Fini, M.; Salvage, J.; Nicoli Aldini, N.; Giardino, R.; Ambrosio, L.; Nicolais, L.; Santin, M. Bone regeneration potential of a soybean-based filler: Experimental study in a rabbit cancellous bone defects. *J. Mater. Sci. Mater. Med.* **2009**, *21*, 615–626. [CrossRef] [PubMed]

149. Tang, J.; Li, X.; Bao, L.; Chen, L.; Hong, F.F. Comparison of two types of bioreactors for synthesis of bacterial nanocellulose tubes as potential medical prostheses including artificial blood vessels. *J. Chem. Technol. Biotechnol.* **2017**, *92*, 1218–1228. [CrossRef]

150. McCullen, S.D.; Ramaswamy, S.; Clarke, L.I.; Gorga, R.E. Nanofibrous composites for tissue engineering applications. *Wiley Interdiscip. Rev. Nanomed. Nanobiotechnol.* **2009**, *1*, 369–390. [CrossRef] [PubMed]

151. Bodin, A.; Bäckdahl, H.; Fink, H.; Gustafsson, L.; Risberg, B.; Gatenholm, P. Influence of cultivation conditions on mechanical and morphological properties of bacterial cellulose tubes. *Biotechnol. Bioeng.* **2007**, *97*, 425–434. [CrossRef] [PubMed]

152. Feng, C.; Khulbe, K.C.; Matsuura, T. Recent progress in the preparation, characterization, and applications of nanofibers and nanofiber membranes via electrospinning/interfacial polymerization. *J. Appl. Polym. Sci.* **2010**, *115*, 756–776. [CrossRef]

153. Shi, M.; Yang, R.; Li, Q.; Lv, K.; Miron, R.J.; Sun, J.; Li, M.; Zhang, Y. Inorganic Self-Assembled Bioactive Artificial Proto-Osteocells Inducing Bone Regeneration. *ACS Appl. Mater. Interfaces* **2018**, *10*, 10718–10728. [CrossRef] [PubMed]

© 2018 by the authors. Licensee MDPI, Basel, Switzerland. This article is an open access article distributed under the terms and conditions of the Creative Commons Attribution (CC BY) license (http://creativecommons.org/licenses/by/4.0/).

polymers

MDPI

Article

Furanoate-Based Nanocomposites: A Case Study Using Poly(Butylene 2,5-Furanoate) and Poly(Butylene 2,5-Furanoate)-*co*-(Butylene Diglycolate) and Bacterial Cellulose

Marina Matos [1], Andreia F. Sousa [1,*], Nuno H. C. S. Silva [1], Carmen S. R. Freire [1], Márcia Andrade [2], Adélio Mendes [2] and Armando J. D. Silvestre [1]

[1] CICECO—Aveiro Institute of Materials, Departmento de Química, Universidade de Aveiro, 3810-193 Aveiro, Portugal; marina.matos@ua.pt (M.M.); nhsilva@ua.pt (N.H.C.S.S.); cfreire@ua.pt (C.S.R.F.); armsil@ua.pt (A.J.D.S.)

[2] Laboratory for Process Engineering, Environment, Biotechnology and Energy (LEPABE), Faculdade de Engenharia da Universidade do Porto, Rua Dr. Roberto Frias, 4200-465 Porto, Portugal; mrsa@fe.up.pt (M.A.); mendes@fe.up.pt (A.M.)

* Correspondence: andreiafs@ua.pt; Tel.: +351-234-370-200

Received: 10 July 2018; Accepted: 22 July 2018; Published: 24 July 2018

Abstract: Polyesters made from 2,5-furandicarboxylic acid (FDCA) have been in the spotlight due to their renewable origins, together with the promising thermal, mechanical, and/or barrier properties. Following the same trend, (nano)composite materials based on FDCA could also generate similar interest, especially because novel materials with enhanced or refined properties could be obtained. This paper presents a case study on the use of furanoate-based polyesters and bacterial cellulose to prepare nanocomposites, namely acetylated bacterial cellulose/poly(butylene 2,5-furandicarboxylate) and acetylated bacterial cellulose/poly(butylene 2,5-furandicarboxylate)-*co*-(butylene diglycolate)s. The balance between flexibility, prompted by the furanoate-diglycolate polymeric matrix; and the high strength prompted by the bacterial cellulose fibres, enabled the preparation of a wide range of new nanocomposite materials. The new nanocomposites had a glass transition between −25–46 °C and a melting temperature of 61–174 °C; and they were thermally stable up to 239–324 °C. Furthermore, these materials were highly reinforced materials with an enhanced Young's modulus (up to 1239 MPa) compared to their neat copolyester counterparts. This was associated with both the reinforcing action of the cellulose fibres and the degree of crystallinity of the nanocomposites. In terms of elongation at break, the nanocomposites prepared from copolyesters with higher amounts of diglycolate moieties displayed higher elongations due to the soft nature of these segments.

Keywords: 2,5-furandicarboxylic acid; poly(1,4-butylene 2,5-furandicarboxylate); biobased materials; bacterial cellulose; nanocomposites; mechanical properties

1. Introduction

The last decades have seen a burgeoning search for more sustainable chemicals, polymers, and materials due to severe environmental concerns and to the announced depletion of fossil resources [1]. In this context, renewable-based chemicals, such as those derived from C5 and C6 biomass sugars, namely the 2,5-furandicarboxylic acid (FDCA), and the polyesters thereof, have been in the spotlight [2]. Some of the most successful examples, due to their promising properties, comparable to fossil-based terephthalate homologues, including poly(ethylene 2,5-furandicarboxylate) (PEF) [3,4], and poly(1,4-butylene 2,5-furandicarboxylate) (PBF) [5–14], also known as poly(ethylene 2,5-furanoate) and poly(1,4-butylene 2,5-furanoate), respectively. They are expected to replace poly(ethylene terephthalate) (PET) and poly(1,4-butylene terephthalate) (PBT), respectively, on various conventional applications of thermoplastics, such as, for example in packaging materials in the case of PEF or in electronic applications in the case of PBF [2].

Furthermore, FDCA-derived copolyesters have also been extensively studied, with the aim of expanding or refining even further the properties and/or potential applications of their parent homopolymers [7,15–22]. Amongst the wide library of these copolymers, furanoate-aliphatic copolyesters were the most studied [2,15–17,21–30], and those incorporating ether linkages, such as the work of Lotti et al. based on diglycolic acid [25] or of Sousa et al. using poly(ethylene glycol) [16], are particularly interesting. For example, the 100% renewable poly(butylene 2,5-furanoate)-*co*-(butylene diglycolate)s (with 60 to 90 mol % of furanoate moieties), henceforth designated by PBF-*co*-PBDG, are biodegradable and could have an elongation at break of up to four times higher than PBF [25]. In fact, the incorporation of high quantities of soft butylene diglycolate units brings significant improvement in the elongation, but at the expense of the Young's Modulus (roughly 10 times lower compared to PBF). In terms of gas barrier properties, PBF-*co*-PBDG can exhibit adequate behaviour for packaging materials applications. The oxygen gas transmission rate (GTR) varied between 111–193 $cm^3 \, m^{-2} \, d^{-1} \, bar^{-1}$ [25].

More recently, (nano)composite and hybrid materials based on furanoate-based polymeric matrices have also been developed [31–36], although still they are modestly and mostly restricted to PEF. However, the significant properties improvement of the ensuing materials, relevant to their processing and/or application (e.g., crystallisation rate improvement), will predictably foster their rapid development in the near future. PEF-derived hybrid materials were prepared by compounding PEF with inorganic fillers, added during the synthesis of the polymer. For example, Bikiaris and co-workers [32] demonstrated that the in situ preparation of PEF/SiO_2 and PEF/TiO_2 hybrid materials, during solid state polymerization, lead to slightly higher molecular weight PEF due to the presence of the SiO_2 or TiO_2 nanoparticles. Lotti et al. [35] also synthesised hybrid materials based on PEF containing either graphene oxide or multi walled carbon nanotubes (non-functionalised, or functionalised with –COOH or –NH_2 groups). Differential scanning calorimetry analysis indicated that all the fillers acted as nucleating agents for the PEF crystallisation, albeit in a different extent. Other works are focused on PEF-derived (nano)composites with cellulose fibres [31,36], organically modified montmorillonite clays and sepiolite clays [33,34]. Of particular interest is the work carried out by the Guigo and Sbirrazzuoli group on PEF composites using small quantities of nanocrystalline cellulose (around 4 wt %) and prepared via twin screw extrusion [31] or solvent casting [36]. These composites have enhanced crystallisation properties in the presence of the fibres, namely faster crystallisation [31] and nucleating effects [36], despite some compatibility problems associated with the hydrophilic nature of pristine cellulose compared to PEF homopolyester [36]. Adding to this, nanocellulose fibres, in particular bacterial cellulose (BC) produced by *Gluconoacetobacter sacchari* bacterial strain at high purity, due to its nanofibrillar structure having unique physical and chemical properties as a nanocomposite [37,38], including optically transparency and high mechanical strength [39]. However, to the best of our knowledge, BC has never before been used in the preparation of furanoate-based nanocomposites. In this vein, this study presents a new family of fully bio-based nanocomposites, prepared from a series of PBF-*co*-PBDG copolyesters, or PBF, and modified bacterial cellulose previously treated with heterogeneous acetylation (to improve compatibility with the thermoplastic matrices). These PBF-*co*-PBDG and PBF-acetylated-BC nanocomposites were chosen as a case study for the broader development of furanoate-based nanocomposites, and in particular for their potential to enhance their mechanical properties. The newly prepared nanomaterials were fully characterised through several structural, thermal, and mechanical techniques, as well as in terms of gas permeability, aiming to access their potential use for packaging applications.

Polymers **2018**, *10*, 810

2. Experimental

2.1. Materials

Bacterial cellulose in the form of wet membranes was produced using the *Gluconoacetobacter sacchari* bacterial strain and conventional culture medium conditions, as described elsewhere [40]. 2,5-Furandicarboxylic acid (FDCA, >98%) and 1,1,1,3,3,3-hexafluoro-2-propanol (HFP, >99%) were purchased from TCI Europe NV. Diglycolic acid (DGA, 98%), 1,4-butanediol (BD, 99%), titanium (IV) tert-butoxide (Ti(OBu)$_4$, pro-analysis), trifluoroacetic acid, (TFA, 99%) and deuterated trifluoroacetic acid (TFA-d, 99 atom % D) were supplied by Sigma-Aldrich Chemicals Corporation (Sintra, Portugal). Sulfuric acid (H$_2$SO$_4$, 96%) was supplied by Acros Organic (Geel, Belgium). All chemicals were used as received.

2.2. Heterogeneous Acetylation of Bacterial Cellulose

Prior to heterogeneous acetylation, the BC wet membrane was disintegrated using a blender and an Ultra-Turrax equipment (15 min at 20,500 rpm), and solvent exchanged with ethanol and acetone (in triplicate). Heterogeneous acetylation of BC fibres was then carried out following a well-established protocol described elsewhere [39]. Briefly, acetic anhydride (225 mL) was placed in a 500 mL round flask into an ice bath for 20 min, then 1 mL of H$_2$SO$_4$ was added, and finally the wet BC fibres ($\approx$40 g) were added to the mixture. The reaction was allowed to proceed under stirring for 4 h at 30 °C. The ensuing BC-acetylated fibres (Ac-BC) were filtered and sequentially washed with water, acetone, ethanol, water, and again with ethanol. Finally, Ac-BC nanofibres were Soxhlet-extracted with ethanol for 12 h to remove any residual trace of acetic anhydride or other impurities, and solvent exchanged with acetone followed by chloroform.

2.3. Preparation of the Acetylated BC/Poly(Butylene Furandicarboxylate-co-Butylene Diglycolate) Nanocomposites (Ac-BC/PBF-co-PBDG)

2.3.1. Synthesis of PBF-*co*-PBDG Copolyesters and Corresponding Homopolyesters

The polyesters were synthesized via a procedure described elsewhere [18,41]: Fisher esterification of FDCA, esterification, and finally polycondensation reaction. In brief, dimethyl 2,5-dimethylfurandicarboxylate (DMFDC) was first prepared by reacting FDCA (192.2 mmol) with an excess of methanol (364 mL), under acidic conditions (HCl, 15 mL), at 80 °C for 15 h. The reaction mixture was allowed to cool down, and the ensuing white precipitate was isolated by filtration in 70% yield. Secondly, DMFDC and DGA (mol % DMFDC/mol % DGA $\approx$ 90/10, 75/25, 50/50, 25/75 and 10/90) were mixed with an excess of BD (1.5 mol per 1 mol of DMFDC and DGA) under a nitrogen atmosphere. The temperature was then raised to 110 °C, Ti(OBu)$_4$ catalyst (1.4 mmol) added and the temperature was again progressively raised to 200 °C. Here, a slightly different procedure was followed, depending on the polyester being synthesized.

In the case of PBF, PBF-*co*-PBDG-90/10, 75/25, and 50/50 (i.e., those polyesters prepared from higher amounts of DMFDC) the reaction mixture was kept at 200 °C for 4 h. Then, the reaction proceeded by applying a vacuum (ca. 10^{-3} bar) for 1 h. Subsequently, the temperature was raised again to approximately 210 °C and kept at that temperature for more than 4 h. In the other cases of PBDG, PBF-*co*-PBDG-25/75 and 10/90, the period at 200 °C was only 2 h, followed by an additional 2 h period, at 210 °C. In the third-step, the reaction proceeded at 210 °C under vacuum, and then the temperature was raised to 220 °C for 4 h.

Then, the mixture was purified by dissolving the polymers in TFA (20 mL), pouring in an excess of cold methanol (ca. 1 L), separation by filtration, and drying. The isolation yields of the polymers were ca. 60%, which was in accordance with previous results [16].

2.3.2. Preparation of Ac-BC/PBF-*co*-PBDG Nanocomposites, and Corresponding Homopolyesters Nanocomposites

The nanocomposites were prepared by a well-known solvent casting approach. The polyesters (0.21 g) were mixed with a BC or Ac-BC chloroform dispersion (0.0045 g/mL, 20 mL) under magnetic stirring for 3 h. The mixture was then deposited into a square Teflon mould (6.5 cm^2) and the films were cast, at room temperature, for a minimum of 15 h, and finally heated at 30 °C, under vacuum, for 12 h to remove any remaining solvent. The ensuing films had a thickness of approximately 0.098 ± 0.001 mm.

2.4. Characterisation Techniques

Attenuated total reflectance Fourier transform infrared (ATR FTIR) spectra were obtained using a PARAGON 1000 Perkin-Elmer FTIR spectrometer equipped with a single-horizontal Golden Gate ATR cell. The spectra were recorded after 128 scans, at a resolution of 4 cm^{-1}, within a range of 500 to 4000 cm^{-1}. ^{1}H and ^{13}C nuclear magnetic resonance (NMR) spectra were recorded using a Bruker AMX 300 spectrometer, operating at 300 or 75 MHz, respectively. All chemical shifts (δ) were expressed as parts per million, downfield from tetramethylsilane (used as the internal standard). Elemental analyses (C and H) were conducted in triplicate using a LECO TruSpec analyser. The degree of substitution (DS) was estimated through the approach of Vaca-Garcia et al. [42]: DS= (5.13766 − 11.5592 × C)/(0.996863 × C − 0.856277 × n + n × C) where n and C stand for the number of carbon atoms in the acyl group and for the carbon contents, respectively.

Scanning electron microscopy (SEM) images of the surface and cross-sections of films were acquired using a field emission gun-SEM Hitachi SU70 microscope operating at 4 kV. Samples were deposited onto a sample holder and coated with carbon twice.

X-ray diffraction (XRD) analyses were performed using a Philips X'pert MPD diffractometer operating with CuKα radiation (λ = 1.5405980 Å) at 40 kV and 50 mA. Samples were scanned in the 2θ range of 5° to 50°, with a step size of 0.04°, and a time per step of 50 s.

Differential scanning calorimetry (DSC) thermograms were obtained with a DSC Q100 V9.9 Build 303 (Universal V4.5A) calorimeter from Texas Instruments, using steel DSC pans. Scans were carried out under nitrogen with a heating rate of 10 °C min^{-1} in the temperature range of −90 to 250 °C. Two heating/cooling cycles were repeated. Glass transition (T$_g$) was determined using the midpoint approach (second heating trace); melting (T$_m$) and crystallisation (T$_{cc}$) temperatures were determined as the maximum of the exothermic crystallisation peak, and the minimum of the melting endothermic peak during the second heating cycle, respectively.

Thermogravimetric analyses (TGA) were carried out with a Setaram SETSYS analyser equipped with an alumina plate. Thermograms were recorded under a nitrogen flow of 20 mL min^{-1} and heated at a constant rate of 10 °C min^{-1} from room temperature up to 800 °C. Thermal decomposition temperatures were taken at the onset of significant weight loss (5%) and at maximum decomposition temperatures from the heated samples (T$_{d,5\%}$ and T$_d$, respectively).

Tensile tests were obtained with an Instron 5564 tensile testing machine at a cross-head speed of 10 mm min^{-1} using a 500 N static load cell. The tensile test specimens were rectangular strips (50 mm × 10 mm) pre-conditioned for 72 h at 50% humidity and 30 °C. Each measurement was repeated at least five times.

Contact angle (CA$_{water}$) measurements with water were carried out using a Contact Angle System OCA20 goniometer (DataPhysics, Filderstadt, Germany) with SCA20 software using the sessile drop approach, and recorded during 40 s. Water was used as probe liquid, and for each specimen, drops of 3 μL were deposited using a syringe (50 μL) onto the nanocomposite film surface. The error analysis was obtained by the standard deviation of at least five independent determinations.

Permeation measurements were performed in a system that included a membrane cell connected to a tank with a calibrated volume (at the permeate side) and to a gas cylinder (at the feed side). Prior to permeation tests, the films were glued to steel O-rings with an epoxy glue (Araldite® Standard, Huntsman Advanced Materials, Basel, Switzerland); the glue was also applied along the interface of the steel O-ring and the film, as described elsewhere [43]. A sintered metal disc covered with a filter paper was used as support for the film in the test cell. Single gases were tested at 30 °C, where the feed pressure was 1 bar and the permeate pressure was ca. 0.03 bar. The tests were performed in a standard pressure-rise setup using an acquisition program based on LabView® platform (National Instruments, Austin, TX, USA). The permeability towards a pure component *i* was determined accordingly to: $L_i = \frac{F_i}{\Delta P_i/l}$, where F_i is the flux of species *i*, ΔP_i is the partial pressure difference of species between the two sides of the membrane, and *l* is the film thickness. The permeability to the pure component was computed from the experimental data as follows: $L_i = \frac{lV_p v_M}{RTA(P_f - P_p)}\frac{\Delta P_p}{\Delta t}$, where V_p is the volume of the permeate tank, v_M is the molar volume of the gas at normal conditions, R is the gas constant, T is the absolute temperature, t is the time, A is the effective permeating area of the film, and P_f and P_p are the feed pressure and permeate pressure, respectively, and ΔP_p is the permeate pressure increment for the elapsed time Δt.

3. Results

3.1. From Furanoate-Glycolate Copolyesters to Acetylated Bacterial Cellulose-Based Nanocomposites

A series of Ac-BC/PBF-*co*-PBDGs, Ac-BC/PBF, and Ac-BC/PBDG nanocomposites were developed for the first time following a three-step procedure (Scheme 1).

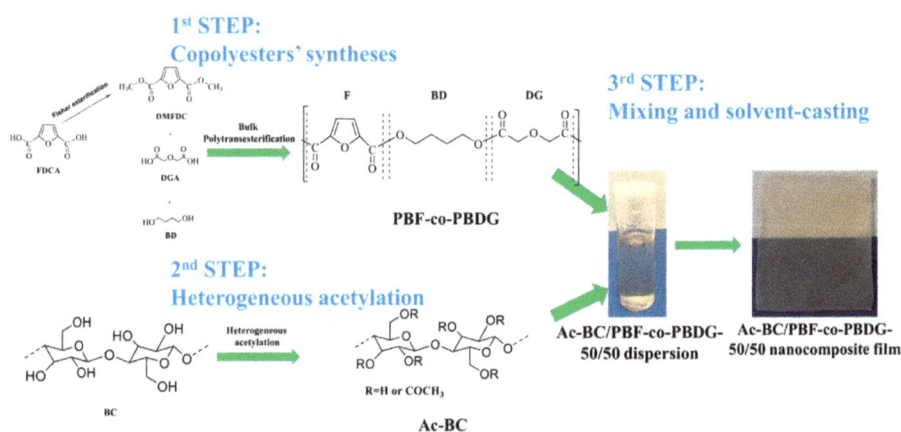

Scheme 1. The Ac-BC/PBF-*co*-PBDG composite preparation approach.

In the first step, the (co)polyesters were prepared by a conventional bulk polyesterification approach [18,41]. These (co)polyesters spanned from the neat PBF to neat PBDG, and encompassed their copolyesters with different relative furanoate/digycolate amounts (90/10, 75/25, and the never reported 50/50, 25/75, and 10/90 mol %). The selection of these (co)polyesters was based on their promising properties, notably biodegradability and high elongation at break [25], and aiming to further improve their mechanical properties.

In the second step, heterogeneous acetylation of the BC fibres were performed using acetic anhydride. The degree of acetylation (DS) of the Ac-BC was determined using elemental analysis, by the Vaca-Garcia et al. [42] approach, and the resulting value was 0.87.

In the third step, nanocomposite films of each (co)polyester and Ac-BC were obtained by solvent-casting aiming to obtain novel nanomaterials with enhanced mechanical properties. Importantly, this approach could be generally applied to other furanoate thermoplastics as a strategy to improve their mechanical performance, namely to recycled PEF. Recycled thermoplastics lose some of their high-performance mechanical properties, mostly due to a reduction of the molecular weight. However, compounding these thermoplastics with the Ac-BC nanofibres could play a reinforcing role.

The relative amount of (co)polyester/Ac-BC used in this work in the nanocomposites preparation was approximately equal to 70/30 wt %, based on the fact that a minimum of 30 wt % of Ac-BC was required to form the films. For comparison reasons, films of each individual component of the nanocomposites were additionally prepared. Pure Ac-Bc generates a white thin film, but the neat copolyesters did not form films by solvent-casting; in fact, this was consistent with the fact that a minimum of 30 wt % of Ac-BC was needed in order to obtain the nanocomposites films.

For comparison reasons, nanocomposites of non-acetylated BC/PBF-*co*-PBDGs were also prepared following a similar approach. However, these materials were shown to be very heterogeneous; hence they were no further investigated. On the contrary the nanocomposite materials prepared using the Ac-BC fibres were homogeneous and translucent, indicating a good dispersion of the modified BC in the thermoplastic polymeric matrices.

3.2. Structure and Morphology

The starting (co)polymers components were studied ^{1}H and ^{13}C NMR analysis. The main results are recorded in the Supplementary data (Figure S1, and also Tables S1 and S2), and were consistent with previously published data [25]. One important aspect studied, due to the influence on the final properties of the (co)polyesters and consequently also on the related nanocomposites, were the assessment of the real furanoate/diglycolate incorporation (Table S2). Results indicated a trend towards incorporating slightly more diglycolate moieties in the copolymer back-bone than in the initial feed ratio, except for PBF-*co*-PBDG-50/50 copolyester (7 mol % higher than expected).

All furanoate-based nanocomposites and corresponding components (Ac-BC, PBF-*co*-PBDG, PBF, and PBDG polyesters) were also thoroughly characterised by means of ATR FTIR spectroscopy (Figure 1 and Figures S2–S4 of Supplementary data).

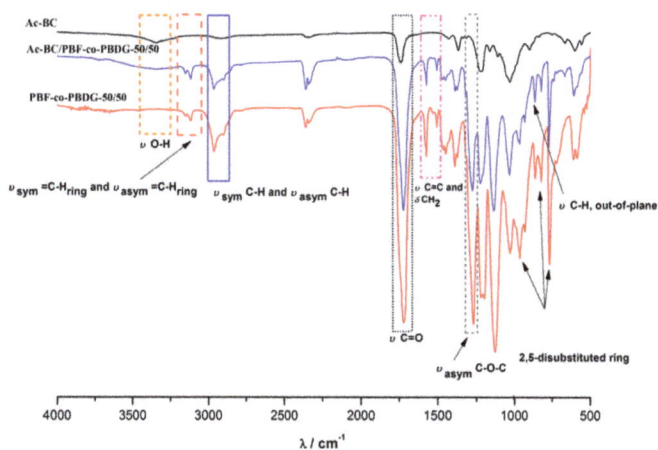

Figure 1. Attenuated total reflectance Fourier transform infrared (ATR FTIR) spectra of Ac-BC/PBF-*co*-PBDG-50/50 nanocomposite and corresponding Ac-BC and PFB-*co*-PBDG-50/50 components.

The Ac-BC/PBF-*co*-PBDG nanocomposites displayed the typical vibration modes of furanoate-based polyesters [44] and in particular of PBF-*co*-PBDGs: two week bands centred at 3150 and 3115 cm^{-1} arising from the symmetrical and asymmetrical C–H stretching of the furanic ring (ν_{sym} = C–H$_{ring}$ and ν_{asym} = C–H$_{ring}$), two other weak bands near 2962 and 2890 cm^{-1} arising from the symmetrical and asymmetrical C–H stretching characteristics of the methylene groups of the BD and diglycolate moieties (ν_{sym} C–H and ν_{asym} C–H), and a very intense band centred at 1720 cm^{-1}, arising from the carbonyl stretching vibration, typical of ester groups (ν C=O). In addition, these spectra also showed a band near 1506 cm^{-1}, assigned to both the C=C stretching and CH$_2$ in plane deformation (ν C=C, δ CH$_2$, respectively), a band near 1263 cm^{-1} arising from the ν_{asym} C–O–C stretching, and several vibrations in the finger print region related to the 2,5-disubstitued ring. The vibrational modes of acetylated BC and those of the polyesters were partially overlapped, as can be confirmed by inspection of the corresponding spectra of Figure 1. However, a distinct feature of the nanocomposites spectra due to the cellulose incorporation was the broad band detected near the 3351 cm^{-1} characteristics of the ν O–H. All of the characteristic vibrational features of Ac-BC and BC precursors are summarised in the Supplementary data (Figure S2).

With regards to the morphology, SEM micrographs of the nanocomposites with higher content of diglycolate units ($\geq$50 mol %), collected at different magnifications (Figure 2 and Figure S5 of Supplementary data), showed a smoother and uniform surface, thus indicating enhanced compatibility between the fibres and those polymeric matrices. From this perspective, the amount of Ac-BC used (around 30 wt %) and the heterogeneous acetylation of the cellulose fibres carried out in order to increase the cellulose hydrophobicity (DS $\approx$ 0.87) and, thus, the compatibility between the modified fibres and the polyesters, was shown to be an adequate approach for obtaining homogeneous nanocomposites, especially in the case of Ac-Bc/PBF-*co*-PBDG-10/90 and -25/75, and Ac-BC/PBDG. A more extensive acetylation of the fibres could, in principle, increase the compatibility of the more hydrophobic furanoate polyesters (such as, PBF, PBF-*co*-PBDG-90/10) [25], but this would also have disrupted the characteristic BC nanostructure and thus would have extensively affected the properties of the ensuing materials. Another, possibility, worth considering in future work, will be the addition of an extra compatibility agent, or even a plasticiser acting also as a compatibility agent. Nevertheless, this would have brought an extra complexity to the data interpretation of the nanocomposite-systems, deviating from the present study as a more in-depth analysis of the basic principles governing cellulose/PBF-*co*-PBDG properties.

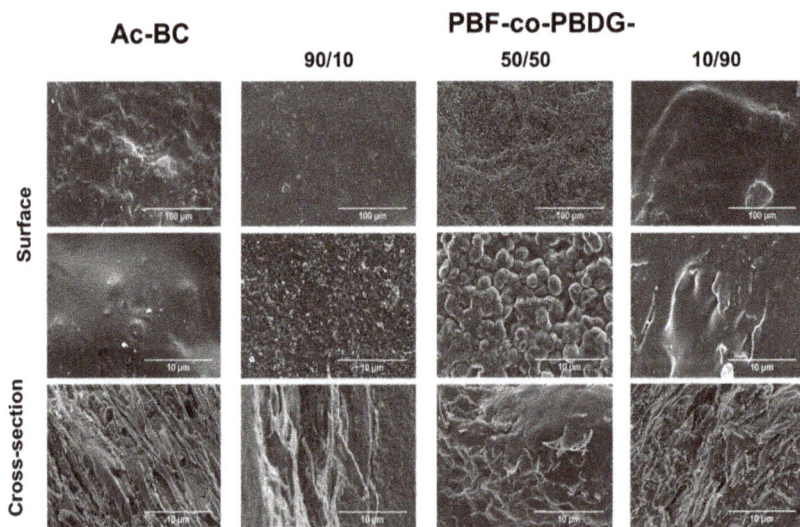

Figure 2. Surface (**top**) and cross-section (**bottom**) SEM micrographs of Ac-BC film and of selected nanocomposite films.

It is evident from the cross-section pictures (Figure 2 and Figure S5 of Supplementary data), the presence of Ac-BC nanofibres embedded within the polymeric matrix. Further, these results confirmed that the interfacial adhesion between the Ac-BC fibres and the polymeric matrices was particularly good for the high diglycolate content polyesters, namely PBF-*co*-PBDG-10/90, -25/75, and PBDG.

The nanocomposite hydrophobicity was evaluated through water contact angle (CA$_{water}$) measurements at several points in time for 40 s after the water droplet deposition. The main results are displayed in Figure 3 and summarised in Table S3 of Supplementary data.

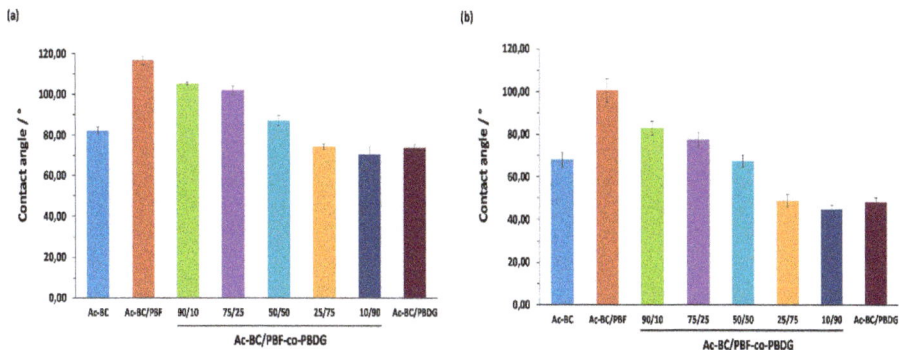

Figure 3. Water contact angles at (**a**) 0 and (**b**) 15 s.

The CA$_{water}$ decreased drastically over the initial 5 s, and then roughly maintained constant. This behaviour was due to the initial re-orientation of the functional groups at surface of the films, allowing the water drops to spread more easily [45]. Among different nanocomposites, the CA$_{water}$, after 15 s, increased with the increasing furanoate content in the copolyester (from 45 to 100°), mostly due to the hydrophobic character of the furanoate-based polyesters [41]. The Ac-BC film showed an intermediate CA$_{water}$ of approximately 67.9° after 15 s, in accordance with nanofibre affinity to the polyesters, and consequently good dispersion in the thermoplastic matrices, especially PBF-*co*-PBDG-50/50 to -10/90, and PBDG. The wide range of water contact angles covered by these nanomaterials, from highly hydrophobic (ca. 100.57°) to moderate hydrophilic (ca. 44.81°) was an interesting feature worth exploiting in different applications, such as, for example in packaging [25] or textiles.

3.3. Cristallinity and Thermal Behaviour

The nature of the crystalline domains of the nanocomposites prepared with a wide range of furanoate/diglycolate copolyesters and with Ac-BC fibres was evaluated by XRD (Figure 4 and Figure S6 of Supplementary Materials).

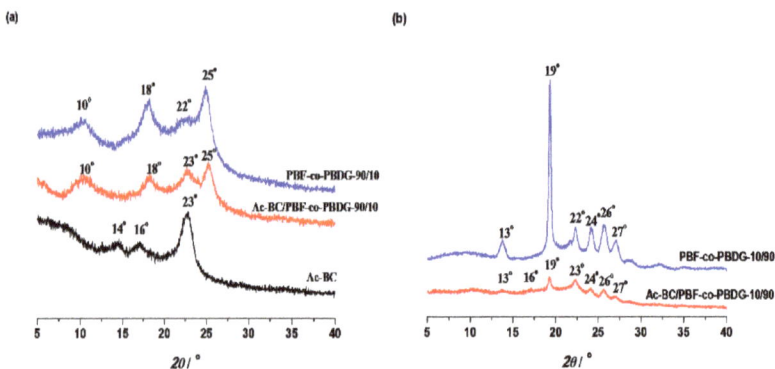

Figure 4. X-Ray diffractograms of: (**a**) Ac-BC/PBF-*co*-PBDG-90/10 nanocomposite film and corresponding Ac-BC film and PBF-*co*-PBDG-90/10 components, and (**b**) Ac-BC/PBF-*co*-PBDG-10/90 nanocomposite film and corresponding Ac-BC film and PBF-*co*-PBDG-10/90 components.

The nanocomposites prepared with the copolyesters containing a higher amount of furanoate moieties (i.e., PBF-*co*-PBDG-90/10 to 50/50) roughly displayed the typical diffraction pattern of PBF, with strong reflections at $2\theta \approx 18$ and $25°$, and smaller peaks at $2\theta \approx 10$ and $22°$ [8]. In the case of the diffractogram of the nanocomposite prepared with the copolyester containing the lowest amount of furanoate moieties (Ac-BC/PBF-*co*-PBDG-10/90), the main peaks observed were those typical of PBDG precursors, viz: $2\theta \approx 14$, 19, 22, 24, 26, and $27°$ [25]. These results allowed one to associate the crystalline domain of Ac-BC/PBF-*co*-PBDG-90/10, 75/25, and 50/50 nanocomposites to PBF, whereas in the case of Ac-BC/PBF-*co*-PBDG-10/90, it was essentially related to PBDG. In addition, the XRD diffraction patterns of the nanocomposite films were all naturally related with the copolyesters counterparts (in the form of powder), despite some differences in the sharpness of the reflection peaks, as easily attested by comparing both (Figure S6 of Supplementary data). These results could be associated with the incorporation of Ac-BC fibres into the polymeric matrix and/or due to solvent casting film formation conditions.

In the particular case of Ac-BC/PBF-*co*-PBDG-25/75, a more pronounced effect was noted; indeed, this nanocomposite was amorphous, displaying accordingly on its diffractogram a pronounced amorphous halo cantered at $22°$, despite its precursor displaying some crystallinity (see Figure S6 of Supplementary Materials).

Importantly, the thermal and mechanical behaviour of all nanocomposites were influenced by their degree of crystallinity and also by the nature of this domain, as discussed below.

All Ac-BC/PBF-*co*-PBDG nanocomposites and the corresponding individual components precursors were further characterised in terms of their thermal behaviour through DSC and TGA analyses (Table 1, Figure 5 and, Table S4 and Figures S7–S9 of Supplementary Materials).

Table 1. Important thermal values obtained from differential scanning calorimetry (DSC) and thermogravimetric analysis (TGA) analyses.

Sample	T_{cc} [1]/°C	T_g [1]/°C	T_m [1]/°C [1]	$T_{d,5\%}$ [2]/°C	$T_{d,max}$ [2]/°C
Ac-BC/PBF	86.5	46.1	173.5	323.8	354.7; 384.2
Ac-BC/PBDG	-	−24.9	66.1 [3]	284.0	362.1; 384.0
		Ac-BC/PBF-*co*-PBDG-			
90/10	76.3	25.8	162.9	305.8	354.9; 383.0
75/25	60.3	15.2	144.8	300.2	353.9; 376.6
50/50	-	−1.8	94.6	297.9	348.2; 380.7
25/75	-	−12.6	-	238.8	362.3; 378.6
10/90	-	−20.4	61.4 [3]	293.6	359.8; 384.8

[1] Determined by DSC from the second heating scan at 10 °C min^{-1}. [2] Determined by TGA at 20 °C min^{-1}.
[3] Determined by DSC from the first heating scan at 10 °C min^{-1}.

The DSC traces of the nanocomposites (Table 1 and Figure 5) were in accordance with the semi-crystalline nature of the nanocomposites or instead with the amorphous character of one of these materials, as XRD results indicated. Therefore, the DSC traces of Ac-BC/PBF, Ac-BC/PBF-*co*-PBDG-90/10, -50/50 and -10/90, and Ac-Bc/PBDG displayed a glass transition (T_g), followed by a melting (T_m) event at 46.1 to −24.9 °C, and 173.5 to 61.4 °C, respectively. An additional cold crystallisation (T_{cc}) event was also observed after the T_g in the cases of Ac-BC/PBF, PBF-*co*-PBDG-90/10 and -75/25, which might be associated to an additional nucleation effect of Ac-BC fibres [46]. The corresponding traces of neat PBF and PBF-*co*-PBDG-90/10 (co)polyesters (Figure S7 and Table S4 of Supplementary Material) did not showed a cold crystallisation event.

In regard to Ac-BC/PBF-*co*-PBDG-25/75, the corresponding thermogram displayed only a step in the baseline at ca. −12.6 °C, attributed to the glass transition temperature, due to its essentially amorphous nature, in agreement with the XRD results.

For all nanocomposites, T_g decreased (around 13 °C) with an increased amount of diglycolate units in the copolyester. In the same vein, the T_m of the nanocomposites also decreased from 173.5 to 61.4 °C with an increasing amount of soft diglycolate segments. This trend was also observed in the case of neat (co)polyesters prepared in this work (Table S4 of Supplementary Material) and reported elsewhere [25].

In addition, the T_g of the nanocomposite films tended to be higher than those obtained for the corresponding (co)polyester component, in agreement with the higher stiffness of the nanocomposites. For example, Ac-Bc/PBF-*co*-PBDG-75/25 had a T_g of 25.6 °C, whereas the same parameter for PBF-*co*-PBDG-75/25 was 13.8 °C. In regard to the T_m, the nanocomposites (Table 1) and the corresponding copolyesters synthesised in this work (Table S4 of Supplementary Material) had very similar results, but they were higher than literature values [25].

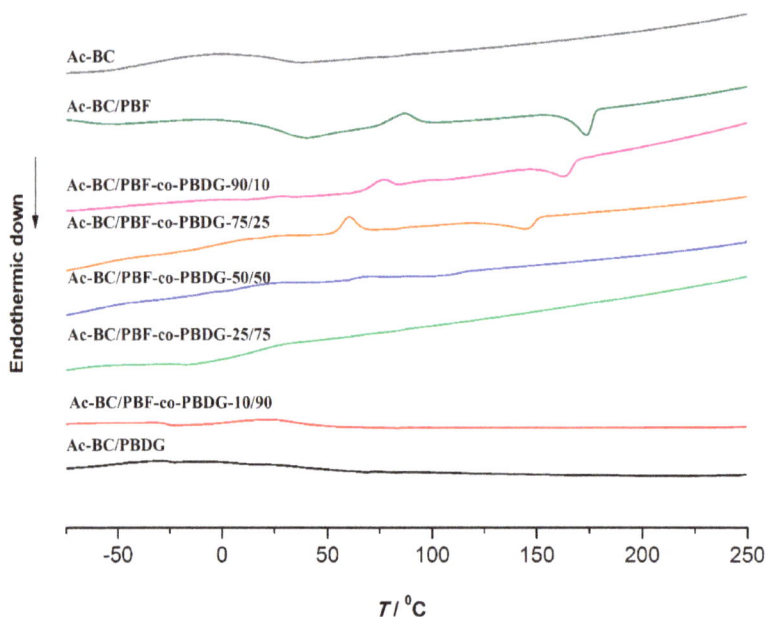

Figure 5. DSC traces of the nanocomposites and Ac-BC.

The typical TGA thermograms of the newly prepared furanoate nanocomposites (Table 1 and Figure S9 of Supplementary Material) displayed two major characteristic steps at maximum decomposition temperatures ($T_{d,max}$) of 348–362 °C and 376–385 °C. The first step was due to the Ac-BC decomposition and was quite close to that observed for the neat Ac-BC fibres (363.0 °C) and comparable to previously reported results [39]. The other decomposition step was associated with the polyesters enriched fraction and was observed at higher temperatures than the related polyester precursor. For example, in the case of Ac-BC/PBF-*co*-PBDG-50/50 the second $T_{d,max}$ was equal to 380.7 °C, whereas the same parameter was equal to 365.1 °C for the PBF-*co*-PBDG-50/50 copolyester. These results were comparable to those reported for other nanocomposites of Ac-BC and poly(lactic acid) [39].

The nanocomposites were thermally stable up to 238–306 °C (Table 1), indicating a decrease of the stability compared to the related polyesters (360–380 °C) (Table S4 of Supplementary Materials). The same effect was previously reported with PEF/cellulose materials [31]. Additionally, one can notice, on both nanocomposites and the corresponding polyester component, an increase of the $T_{d,5\%}$ with the amount of furanoate incorporated into the polymeric matrix backbone. These thermal features enabled the establishment a maximum working temperature of up to 306 °C for the novel nanocomposites.

3.4. Mechanical Properties and Permeability Assays for Oxygen

Tensile tests were performed to assess the mechanical performance of these novel furanoate-based nanocomposites and in particular to evaluate the effect of compounding cellulose nanofibres (Ac-BC) with PBF-*co*-PBDG copolyesters. The stress-strain behaviour of Ac-BC/PBF-*co*-PBDGs was revealed to be dependent on a complex interplay of factors, namely the chemical composition of the related (co)polyesters and the presence of nanofibres, as well as the crystallinity of the new materials. The main results are displayed in Figure 6 and summarised in Table S5.

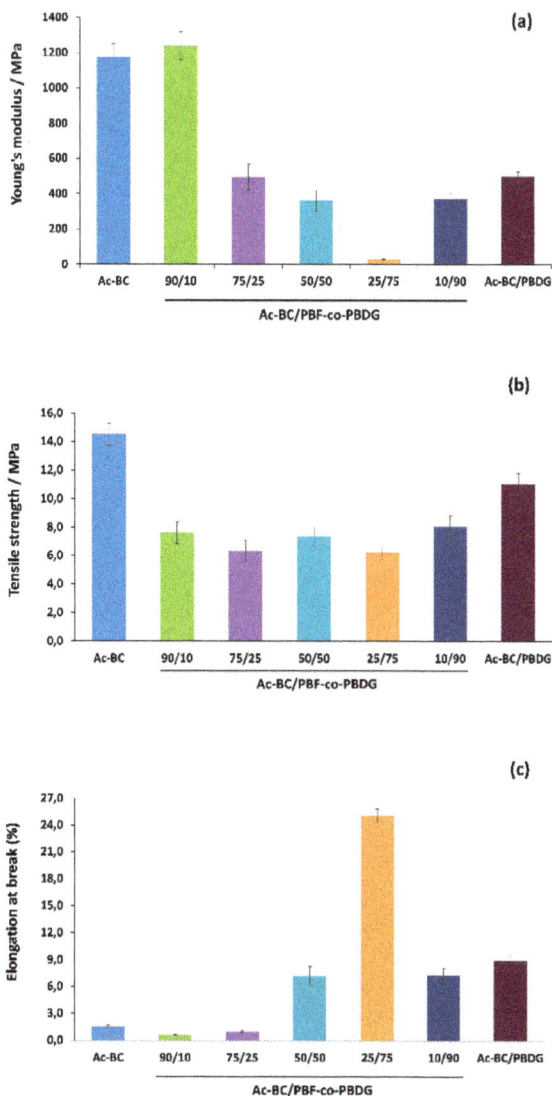

Figure 6. (a) Young's modulus, (b) tensile strength and (c) elongation at break of the nanocomposites and of the Ac-BC component.

Ac-BC/PBF-*co*-PBDG-90/10 exhibited the highest Young's modulus, approximately 1239.3 MPa, in accordance with this nanocomposite having the highest amount of rigid furanoate moieties. This result was significantly higher than that previous reported to the related film of the neat copolyester (ca. 373 MPa) [25]. The cellulose fibres played here had a reinforcing role in the nanocomposites, as well as due to the higher crystallinity of the copolyester prepared in this work, thus explaining the Young's modulus increase.

Importantly, the Young's modulus of Ac-BC/PBF-*co*-PBDG-90/10 (ca. 1239.3 MPa) was very close to that routinely reported for PBF [11,25,28], and also very near to neat Ac-BC film (ca. 1172.8 MPa). This was a very interesting result because this nanocomposite had good mechanical properties, comparable to those of PBF; but had the advantage of being biodegradable oppositely to PBF [25].

The other nanocomposites showed lower Young's modulus (ca. 499.9–30.3 MPa), decreasing with decreasing amounts of furanoate units from 75/25 to 25/75, but slightly increasing again to Ac-BC/PBF-*co*-PBDG-10/90 and PBDG. This inverted bell-shape trend behaviour for the first decreasing trend was in agreement with the decrease of stiff furanoate moieties; and for the second increasing trend it was most probably associated with an increase in crystallinity as prompted by the substantial amount of diglycolate segments, and thus a lower degree of randomness (as well as disclosed by XRD analysis). In addition, the nanocomposites Young's modulus results were typically much higher than those reported for the neat copolyesters films produced by Soccio et al. [25] due to the expectable reinforcing role of the cellulose fibres [39], and due to a higher crystallinity of the herein prepared nanocomposites.

In terms of elongation at break, the nanocomposites prepared from copolyesters with higher amounts of diglycolate moieties displayed higher elongations due to the soft nature of these segments. The highest result was obtained with Ac-BC/PBF-*co*-PBDG-25/75 (ca. 25.02%). Despite this composite having a huge gain in elasticity, especially compared to cellulose fibres, the elongation at break was still lower than those of neat copolyesters [25].

The nanocomposites barrier properties were evaluated in terms of permeability to oxygen, and preliminary results indicated that the nanocomposites and the corresponding (co)polyesters films prepared elsewhere [25] had similar permeabilities towards oxygen. Ac-BC/PBF-*co*-PBDG-90/10 showed to have a permeability to O_2 that was equal to 3.49×10^2 Barrer, whereas Ac-BC had 1.75×10^5, in accordance with the well-documented [44,47] superior barrier properties of furanoate-based polymers. These were attractive properties worth exploiting for applications within packaging.

4. Conclusions

Furanoate-based (nano)composites using bacterial nano-cellulose was here reported for the first time, revealing great potential to broaden the properties of this material. In the present case study involving Ac-BC/PBF and Ac-BC/PBF-*co*-PBDGs nanocomposites, they were shown to have high stiffness (evaluated by the Young's modulus, from 30.3 to 1239 MPa) compared to the neat (co)polyesters counterparts. Concomitantly, these nanocomposites still displayed reasonable elasticity (elongation at break) compared e.g., to cellulose or PBF. These properties were only possible by judiciously tailoring the composition of the nanocomposites, especially the critical diglycolate/furanoate amount in the copolyester, as well as by compounding the (co)polyesters with acetylated cellulose (tailoring crystallinity, homogeneity, among other properties). Moreover, the permeabilities to oxygen results were quite attractive, being in the order of magnitude of PBF, justifying further exploitation of these nanocomposites for applications within packaging.

Supplementary Materials: The following are available online at http://www.mdpi.com/2073-4360/10/8/810/s1, Scheme S1: Chemical structures of the triad units of the PBF-*co*-PBDG copolyesters, Figure S1: ^{1}H NMR spectra in TFA-d of PBF-*co*-PBDG copolyesters and related PBF and PBDG homopolyesters, Table S1: Main ^{1}H NMR resonances of PBF-*co*-PBDG copolyesters and related PBF and PBDG homopolyesters, Table S2: Comparison between the initial molar feed percentage and the real molar percentage of furanoate and diglycolate moieties, Figure S2: ATR FTIR spectra of the acetylated bacterial cellulose (Ac-BC) and of the unmodified bacterial cellulose (BC) fibres, Figure S3: ATR FTIR spectra of PBF-*co*-PBDG copolyesters and of PBF- and PBDG-related homopolyesters, Figure S4: ATR FTIR spectra of all Ac-BC/PBF-*co*-PBDG nanocomposites, Figure S5: SEM micrographs of Ac-BC film and of the nanocomposites of the (a) surface (500 X and 5.0 kX) and (b) cross-section (500 X and 5.0 kX), Table S3: Water contact angles of the composite films measured at several points in time for 40 s, Figure S6: X-ray diffractograms of the (a) neat (co)polyesters and (b) corresponding nanocomposites, Table S4: Important thermal values of the (co)polyesters and Ac-BC obtained by DSC and TGA analyses, Figure S7: DSC traces of the PBF-*co*-PBDGs and related PBF and PBDG homopolyesters.

Author Contributions: This study conceptualisation was carried out by A.F.S. and supervision was performed by A.F.S. and A.J.D.S. The experiments were carried out by M.M.; the bacterial cellulose was produced by N.H.C.S.S. and supervision of C.S.R.F., the permeation measurements and their supervision were performed by M.A. and A.M.; and the manuscript was written and revised with the contribution of all authors.

Acknowledgments: FCT and POPH/FSE are gratefully acknowledged for funding a doctoral grant to M.M. (PD/BD/52501/2014), post-doctoral grant to A.F.S. (SFRH/BPD/73383/2010) and a Researcher Contract to C.F. (IF)/01407/2012). M.A. is thankful to POCI-01-0145-FEDER-006939 (Laboratory for Process Engineering, Environment, Biotechnology and Energy—UID/EQU/00511/2013) funded by the European Regional Development Fund (ERDF), through COMPETE2020—Programa Operacional Competitividade e Internacionalização (POCI), and by national funds, through FCT—Fundação para a Ciência e a Tecnologia and NORTE-01-0145-FEDER-000005—LEPABE-2-ECO-INNOVATION, supported by the North Portugal Regional Operational Programme (NORTE 2020), under the Portugal 2020 Partnership Agreement, through the European Regional Development Fund (ERDF) for the fellowship grant. This work was developed within the scope of the project CICECO-Aveiro Institute of Materials, POCI-01-0145-FEDER-007679 (FCT Ref. UID/CTM/50011/2013), financed by national funds through the FCT/MEC and when appropriate co-financed by FEDER under the PT2020 Partnership Agreement.

Conflicts of Interest: The authors declare no conflict of interest.

References

1. Vilela, C.; Sousa, A.F.; Fonseca, A.C.; Serra, A.C.; Coelho, J.F.J.; Freire, C.S.R.; Silvestre, A.J.D. The quest for sustainable polyesters–insights into the future. *Polym. Chem.* **2014**, *5*, 3119–3141. [CrossRef]
2. Sousa, A.F.; Vilela, C.; Fonseca, A.C.; Matos, M.; Freire, C.S.R.; Gruter, G.J.M.; Coelho, J.F.J.; Silvestre, A.J.D. Biobased polyesters and other polymers from 2,5-furandicarboxylic acid: A tribute to furan excellency. *Polym. Chem.* **2015**, *6*, 5961–5983. [CrossRef]
3. Drewitt, J.G.N.; Lincocoln, J. Improvements in Polymers. UK Patent GB621971-A, 12 November 1946.
4. Gandini, A.; Silvestre, A.J.D.; Neto, C.P.; Sousa, A.F.; Gomes, M.M. The furan counterpart of poly(ethylene terephthalate): An alternative material based on renewable resources. *J. Polym. Sci. Part. A Polym. Chem.* **2009**, *47*, 295–298. [CrossRef]
5. Papageorgiou, G.Z.; Tsanaktsis, V.; Papageorgiou, D.G.; Exarhopoulos, S.; Papageorgiou, M.; Bikiaris, D.N. Evaluation of polyesters from renewable resources as alternatives to the current fossil-based polymers. Phase transitions of poly(butylene 2,5-furan-dicarboxylate). *Polymer* **2014**, *55*, 3846–3858. [CrossRef]
6. Jiang, M.; Liu, Q.; Zhang, Q.; Ye, C.; Zhou, G. A series of furan-aromatic polyesters synthesized via direct esterification method based on renewable resources. *J. Polym. Sci. Part. A Polym. Chem.* **2012**, *50*, 1026–1036. [CrossRef]
7. Wu, B.; Xu, Y.; Bu, Z.; Wu, L.; Li, B.G.; Dubois, P. Biobased poly(butylene 2,5-furandicarboxylate) and poly(butylene adipate-co-butylene 2,5-furandicarboxylate)s: From synthesis using highly purified 2,5-furandicarboxylic acid to thermo-mechanical properties. *Polymer* **2014**, *55*, 3648–3655. [CrossRef]
8. Ma, J.; Yu, X.; Xu, J.; Pang, Y. Synthesis and crystallinity of poly(butylene 2,5-furandicarboxylate). *Polymer* **2012**, *53*, 4145–4151. [CrossRef]
9. Jiang, Y.; Woortman, A.J.J.; van Ekenstein, G.O.R.A.; Loos, K. A biocatalytic approach towards sustainable furanic-aliphatic polyesters. *Polym. Chem.* **2015**, *6*, 5198–5211. [CrossRef]
10. Tsanaktsis, V.; Vouvoudi, E.; Papageorgiou, G.Z.; Papageorgiou, D.G.; Chrissafis, K.; Bikiaris, D.N. Thermal degradation kinetics and decomposition mechanism of polyesters based on 2,5-furandicarboxylic acid and low molecular weight aliphatic diols. *J. Anal. Appl. Pyrolysis* **2015**, *112*, 369–378. [CrossRef]
11. Zhu, J.; Cai, J.; Xie, W.; Chen, P.; Gazzano, M.; Scandola, M.; Gross, R.A. Poly(butylene 2,5-furandicarboxylate), a biobased alternative to PBT: Synthesis, physical properties, and crystal structure. *Macromolecules* **2013**, *46*, 796–804. [CrossRef]
12. Thiyagarajan, S.; Vogelzang, W.; Knoop, R.J.; Frissen, A.E.; Van Haveren, J.; Van Es, D.S. Biobased furandicarboxylic acids (FDCAs): Effects of isomeric substitution on polyester synthesis and properties. *Green Chem.* **2014**, *16*, 1957–1966. [CrossRef]
13. Ma, J.; Pang, Y.; Wang, M.; Xu, J.; Ma, H.; Nie, X. The copolymerization reactivity of diols with 2,5-furandicarboxylic acid for furan-based copolyester materials. *J. Mater. Chem.* **2012**, *22*, 3457–3461. [CrossRef]
14. Morales-Huerta, J.C.; de Ilarduya, A.M.; Muñoz-Guerra, S. Poly(alkylene 2,5-furandicarboxylate)s (PEF and PBF) by ring opening polymerization. *Polymer* **2016**, *87*, 148–158. [CrossRef]
15. Matos, M.; Sousa, A.F.; Fonseca, A.C.; Freire, C.S.R.; Coelho, J.F.J.; Silvestre, A.J.D. A new generation of furanic copolyesters with enhanced degradability: Poly(ethylene 2,5-furandicarboxylate)-*co*-poly(lactic acid) copolyesters. *Macromol. Chem. Phys.* **2014**, *215*, 2175–2184. [CrossRef]

16. Sousa, A.F.; Guigo, N.; Pozycka, M.; Delgado, M.; Soares, J.; Mendonça, P.V.; Coelho, J.F.J.; Sbirrazzuoli, N.; Silvestre, A.J.D. Tailored design of renewable copolymers based on poly(1,4-butylene 2,5-furandicarboxylate) and poly(ethylene glycol) with refined thermal properties. *Polym. Chem.* **2018**, *9*, 722–731. [CrossRef]

17. Morales-Huerta, J.C.; Ciulik, C.B.; De Ilarduya, A.M.; Muñoz-Guerra, S. Fully bio-based aromatic-aliphatic copolyesters: Poly(butylene furandicarboxylate-*co*-succinate)s obtained by ring opening polymerization. *Polym. Chem.* **2017**, *8*, 748–760. [CrossRef]

18. Wu, L.; Mincheva, R.; Xu, Y.; Raquez, J.M.; Dubois, P. High molecular weight poly(butylene succinate-*co*-butylene furandicarboxylate) copolyesters: From catalyzed polycondensation reaction to thermomechanical properties. *Biomacromolecules* **2012**, *13*, 2973–2981. [CrossRef] [PubMed]

19. Yu, Z.; Zhou, J.; Cao, F.; Wen, B.; Zhu, X.; Wei, P. Chemosynthesis and characterization of fully biomass-based copolymers of ethylene glycol, 2,5-furandicarboxylic acid, and succinic acid. *J. Appl. Polym. Sci.* **2013**, *130*, 1415–1420. [CrossRef]

20. Zhou, W.; Wang, X.; Yang, B.; Xu, Y.; Zhang, W.; Zhang, Y.; Ji, J. Synthesis, physical properties and enzymatic degradation of bio-based poly(butylene adipate-co-butylene furandicarboxylate) copolyesters. *Polym. Degrad. Stab.* **2013**, *98*, 2177–2183. [CrossRef]

21. Zheng, M.Y.; Zang, X.L.; Wang, G.X.; Wang, P.L.; Lu, B.; Ji, J.H. Poly(butylene 2,5-furandicarboxylate-ε-caprolactone): A new bio-based elastomer with high strength and biodegradability. *Express Polym. Lett.* **2017**, *11*, 611–621. [CrossRef]

22. Papageorgiou, G.Z.; Papageorgiou, D.G. Solid-state structure and thermal characteristics of a sustainable biobased copolymer: Poly(butylene succinate-*co*-furanoate). *Thermochim. Acta* **2017**, *656*, 112–122. [CrossRef]

23. Peng, S.; Bu, Z.; Wu, L.; Li, B.G.; Dubois, P. High molecular weight poly(butylene succinate-co-furandicarboxylate) with 10 mol % of BF unit: Synthesis, crystallization-melting behavior and mechanical properties. *Eur. Polym. J.* **2017**, *96*, 248–255. [CrossRef]

24. Hu, H.; Zhang, R.; Wang, J.; Ying, W.B.; Zhu, J. Synthesis and structure-property relationship of bio-based biodegradable poly(butylene carbonate-*co*-furandicarboxylate). *ACS Sustain. Chem. Eng.* **2018**, *6*, 7488–7498. [CrossRef]

25. Soccio, M.; Costa, M.; Lotti, N.; Gazzano, M.; Siracusa, V.; Salatelli, E.; Manaresi, P.; Munari, A. Novel fully biobased poly(butylene 2,5-furanoate/diglycolate) copolymers containing ether linkages: Structure-property relationships. *Eur. Polym. J.* **2016**, *81*, 397–412. [CrossRef]

26. Hu, H.; Zhang, R.; Wang, J.; Ying, W.B.; Zhu, J. Fully bio-based poly(propylene succinate-co-propylene furandicarboxylate) copolyesters with proper mechanical, degradation and barrier properties for green packaging applications. *Eur. Polym. J.* **2018**, *102*, 101–110. [CrossRef]

27. Kasmi, N.; Majdoub, M.; Papageorgiou, G.Z.; Bikiaris, D.N. Synthesis and crystallization of new fully renewable resources-based copolyesters: Poly(1,4-cyclohexanedimethanol-co-isosorbide 2,5-furandicarboxylate). *Polym. Degrad. Stab.* **2018**, *152*, 177–190. [CrossRef]

28. Cai, X.; Yang, X.; Zhang, H.; Wang, G. Aliphatic-aromatic poly(carbonate-*co*-ester)s containing biobased furan monomer: Synthesis and thermo-mechanical properties. *Polymer* **2018**, *134*, 63–70. [CrossRef]

29. Wang, X.; Wang, Q.; Liu, S.; Wang, G. Biobased copolyesters: Synthesis, structure, thermal and mechanical properties of poly(ethylene 2,5-furandicarboxylate-co-ethylene 1,4-cyclohexanedicarboxylate). *Polym. Degrad. Stab.* **2018**, *154*, 96–102. [CrossRef]

30. Wang, J.; Liu, X.; Jia, Z.; Sun, L.; Zhang, Y.; Zhu, J. Modification of poly(ethylene 2,5-furandicarboxylate) (PEF) with 1,4-cyclohexanedimethanol: Influence of stereochemistry of 1,4-cyclohexylene units. *Polymer* **2018**, *137*, 173–185. [CrossRef]

31. Codou, A.; Guigo, N.; van Berkel, J.G.; de Jong, E.; Sbirrazzuoli, N. Preparation and crystallization behavior of poly(ethylene 2,5-furandicarboxylate)/cellulose composites by twin screw extrusion. *Carbohyd. Polym.* **2017**, *174*, 1026–1033. [CrossRef] [PubMed]

32. Achilias, D.S.; Chondroyiannis, A.; Nerantzaki, M.; Adam, K.V.; Terzopoulou, Z.; Papageorgiou, G.Z.; Bikiaris, D.N. Solid state polymerization of poly(ethylene furanoate) and its nanocomposites with SiO$_2$ and TiO$_2$. *Macromol. Mater. Eng.* **2017**, *302*, 1–15. [CrossRef]

33. Martino, L.; Guigo, N.; van Berkel, J.G.; Sbirrazzuoli, N. Influence of organically modified montmorillonite and sepiolite clays on the physical properties of bio-based poly(ethylene 2,5-furandicarboxylate). *Compos. Part. B-Eng.* **2017**, *110*, 96–105. [CrossRef]

34. Martino, L.; Niknam, V.; Guigo, N.; van Berkel, J.G.; Sbirrazzuoli, N. Morphology and thermal properties of novel clay-based poly(ethylene 2,5-furandicarboxylate) (PEF) nanocomposites. *RSC Adv.* **2016**, *6*, 59800–59807. [CrossRef]

35. Lotti, N.; Munari, A.; Gigli, M.; Gazzano, M.; Tsanaktsis, V.; Bikiaris, D.N.; Papageorgiou, G.Z. Thermal and structural response of in situ prepared biobased poly(ethylene 2,5-furan dicarboxylate) nanocomposites. *Polymer* **2016**, *103*, 288–298. [CrossRef]

36. Codou, A.; Guigo, N.; Van Berkel, J.G.; De Jong, E.; Sbirrazzuoli, N. Preparation and characterization of poly(ethylene 2,5-furandicarboxylate)/nanocrystalline cellulose composites via solvent casting. *J. Polym. Eng.* **2017**, *37*, 869–878. [CrossRef]

37. Tomé, L.C.; Gonçalves, C.M.B.; Boaventura, M.; Brandão, L.; Mendes, A.M.; Silvestre, A.J.D.; Neto, C.P.; Gandini, A.; Freire, C.S.R.; Marrucho, I.M. Preparation and evaluation of the barrier properties of cellophane membranes modified with fatty acids. *Carbohydr. Polym.* **2011**, *83*, 836–842. [CrossRef]

38. Sousa, A.F.; Vilela, C.; Matos, M.; Freire, C.S.R.; Silvestre, A.J.D.; Coelho, J.F.J. Polyethylene terephthalate: Copolyesters, composites, and renewable alternatives. In *Poly(Ethylene Terephthalate) Based Blends, Composites and Nanocomposites*; Visakh, P.M., Liang, M., Eds.; Elsevier: New York, NY, USA, 2015; pp. 113–141.

39. Tomé, L.C.; Pinto, R.J.B.; Trovatti, E.; Freire, C.S.R.; Silvestre, A.J.D.; Gandini, A.; Neto, C.P. Transparent bionanocomposites with improved properties prepared from acetylated bacterial cellulose and poly(lactic acid) through a simple approach. *Green Chem.* **2011**, *13*, 419–427. [CrossRef]

40. Gomes, F.P.; Silva, N.H.C. S.; Trovatti, E.; Serafim, L.S.; Duarte, M.F.; Silvestre, A.J.D.; Neto, C.P.; Freire, C.S.R. Production of bacterial cellulose by Gluconacetobacter sacchari using dry olive mill residue. *Biomass Bioenergy* **2013**, *55*, 205–211. [CrossRef]

41. Soares, M.J.; Dannecker, P.K.; Vilela, C.; Bastos, J.; Meier, M.A.R.; Sousa, A.F. Poly(1,20-eicosanediyl 2,5-furandicarboxylate), a biodegradable polyester from renewable resources. *Eur. Polym. J.* **2017**, *90*, 301–311. [CrossRef]

42. Vaca-Garcia, C.; Borredon, M.E.A. Gaseta Determination of the degree of substitution (DS) of mixed cellulose esters by elemental analysis. *Cellulose* **2001**, *8*, 225–231. [CrossRef]

43. Campo, M.C.; Magalhães, F.D.; Mendes, A. Carbon molecular sieve membranes from cellophane paper. *J. Membr. Sci.* **2010**, *350*, 180–188. [CrossRef]

44. Araujo, C.F.; Nolasco, M.M.; Ribeiro-Claro, P.J.A.; Rudić, S.; Silvestre, A.J.D.; Vaz, P.D.; Sousa, A.F. Inside PEF: Chain conformation and dynamics in crystalline and amorphous domains. *Macromolecules* **2018**, *51*, 3515–3526. [CrossRef]

45. Liukkonen, A. Contact angle of water on paper components: Sessile drops versus environmental scanning electron microscope measurements. *Scanning* **1997**, *19*, 411–415. [CrossRef]

46. Zhang, X.; Li, W.; Ye, B.; Lin, Z.; Rong, J. Studies on confined crystallization behavior of nanobiocomposites consisting of acetylated bacterial cellulose and poly(lactic acid). *J. Thermoplast. Compos. Mater.* **2013**, *26*, 346–361. [CrossRef]

47. Burgess, S.K.; Leisen, J.E.; Kraftschik, B.E.; Mubarak, C.R.; Kriegel, R.M.; Koros, W.J. Chain mobility, thermal, and mechanical properties of poly(ethylene furanoate) compared to poly(ethylene terephthalate). *Macromolecules* **2014**, *47*, 1383–1391. [CrossRef]

© 2018 by the authors. Licensee MDPI, Basel, Switzerland. This article is an open access article distributed under the terms and conditions of the Creative Commons Attribution (CC BY) license (http://creativecommons.org/licenses/by/4.0/).

![polymers logo] *polymers*

MDPI

Article

Partially Renewable Poly(butylene 2,5-furandicarboxylate-*co*-isophthalate) Copolyesters Obtained by ROP

Juan Carlos Morales-Huerta, Antxon Martínez de Ilarduya * and Sebastián Muñoz-Guerra *

Department d'Enginyeria Química, Universitat Politècnica de Catalunya, ETSEIB, Diagonal 647, 08028 Barcelona, Spain; juan.carlos.morales.huerta@upc.edu
* Correspondence: antxon.martinez.de.ilarduia@upc.edu (A.M.d.I.); sebastian.munoz@upc.edu (S.M.-G.);
  Tel.: +34-93-401-0910 (A.M.d.I.); +34-93-401-6680 (S.M.-G.)

Received: 5 April 2018; Accepted: 25 April 2018; Published: 28 April 2018

Abstract: Cyclic butylene furandicarboxylate ($c(BF)_n$) and butylene isophthalate ($c(BI)_n$) oligomers obtained by high dilution condensation reaction were polymerized in bulk at 200 °C with $Sn(Oct)_2$ catalyst via ring opening polymerization to give homopolyesters and copolyesters ($coPBF_xI_y$) with weight average molar masses in the 60,000–70,000 g·mol^{-1} range and dispersities between 1.3 and 1.9. The composition of the copolyesters as determined by NMR was practically the same as that of the feed, and they all showed an almost random microstructure. The copolyesters were thermally stable up to 300 °C and crystalline for all compositions, and have T_g in the 40–20 °C range with values decreasing almost linearly with their content in isophthalate units in the copolyester. Both melting temperature and enthalpy of the copolyesters decreased as the content in butylene isophthalate units increased up to a composition 30/70 (BF/BI), at which the triclinic crystal phase made exclusively of butylene furanoate units changed to the crystal structure of PBI. The partial replacement of furanoate by isophthalate units decreased substantially the crystallizability of PBF.

Keywords: PBF; PBI; copolyesters; ROP; cyclic oligomers; thermal properties; crystallization

1. Introduction

Due to popular awareness of sustainability, polymers obtained from renewable sources have been developed in the last decade, with the purpose of replacing those obtained from fossil resources [1–5].

One renewable monomer that has attracted much attention is 2,5-furandicarboxylic acid (FDCA), an aromatic building block obtained from C5 and C6 sugars, that is able to replace terephthalic acid, a petrochemical compound widely used for the preparation of aromatic polyesters such as PET or PBT [6–8]. Poly(ethylene furanoate) (PEF) has been extensively studied because it has not only similar properties to PET, but also improved gas barrier properties, which make it a serious alternative for applications in soft drink bottles. In contrast, poly(butylene furanoate) (PBF) has been much less studied; as such, the knowledge available on this polyester is relatively scarce. PBF is a semicrystalline polymer with a melting temperature of 172 °C and a glass transition temperature of 39 °C [9]. As with PBT, the presence of the butylene segment in the repeating unit of PBF confers to this polymer a strong propensity for rapid crystallization, which is inconvenient for some injection molding processes due to the excessive mold shrinkage. In order to overcome these problems, one solution is copolymerization. The insertion of either a diol or diacid comonomeric unit in small quantities in the PBF chain decreases both the melting temperature and enthalpy, therefore reducing processing costs. Copolymerization has been applied to various technical polyesters in order to tune their thermal properties, such as crystallizability, melting or glass transition temperatures [10]. PBF copolyesters with enhanced biodegradability have already been prepared by either melt polycondensation [11–14] or ring opening polymerization (ROP) of cyclic oligomers [15,16]. This last method, which uses cyclic oligomers for the synthesis, has some advantages because it does not require by-product removal during reaction, implies small or no heat exchange, and attains very high-molecular-weight polymers in reaction times of minutes. ROP for these systems was first examined in detail by Brunelle [17] and recently reviewed by Hodge [18] and Strandman et al. [19]. The technique has been successfully used by us to prepare various PEF and PBF copolyesters with enhanced properties [20,21].

In this work we would like to report on the synthesis and characterization, evaluation of thermal properties, and crystallization behavior of new, partially renewable PBF copolyesters containing isophthalate units that are prepared by ROP of mixtures of cyclic butylene furanoate and butylene isophthalate oligomers. The 3D chemical structures of FDCA and isophthalic acid (IPA) are depicted in Scheme 1.

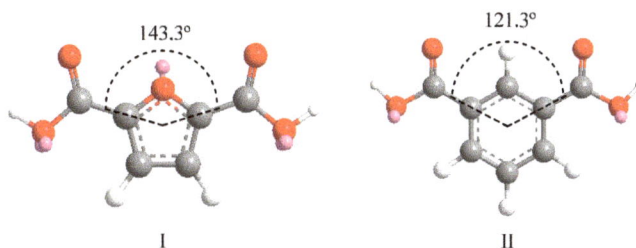

Scheme 1. 3D models of 2,5-furandicarboxylic acid (I) and isophthalic acid molecules (II).

2. Materials and Methods

2.1. Materials

2,5-Furandicarboxylic acid (FDCA, >98% purity) was purchased from Satachem (Shanghai, China). Isophthalic acid (IPA, 99%), 1,4-butanediol (BD), thionyl chloride (SOCl$_2$, 99%), and 1,4-di-azabicyclo[2.2.2]octane (DABCO, 99%) and tin(II) ethylhexanoate (Sn(Oct)$_2$, 99%) catalysts were purchased from Sigma-Aldrich Co. Triethylamine (Et$_3$N, 98%) was purchased from Panreac. Solvents used for reaction, isolation and purification were of high-purity grade and used as received except tetrahydrofuran (THF) that was dried on 3 Å-molecular sieves. The DABCO catalyst was purified by sublimation.

2.2. Methods

^{1}H- and ^{13}C-NMR spectra were recorded on a Bruker AMX-300 spectrometer (Billerica, MA, USA) at 25 °C, operating at 300.1 and 75.5 MHz, respectively. For NMR analysis, monomers, cyclic oligomers and intermediate compounds were dissolved in deuterated chloroform (CDCl$_3$) and polymers in pure CDCl$_3$ or in a mixture of trifluoroacetic acid (TFA) and CDCl$_3$ (1:8). About 10 and 50 mg of sample in 1 mL of solvent were used for ^{1}H- and ^{13}C-NMR, respectively. Sixty-four scans were recorded for ^{1}H, and between 1000 and 10,000 scans for ^{13}C-NMR. Spectra were internally referenced to tetramethylsilane (TMS).

High-performance liquid chromatography (HPLC) analysis was performed at 25 °C in a Waters apparatus equipped with a UV detector of Applied Biosystems operating at 254 nm wavelength, and a Scharlau Science column (Si60, 5 μm; 250 × 4.6 mm). Cyclic oligomers (1 mg) were dissolved in chloroform (1 mL) and eluted with hexane/1,4-dioxane 70/30 (v/v) at a flow rate of 1.0 mL·min^{-1}. Molecular weight analysis was performed by GPC on a Waters equipment provided with RI and UV detectors. 100 μL of 0.1% (w/v) sample solution were injected and chromatographed with a flow of 0.5 mL·min^{-1} of 1,1,1,3,3,3-hexafluoroisopropanol (HFIP). HR5E and HR2 Waters linear Styragel columns (7.8 mm × 300 mm, pore size 103–104 Å) were packed with crosslinked polystyrene and protected with a precolumn. Molar mass average and distributions were calculated against PMMA standards.

Matrix-assisted laser desorption/ionization time of flight (MALDI-TOF) mass spectra were recorded in a 4700 Proteomics Analyzer instrument (Applied Biosystems, Foster City, CA, USA) at the Proteomics Platform of Barcelona Science Park, University of Barcelona. Spectra acquisition was performed in the MS reflector positive-ion mode. About 0.1 mg of sample was dissolved in 50 μL of DCM and 2 μL of this solution was mixed with an equal volume of DCM solution of anthracene (10 mg·mL^{-1}); the mixture was then left to evaporate to dryness onto the stainless steel plate of the analyzer. The residue was then covered with 2 μL of a solution of 2,5-dihydroxibenzoic acid in acetonitrile/H$_2$O (1/1) containing 0.1% TFA, and the mixture was left to dry prior to exposition to the laser beam.

The thermal behavior of cyclic compounds and polymers were examined by differential scanning calorimetry (DSC), using a Perkin-Elmer Pyris 1 apparatus (Waltam, MA, USA). The thermograms were recorded from 3 to 6 mg samples at heating and cooling rates of 10 °C·min^{-1} under a nitrogen flow of 20 mL·min^{-1}. Indium and zinc were used as standards for temperature and enthalpy calibration. The glass transition temperature (T_g) was taken as the inflection point of the heating DSC traces recorded at 20 °C·min^{-1} from melt-quenched samples, and the melting temperature (T_m) was taken as the maximum of the endothermic peak appearing on heating traces. Thermogravimetric analyses were performed on a Mettler-Toledo TGA/DSC 1 Star System under a nitrogen flow of 20 mL·min^{-1} at a heating rate of 10 °C·min^{-1} and within a temperature range of 30 to 600 °C. X-ray diffraction patterns from powdered samples coming directly from synthesis were recorded on a PANalytical X'Pert PRO MPD θ/θ diffractometer using the CuKα radiation of wavelength 0.1542 nm.

2.3. Synthesis

2.3.1. Synthesis of Cyclic Oligomers

Cyclic oligomers of butylene 2,5-furandicarboxylate $c(BF)_n$ and butylene isophthalate $c(BI)_n$ were synthesized by high dilution condensation (HDC) from equimolar mixtures of BD and furandicarboxylic dichloride (FDCA-Cl$_2$) and isophthaloyl chloride (IPA-Cl$_2$), respectively, as previously reported Brunelle et al. [22], and more recently by us [23]. Briefly, a three necked round bottom flask charged with 250 mL of THF was cooled to 0 °C; 12.5 mmol (1.40 g) of DABCO was then added under stirring. 5 mmol (0.96 g) of FDCA-Cl$_2$ or 5 mmol (1.01 g) of IPA-Cl$_2$ in 10 mL and 5 mmol (0.46 g) of BD in THF were drop-wise added simultaneously for 40 min using two addition funnels, in order to maintain the reagents equimolarity in the reaction mixture. The reaction was finished by adding 1 mL of water, followed by 5 mL of 1M HCl; after stirring for 5 min, the mixture was diluted with DCM and filtered. The filtrate was washed with 0.1M HCl, dried on MgSO$_4$, and evaporated to dryness to render a mixture of linear and cyclic oligomers. Linear oligomers were removed by chromatography through a short column of silica gel using a cold mixture of DCM/diethyl ether 90/10 (v/v) as eluent. $c(BF)_n$: ^{1}H-NMR (δ ppm, CDCl$_3$, 300 MHz): 7.23, 7.24, 7.25 (3s, 2H), 4.40 (m, 4H), 1.92, 1.99 (2m, 4H), ^{13}C-NMR (δ ppm, CDCl$_3$, 75.5 MHz): 158.1, 157.9, 146.7, 146.5, 118.7, 118.6, 118.5, 65.0, 64.8, 25.4. $c(BI)_n$: ^{1}H-NMR (δ ppm, CDCl$_3$, 300 MHz): 8.62, 8.60 (2m, 1H), 8.26, 8.21 (2m, 2H), 7.56, 7.50 (2m, 1H), 4.45 (m, 4H), 2.01, 1.97 (2m, 4H). ^{13}C-NMR (δ ppm, CDCl$_3$, 75.5 MHz): 165.6, 165.5, 134.2, 133.8, 130.6, 130.5, 130.3 129.7, 128.8, 128.6, 64.8, 64.7, 64.6, 25.5, 25.4.

2.3.2. Synthesis of Polymers

Mixtures of cyclic oligoesters ($c(BF)_n$ and $c(BI)_n$) at different molar ratios were polymerized following the procedure used by us for the synthesis of other PBF copolyesters [15,21]. A total of 47 mmol of the mixture of cyclic oligomers with the selected composition together with 0.5 mol % of Sn(Oct)$_2$ were dissolved in 10 mL of CHCl$_3$, the solution evaporated, and the remaining solid dried under vacuum at room temperature for 24 h. Subsequently, the mixture was left to react in a three necked round bottom flask for 6 h at 200 °C under a flow of N_2. For optimization of the reaction, the 50:50 mixture was polymerized at 180, 200 and 230 °C. The evolution in the molecular weight of all reactions was monitored by drawing aliquots at different times and analyzing them by GPC. The resulting polymers—without further treatment—were analyzed by NMR, GPC, TGA, DSC and WAXS.

3. Results and Discussion

The synthetic route used for the preparation of $c(BF)_n$ and $c(BI)_n$ cyclic oligomers and their ROP is represented in Figure 1. In a first step, the cyclic oligomers were obtained by HDC of BD and either FDCA-Cl_2 or IPA-Cl_2. The mixture of cyclic oligomers and linear species were separated by column chromatography, and the purity of cyclic oligomers was ascertained by HPLC, NMR and MALDI-TOF mass spectra (Figures S2 and S3).

The ^{1}H-NMR spectra show the absence of any peaks at around 3.8 ppm due to the presence of CH_2OH groups, which indicates that only cyclic oligomers are present in the purified fractions. Some signals in both ^{1}H- and ^{13}C-NMR spectra are split due to the sensitivity of these nuclei to the size of the oligomeric cycle. On the other hand, MALDI-TOF MS spectra allowed determining the molar mass of the different cyclic species.

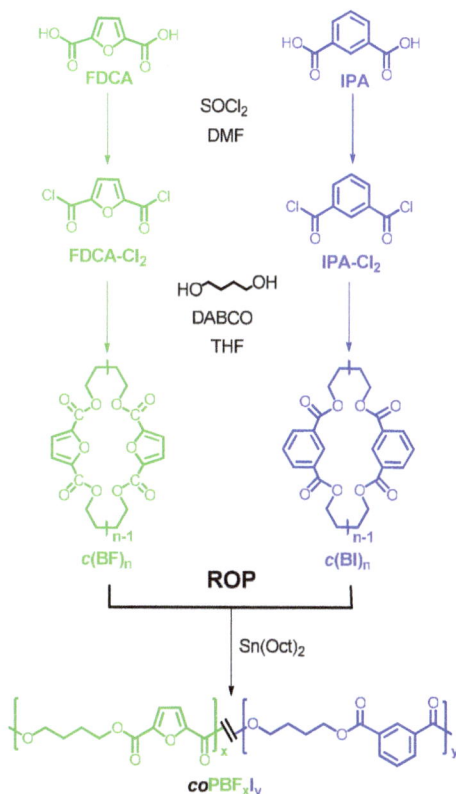

Figure 1. Synthesis route to poly(butylene 2,5-furandicarboxylate-*co*-isophthalate) (*co*PBF$_x$I$_y$) via ROP.

Table 1 shows the composition of the different cycles for the two oligomeric fractions as determined by HPLC. A mixture of cyclic oligomers, mainly from dimer to tetramer, were obtained for both $c(BF)_n$ and $c(BI)_n$ being the dimer the predominant cycle size. Both the flexibility of the butylene unit and the 1,3- or 2,5-substitution in benzene or furan respectively, favor the cyclization reaction due to the probable low ring strain of the cycles made of two repeating units.

Table 1. Cyclization reaction results.

	Yield (%)	Composition[1] (2/3/4)	$T_m{}^2$ (°C)	$^\circ T_{5\%}{}^2$ (°C)	$T_d{}^2$ (°C)
$c(BF)_n$	67	61/31/8	147	276	387
$c(BI)_n$	70	75/15/10	149	330	399

[1] Relative content (w/w) of the reaction product in cyclic dimer, trimer and tetramer as measured by HPLC. [2] Melting and decomposition temperatures measured by DSC and TGA.

The thermal properties of these cycles were evaluated by DSC and TGA (Figure S4 and Table 1). Both $c(BF)_n$ and $c(BI)_n$ showed melting peaks at around 150 °C, and it was observed that they were thermally stable up to 276 °C and 330 °C, respectively, which allowed their thermal polymerization at the temperature used for reaction (200 °C) without perceiving degradation.

These cycles were then polymerized via ROP in bulk. First, an equimolar mixture of the two cyclic fractions was made to react in order to test the effect of time and temperature on the polymerization results. Three different temperatures above the melting point of the cycles were chosen, i.e., 180, 200 and 230 °C, and sample aliquots were drawn at scheduled periods to determine the evolution of the molar mass of the copolymer produced under different conditions. It was found that the molar mass of the polymer did not increase after six hours of reaction at above 200 °C, (Figure 2a); such a temperature was then chosen for carrying out all copolyesters synthesis. The evolution of the molar mass of the copolyesters with time of reaction is depicted in Figure 2b, where a similar tendency is observed for all the series. However the maximum molar mass attained was observed to increase slightly with the content in furanoate units in the copolyester (Figure 2c and Table 2), which can be due to the higher reactivity of the furanoate over isophthalate cyclic oligomers or the higher thermal stability of the former. The dispersities of the obtained copolyesters oscillated between 1.30 and 1.78 values, which are in accordance with those obtained by entropically driven ROP [18,19].

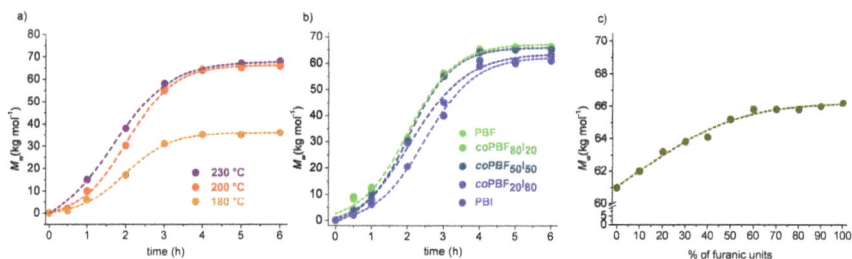

Figure 2. (**a**) Evolution of M_w of $coPBF_{50}I_{50}$ with reaction time at different temperatures; (**b**) Evolution of M_w of $coPBF_xI_y$ with reaction time at 200 °C for different compositions; (**c**) Effect of composition of $coPBF_xI_y$ on the M_w of the copolyester produced at 200 °C after 6 h of reaction.

Table 2. Results of molecular weight and microstructure analysis for *co*PBF$_x$I$_y$ copolyesters obtained via ROP.

Copolyester	Yield (%)	x$_{BF}$/y$_{BI}$ [1] (mol/mol)	Molecular Weight [2]		Dyad Content (mol %) [3]			Sequence Length [4]		R [4]
			M$_w$	Đ	FBF	FBI + IBF	IBI	n$_{BF}$	n$_{BI}$	
PBF	90	100/0	66,200	1.65	-	-	-	-	-	-
*co*PBF$_{90}$I$_{10}$	88	89/11	66,000	1.50	79.7	18.4	1.7	11.30	1.18	0.94
*co*PBF$_{80}$I$_{20}$	85	81/19	65,800	1.45	72.7	14.1	13.2	9.57	1.50	0.94
*co*PBF$_{70}$I$_{30}$	86	70/30	65,800	1.28	43.1	41.7	14.8	3.06	1.71	0.91
*co*PBF$_{60}$I$_{40}$	85	64/36	65,800	1.30	43.3	42.9	14.1	3.02	1.65	0.93
*co*PBF$_{50}$I$_{50}$	86	48/52	65,200	1.45	25.8	36.8	37.5	2.41	3.04	0.80
*co*PBF$_{40}$I$_{60}$	89	40/60	64,100	1.62	22.1	32.5	45.4	2.36	3.79	0.72
*co*PBF$_{30}$I$_{70}$	91	31/69	63,800	1.78	17.9	27.5	55.4	2.24	5.04	0.75
*co*PBF$_{20}$I$_{80}$	88	18/82	63,200	1.60	5.6	24.2	70.2	1.46	6.79	0.83
*co*PBF$_{10}$I$_{90}$	87	10/90	62,000	1.45	1.7	16.2	82.2	1.21	11.17	0.91
PBI	93	0/100	61,000	1.50	-	-	-	-	-	-

[1] Determined by ^{1}H-NMR. [2] Weight-average molar masses in g·mol^{-1} and dispersities determined by GPC. [3] Determined by deconvolution of the ^{13}C-NMR peaks appearing in the 64.6–65.2 ppm region. [4] Number average sequence lengths and degree of randomness (R), calculated using the expressions mentioned in the text.

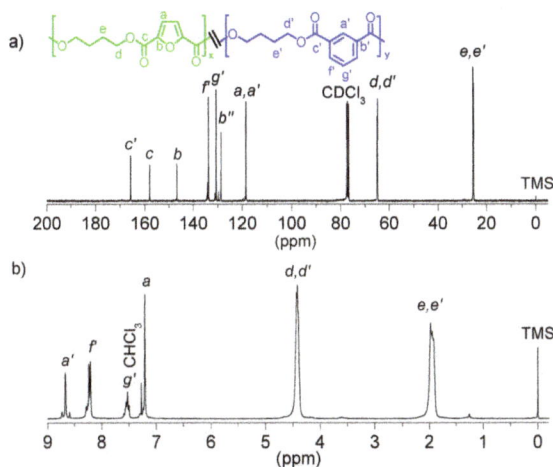

Figure 3. (a) ^{13}C- and (b) ^{1}H-NMR of *co*PBF$_{50}$I$_{50}$ with peak assignments.

The polyesters were obtained in good yields (85–93%). The chemical structure and composition of *co*PBF$_x$I$_y$ copolymers were determined by NMR. Figure 3 shows both ^{1}H- and ^{13}C-NMR spectra of *co*PBF$_{50}$I$_{50}$ with peak assignments as a representative of the series. NMR spectra for all series are depicted in Figure S5 of SI document. Signals due to the furanic proton *a* and isophthalic protons *f'* were chosen for the determination of the copolyester composition. In general, a good correlation between the feed and the final copolyester composition determined by ^{1}H-NMR was found, with slight fluctuations probably due to uncontrolled cycles volatilization (Table 2).

In contrast, ^{13}C-NMR spectra were used for the determination of the copolymer microstructure. Each carbon signal of the butylene segment split into four peaks due to its sensitivity to sequence distribution at the level of dyads (Figure 4). The assignment of the peaks contained in the different dyads was straightforward by comparison to those appearing in both PBF and PBI homopolyesters. By deconvolution of these signals, the dyad content (FBF, FBI + IBF, IBI) could be obtained and the number average sequence length and degree of randomness (R) could be determined for each copolymer by applying the following expressions [24]:

$$\bar{n}_{BF} = \frac{FBF + \frac{1}{2}(FBI + IBF)}{\frac{1}{2}(FBI + IBF)} ; \; \bar{n}_{BI} = \frac{IBI + \frac{1}{2}(FBI + IBF)}{\frac{1}{2}(FBI + IBF)} ; R = \frac{1}{\bar{n}_{BF}} + \frac{1}{\bar{n}_{BI}}$$

The degree of randomness was near to one with lower values for copolymers with contents in isophthalate units of between 50 mol % and 80 mol %. These values are higher than those that should be expected when only the ROP reaction takes place. In such cases, blocky copolymers and lower values of R should be obtained [25]. The observed values indicate that extensive transesterifications took place during polymerization, leading to nearly statistical copolymers (Table 2).

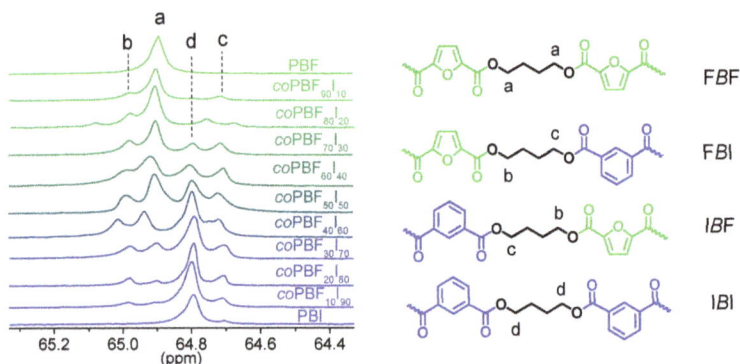

Figure 4. ^{13}C-NMR spectra of *co*PBF$_x$I$_y$ copolyesters in the region of the first methylene of the oxybutylene segment.

The thermal stability of PBF, PBI and their copolyesters was evaluated by TGA under an inert atmosphere. Both, polyesters, and copolyesters were observed to be thermally stable up to 300 °C with onset temperatures above 330 °C, a temperature of maximum decomposition rate close to 400 °C (Figure 5 and Table 3), and remaining weights at 600 °C between 7% and 11%. These values indicate that both homopolyesters and copolyesters have a good thermal stability, and may be processed above their melting temperature without suffering significant thermal degradation.

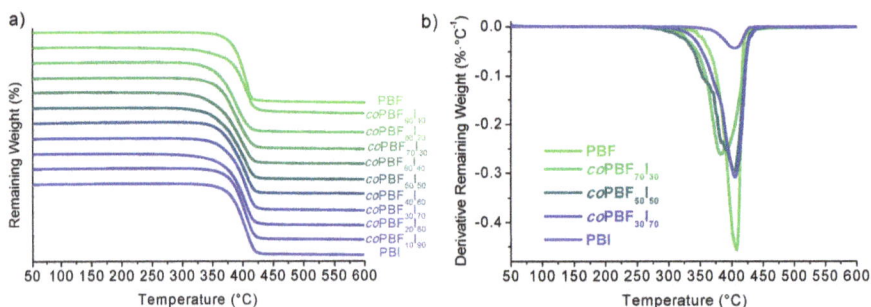

Figure 5. TGA analysis of *co*PBF$_x$I$_y$. (**a**) Weight loss vs. temperature traces; (**b**) Derivative curves.

Thermal properties of the copolyesters, such as melting and glass transition temperatures, have been evaluated by DSC. The DSC traces obtained at heating from samples coming directly from synthesis are depicted in Figure 6a–c and data taken from these thermograms are collected in Table 3.

Table 3. Thermal properties of $coPBF_xI_y$ copolyesters prepared via ROP.

Copolyester	TGA			DSC					Crystallization Kinetics		
				First Heating			Second Heating				
	^oT_d [1] (°C)	$^{max}T_d$ (°C)	R_w (%)	T_g (°C)	T_m (°C)	ΔH (J·mol^{-1})	T_m (°C)	ΔH (J·mol^{-1})	n^2	lnK^2	$t_{1/2}$ (min)
PBF	364	407	7	41	173	45	173	39	2.2	−4.4	6.9
coPBF$_{90}$I$_{10}$	340	404	11	35	164	41	163	35	2.5	−8.6	27.8
coPBF$_{80}$I$_{20}$	338	390	9	30	158	29	156	26	-	-	-
coPBF$_{70}$I$_{30}$	342	396	7	28	143	21	142	15	-	-	-
coPBF$_{60}$I$_{40}$	330	396	7	28	137	12	137	4	-	-	-
coPBF$_{50}$I$_{50}$	334	403	7	27	126	10	-	-	-	-	-
coPBF$_{40}$I$_{60}$	355	404	8	27	122	3	-	-	-	-	-
coPBF$_{30}$I$_{70}$	352	405	7	27	118	1	-	-	-	-	-
coPBF$_{20}$I$_{80}$	336	403	8	26	106	2	-	-	-	-	-
coPBF$_{10}$I$_{90}$	363	403	8	24	130	5	129	1	-	-	-
PBI	364	404	8	21	142	31	141	1	-	-	-

[1] ^oT_d obtained at 5% of weight lost. [2] Avrami parameters obtained from isothermal crystallization at 146 °C.

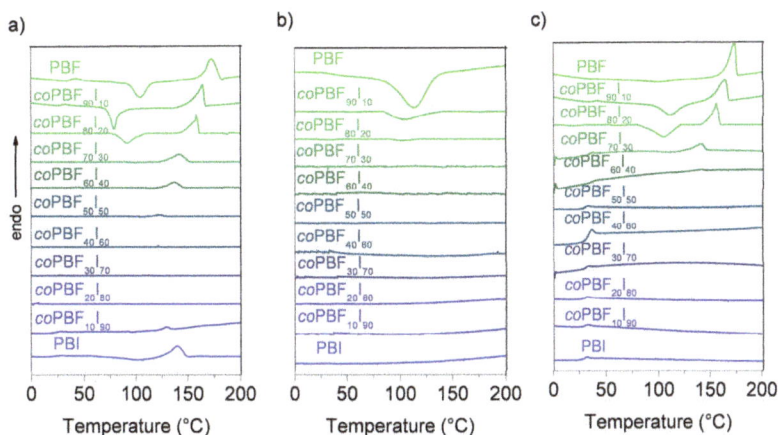

Figure 6. DSC analysis of $coPBF_xI_y$ copolyesters. (**a**) First heating; (**b**) cooling from the melt; and (**c**) second heating.

DSC traces recorded at heating from $coPBF_xI_y$ samples quenched from the melt showed a single T_g intermediate between the two homopolyesters, with a value that decreased continuously from 41 to 21 °C with the content of isophthalate units in the copolyester. This feature is usually taken as an indication of the presence of random copolymers or miscible polymers blends. On the other hand, the replacement of furanoate by terephthalate units showed the opposite effect in the copolyesters, with a small increase of T_g with the content of terephthalate units [21].

The DSC analysis showed that, according to expectations, both PBF and PBI are semicrystalline polyesters with melting temperatures of 173 and 141 °C, and melting enthalpies of 45 and 31 kJ·mol^{-1}, respectively. The insertion of butylene isophthalate units in PBF restricted its crystallinity, reducing gradually both the melting temperature and enthalpy as their content increased. In fact, copolyesters with contents between 60 mol % and 80 mol % of isophthalate units showed very low melting enthalpy values at the first heating (≤ 3 kJ·mol^{-1}), and were unable to crystallize upon cooling from the melt. This last effect was not observed for PBF copolyester series containing terephthalate units in the copolyester, and all copolyesters were observed to be semicrystalline [21].

The intensity profiles of the X-rays scattered by powder pristine samples of representative $coPBF_xI_y$ copolyesters and the homopolyesters examined in the present work are compared in Figure 7.

In agreement with DSC results, discrete scattering was observed for copolyesters containing up to 50 mol % of isophthalate units, with spacings at around 4.9, 3.9 and 3.5 Å. Peaks broadening with the content of isophthalate units in the copolyester agree with should be expected for the impoverishment of the polymer crystallites, and their slight displacement upwards is probably due to the crystal lattice strain caused by the isophthalate units placed at the crystalline-amorphous interphase. From a comparison of these profiles with those produced by PBF, it can be inferred that semicrystalline copolyesters share the triclinic structure of the homopolyester, a fact that implies the exclusion of the isophthalate units from the crystal lattice [9]. Over 50 mol % of isophthalate units, the copolyesters produced amorphous scattering up to 90 mol %, where the profiles showed weak reflections at 5.3 and 3.6 Å, characteristic of the crystal structure of PBI [26].

In order to better understand the effect of the copolymerization on the crystallization behavior, a preliminary isothermal crystallization study has been carried out on PBF homopolymer and a copolymer containing 10 mol % of isophthalate units. Samples that were melted and quenched to 146 °C were isothermally crystallized for one hour at this temperature, and the crystallization enthalpy values generated over time were registered by DSC. It was found that the relative crystallinity (X_t) increased following a sigmoidal trend in both cases (Figure 8), but that the crystallization rate decreased substantially in the $coPBF_{90}I_{10}$ copolyester.

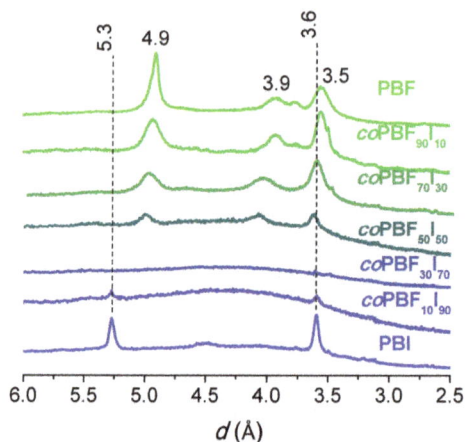

Figure 7. WAXS diffractograms of $coPBF_xI_y$.

The crystallization kinetics was analyzed by means of the Avrami approach. Taking the logarithm of both sides of the Avrami equation gives following equation:

$$\log(-\ln(1 - X_t)) = \log(K) + n \log(t - t_o)$$

where X_t is the fraction of crystallized material, K is the temperature-dependent rate constant, t and t_o are the elapsed and the onset times respectively, and n is the Avrami exponent, indicative of the type of nucleation and dimensionality of crystal growth. Both, n and $\log(K)$ were determined from the slope and the intercept of the linear plot of $\log(-\ln(1 - X_t))$ against $\log(t - t_o)$, respectively, and the resulting values are compared in Table 3 (Figure S6). These results led us to conclude that a similar nucleation/growing mechanism was operating in the crystallization of the two samples, since close values were obtained for n in both cases. In contrast, the crystallization half time increased by a factor of around four in the copolyester, revealing that crystallizability of PBF becomes severely hindered by copolymerization, even for small comonomer contents.

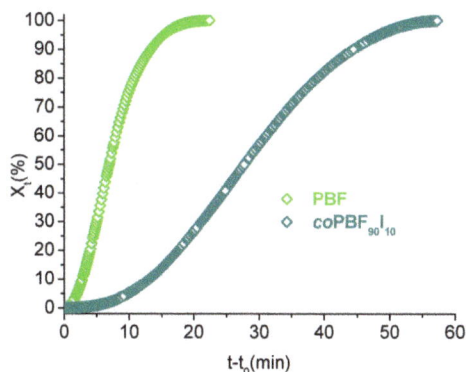

Figure 8. Evolution of the relative crystallinity as a function of time in the isothermal crystallization of PBF and *co*PBF$_{90}$I$_{10}$ at 146 °C.

4. Conclusions

Partially renewable random poly(butylene furoate) copolyesters (PBF-PBI) containing butylene isophthalate units have been synthesized by ROP of cyclic oligoesters. They were obtained with high molecular weights in good yields, and apparently free from impurities. These results are comparable to those obtained in the preparation of PBF-PBT copolyesters, and proove the suitability of the ROP technique as a general tool to synthesize PBF copolyesters. The crystallinity of the PBF-PBI copolyesters was drastically repressed by the insertion of the isophthalate units. This effect was found to be more efficient than that observed for PBF-PBT copolyesters. The crystallizability, even for small contents of isophthalate units, was also drastically reduced, which opens the possibility to use them, provided that they display good mechanical properties, for applications where a more precise control of the crystallization rate is required.

Supplementary Materials: The following are available online at line at www.mdpi.com/xxx/s1, Figure S1: (a) Figure S1: (a) ^{13}C-NMR, (b) ^{1}H-NMR spectra of isophthaloyl chloride, Figure S2: (a) ^{1}H-NMR, (b) HPLC and (c) MALDI-ToF of c(BF)$_n$, Figure S3: (a) ^{1}H-NMR, (b) HPLC and (c) MALDI-ToF of c(BI)$_n$, Figure S4: (a) DSC and (b) TGA analysis of c(BF)$_n$ and c(BI)$_n$, Figure S5: (a) ^{13}C- and (b) ^{1}H-NMR of coPBF$_x$I$_y$, Figure S6: Double logarithmic plot of the Avrami equation for experimental data recorded from the isothermal crystallization of coPBF$_{90}$I$_{10}$ and PBF.

Author Contributions: The manuscript was completed through contributions of all authors. S.M.-G. and A.M.d.I. conceived and designed the experiments; J.C.M.-H. performed the experiments; all three authors analyzed the data; A.M.d.I. and S.M.-G. wrote the paper.

Acknowledgments: Financial support for this research was afforded by MINECO with grants MAT-2012-38044-CO3-03 and MAT-2016-77345-CO3-03. J.C.M.-H. thanks to CONACYT (Mexico) for the Ph.D. grant awarded.

Conflicts of Interest: The authors declare no conflict of interest.

References

1. Miller, S.A. Sustainable polymers: Replacing polymers derived from fossil fuels. *Polym. Chem.* **2014**, *5*, 3117–3118. [CrossRef]
2. Gandini, A. The irruption of polymers from renewable resources on the scene of macromolecular science and technology. *Green Chem.* **2011**, *13*, 1061–1083. [CrossRef]
3. Gandini, A.; Lacerda, T.M. From monomers to polymers from renewable resources: Recent advances. *Prog. Polym. Sci.* **2015**, *48*, 1–39. [CrossRef]
4. Corma, A.; Iborra, S.; Velty, A. Chemical routes for the transformation of biomass into chemicals. *Chem. Rev.* **2007**, *107*, 2411–2502. [CrossRef] [PubMed]

5. Coates, G.W.; Hillmyer, M.A. A virtual issue of macromolecules: Polymers from renewable resources. *Macromolecules* **2009**, *42*, 7987–7989. [CrossRef]

6. Sousa, A.F.; Vilela, C.; Fonseca, A.C.; Matos, M.; Freire, C.S.R.; Gruter, G.J.M.; Coelhob, J.F.J.; Silvestre, A.J.D. Biobased polyesters and other polymers from 2,5-furandicarboxylic acid: A tribute to furan excellency. *Polym. Chem.* **2015**, *6*, 5961–5983. [CrossRef]

7. Gandini, A.; Lacerda, T.M.; Carvalho, A.J.F.; Trovatti, E. Progress of polymers from renewable resources: Furans, vegetable oils, and polysaccharides. *Chem. Rev.* **2016**, *116*, 1637–1669. [CrossRef] [PubMed]

8. Papageorgiou, G.Z.; Papageorgiou, D.G.; Terzopoulou, Z.; Bikiaris, D.N. Production of bio-based 2,5-furan dicarboxylate polyesters: Recent progress and critical aspects in their synthesis and thermal properties. *Eur. Polym. J.* **2016**, *83*, 202–229. [CrossRef]

9. Zhu, J.H.; Cai, J.L.; Xie, W.C.; Chen, P.H.; Gazzano, M.; Scandola, M.; Gross, R.A. Poly(butylene 2,5-furan dicarboxylate), a biobased alternative to PBT: Synthesis, physical properties, and crystal structure. *Macromolecules* **2013**, *46*, 796–804. [CrossRef]

10. Kint, D.P.R.; Muñoz-Guerra, S. Modification of the thermal properties and crystallization behaviour of poly(ethylene terephthalate) by copolymerization. *Polym. Int.* **2003**, *52*, 321–336. [CrossRef]

11. Zheng, M.Y.; Zang, X.L.; Wang, G.X.; Wang, P.L.; Lu, B.; Ji, J.H. Poly(butylene 2,5-furandicarboxylate-epsilon-caprolactone): A new bio-based elastomer with high strength and biodegradability. *Express Polym. Lett.* **2017**, *11*, 611–621. [CrossRef]

12. Wu, B.S.; Xu, Y.T.; Bu, Z.Y.; Wu, L.B.; Li, B.G.; Dubois, P. Biobased poly(butylene 2,5-furandicarboxylate) and poly(butylene adipate-co-butylene 2,5-furandicarboxylate)s: From synthesis using highly purified 2,5-furandicarboxylic acid to thermo-mechanical properties. *Polymer* **2014**, *55*, 3648–3655. [CrossRef]

13. Oishi, A.; Iida, H.; Taguchi, Y. Synthesis of poly(butylene succinate) copolymer including 2,5-furandicarboxylate. *Kobunshi Ronbunshu* **2010**, *67*, 541–543. [CrossRef]

14. Papageorgiou, G.Z.; Papageorgiou, D.G. Solid-state structure and thermal characteristics of a sustainable biobased copolymer: Poly(butylene succinate-co-furanoate). *Thermochim. Acta* **2017**, *656*, 112–122. [CrossRef]

15. Morales-Huerta, J.C.; Ciulik, C.B.; Martínez de Ilarduya, A.; Muñoz-Guerra, S. Fully bio-based aromatic-aliphatic copolyesters: Poly(butylene furandicarboxylate-co-succinate)s obtained by ring opening polymerization. *Polym. Chem.* **2017**, *8*, 748–760. [CrossRef]

16. Morales-Huerta, J.C.; Martínez de Ilarduya, A.; Muñoz-Guerra, S. Blocky poly(ε-caprolactone-co-butylene 2,5-furandicarboxylate) copolyesters via enzymatic ring opening polymerization. *J. Polym. Sci. Pol. Chem.* **2018**, *56*, 290–299. [CrossRef]

17. Brunelle, D.J. Synthesis and polymerization of cyclic polyester oligomers. In *Modern Polyesters: Chemistry and Technology of Polyesters and Copolyesters*; John Wiley & Sons, Ltd.: Hoboken, NJ, USA, 2004; pp. 117–142. [CrossRef]

18. Hodge, P. Entropically driven ring-opening polymerization of strainless organic macrocycles. *Chem. Rev.* **2014**, *114*, 2278–2312. [CrossRef] [PubMed]

19. Strandman, S.; Gautrot, J.E.; Zhu, X.X. Recent advances in entropy-driven ring-opening polymerizations. *Polym. Chem.* **2011**, *2*, 791–799. [CrossRef]

20. Morales-Huerta, J.C.; Martínez de Ilarduya, A.; Muñoz-Guerra, S. A green strategy for the synthesis of poly(ethylene succinate) and its copolyesters via enzymatic ring opening polymerization. *Eur. Polym. J.* **2017**, *95*, 514–519. [CrossRef]

21. Morales-Huerta, J.C.; Martínez de Ilarduya, A.; Muñoz-Guerra, S. Sustainable aromatic copolyesters via ring opening polymerization: Poly(butylene 2,5-furandicarboxylate-co-terephthalate)s. *ACS Sustain. Chem. Eng.* **2016**, *4*, 4965–4973. [CrossRef]

22. Brunelle, D.J.; Bradt, J.E.; Serth-Guzzo, J.; Takekoshi, T.; Evans, T.L.; Pearce, E.J.; Wilson, P.R. Semicrystalline polymers via ring-opening polymerization: Preparation and polymerization of alkylene phthalate cyclic oligomers. *Macromolecules* **1998**, *31*, 4782–4790. [CrossRef] [PubMed]

23. Morales-Huerta, J.C.; Martínez de Ilarduya, A.; Muñoz-Guerra, S. Poly(alkylene 2,5-furandicarboxylate)s (PEF and PBF) by ring opening polymerization. *Polymer* **2016**, *87*, 148–158. [CrossRef]

24. Randall, J. *Polymer Sequence Determination: Carbon-13 NMR Method*; Elsevier Science: New York, NY, USA, 2012.

25. Kamau, S.D.; Hodge, P.; Williams, R.T.; Stagnaro, P.; Conzatti, L. High throughput synthesis of polyesters using entropically driven ring-opening polymerizations. *J. Comb. Chem.* **2008**, *10*, 644–654. [CrossRef] [PubMed]

26. Sanz, A.; Nogales, A.; Ezquerra, T.A.; Lotti, N.; Munari, A.; Funari, S.S. Order and segmental mobility during polymer crystallization: Poly(butylene isophthalate). *Polymer* **2006**, *47*, 1281–1290. [CrossRef]

© 2018 by the authors. Licensee MDPI, Basel, Switzerland. This article is an open access article distributed under the terms and conditions of the Creative Commons Attribution (CC BY) license (http://creativecommons.org/licenses/by/4.0/).

polymers

Article

Synthesis and Characterization of Renewable Polyester Coil Coatings from Biomass-Derived Isosorbide, FDCA, 1,5-Pentanediol, Succinic Acid, and 1,3-Propanediol

Mónica Lomelí-Rodríguez [1,*], José Raúl Corpas-Martínez [1], Susan Willis [2], Robert Mulholland [2] and Jose Antonio Lopez-Sanchez [1,*]

[1] Stephenson Institute for Renewable Energy, Department of Chemistry, University of Liverpool, Crown Street, Liverpool L69 7ZD, UK; Jose.Raul.Corpas-Martinez@liverpool.ac.uk
[2] Becker Industrial Coatings Ltd, Goodlass Road, Speke, Liverpool L24 9HJ, UK; susan.willis@beckers-group.com (S.W.); Robert.Mulholland@beckers-group.com (R.M.)
*   Correspondence: monica.lomeli.r@gmail.com (M.L.-R.); jals@liverpool.ac.uk (J.A.L.-S.); Tel.: +44-(0)151-794-3535 (J.A.L.-S.)

Received: 20 April 2018; Accepted: 22 May 2018; Published: 29 May 2018

Abstract: Biomass-derived polyester coatings for coil applications have been successfully developed and characterized. The coatings were constituted by carbohydrate-derived monomers, namely 2,5-furan dicarboxylic acid, isosorbide, succinic acid, 1,3-propanediol, and 1,5-pentanediol, the latter having previously been used as a plasticizer rather than a structural building unit. The effect of isosorbide on the coatings is widely studied. The inclusion of these monomers diversified the mechanical properties of the coatings, and showed an improved performance against common petrochemical derived coatings. This research study provides a range of fully bio-derived polyester coil coatings with tunable properties of industrial interest, highlighting the importance of renewable polymers towards a successful bioeconomy.

Keywords: biomass; coatings; isosorbide; FDCA; polyester; biopolymer; 1-5-pentanediol

---

1. Introduction

The environmental and sustainability problems currently faced worldwide call for the implementation and usage of renewable materials across industries. Within polymers, the future in polyesters relies on the inclusion of monomers sourced from bioderived feedstocks [1]. In this vein, our group has previously developed the synthesis, kinetic modelling, and process optimization and intensification of bioderived polyesters based on furan 2,5-dicarboxylic acid (FDCA), succinic acid, 1,3-propanediol, and 1,5-pentanediol [2–4]. Common polyester coatings are normally prepared from diacids such as terephthalic acid, isophthalic acid, phthalic anhydride, and adipic acid. The polyalcohols could include difunctional monomers such as neopentyl alcohol, ethylene glycol, or polyfunctional compounds, including trimethylol propane, among other compounds exceeding two functionalities [5]. In this regard, carbohydrates are a vast source of renewable, biomass-derived diols suitable for coil coatings, such as 1,4:3,6-dianhydrohexitols [6]. Dianhydrohexitols are a by-product of the starch industry obtained by the reduction of hexose sugars followed by dehydration. 1,4:3,6-dianhydro-D-glucitol (isosorbide), 1,4:3,6-dianhydrohexitol-D-mannitol (isomannide), and 1,4:3,6-dianhydro-L-iditol (isoidide) are the three main diastereoisomers derived from D-glucose, D-mannose, and L-fructose [7], respectively, and are shown in Figure 1. The reader is encouraged to refer to additional sources for detailed information on the synthesis, chemistry, and properties of 1,4:3,6-dianhydrohexitols [8–10]. Isosorbide bears a considerable potential for the production of new tailored chemicals from renewable resources as it is conformed by two *cis*-connected tetrahydrofuran rings with secondary hydroxyl groups in the 2- (*endo*) and 5- (*exo*) positions, which allow for further functionalization or direct processing [9]. The use of isosorbide in polyesters can be motivated by several features: rigidity, chirality, non-toxicity [8], and recently, its use as a monomer for the preparation of UV-cured coatings has been highlighted [11]. Isosorbide has a relatively high thermostability and low segmental mobility, and can be used to improve the glass transition temperatures of polyesters [12]. However, the hydroxyl group in *endo* position easily forms intra-molecular hydrogen bonds with the oxygen in the other ring, which leads to the poor reactivity of the secondary hydroxyl group and the low number average molecular weights of copolyesters [13].

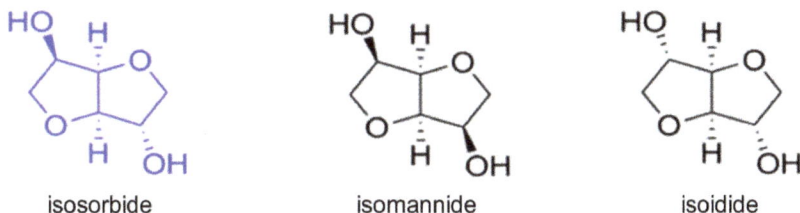

Figure 1. Molecular structures of isosorbide, isomannide, and isoidide.

Previous work on isosorbide-based polyesters has been reported, mainly based on a variety of biomass-derived monomers, mainly citric acid [14], lactic acid [15], succinic acid [16–19], sebacic acid [17,20], itaconic acid [18], 1,4-cyclohexanedimethanol [21], and dimethyl-2,5-furan dicarboxylate [21,22]. Goerz et al. [18] studied the synthesis of polyesters from isosorbide, itaconic acid and succinic acid. The obtained polyesters had glass transition temperatures (T_g) from 57 °C to 65 °C and molecular weights from 1200 Da up to 3500 Da, depending on the molar ratio of the monomers. In the field of coatings, Noordover et al. [14,19,23] reported the synthesis of terpolyesters for powder coatings based on isosorbide, succinic acid, citric acid, and aliphatic diols such as 1,3-propanediol, 1,4-butanediol, and neopentyl glycol, showing number average molecular weights (M_n) from 2700 up to 4600 Da, and highlighting the effect of isosorbide content on the glass transition temperature [19]. Gioia et al. [24] synthesized polyesters based on recycled PET, succinic acid, and isosorbide for powder coating applications

Jacquel et al. [25] prepared bioderived copolyesters of succinic acid and isosorbide by varying the mol % isosorbide from 5 to 20%. The polyesters had T_g from −28 °C to −11 °C, which increased by increasing the mol % isosorbide, while the esterification yield decreased. Zhou et al. [12,13,20] studied the properties and crystallization kinetics of copolyesters based on isosorbide, sebacic acid, and either 1,10-decanediol [20] or 1,3-propanediol [12]. The authors varied the mol % isosorbide from 5.3 mol % to 66.2%, reporting T_g ranging from −26 °C to −5 °C, although no glass transition was observed when the mol % isosorbide was below 30%. The M_n range was broad, with the polyester of isosorbide and sebacic acid showing the lowest M_n (2800 Da), whereas the copolyester with 25.4 mol % isosorbide had the highest M_n, 17000 Da. No relationship was found between the M_n and the amount of incorporated isosorbide [20].

Besides the influence of isosorbide on the final polymer, the effect of catalysts has been studied. For instance, during the synthesis of poly(ethylene terephthalate-co-isosorbide terephthalate), the authors found that combinations of antimony oxide with lithium, magnesium or aluminum based salts successfully increased the efficiency of the transesterification step [26].

Furan 2,5-dicarboxylic acid has been subject to extensive research over the last years, with a drive to develop it as a green chemical building block for polyesters [27–33]. Furan 2,5-dicarboxylic acid is a versatile, bioderived carboxylic acid which was envisioned as a replacement for terephthalic acid in the synthesis of poly(ethylene terephthalate) (PET) and poly(butylene terephthalate) (PBT), although many other polyfuronoates have been developed. Poly(ethylene 2,5-furandicarboxylate) (PEF) shows greatly improved barrier and mechanical properties, higher glass transition temperature, reduced oxygen permeability, and slower chain mobility than its terephthalic acid counterpart [34]. The reactivity and kinetic modelling of the catalyst effect on the PEF polycondensation reaction was also recently reported [35]. The synthesis of potentially 100% renewable FDCA copolyesters with different biomonomers has been explored as well, namely with succinic acid [2,3,30,33,36–38], 1,3-propanediol [2,39], along with some work done with copolyesters of FDCA with lactic acid [40], and isosorbide or its derivatives [21,22,25,41,42]. Tsanaktsis et al. [32] reported for the first time the synthesis of poly(pentylene furonoate) (PPeF) along with poly(heptylene furonoate) (PHepF). Poly(pentylene furonoate) was identified as a semicrystalline polyester with a melting point at 94 °C, T_g at 19 °C, and a maximum decomposition temperature at 394 °C. The same research group studied the thermal decomposition mechanism of PPeF, PHepF, and poly(nonylene furanoate) (PNF) [43]. The authors found that the decomposition of PPeF released gases, such as CO and CO_2, along other degradation products such as dienes, and vinyl- and carboxyl-terminated molecules.

Lately, the thermal properties of FDCA polyesters have been extensively studied, such as poly(octylene furanoate) [44], along with thermal degradation of different polyfuronoates [45,46]. The potential of polyesters based on FDCA is steadily increasing and their industrialization and commercialization will eventually become a reality. Papageorgiou et al. [47] recently reviewed the current status and latest progress of polyfuranoates.

This research study presents the development and synthesis of new polyester resins based on isosorbide, FDCA, succinic acid, and 1,5-pentanediol to be used in coil coating applications. 1,5-pentanediol has normally been used as a polymerization additive (plasticizer) [48] rather than a main diol during polyesterification; however, we have considered it since its bioderived synthesis is playing a major role within the biorefinery concept, with promising prospects [49–54]. Resins with 1,3-propanediol are included as comparison. The paper focuses on the real applicability of the bioderived coatings instead of providing a detailed description on their compositions or chemical characterization. Instead, common mechanical testing analyses for coatings are included. Our objective is therefore to compel a fully biomass-derived, coil coatings library not only of the base polyester resins, but also of the final coatings. We believe our overall work delivers a good basis for the implementation of biomass-derived polymers in large scale and aims to become a strong industrial reference to embrace renewable feedstocks.

2. Materials and Methods

2.1. Materials

Furan 2,5-dicarboxylic acid (>98%) was purchased from Manchester Organics Ltd. (Runcorn, UK) Isosorbide (>98%), succinic acid (>99%), and 1,5-pentanediol (>99%) were purchased from Acros Organics (England, UK). $SnCl_2$ (>98%) was purchased from Alfa Aesar (Haverhill, MA, USA). All other chemicals were of analytical grade and obtained either from Sigma Aldrich (St. Louis, MO, USA) or Fisher Scientific (Hampton, NH, USA).

2.2. Synthesis of Isosorbide-Based Renewable Polyesters

The structures and the synthetic procedure of the isosorbide polyesters with 1,3-propanediol (PPFIS) and 1,5-pentanediol (PPeFIS) are depicted in Scheme 1, whereas Tables 1 and 2 summarize the polyesters synthesized, along with their M_n, weight-average molecular weight (M_w), and dispersity (Đ), which were calculated by gel permeation chromatography (GPC), following the method described later on in Section 2. The polyesters' nomenclature is based on the diol used, the diacid molar ratio, and the molar proportion of isosorbide present. For a typical polyester, the first *P* refers to the suffix *poly*, followed either by *P* or *Pe*, if synthesized with either 1,3-propanediol or 1,5-pentanediol, respectively.

Next, the notation F_x, indicates the presence of FDCA in x mol %. Similarly, S_y refers to the succinic acid present in y mol % whilst I_z corresponds to the z mol % isosorbide.

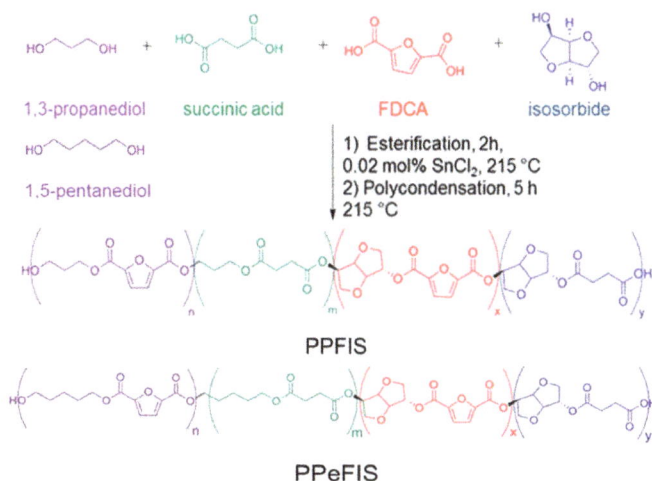

Scheme 1. Synthesis of isosorbide-based polyesters with succinic acid, furan 2,5-dicarboxylic acid (FDCA), 1,3-propanediol (PPFIS), and 1,5-pentanediol (PPeFIS).

Table 1. Synthesized biomass-derived polyesters with 1,3-propanediol and isosorbide (PPFIS).

Polymer [a]	Temperature, °C	Mol% FDCA	Mol% Isosorbide	R [b]	M_n, Da	M_w, Da	Đ
$PPF_{15}I_{30}S_{85}$			30		700	1100	1.6
$PPF_{15}I_{60}S_{85}$	215	15	60	1.5	650	1000	1.5
$PPF_{15}I_{70}S_{85}$			70		600	1000	1.5
$PPF_{30}I_{30}S_{70}$			30		900	1500	1.7
$PPF_{30}I_{60}S_{70}$	215	30	60	1.5	600	1000	1.6
$PPF_{30}I_{70}S_{70}$			70		500	700	1.3
$PPF_{70}I_{10}S_{30}$			10		- [c]		
$PPF_{70}I_{30}S_{30}$	215	70	30	1.5	- [c]		
$PPF_{70}I_{50}S_{30}$			50		-		
$PPF_{85}I_{10}S_{15}$			10		- [c]		
$PPF_{85}I_{30}S_{15}$	215	85	30	1.5	-		
$PPF_{85}I_{30}S_{15}$			50		-		

[a] Catalyst: 0.02 mol % $SnCl_2$/mol diacids, [b] Molar ratio diols:diacids. [c] Polymer insoluble in tetrahydrofuran.

Polyesterification reactions were performed at two different scales: 250 mL and 500 mL. The choice of scale depended on the mol % FDCA in the feed, because the bulk viscosity of the system increased as the mol % FDCA was increased. Hence, all the polyesters bearing furanic content above 50 mol % were synthesized at the 500 mL scale. The different scales also facilitated cleaning and recovery of the product, while minimizing the use of solvents. Despite working with two different volumes, the geometry of the stirrer and the shape of the reactor (round-bottom) were the same in both configurations, as well as the nitrogen flow rate. (No flowmeter was in place but the nitrogen flow was set to 2 bubbles·s^{-1}). The general polymerization reactors set-up and procedure are available in our previous publications [2–4].

The experimental synthesis was followed for all polyesters by adjusting the ratio of the monomers accordingly. The exact quantities for each polyester are available in Supplementary Material Tables S1 and S2, along with the purification polyester method. In a typical polymerization to synthesize $PPF_{15}I_{30}S_{85}$ to a 250 mL 4-neck-round bottom flask equipped with an overhead stirrer, was added 59 g (0.77 mol) of 1,3-propanediol, 48 g (0.33 mol) of isosorbide, and 17 g (0.11 mol) of FDCA. Secondly, the reactor was heated up to 150 °C and 74 g (0.62 mol) of succinic acid and $SnCl_2$ were added. The temperature was increased to 215 °C and was continuously stirred at 350 ppm. The esterification stage was completed after 2 h when all the water had been released and the head temperature on top of the distillation column was back to ambient temperature. Then, the polycondensation reaction was carried out by azeotropic distillation by changing the packed column to a Dean Stark trap, adding 3 wt % xylene as azeotropic agent under atmospheric pressure for 5 hours. The reaction was finally cooled down and the polymer was collected and purified for characterization. The polyesters' color depended on the amount of FDCA and isosorbide in the mixture. Although $SnCl_2$ was the only catalyst used, tin catalysts showed less intense coloration in a comparative polycondensation study using titanium catalysts with furanoates [35]. The intensity color range of the polyesters went from a light yellow to brown with increasing FDCA, although no color space system measurements were undertaken, as our coatings did not require any color specification.

Table 2. Synthesized biomass-derived polyesters with 1,5-pentanediol and isosorbide (PPeFIS).

Polymer [a]	Temperature, C	mol % FDCA	mol % Isosorbide	r [b]	M_n, Da	M_w, Da	Đ
$PPeF_{15}I_{10}S_{85}$			10		1500	3100	2.1
$PPeF_{15}I_{30}S_{85}$			30		1200	2500	2.2
$PPeF_{15}I_{50}S_{85}$	215	15	50	1.5	800	1500	1.8
$PPeF_{15}I_{60}S_{85}$			60		500	1100	2.1
$PPeF_{15}I_{70}S_{85}$			70		700	1100	1.7
$PPeF_{30}I_{10}S_{70}$			10		1300	2700	2.0
$PPeF_{30}I_{30}S_{70}$			30		1100	2800	2.5
$PPeF_{30}I_{50}S_{70}$	215	30	50	1.5	900	1600	1.9
$PPeF_{30}I_{60}S_{70}$			60		500	700	1.5
$PPeF_{30}I_{70}S_{70}$			70		600	1000	1.6
$PPeF_{70}I_{10}S_{30}$			10		1800	3800	2.1
$PPeF_{70}I_{30}S_{30}$	215	70	30	1.5	1300	3200	2.4
$PPeF_{70}I_{50}S_{30}$			50		1000	1900	2.0
$PPeF_{85}I_{10}S_{15}$			10		2300	5400	2.3
$PPeF_{85}I_{30}S_{15}$	215	85	30	1.5	1300	2800	2.1
$PPeF_{85}I_{30}S_{15}$			50		1000	1800	1.8

[a] Catalyst: 0.02 mol % $SnCl_2$/mol diacids, [b] Molar ratio diols:diacids.

2.3. Characterization Methods

2.3.1. Nuclear Magnetic Resonance Spectroscopy (^{1}H NMR)

^{1}H NMR measurements were performed on a Brucker NMR spectrometer (400 MHz). Deuterated chloroform ($CDCl_3$) was used as solvent for all samples.

2.3.2. Gel Permeation Chromatography

Gel permeation chromatography was carried out on an Agilent 1260 Infinity with two Agilent ResiPore Organic 250 × 4.6 mm columns, a guard column, and a refractive index detector. The eluent was tetrahydrofuran (THF) at a flow rate of 0.3 mL/min. Molecular weights were calculated using a conventional calibration with polystyrene standards.

2.3.3. Differential Scanning Calorimetry (DSC)

Differential Scanning Calorimetry measurements were performed using a TA Instruments Q2000 analyzer with a RC590 cooling system using a standard heat-cool-heat method. The temperature range was −50 °C to 200 °C. Both the heating and cooling rates were 10 °C/min in N_2. The amount of sample was approximately 6 ± 0.1 mg. The samples were deposited in Tzero aluminum pans.

2.3.4. Thermal Gravimetric Analysis (TGA)

The thermal stability of the polyesters was determined by TGA using TA Instruments Q5000 equipment under N_2 atmosphere. The samples were placed in aluminum pans and heated from room temperature to 550 °C at a rate of 10 °C/min.

2.3.5. Coatings Characterization

Coatings were applied at 18–20 microns directly onto a smooth polyester pre-primed steel substrate and cured for 40 s to reach a peak metal temperature of 224–232 °C. The physical testing methods carried out on the metal panels are described in Supplementary Material, Section 2 [55–61].

3. Results and Discussion

3.1. ^1H NMR

The intention of the present ^1H NMR analysis was to solely provide a general insight into the regions that identify the carbohydrate-derived polyesters. Furthermore, no quantitative analysis or precise definition of every signal was undertaken. The analysis by ^{13}C NMR and 2D NMR could facilitate the study. The NMR spectra of the isosorbide monomer is included in Supplementary Material, Figure S1, whilst illustrative ^{13}C NMR and 2D NMR spectra of a representative PPeFIS polyester are depicted in Figures S2 and S3.

The chemical shifts and assignments are summarized in Table 3. Figure 2 shows the NMR spectra for 1,5-pentanediol copolyesters $PPeF_{85}I_{10}S_{15}$, $PPeF_{30}I_{30}S_{70}$, and $PPeF_{70}I_{50}S_{30}$, which contain 10, 30, and 50 mol % isosorbide, respectively. The identification of single peaks was a complex process since the incorporation of isosorbide could possibly facilitate the formation of cyclic structures and short chain oligomers, along with the probable presence of unreacted isosorbide (potentially between 3.5–4.5 ppm, Figure 2) at the end of the chains, leaving the secondary hydroxyl unreacted. The presence of impurities in the isosorbide monomer was a possibility as well (>98% purity), as it is one of the limitations of industrially-sourced isosorbide, along with residual monomer.

Table 3. Assignment of chemical shifts of PPeFIS and integrations for $PPeF_{85}I_{10}S_{15}$.

Polyester	Assignment of Chemical Shifts (CDCl₃, δ/ppm)						
	a	b,b′	c	d	e	f	g,j
PPeFIS	7.20	2.62, 2.69	3.68	4.62–4.67	3.80–3.97	4.09	1.80
	h,k	i	l	m	n	o	p
	1.66	4.34	1.42–1.48	5.21–5.25	4.84–4.97	4.48	5.40–5.46
Integrations for $PPeF_{85}I_{10}S_{15}$							
	a	b,b′	c	d	e	f	g,j
	2.03	2.00, 1.13	0.47	0.53	1.17	2.24	4.05
	h,k	i	l	m	n	o	p
	1.97	4.57	2.95	0.14	0.21	0.21	0.47

The formation of the FDCA and succinic acid esters is confirmed by the shifts at 4.34 (*i*) and 4.09 (*f*), respectively, as the protons of the FDCA esters tend to shift to higher ppm values [29–31]. The assignment *a* (7.2 ppm) corresponds to the protons of the furan ring of FDCA. The signals between 3.8 ppm and 5.4 ppm (*e,d,m,n,o,p*) are attributed to the protons of isosorbide, which is in good agreement with previous literature of different isosorbide-based polyesters: 3.9–5.6 ppm [24], 3.7–5.2 ppm [17], 3.8–5.4 ppm [20], and 3.8–5.15 ppm [19]. The differences in shifts between peaks *p* (5.40–5.46 ppm) and *m* (5.21–5.25ppm) are the suggested result of the *endo* (*p*) and *exo* (*m*) stereochemistry of the two hydroxyl groups of isosorbide, as well as between peaks *d* (4.62–4.67 ppm) and *n* (4.84–4.97 ppm) [19]. Gioia et al. [24] identified the *endo* and *exo* OH groups at 5.2 ppm and 5.5 ppm, respectively.

The broadening of the CH_2-signal of succinic acid (*b* and *b'*, 2.62 and 2.69 ppm) was derived from the presence of two diols, 1,5-pentanediol and isosorbide, coupled with the *endo* and *exo* stereochemistry of isosorbide. A similar behavior was reported for polyesters conformed by 1,3-propanediol and isosorbide with succinic acid, where the succinic acid shifts were observable between 2.6 and 2.8 ppm [19]. Apparently, from $PPeF_{85}I_{10}S_{15}$ down to $PPeF_{30}I_{50}S_{70}$ with 50 mol % isosorbide, the characteristic peaks of 1,5-pentanediol (1.42–1.80 ppm) decrease as the isosorbide concentration increases [20], although a quantitative analysis should be performed to confirm the assumption.

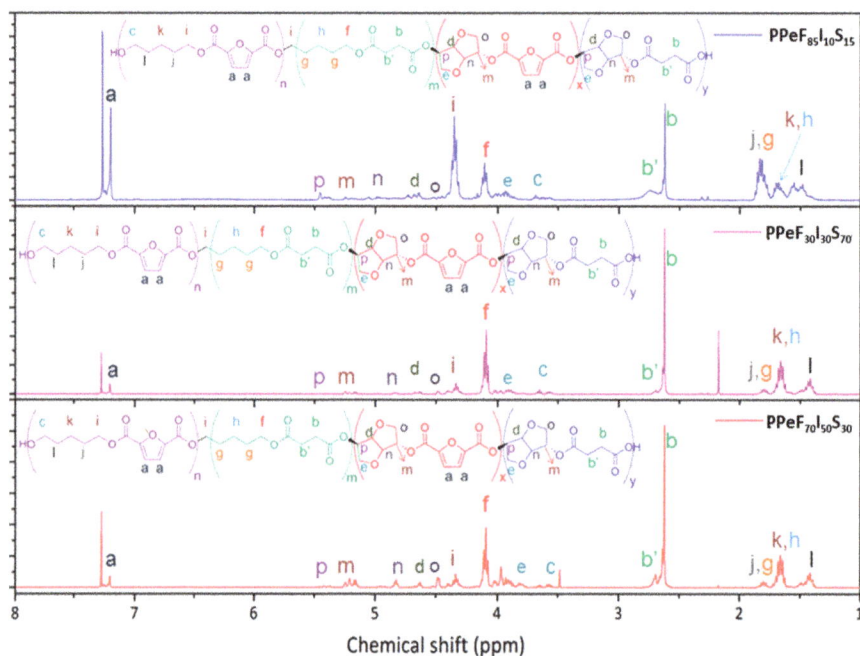

Figure 2. ^1H NMR spectra of $PPeF_{85}I_{10}S_{15}$, $PPeF_{30}I_{30}S_{70}$, and $PPeF_{70}I_{50}S_{30}$. The mol % isosorbide increases from top to bottom.

3.2. GPC

M_n, M_w, and Đ were measured by GPC. Tables 1 and 2 show the results obtained for the isosorbide-based polyesters, whilst Figure 3 shows the chromatographs for the $PPEF_{15}IS_{85}$ and $PPeF_{85}IS_{15}$ families, respectively.

The range of M_w for the polyesters was between 700 and 10200 Da, whereas M_n fell within 500 and 3100 Da, indicating the great influence that the addition of isosorbide imparts, which allows a great versatility within the properties of these biomass-derived polyesters. The results obtained suggest that in general, M_n and M_w decreased as the isosorbide content increased. The dispersity of PPeFIS fell within 1.5/2.4 which was close to the expected value of two for polyesters [62], whilst PPFIS was between 1.3–1.6, suggesting moderate polydisperse samples. The narrowness of the distributions with isosorbide concentrations from 10% to 30% (Figures 3 and 4) were an indication of the strength and toughness of the polyesters. On the other hand, the narrow peaks at higher reaction times were an indication of the presence of short, oligomeric species, and potentially residual isosorbide, as previously seen in the NMR spectra. The highest M_n and M_w were achieved with the furan rich polyesters $PPeF_{85}IS_{15}$ and $PPeF_{70}IS_{30}$ and the trend is followed as the FDCA content decreased, although $PPeF_{70}IS_{30}$ had slightly higher molecular weight when the mol % isosorbide was between

30 and 50%. The incorporation of isosorbide was limited to 50 mol % for these two polyester families, as the mixture becomes extremely viscous and highly unprocessable above that concentration.

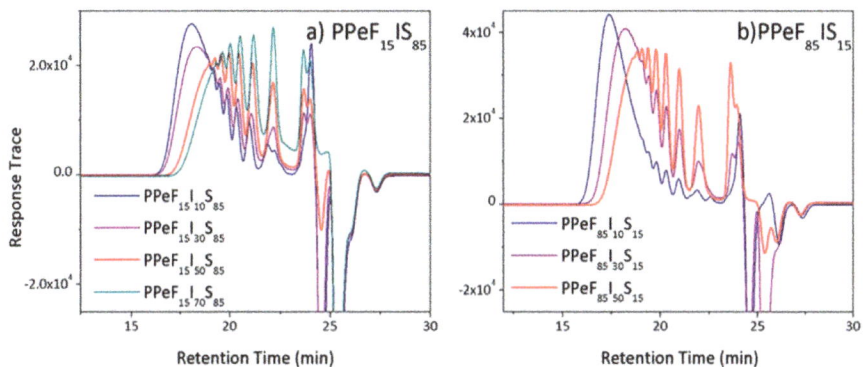

Figure 3. Gel permeation chromatography (GPC) chromatographs of (**a**) PPeF$_{15}$IS$_{85}$ and (**b**) PPeF$_{85}$IS$_{15}$.

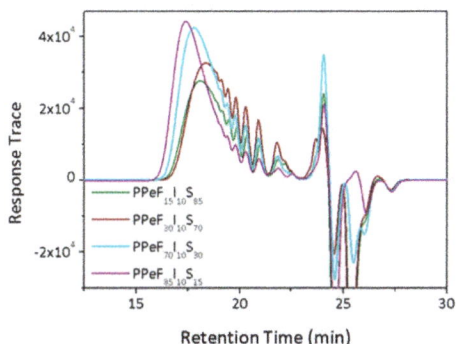

Figure 4. GPC chromatograms of polyesters PPeFIS with 10 mol % isosorbide.

The molecular weight as a function of FDCA/succinic acid composition is depicted in Figure 4 for copolyesters bearing 10 mol % isosorbide. The chromatograms for 30 and 50 mol % isosorbide are available in Supplementary Material Figures S4 and S5.

The copolyesters PPeF$_{70}$I$_{10}$S$_{30}$ and PPeF$_{85}$I$_{10}$S$_{15}$ with 10 mol % isosorbide presented M$_w$ of 3800 Da and 5400 Da, respectively and the dispersities of both compositions with different mol % isosorbide were above two. These values represent the highest molecular weights among the copolyesters of FDCA and succinic acid with isosorbide, whereas the lowest M$_w$ figures correspond to PPeF$_{15}$IS$_{85}$ and PPeF$_{30}$IS$_{70}$ with either 60 or 70 mol % isosorbide (1000 Da). In the case of PPeF$_{15}$IS$_{85}$ and PPeF$_{30}$IS$_{70}$, dispersities of two and M$_w$ above 2000 Da were obtained when the isosorbide content was limited to 30%.

The results for PPFIS resembled our findings with PPeFIS copolyesters. It was observed that M$_n$ decreases as isosorbide content increases, for all compositions. The chromatograms of PPF$_{15}$IS$_{85}$ and PPF$_{30}$IS$_{70}$ are available in Supplementary Material Figure S6. Unfortunately, the results for PPF$_{70}$IS$_{30}$ and PPF$_{85}$IS$_{15}$ are not available since the samples were insoluble in THF.

The decrease in M$_n$ as the isosorbide content increases has been reported in the literature [12,15,19,20,63]. One of the possible explanations was the decreased reactivity of the secondary OH groups when compared with the primary OH groups present in aliphatic linear diols, possibly corresponding to a lower acidic character [24], and most importantly, nucleophilicity. Nucleophilicity is affected by steric hindrance, since the bulkier a given nucleophile is, the slower the rate of its reactions and therefore the lower its nucleophilicity. In the case of isosorbide, although the nucleophilicity of the *endo* hydroxyl group is increased, the steric hindrance caused by hydrogen bonding makes the *exo* hydroxyl group more reactive [64]. Consequently, the difference in reactivity of the OH groups present in isosorbide was mainly due to their stereochemical nature—*endo* and *exo*—where the steric hindrance of the *endo* hydroxyl group is known to decrease the whole reactivity of the system [24]. The OH in *endo* position is more likely to form intra-molecular hydrogen bonding with the oxygen in main chains, while the other in *exo* position is more reactive in polycondensation reactions [65]. In addition, the *endo* hydroxyl is protected by the steric bulk of the rest of the molecule [66]. It might be worthwhile to do a kinetic study to explore different catalysts and reaction times and their influence on the final molecular weight of the polyesters. Wei et al. [20] showed that increasing the isosorbide content above 30% resulted in a significant decrease in the number average molecular weights of the polyesters. In fact, the M_n of the homopolymer poly(isosorbide sebacate) was lower than 3000 Da. Sadler et al. [63] synthesized unsaturated polyesters of phthalic anhydride, maleic anhydride, ethylene glycol, and isosorbide. The polyesters had isosorbide contents from 10 to 25 mol %, and the molecular weight decreased accordingly from 7000 to 3500 Da. In the same vein, Noordover et al. [19] reported M_n from 2000 to 3100 Da for polyesters of succinic acid and isosorbide, and M_n from 2700 to 4600 Da with the addition of 1,4-butanediol or neopentyl glycol as the second diol monomer. Our results suggest that the incorporation of more than 50 mol % isosorbide when processing via azeotropic distillation is considerably detrimental to the molecular weight of the polyesters, even though a catalyst is added to the reaction mixture. This could be due either to changes in stoichiometry, so more hydroxyl functionality will reduce the molecular weight, or due to the isosorbide reacting as a mono-functional monomer. If the isosorbide content is kept between 10 and 30 mol %, the molecular weights are in the desirable range for coatings (i.e., 2000 to 6000 Da) [5,19].

3.3. DSC

The DSC scans for PPeF$_{30}$IS$_{70}$ and PPeF$_{70}$IS$_{30}$ are shown in Figure 5, whilst the T_g, melting temperatures (T_m), and the melting enthalpies (ΔH_m) are summarized in Table 4. The T_g of the parent resins with no isosorbide is included as reference. The isosorbide content and glass transition temperature kept a linear relationship as expected. In all the different compositions, the T_g increased as a function of the mol % isosorbide. Sadler et al. [63] demonstrated that adding as little as 10 mol % isosorbide to the reaction mixture in place of an equivalent amount of a linear diol resulted in a significant increase in T_g. Moreover, adding 60 and 80 mol % isosorbide to resins based on succinic acid and neopentyl glycol achieved final glass transition temperatures from 30.5 °C and 47.1 °C, respectively [19].

Within the range of molecular weights of the resins and taking into account the diacid composition of the base polyester, a minimum content of 50 mol % isosorbide was needed in order to achieve glass transition temperatures above 0 °C and above room temperature. The copolyesters PPeF$_{70}$IS$_{30}$ and PPeF$_{85}$IS$_{15}$ exhibited the highest T_g among the 1,5-pentanediol polyesters, achieving values of 31 °C and 22 °C, respectively. Surprisingly, PPeF$_{70}$I$_{50}$S$_{30}$ had higher T_g (31 °C) than PPeF$_{85}$I$_{50}$S$_{15}$ (22 °C), with 50 mol % isosorbide, but when only 10 mol % was added, the difference in respect to the base polyester is either negligible (−11 °C for PPeF$_{70}$I$_{10}$S$_{30}$ and −10 °C for PPeF$_{70}$S$_{30}$) or very little as per 85 mol % FDCA (4 °C, PPeF$_{85}$I$_{10}$S$_{15}$ and 1 °C, PPeF$_{85}$S$_{15}$). Nonetheless, the slight difference might not be significant and could come down to the acid value that was processed.

For the succinic acid rich polyesters PPeF$_{15}$IS$_{85}$ and PPeF$_{30}$IS$_{70}$, there is an abrupt increase in the glass transition temperature when the isosorbide content is 60% or above, going from values of approximately −40 °C up to 30 °C. This suggests that our library of polyester resins, based in three main monomers, would have different end properties just by tuning the desired content of isosorbide, expanding the potential applications of our renewable coatings. Figure 6 shows T_g and M_n as a function of isosorbide content for $PPeF_{70}IS_{30}$ and $PPeF_{85}IS_{15}$. The corresponding M_n-T_g-isosorbide relationships for $PPeF_{15}IS_{85}$ and $PPeF_{30}IS_{70}$ are depicted in Figures S10 and S11, Supplementary Material. It is observed how M_n decreases as the isosorbide content increases, as explained in the GPC section.

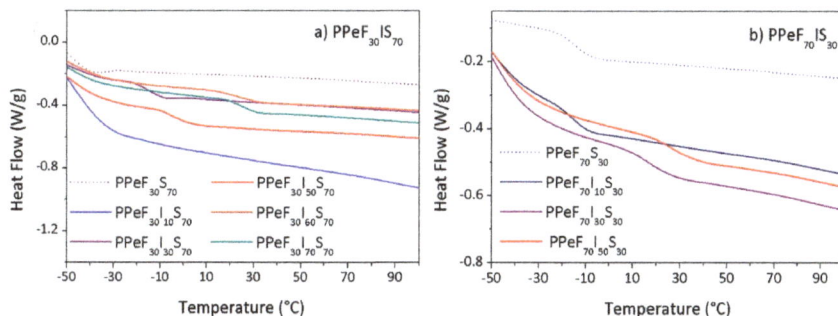

Figure 5. Second heating scan at 10 °C/min for polyesters (**a**) $PPeF_{30}IS_{70}$ including $PPeF_{30}S_{70}$ as the no-isosorbide reference and (**b**) $PPeF_{70}IS_{30}$ including $PPeF_{70}S_{30}$ as the no-isosorbide reference.

Table 4. Thermal transitions of PPeFIS polyesters measured by differential scanning calorimetry (DSC).

Polyester	Mol% Isosorbide	M_w, Da	T_g, °C	T_m, °C	ΔH_m, J/g
$PPeF_{15}S_{85}$	0	2700	−46	-	-
$PPeF_{15}I_{10}S_{85}$	10	3100	−43	-	-
$PPeF_{15}I_{30}S_{85}$	30	2500	−26	-	-
$PPeF_{15}I_{50}S_{85}$	50	1500	−12	-	-
$PPeF_{15}I_{60}S_{85}$	60	1100	7	-	-
$PPeF_{15}I_{70}S_{85}$	70	1100	35	117	2.6
$PPeF_{30}S_{70}$	0	2800	−39	-	-
$PPeF_{30}I_{10}S_{70}$	10	2600	−39	-	-
$PPeF_{30}I_{30}S_{70}$	30	2800	−12	155	3.5
$PPeF_{30}I_{50}S_{70}$	50	1600	0.7	139	1.3
$PPeF_{30}I_{60}S_{70}$	60	700	23	105	0.1
$PPeF_{30}I_{70}S_{70}$	70	1000	24	77	-
$PPeF_{70}S_{30}$	0	5900	−10	-	-
$PPeF_{70}I_{10}S_{30}$	10	3800	−11	-	-
$PPeF_{70}I_{30}S_{30}$	30	3200	19	94	0.4
$PPeF_{70}I_{50}S_{30}$	50	1900	31	105	1.2
$PPeF_{85}S_{15}$	0	4100	1	50	-
$PPeF_{85}I_{10}S_{15}$	10	5400	4	119	0.1
$PPeF_{85}I_{30}S_{15}$	30	2800	20	101	1.0
$PPeF_{85}I_{50}S_{15}$	50	1800	22	139	1.4

Nevertheless, some other factors, such as impurities, chemical conformation, degree of crystallinity, and molecular weight could be relevant for the analysis of thermal transitions. Consequently, further work could be focused on the particular effect of any of the above factors for selected polyesters, potentially those with closer molar masses or bearing the greatest industrial feasibility.

The glass transition temperatures of the PPFIS family, prepared with 1,3-propanediol, are slightly higher since they present lower chain flexibility than their 1,5-pentanediol counterparts. The results are available in Supplementary Material Table S3. The inclusion of 30 mol % isosorbide leads to an increase of T_g of about 30 °C (−45 °C to −14 °C) achieving a T_g at 6.2 °C with 70 mol % isosorbide. The copolyesters $PPF_{70}I_{50}S_{30}$ and $PPF_{85}I_{50}S_{15}$ with 50 mol % isosorbide exhibited the highest T_g among all the polyesters, achieving values of 29.2 and 53.2 °C, respectively. Representative DSC thermograms of PPFIS polyesters are available in Supplementary Material Figures S7 and S8.

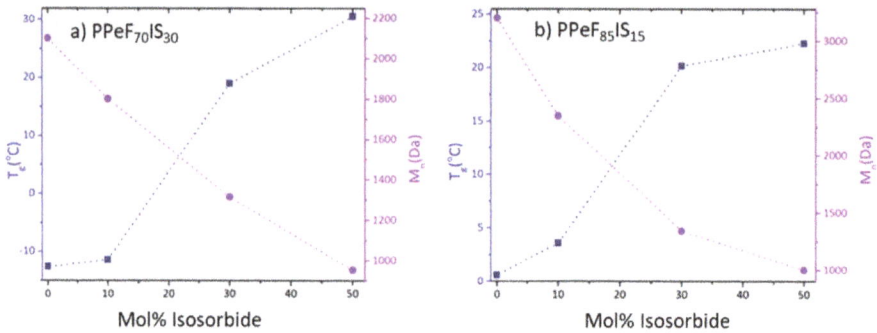

Figure 6. T_g and M_n as a function of mol % isosorbide for (**a**) PPeF$_{70}$IS$_{30}$ and (**b**) PPeF$_{85}$IS$_{15}$.

Tailoring the different ratios of diacids and diols allowed to synthesize amorphous and semicrystalline polyesters. Regarding melting temperatures (T_m) 1,5-pentanediol polyesters with FDCA contents of 30% and above showed melting endotherms from 77.3 °C to 154.7 °C, whereas for 1,3-propanediol the only observable T_m's (97–131 °C) correspond to those containing 70 and 85 mol % FDCA. In the case of PPF$_{85}$IS$_{15}$ and PPeF$_{30}$IS$_{70}$, T_m decreased with an increase of the isosorbide content, a phenomenon that was previously observed in copolyesters of sebacic acid, 1,3-propanediol, and varying isosorbide content [12], whereas the other compositions exhibit the opposite trend.

Particularly, polyesters PPF$_{70}$IS$_{30}$ and PPF$_{85}$IS$_{15}$ did not show a single melting peak in the DSC heating scan, as observed in Figure 7 for PPF$_{85}$IS$_{15}$ and Figure S9 for PPF$_{70}$IS$_{30}$. This behavior has been reported in the literature [12,67–69] for different copolyesters, and it is thought to be potentially the result of the existence of different crystal types in the same polymer sample [67], melting-recrystallization-remelting processes [68,70,71], or the presence of different molecular weight species [12]. It is necessary to confirm the assumptions regarding crystallinity features by performing an X-Ray Diffraction (XRD) analysis of the polyesters. There is a strong possibility that the polyesters are conformed by different oligomers with varying molecular weights, as the presence of isosorbide tends to form cyclic oligomeric structures [19,72]. Again, the confirmation of these structures requires further robust analysis, such as MALDI-TOF-MS, although this is outside our current scope.

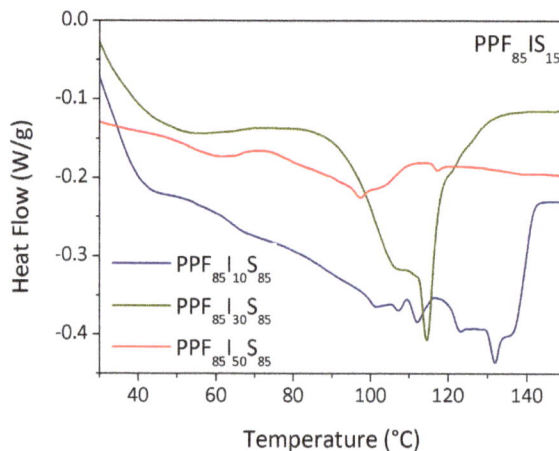

Figure 7. First heating scan at 10 °C/min for polyesters PPF$_{85}$I$_{10}$S$_{15}$, PPF$_{85}$I$_{30}$S$_{15}$, and PPF$_{85}$I$_{50}$S$_{15}$.

Previous syntheses of isosorbide-derived polyesters have reported glass transition temperatures in the range presented herein or below. For instance, the T_g's of poly(isosorbide sebacate) and poly(isosorbide itaconate) were 6.4 °C (M_n = 19200 Da) and 34.5 °C (M_n = 8700 Da), respectively [17]. Wei et al. [20] synthesized polyesters of 1,10-decanediol, sebacic acid and isosorbide with different mol % isosorbide, namely, 43.9%, 66.2%, and 100%. The polyesters showed T_g's at −26 °C, −18 °C, and −5 °C, with M_w = 8900, 8699, and 2800 Da, respectively. The authors did not observe glass transitions for 5.3 or 16.7 mol % isosorbide, which they attributed to the fact that long chain aliphatic polyesters have a strong crystallization capacity and crystallize very fast. Terzopoulou and coworkers [22] showed that poly(isosorbide-2,5-furanoate) presented a Tg at 157 °C, whilst the Tg's for copolyesters of FDCA, isosorbide, and 1,4-cyclohexane-dimethanol (CHDM) increased from 75.3 to 103.5 °C, with decreasing CHDM/isosorbide ratios from 95/5 to 60/40 [21].

3.4. TGA

Figure 8 shows the thermograms for the polyesters PPeF$_{70}$IS$_{30}$ and PPeF$_{85}$IS$_{15}$. Thermal gravimetric analysis indicates that the thermal stability ranges from 308 °C to 371 °C for the 1,5-pentanediol library. Expectedly, PPFIS polyesters showed a slightly higher and narrower T_d range between 362 and 372 °C, suggesting that the diol is a fundamental factor for thermal stability. Previous studies have shown that the thermal stability decreases with increasing methylene units. For example, the Bikiaris group [43] showed that poly(1,4-butylene furanoate) (PBF) was less thermally stable than the furanoates synthesized with shorter chain diols, such as ethylene glycol and 1,3-propanediol. The thermal decomposition temperatures (T_d) for PPeFIS and PPFIS are summarized in Tables S4 and S5 in Supplementary Material, along with the TGA thermograms of PPeF$_{15}$IS$_{85}$ and PPeF$_{30}$IS$_{70}$, which correspond to Figure S12.

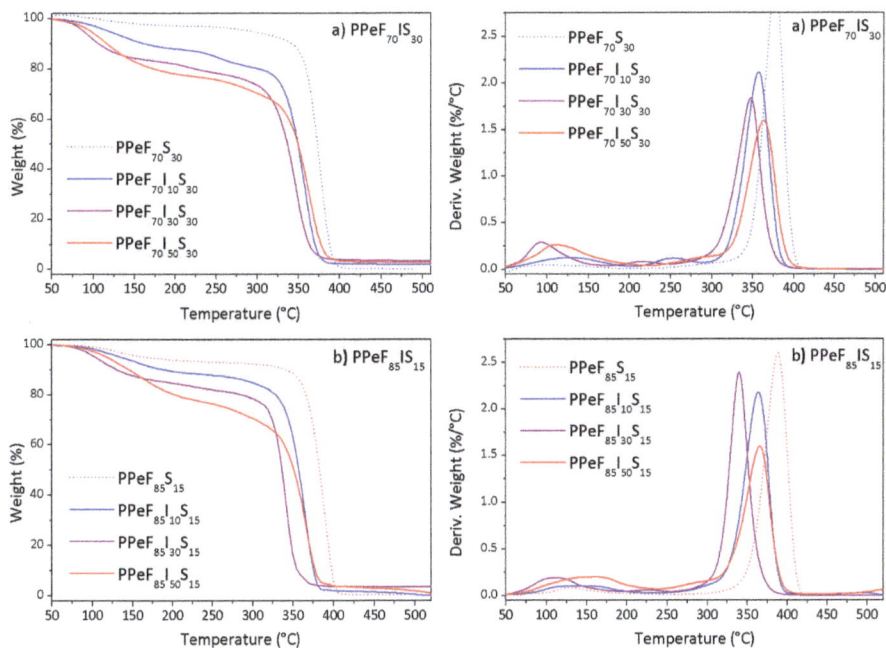

Figure 8. Weight % and derivative weight thermograms of polyesters (**a**) PPeF$_{70}$IS$_{30}$ and (**b**) PPeF$_{85}$IS$_{15}$ (N$_2$ flow, 10 °C·min^{-1}).

For both linear aliphatic diols, the incorporation of isosorbide to the main polymers resulted in a decrease of approximately 30–40 °C in the maximum T_d of the resins, which were close (~385 °C) or above 400 °C. Within each diacidcomposition, the results suggest that the PPeFIS polyesters with highest mol % isosorbide present the highest decomposition temperature. However, the same trend was not observed for PPFIS, where all the polyesters had similar T_d values.

Specifically, for 1,3-propanediol, the thermogravimetric data suggests the presence of diverse species within the polymer matrix, as three different decomposition temperatures (T_{d1}, T_{d2}, and T_{dmax}, Table S5) are observed. All the diacid compositions present the first decomposition temperature between 98.7 °C and 194 °C, followed by a second transition at 180–290 °C. The lowest T_d's appear to be prompted by isosorbide concentrations of above 50 mol %. The nature of these species has not been determined; however, previous research has shown that polyesters undergo decomposition mechanisms initiated by scission of an alkyl-oxygen bond, suggesting a random-chain scission [73]. Also, the decomposition of these polyesters is suggested to be dominated by cyclic or open chain oligomers with carboxylic-end groups. The formation of these cyclic oligomers is done through an intramolecular exchange reaction which happens below 300 °C [74] which could be the same process that took place in our polyesters. Cyclization of isosorbide-based polyesters has indeed been reported by other authors [15,19]. In the case of polyesters with low FDCA and rich isosorbide contents, data suggests the formation of more volatile products, which some authors have determined to arise from the secondary breakdown of end-groups, which follows the primary step of cyclic concerted decomposition [75].

3.5. Coatings

Tables 5 and 6 summarize the results obtained. For succinic acid-rich polyesters, the chosen samples were those with higher isosorbide concentrations. A reference, petro-derived coating *R* has been included for comparison. Specifically, in the case of 1,3-propanediol resins, it was not possible to make paints with PPF$_{15}$I$_{70}$S$_{85}$ and PPF$_{30}$I$_{70}$S$_{70}$ (70 mol % isosorbide) as they were immiscible with the paint solvent system, which consisted of a 50:50 mixture of butanol and dibasic ester. In the case of PPF$_{70}$I$_{50}$S$_{30}$, the resin was a brittle solid, which prevented solution in the same paint solvents. As similarly reported for some coating applications [14,19,24,63], the introduction of isosorbide into the synthesis of the polyester resins improved the thermomechanical properties of the resulting paints, compared to the parent polyesters. The no-isosorbide parent polyesters, PPeF1$_5$S$_{85}$ and PPeF$_{30}$S$_{70}$, had very low microhardness (11 N·m^{-2} and 12 N·m^{-2}, respectively) and low T_g (−28 °C and −17 °C, respectively), but when 70 mol % isosorbide was added (PPeF$_{15}$I$_{70}$S$_{85}$ and PPeF$_{30}$I$_{70}$S$_{70}$) the microhardness abruptly increased to 237 N·m^{-2} and 287 N·m^{-2}, with T_g at 66 °C and 53 °C. The resins were not very flexible as T-Bend was 5T and 6T and both presented severe cracking after the Erichsen testing, which suggested that the concentration of FDCA should be increased. We previously reported the effect of the FDCA on the final polyester resin [2]. PPeF$_{70}$I$_{30}$S$_{30}$ with an isosorbide concentration of 30 mol % and an increased FDCA concentration of 70% showed an improvement in flexibility (2.5T) respect to the reference resin (3T), T_g on specification at 34 °C and better impact resistance as it presented a slight cracking. Moreover, PPeF$_{70}$I$_{50}$S$_{30}$ had above-specification Erichsen (No cracking), T_g (67 °C) and microhardness (270 N·m^{-2}) than the reference resin. Likewise, comparing PPeF$_{70}$I$_{50}$S$_{30}$ with its parent resin with no isosorbide, the performance was considerably improved, as the properties recorded for the latter were T_g = 6 °C, Microhardness = 20 N·m^{-2} and T-Bend no cracking of 1.5T.

PPeF$_{85}$IS$_{15}$ resins again showed better microhardness and higher T_g than PPeF$_{85}$S$_{15}$, as the results were 21 N·m^{-2} and 15 °C, respectively. PPeF$_{85}$I$_{50}$S$_{15}$ (Table 6), despite having the highest T_g (69 °C) and enhanced microhardness (299 N·m^{-2}) was slightly harder (Pencil Hardness 2H) and presents poor flexibility measured by T-Bend no cracking (5.5T). PPeF$_{85}$I$_{30}$S$_{15}$ therefore presented the best overall properties (Table 6).

Table 5. Physical test results on white paints based on polyesters PPeF$_{15}$IS$_{85}$-PPeF$_{70}$IS$_{30}$ and the standard reference resin R.

Test	Standard	R	PPeF$_{15}$I$_{70}$S$_{85}$	PPeF$_{30}$I$_{70}$S$_{70}$	PPeF$_{70}$I$_{30}$S$_{30}$	PPeF$_{70}$I$_{50}$S$_{30}$
Pencil hardness	EN13523-4	H	2H	3H	H	H
Gloss top coat	13523-2	35	39	35	40	39
Reverse impact 80″ lb	13523-5	Moderate cracking	Moderate cracking	Moderate cracking	Slight cracking	Moderate cracking
Erichsen 7.5 mm	13523-6	Moderate cracking	Severe cracking	Severe cracking	Slight cracking	No cracking
T-bend (no tape pick off)	13523-7	0.5T	1T	1.5T	0T	1T
T-Bend no cracking	13523-7	3T	5T	6T	2.5T	5T
MEK^a rubs primer	BSSP 3.522.11	110	110	110	110	110
T_g, °C (onset/midpoint)		28/35	53/66	51/53	27/34	53/67
Microhardness, N·m^{-2}		216	237	287	190	270

^{a} Methyl ethyl ketone.

Table 6. Physical test results on white paints based on polyesters PPeF$_{85}$I$_{10}$S$_{15}$, PPeF$_{85}$I$_{30}$S$_{15}$, PPeF$_{85}$I$_{50}$S$_{15}$, PPF$_{85}$I$_{30}$S$_{15}$ and the standard reference resin R.

Test	Standard	R	PPeF$_{85}$I$_{10}$S$_{15}$	PPeF$_{85}$I$_{30}$S$_{15}$	PPeF$_{85}$I$_{50}$S$_{15}$	PPF$_{85}$I$_{30}$S$_{15}$
Pencil hardness	EN13523-4	H	H	F	2H	H
Gloss top coat	13523-2	35	40	43	38	39
Reverse impact 80″ lb	13523-5	Moderate cracking	No cracking	Moderate cracking	Severe cracking	Slight cracking
Erichsen 7.5 mm	13523-6	Moderate cracking	No cracking	No cracking	Severe cracking	Slight cracking
T-bend (no tape pick off)	13523-7	0.5T	1.5T	1.5T	1T	1.5T
T-Bend no cracking	13523-7	3T	1T	2T	5.5T	2.5T
MEK rubs primer	BSSP 3.522.11	110	110	110	110	110
T_g, °C (onset/midpoint)		28/35	26/32	34/42	58/69	29/34
Microhardness, N·m^{-2}		216	174	287	299	169

4. Conclusions

We successfully diversified and enhanced the properties of novel biomass-derived polyester coatings based on FDCA, succinic acid, and either 1,3-propanediol or 1,5-pentanediol by incorporating isosorbide into the polyesters' backbone. Depending on the concentration of isosorbide, the molecular weight of the polyesters could be tuned from 700 to 10,200 Da. In general, the molecular weight decreased as the isosorbide content increased. This behavior could be attributed to the difference in reactivity of the OH groups present in isosorbide. The thermal results suggested that a minimum 50 mol % isosorbide is needed to achieve glass transition temperatures above 0 °C and above room temperature. The T_g could vary approximately 40 degrees by solely adjusting the molar isosorbide concentration.

Paint testing results showed that the best isosorbide resins for coil coating applications were PPeF$_{85}$I$_{30}$S$_{15}$ and PPeF$_{70}$I$_{30}$S$_{30}$, along with PPF$_{85}$I$_{30}$S$_{15}$. The consideration of bioderived 1,5-pentanediol as a main building block opens the possibility towards the development of flexible coatings but with better hardness than currently used diols, such as 1,6-hexanediol, enhanced by the presence of FDCA and isosorbide.

The inclusion of the carbohydrate-derived diol isosorbide to our coatings did promote better performance in terms of thermomechanical properties, and allowed for the tuning of fully biomass-derived resins that could easily replace the current petrochemical-derived ones in different applications, with controlled molecular weights and glass transition temperatures. Exterior durability testing for top coats would still be needed, but our bio-derived coatings could be used as backing coatings and interior finishes. Future work however needs to be done in terms of the variables effects on the T_g, the identification of the nature of the different oligomers or cyclic structures formed during the polymerization, as well as a kinetic study to optimize the reaction conditions to overcome the isosorbide's low-reactive nature.

Polymers **2018**, *10*, 600

Supplementary Materials: The following are available online at http://www.mdpi.com/2073-4360/10/6/600/s1, Figure S1: ^{1}H NMR of isosorbide monomer, Table S1: Monomer Charge for PPFIS Polyesters, Table S2: Monomer Charge for PPeFIS polyesters, Figure S2: ^{13}C NMR of PPeF$_{15}$I$_{50}$S$_{85}$, Figure S3: HSQC of PPeF$_{15}$I$_{50}$S$_{85}$, Figure S4: GPC chromatogram of polyesters PPeFIS with 30 mol % isosorbide, Figure S5: GPC chromatogram of polyesters PPeFIS with 50 mol % isosorbide, Figure S6: GPC chromatogram of polyesters PPF$_{15}$IS$_{85}$ and PPF$_{30}$IS$_{70}$, Table S3: Thermal transitions of PPFIS measured by DSC, Figure S7: DSC thermogram of PPeF$_{15}$IS$_{85}$, Figure S8: DSC thermogram of PPF$_{30}$IS$_{70}$, Figure S9: First heating scan at 10 °C/min for polyesters PPF$_{70}$IS$_{30}$, Figure S10: T$_g$-M$_n$-mol % isosorbide relationship for PPeF$_{15}$IS$_{85}$, Figure S11: T$_g$-M$_n$-mol % isosorbide relationship for PPeF1$_{30}$IS$_{70}$, Table S4: Characteristic decomposition temperatures T$_{d1}$, T$_{dmax}$ and weight loss% of PPeFIS, Table S5: Characteristic decomposition temperatures T$_{d1}$, T$_{d2}$ and T$_{dmax}$ and weight loss% of PPFIS, Figure S12: TGA thermograms for PPeF$_{15}$IS$_{85}$ and PPeF$_{30}$IS$_{70}$.

Author Contributions: M.L.-R. conceived and designed the experiments; M.L.-R. and J.R.C.-M. performed the experiments; M.L.-R., J.R.C.-M., R.M., and S.W. analyzed the data; R.M. and S.W. contributed reagents/materials/analysis tools; M.L.-R. and J.A.L.-S wrote the paper; J.A.L.-S. reviewed and approved the paper.

Acknowledgments: All authors are grateful to the EPSRC for funding support. Mónica Lomelí-Rodríguez is grateful to Consejo Nacional de Ciencia y Tecnología (CONACYT) for funding support and to Becker Industrial Coatings Ltd. in Liverpool for the valuable discussions and the facilities provided.

Conflicts of Interest: The authors declare no conflict of interest. The founding sponsors had no role in the design of the study; in the collection, analyses, or interpretation of data; in the writing of the manuscript, and in the decision to publish the results.

References

1. Siyab, N.; Tenbusch, S.; Willis, S.; Lowe, C.; Maxted, J. Going Green: Making reality match ambition for sustainable coil coatings. *J. Coat. Technol. Res.* **2016**, *13*, 629–643. [CrossRef]

2. Lomelí-Rodríguez, M.; Martín-Molina, M.; Jiménez-Pardo, M.; Nasim-Afzal, Z.; Cauët, S.I.; Davies, T.E.; Rivera-Toledo, M.; Lopez-Sanchez, J.A. Synthesis and kinetic modeling of biomass-derived renewable polyesters. *J. Polym. Sci. Part A Polym. Chem.* **2016**, *54*, 2876–2887. [CrossRef]

3. Lomelí-Rodríguez, M.; Rivera-Toledo, M.; López-Sánchez, J.A. Optimum Batch-Reactor Operation for the Synthesis of Biomass-Derived Renewable Polyesters. *Ind. Eng. Chem. Res.* **2017**, *56*, 549–559. [CrossRef]

4. Lomelí-Rodríguez, M.; Rivera-Toledo, M.; López-Sánchez, J.A. Process Intensification of the Synthesis of Biomass-Derived Renewable Polyesters: Reactive Distillation and Divided Wall Column Polyesterification. *Ind. Eng. Chem. Res.* **2017**, *56*, 3017–3032. [CrossRef]

5. Gubbels, E.; Drijfhout, J.P.; Posthuma-van Tent, C.; Jasinska-Walc, L.; Noordover, B.A.J.; Koning, C.E. Bio-based semi-aromatic polyesters for coating applications. *Prog. Organ. Coat.* **2014**, *77*, 277–284. [CrossRef]

6. Naves, A.F.; Fernandes, H.T.C.; Immich, A.P.S.; Catalani, L.H. Enzymatic syntheses of unsaturated polyesters based on isosorbide and isomannide. *J. Polym. Sci. Part A Polym. Chem.* **2013**, *51*, 3881–3891. [CrossRef]

7. Feng, X.; East, A.J.; Hammond, W.B.; Zhang, Y.; Jaffe, M. Overview of advances in sugar-based polymers. *Polym. Adv. Technol.* **2011**, *22*, 139–150. [CrossRef]

8. Fenouillot, F.; Rousseau, A.; Colomines, G.; Saint-Loup, R.; Pascault, J.-P. Polymers from renewable 1, 4: 3, 6-dianhydrohexitols (isosorbide, isomannide and isoidide): A review. *Prog. Polym. Sci.* **2010**, *35*, 578–622. [CrossRef]

9. Rose, M.; Palkovits, R. Isosorbide as a renewable platform chemical for versatile applications—Quo Vadis? *ChemSusChem* **2012**, *5*, 167–176. [CrossRef] [PubMed]

10. Flèche, G.; Huchette, M. Isosorbide. Preparation, Properties and Chemistry. *Starch Stärke* **1986**, *38*, 26–30. [CrossRef]

11. Fertier, L.; Ibert, M.; Buffe, C.; Saint-Loup, R.; Joly-Duhamel, C.; Robin, J.J.; Giani, O. New biosourced UV curable coatings based on isosorbide. *Prog. Organ. Coat.* **2016**, *99*, 393–399. [CrossRef]

12. Zhou, C.; Wei, Z.; Yu, Y.; Wang, Y.; Li, Y. Biobased copolyesters from renewable resources: Synthesis and crystallization kinetics of poly(propylene sebacate-*co*-isosorbide sebacate). *RSC Adv.* **2015**, *5*, 68688–68699. [CrossRef]

13. Wang, G.; Jiang, M.; Zhang, Q.; Wang, R.; Zhou, G. Biobased copolyesters: Synthesis, crystallization behavior, thermal and mechanical properties of poly(ethylene glycol sebacate-*co*-ethylene glycol 2,5-furan dicarboxylate). *RSC Adv.* **2017**, *7*, 13798–13807. [CrossRef]

14. Noordover, B.A.; Duchateau, R.; van Benthem, R.A.; Ming, W.; Koning, C.E. Enhancing the functionality of biobased polyester coating resins through modification with citric acid. *Biomacromolecules* **2007**, *8*, 3860–3870. [CrossRef] [PubMed]

15. Kricheldorf, H.R.; Weidner, S.M. High T g copolyesters of lactide, isosorbide and isophthalic acid. *Eur. Polym. J.* **2013**, *49*, 2293–2302. [CrossRef]

16. Garaleh, M.; Yashiro, T.; Kricheldorf, H.R.; Simon, P.; Chatti, S. (Co-) Polyesters Derived from Isosorbide and 1, 4-Cyclohexane Dicarboxylic Acid and Succinic Acid. *Macromol. Chem. Phys.* **2010**, *211*, 1206–1214. [CrossRef]

17. Park, H.-S.; Gong, M.-S.; Knowles, J.C. Synthesis and biocompatibility properties of polyester containing various diacid based on isosorbide. *J. Biomater. Appl.* **2012**, *27*, 99–109. [CrossRef] [PubMed]

18. Goerz, O.; Ritter, H. Polymers with shape memory effect from renewable resources: Crosslinking of polyesters based on isosorbide, itaconic acid and succinic acid. *Polym. Int.* **2013**, *62*, 709–712. [CrossRef]

19. Noordover, B.A.; van Staalduinen, V.G.; Duchateau, R.; Koning, C.E.; van Benthem, R.A.; Mak, M.; Heise, A.; Frissen, A.E.; van Haveren, J. Co-and terpolyesters based on isosorbide and succinic acid for coating applications: Synthesis and characterization. *Biomacromolecules* **2006**, *7*, 3406–3416. [CrossRef] [PubMed]

20. Wei, Z.; Zhou, C.; Yu, Y.; Li, Y. Biobased copolyesters from renewable resources: Synthesis and crystallization behavior of poly (decamethylene sebacate-*co*-isosorbide sebacate). *RSC Adv.* **2015**, *5*, 42777–42788. [CrossRef]

21. Kasmi, N.; Majdoub, M.; Papageorgiou, G.Z.; Bikiaris, D.N. Synthesis and crystallization of new fully renewable resources-based copolyesters: Poly(1,4-cyclohexanedimethanol-*co*-isosorbide 2,5-furandicarboxylate). *Polym. Degrad. Stab.* **2018**, *152*, 177–190. [CrossRef]

22. Terzopoulou, Z.; Kasmi, N.; Tsanaktsis, V.; Doulakas, N.; Bikiaris, D.N.; Achilias, D.S.; Papageorgiou, G.Z. Synthesis and Characterization of Bio-Based Polyesters: Poly(2-methyl-1,3-propylene-2,5-furanoate), Poly(isosorbide-2,5-furanoate), Poly(1,4-cyclohexanedimethylene-2,5-furanoate). *Materials* **2017**, *10*, 801. [CrossRef] [PubMed]

23. Van Haveren, J.; Oostveen, E.A.; Miccichè, F.; Noordover, B.A.J.; Koning, C.E.; van Benthem, R.A.T.M.; Frissen, A.E.; Weijnen, J.G.J. Resins and additives for powder coatings and alkyd paints, based on renewable resources. *J. Coat. Technol. Res.* **2007**, *4*, 177–186. [CrossRef]

24. Gioia, C.; Vannini, M.; Marchese, P.; Minesso, A.; Cavalieri, R.; Colonna, M.; Celli, A. Sustainable polyesters for powder coating applications from recycled PET, isosorbide and succinic acid. *Green Chem.* **2014**, *16*, 1807–1815. [CrossRef]

25. Jacquel, N.; Saint-Loup, R.; Pascault, J.-P.; Rousseau, A.; Fenouillot, F. Bio-based alternatives in the synthesis of aliphatic–aromatic polyesters dedicated to biodegradable film applications. *Polymer* **2015**, *59*, 234–242. [CrossRef]

26. Bersot, J.C.; Jacquel, N.; Saint-Loup, R.; Fuertes, P.; Rousseau, A.; Pascault, J.P.; Spitz, R.; Fenouillot, F.; Monteil, V. Efficiency Increase of Poly (ethylene terephthalate-co-isosorbide terephthalate) Synthesis using Bimetallic Catalytic Systems. *Macromol. Chem. Phys.* **2011**, *212*, 2114–2120. [CrossRef]

27. Gandini, A.; Silvestre, A.J.; Neto, C.P.; Sousa, A.F.; Gomes, M. The furan counterpart of poly (ethylene terephthalate): An alternative material based on renewable resources. *J. Polym. Sci. Part A Polym. Chem.* **2009**, *47*, 295–298. [CrossRef]

28. Jiang, M.; Liu, Q.; Zhang, Q.; Ye, C.; Zhou, G. A series of furan-aromatic polyesters synthesized via direct esterification method based on renewable resources. *J. Polym. Sci. Part A Polym. Chem.* **2012**, *50*, 1026–1036. [CrossRef]

29. Papageorgiou, G.Z.; Tsanaktsis, V.; Papageorgiou, D.G.; Exarhopoulos, S.; Papageorgiou, M.; Bikiaris, D.N. Evaluation of polyesters from renewable resources as alternatives to the current fossil-based polymers. Phase transitions of poly (butylene 2, 5-furan-dicarboxylate). *Polymer* **2014**, *55*, 3846–3858. [CrossRef]

30. Sousa, A.; Fonseca, A.; Serra, A.; Freire, C.; Silvestre, A.; Coelho, J. New unsaturated copolyesters based on 2, 5-furandicarboxylic acid and their crosslinked derivatives. *Polym. Chem.* **2016**, *7*, 1049–1058. [CrossRef]

31. Vannini, M.; Marchese, P.; Celli, A.; Lorenzetti, C. Fully biobased poly (propylene 2, 5-furandicarboxylate) for packaging applications: Excellent barrier properties as a function of crystallinity. *Green Chem.* **2015**, *17*, 4162–4166. [CrossRef]

32. Tsanaktsis, V.; Terzopoulou, Z.; Nerantzaki, M.; Papageorgiou, G.Z.; Bikiaris, D.N. New poly (pentylene furanoate) and poly (heptylene furanoate) sustainable polyesters from diols with odd methylene groups. *Mater. Lett.* **2016**, *178*, 64–67. [CrossRef]

33. Terzopoulou, Z.; Tsanaktsis, V.; Bikiaris, D.N.; Exarhopoulos, S.; Papageorgiou, D.G.; Papageorgiou, G.Z. Biobased poly (ethylene furanoate-co-ethylene succinate) copolyesters: Solid state structure, melting point depression and biodegradability. *RSC Adv.* **2016**, *6*, 84003–84015. [CrossRef]

34. Burgess, S.K.; Leisen, J.E.; Kraftschik, B.E.; Mubarak, C.R.; Kriegel, R.M.; Koros, W.J. Chain mobility, thermal, and mechanical properties of poly (ethylene furanoate) compared to poly (ethylene terephthalate). *Macromolecules* **2014**, *47*, 1383–1391. [CrossRef]

35. Terzopoulou, Z.; Karakatsianopoulou, E.; Kasmi, N.; Tsanaktsis, V.; Nikolaidis, N.; Kostoglou, M.; Papageorgiou, G.Z.; Lambropoulou, D.A.; Bikiaris, D.N. Effect of catalyst type on molecular weight increase and coloration of poly(ethylene furanoate) biobased polyester during melt polycondensation. *Polym. Chem.* **2017**, *8*, 6895–6908. [CrossRef]

36. Wu, L.; Mincheva, R.; Xu, Y.; Raquez, J.-M.; Dubois, P. High molecular weight poly (butylene succinate-co-butylene furandicarboxylate) copolyesters: From catalyzed polycondensation reaction to thermomechanical properties. *Biomacromolecules* **2012**, *13*, 2973–2981. [CrossRef] [PubMed]

37. Hbaieb, S.; Kammoun, W.; Delaite, C.; Abid, M.; Abid, S.; El Gharbi, R. New copolyesters containing aliphatic and bio-based furanic units by bulk copolycondensation. *J. Macromol. Sci. Part A* **2015**, *52*, 365–373. [CrossRef]

38. Yu, Z.; Zhou, J.; Cao, F.; Wen, B.; Zhu, X.; Wei, P. Chemosynthesis and characterization of fully biomass-based copolymers of ethylene glycol, 2, 5-furandicarboxylic acid, and succinic acid. *J. Appl. Polym. Sci.* **2013**, *130*, 1415–1420. [CrossRef]

39. Papageorgiou, G.Z.; Papageorgiou, D.G.; Tsanaktsis, V.; Bikiaris, D.N. Synthesis of the bio-based polyester poly (propylene 2, 5-furan dicarboxylate). Comparison of thermal behavior and solid state structure with its terephthalate and naphthalate homologues. *Polymer* **2015**, *62*, 28–38. [CrossRef]

40. Matos, M.; Sousa, A.F.; Fonseca, A.C.; Freire, C.S.; Coelho, J.F.; Silvestre, A.J. A New Generation of Furanic Copolyesters with Enhanced Degradability: Poly (ethylene 2, 5-furandicarboxylate)-co-poly (lactic acid) Copolyesters. *Macromol. Chem. Phys.* **2014**, *215*, 2175–2184. [CrossRef]

41. Storbeck, R.; Ballauff, M. Synthesis and properties of polyesters based on 2, 5-furandicarboxylic acid and 1, 4: 3, 6-dianhydrohexitols. *Polymer* **1993**, *34*, 5003–5006. [CrossRef]

42. Wu, J.; Eduard, P.; Thiyagarajan, S.; Noordover, B.A.; van Es, D.S.; Koning, C.E. Semi-Aromatic Polyesters Based on a Carbohydrate-Derived Rigid Diol for Engineering Plastics. *ChemSusChem* **2015**, *8*, 67–72. [CrossRef] [PubMed]

43. Terzopoulou, Z.; Tsanaktsis, V.; Nerantzaki, M.; Papageorgiou, G.Z.; Bikiaris, D.N. Decomposition mechanism of polyesters based on 2,5-furandicarboxylic acid and aliphatic diols with medium and long chain methylene groups. *Polym. Degrad. Stab.* **2016**, *132*, 127–136. [CrossRef]

44. Papageorgiou, G.Z.; Guigo, N.; Tsanaktsis, V.; Papageorgiou, D.G.; Exarhopoulos, S.; Sbirrazzuoli, N.; Bikiaris, D.N. On the bio-based furanic polyesters: Synthesis and thermal behavior study of Poly (octylene furanoate) using Fast and Temperature Modulated Scanning Calorimetry. *Eur. Polym. J.* **2015**, *68*, 115–127. [CrossRef]

45. Tsanaktsis, V.; Vouvoudi, E.; Papageorgiou, G.Z.; Papageorgiou, D.G.; Chrissafis, K.; Bikiaris, D.N. Thermal degradation kinetics and decomposition mechanism of polyesters based on 2, 5-furandicarboxylic acid and low molecular weight aliphatic diols. *J. Anal. Appl. Pyrolysis* **2015**, *112*, 369–378. [CrossRef]

46. Terzopoulou, Z.; Tsanaktsis, V.; Nerantzaki, M.; Achilias, D.S.; Vaimakis, T.; Papageorgiou, G.Z.; Bikiaris, D.N. Thermal degradation of biobased polyesters: Kinetics and decomposition mechanism of polyesters from 2, 5-furandicarboxylic acid and long-chain aliphatic diols. *J. Anal. Appl. Pyrolysis* **2016**, *117*, 162–175. [CrossRef]

47. Papageorgiou, G.Z.; Papageorgiou, D.G.; Terzopoulou, Z.; Bikiaris, D.N. Production of bio-based 2, 5-furan dicarboxylate polyesters: Recent progress and critical aspects in their synthesis and thermal properties. *Eur. Polym. J.* **2016**, *83*, 202–229. [CrossRef]

48. Oldring, P.K.; Tuck, N. *Resins for Surface Coatings, Alkyds & Polyesters*; John Wiley & Sons: Hoboken, NJ, USA, 2000; Volume 2.

49. Brentzel, Z.J.; Barnett, K.J.; Huang, K.; Maravelias, C.T.; Dumesic, J.A.; Huber, G.W. Chemicals from Biomass: Combining Ring-Opening Tautomerization and Hydrogenation Reactions to Produce 1,5-Pentanediol from Furfural. *ChemSusChem* **2017**, *10*, 1351–1355. [CrossRef] [PubMed]

50. Sun, D.; Sato, S.; Ueda, W.; Primo, A.; Garcia, H.; Corma, A. Production of C4 and C5 alcohols from biomass-derived materials. *Green Chem.* **2016**, *18*, 2579–2597. [CrossRef]

51. Huang, K.; Brentzel, Z.J.; Barnett, K.J.; Dumesic, J.A.; Huber, G.W.; Maravelias, C.T. Conversion of Furfural to 1,5-Pentanediol: Process Synthesis and Analysis. *ACS Sust. Chem. Eng.* **2017**, *5*, 4699–4706. [CrossRef]

52. Chatterjee, M.; Ishizaka, T.; Kawanami, H. Hydrogenation of 5-hydroxymethylfurfural in supercritical carbon dioxide-water: A tunable approach to dimethylfuran selectivity. *Green Chem.* **2014**, *16*, 1543–1551. [CrossRef]

53. Koso, S.; Furikado, I.; Shimao, A.; Miyazawa, T.; Kunimori, K.; Tomishige, K. Chemoselective hydrogenolysis of tetrahydrofurfuryl alcohol to 1, 5-pentanediol. *Chem. Commun.* **2009**, 2035–2037. [CrossRef] [PubMed]

54. Liu, S.; Amada, Y.; Tamura, M.; Nakagawa, Y.; Tomishige, K. One-pot selective conversion of furfural into 1, 5-pentanediol over a Pd-added Ir–ReO x/SiO 2 bifunctional catalyst. *Green Chem.* **2014**, *16*, 617–626. [CrossRef]

55. Standard Test Methods for Indentation Hardness of Organic Coatings. Available online: https://compass. astm.org/download/D1474D1474M.4895.pdf (accessed on 15 February 2017).

56. Coil Coated Metals. Test Methods. Pencil Hardness. Available online: http://shop.bsigroup.com/ ProductDetail/?pid=000000000030268977 (accessed on 14 February 2017).

57. Standard Test Method for Specular Gloss. Available online: https://compass.astm.org/Standards/ HISTORICAL/D523-89R99.htm (accessed on 14 February 2017).

58. Standard Practice for Assessing the Solvent Resistance of Organic Coatings Using Solvent Rubs. Available online: https://compass.astm.org/download/D5402-93R99.15287.pdf (accessed on 15 February 2015).

59. ISO 14577-1:2015 Metallic Materials—Instrumented Indentation Test for Hardness and Materials parameters —Part 1: Test Method. Available online: https://www.iso.org/obp/ui/#iso:std:iso:14577:-1:ed-2:v1:en (accessed on 15 February 2017).

60. Standard Test Method for Assignment of the Glass Transition Temperatures by Differential Scanning Calorimetry or Differential Thermal Analysis. Available online: https://compass.astm.org/download/ E1356-98.33431.pdf (accessed on 15 February 2017).

61. Standard Test Method for Coating Flexibility of Prepainted Sheet. Available online: https://compass.astm. org/download/D4145.28079.pdf (accessed on 15 February 2017).

62. Allcock, H.R.; Lampe, F.W.; Mark, J.E.; Allcock, H. *Contemporary Polymer Chemistry*; Pearson/Prentice Hall: Upper Saddle River, NJ, USA, 2003.

63. Sadler, J.M.; Toulan, F.R.; Palmese, G.R.; La Scala, J.J. Unsaturated polyester resins for thermoset applications using renewable isosorbide as a component for property improvement. *J. Appl. Polym. Sci.* **2015**, *132*, 42315. [CrossRef]

64. Koo, J.M.; Hwang, S.Y.; Yoon, W.J.; Lee, Y.G.; Kim, S.H.; Im, S.S. Structural and thermal properties of poly(1,4-cyclohexane dimethylene terephthalate) containing isosorbide. *Polym. Chem.* **2015**, *6*, 6973–6986. [CrossRef]

65. Łukaszczyk, J.; Janicki, B.; Kaczmarek, M. Synthesis and properties of isosorbide based epoxy resin. *Eur. Polym. J.* **2011**, *47*, 1601–1606. [CrossRef]

66. Zhu, Y.; Molinier, V.; Durand, M.; Lavergne, A.; Aubry, J.-M. Amphiphilic Properties of Hydrotropes Derived from Isosorbide: Endo/Exo Isomeric Effects and Temperature Dependence. *Langmuir* **2009**, *25*, 13419–13425. [CrossRef] [PubMed]

67. Papageorgiou, G.Z.; Achilias, D.S.; Bikiaris, D.N. Crystallization Kinetics and Melting Behaviour of the Novel Biodegradable Polyesters Poly(propylene azelate) and Poly(propylene sebacate). *Macromol. Chem. Phys.* **2009**, *210*, 90–107. [CrossRef]

68. Papageorgiou, G.Z.; Bikiaris, D.N. Crystallization and melting behavior of three biodegradable poly(alkylene succinates). A comparative study. *Polymer* **2005**, *46*, 12081–12092. [CrossRef]

69. Wang, X.; Zhou, J.; Li, L. Multiple melting behavior of poly(butylene succinate). *Eur. Polym. J.* **2007**, *43*, 3163–3170. [CrossRef]

70. Gunaratne, L.M.W.K.; Shanks, R.A. Multiple melting behaviour of poly(3-hydroxybutyrate-co-hydroxyvalerate) using step-scan DSC. *Eur. Polym. J.* **2005**, *41*, 2980–2988. [CrossRef]

71. Song, P.; Chen, G.; Wei, Z.; Zhang, W.; Liang, J. Calorimetric analysis of the multiple melting behavior of melt-crystallized poly(L-lactic acid) with a low optical purity. *J. Therm. Anal. Calorim.* **2013**, *111*, 1507–1514. [CrossRef]

72. Chatti, S.; Weidner, S.M.; Fildier, A.; Kricheldorf, H.R. Copolyesters of isosorbide, succinic acid, and isophthalic acid: Biodegradable, high Tg engineering plastics. *J. Polym. Sci. Part A Polym. Chem.* **2013**, *51*, 2464–2471. [CrossRef]
73. Beyler, C.L.; Hirschler, M.M. Thermal decomposition of polymers. In *SFPE Handbook of Fire Protection Engineering*; Springer: Berlin, Germany, 2002; p. 32.
74. Montaudo, G.; Puglisi, C.; Samperi, F. Primary thermal degradation mechanisms of PET and PBT. *Polym. Degrad. Stab.* **1993**, *42*, 13–28. [CrossRef]
75. Goldfarb, I.J.; McGuchan, R. *Thermal Degradation of Polyesters. 1. Aliphatic Polymers*; DTIC Document; DTIC: Fort Belvoir, VA, USA, 1968.

© 2018 by the authors. Licensee MDPI, Basel, Switzerland. This article is an open access article distributed under the terms and conditions of the Creative Commons Attribution (CC BY) license (http://creativecommons.org/licenses/by/4.0/).

![polymers logo]

polymers

MDPI

Article

FA Polymerization Disruption by Protic Polar Solvents

Guillaume Falco, Nathanaël Guigo *, Luc Vincent and Nicolas Sbirrazzuoli *

Institut de Chimie de Nice, Université Nice-Sophia Antipolis, Université Côte d'Azur, UMR CNRS 7272, Nice CEDEX 06108, France; guillaume.falco@insa-lyon.fr (G.F.); lvincent@unice.fr (L.V.)
* Correspondence: Nathanael.Guigo@unice.fr (N.G.); Nicolas.Sbirrazzuoli@unice.fr (N.S.); Tel.: +33-049-207-6179

Received: 20 April 2018; Accepted: 8 May 2018; Published: 15 May 2018

Abstract: Furfuryl alcohol (FA) is a biobased monomer derived from lignocellulosic biomass. The present work describes its polymerization in the presence of protic polar solvents, i.e., water or isopropyl alcohol (IPA), using maleic anhydride (MA) as an acidic initiator. The polymerization was followed from the liquid to the rubbery state by combining DSC and DMA data. In the liquid state, IPA disrupts the expected reactions during the FA polymerization due to a stabilization of the furfuryl carbenium center. This causes the initiation of the polymerization at a higher temperature, which is also reflected by a higher activation energy. In the water system, the MA opening allows the reaction to start at a lower temperature. A higher pre-exponential factor value is obtained in that case. The DMA study of the final branching reaction occurring in the rubbery state has highlighted a continuous increase of elastic modulus until 290 °C. This increasing tendency of modulus was exploited to obtain activation energy dependences (E_α) of FA polymerization in the rubbery state.

Keywords: renewable resources; lignocellulosic biomass; polymerization; reaction mechanisms; furfuryl alcohol

1. Introduction

Lignocellulosic biomass appears as an important renewable source for the fabrication of organic polymeric materials [1–3]. In this line, furan derivatives [4] such as 5-hydroxymethylfurfural [5] or furfural [6] are respectively obtained from C_6 or C_5 sugar dehydrations [7] in the bio-refinery processes. These two compounds are among the most interesting biobased building blocks to design a sustainable chemistry [8] and they can be further transformed [9] in, for instance, 2,5-furandicarboxylic acid and furfuryl alcohol (FA), respectively. These latter compounds are indeed the top two of the class of biobased precursors used for their polymerization abilities.

FA is an attractive biobased compound which has the particularity to homopolymerize in poly(furfuryl alcohol) (PFA) under an acid catalyzed condition, leading to a strong reticulated network with high thermomechanical properties [10]. These advantages make the PFA an excellent biobased alternative in industrial processes such as for foundry molds [11] or wood reinforcement [12]. Other studies have demonstrated several applications for this eco-friendly polymer as the fabrication of carbon nanospheres [13], hybrid materials with silica [14,15], fully biobased composites with several natural fibers [16,17], or copolymers with a combination of vegetable oils [18,19].

As shown in Figure 1, the FA polymerization mechanism [10,20,21] is complex and can be divided into two steps. The first polymerization step occurs in the liquid phase with the influence of the acid initiator. This step consists of the formation of FA oligomers by polycondensation from an active furfuryl carbenium center. The second step leads to a three-dimensional branched polymer through Diels-Alder cycloadditions between the formed oligomers. Furthermore, side reactions can also occur. First, an electrophilic addition [22] of the active furfuryl carbenium center on the conjugate form of the oligomer can occur, increasing the heterogeneity of the final 3D network. Another reaction that was reported in the literature is FA furan ring opening [20], which preferably starts via C–O bond breaking [23]. Although spectroscopic [10,24] and theoretical [25] studies have highlighted the low percentage of furan ring-opening during the FA polymerization, these side reactions may have a significant impact on the final polymer structure. Indeed, furan ring openings lead to a lower final crosslinking density, impacting the mechanical properties of the final polymer [26]. Furthermore, these opened furan structures are amplified by the addition of protic polar solvents, suggesting disruption mechanisms of FA polymerization.

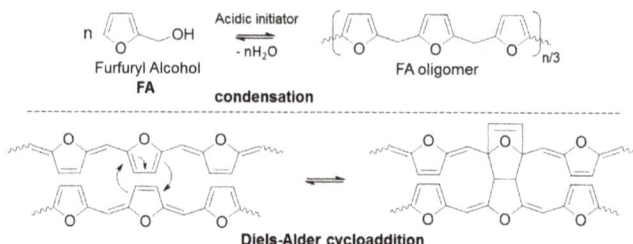

Figure 1. Polymerization mechanisms of FA.

This paper aims to study the disruption effect of protic polar solvents during the polymerization of FA. Both water and isopropyl alcohol (IPA) were added as protic polar solvents in FA formulations with maleic anhydride (MA) as an acidic initiator. The polymerization evolution was followed from the liquid state by Differential Scanning Calorimetry (DSC) and additionally from the rubbery state by Dynamic Mechanical Analysis (DMA). The study provides, for the first time, an advanced isoconversional analysis of FA polymerization kinetics over a wide temperature range and including two stages of polymerization, i.e., an early liquid state and final rubbery state. The obtained variations of effective activation energy, E_α, are interpreted as a function of the extent of conversion, α, and of the temperature. Important changes in the polymerization mechanism are discussed.

2. Materials and Methods

2.1. Materials

Furfuryl alcohol (FA) (M_w = 98.10 g·mol^{-1}, b.p. = 170 °C, purity 99%), isopropyl alcohol (IPA) (M_w = 60.10 g·mol^{-1}, b.p. = 82 °C, purity > 99.7%), and maleic anhydride (MA) (M_w = 98.06 g·mol^{-1}, m.p. = 51–56 °C, purity > 99%) were obtained from Aldrich Chemical Co (Milwaukee, WI, USA).

2.2. Preparation of Liquid FA/solvent Mixtures

Three different liquid formulations containing FA, maleic anhydride (MA), and protic polar solvents were prepared for DSC investigation, namely: FA/MA (100/2); FA/MA/water; and FA/MA/IPA (50/1/50). MA was used as an acidic initiator of FA homopolymerization and added with 2% *w/w* compared to FA. For mixtures containing solvents, IPA or ultra-pure water (conductivity 2 µS·cm^{-1}) was added to an equivalent quantity of FA (50/50 *w/w*) with 1% of MA to maintain the same initiator ratio as FA/MA.

2.3. Preparation of Cured PFA Materials

Three different reticulated resins were prepared from liquid formulations used in the DSC study, i.e., FA/MA (100/2); FA/MA/water; and FA/MA/IPA (50/1/50). Pre-polymers were prepared in a PTFE round-bottom flask by heating approximatively 15 grams of liquid formulations. The mixtures were vigorously stirred during the overall process to lead to homogenous final polymers. The synthesis of each furanic resin was realized in two steps. The first pre-polymerization step was the same for the three resins. Blends were heated at around 85 °C for one hour in a round-bottom flask. A condenser was used to avoid the evaporation of solvent or FA and to induce sufficient interactions between FA (or FA oligomers) and the protic polar solvent. For the second step, the condenser was removed and the resulting mixtures were placed at 100 °C. The temperature was increased by 10 °C every 30 min to obtain a highly viscous resin (~10^3 Pa·s^{-1}). To reach the desired viscosity, FA/MA and FA/MA/water were heated to 120 °C, while FA/MA/IPA was heated to 140 °C.

Then, the pre-polymers obtained were introduced in a silicon mold and were cured for two hours under pressure (~10 bars), respectively, at 160 °C for FA/MA and FA/MA/water and at 180 °C for FA/MA/IPA. This step was conducted to obtain rigid materials. The second curing step under pressure was essential to avoid the formation of holes in the PFA material (due to water evaporation by the polycondensation of FA), which can modify the mechanical or the thermal performances of the final materials.

After slow cooling to room temperature, polymers were unmolded and a post-curing step with a duration of one hour at 180 °C for FA/MA and FA/MA/water and at 220 °C for FA/MA/IPA was conducted in order to evaporate unreacted FA monomers and avoid possible trapped solvent. The three materials are labelled in the DMA study as follows: reference PFA (for PFA cured with pure FA), PFA/water (for PFA obtained via the water route), and PFA/IPA (for PFA obtained via the IPA route).

2.4. Analytical Techniques

Differential scanning calorimetry (DSC) measurements were performed on a Mettler-Toledo DSC-1(Mettler-Toledo GmbH, Schwerzenbach, Switzerland) equipped with a FRS5 sensor (with 56 thermocouples Au-Au/Pd, Mettler-Toledo GmbH, Schwerzenbach, Switzerland) and STAR© software (Mettler-Toledo GmbH, Schwerzenbach, Switzerland) for data analysis. Temperature and enthalpy calibrations were performed using indium and zinc standards. Samples of about 10 mg were placed into a sealed 30 µL high-pressure crucible. The DSC measurements of FA polymerization were conducted at the heating rates of 1, 2, 4, and 6 °C·min^{-1}.

Dynamic mechanical properties were carried out on a Mettler-Toledo DMA 1 (Mettler-Toledo GmbH, Schwerzenbach, Switzerland) in tensile mode equipped with STAR© software for curves analysis. The sample dimensions were 15.00 (length), 4.50 mm (width), and 1.50 mm (thickness) (±0.01 mm). The experiments were performed from 25 to 350 °C, at a frequency of 1 Hz and with a heating of 1, 2, and 4 °C·min^{-1}.

The DSC and DMA data were treated with an advanced isoconversional method to realize a kinetic study and to compute the activation energy dependency (E_α) of FA polymerization. These computations were realized with internal software [27,28].

2.5. Theoretical Approaches

Isoconversional methods are amongst the more reliable kinetic methods for the treatment of thermoanalytical data, see, for example, [29–32]. The ICTAC Kinetics Committee has recommended the use of multiple temperature programs for the evaluation of reliable kinetic parameters [30]. The main advantages of isoconversional methods are that they afford an evaluation of the activation energy, E_α, without assuming any particular form of the reaction model, $f(\alpha)$ or $g(\alpha)$, and that a change in the E_α variation, called E_α-dependency, can generally be associated with a change in the reaction mechanism or in the rate-limiting step of the overall reaction rate, as measured with thermoanalytical techniques.

Polymerizations are frequently accompanied by a significant amount of heat released, thus cure kinetics can be easily monitored by DSC. It is generally assumed that the heat flow measured by calorimetry is proportional to the process rate [33]. Thus, the extent of conversion at time t, α_t, is computed according to Equation (1), as follows:

$$\alpha_t = \frac{\int_{t_i}^{t}(dQ/dt)dt}{\int_{t_i}^{t_f}(dQ/dt)dt} \tag{1}$$

where dQ/dt is the heat flow; t_i is the time at which the process initiates (i.e., the respective heat flow becomes detectable); and t_f is the time at which the process finishes (i.e., the heat flow falls below the detection limit). The denominator represents the total transformation heat released during curing (Q).

In a more general sense, the extent of conversion can be determined as a change of any physical property associated with the reaction progress. For this, the physical property has to be normalized to lay between 0 and 1. If the shear modulus changes with the reaction progress, an extent of conversion can be defined as follows:

$$\alpha_t = \frac{G'_t - G'_{t_i}}{G'_{t_f} - G'_{t_i}} \tag{2}$$

where G' is the shear modulus measured by DMA (or dynamic rheometric) experiments at time t, and where t_i and t_f have the same meaning as in Equation (1).

The general form of the basic rate equation is usually written as [33,34]:

$$\frac{d\alpha}{dt} = A \exp\left(\frac{-E}{RT}\right) f(\alpha) \tag{3}$$

where T is the temperature, $f(\alpha)$ is the differential form of the reaction model that represents the reaction mechanism, E is the activation energy, and A is the pre-exponential factor.

The advanced isoconversional method [27,33,35,36] used in this study is presented in Equations (4) and (5) and has been derived from Equation (3), as follows:

$$\Phi(E_\alpha) = \sum_{i=1}^{n}\sum_{j\neq1}^{n} \frac{J[E_\alpha, T_i(t_\alpha)]}{J[E_\alpha, T_j(t_\alpha)]} \tag{4}$$

$$J[E_\alpha, T_i(t_\alpha)] = \int_{t_\alpha-\Delta\alpha}^{t_\alpha} \exp\left[\frac{-E_\alpha}{RT(t)}\right] dt \tag{5}$$

where E_α is the effective activation energy. The E_α value is determined as the value that minimizes the function Φ (E_α). This method is applicable to any arbitrary temperature program $T_i(t)$ and uses a numerical integration of the integral with respect to the time. E_α is computed for each value of α generally in the range 0.02–0.98 with a step of 0.02. For each i-th temperature program, the time $t_{\alpha,i}$ and temperature $T_{\alpha,i}$ related to selected values of α are determined by an accurate interpolation [27,28]. The software developed can treat any kind of isothermal or non-isothermal data from DSC, calorimetry (C80), TGA, DMA, or rheometry [27,28,37,38].

3. Results and Discussion

3.1. FA Polymerization Evolution from the Liquid State

3.1.1. Non-Isothermal DSC Investigation

Figure 2 shows non-isothermal DSC data obtained during FA polymerization and in the presence of protic polar solvents such as water and IPA. The data of polymerization without solvent are presented in the insert of Figure 2 for comparison (reference system). The reference system shows a single asymmetric thermal event, while the addition of solvents significantly modifies the heat flow curve shapes, with the appearance of a shoulder or a second peak. This observation leads to the hypothesis of a change in the polymerization pathways. In the FA/MA/water mix, the first peak is predominant. It is followed by a broad thermal event that could correspond to secondary reactions or residual cross-links between FA oligomers. For the FA/MA/IPA mix, the polymerization pathway seems different from the other two systems due to a second and predominant thermal event. Moreover, the FA polymerization reactions are shifted to a higher temperature. The protic character of IPA due to its high dipolar moment (μ_{IPA} = 1.70 D) can explain this. The first step of FA polymerization starts with the formation of an active furfuryl carbenium center [10]. Thus, in this system, IPA slows down the formation rate of the carbenium center by forming hydrogen bonds and a solvation sphere with the hydroxyl groups of FA [39]. These effects could also occur in the presence of water, which also presents a high dipolar moment (μ_{water} = 1.85 D). However, the fact that MA is opened in maleic acid in the presence of water should also be taken into account. The pKa$_1$ of maleic acid is very low (~1.8) and decreases the pH of the mix. When the FA polymerization is initiated in an acid catalyzed condition, the MA opening releases protons that allow the condensations to start at a lower temperature. Contrary to IPA, the slowdown effect due to solvation is counterbalanced by the formation of H$^+$ due to the MA hydrolysis, which results in earlier initiation of FA polymerization with water.

These curves have been used to estimate the reaction heat released during the reaction (Q) by the integration of DSC peaks. The reaction heat values obtained for FA/MA/water and FA/MA/IPA systems are summarized and compared to the FA reference system [40] (Table 1). In order to compare values obtained from each formulation, the data were reported as the mass of FA. FA/MA/IPA systems show decreasing reaction heat values with an increasing heating rate. The values of the reaction heat are 2.4 to 3.5 times smaller than the reference values. Moreover, the reaction heat (Q) depends on the amplitude of the first thermal event, which is rather associated with initiation reactions. Indeed, increasing the heating rate allows less time for initiation reactions to take place and leads to a decrease of the reaction heat. These decreasing values, combined with the particular thermal events of this mix (two distinct peaks), suggest possible interactions between FA and IPA.

In the case of the FA/MA/water system, the opposite tendency occurs. The reaction heat increases with the heating rate. On the other hand, the highest reaction heat value of the FA/MA/water mix (712 g·mol^{-1}) obtained for the fastest rate is almost the same as the reaction heat value of the FA/MA reference mix at the slower rate (709 g·mol^{-1}). Thus, the heating rate considerably affects the reaction mechanism, which confirms the assumption of a complex multi-step polymerization mechanism. This also indicates that the final properties of the materials will be completely different according to the temperature domain used for curing the system.

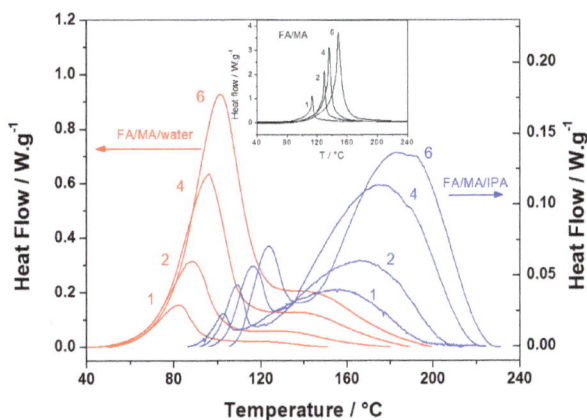

Figure 2. Non-isothermal DSC curves for the curing of FA/MA/water (red lines) and FA/MA/IPA (blue lines). Insert: FA/MA cure without solvent. The heating rates in °C·min^{-1} are indicated by the curves.

Table 1. Reaction heat (Q) reported to the mass of FA for the three different systems: FA/MA/water, FA/MA/IPA, and FA/MA.

	FA/MA/Water	FA/MA/IPA	FA/MA (Reference)
$\beta/°C \cdot min^{-1}$		$Q/J \cdot g^{-1}$ of FA	
1	490 ± 20	300 ± 30	709 ± 30
2	632 ± 30	230 ± 20	685 ± 30
4	678 ± 30	216 ± 20	620 ± 30
6	712 ± 30	168 ± 20	593 ± 30

3.1.2. E_α vs. α-Dependence

Model-free advanced isoconversional analysis was employed to highlight new insights on the complex polymerization mechanism and kinetics in the presence of solvents. Figure 3 represents the E_α dependencies with the extent of conversion (α). Analysis of the E_α-dependencies clearly indicates a complex mechanism that involves several chemical steps, each of them having their own activation energy. As a result, each increasing and decreasing part of the effective activation energy (E_α) can be associated with changes in the rate-limiting steps of the overall polymerization.

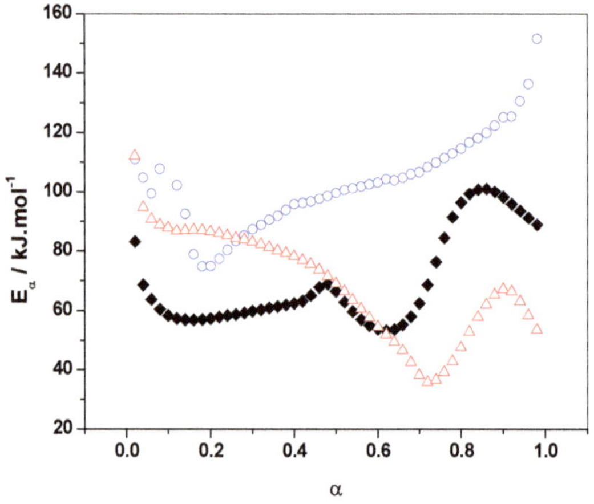

Figure 3. Dependence of the effective activation energy (E_α) on extent of conversion of FA/MA/water (open red triangles), FA/MA/IPA (open blue circles), and FA/MA (solid black lozenges).

The three systems exhibit decreasing E_α values in the initial stages of the reaction (for α values, until 0.10 for reference and water systems, and 0.20 for the IPA system) that can be attributed to an autocatalytic step [41]. This initial step corresponds to the formation of an active furfuryl carbenium center that will induce the polymerization. The longer decay of FA/MA/IPA E_α values to a higher extent of conversion (0.20) confirms the hypothesis of interactions between FA and IPA by hydrogen bonds or solvation spheres, which slow down the autocatalytic step.

The FA disruption polymerization by IPA is also confirmed by the completely different activation energy dependency obtained compared to the two other systems. Indeed, while alternating decreasing and increasing values are obtained for FA/MA and FA/MA/water, the FA/MA/IPA system reveals a continuous activation energy increase for α values from 0.20 to 1. The progressive increase observed might correspond to competitive or consecutive reactions (i.e., condensation, Diels-Alder and possible side reactions) during the polymerization [27]. On the other hand, E_α-dependencies of FA/MA/water and reference systems show several similarities. The polycondensation of FA occurs for $\alpha \approx 0.10$ to $\alpha \approx 0.40$ for the two systems. A slight E_α decrease for FA/MA/water and a slight increase for the reference system characterize this step. The presence of water decreases the viscosity, which facilitates the polymerization of FA.

For $\alpha \approx 0.48$, E_α values of both systems are very close and a slight increase in the activation energy is observed for the reference system. It has been demonstrated that this increasing tendency for the reference system is correlated with the high viscosity increase due to the formation of crosslinks [40]. This phenomenon is not visible for FA/MA/water due to the presence of water, which still induces a lower viscosity at the same extent of conversion.

For $\alpha > 0.48$, the decreasing tendency for the two curves is still the same, with a slight shift to a higher extent of conversion for the FA/MA/water system. At this stage of the reaction, the molecular mobility strongly decreases, which induces a decrease of the reaction rate. Thus, the overall polymerization becomes controlled by the diffusion of short linear chains. This is mainly due to an increase of the viscosity of the system, which reduces the chemical reactions rates and leads to a decrease of E_α [41]. This decrease is observed for both systems and is characteristic of a transition from a kinetic to diffusion regime [40]. Diffusion control generally becomes rate limiting when the characteristic relaxation time of the reaction medium markedly exceeds the characteristic time of the reaction itself [29]. This decrease to low E_α values is less marked, but is also present for the FA/MA system (E_α decreasing from 69 to 53 kJ·mol^{-1} for α from 0.48 to 0.63) than FA/MA/water (E_α decreasing from 71 to 36 kJ·mol^{-1} for α from 0.48 to 0.73). This indicates that diffusion control is more important in the presence of water. This will be explained by an analysis of the dependence of the effective activation energy (E_α) on temperature (T).

Following this decrease, an increasing E_α-dependency is observed for both FA/MA (until $\alpha \approx 0.90$) and FA/MA/water (until $\alpha \approx 0.85$) systems and corresponds to an increase in molecular mobility due to temperature increase. This mobility increase permits the chemical reaction to be reactivated and corresponds to the formation of chemical bonds in the gelled state by Diels-Alder cycloadditions. Finally, the last decrease is attributed to the diffusion of unreacted FA monomers [40].

According to Figure 3, E_α for FA/MA/water is always higher than E_α for FA/MA for $0 < \alpha < 0.60$, while the reaction in the presence of water is shifted to lower temperatures (Figure 2). Generally, a higher activation energy shifts the reaction to a higher temperature. Thus, in this case, it seems that the opposite effect is obtained. Because E and A have opposite effects regarding the shift of a reaction to a lower or higher temperature, this observation could be explained by a higher value of the pre-exponential factor for the FA/MA/water system. In order to verify this, pre-exponential factors were computed for FA/MA and FA/MA/water systems using the model-free method explained in detail in ref. [27]. The method uses the so-called false compensation effect that allows one to establish a relationship between E_α and $\ln A_\alpha$ in the form: $\ln A_\alpha = aE_\alpha + b$ and is based on the practical observation that for complex (multi-step) processes, the same experimental curve can be described by several reaction models. Once this relation has been established, it is possible to compute $\ln A_\alpha$ in a model-free way using the value of E_α obtained for each α value with an advanced isoconversional method. The computations were realized for a heating rate of 2 °C·min^{-1} and an extent of conversion $0.05 < \alpha < 0.15$. This range was selected in order to compute the pre-exponential factors at the very beginning of the reaction. The models that lead to the best fit ($r^2 > 0.9995$) were the model numbers 4, 5, 6, 10, 11, 12, and 13 of ref. [27]. The parameters found for FA/MA are $a = 0.29475$ mol·kJ^{-1} and $b = -6.47597$ ($r^2 = 0.9984$), and for FA/MA/water, they are $a = 0.32955$ mol·kJ^{-1} and $b = -6.23381$ ($r^2 = 0.9987$). These values confirm our hypothesis that much higher values for the pre-exponential factor are obtained in the presence of water. As an example, the pre-exponential factors obtained for FA/MA and FA/MA/water for $\alpha = 0.10$ are 4.48×10^4 s^{-1} and 7.23×10^9 s^{-1}, respectively.

3.1.3. Additionnal Kinetic Computation of FA/MA/IPA

To better understand the complex reactivity of FA/MA/IPA, additional computations were performed for each thermal event of this system (see Figure 2). Thus, the E_α calculated from the first thermal event (before 120/140 °C) is shown in Figure 4a, while Figure 4b shows the E_α calculated from the second thermal event (after 120/140 °C). It can be seen that the activation energy of Figure 4a has the same tendency as the curve of Figure 3 from $\alpha = 0$ to $\alpha = 0.20$, with a continuous E_α decrease. Therefore, the first thermal event of the FA/MA/IPA thermograms of Figure 2 is the result of the autocatalytic step. The second thermal event could be related to competitive reactions during the polymerization. These competitive reactions are displayed in Figure 4b, where a continuous increase of E_α for α from 0.20 to 1 is observed.

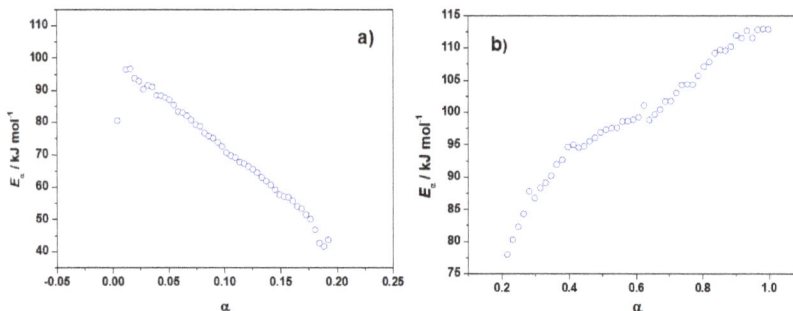

Figure 4. Partial dependence of the effective activation energy (E_α) calculated from the non-isothermal DSC curves of FA/MA/IPA mix of Figure 2. (**a**) from the first thermal event (before 120/140 °C) and (**b**) from the second thermal event (after 120/140 °C).

3.1.4. E_α vs. T-Dependence

The apparent activation energy can also be computed as a function of temperature (E_α–T dependency), by taking an average temperature associated with each value of the extent of conversion. Figure 5 represents the E_α-dependencies with temperature. Analysis of E_α–T values shows that the various rate-limiting steps are extended over a higher temperature range for the solvent systems and each of these steps takes place at different temperatures depending on the polymerization environment of FA. More precisely, Figure 5 highlights that the reaction of the FA/MA/water system is shifted to lower temperatures and the reaction of FA/MA/IPA to higher temperatures compared to the reference.

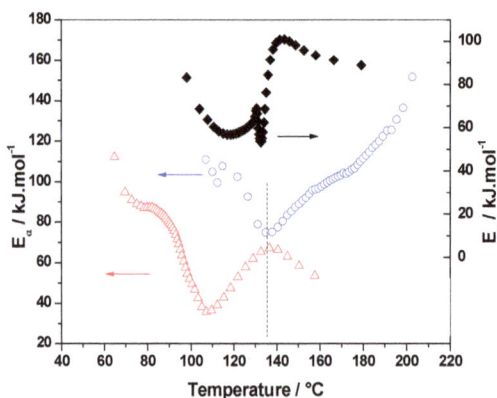

Figure 5. Dependence of the effective activation energy (E_α) on temperature (T) for non-isothermal curing of FA/MA/water (open red triangles), FA/MA (solid black lozenges), and FA/MA/IPA (open blue circles) systems.

Indeed, the first E_α decrease associated with the autocatalytic step occurs between 100–115 °C for FA/MA, 60–75 °C for FA/MA/water, and 120–140 °C for FA/MA/IPA system. The shift to lower temperatures observed for the water mix system is explained by the easier MA opening into maleic acid due to the presence of water and correlates with the higher pre-exponential factor values previously reported. For FA/MA/IPA, analysis of the E_α–T dependency confirms the higher activation energy barrier for the initial stages of the polymerization of FA in the presence of IPA instead of water. Furthermore, the polymerization reactions (i.e., condensation and Diels-Alder reactions) of this mix begin at 135 °C, i.e., when the polymerization is finished or almost finished for the two other formulations. This confirms our previous hypothesis that the solvation sphere formed between FA and IPA is very stable and hinders the reactions.

Another significant difference highlighted by the analysis of the E_α–T dependency is the secondary E_α decreasing values due to the diffusion regime. This stage is observed in a very short temperature range for FA/MA (131–133 °C), while this step takes place over a wider temperature range for FA/MA/water (90–105 °C). For the FA/MA/IPA system, this stage is absent and E_α values exhibit the same tendency as in Figure 3, where the major part of the FA/MA/IPA dependency demonstrates increasing E_α values. This E_α decrease characteristic of the diffusion control of small molecules is more pronounced for the FA/MA/water system (86 to 36 kJ·mol^{-1}) than for the FA/MA system (69 to 54 kJ·mol^{-1}), in agreement with the results of Figure 3. This could seem to be, a priori, contradictory. Nevertheless, analysis of the E_α–T dependency shows that this decrease occurs at much lower temperatures for the FA/MA/water system. This explains that diffusion control is more pronounced in the presence of water due to the inferior molecular mobility at a lower temperature.

After the second decreasing step observed for FA/MA and FA/MA/water, the E_α values re-increase significantly in a very sharp temperature interval (133–141 °C) for the reference system. This is attributed to the reactivation of chemical reactions and diffusion of long segments of the polymer chains. A similar increase is observed for the water system, but at a lower temperature and in a wider temperature interval (107–136 °C). This final re-increase was attributed to the cross-links formation in the gelled state due to Diels-Alder cycloadditions [40]. Thus, the addition of water allows for the re-activation of chemical reactions at a lower temperature, probably because of the higher mobility of the system.

3.2. Residual Cross-Linking Reactions in Rubbery State

3.2.1. Non-Isothermal DMA Investigation

Once polymerized, the mechanical properties of the three PFAs were evaluated by DMA. Elastic moduli of the three materials measured during the first heating scan or during the second scan (after a first heating to 250 °C) are shown in Figure 6. The decreasing moduli of about two decades between 30 and 130 °C for all the systems during the first scan are unambiguously attributed to the cooperative α-relaxation process of PFA chains commonly associated with the glass transition. Above 170 °C, i.e., from temperature range corresponding to the post-curing treatment, the elastic modulus increases for the three samples. This increase may correspond to residual cross-links occurring in solid PFA resins. As shown in Figure 6, the increase is more important when the PFA has been cured in the presence of protic polar solvents (e.g. water or IPA). In particular, it can be deduced that polymerization via IPA is less complete since the re-increase is about one decade and accordingly the amount of residual cross-links is more important compared to the other systems. Compared to the first scans, the second DMA scans measured after a first heating to 250 °C both show higher moduli and higher glass transition temperatures for all the samples. This indicates that the re-increase of moduli above 170 °C during the first scan can be attributed to residual crosslinks, thus increasing the crosslink density and consequently the chain mobility in PFA. It is worth noting that PFA/IPA and PFA/water shows lower glass transition temperatures and lower values of moduli compared to PFA. These features were attributed to furan ring opening reactions that are exacerbated during polymerization with protic polar solvents, thus leading to a lower cross-link density due to these more mobile entities [26].

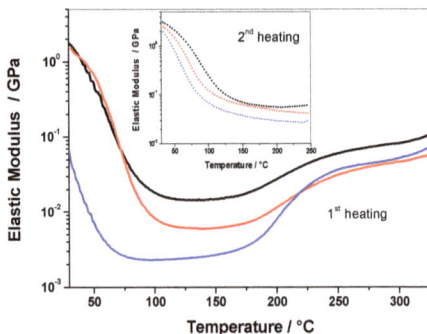

Figure 6. Elastic modulus vs. temperature obtained during the first heating (line) or the second heating (dot) at 2 °C·min^{-1} of reference PFA (black line), PFA/water (red line), and PFA/IPA (blue line).

Additional dynamic mechanical measurements were carried out at 1 and 4 °C·min^{-1} on each polymer from 25 to 350 °C (Figure 7) in order to further understand the kinetic of residual cross-links occurring in solid PFA resins. These DMA measurements performed at various heating rates were conducted to show that this increase is not due to the volatilization of either protic polar solvent added in formulations or water from FA polycondensations, which could remain trapped within polymer chains. These curves show that the moduli increase depends on the heating rate. Therefore, it can be deduced that it should be a kinetic phenomenon and not a thermodynamic phenomenon, such as a solvent evaporation. Indeed, first-order thermodynamic transitions such as vaporization should be independent of the heating rate. It is interesting to note that the modulus increase is observed from 170°C to 290 °C for the three systems in DMA, while no thermal event was recorded at a temperature higher than 220 °C on the first DSC heating curves. This clearly highlights the interest of using DMA measurements to identify residual crosslinks occurring in the rubbery state after intensive post-curing.

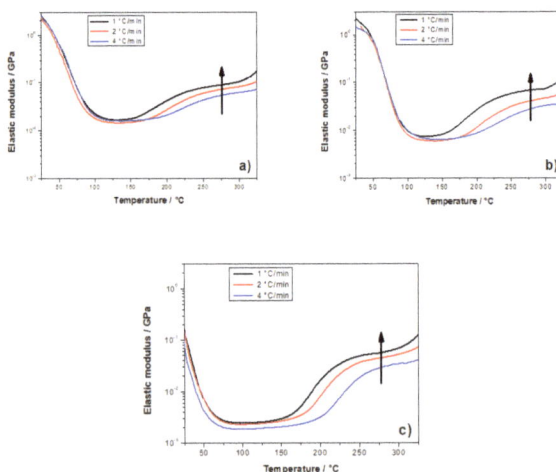

Figure 7. (a) Elastic modulus (*E'*) of three PFA samples obtained by DMA at 1, 2, and 4 °C. **(a)** reference PFA, **(b)** PFA/water, and **(c)** PFA/IPA.

Then, the modulus increase above 150 °C clearly indicates that chemical reactions still occur in the rubbery state in the different PFA samples. The modulus increase is more important when the samples have been cured in the presence of protic polar solvents (e.g. water or IPA). More precisely, the polymerization in the presence of IPA is less complete since the re-increase and accordingly the amount of residual cross-links is more important compared to the other systems.

3.2.2. E_α vs. T-Dependence

The kinetic of formation of these residual cross-links in the rubbery state was evaluated from the treatment of the DMA curves of Figure 7. For that purpose, the elastic modulus was normalized between 0 and 1 according to Equation (2), in order to get an extent of conversion, α, for the cross-links in the rubbery state. These data, obtained on the three PFA samples, were computed with an advanced isoconversional method (Equations (5) and (6)).

Figure 8a shows the E_α dependencies computed as a function of temperature. The three samples demonstrate increasing E_α values, which could correspond to a kinetic control by the diffusion of long polymeric chains. As the PFA material progressively cross-links in the solid state, the molecular mobility becomes more restricted, which leads to an increase in the activation energy (E_α). The PFA cured without solvent (i.e., PFA/MA, black lozenges) has the highest E_α values, which indicates that the long chains are more constrained in comparison to the PFA samples cured in the presence of solvents.

The E_α–dependencies of the reference system obtained for polymerization from the liquid state (DSC measurements) and for additional cross-links occurring in the rubbery state (DMA measurements) were plotted in Figure 8b. As can be seen, there is a perfect continuity between the two dependencies, with the final values of the reaction starting from the liquid state being in good agreement with the first values of the reaction starting from the rubbery state. Thus, the combination of DSC and DMA data allows one to study the polymerization kinetics and to get mechanistic information over a wide temperature range, starting from the liquid state and ending in the solid state.

The modification of FA polymerization by the protic polar solvents (see Figure 8a) gives lower E_α values for reactions continuing in the rubbery state. The occurrence of ring opening reactions reduces the cross-link density [26] and thus, molecular mobility in the rubbery state is promoted. This result is in agreement with the lower values of the elastic modulus (E') in the rubbery state for the PFA samples prepared with solvents (Figure 6). Absolute E_α values of PFA/water and PFA/IPA cannot be directly compared because these two samples were subjected to different post-curing treatments in order to obtain a sufficiently cohesive material for permitting DMA measurements. Nevertheless, the values are in the same order of magnitude.

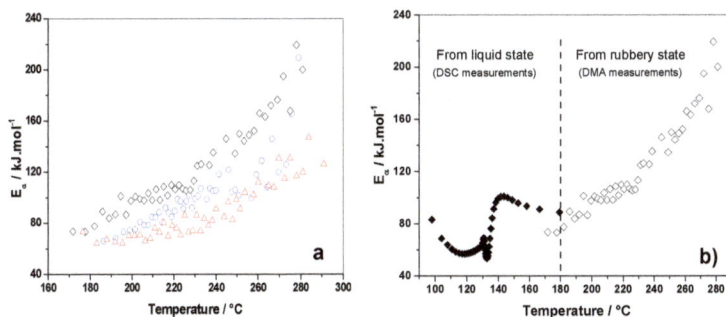

Figure 8. (a) Effective activation energy dependency (E_α) as a function of temperature obtained from DMA measurements of Figure 7 (black open lozenges: PFA/MA, red open triangles: PFA/MA/water, blue open circles: PFA/MA/IPA); **(b)** Continuous E_α from liquid state (DSC measurements) to solid state (DMA measurements) for the FA/MA system.

4. Conclusions

The presence of solvent significantly disturbs the polymerization kinetics of FA. The autocatalytic stage is modified, likely due to a stabilization of the furfuryl carbenium center by either water or IPA. In the presence of water, the autocatalytic stage is less marked as the initiator, maleic anhydride, is rapidly hydrolyzed into maleic acid, which induces the onset of polymerization reactions at lower temperatures. In this mix, we note that FA polymerization rate-limiting steps are similar to the reference system, with only few variations of E_α values and temperatures of reaction. However, the presence of IPA shifts the beginning of the polymerization to a higher temperature. Furthermore, E_α values of FA/MA/IPA clearly indicate that main reactions and other side reactions seem to occur in parallel and throughout the whole polymerization. The only similarity between the E_α dependency of FA/MA/IPA and the two other systems is the autocatalytic step, which is characterized by a separate thermal event.

The DMA study has highlighted a continuous increase of the elastic modulus attributed to further cross-links in the rubbery state until 290 °C. These additional reactions occurring in the rubbery plateau cannot be highlighted by DSC since the potential exothermic variation is too small to be detected. The E' variations obtained for different heating rates permitted us to obtain E_α-dependency in the rubbery state. It was shown that the modification of FA polymerization by protic polar solvents slightly decreases E_α values in the rubbery state due to the increase of molecular mobility. For reference PFA, the E_α-dependencies obtained from the liquid state and from the rubbery state are in perfect continuity, which prove the validity of the new approach proposed to obtain mechanistic information over a very wide temperature range for complex polymerization.

Author Contributions: N.G. and N.S.; Methodology, G.F. L.V. and N.G.; Software, N.S. internal software; Validation, G.F. N.G. and L.V.; Formal Analysis, N.G. and N.S.; Investigation, G.F. and L.V.; Resources, L.V. and N.S.; Data Curation, G.F. N.G. and N.S.; Writing-Original Draft Preparation, G.F. N.G.; Writing-Review & Editing, G.F. N.G. and N.S.; Visualization, G.F. N.G. and N.S.; Supervision, N.G. and N.S..; Project Administration, N.G. and N.S.; Funding Acquisition, N.S.", please turn to the CRediT taxonomy for the term explanation. Authorship must be limited to those who have contributed substantially to the work reported.

Funding: European Project POLYWOOD, <Wood-polymer composites for use in marine environments> No. 219294/O30, The Research Council of Norway RCN, Région PACA (France) project ECOMOBIL

Acknowledgments: The authors wish to thank Mettler-Toledo Inc. for fruitful collaboration and scientific exchanges. The company Kebony (Norway) are also gratefully acknowledged for the European Project POLYWOOD, <Wood-polymer composites for use in marine environments>. The authors also gratefully acknowledge Région PACA (France) for financial support of the project ECOMOBIL.

Conflicts of Interest: The authors declare no conflict of interest.

References

1. Corma, A.; Iborra, S.; Velty, A. Chemical routes for the transformation of biomass into chemicals. *Chem. Rev.* **2007**, *107*, 2411–2502. [CrossRef] [PubMed]
2. Isikgor, F.H.; Becer, C.R. Lignocellulosic biomass: A sustainable platform for the production of bio-based chemicals and polymers. *Polym. Chem.* **2015**, *6*, 4497–4559. [CrossRef]
3. Kamm, B.; Kamm, M.; Schmidt, M.; Hirth, T.; Schulze, M. Lignocellulose-based chemical products and product family trees. In *Biorefineries–Industrial Processes and Products*; Kamm, B., Gruber, P.R., Kamm, M., Eds.; WILEY-VCH Verlag GmbH & Co.: Weinheim, Germany, 2006.
4. Chheda, J.N.; Román-Leshkov, Y.; Dumesic, J.A. Production of 5-hydroxymethylfurfural and furfural by dehydration of biomass-derived mono- and poly-saccharides. *Green Chem.* **2007**, *9*, 342–350. [CrossRef]
5. Rosatella, A.A.; Simeonov, S.P.; Frade, R.F.M.; Afonso, C.A.M. 5-hydroxymethylfurfural (HMF) as a building block platform: Biological properties, synthesis and synthetic applications. *Green Chem.* **2011**, *13*, 754–793. [CrossRef]
6. Mariscal, R.; Maireles-Torres, P.; Ojeda, M.; Sadaba, I.; Lopez Granados, M. Furfural: A renewable and versatile platform molecule for the synthesis of chemicals and fuels. *Energy Environ. Sci.* **2016**, *9*, 1144–1189. [CrossRef]

7. Gandini, A. The irruption of polymers from renewable resources on the scene of macromolecular science and technology. *Green Chem.* **2011**, *13*, 1061–1083. [CrossRef]
8. Bozell, J.J.; Petersen, G.R. Technology development for the production of biobased products from biorefinery carbohydrates—the us department of energy's "top 10" revisited. *Green Chem.* **2010**, *12*, 539–554. [CrossRef]
9. Gandini, A.; Lacerda, T.M.; Carvalho, A.J.F.; Trovatti, E. Progress of polymers from renewable resources: Furans, vegetable oils, and polysaccharides. *Chem. Rev.* **2016**, *116*, 1637–1669. [CrossRef] [PubMed]
10. Choura, M.; Belgacem, N.M.; Gandini, A. Acid-catalyzed polycondensation of furfuryl alcohol: Mechanisms of chromophore formation and cross-linking. *Macromolecules* **1996**, *29*, 3839–3850. [CrossRef]
11. Oliva-Teles, M.T.; Delerue-Matos, C.; Alvim-Ferraz, M.C.M. Determination of free furfuryl alcohol in foundry resins by chromatographic techniques. *Anal. Chim. Acta* **2005**, *537*, 47–51. [CrossRef]
12. Lande, S.; Westin, M.; Schneider, M.H. Eco-efficient wood protection: Furfurylated wood as alternative to traditional wood preservation. *Manag. Environ. Qual. Int. J.* **2004**, *15*, 529–540. [CrossRef]
13. Ju, M.; Zeng, C.; Wang, C.; Zhang, L. Preparation of ultrafine carbon spheres by controlled polymerization of furfuryl alcohol in microdroplets. *Ind. Eng. Chem. Res.* **2014**, *53*, 3084–3090. [CrossRef]
14. Spange, S.; Grund, S. Nanostructured organic–inorganic composite materials by twin polymerization of hybrid monomers. *Adv. Mater.* **2009**, *21*, 2111–2116. [CrossRef]
15. Bosq, N.; Guigo, N.; Falco, G.; Persello, J.; Sbirrazzuoli, N. Impact of silica nanoclusters on furfuryl alcohol polymerization and molecular mobility. *J. Phys. Chem. C* **2017**, *121*, 7485–7494. [CrossRef]
16. Deka, H.; Misra, M.; Mohanty, A. Renewable resource based "all green composites" from kenaf biofiber and poly(furfuryl alcohol) bioresin. *Ind. Crops Prod.* **2013**, *41*, 94–101. [CrossRef]
17. Guigo, N.; Mija, A.; Vincent, L.; Sbirrazzuoli, N. Eco-friendly composite resins based on renewable biomass resources: Polyfurfuryl alcohol/lignin thermosets. *Eur. Pol. J.* **2010**, *46*, 1016–1023. [CrossRef]
18. Roudsari, G.M.; Misra, M.; Mohanty, A.K. A study of mechanical properties of biobased epoxy network: Effect of addition of epoxidized soybean oil and poly(furfuryl alcohol). *J. Appl. Polym. Sci.* **2017**, *134*. [CrossRef]
19. Pin, J.M.; Guigo, N.; Vincent, L.; Sbirrazzuoli, N.; Mija, A. Copolymerization as a strategy to combine epoxidized linseed oil and furfuryl alcohol: The design of a fully bio-based thermoset. *ChemSusChem* **2015**, *8*, 4149–4161. [CrossRef] [PubMed]
20. Conley, T.; Metil, I. An investigation of the structure of furfuryl alcohol polycondensatee with infrared spectroscopy. *J. Appl. Polym. Sci.* **1963**, *7*, 37–52. [CrossRef]
21. Dunlop, A.; Peters, F. *The Furans*; Reinhold Publishing Corporation: New York, NY, USA, 1953.
22. Montero, A.L.; Montero, L.A.; Martínez, R.; Spange, S. Ab initio modelling of crosslinking in polymers. A case of chains with furan rings. *J. Mol. Struct. THEOCHEM* **2006**, *770*, 99–106. [CrossRef]
23. Wang, S.; Vorotnikov, V.; Vlachos, D.G. A DFT study of furan hydrogenation and ring opening on Pd (111). *Green Chem.* **2014**, *16*, 736–747. [CrossRef]
24. Kim, T.; Jeong, J.; Rahman, M.; Zhu, E.; Mahajan, D. Characterizations of furfuryl alcohol oligomer/polymerization catalyzed by homogeneous and heterogeneous acid catalysts. *Korean J. Chem. Eng.* **2014**, *31*, 2124–2129. [CrossRef]
25. Kim, T.; Assary, R.S.; Marshall, C.L.; Gosztola, D.J.; Curtiss, L.A.; Stair, P.C. Acid-catalyzed furfuryl alcohol polymerization: Characterizations of molecular structure and thermodynamic properties. *ChemCatChem* **2011**, *3*, 1451–1458. [CrossRef]
26. Falco, G.; Guigo, N.; Vincent, L.; Sbirrazzuoli, N. Opening furan for tailoring properties of biobased poly(furfuryl alcohol) thermoset. *ChemSusChem* **2018**. [CrossRef] [PubMed]
27. Sbirrazzuoli, N. Determination of pre-exponential factors and of the mathematical functions $f(\alpha)$ or $g(\alpha)$ that describe the reaction mechanism in a model-free way. *Thermochim. Acta* **2013**, *564*, 59–69. [CrossRef]
28. Sbirrazzuoli, N. Is the friedman method applicable to transformations with temperature dependent reaction heat? *Macromol. Chem. Phys.* **2007**, *208*, 1592–1597. [CrossRef]
29. Vyazovkin, S. *Isoconversional Kinetics of Thermally Stimulated Processes*; Springer: Berlin, Germany, 2015.
30. Vyazovkin, S.; Burnham, A.K.; Criado, J.M.; Pérez-Maqueda, L.A.; Popescu, C.; Sbirrazzuoli, N. Ictac kinetics committee recommendations for performing kinetic computations on thermal analysis data. *Thermochim. Acta* **2011**, *520*, 1–19. [CrossRef]
31. Vyazovkin, S.; Sbirrazzuoli, N. Isoconversional kinetic analysis of thermally stimulated processes in polymers. *Macromol. Rapid Commun.* **2006**, *27*, 1515–1532. [CrossRef]

32. Papageorgiou, G.Z.; Achilias, D.S.; Karayannidis, G.P. Estimation of thermal transitions in poly(ethylene naphthalate): Experiments and modeling using isoconversional methods. *Polymer* **2010**, *51*, 2565–2575. [CrossRef]

33. Sbirrazzuoli, N.; Vincent, L.; Vyazovkin, S. Comparison of several computational procedures for evaluating the kinetics of thermally stimulated condensed phase reactions. *Chemom. Intell. Lab. Syst.* **2000**, *54*, 53–60. [CrossRef]

34. Atkins, P.; Paula, J.D. *Physical Chemistry*, 9th ed.; W.H Freeman: New York, NY, USA, 2010.

35. Vyazovkin, S. Evaluation of activation energy of thermally stimulated solid-state reactions under arbitrary variation of temperature. *J. Comput. Chem.* **1997**, *18*, 393–402. [CrossRef]

36. Vyazovkin, S. Modification of the integral isoconversional method to account for variation in the activation energy. *J. Comput. Chem.* **2001**, *22*, 178–183. [CrossRef]

37. Sbirrazzuoli, N.; Brunel, D.; Elegant, L. Different kinetic equations analysis. *J. Therm. Anal. Calorim.* **1992**, *38*, 1509–1524. [CrossRef]

38. Sbirrazzuoli, N.; Girault, Y.; Elegant, L. Simulations for evaluation of kinetic methods in differential scanning calorimetry. Part 3—Peak maximum evolution methods and isoconversional methods. *Thermochim. Acta* **1997**, *293*, 25–37. [CrossRef]

39. Kim, T.; Assary, R.S.; Kim, H.; Marshall, C.L.; Gosztola, D.J.; Curtiss, L.A.; Stair, P.C. Effects of solvent on the furfuryl alcohol polymerization reaction: Uv raman spectroscopy study. *Catal. Today* **2013**, *205*, 60–66. [CrossRef]

40. Guigo, N.; Mija, A.; Vincent, L.; Sbirrazzuoli, N. Chemorheological analysis and model-free kinetics of acid catalysed furfuryl alcohol polymerization. *Phys. Chem. Chem. Phys.* **2007**, *9*, 5359–5366. [CrossRef] [PubMed]

41. Vyazovkin, S.; Sbirrazzuoli, N. Mechanism and kinetics of epoxy−amine cure studied by differential scanning calorimetry. *Macromolecules* **1996**, *29*, 1867–1873. [CrossRef]

© 2018 by the authors. Licensee MDPI, Basel, Switzerland. This article is an open access article distributed under the terms and conditions of the Creative Commons Attribution (CC BY) license (http://creativecommons.org/licenses/by/4.0/).

![polymers logo] *polymers*

MDPI

Article

Barrier Properties of Poly(Propylene Cyclohexanedicarboxylate) Random Eco-Friendly Copolyesters

Valentina Siracusa [1,*], Laura Genovese [2], Carlo Ingrao [1], Andrea Munari [2] and Nadia Lotti [2]

[1] Department of Chemical Science, University of Catania, Viale A. Doria 6, 95125 Catania (CT), Italy; ing.carloingrao@gmail.com

[2] Department of Civil, Chemical, Environmental and Materials Engineering, University of Bologna, Via Terracini 28, 40131 Bologna (BO), Italy; laura.genovese@unibo.it (L.G.); andrea.munari@unibo.it (A.M.); nadia.lotti@unibo.it (N.L.)

* Correspondence: vsiracus@dmfci.unict.it; Tel.: +39-338-727-5526

Received: 12 April 2018; Accepted: 3 May 2018; Published: 5 May 2018

Abstract: Random copolymers of poly(propylene 1,4-cyclohexanedicarboxylate) containing different amounts of neopentyl glycol sub-unit were investigated from the gas barrier point of view at the standard temperature of analysis (23 °C) with respect to the three main gases used in food packaging field: N_2, O_2, and CO_2. The effect of temperature was also evaluated, considering two temperatures close to the T_g sample (8 and 15 °C) and two above T_g (30 and 38 °C). Barrier performances were checked after food contact simulants and in different relative humidity (RH) environments obtained with two saturated saline solutions (Standard Atmosphere, 23 °C, 85% of RH, with saturated KCl solution; Tropical Climate, 38 °C, 90% RH, with saturated KNO_3 solution). The results obtained were compared to those of untreated film, which was used as a reference. The relationships between the gas transmission rate, the diffusion coefficients, the solubility, and the copolymer composition were established. The results highlighted a correlation between barrier performance and copolymer composition and the applied treatment. In particular, copolymerization did not cause a worsening of the barrier properties, whereas the different treatments differently influenced the gas barrier behavior, depending on the chemical polymer structure. After treatment, Fourier transform infrared analysis confirmed the chemical stability of these copolymers. Films were transparent, with a light yellowish color, slightly more intense after all treatments.

Keywords: biodegradable polymers; Poly(propylene 1,4-cyclohexanedicarboxylate); random copolymers; gas barrier properties; food packaging; eco-friendly copolyesters; food simulants; relative humidity

---

1. Introduction

Plastic packaging, particularly that used for food packaging, accounts for a large proportion of the total polymer production due to a combination of several favorable factors such as being lightweight, flexible, strong, stable, impermeable, and easy to sterilize. Due to their high versatility and safety, plastics are used for fresh meat, beverages, oils and sauces, fruit and vegetables, yoghurt, fish, essentially for all kinds of food. Most importantly, food packaging must guarantee food conservation and preservation for long periods, simultaneously reducing time wastage and the use of preservatives. Prolongation of the shelf-life results in considerable savings in terms of money, material consumption, and food waste.

Shelf-life has been defined many times, but no definition has been regulated. The European Commission Regulation (EC) No. 2073/2005 provided the following definition: shelf-life is the time frame corresponding to the period preceding the "use by" or the "minimum durability date", with "use by" and "minimum durability date" taken from Art. 9 and 10 of Directive 200/13/EC). Food shelf-life is governed by several factors: (1) the intrinsic food characteristics, like pH, water activity, fat content, nutrient, respiration rate, and biological structure; (2) environmental influences such as temperature, relative humidity, and gas surrounding (permeability factor); and (3) the type of packaging. Control over these three parameters contributes to prolonging, or at least maintaining food quality during the "use by" time.

Moreover, food deterioration can occur via chemical, biochemical, physical, and microbiological attack. Several materials, such as paper, glass, or metal, have been used for food preservation. Only in the last century have plastics appeared, being the current most used material for food packaging. Plastics must fulfil strict requirements to be used for this application. In particular, they have to be "passive", "active", and/or "intelligent". "Passive" means that packaging must not interact with food; plastic packaging must maintain the same physical, chemical, mechanical, and permeability behavior during the food's entire shelf life, and no migration of monomers or additives can occur from packaging to food. Food organoleptic properties, including smell, taste, color, and texture, must also remain unaltered. "Active" implies that packaging should perform an active role when used for food preservation (modify its properties), remaining "passive" in order to ensure food shelf life extension, to improve safety and sensory properties, and to maintain the food quality [1]. To this purpose, the incorporation of additives could represent a solution to create an "active packaging". So far, many additives have been introduced in the market: oxygen scavengers, CO_2, O_2, C_2H_4, and moisture absorbers, antimicrobial agents, etc. Lastly, "intelligent" means that plastic has to be able to communicate with food inside the package and the surroundings outside the package. Although "passive" and "active" represent intrinsic material properties, "intelligence" is an added characteristic, readable by the consumer [2].

The huge amount of plastic packaging produced annually quickly becomes, significantly contributing to the pollution of aquatic and terrestrial environments. Therefore, growing environmental awareness is creating a need for package film to be eco-friendly [3]. In addition to recycling used packages, the development of materials with biodegradability and/or compostability attributes would reduce municipal solid waste [4]. As a result, biodegradability is not only a functional requirement, but is also crucial from an environmental point of view [3], especially considering that the bioplastics industry is growing at a rate of more than 20% per year [5].

With the aim of broadening the spectrum of bioplastics used in this field, several research groups have been focusing their efforts on developing new bioplastics, compounds, and master batches. Until now, only a limited number of these materials have been made available on the market for food packaging applications, with most common being aliphatic polyesters, above all Poly(lactic acid) (PLA), starch, and cellulose [4,6,7]. Aliphatic polyesters can be considered very competitive, as most are biodegradable and bio-based. Some of the monomers used for their production, such as succinic acid, adipic acid, 1,3-propanediol, 1,4-butanediol, lactic acid, and γ-butyrolactone, can be obtained from both petroleum resources and renewable resources [8]. Within this class, polyesters with cycloaliphatic units like poly(alkylene 1,4-cyclohexanedicarboxylate)s offer several advantages: excellent tensile strength, stiffness and impact properties, high thermal stability, and easily attackable by microorganisms that allow these materials to be biodegradable. Furthermore, most of poly(alkylene 1,4-cyclohexanedicarboxylate)s are biodegradable and bio-based materials. The monomer 1,4-cyclohexane dicarboxylic acid can be obtained from bio-based terephthalic acid starting from limonene and other terpenes [8]. Our research group synthesized a poly(butylene 1,4-cyclohexane dicarboxylate) homopolymer as well as some of its random and block copolymers with the aim of evaluating the potential of these new materials to be used for eco-friendly food packaging. The barrier performances appeared to be promising, being superior compared with some traditional fossil-based plastics, strictly correlated to material chemical structure [9–11].

Polymers **2018**, *10*, 502

Poly(propylene 1,4-cyclohexanedicarboxylate) (PPCE) and its random copolymers containing different amounts of neopentyl glycol sub-units (P(PCExNCEy) have been synthesized and characterized from the molecular, thermal, structural, and mechanical points of view [12]. These new polyesters represent a new class of bio-based and biodegradable ecofriendly materials with the potential for use in food packaging applications.

Copolymerization is a useful tool for obtaining materials with well-tailored properties for the final desired application [13,14]. Using this strategy, synthetizing a new class of materials with improved characteristics is possible without compromising the pre-existing satisfying characteristics.

As the gas barrier behavior is fundamental in food packaging applications in order to select the best material for prolonging food shelf life, while maintaining the safety and quality of the packed food throughout the storage period, our research work focused on the study of such properties in different situations for PPCE and P(PCExNCEy) random copolymers. The gas barrier behavior was first analyzed in standard conditions with the three main gases, nitrogen, oxygen, and carbon dioxide (N_2, O_2 and CO_2), used with the modified atmosphere packaging technique (MAP). The permeation mechanism was studied in the temperature range of 8 to 38 °C in order to understand the influence of temperature and to calculate the activation energy of the permeation process. Then, a different moisture environment was considered, simulating storage under standard atmosphere (85% relative humidity) and in tropical atmosphere (90% relative humidity). Lastly, food contact was mimicked, with the use of four food simulant liquids, following the guideline provided by the European Regulation for packaging in contact with food. The correlation between chemical polymer structure and barrier properties was determined to establish structure-property relationships, which are of fundamental importance for evaluating the suitability of a certain material for a specific application.

2. Materials and Methods

2.1. Materials

1,4-dimethylcyclohexanedicarboxylate (DMCE), containing 99% trans isomer (TCI Europe, Zwijndrecht, Belgium), 1,3-propanediol (1,3-PD), neopentyl glycol (NPG), and titanium tetrabutoxide (Ti(OBu)$_4$) (Sigma Aldrich, Milan, Italy) were used as supplied, without any preliminary purification. Only the catalyst Ti(OBu)$_4$ was distilled before use.

2.2. Polymer Synthesis, Film Preparation, and Thickness Determination

Poly(propylene cyclohexanedicarboxylate) (PPCE) and poly(propylene/neopenthyl glycol cyclohexanedicarboxylate) random copolymers (P(PCExNCEy)) were synthesized according to the procedure reported by Genovese et al. [12]. The polymerization was performed starting from DMCE and 1,3-PD for the homopolymer, and different ratios of PD/NPG for the copolymers, with 40% mol excess glycol with respect to dimethylester. A total of 150 ppm of Ti(OBu)$_4$ per g of polymer were used. The polymers were prepared according to the two-stage polymerization procedure. In the first step, the temperature was raised to 180 °C and kept constant for about 120 min, until more than 90% of the methanol was distilled off. In the second step, the pressure was reduced to about 0.1 mbar to facilitate the removal of the residual methanol and the excess glycol. The temperature was increased to 240 °C and the polymerization was performed for about 180 min. Temperature and torque were continuously recorded during the polymerization. The syntheses were performed in a 250 mL glass reactor under continuous stirring in a thermostated silicon oil bath.

The obtained copolymers were indicated as P(PCExNCEy) with x and y of the mol % of propylene 1,4-cyclohexanedicarboxylate (PCE) and neopenthyl glycol 1,4-cyclohexanedicarboxylate (NCE) co-monomeric units, respectively. For simplicity, the chemical formula of the synthesized copolyesters is shown in Figure 1.

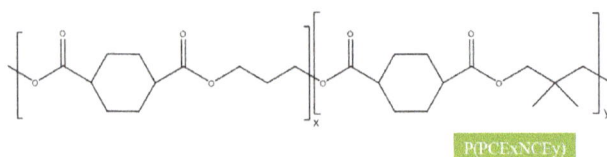

P(PCExNCEy)

Figure 1. Chemical formula of neopentyl glycol sub-units P(PCExNCEy) random copolyesters.

Films of P(PCExNCEy) were obtained by hot pressing in a hydraulic press (Carver Inc., Wabash, IN, USA). The powder was placed between two sheets of Teflon for 2 min at a temperature T equal to $T_m + 40\,°C$, with T_m being the fusion temperature determined by calorimetric experiments. The films were cooled directly in the press until reaching room temperature by running water.

Before completing the analyses, the films were maintained at room temperature for at least three weeks to reach crystallinity equilibrium.

The film thickness was determined using the Sample Thickness Tester DM-G (Brugger Feinmechanik GmbH, Munich, Germany), consisting of a digital indicator (Digital Dial Indicator) connected to a PC. The reading was made twice per second (the tool automatically performs at least three readings), measuring a minimum, a maximum, and an average value. The thickness value is expressed in µm and the measured values range from 240 to 310 µm, with a resolution of 0.001 µm. The reported results represent the mean value thickness of the three experimental tests run at 10 different points on the polymer film surface at room temperature.

2.3. Gas Transport Measurements

The determination of the gas barrier behavior was performed using a manometric method, using a Permeance Testing Device, type GDP-C (Brugger Feinmechanik GmbH, Munich, Germany), according to ASTM 1434-82 (Standard test Method for Determining Gas Permeability Characteristics of Plastic Film and Sheeting), DIN 53 536 in compliance with ISO/DIS 15 105-1, and according to the Gas Permeability Testing Manual [15].

After a preliminary high vacuum desorption of the system, the upper chamber was filled with the gas test at ambient pressure. A pressure transducer, set in the chamber below the film, recorded the increasing gas pressure as a function of time. The gas transmission rate (GTR, expressed in $cm^3 \cdot m^{-2} \cdot d^{-1} \cdot bar^{-1}$) was determined by considering the increase in pressure in relation to the time and the volume of the device. The pressure was given by the instrument in bar units. To obtain the data value in kPa, the primary SI units, we used the following correction factor: 1 bar = 10 kPa, according to NIST special publication 811 [16]. Films were analyzed at 23 °C (the standard temperature of analysis), at 8 and 15 °C (around T_g), and 30 and 38 °C (above T_g), with a food grade gas stream of 100 $cm^3 \cdot min^{-1}$ and a 0% gas RH. Gas transmission measurements were performed at least in triplicate and the mean value is presented. Method A was used for the analysis, as previously reported in the literature [15,17] with the evacuation of the top and bottom chambers. The sample temperature was set by an external thermostat HAAKE-Circulator DC10-K15 type (Thermo Fisher Scientific, Waltham, MA, USA).

The experiments were performed in triplicate and the results are presented as the average ± standard deviation.

The transport phenomena background is well described in the literature, with a full description of the mathematical equation and interpretation [18,19].

2.4. Relative Humidity Solution

According to the procedure reported in the Gas Permeability Testing Manual [15], the analyses were performed at different relative humidity (RH) obtained with several saturated saline solutions. In particular, we performed the analyses at Standard Atmosphere, which is 23 °C, 85% RH, with a saturated KCl solution, and at Tropical Climate, which is 38 °C, 90% RH, with a saturated KNO_3 solution.

The values of the relative humidity for the saline solutions were obtained from DIN 53 122 part 2. A glass-fiber round filter humidified with the desired saturated saline solution was inserted in the humid part of the top permeation cell. Method C was used, with gas flow blocked from the test specimen during evacuation. Using this method, the test gas was humidified inside the permeation cell. This method evacuates only the area of the bottom part of the sample. On the top part of the test specimen, filled with the humidified gas, normal ambient pressure was applied.

2.5. Simulant Liquids

The food contact simulation was performed in accordance with *EU Regulations No. 10/2011 on plastic materials and articles intended to come into contact with food* [20]. Four solutions were used as food simulants: (1) Simulant A, Ethanol 10% (*v*/*v*), 10 days, 40 °C; (2) Simulant B, Acetic acid 3% (*v*/*v*), 10 days, 40 °C; (3) Simulant C, Ethanol 20% (*v*/*v*), 10 days, 40 °C; and (4) Simulant D1, Ethanol 50% (*v*/*v*), 10 days, 40 °C.

Measurements were collected from a completely immersed 12 cm × 12 cm film specimen. A total of 200 mL of simulant were placed into 400 mL glass flasks containing the film sample and the flasks were then covered with caps. Samples were placed in a stove (Universalschrank UF110, Memmert GmbH + Co. KG, Schwabach, Germany). After the assay time elapsed, the specimens were removed from the flasks, washed with distilled water twice, and dried with blotting paper. Before analysis, the films were kept at room temperature, in dry ambient conditions for at least two weeks. The samples were tested in triplicate.

2.6. FTIR Spectroscopic Analysis

The Fourier transform infrared (FTIR)/ATR spectra were recorded on sample films by a Perkin-Elmer-1725-X Spectrophotometer (Labexchange Group, Burlandingen, Germany). Spectra were recorded from 4000 to 600 cm^{-1} with a resolution of 4.0 cm^{-1}. The results are an average of 10 experimental tests, run on 10 different sample points to test the results' reproducibility. Sixty-four scans were recorded on each sample. The experiments were performed at room temperature, directly on the samples, without any preliminary treatments.

2.7. Color Evaluation

The color of the film samples was measured using a HunterLab ColorFlex EZ 45/0° color spectro-photometer (Reston, VA, USA) with D65 illuminant and 10° observer according to ASTM E308. Measurements were recorded using CIE Lab scale. The instrument was calibrated with a black and white tile before the measurements. Results were expressed as L^* (lightness), a^* (red/green), and b^*(yellow/blue) parameters. The total color difference (ΔE) was calculated using the following equation: $\Delta E = [(\Delta L)^2 + (\Delta a)^2 + (\Delta b)^2]^{0.5}$, where ΔL, Δa, and Δb are the differentials between a sample color parameter (L^*, a^*, and b^*) and the color parameter of a standard white plate used as the film background ($L' = 66.39$, $a' = -0.74$, and $b' = 1.25$). Chromaticity $C^* = [(a^*)^2 + (b^*)^2]^{0.5}$ and hue angle $h_{ab} = [\arctan (b^*/a^*)/2\pi]$ 360, were calculated, as previously reported in the literature [21–23]. Measurements were recorded in triplicate at random positions over the film surface. Average values are reported.

2.8. Molecular Weight Determination

Molecular weights were evaluated by gel-permeation chromatography (GPC) at 30 °C using a 1100 high performance liquid chromatography (HPLC) system (Agilent Technologies, Santa Clara, CA, U.S.) equipped with PLgel μmeter MiniMIX-C column (Agilent Technologies). A refractive index was used as the detector. Chloroform was used as the eluent with a 0.3 mL/min flow and sample concentrations of about 2 mg/mL. A molecular weight calibration curve was obtained with polystyrene standards in the range of 2000 to 100,000 g/mol.

2.9. Statistical Analysis

One-way analysis of variance (ANOVA) and testing of the mean comparisons according to Fisher's least significant differences (LSDs) was applied to the obtained results, with a level of significance of 0.05. The statistical package STS Statistical for Windows, version 6.0 (Statsoft Inc., Tulsa, OK, USA) was used. Values are given as mean $\pm$ SD of 3 replicates.

3. Results and Discussion

3.1. Molecular Characterization

All samples were synthesized and characterized as previously reported [12]. Some of the characterization data are provided in Table 1. These data are useful for the future interpretation of gas barrier behavior.

Table 1. Molecular, thermal, diffractometric, and mechanical characterization data of poly(propylene 1,4-cyclohexanedicarboxylate) (PPCE) and neopentyl glycol sub-units (P(PCExNCEy)) random copolyesters [12].

Polymer	M_n [a]	D_{index} [b]	NCE [c] (mol %)	Thickness (μm)	T_m [d] (°C)	T_g [e] (°C)	X_c [d] (%)
PPCE	36,398	2.2	0	246 $\pm$ 22	148	9	29 $\pm$ 4
P(PCE95NCE5)	29,549	2.9	5	292 $\pm$ 31	142	11	26 $\pm$ 3
P(PCE90NCE10)	31,124	2.2	10	268 $\pm$ 18	135	12	25 $\pm$ 2
P(PCE85NCE15)	27,522	2.6	15	238 $\pm$ 33	125	13	25 $\pm$ 2
P(PCE80NCE20)	25,386	2.4	20	308 $\pm$ 10	119	13	24 $\pm$ 2

Note: [a] number average molecular weight calculated by GPC analysis; [b] polydispersity index (D_{index}) calculated by GPC analysis; [c] experimental copolymer composition calculated by ^{1}H NMR; [d] from differential scanning calorimetry, first scan; and [e] from differential scanning calorimetry, second scan.

The synthetized polyesters appeared as opaque and slightly yellow powders, and were characterized by high and comparable molecular weight and by a real copolymer molar composition very close to the feed copolymer. The chemical structure and the real copolymer composition were determined by ^{1}H NMR analysis. By NMR, it was possible to calculate the ratio between the *trans* and *cis* forms of the 1,4-cyclohexylene ring present in the DMCE molecule. In particular, less than 5% of the *cis* form was evicted after the polymerization process. All the copolymers showed a high thermal stability, comparable to that of the PPCE homopolymer, due to the presence of a cycloaliphatic ring, which is more thermally stable than a benzene ring [12]. The glass transition temperature was not always evident. A slight increase in T_g as the amount of NCE co-units increased was observed, in agreement with previously obtained results [24–26]. However, all the studied polymers were in the rubbery state at room temperature. Both calorimetric and diffractometric analyses showed a modest reduction in the degree of polymer crystallinity by copolymerization, suggesting the partial inclusion of NCE co-units into the PPCE crystal lattice (partial co-crystallization) [12].

3.2. Barrier Properties

Carbon dioxide, oxygen, and nitrogen are the main gases used in the food packaging field, especially in Modified Atmosphere Packaging (MAP) technology. These gases may transfer through the packaging wall, continuously influencing the food shelf life as well as food safety and quality. Therefore, gas permeation studies are fundamental to understand and find the best packaging solution, avoiding food damage and losses. As reported in the literature, several parameters can be considered that affect the final barrier performance of a polymeric film, such as temperature, thickness, polymer chemical structure, moisture environment, kind of food in contact with the package, etc. [2].

The samples were analyzed at 23 °C (standard condition) in the range of 8 to 38 °C, in order to study the temperature-permeability dependence, in two different moisture ambient environments, simulating the standard atmosphere (23 °C, 85% RH obtained with KCl saturated saline solution) and a tropical atmosphere (at 38 °C, 90% of RH obtained with KNO_3 saturated saline solution). The barrier performances were also evaluated after food simulant contact. The results were compared to untreated samples to study the effect of these treatments.

3.2.1. Barrier Properties under the Standard Condition

Permeability was expressed as GTR ($cm^3 \cdot cm/m^2 \cdot d \cdot atm$), normalized for sample thickness. To convert this unit to others reported in the literature, the converting factors reported by Robertson were used [2]. The theoretical models, the permeation process mechanism, and the permeability coefficients were previously described [18,27].

The gas transmission rate data, recorded for all the samples under study, are reported in Figure 2. Measures were recorded at 23 °C with the three main gases, N_2, O_2, and CO_2, used for food packaging, especially for the modified atmosphere packaging technique (MAP). As previously reported [28], O_2 is responsible for the food respiration rate. By decreasing O_2, it is possible to reduce the enzymatic degradation, extending the food shelf-life. However, if the O_2 amount becomes too low, off-flower and off-odors could be produced, leading to food tissue deterioration. CO_2, as explained by Farber [29], confers antimicrobial behavior to the food packed, whereas N_2 is used as an inert gas to complete the inside package atmosphere and to prevent film collapse. The best mix of these gases is fundamental to preserving food quality and safety during the entire storage time. In Table 2, the *S*, *D*, and t_L data recorded with the CO_2 gas test are reported. It was not possible to detect the same parameters with the N_2 and O_2 gas tests because of the inability to fit the slope of the linear portion of the GTR curves [30]. Perm-selectivity ratios are also reported in Table 2.

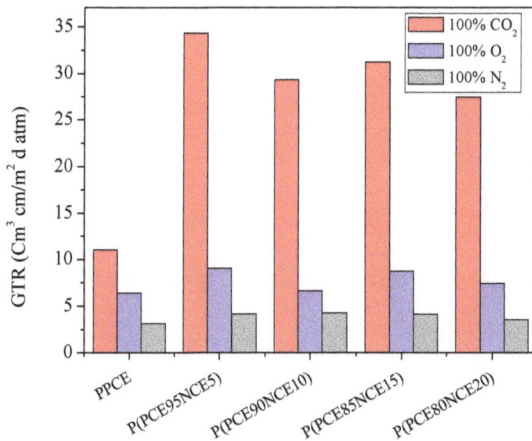

Figure 2. Gas transmission rate at 23 °C for the nitrogen (N_2), oxygen (O_2), and carbon dioxide (CO_2) gas tests.

Table 2. GTR, S, D, and t_L data for the carbon dioxide (CO_2) gas test and perm-selectivity ratio of the films.

Sample	GTR ($cm^3 \cdot cm/m^2 \cdot d \cdot atm$)	S ($cm^3/cm^2 \cdot atm$)	D (cm^3/s)	t_L (s)	CO_2/O_2	CO_2/N_2	O_2/N_2
PPCE	11.04	$3.02 \pm 4.04 \cdot E^{-02}$	$4.18 \cdot E^{-09} \pm 0.16 \cdot E^{-11}$	8513 ± 86	1.73	3.53	2.04
P(PCE95NCE5)	34.35	$8.44 \cdot E^{-01} \pm 4.04 \cdot E^{-03}$	$4.64 \cdot E^{-08} \pm 1.15 \cdot E^{-10}$	4623 ± 15	3.80	8.26	2.17
P(PCE90NCE10)	29.36	$1.69 \pm 1.0 \cdot E^{-02}$	$1.98 \cdot E^{-08} \pm 1.15 \cdot E^{-10}$	3464 ± 18	4.41	6.88	1.56
P(PCE85NCE15)	31.25	$7.48 \cdot E^{-01} \pm 1.60 \cdot E^{-02}$	$4.77 \cdot E^{-08} \pm 1.05 \cdot E^{-09}$	3312 ± 73	3.58	7.59	2.12
P(PCE80NCE20)	27.47	$1.22 \pm 3.51 \cdot E^{-02}$	$2.57 \cdot E^{-08} \pm 6.51 \cdot E^{-10}$	6043 ± 145	3.70	7.78	2.10

As observed from the experimental data, CO_2 was more permeable than O_2 and N_2 for all the samples under study, as reported for other similar polymers previously investigated [28,31], due to the diffusivity drop and solubility increase with decreasing permeant size (molecular diameter of CO_2 3.4 Å, oxygen molecular diameter 3.1 Å, and nitrogen molecule diameter 2.0 Å, respectively) [2]. Copolymerization appeared to affect slightly the permeation behavior, which changed with copolymer composition. As previously reported [12], an insertion of bigger NCE units into the PPCE crystal cell was observed by X-ray diffraction analysis. Consequently, the cocrystallization supported the modest decrease in the degree of crystallinity, as well as the reduction in the melting temperature. The introduction of NCE co-units along PPCE macromolecular chains caused a worsening of the barrier performance, particularly pronounced in the case of the most permeable gas, carbon dioxide. As an example, for a copolymer containing the highest amount of comonomeric units (P(PCE80NCE20)), the ratio of GTR to carbon dioxide is 2.5 times higher than that of the PPCE homopolymer, whereas GTR to oxygen is of the same order of magnitude. In general, in the case of the oxygen and nitrogen gas tests, the GTRs of the copolymers are only slightly higher than that of the PPCE homopolymer. This result is analogous to the one we previously found by investigating Poly(L-lactic acid) (PLLA)-based triblock copolymers, containing poly(butylene/neopentyl glycol succinate) random copolymers as central soft block [32]. The short ramifications (methyl groups) exert an obstacle effect toward the small permeant molecules, such as oxygen and nitrogen; their effect being practically negligible in the case of large CO_2 molecules. Hydrophobic side alkyl groups contribute to reducing the solubility of carbon dioxide in the polymer matrix, with a consequent increase in GTR (see value reported in Table 2). On the contrary, the diffusion coefficient of copolymers appeared to be higher, explaining the higher values of GTR in the CO_2 gas test. The effect of copolymer composition is not as clear; in fact, an alternating trend was observed with increasing the NCE co-unit amount. As mentioned above, all the copolymers at room temperature were above their glass transition temperature and semicrystalline with a similar crystallinity degree. The polymer polydispersity index (D_{index}) is among the factors affecting permeability behavior: the lower the D_{index}, the better the barrier performances. As seen from the data reported in Table 1, P(PCE95NCE5) and P(PCE85NCE15) had higher D_{index} values with respect to P(PCE90NCE10) and P(PCE80NCE10) samples, explaining the observed GTR trend (Figure 2).

The D, S, and t_L parameters were previously fully described by Siracusa et al. [18]. As can be observed from the data reported in Table 2, with respect to the PPCE homopolymer, the S value decreased, meaning a lower CO_2 solubility within the matrix, whereas the D value increased, meaning more diffusivity of the gas molecules through the films. Consequently, the t_L value was lower than the homopolymer value, due to the less time being needed to reach the permeability steady-state. The GTR rate increase recorded for the copolymers is a direct consequence of this behavior.

Perm-selectivity values, which represent the permeability ratio of different permeants, are also reported in Table 2. Those values are useful because they allow the calculation of the unknown GTR date, knowing the GTR value of another gas. As reported in the literature [33], for many polymers the $N_2:O_2:CO_2$ is in the range of 1:4:16. Our results are very different than those tabulated in the literature, demonstrating that using these values for calculation would not be appropriate. As previously demonstrated [11,31], those ratios are not constant, but depend on the chemical structure, temperature, moisture, and kind of simulant in contact with food.

Lastly, PPCE homopolymer and P(PCE80NCE20) copolymer were compared with some common petrochemical-based polymeric packaging materials (Figure 3). Both samples exhibit very good performance in terms of barrier properties against CO_2 and O_2, being worse than Nylon6 and polyethylene terephthalate (PET). Whilst this comparison is far from being exhaustive, it can be considered meaningful to highlight the potential of PPCE and PPCE-based copolymers for use as high barrier films.

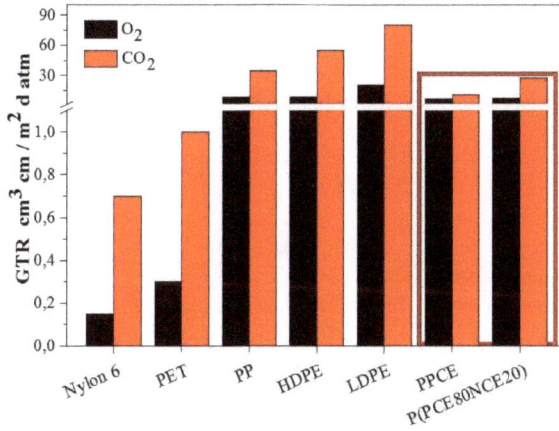

Figure 3. Gas transmission rate of O_2 and CO_2 gases for PPCE and P(PCE80NCE20) and some common petrochemical-based polymeric packaging materials [13].

3.2.2. Activation Energy of Gas Transport Process

Temperature is one of the most important parameters affecting food respiration rate. To understand the behavior of polymer membranes in terms of gas permeation, the effect of temperature on permeation behavior has to be evaluated [26]. To establish a correlation with temperature and calculate the activation energies of the permeation processes, the barrier behavior was investigated in the range of 8 to 38 °C. This temperature range was chosen considering all possible temperature scenarios, from food preservation (lower temperatures) to food handling (higher temperatures). Limited information is available about film permeability properties at different storage temperatures. As can be observed in Figure 4, an increase in the GTR was recorded for all samples, being less evident for the PPCE homopolymer.

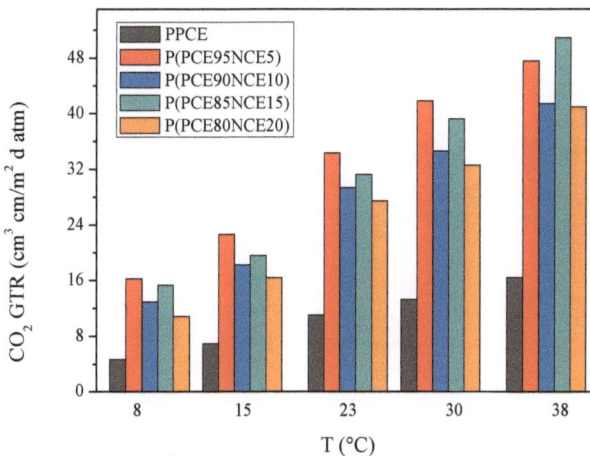

Figure 4. CO_2-GTR at different temperatures for PPCE and P(PCExNCEy) copolymers.

The CO_2 gas transfer rate was still higher than those of O_2 and N_2 (data not reported). The behavior confirmed the previously observed gas barrier trend. S, D, and t_L followed the theoretical behavior. By increasing the temperature, S decreased, D increased, and t_L decreased. The S value, correlated with gas solubility, indicated a reduction in CO_2 interaction with the polymer matrix. Consequently, D, which is correlated to the kinetic parameter, increased and the t_L decreased with increasing temperature. To describe the permeation dependence on the temperature, the Arrehenius model was used. The activation energy for the gas transmission rate (E_{GTR}), the specific heat of gas solution (H_S), and gas diffusivity (E_D) processes were calculated using the mathematical relationships reported in the literature [17,33,34]. The activation energy was calculated from the slope ($-E_a/R$) of the straight line obtained plotting ln(GTR) as a function of 1/T for all the samples under investigation (Figure 5), where R is the universal gas constant, equal to 8.314 J/mol K. From Figure 5, the experimental data are well-fitted by an Arrhenius-type equation; a high value of the regression coefficient (R^2) was obtained from the fitting. In Table 3, the gas transmission rate data in the temperature range of 8 to 38 °C with CO_2 gas test are reported. No data were obtained for the O_2 and N_2 gas tests because it was not possible to fit the slope of the linear portion of the GTR curves [30].

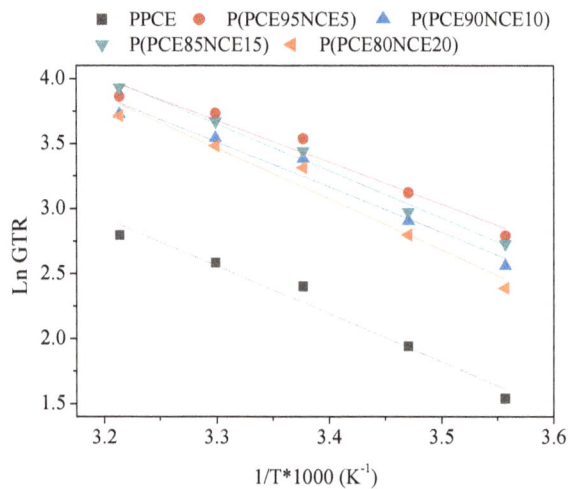

Figure 5. Arrhenius plot of GTR for PPCE and P(PCExNCEy) copolymers.

Table 3. E_{GTR}, H_S, and E_D data for CO_2 pure gas, in the 8–38 °C temperature range, with the linear regression coefficients R^2 provided in brackets.

Sample	PPCE	P(PCE95NCE5)	P(PCE90NCE10)	P(PCE85NCE15)	P(PCE80NCE20)
E_{GTR} (KJ/mol)	30.7 ± 0.13 (0.97)	26.8 ± 0.18 (0.97)	28.9 ± 0.11 (0.97)	30.1 ± 0.14 (0.99)	32.5 ± 0.21 (0.97)
H_S (KJ/mol)	-	-15.4 ± 0.11 (0.60)	-35.8 ± 0.16 (0.18)	5.7 ± 0.11 (0.02)	-7.24 ± 0.12 (0.01)
E_D (KJ/mol)	-	553 ± 0.18 (0.88)	8.32 ± 0.12 (0.00)	95.1 ± 0.15 (0.97)	388 ± 0.11 (0.15)

In general, the activation energy values for gases that migrate through a polymer membrane range from 12 to 63 KJ/mol [34]. The activation energies for the samples under investigation ranged from 27 to 33 KJ/mol, being very similar to those reported by other authors PET, Polyethylene furanoate (PEF), and poly(neopentyl glycol furanoate) (PNF) [19,26,35]. As reported in the literature [36–38], high activation energy implies more sensitivity to temperature deviation. Whereas the permeation process is characterized by a good correlation with the temperature variation, the sorption and diffusion process shows consistent deviation due to the chemical composition of the polymers. In general, the solubility decreased with temperature and it is the parameter correlated with the polymer composition.

The recorded trend is a confirmation that the gas interacts differently with the polymer matrix. From low to high temperatures, the ln$S$ value fluctuated as did the ln$D$ values.

3.2.3. Barrier Properties at Different Relative Humidity

Materials characterized by good barrier properties in dry ambient conditions can perform differently when tested in different environments, for example in water. As reported in the literature [39], in the case of low barrier films, the medium reduces the gas permeation rate, whereas for higher barrier materials, like poly(vinylidene-chloride) (PVDG), the influence of the environment on the permeation process is almost undetectable. The GTR results for the PPCE and P(PCExNCEy) samples at different relative humidity are reported in Figure 6.

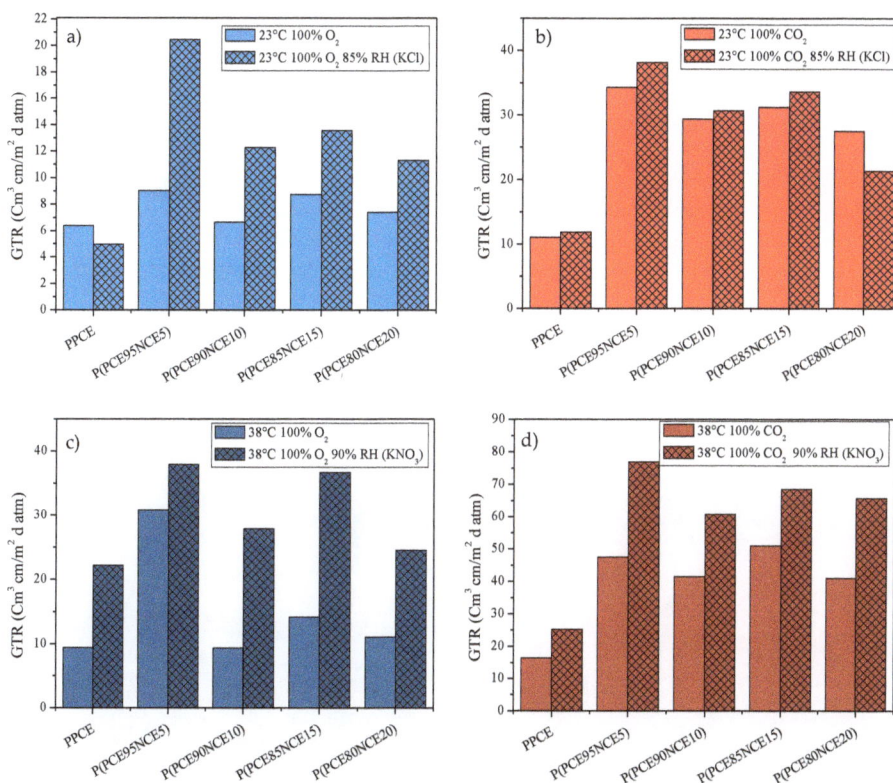

Figure 6. (**a**) O_2-GTR and (**b**) CO_2-GTR at 23 °C with 85% relative humidity (Standard Atmosphere). (**c**) O_2-GTR and (**d**) CO_2-GTR at 38 °C, 90% relative humidity (Tropical Atmosphere).

As reported in the literature [40–42], plasticization and swelling phenomena can occur in moist ambient environments. The hydrogen bonds and/or dipole–dipole interactions between water and the polar side of the polymer chains are responsible for this behavior. In particular, due to the water interaction, small network fragments are lost, promoting the transfer of the gas throughout the film. The effects of these phenomena become more intense as the relative humidity and temperature increase. As shown in Figure 6, an increase in the GTR was recorded with increasing RH due to the presence of ester polar groups. In particular, for the sake of simplicity, Table 4 reports the percentage of increase or decrease with respect to the results recorded on the samples without treatments.

Table 4. Percentage increase (+) or decrease (−) in GTR for PPCE and P(PCExNCEy) samples under different ambient moistures.

Sample	CO_2			
	23 °C 85% RH (KCl)	38 °C 90% RH (KNO$_3$)	23 °C 85% RH (KCl)	38 °C 90% RH (KNO$_3$)
PPCE	−22%	+136%	+7%	+54%
P(PCE95NCE5)	+126%	+23%	+11%	+62%
P(PCE90NCE10)	+85%	+197%	+5%	+47%
P(PCE85NCE15)	+55%	+159%	+8%	+35%
P(PCE80NCE20)	+53%	+121%	-22%	+60%

A progressive increase in the gas transmission rate was recorded at increasing RH and temperature. In particular, a considerable increase was recorded at 38 °C from 0% to 90% RH for both gases. In general, barrier properties worsened at higher relative humidity, highlighting that the water played an important role in the gas transport process in a wet polymer matrix. The permeability of wet polymers did not follow the same trend as the dry samples. In some cases, the O_2-GTR was higher than the CO_2-GTR and vice versa, highlighting that several different factors play a role in the gas barrier behavior. The concomitant action of these factors prevents the establishment of a clear correlation between the chemical structure and barrier behavior under different environmental conditions.

Regardless, these results are important because they account for the strong gas-water-polymer matrix interactions always present for the real conditions during the use of the packaging.

3.2.4. Barrier after Food Simulant Contact

The GTR data recorded after food simulant contact with the CO_2 and O_2 gas tests are reported in Figure 7, together with those of untreated samples added for comparison.

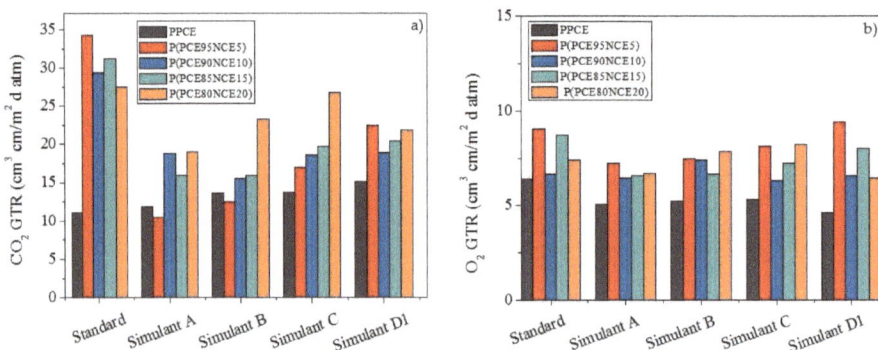

Figure 7. (a) CO_2-GTR and (b) O_2-GTR after food simulant contact for PPCE homopolymer and P(PCE$_x$NCE$_y$) copolymers.

When polymer matrixes are used for food packaging applications, and the polymer packages are consequently placed in contact with food, several scenarios must be considered due to the different characteristics of food.

In particular, as reported in the literature [43,44], food can be aqueous, acid, containing oil/fat, oily or fatty, alcoholic, or low moisture content solid food. To understand the behavior of the materials under study, the worst of the foreseeable conditions were chosen in terms of contact time and temperature (Tables 1–3 of the E.U. Regulation) [20]. To perform the analysis, the test number OM2 was chosen in order to analyze a large spectra of food packaging scenarios with a contact time of 10 days at 40 °C. This test is used to analyze the packaging behavior for any long-term food storage at room temperature or below, at heating up to 70 °C for 2 h, and at heating up to 100 °C for up to 15 min. This test also covers the food contact conditions covered by test numbers OM1 and OM3. In particular, food simulants A, B, C, and D1 were chosen. According to the E.U. regulation: (1) simulant A, B, and C simulate the contact of packaging plastic material with food characterized by a hydrophilic character, able to extract hydrophilic substances; (2) simulant B is indicated for food with a pH below 4.5; (3) simulant C is indicated for alcoholic food with an alcohol content up to 20% and for food containing relevant amounts of organic ingredients that render the food more lipophilic; and (4) simulant D1 is indicated for food with a lipophilic character and able to extract lipophilic substances, that mimic alcoholic food with an alcohol content above 20%, and oil in water solution.

For simplicity, the percentages of GTR increase or decrease (±, %) with respect to the results recorded on the untreated samples are reported in Table 5.

Table 5. Percentage (%) of CO_2-GTR/O_2-GTR increase (+) or decrease (−) for PPCE and P(PCExNCEy) copolymers after food simulant contact.

Sample/Simulant	Simulant A	Simulant B	Simulant C	Simulant D1
	CO_2/O_2			
PPCE	+8%/−21%	+24%/−18%	+25%/−17%	+37%/−28%
P(PCE95NCE5)	−70%/−20%	−64%/−17%	−50%/−10%	−34%/+4%
P(PCE90NCE10)	−36%/−3%	−47%/+11%	−36%/−5%	−36%/−1%
P(PCE85NCE15)	−49%/−25%	−49%/−24%	−37%/−17%	−34%/−8%
P(PCE80NCE20)	−31%/−10%	−15%/+6%	−3%/+11%	−20%/−13%

When CO_2 was used as the test gas, an increase in the GTR value was recorded for the PPCE homopolymer, ranging from 8 to 37%, with the highest value for simulant D1. Interestingly, for all the copolymers investigated, a consistent decrease in the GTR values was recorded, indicating the great stability of these materials when in contact with the food simulants under the worst conditions. In the case of the O_2 gas test, a general stability, with a light decrease in the GTR values, was also recorded. This behavior was supported by the D, S, and t_L data (data not reported). The S value increased after food simulant contact, indicating a higher compatibility of the gas with the polymer. Consequently, the D value decreased, due to the lower speed of the gas molecules moving through the polymer membrane. Thus, the time to attain the steady-state was longer, due to requiring more time for the gas molecules to homogeneously arrange inside the polymer. Due to the influence of several parameters in the permeability process, clear correlation with polymer chemical structure could not be confirmed [2]. Further analyses are in progress to correlate the changes in gas permeation behavior to the chemical and morphological characteristics (change in the crystalline/amorphous ratio after treatments).

3.3. FTIR Characterization, Molecular Weigt Determination and Color Evaluation

3.3.1. FTIR Characterization and Molecular Weight Determination

FTIR spectra were recorded for each sample to investigate the change in the chemical structure due to the different treatments. The principal absorption bands for all films are summarized in Table 6. From the spectra, no substantial changes were recorded after each treatment. The main peaks were still present with a small change in the band intensity by increasing the NCE co-unit content due to the presence of –CH_3 pendant groups. A shift of no more than ±10 cm^{-1} was recorded with respect to the untreated samples. The shift was more evident after food contact with simulant liquids, as also seen for gas barrier permeability.

The interaction with polar liquids was evidenced by a slight increase in the –OH band intensity that was similar in all samples. These results highlighted the suitability of these materials to be used in humid environments as well as in contact with food. Furthermore, a molecular weight decrease of no more than 1–2% was recorded for all samples after stressed treatment, confirming the stability of such materials.

Migration tests need to be performed to evaluate the effectiveness of such materials to be placed in contact with real food.

Table 6. Fourier transform infrared (FTIR) data for PPCE and P(PCExNCEy) films.

Chemical Group	Peak Position (cm^{-1})
–OH stretch (free)	3578
CH-stretch (of CH$_2$)	2916 (ν_{as} CH$_2$), 2853 (ν_s CH$_2$)
–CH$_3$ (pendant group)	2871 (ν_s)
–C=O normal carbonyl stretch	1712
–CH-deformation symmetric and asymmetric bending	1472 (δ_s CH$_2$)
C–O–H in-plane bend	1424
–CH$_3$	1451 (δ_{as}), 1378 (δ_s)
–CH$_2$-scissoring	1438
–C=O bending	1245
–C–O stretching	1178, 1153
–OH bending	1046
–CH$_2$ wagging and twisting	1243, 1180
–CH$_2$ rocking	731
O–H out-of-plane	992 (as), 945(s)
C–C stretch	920, 809

3.3.2. Color Evaluation

Film transparency and color are important requisites, especially if the material is used for food packaging application. Food color, associated with a high amount of naturally present pigments, has been always considered one of the key factors for evaluating food quality and taste, especially from the final consumers. Therefore, as previously reported [44], packaging should interfere as little as possible with the color of the food product, in order to preserve the consumer attractiveness. In Table 7, the film surface color determination for PPCE and P(PCExNCEy) samples are reported and compared to a white standard.

Table 7. Lightness coefficient (*L**), *a**, and *b**, total color difference (Δ*E*), *C** and h_{ab} of PPCE film and P(PCExNCEy) films.

Sample	*L**	*a**	*b**	Δ*E*	*C**	h_{ab}
White standard	66.47 ± 0.01	−0.73 ± 0.01	1.22 ± 0	-	1.42	121
PPCE	63.67 ± 0.14	−0.89 ± 0.03	1.78 ± 0.13	2.85	1.99	139
P(PCE95NCE5)	61.86 ± 0.69	−0.88 ± 0.02	2.93 ± 0.52	4.92	3.06	107
P(PCE90NCE10)	63.49 ± 0.60	−0.87 ± 0.02	1.85 ± 0.28	3.05	2.04	115
P(PCE85NCE15)	62.92 ± 0.32	−0.95 ± 0.05	2.52 ± 0.27	3.79	2.69	111
P(PCE80NCE20)	61.96 ± 0.43	−0.98 ± 0.02	2.88 ± 0.24	4.81	3.04	109

h_{ab} = 0°, red-purple; h_{ab} = 90°, yellow; h_{ab} = 180°, green; h_{ab} = 270°, blue.

On the CIE Lab Color scale, the lightness coefficient (*L**) ranges from black (0) to white (100). For any *L** value, the coordinates *a** and *b** situate the color on a rectangular coordinate grid perpendicular to the *L** axis. At the origin (*a** = 0 and *b** = 0) the color is achromatic (gray). Moving on the horizontal axis, a positive *a** value indicates a hue of red-purple and a negative *a** value indicates a green hue. Moving on the vertical axis, a positive *b** value indicates a yellow hue and a negative *b** value indicates a blue hue [45]. P(PCExNCEy) films showed an L* closer to white, whereas *a** and *b** indicated a faint tendency toward a yellowish color (h_{ab} over 90°), like the as-synthesized polymer powder. A very low *C** was recorded, meaning low color saturation and consequently a good transparency of the film, despite some differences being recorded related to the copolymer composition. The same characteristics were observed after treatments, indicating the good stability of the samples.

4. Conclusions

A new class of aliphatic biobased polyesters, previously synthesized and characterized from the thermal and mechanical points of view, was subjected to studies aiming to evaluate their barrier performances.

The results obtained are extremely interesting as the copolymers under investigation could be considered good candidates for food packaging application using the modified atmosphere packaging technique (MAP). The introduction of a neopentyl glycol unit into the PPCE did not result in a significant worsening of barrier performance with respect oxygen and nitrogen. As oxygen promotes the oxidation process, with subsequent deterioration of the chemical-physical, organoleptic, and quality properties of the packed food, a low oxygen permeation value can be considered a good result. Conversely, low permeation of nitrogen is a guarantee of package stability, avoiding bag collapses. In the case of the larger and polar CO_2 molecules, a worsening in barrier performance due to copolymerization was found; the two side methyl groups present in the macromolecular chains rendered the polymer less polar and therefore decreased the solubility of carbon dioxide in the polymer matrix. However, an atmosphere poor in oxygen and rich in carbon dioxide decreases the metabolism of packed products or the spoilage activity, maintaining and/or prolonging the desired food shelf-life.

A general worsening in the gas barrier properties after measurement in different moisture environments was recorded, showing an important interaction between the polymer matrix and water. On the contrary, all the samples under investigation showed good stability after food simulant contact.

In conclusion, due to their bio-based and biodegradable nature, the new investigated polyesters can be considered good candidates for substitution of the traditional petroleum-based polymers for packaging application.

Author Contributions: V.S. and N.L. conceived and designed the experiments; V.S. performed the experiments; V.S., N.L. and L.G. analyzed the data; V.S., N.L., A.M. contributed to reagents/materials/analysis tools; V.S. wrote the paper; N.L. and C.I. performed the final revision of the paper.

Conflicts of Interest: The authors declare no conflict of interest. The founding sponsors had no role in the design of the study; in the collection, analyses, or interpretation of data; in the writing of the manuscript, and in the decision to publish the results.

References

1. Vermeiren, L.; Devlieghere, F.; Beest, V.; Kruijf, N.D.; Debevere, J. Developments in the active food packaging of foods. *Trend Food Sci. Technol.* **1999**, *10*, 77–86. [CrossRef]
2. Robertson, G.L. Chapter 4: Optical, Mechanical and Barrier Properties of Thermoplastic Polymers. In *Food Packaging—Principles and Practice*, 3rd ed.; Taylor & Francis Group, CRC Press: Boca Raton, FL, USA, 2013; pp. 91–130. ISBN 978-1-4398-6242-1.
3. Siracusa, V.; Rocculi, P.; Romani, S.; Dalla Rosa, M. Biodegradable polymer for food packaging: A review. *Trend Food Sci. Technol.* **2008**, *19*, 634–643. [CrossRef]
4. Peelman, N.; Ragaert, P.; De Meulenaer, B.; Adons, D.; Peeters, R.; Cardon, L.; Van Impe, F.; Devlieghere, F. Application of bioplastics for food packaging. *Trends Food Sci. Technol.* **2013**, *32*, 128–141. [CrossRef]
5. European Bioplastics–Bioplastics, Facts and Figures 2017. Available online: http://en.european-bioplastics.org (accessed on 28 July 2017).
6. Rabnawaz, M.; Wyman, I.; Auras, R.; Cheng, S. A roadmap towards green packaging: The current status and future outlook for polyesters in the packaging industry. *Green Chem.* **2017**. [CrossRef]
7. Nakajima, H.; Dijkstra, P.; Loos, K. The Recent Developments in Biobased Polymers toward General and Engineering Applications: Polymers that are Upgraded from Biodegradable Polymers, analogous to Petroleum-Derived Polymers, and New Developed. *Polymers* **2017**, *9*, 523. [CrossRef]
8. Chen, G.-Q.; Patel, M.K. Plastics Derived from Biological Sources: Present and Future: A technical and Environmental Review. *Chem. Rev.* **2012**, *112*, 2082–2099. [CrossRef] [PubMed]

9. Fabbri, M.; Soccio, M.; Gigli, M.; Guidotti, G.; Gamberini, R.; Gazzano, M.; Siracusa, V.; Rimini, B.; Lotti, N.; Munari, A. Design of fully aliphatic multiblock poly(ester urethane)s displaying thermoplastic elastomeric properties. *Polymer* **2016**, *83*, 154–161. [CrossRef]

10. Gigli, M.; Lotti, N.; Siracusa, V.; Gazzano, M.; Munari, A.; Dalla Rosa, M. Effect of molecular architecture and chemical structure on solid-state and barrier properties of heteroatom-containing aliphatic polyesters. *Eur. Polym. J.* **2016**, *78*, 314–325. [CrossRef]

11. Gigli, M.; Lotti, N.; Gazzano, M.; Siracusa, V.; Finelli, L.; Munari, A.; Dalla Rosa, M. Biodegradable aliphatic copolyesters containing PEG-like sequences for sustainable food packaging applications. *Polym. Degrad. Stab.* **2014**, *105*, 96–106. [CrossRef]

12. Genovese, L.; Lotti, N.; Gazzano, M.; Finelli, L.; Munari, A. New eco-friendly random copolyesters based on poly(propylene cyclohexanedicarboxylate): Structure-properties relationships. *eXPRESS Polym. Lett.* **2015**, *9*, 972–983. [CrossRef]

13. Mensitieri, G.; Di Maio, E.; Buonocore, G.G.; Nedi, I.; Oliviero, M.; Sansone, L.; Iannace, S. Processing and shelf life issues of selected food packaging materials and structures from renewable resource. *Trends Food Sci. Technol.* **2011**, *22*, 72–80. [CrossRef]

14. Genovese, L.; Soccio, M.; Gigli, M.; Lotti, N.; Gazzano, M.; Siracusa, V.; Munari, A. Gas permeability, mechanical behaviour and compostability of fully-aliphatic bio-based multiblock poly(ester urethane)s. *RSC Adv.* **2016**, *6*, 55331–55342. [CrossRef]

15. Brugger Feinmechanik GmbH. *Gas Permeability Testing Manual*; Brugger Feinmechanik GmbH: Munchen, Germany, 2008.

16. NIST-National Instition of Standards and Technology. *Guide for the Use of the International System of Units (SI)*; Special Publication 811; Thompson, A., Taylor, B.N., Eds.; U.S. Department of Commerce: Washington, DC, USA, 2008.

17. Siracusa, V. Food packaging permeability behaviour: A report. *Int. J. Polym. Sci.* **2012**, *1*, 1–11. [CrossRef]

18. Siracusa, V.; Dalla Rosa, M.; Iordanskii, A. Performance of poly(lactic acid) surface modified films for food packaging application. *Materials* **2017**, *10*, 850. [CrossRef] [PubMed]

19. Burgess, S.K.; Kriegel, R.M.; Koros, W.J. Carbon Dioxide Sorption in Amorphous Poly(ethylene furanoate). *Macromolecules* **2015**, *48*, 2184–2193. [CrossRef]

20. European Union (EU). Regulation No. 10/2011 on Plastic Materials and Articles Intended to Come into Contact with Food. Available online: http://eur-lex.europa.eu/legal-content/EN/ALL/?uri=CELEX%3A32011R0010 (accessed on 4 February 2011).

21. Glicerina, V.; Balestra, F.; Dalla Rosa, M.; Bergenhstal, B.; Tornberg, E.; Romani, S. The Influence of Different Processing Stages on particle size, microstructure and appearance of dark chocolate. *J. Food Sci.* **2014**, *79*, E1359–E1365. [CrossRef] [PubMed]

22. Galus, S.; Lenart, A. Development and characterization of composite edible films based on sodium alginate and pectin. *J. Food Eng.* **2013**, *115*, 459–465. [CrossRef]

23. Syahidad, K.; Rosnah, S.; Noranizan, M.A.; Zaulia, O.; Anvarjon, A. Quality change of fresh cut cantaloupe (Cucumis melo L. var Reticulatus cv. Glamour) in different types of polypropylene packaging. *Int. J. Res.* **2015**, *22*, 753–760.

24. Soccio, M.; Lotti, N.; Finelli, L.; Gazzano, M.; Munari, A. Neopenthyl glycol containing poly(propylene azelate)s: Synthesis and thermal properties. *Eur. Polym. J.* **2007**, *43*, 3301–3313. [CrossRef]

25. Soccio, M.; Lotti, N.; Finelli, L.; Gazzano, M.; Munari, A. Neopenthyl glycol containing poly(propylene terephthalate)s: Structure-properties relationships. *J. Polym. Sci. Part B Polym. Phys.* **2008**, *46*, 170–181. [CrossRef]

26. Genovese, L.; Lotti, N.; Siracusa, V.; Munari, A. Poly(Neopentyl Glycol Furanoate): A Member of the Furan-Based Polyester Family with Smart Barrier Performances for Sustainable Food Packaging Applications. *Materials* **2017**, *10*, 1028. [CrossRef] [PubMed]

27. Genovese, L.; Lotti, N.; Gazzano, M.; Siracusa, V.; Dalla Rosa, M.; Munari, A. Novel biodegradable aliphatic copolyesters based on poly(butylene succinate) containing thioether-linkages for sustainable food packaging applications. *Polym. Degrad. Stab.* **2016**, *132*, 191–201. [CrossRef]

28. Guidotti, G.; Soccio, M.; Siracusa, V.; Gazzano, M.; Salatelli, E.; Munari, A.; Lotti, N. Novel Random PBS-Based Copolymers Containing Aliphatic Side Chains for Sustainable Flexible Food Packaging. *Polymers* **2017**, *9*, 724. [CrossRef]

29. Farber, J.M. Microbiological aspects of modified—Atmosphere packaging technology—A review. *J. Food Prot.* **1991**, *54*, 58–70. [CrossRef]

30. Alavi, S.; Thomas, S.; Sandeep, K.P.; Kalarikkal, N.; Varghese, J.; Yaragalla, S. *Polymer for Packaging Application*; CRC Press: Boca Raton, FL, USA, 2014; Volume 2, pp. 39–52.

31. Genovese, L.; Gigli, M.; Lotti, N.; Gazzano, M.; Siracusa, V.; Munari, A.; Dalla Rosa, M. Biodegradable Long Chain Aliphatic Polyesters Containing Ether-Linkages: Synthesis, Solid-State, and Barrier Properties. *Ind. Eng. Chem. Res.* **2014**, *53*, 10965–10973. [CrossRef]

32. Genovese, L.; Soccio, M.; Lotti, N.; Gazzano, M.; Siracusa, V.; Salatelli, E.; Balestra, F.; Munari, A. Design of biobased PLLA triblock copolymers for sustainable food packaging: Thermo-mechanical properties, gas barrier ability and compostability. *Eur. Polym. J.* **2017**, *95*, 289–303. [CrossRef]

33. Shmid, M.; Zillinger, W.; Muller, K.; Sangerlaub, S. Permeation of water vapour, nitrogen, oxygen and carbon dioxide through whey protein isolated based films and coatings—Permselectivity and activation energy. *Food Packag. Shelf Life* **2015**, *6*, 21–29. [CrossRef]

34. Siracusa, V.; Ingrao, C. Correlation amongst gas barrier behavior, temperature and thickness in BOPP films for food packaging usage: A lab-scale testing experience. *Polym. Test.* **2017**, *59*, 277–289. [CrossRef]

35. Burgess, S.K.; Karvan, O.; Johnson, J.R.; Kriegel, R.M.; Koros, W.J. Oxygen sorption and transport in amorphous poly(ethylene furanoate). *Polymer* **2014**, *55*, 4748–4756. [CrossRef]

36. Atkins, P.; Jones, L. *Chemical Principles: The Quest for Insight*, 5th ed.; Freeman WH & Co.: New York, NY, USA, 2012.

37. Auras, R.A.; Harte, B.; Selke, S.; Hernandez, R. Mechanical, Physical and Barrier Properties of Poly(lactide) Films. *J. Plast. Film Sheeting* **2003**, *19*, 123–135. [CrossRef]

38. Kim, S.W.; Choi, H.M. Morphology, thermal, mechanical and barrier properties of grapheme oxide/poly(lactic acid) nanocomposite films. *Korean J. Chem. Eng.* **2016**, *33*, 330–336. [CrossRef]

39. Galić, K.; Ciković, N. Permeability characterization of solvent treated polymer materials. *Polym. Test.* **2001**, *20*, 599–606. [CrossRef]

40. Abenojar, J.; Pantoja, M.; Matinez, M.A.; Del Real, J.C. Aging by mixture and/or temperature of epoxy/SiC composites: Thermal and mechanical properties. *J. Comp. Mater.* **2015**, *49*, 2963–2975. [CrossRef]

41. Meisrr, A.; Willstrand, K.; Possart, W. Influence of composition, humidity and temperature on chemical aging in epoxies: A local study of the interphase with air. *J. Adhes.* **2010**, *86*, 222–243. [CrossRef]

42. Lawton, J.W.; Doane, W.M.; Willett, J.L. Aging and moisture effects on the tensile properties of Starch/Poly(hydroxyester ether) composites. *J. Appl. Polym. Sci.* **2006**, *100*, 3332–3339. [CrossRef]

43. Siracusa, V.; Lotti, N.; Munari, A.; Dalla Rosa, M. Poly(butylene succinate) and poly(butylene-succinate-co-adipate) for food packaging application: Gas barrier properties after stressed treatments. *Polym. Degrad. Stab.* **2015**, *119*, 35–45. [CrossRef]

44. Guidotti, G.; Gigli, M.; Soccio, M.; Lotti, N.; Gazzano, M.; Siracusa, V.; Munari, A. Poly(butylene 2,5-thiophenedicarboxylate): An Added Value to the Class of High Gas Barrier Biopolyesters. *Polymer* **2018**, *9*, 167. [CrossRef]

45. McGuire, R.G. Reporting of objective color measurements. *HortScience* **1992**, *27*, 1254–1255.

© 2018 by the authors. Licensee MDPI, Basel, Switzerland. This article is an open access article distributed under the terms and conditions of the Creative Commons Attribution (CC BY) license (http://creativecommons.org/licenses/by/4.0/).

MDPI

Article

Solid-State Polymerization of Poly(Ethylene Furanoate) Biobased Polyester, II: An Efficient and Facile Method to Synthesize High Molecular Weight Polyester Appropriate for Food Packaging Applications

Nejib Kasmi [1], George Z. Papageorgiou [2],*, Dimitris S. Achilias [1] and Dimitrios N. Bikiaris [1],*

[1] Laboratory of Polymer Chemistry and Technology, Department of Chemistry,
     Aristotle University of Thessaloniki, GR-541 24 Thessaloniki, Macedonia, Greece;
     nejibkasmi@gmail.com (N.K.); axilias@chem.auth.gr (D.S.A.)
[2] Chemistry Department, University of Ioannina, P.O. Box 1186, 45110 Ioannina, Greece
*   Correspondence: gzpap@cc.uoi.gr (G.Z.P.); dbic@chem.auth.gr (D.N.B.);
     Tel.: +30-265-1008354 (G.Z.P.); +30-231-0997812 (D.N.B.)

Received: 10 April 2018; Accepted: 23 April 2018; Published: 25 April 2018

Abstract: The goal of this study was to synthesize, through a facile strategy, high molecular weight poly(ethylene furanoate) (PEF), which could be applicable in food packaging applications. The efficient method to generate PEF with high molecular weight consists of carrying out a first solid-state polycondensation under vacuum for 6 h reaction time at 205 °C for the resulting polymer from two-step melt polycondensation process, which is catalyzed by tetrabutyl titanate (TBT). A remelting step was thereafter applied for 15 min at 250 °C for the obtained polyester. Thus, the PEF sample was ground into powder, and was then crystallized for 6 h at 170 °C. This polyester is then submitted to a second solid-state polycondensation (SSP) carried out at different reaction times (1, 2, 3.5, and 5 h) and temperatures 190, 200, and 205 °C, under vacuum. Ultimately, a significant increase in intrinsic viscosity is observed with only 5 h reaction time at 205 °C during the second SSP being needed to obtain very high molecular weight PEF polymer greater than 1 dL/g, which sufficient for manufacturing purposes. Intrinsic viscosity (IV), carboxyl end-group content (–COOH), and thermal properties, via differential scanning calorimetry (DSC), were measured for all resultant polyesters. Thanks to the post-polymerization process, DSC results showed that the melting temperatures of the prepared PEF samples were steadily enhanced in an obvious way as a function of reaction time and temperature increase. It was revealed, as was expected for all SSP samples, that the intrinsic viscosity and the average molecular weight of PEF polyester increased with increasing SSP time and temperature, whereas the number of carboxyl end-group concentration was decreased. A simple kinetic model was also developed and used to predict the time evolution of polyesters IV, as well as the carboxyl and hydroxyl end-groups of PEF during the SSP.

Keywords: poly(ethylene furanoate); solid-state polymerization; high molecular weight; thermal properties; polyester; remelting process

---

1. Introduction

The search for sustainable biobased alternatives for polymer production has dramatically intensified in recent years, due to an increasing awareness of finite fossil fuel resources and the disrupting climatic effects of greenhouse gas emissions [1–3]. For this reason, the interest in biomass has rapidly emerged as a renewable source of chemicals and mainly monomers for bio-based polymers

Polymers **2018**, *10*, 471

production [4,5]. The demand for this attractive feedstock in nature is motivating both industrial and scientific communities to innovate a new generation of polymers, and therefore, an opening of the way to an all-inclusive sustainability [6]. Extensive research has escalated in recent years into biorefineries technology development, which has shown a burgeoning surge for producing green monomers [7]. 2,5-Furandicarboxylic acid (FDCA) is the most promising rigid bio-based building block, which has been recognized as one of the twelve most important renewable-based monomers [8]. This furan derivative, which may provide a suitable alternative for terephthalic acid, can be prepared by catalytic oxidation of 5-hydroxymethylfurfural (HMF) derived from C6 sugars or polysaccharides [9,10]. In fact, extensive efforts made up to date on the synthesis of different homopolyesters derived from the renewable-based monomer (FDCA) and various diols [11–19]. The most successful furanic biobased polyester is poly(ethylene furan dicarboxylate) (PEF), which is produced from 2,5-furandicarboxylic acid (2,5-FDCA) and ethylene glycol. It is a fully biosourced alternative to its commercial analogue polyethylene terephthalate (PET), produced from petroleum-derived terephthalic acid [20]. Extensive research efforts were triggered since the last decade towards PEF, and its historical progress has been extensively described in two recent extended reports [21,22].

Recently, intensive investigations have been conducted on the commercial polyester (PEF) and its interest is increasing in a spectacular way, due to its renewable nature and promising features. Compared to PET, PEF showed excellent thermal properties, i.e., its processability at lower temperatures due to a lower melting temperature (T_m), and the ability to withstand high temperatures thanks to a higher glass transition temperature (T_g) [23], as well as it is characterized by greatly improved thermal stability up to 320 °C [24]. An impressive 10–27-fold and 19-fold reduction has been reported in oxygen and carbon dioxide permeability, respectively, for PEF compared to PET [25]. Other attractive properties, such as excellent mechanical properties [26], reduced carbon footprint [27], and ability to formulate in films, fibers, and mostly bottles make PEF an appealing substituent to PET [28]. The combination of all PEF features aforementioned are suitable for use as bottles in food and beverage packaging. In 2010, manufacturing of PEF using Avantium's YXY technology has been started by Avantium in Netherlands [29] for typical applications for packaging of water and fibers, alcoholic beverages, and soft drinks, among others.

Apart from several reports dealing with the emerging topic of PEF by highlighting on its attractive features compared with its analog PET, a drawback still currently an obstacle of interest for researchers, besides the undesired yellowing of the final polyester, as well as its high brittleness, is the production of high molecular weight PEF, which ensures, therefore, the resulting polymer processing in a large safety without any deterioration of its mechanical properties. The decomposition of 2,5-FDCA at high reaction temperatures during melt polycondensation reactions, as well as the important role in the molecular weight increase of the catalyst type used could be the main cause for the emergence of the low molecular weight defect for PEF.

Solid-state polymerization (SSP), performed under mild conditions, has numerous potential advantages and a strong industrial interest to overcome the polymers' molecular weight limitations obtained from the melt methods [30]. This well-known technique is extensively employed in industry as an extension of the melt polycondensation to produce high molecular weight polyesters with improved properties suitable for wide range of applications (e.g., bottles, films, and fiber production). This competitive process to conventional melt polycondensation, involving heating of the starting partially crystalline polyester at a temperature between its glass transition temperature (T_g) and its melting point (T_m), is used mainly for PET manufacturing to get over its relatively low molecular weight [31–38].

PEF has been the topic of a significant number of publications addressed on its biaxial orientation [39], thermal properties, glass transition, mechanical properties, and isothermal or non-isothermal crystallization [40–49], as well as investigations on the synthesis and full characterization of this biobased polyester have been well discussed [50–59].

Surprisingly, although there are numerous studies on PEF synthesis and characterization, only very few publications deal with the industrially relevant process (SSP) of PEF, which leads to manufacturing of high-molecular weight polyester. To date, only three preceding reports [60,61] have been addressed on SSP of PEF as a third stage after two-step melt polycondensation process.

In this context, Knoop et al. [62] have managed to apply SSP method under reduced pressure for PEF, using Ti(IV)-isopropoxide as catalyst. This study, which aims to increase the polymer degree of polymerization during several hours, revealed that PEF with molecular weight of 25.000 g·mol^{-1} was obtained after 24 h SSP, and it was increased to 83.000 g·mol^{-1} after 72 h SSP of heating at 180 °C. The goal of this report was chiefly focused on the crystallization investigation, and its influence on the mechanical properties of high molecular weight polyester PEF. As it was presented by Hong [63], SSP has been recently carried out for PEF. It was found that the IV increased from 0.6 to 0.64 and 0.72 dL/g after 24 and 48 h of SSP reaction time, respectively.

The current report extends the very limited studies, available nowadays in literature, regarding the synthesis via SSP of high molecular weight PEF. Such work is necessary to enable large-scale industrial applications in PEF-market. In the present study, a facile and efficient method, which is never applied to PEF, has been revealed to effectively circumvent the limited molecular weight of this biobased polyester, and thereafter, the access to very high intrinsic viscosity (IV), up to 1.02 dL/g, appropriate to several applications, such as frozen food trays, tire-cord applications, and principally, for bottle manufacturing [64–67].

The efficiency of the developed method herein was compared, regarding molecular weight increase, with resulting PEF samples from the application of one SSP reaction to two prepolymers having different initial IV values.

The effect of the temperature and reaction time on the molecular weight increase of the obtained polyester PEF was studied in detail using both experimental measurements and a simple kinetic theoretical model.

2. Experimental

2.1. Materials

2,5-Furan dicarboxylic acid (2,5-FDCA, purum 97%), ethylene glycol (99.8%), and tetrabutyl titanate (TBT) (97%) catalyst was purchased from Aldrich Co. (Chemie GmbH, Unna, Germany). All other materials and solvents used were of analytical grade.

2.2. Synthesis of 2,5-Dimethylfuran-Dicarboxylate(DMFD)

DMFD was prepared as described in the reported procedure [49], whereby a reaction was performed into a round bottom flask (500 mL) in presence of 2,5-furandicarboxylic acid (15.6 g), 200 mL of anhydrous methanol, and 2 mL of concentrated sulfuric acid. The mixture was refluxed for 5 h. The excess of methanol was removed by distillation and filtration was performed out through a disposable Teflon membrane filter (Chemie GmbH, Unna, Germany). During filtration, DMFD was precipitated as white powder, and then 100 mL of distilled water was added, after cooling. Na$_2$CO$_3$ 5% *w*/*v* was added during stirring while pH was measured continuously to neutralize, partially, the dispersion. DMFD was recuperated as white powder, which was collected by filtration and washed with distilled water, and after drying, was recrystallized with a mixture of 50/50 *v*/*v* methanol/water. According to this procedure, white needles of DMFD were prepared (yield about 83%) with melting point 115 °C, and purity measured by hydrogen nuclear magnetic resonance (^{1}H NMR) 99.5% (Bruker spectrometer, Bremen, Germany).

2.3. Polyester Synthesis

The PEF polyesters were prepared through the two-stage melt polycondensation method (esterification and polycondensation) in a glass batch reactor as described in our previous work [13]. DMFD and ethylene glycol at a molar ratio of diester/diol = 1:2 were charged with 400 ppm of TBT as catalyst into the reaction tube of the polyesterification apparatus. The reaction mixture was heated under controlled argon flow for 2 h at a temperature of 160 °C, for an additional 1 h at 170 °C, and finally, at 180–190 °C for 1 h. The transesterification stage (first step) was considered complete after the collection of almost all the theoretical amount of methanol, which was removed from the reaction mixture by distillation and collected in a graduated cylinder. In the second stage of polycondensation, the vacuum was gradually reduced to 5.0 Pa over a time of about 30 min, to remove the excess diol, to avoid excessive foaming, and furthermore, to minimize oligomer sublimation, which is a potential problem during melt polycondensation. The temperature was gradually increased, during this time interval, to 230 °C, while stirring speed was also increased to 720 rpm. The reaction was maintained for 2 h at this temperature. After the polycondensation reaction was completed, PEF sample was removed from the reactor, milled, and washed with methanol.

2.4. Solid-State Polycondensation

Solid-state polymerization (SSP) was performed using an apparatus involving five volumetric flasks (100 mL) which were connected to a vacuum line, and were immersed in a potassium nitrate/sodium nitrite thermostated bath, having a precision within ±0.5 °C. Crystallized PEF (2 g) with a particle size fraction of −0.40 + 0.16 mm was charged in each one of the volumetric flasks under vacuum, stabilized beneath 3 and 4 Pa. The reaction temperature was kept constant at 190, 200, or 205 °C. The reaction flasks were withdrawn from the bath after 1, 2, 3.5, and 5 h for analysis of the PEF sample's intrinsic viscosity (IV), to identify the molecular weight of the resulting PEF samples, as well as measuring the carboxyl end-group concentration (COOH). The protocol mentioned above was conducted for two prepolymer PEF samples obtained from melt polycondensation procedure with different initial IV values (0.28 and 0.38 dL/g). The resulting PEF polyesters from SSP are respectively named PEF/TBT.1 and PEF/TBT.2.

The effective method used herein to increase the molecular weight of PEF involves the application of one SSP for 6 h at 205 °C, followed by a remelting step of the obtained polyester for 15 min at 250 °C, and thereafter, a second SSP reaction was applied at different reaction times (1, 2, 3.5, and 5 h) and temperatures 190, 200, and 205 °C, under vacuum. The resulting PEF polyesters are labeled as PEF/TBT.3.

2.5. Polyester Characterization

2.5.1. Intrinsic Viscosity Measurement

For intrinsic viscosity [η] measurements, PEF samples (1 wt %) have been dissolved in a mixture of phenol/tetrachloroethane (60:40 w/w) at 90 °C, and their flow time was measured using an Ubbelohde viscometer (Schott Gerate GMBH, Hofheim, Germany) at 25 °C. The [η] value of each sample was calculated using the following Solomon-Ciuta equation:

$$[\eta] = [2\{t/t_0 - \ln(t/t_0) - 1\}]^{1/2}/c \tag{1}$$

where c is the concentration of the solution; t_0 the flow time of pure solvent; and t, the flow time of solution. Three different measurements were repeated for each sample to ensure the accuracy of the results, and the average value was calculated.

2.5.2. Molecular Weight

The number average molecular weight ($\overline{M}_n$) of the PEF polyester samples was calculated from the intrinsic viscosity [η] values, using the Berkowitz equation [68], as was modified in our previous work [69]:

$$\overline{M}_n = 3.29 \times 10^4 [\eta]^{1.54} \tag{2}$$

2.5.3. End-Group Analysis

Carboxyl end-group content (C.C.) of the PEF polyesters was determined according to Pohl's method, by titrating a solution of the polyester in benzyl alcohol/chloroform mixture. NaOH solution in benzyl alcohol was used as standard solution, and phenol red as indicator [70]. Three different measurements were performed for each sample, and the average value was calculated.

2.5.4. Differential Scanning Calorimetry (DSC)

Differential scanning calorimetry (DSC) study of PEF was carried out on a Perkin-Elmer, Pyris Diamond DSC differential scanning calorimeter (Perkin-Elmer, Waltham, MA, USA), calibrated with high purity indium and zinc standards. For each measurement, a sample of 7 ± 0.1 mg was sealed in aluminum pans, and was then scanned in the instrument from 30 to 240 °C at a heating rate of 20 °C/min under nitrogen flow (50 mL/min). The melting temperature (T_m), the heat of fusion (ΔH_m), and the glass transition temperature (T_g) the of the PEF samples were determined from these scans.

3. Modeling the PEF SSP Kinetics

3.1. Reaction Mechanism

The reactions taking place during SSP of PEF include polycondensation/transesterification, esterification, thermal degradation, and side reactions of vinyl end-groups [31], and they are illustrated in the following Equations (3)–(6). In these equations, k_1, K_1 and k_2, K_2 denote the forward and equilibrium rate constants of transesterification and esterification reactions, respectively, k_d and k_v refer to the kinetic rate constants of the degradation and polycondensation of vinyl end-group reactions, which are considered one way.

Polycondensation/transesterification

$$k_1 \quad k_1/K_1 \tag{3}$$

Esterification

$$k_2 \quad k_2/K_2 \tag{4}$$

Thermal degradation

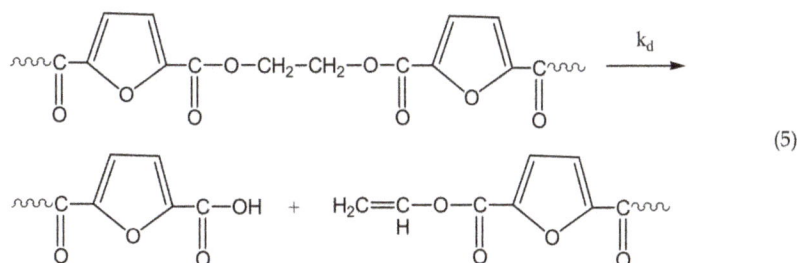

$$k_d \tag{5}$$

Polycondensation of vinyl end-groups

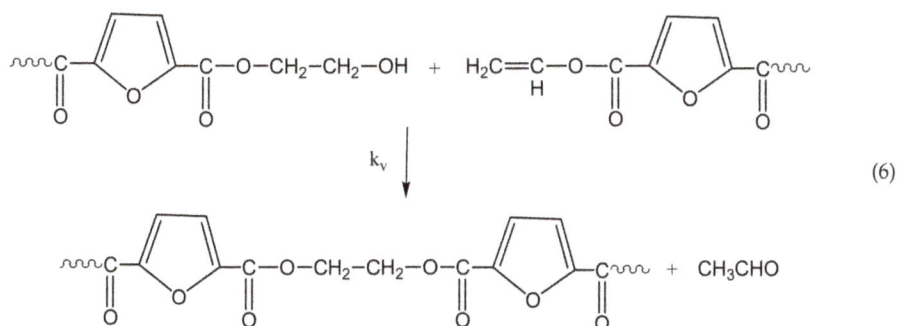

$$k_v \tag{6}$$

$$\cdots + CH_3CHO$$

The molecular weight of the polymer is increased by two reactions: in the first, Equation (3), two hydroxyl end-groups react, and ethylene glycol is produced. In the second, Equation (4), a carboxyl end-group reacts with a hydroxyl, and water is released as byproduct. By contrast, when thermal degradation takes place (Equation (5)), the molecular weight of the polyester can be decreased from the cleavage of an ester bond in the macromolecular chain, generating a vinyl ester end-group and a carboxyl end-group. In addition, acetaldehyde may be released from the side reaction of a vinyl ester end-group with a hydroxyl end-group, resulting also in an increase of the molecular weight (Equation (6)). The overall reaction rate is influenced by a combination of several factors, including intrinsic reaction kinetics, change of polymer degree of crystallization, and diffusional limitations of the reactive end-groups, and of the desorbing volatile byproducts (i.e., glycol and water) [71].

3.2. Simplified Mathematical Model

The problem of modelling the SSP kinetics is complicated, since, besides chemical kinetics, describing the rate of change of the concentration of the species as a function of time, diffusion phenomena should be incorporated, which results in additional variation with the distance from the interface [31]. Thus, two independent variables are introduced, resulting in a set of partial differential equations that should be solved, including several kinetic, diffusional, and crystallization parameters [72]. Using such complicated models to simulate a few experimental data points is out of any physical meaning. Since in this investigation, only five data points were measured at each experimental condition, a simple kinetic model was adopted afterAgarwal and co-workers [73,74]. This approach was originally developed for the solid-state polycondensation of PET, and successfully applied by our group in modelling the SSP of PET with several nanoadditives, as well as of PEF with nanoadditives [75–77].

In order to develop the mathematical model, several assumptions were made, including the following:

- All kinetic rate constants are considered independent of polymer chain length (only end-group reactivity is considered).
- Backward reactions in Equations (3) and (4) are eliminated, due to the fast removal of the water and ethylene glycol, produced in the reaction mixture, caused by the application of high vacuum (beneath 3 and 4 Pa).
- Due to the performance of the polycondensation at relatively low temperatures (i.e., 190–205 °C), no side reactions for the formation of acetaldehyde or thermal degradation are considered (Equations (5) and (6) are eliminated).
- Diffusional limitations on account of desorbing volatile species are neglected.
- Then, the rate of change of hydroxyl [OH] and carboxyl [COOH] end-groups can be described by the following expressions [73,74]:

$$\frac{d[OH]_t}{dt} = -2k_1[OH]_t^2 - [COOH]_t[OH]_t \tag{7}$$

$$\frac{d[COOH]_t}{dt} = -k_2[COOH]_t[OH]_t \tag{8}$$

where $[OH]_t$ and $[COOH]_t$ denote the actual "true" hydroxyl and carboxyl end-group concentrations, respectively.

The term "actual hydroxyl and carboxyl end-groups" was introduced by Ma and Agarwal [73,74], in order to account for the slowdown in SSP kinetics at high [η] values. Accordingly, a part of the carboxyl ([COOH]) and hydroxyl end-groups ([OH]) were considered to be rendered temporarily inactive (denoted as $[COOH]_i$ and $[OH]_i$) and the actual concentration of OH and COOH in Equations (7) and (8) can be expressed as

$$[OH]_t = [OH] - [OH]_i \tag{9}$$

$$[COOH]_t = [COOH] - [COOH]_i \tag{10}$$

where [OH], [COOH] and $[OH]_i$, $[COOH]_i$ denote the concentration of the total and temporarily inactivated OH and COOH end-groups, respectively.

Moreover, the number average molecular weight is expressed as

$$\overline{M}_n = \frac{2}{[COOH] + [OH]} \tag{11}$$

Equations (7) and (8), together with Equations (2) and (9)–(11), constitute a set of ordinary differential equations which can be easily solved numerically using a varying step-size Runge-Kutta method, to give results on the variation of the intrinsic viscosity and the concentration of –COOH and –OH end-groups as a function of time. Four adjustable parameters, namely k_1, k_2, $[OH]_i$, and $[COOH]_i$, are estimated at each experimental condition by simultaneous fitting of the values of [OH], [COOH], and IV to the experimental data points as a function of time.

4. Results and Discussion

The chemical structure of the initial polyesters before carrying out the SSP procedure (PEF/TBT.1, PEF/TBT.2, and PEF/TBT.3) was confirmed by ^{1}H NMR spectroscopy, as shown in Figure 1. The resonances appeared (b) at 5.4 for PEF/TBT.1, 5.32 for PEF/TBT.2, and 5.31 ppm for PEF/TBT.3 correspond to the methylene protons of the ethylene group. The peaks labelled (a) at 7.98, 7.91, and 7.90 ppm were respectively assigned to the ring protons (2 H, s) of PEF/TBT.1, PEF/TBT.2, and PEF/TBT.3.

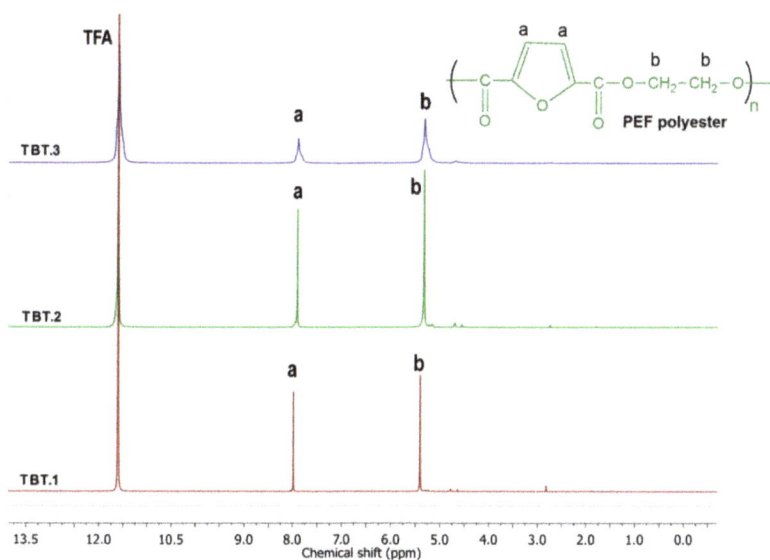

Figure 1. ¹H NMR spectra of poly(ethylene furanoate) (PEF)/ tetrabutyl titanate (TBT) samples.

4.1. Kinetic Study of the Solid-State Polymerization of PEF

As it was also pointed in part I of this research [58], SSP of PEF results in polyesters having increased average molecular weights. The effect of temperature and time on the molecular weight increase during SSP of PEF with TBT catalyst was also investigated here. Two initial PEF samples were employed, having initial IV equal to 0.27 and 0.38 dL/g, and given the names PEF/TBT.1 and PEF/TBT.2, respectively. These IV values correspond to polyesters having initial average degree of polymerization equal to 24 and 40, respectively. As it can be seen in Figure 2, the IV of neat PEF starting at 0.27 dL/g, increases to 0.47, 0.50, and 0.53 dL/g after 5 h of SSP at 190, 200, and 205 °C, respectively. Similar final IV values (i.e., 0.47, 0.50, and 0.54 dL/g after 5 h of SSP at 190, 200, and 205 °C) were measured for the polyester starting at 0.38 dL/g. Increased temperatures favor the increase in the IV values, since both esterification and transesterification reactions are accelerated. Furthermore, diffusion of byproducts produced (such as water and ethylene glycol) is much slower at low SSP temperatures. For these reasons, by increasing the SSP temperature to 205 °C, the IV increase is much higher compared to that at 190 °C (Figure 2a,b). At temperatures close to the melting point of PEF, the macromolecular chains have higher mobility and thus, hydroxyl end-groups react with carboxyl end-groups more easily, joining the macromolecular chains and increasing the molecular weight of PEF. Crystallinity plays also an important role since as higher it is as lower will be the diffusion rate of formed byproducts.

PEF samples, described above, at the end of the SSP, show a relatively low intrinsic viscosity, ranging from 0.47 to 0.54 dL/g. These low values also correspond to low molecular weights, resulting in polyesters having inferior mechanical performance, which is not appropriate for several promising applications. One way to increase the average molecular weight of a polymer conducted after SSP is to employ the so-called remelting process. This is widely applied in the industrial production of polyamides [78–81], and to a lesser extent, in PET [82,83]. Although many papers have been reported on the advantageous features of the remelting process with respect to the MW increase, to the best of our knowledge, there are no such studies in applying this technique to PEF. In this context, this is the first time that synthesis of PEF is undertaken via this method, in order to achieve a high molecular weight, and consequently, to overcome the inferior properties of the resulting polyester. Therefore, this procedure was applied on PEF/TBT sample for 30 min at 240 °C under argon atmosphere, preceded by SSP involving the same PEF polyester (initial sample with IV = 0.38 dL/g) at 205 °C for 6 h. Once the reaction was over, the polymer was cooled at room temperature, and then, a second SSP reaction was conducted at time intervals 1, 2, 3.5, and 5 h at constant temperatures 190, 200, and 205 °C under vacuum. The evolution of the intrinsic value of the resulting PEF/TBT.3 polyester is depicted in Figure 2c. The IV of PEF starts at 0.61 dL/g, and increases to 0.76, 0.86, and 1.02 dL/g after 5 h of SSP at 190, 200, and 205 °C, respectively. As can be seen, a higher increase of the polymerization rate has been proved, especially at SSP temperatures of 205 °C. This effect is due to the remelting features, which begets a redistribution/homogenization of the reactive end-group separation, and on the other hand, it reduces the polyester water content, which occurs during remelting, and afterwards, making end-group diffusion much easier [30,81].

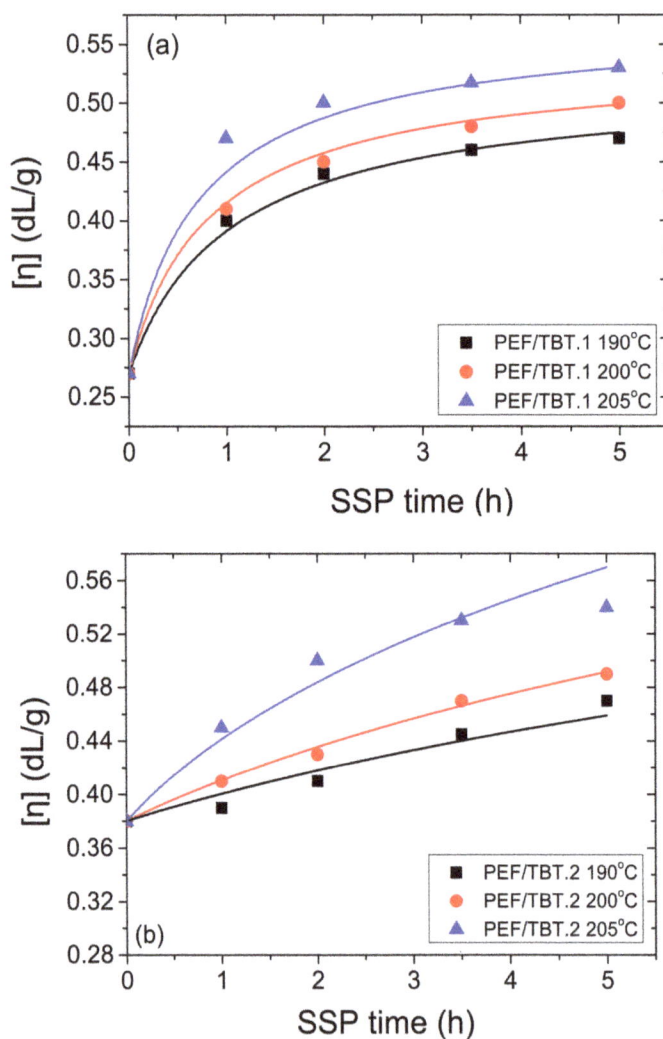

Figure 2. *Cont.*

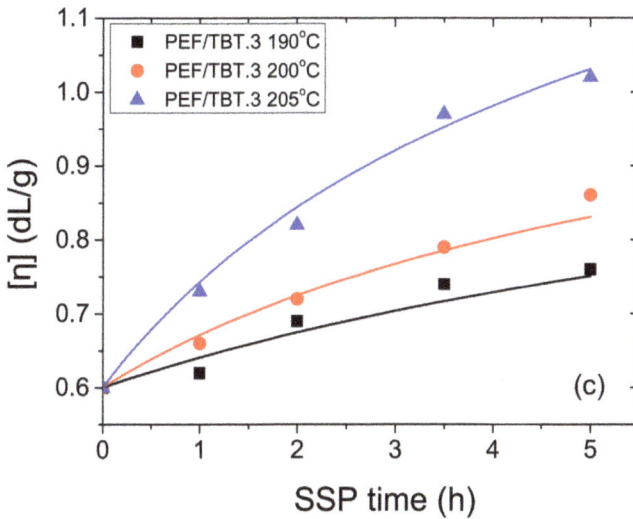

Figure 2. Variation of the intrinsic viscosity with time during solid-state polymerization (SSP) of PEF using TBT catalyst and three different initial IV values; PEF/TBT.1 (**a**), PEF/TBT.2 (**b**), and PEF/TBT.3 (**c**) at different temperatures. Continuous lines represent the theoretical kinetic model simulation results.

The calculated number average molecular weight, M_n, as well as the corresponding number-average degree of polymerization from the experimentally measured IV of all prepared samples are summarized in Table 1. The final PEF polyesters obtained after 2, 3.5, and 5 h of SSP at 205 °C for the PEF/TBT.3 sample are characterized by high IV values of 0.82, 0.97, and 1.02 dL/g, corresponding to number average molecular weights of 24,236; 31,392; and 33,920 g·mol^{-1}, respectively. These values are much higher compared to the corresponding values obtained for the PEF/TBT.1 and PEF/TBT.2 samples, which are similar at 205 °C, and range from 11,300 to 12,700 g·mol^{-1}. It can be stated that the reached IV range is enough to meet specific end-use requirements. It is suitable for packaging applications, such as carbonated beverage bottles and manufacturing of sheet grades for thermoforming [66].

Table 1. Number average molecular weights ($\overline{M}_n$, g·mol^{-1}) of PEF polyester using TBT catalyst obtained after SSP at different temperatures and times. The value includes, in parentheses, the corresponding number average degree of polymerization.

Temperature (°C)	SSP time (h)	PEF/TBT.1	PEF/TBT.2	PEF/TBT.3
	0	4400 (24)	7400 (40)	15,000 (81)
190	1	8000 (44)	7700 (42)	15,800 (86)
	2	9300 (51)	8300 (45)	18,600 (101)
	3.5	10,000 (54)	10,000 (54)	20,700 (112)
	5	10,300 (56)	10,600 (58)	21,600 (117)
200	1	8300 (45)	8300 (45)	17,400 (90)
	2	9600 (52)	9000 (49)	19,800 (108)
	3.5	10,600 (58)	10,300 (56)	22,900 (124)
	5	11,300 (62)	11,000 (59)	26,100 (142)
205	1	10,300 (56)	9600 (52)	20,300 (110)
	2	11,300 (62)	11,700 (63)	24,200 (132)
	3.5	11,900 (63)	12,400 (67)	31,400 (171)
	5	12,400 (67)	12,700 (69)	33,900 (184)

Moreover, end-group analysis (–COOH and –OH) was performed on all prepared samples. Figures 3 and 4 show the effect of SSP time and temperature on the –COOH and –OH concentrations of all PEF/TBT samples. It is obvious that the carboxyl and hydroxyl contents decrease with increasing SSP time and temperature, wherein at low SSP temperature (190 and 200 °C), carboxyl end-groups content decreases almost linearly with the SSP time, while it begins to sharply deviate from the linear relationship at 205 °C. Carboxyl end-groups start at approximately 24 eq/10^6 for both PEF/TBT.1 and PEF/TBT.2, and after 5 h of SSP, depending on the temperature, reach values ranging from 9 to 12 eq/10^6 and 11 to 14 eq/10^6 for PEF/TBT.1 and PEF/TBT.2, respectively. Moreover, it was expected that PEF/TBT.3 samples should exhibit a content of carboxyl end-groups much lower compared to those of other PEF/TBT samples, due to their much higher molecular weights. However, this was not the case, as a high COOH concentration of 46.9 eq/10^6 was obtained after remelting. After 5 h of SSP, the carboxyl content was reduced to 32–37 eq/10^6 depending on temperature. The main reason for these high values could be the existence of degradation reactions, which take place principally at high temperatures in the melt phase, wherein a high amount of carboxyl end-groups have been easily created. Accordingly, the major degradation reaction responsible for the generation of carboxyl ends happened mainly via the cleavage of an ester bond in the PEF main chain producing vinyl ester and carboxyl end-groups. More impressive results were observed when calculating the hydroxyl content, which in the case of PEF/TBT.1 and PEF/TBT.2, start at high amounts, i.e., 433 and 245 eq/10^6, respectively, and after 5 h, SSP ranges from 150 to 180 eq/10^6, depending on temperature. By contrast, in the PEF/TBT.3 samples, the hydroxyl content was very much lower, starting at 87 eq/10^6, and after 5 h SSP, decreasing to only 55, 41, and 27 eq/106 at 190, 200, and 205 °C, respectively. Thus, it seems, that in this polyester, in contrast to results obtained for PEF/TBT.1 and PEF/TBT.2, the concentrations of –COOH and –OH are similar.

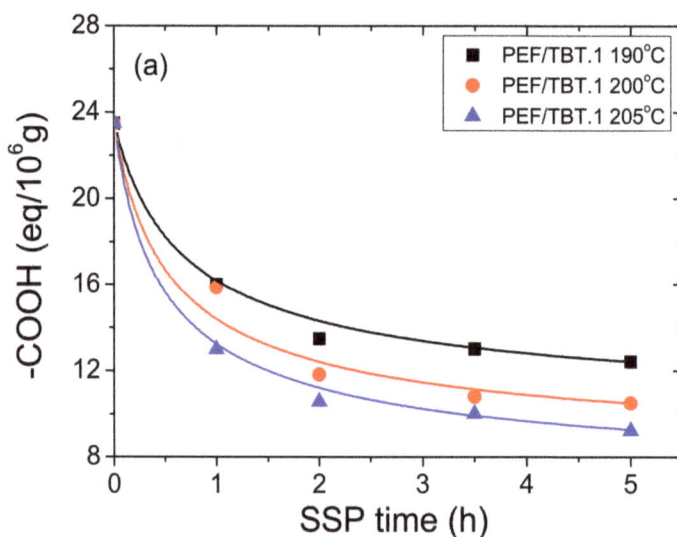

Figure 3. *Cont.*

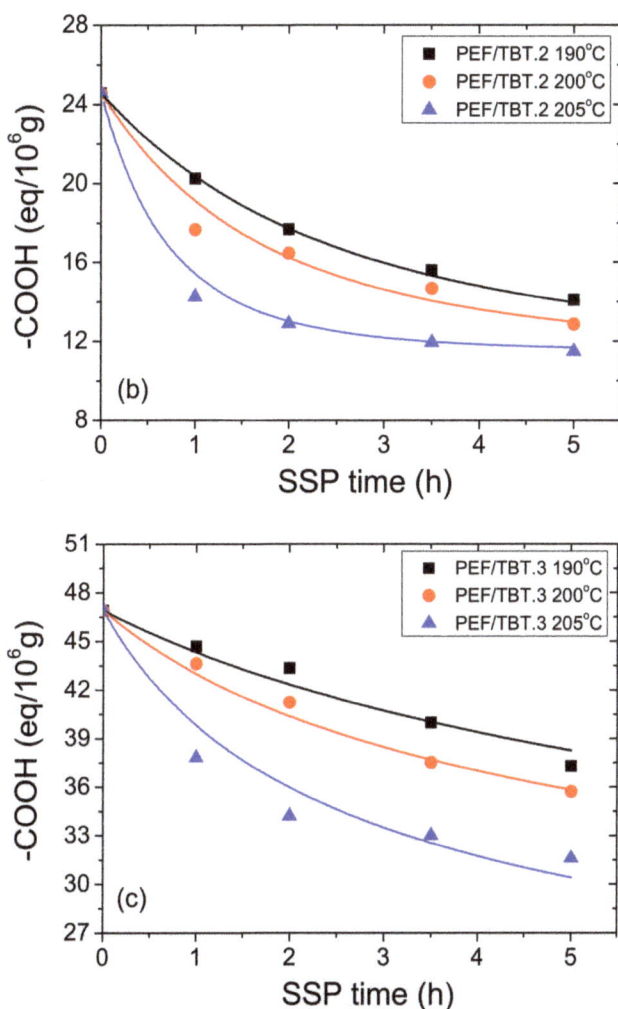

Figure 3. Variation of carboxyl end-groups with time during PEF/TBT.1 (**a**); PEF/TBT.2 (**b**); and PEF/TBT.3 (**c**) SSP at different temperatures. Continuous lines represent the theoretical kinetic model simulation results.

Furthermore, in order to provide results on the kinetic rate constants of the esterification and polycondensation reactions, the theoretical kinetic model presented in Section 3, was employed. Differential Equations (7) and (8) were solved numerically together with Equations (2), (9)–(11), and IV values, as well as the concentration of hydroxyl and carboxyl end-groups were obtained as a function of SSP time. The best-fit values of the parameters k_1, k_2, $[OH]_i$, and $[COOH]_i$ were estimated using the experimental data presented in Figures 2–4 for all PEF/TBT samples at all temperatures. Optimized values are illustrated in Table 2. Results of the theoretical simulation curves are included as continuous lines in the abovementioned figures. As it can be seen, theoretical simulation curves follow satisfactorily the experimental data at all different temperatures. Slight discrepancies could be attributed to the assumptions made, when developing the simple kinetic model. From an inspection of the values reported in Table 2, it seems that the values of the transesterification rate constant, k_1, are lower than that of the esterification rate constant, k_2, for PEF/TBT.1 and PEF/TBT.2, following the concentration of –COOH, which is lower compared to that of –OH for these polyesters. However, in PEF/TBT.3, where both concentrations are similar, the model also resulted in similar k_1 and k_2 values. Decreased k_2 estimated for the PEF/TBT.3 compared to the other polyesters is a direct consequence of the restricted mobility of reactive end-groups (e.g., hydroxyl and carboxyl) to come into close proximity and react, which is facilitated by the polyester's higher molecular weight.

In addition, from Table 2, it was estimated that the best fit value for the hydroxyl inactive groups, $[OH]_i$ (meaning those which are inaccessible to react), are always lower in PEF/TBT.3 compared to the other polyesters, while they are always reduced with increasing temperature. An increase in temperature improves the mobility of the polymer chains, and thus, reduces the number of inactive end species. The lower values in PEF/TBT.3 is a direct consequence of the always lower –OH end-group concentration measured at all reaction times and temperatures compared to other polyesters. Concerning the inactive $[COOH]_i$, the values are always low enough for all polyesters.

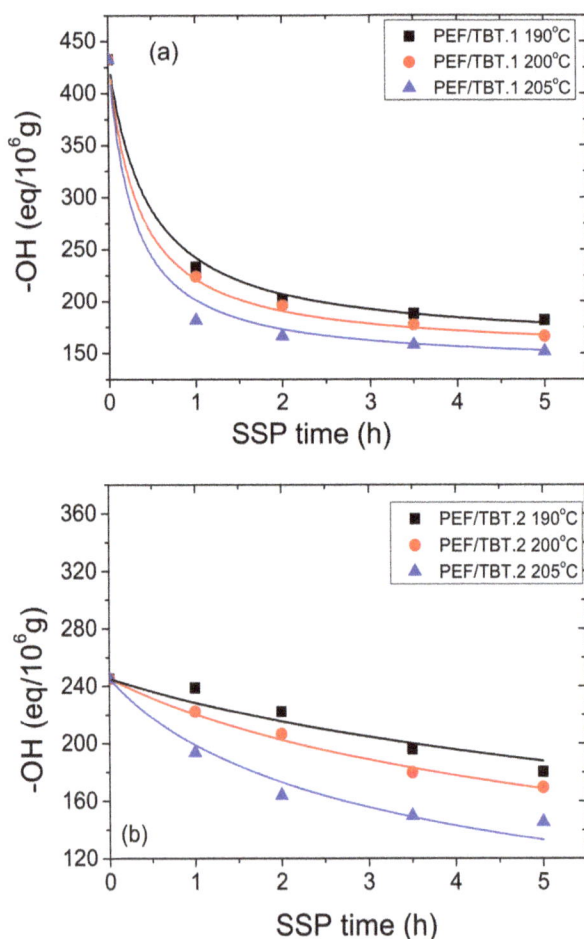

Figure 4. *Cont.*

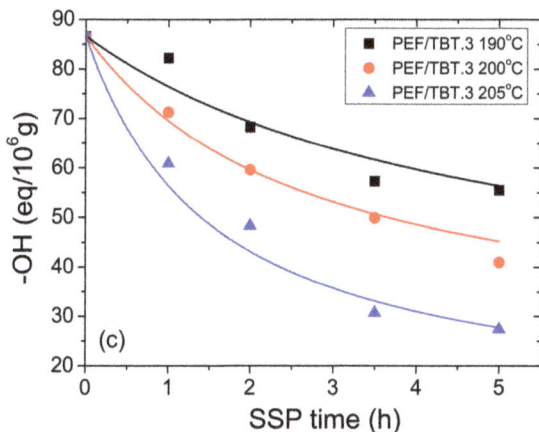

Figure 4. Variation of hydroxyl end-groups with time during PEF/TBT.1 (**a**), PEF/TBT.2 (**b**), and PEF/ TBT.3 (**c**) SSP at different temperatures. Continuous lines represent the theoretical kinetic model simulation results.

Table 2. Kinetic rate constants of the transesterification and esterification reaction and concentration of temporarily inactivated OH and COOH end-groups at different polycondensation temperatures of PEF/TBT.1, PEF/TBT.2, and PEF/TBT.3.

Sample	Temperature (°C)	$k_1 \cdot$(kg/meq)$\cdot$h^{-1}	$k_2 \cdot$(kg/meq)$\cdot$h^{-1}	[OH]$_i$ (meq/kg)	[COOH]$_i$ (meq/kg)
PEF/TBT.1	190	39×10^{-4}	51×10^{-4}	157	9.5
	200	50×10^{-4}	64×10^{-4}	150	7.5
	205	59×10^{-4}	71×10^{-4}	138	6.0
PEF/TBT.2	190	2.0×10^{-4}	22×10^{-4}	60	11.5
	200	3.2×10^{-4}	31×10^{-4}	58	11.5
	205	6.6×10^{-4}	72×10^{-4}	52	11.5
PEF/TBT.3	190	15×10^{-4}	14×10^{-4}	31	9.0
	200	25×10^{-4}	21×10^{-4}	26	9.0
	205	36×10^{-4}	37×10^{-4}	13	9.0

Finally, both kinetic rate constants were correlated with temperature using an Arrhenius-type expression. As expected, the values of all rate constants increase with SSP temperature, in accordance with the mobility and activity of the chain ends. When plotting ln(k) vs 1/T, good straight lines were obtained with a correlation coefficient greater than 0.90. From the slope of these straight lines, the activation energies for the transesterification, E_1, and esterification, E_2, reactions were determined (Table 3). It should be noted that the estimation of the activation energies using only 3 experimental data points (at the three investigated temperatures) results in a somewhat great uncertainty in the values denoted by their high standard deviation. Thus, a safe conclusion concerning the activation energies cannot be set.

Table 3. Activation energies and correlation coefficients of the transesterification and esterification reaction of all PEF/TBT polyesters.

Sample	E_1 (kJ/mol)	R^2	E_2 (kJ/mol)	R^2
PEF/TBT.1	50.0 ± 4.0	0.996	40.7 ± 0.6	0.999
PEF/TBT.2	137.4 ± 11.4	0.950	133.1 ± 15.1	0.907
PEF/TBT.3	105.3 ± 9.7	0.979	112.5 ± 24.1	0.895

4.2. Thermal Analysis of Solid-State Polymerization PEF Polyester Samples

Differential scanning calorimetric (DSC) measurements showed that SSP strongly affects the thermal properties of the PEF samples. The results of the melting behavior of the PEF samples at different temperatures and reaction times are provided in Figure 5 (as well as in Figures S1 and S2 in the Supporting Information). For all samples, increasing either the SSP time or the SSP temperature shifts the endothermic melting peaks to a higher temperature, with an accompanying increase of the crystallinity. The increased molecular weight of the polyester produced during SSP procedure is responsible for the increase in the melting points, as well as the increase in the sharpness of the melting peaks.

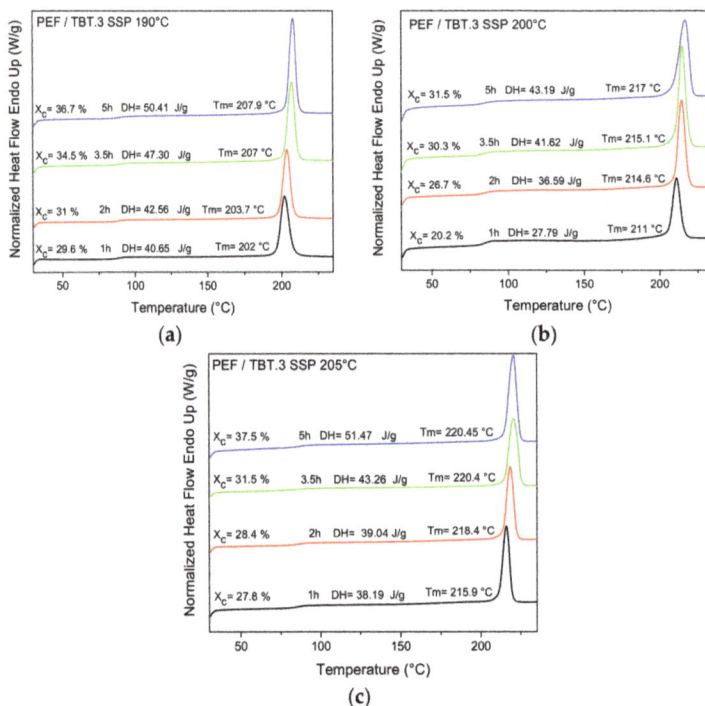

Figure 5. DSC thermograms of PEF/TBT.3 samples prepared after SSP at different temperatures and times (**a**) 190 °C, (**b**) 200 °C, and (**c**) 205 °C.

Table 4 shows the degree of crystallinity values (X_c) for all SSP PEF samples. The latter were calculated from measured melting enthalpy (ΔH_m) using the heat of fusion value for the pure crystalline PEF polyester found in previous report to be about 137 J·g^{-1} [13]. As illustrated in Figure 6, the degree of crystallinity reached the highest value for PEF/TBT.1, and then showed a slightly lower value for PEF/TBT.2, while the polyester PEF/TBT.3 exhibits much lower X_c. Considering though, the fact that the SSP procedure takes place in the amorphous regions, and as the mobility of the diffusion rate and the polymer chain end-groups, which are concentrated only in the amorphous phase of the semi-crystalline polyester, are affected by the crystallinity, it can be inferred that the rapidly increasing molecular weight rate/IV values of PEF/TBT.3, when compared with the other two polyester samples, is due, in fact, to the lowest degree of crystallinity, as presented in Figure 6. This implies that the scape of low molecular weight byproducts got more facile, thus, the increase of molecular weight became faster. This explanation is in good accordance with the M_n/IV values obtained in this study.

Table 4. Degree of crystallinity values (%) of the different SSP PEF/TBT samples.

SSP temperature (°C)	SSP time (h)	PEF/TBT.1	PEF/TBT.2	PEF/TBT.3
	0	23	28.8	2.7
190	1	37.7	38.1	29.6
	2	43.8	38.9	31
	3.5	44	40.9	34.5
	5	47.3	46.2	36.7
200	1	43	39.2	20.2
	2	45.5	40	26.7
	3.5	46.3	40.5	30.3
	5	49.3	48.2	31.5
205	1	46.3	43.3	27.8
	2	48.4	44.2	28.4
	3.5	50.8	44.8	31.5
	5	54.4	45.4	37.5

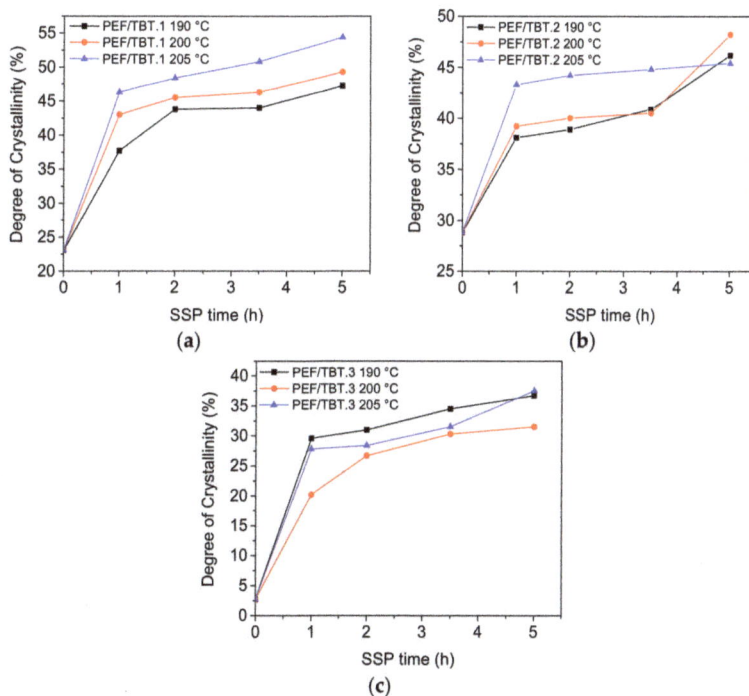

Figure 6. Effect of SSP time and temperature on the evolution of the degree of crystallinity of PEF samples: (**a**) PEF/TBT.1, (**b**) PEF/TBT.2, (**c**) PEF/TBT.3.

5. Conclusions

The present work is, to the best of our knowledge, the first study which investigated the feasibility of SSP after remelting process to synthesize PEF polyesters. Obviously, by introducing a remelting step in the SSP procedure, the latter was found to be a very efficient method that leads to the production of PEF with very high molecular weight appropriate for food packaging applications. The effect of the remelting on the SSP kinetics of PEF was investigated at several temperatures, both experimentally and using a simple kinetic model. As expected, the average molecular weight and the intrinsic viscosity of PEF increased as SSP time and temperature increase. This is because the elimination of formed byproducts during both esterification and transesterification reactions that are occurring during SSP, are diffusion controlled. A simple kinetic model was also developed, and used to predict the time evolution of polyesters' IV, as well as the hydroxyl and carboxyl content during the SSP of PEF. From both the theoretical simulation results and the experimental measurements, it has been demonstrated that the introduction of remelting process in SSP procedure resulted in similar transesterification and esterification kinetic rate constants, as well as in a higher increase of the polymerization rate, and thus, the obtaining of very high molecular weight PEF.

Supplementary Materials: The following are available online at http://www.mdpi.com/2073-4360/10/5/471/s1, Figure S1: DSC thermograms of PEF/TBT.1 samples prepared after SSP at different temperatures and times, Figure S2: DSC thermograms of PEF/TBT.2 samples prepared after SSP at different temperatures and times.

Author Contributions: D.N.B. and G.Z.P. conceived of and designed the experiments and participate to paper writing; N.K. performed the experiments and participate to paper writing; D.S.A. performed all the kinetic model simulations.

Conflicts of Interest: The authors declare no conflict of interest.

References

1. Gandini, A. The irruption of polymers from renewable resources on the scene of macromolecular science and technology. *Green Chem.* **2011**, *13*, 1061–1083. [CrossRef]
2. Meier, M.A.R.; Metzger, J.O.; Schubert, U.S. Plant oil renewable resources as green alternatives in polymer science. *Chem. Soc. Rev.* **2007**, *36*, 1788–1802. [CrossRef] [PubMed]
3. Vilela, C.; Sousa, A.F.; Fonseca, A.C.; Serra, A.C.; Coelho, J.F.J.; Freire, C.S.R.; Silvestre, A.J.D. The quest for sustainable polyesters—Insights into the future. *Polym. Chem.* **2014**, *5*, 3119–3141. [CrossRef]
4. Gandini, A.; Lacerda, T.M. From monomers to polymers from renewable resources: Recent advances. *Prog. Polym. Sci.* **2015**, *48*, 1–39. [CrossRef]
5. Delidovich, I.; Hausoul, P.J.C.; Deng, L.; Pfützenreuter, R.; Rose, M.; Palkovits, R. Alternative Monomers Based on Lignocellulose and Their Use for Polymer Production. *Chem. Rev.* **2016**, *116*, 1540–1599. [CrossRef] [PubMed]
6. Gandini, A.; Lacerda, T.M.; Carvalho, A.J.F.; Trovatti, E. Progress of Polymers from Renewable Resources: Furans, Vegetable Oils, and Polysaccharides. *Chem. Rev.* **2016**, *116*, 1637–1669. [CrossRef] [PubMed]
7. Pellis, A.; Herrero Acero, E.; Gardossi, L.; Ferrario, V.; Guebitz, G.M. Renewable building blocks for sustainable polyesters: New biotechnological routes for greener plastics. *Polym. Int.* **2016**, *65*, 861–871. [CrossRef]
8. Werpy, T.; Petersen, G. *Top Value Added Chemicals from Biomass, Volume I–Results of Screening for Potential Candidates from Sugars and Synthesis Gas*; (DOE/GO-102004–1992); National Renewable Energy Laboratory, US Department of Energy: Springfield, VA, USA, 2004.
9. Thiyagarajan, S.; Genuino, H.C.; van der Waal, J.C.; De Jong, E.; Weckhuysen, B.M.; Van Haveren, J.; Bruijnincx, P.C.; Van Es, D.S. A Facile solid-phase route to renewable aromatic chemicals from biobased furanics. *Angew. Chem. Int. Ed.* **2016**, *55*, 1368–1371. [CrossRef] [PubMed]
10. Esposito, D.; Antonietti, M. Redefining biorefinery: The search for unconventional building blocks for materials. *Chem. Soc. Rev.* **2015**, *44*, 5821–5835. [CrossRef] [PubMed]
11. Vannini, M.; Marchese, P.; Celli, A.; Lorenzetti, C. Fully biobased poly(propylene 2,5-furandicarboxylate) for packaging applications: Excellent barrier properties as a function of crystallinity. *Green Chem.* **2015**, *17*, 4162–4166. [CrossRef]
12. Soares, M.J.; Dannecker, P.-K.; Vilela, C.; Bastos, J.; Meier, M.A.R.; Sousa, A.F. Poly(1,20-eicosanediyl 2,5-furandicarboxylate), a biodegradable polyester from renewable resources. *Eur. Polym. J.* **2017**, *90*, 301–311. [CrossRef]
13. Terzopoulou, Z.; Karakatsianopoulou, E.; Kasmi, N.; Majdoub, M.; Papageorgiou, G.Z.; Bikiaris, D.N. Effect of catalyst type on recyclability and decomposition mechanism of poly(ethylene furanoate) biobased polyester. *J. Anal. Appl. Pyrolysis* **2017**, *126*, 357–370. [CrossRef]

14. Papageorgiou, D.G.; Guigo, N.; Tsanaktsis, V.; Exarhopoulos, S.; Bikiaris, D.N.; Sbirrazzuoli, N.; Papageorgiou, G.Z. Fast crystallization and melting behavior of a long-spaced aliphatic furandicarboxylate bio-based polyester, the poly(dodecylene 2,5-furanoate). *Ind. Eng. Chem. Res.* **2016**, *55*, 5315–5326. [CrossRef]

15. Tsanaktsis, V.; Terzopoulou, Z.; Nerantzaki, M.; Papageorgiou, G.Z.; Bikiaris, D.N. New poly(pentylene furanoate) and poly(heptylene furanoate) sustainable polyesters from diols with odd methylene groups. *Mater. Lett.* **2016**, *178*, 64–67. [CrossRef]

16. Tsanaktsis, V.; Terzopoulou, Z.; Exarhopoulos, S.; Bikiaris, D.N.; Achilias, D.S.; Papageorgiou, D.G.; Papageorgiou, G.Z. Sustainable, eco-friendly polyesters synthesized from renewable resources: Preparation and thermal characteristics of poly(dimethyl-propylene furanoate). *Polym. Chem.* **2015**, *6*, 8284–8296. [CrossRef]

17. Zhu, J.; Cai, J.; Xie, W.; Chen, P.-H.; Gazzano, M.; Scandola, M.; Gross, R.A. Poly(butylene 2,5-furan dicarboxylate), a Biobased Alternative to PBT: Synthesis, Physical Properties, and Crystal Structure. *Macromolecules* **2013**, *46*, 796–804. [CrossRef]

18. Terzopoulou, Z.; Kasmi, N.; Tsanaktsis, V.; Doulakas, N.; Bikiaris, D.N.; Achilias, D.S.; Papageorgiou, G.Z. Synthesis and Characterization of Bio-Based Polyesters: Poly(2-methyl-1,3-propylene-2,5-furanoate), Poly(isosorbide-2,5-furanoate), Poly(1,4-cyclohexane dimethylene-2,5-furanoate). *Materials* **2017**, *10*, 801. [CrossRef] [PubMed]

19. Carlos Morales-Huerta, J.; Martínez De Ilarduya, A.; Muñoz-Guerra, S. Poly(alkylene 2,5-furandicarboxylate)s (PEF and PBF) by ring opening polymerization. *Polymer* **2016**, *87*, 148–158. [CrossRef]

20. De Jong, E.; Dam, M.A.; Sipos, L.; Gruter, G.-J.M. Furandicarboxylic acid (FDCA), A versatile building block for a very interesting class of polyesters. *ACS Symp. Ser.* **2012**, *1105*, 1–13. [CrossRef]

21. Sousa, A.F.; Vilela, C.; Fonseca, A.C.; Matos, M.; Freire, C.S.R.; Gruter, G.-J.M.; Coelho, J.F.J.; Silvestre, A.J.D. Biobased polyesters and other polymers from 2,5-furandicarboxylic acid: A tribute to furan excellence. *Polym. Chem.* **2015**, *6*, 5961–5983. [CrossRef]

22. Papageorgiou, G.Z.; Papageorgiou, D.G.; Terzopoulou, Z.; Bikiaris, D.N. Production of bio-based 2,5-furan dicarboxylate polyesters: Recent progress and critical aspects in their synthesis and thermal properties. *Eur. Polym. J.* **2016**, *83*, 202–229. [CrossRef]

23. Weinberger, S.; Canadell, J.; Quartinello, F.; Yeniad, B.; Arias, A.; Pellis, A.; Guebitz, G.M. Enzymatic Degradation of Poly(ethylene 2,5-furanoate) Powders and Amorphous Films. *Catalysts* **2017**, *7*, 318. [CrossRef]

24. Thiyagarajan, S.; Vogelzang, W.J.I.; Knoop, R.; Frissen, A.E.; Van Haveren, J.; Van Es, D.S. Biobased furandicarboxylic acids (FDCAs): Effects of isomeric substitution on polyester synthesis and properties. *Green Chem.* **2014**, *16*, 1957–1966. [CrossRef]

25. Tsanaktsis, V.; Papageorgiou, D.G.; Exarhopoulos, S.; Bikiaris, D.N.; Papageorgiou, G.Z. Crystallization and Polymorphism of Poly(ethylene furanoate). *Cryst. Growth Des.* **2015**, *15*, 5505–5512. [CrossRef]

26. Burgess, S.K.; Leisen, J.E.; Kraftschik, B.E.; Mubarak, C.R.; Kriegel, R.M.; Koros, W.J. Chain Mobility, Thermal, and Mechanical Properties of Poly(ethylene furanoate) Compared to Poly(ethylene terephthalate). *Macromolecules* **2014**, *47*, 1383–1391. [CrossRef]

27. Eerhart, A.J.J.E.; Faaij, A.P.C.; Patel, M.K. Replacing fossil based PET with biobased PEF; process analysis, energy and GHG balance. *Energy Environ. Sci.* **2012**, *5*, 6407–6422. [CrossRef]

28. Burgess, S.K.; Kriegel, R.M.; Koros, W.J. Carbon dioxide sorption and transport in amorphous poly(ethylene furanoate). *Macromolecules* **2015**, *48*, 2184–2193. [CrossRef]

29. PEF: Game-Changing Plastic. Available online: https://www.avantium.com/yxy/products-applications/ (accessed on 26 March 2015).

30. Steinborn-Rogulska, I.; Rokicki, G. Solid-state polycondensation (SSP) as a method to obtain high molecular weight polymers. Part II. Synthesis of polylactide and polyglycolide via SSP. *Polimery* **2013**, *58*, 85–92. [CrossRef]

31. Vouyiouka, S.N.; Karakatsani, E.K.; Papaspyrides, C.D. Solid state polymerization. *Prog. Polym. Sci.* **2005**, *30*, 10–37. [CrossRef]

32. Papaspyrides, C.D.; Vouyiouka, S.N. *Solid State Polymerization*, 1st ed.; John Wiley & Sons, Inc.: Hoboken, NJ, USA, 2009; pp. 1–294. ISBN 9780470084182.

33. Zhang, J.; Shen, X.-J.; Zhang, J.; Feng, L.-F.; Wang, J.-J. Experimental and modeling study of the solid state polymerization of poly(ethylene terephthalate) over a wide range of temperatures and particle sizes. *J. Appl. Polym. Sci.* **2013**, *127*, 3814–3822. [CrossRef]

34. Li, L.-J.; Duan, R.-T.; Zhang, J.-B.; Wang, X.-L.; Chen, L.; Wang, Y.-Z. Phosphorus-Containing Poly(ethylene terephthalate): Solid-State Polymerization and Its Sequential Distribution. *Ind. Eng. Chem. Res.* **2013**, *52*, 5326–5333. [CrossRef]

35. Gantillon, B.; Spitz, R.; McKenna, T.F. The Solid State Postcondensation of PET, 1: A Review of the Physical and Chemical Processes Taking Place in the Solid State. *Macromol. Mater. Eng.* **2004**, *289*, 88–105. [CrossRef]

36. Karayannidis, G.P.; Kokkalas, D.E.; Bikiaris, D.N. Solid-state polycondensation of poly(ethylene terephthalate) recycled from postconsumer soft-drink bottles. II. *J. Appl. Polym. Sci.* **1995**, *56*, 405–410. [CrossRef]

37. Bikiaris, D.; Karavelidis, V.; Karayannidis, G. A New Approach to Prepare Poly(ethylene terephthalate)/Silica Nanocomposites with Increased Molecular Weight and Fully Adjustable Branching or Crosslinking by SSP. *Macromol. Rapid Commun.* **2006**, *27*, 1199–1205. [CrossRef]

38. Achilias, D.S.; Bikiaris, D.N.; Karavelidis, V.; Karayannidis, G.P. Effect of silica nanoparticles on solid state polymerization of poly(ethylene terephthalate). *Eur. Polym. J.* **2008**, *44*, 3096–3107. [CrossRef]

39. Van Berkel, J.G.; Guigo, N.; Kolstad, J.J.; Sbirrazzuoli, N. Biaxial Orientation of Poly(ethylene 2,5-furandicarboxylate): An Explorative Study. *Macromol. Mater. Eng.* **2018**, *303*, 1700507. [CrossRef]

40. Burgess, S.K.; Mubarak, C.R.; Kriegel, R.M.; Koros, W.J. Physical aging in amorphous poly(ethylene furanoate): Enthalpic recovery, density, and oxygen transport considerations. *J. Polym. Sci. Part B* **2015**, *53*, 389–399. [CrossRef]

41. Guigo, N.; van Berkel, J.; de Jong, E.; Sbirrazzuoli, N. Modelling the non-isothermal crystallization of polymers: Application to poly(ethylene 2,5-furandicarboxylate). *Thermochim. Acta* **2017**, *650*, 66–75. [CrossRef]

42. Codou, A.; Moncel, M.; Van Berkel, J.G.; Guigo, N.; Sbirrazzuoli, N. Glass transition dynamics and cooperativity length of poly(ethylene 2,5-furandicarboxylate) compared to poly(ethylene terephthalate). *Phys. Chem. Chem. Phys.* **2016**, *18*, 16647–16658. [CrossRef] [PubMed]

43. Burgess, S.K.; Mikkilineni, D.S.; Yu, D.B.; Kim, D.J.; Mubarak, C.R.; Kriegel, R.M.; Koros, W.J. Water sorption in poly(ethylene furanoate) compared to poly(ethylene terephthalate). Part 1: Equilibrium sorption. *Polymer* **2014**, *55*, 6861–6869. [CrossRef]

44. Van Berkel, J.G.; Guigo, N.; Kolstad, J.J.; Sipos, L.; Wang, B.; Dam, M.A.; Sbirrazzuoli, N. Isothermal Crystallization Kinetics of Poly(Ethylene 2,5-Furandicarboxylate). *Macromol. Mater. Eng.* **2014**, *300*, 466–474. [CrossRef]

45. Codou, A.; Guigo, N.; Van Berkel, J.; De Jong, E.; Sbirrazzuoli, N. Non-isothermal Crystallization Kinetics of Biobased Poly(ethylene 2,5-furandicarboxylate) Synthesized via the Direct Esterification Process. *Macromol. Chem. Phys.* **2014**, *215*, 2065–2074. [CrossRef]

46. Burgess, S.K.; Mikkilineni, D.S.; Yu, D.B.; Kim, D.J.; Mubarak, C.R.; Kriegel, R.M.; Koros, W.J. Water sorption in poly(ethylene furanoate) compared to poly(ethylene terephthalate). Part 2: Kinetic sorption. *Polymer* **2014**, *55*, 6870–6882. [CrossRef]

47. Stoclet, G.; Gobius Du Sart, G.; Yeniad, B.; De Vos, S.; Lefebvre, J.M. Isothermal crystallization and structural characterization of poly(ethylene-2,5-furanoate). *Polymer* **2015**, *72*, 165–176. [CrossRef]

48. Maini, L.; Gigli, M.; Gazzano, M.; Lotti, N.; Bikiaris, D.N.; Papageorgiou, G.Z. Structural investigation of poly(ethylene furanoate) polymorphs. *Polymers* **2018**, *10*, 296. [CrossRef]

49. Papageorgiou, G.Z.; Tsanaktsis, V.; Bikiaris, D.N. Synthesis of poly(ethylene furandicarboxylate) polyester using monomers derived from renewable resources: Thermal behavior comparison with PET and PEN. *Phys. Chem. Chem. Phys.* **2014**, *16*, 7946–7958. [CrossRef] [PubMed]

50. Pellis, A.; Haernvall, K.; Pichler, C.M.; Ghazaryan, G.; Breinbauer, R.; Guebitz, G.M. Enzymatic hydrolysis of poly(ethylene furanoate). *J. Biotechnol.* **2016**, *235*, 47–53. [CrossRef] [PubMed]

51. Weinberger, S.; Haernvall, K.; Scaini, D.; Ghazaryan, G.; Zumstein, M.T.; Sander, M.; Pellis, A.; Guebitz, G.M. Enzymatic surface hydrolysis of poly(ethylene furanoate) thin films of various crystallinities. *Green Chem.* **2017**, *19*, 5381–5384. [CrossRef]

52. Stoclet, G.; Lefebvre, J.M.; Yeniad, B.; Gobius du Sart, G.; De Vos, S. On the strain-induced structural evolution of Poly(ethylene-2,5-furanoate) upon uniaxial stretching: An in-situ SAXS-WAXS study. *Polymer* **2018**, *134*, 227–241. [CrossRef]

53. Lotti, N.; Munari, A.; Gigli, M.; Gazzano, M.; Tsanaktsis, V.; Bikiaris, D.N.; Papageorgiou, G.Z. Thermal and structural response of in situ prepared biobased poly(ethylene 2,5-furan dicarboxylate) nanocomposites. *Polymer* **2016**, *103*, 288–298. [CrossRef]

54. Jiang, M.; Liu, Q.; Zhang, Q.; Ye, C.; Zhou, G. A Series of Furan-Aromatic Polyesters Synthesized via Direct Esterification Method Based on Renewable Resources. *J. Polym. Sci. Part A* **2012**, *50*, 1026–1036. [CrossRef]

55. Rosenboom, J.-G.; Roo, J.D.; Storti, G.; Morbidelli, M. Diffusion (DOSY) ^{1}H NMR as an Alternative Method for Molecular Weight Determination of Poly(ethylene furanoate) (PEF) Polyesters. *Macromol. Chem. Phys.* **2017**, *218*, 1600436. [CrossRef]

56. Gomes, M.; Gandini, A.; Silvestre, A.J.D.; Reis, B. Synthesis and Characterization of Poly(2,5-furan dicarboxylate)s Based on a Variety of Diols. *J. Polym. Sci. Part A* **2011**, *49*, 3759–3768. [CrossRef]

57. Gruter, G.-J.M.; Sipos, L.; Dam, M.A. Accelerating Research into Bio-Based FDCA-Polyesters by Using Small Scale Parallel Film Reactors. *Comb. Chem. High Throughput Screen.* **2012**, *15*, 180–188. [CrossRef] [PubMed]

58. Kasmi, N.; Majdoub, M.; Papageorgiou, G.Z.; Achilias, D.S.; Bikiaris, D.N. Solid-State Polymerization of Poly(ethylene furanoate) Biobased Polyester, I: Effect of Catalyst Type on Molecular Weight Increase. *Polymers* **2017**, *9*, 607. [CrossRef]

59. Terzopoulou, Z.; Karakatsianopoulou, E.; Kasmi, N.; Tsanaktsis, V.; Nikolaidis, N.; Kostoglou, M.; Papageorgiou, G.Z.; Lambropoulou, D.A.; Bikiaris, D.A. Effect of catalyst type on molecular weight increase and coloration of poly(ethylene furanoate) biobased polyester during melt polycondensation. *Polym. Chem.* **2017**, *8*, 6895–6908. [CrossRef]

60. Achilias, D.S.; Chondroyiannis, A.; Nerantzaki, M.; Adam, K.-V.; Terzopoulou, Z.; Papageorgiou, G.Z.; Bikiaris, D.N. Solid State Polymerization of Poly(Ethylene Furanoate) and Its Nanocomposites with SiO_2 and TiO_2. *Macromol. Mater. Eng.* **2017**, *302*, 1700012. [CrossRef]

61. Sipos, L. A Process for Preparing a Polymer having a 2,5-Furandicarboxylate Moiety within the Polymer Backbone and Such (Co)Polymers, (Furanix Technologies B.V.). W.O. Patent 2010/077133 A1, 8 July 2010.

62. Knoop, R.J.I.; Vogelzang, W.; Van Haveren, J.; Van Es, D.S. High molecular weight poly(ethylene-2,5-furanoate); critical aspects in synthesis and mechanical property determination. *J. Polym. Sci. Part A* **2013**, *51*, 4191–4199. [CrossRef]

63. Hong, S.; Min, K.-D.; Nam, B.-U.; Park, O.O. High molecular weight bio furan-based co-polyesters for food packaging applications: Synthesis, characterization and solid-state polymerization. *Green Chem.* **2016**, *18*, 5142–5150. [CrossRef]

64. Duh, B. Reaction Kinetics for Solid-State Polymerization of Poly(ethylene terephthalate). *J. Appl. Polym. Sci.* **2001**, *81*, 1748–1761. [CrossRef]

65. Weissmann, D. PET Use in Blow Molded Rigid Packaging. In *Applied Plastics Engineering Handbook*, 1st ed.; Kutz, M., Ed.; William Andrew: Waltham, MA, USA, 2011; pp. 603–623. ISBN 978-143773514-7.

66. Gupta, V.B.; Bashir, Z. PET Fibers, Films, and Bottles. In *Handbook of Thermoplastic Polyesters: Homopolymers, Copolymers, Blends and Composites*, 1st ed.; Fakirov, S., Ed.; Wiley-VCH Verlag GmbH & Co. KGaA: Weinheim, Germany, 2002; pp. 317–361, ISBN 9783527301133.

67. Ros-Chumillas, M.; Belissario, Y.; Iguaz, A.; López, A. Quality and shelf life of orange juice aseptically packaged in PET bottles. *J. Food Eng.* **2007**, *79*, 234–242. [CrossRef]

68. Berkowitz, S. Viscosity–molecular weight relationships for poly(ethylene terephthalate) in hexafluorois opropanol–pentafluorophenol using SEC–LALLS. *J. Appl. Polym. Sci.* **1984**, *29*, 4353–4361. [CrossRef]

69. Karayannidis, G.P.; Kokkalas, D.E.; Bikiaris, D.N. Solid-state polycondensation of poly(ethylene terephthalate) recycled from postconsumer soft-drink bottles. I. *J. Appl. Polym. Sci.* **1993**, *50*, 2135–2142. [CrossRef]

70. Pohl, H.A. Determination of carboxyl end groups in a polyester, polyethylene terephthalate. *Anal. Chem.* **1954**, *26*, 1614–1616. [CrossRef]

71. Ravindranath, K.; Mashelkar, R.A. Modeling of Poly(ethylene Terephthalate) Reactors. I. A Semibatch Ester Interchange Reactor. *J. Appl. Polym. Sci.* **1981**, *26*, 3179–3204. [CrossRef]

72. Mallon, F.K.; Ray, W.H. Modeling of solid-state polycondensation. II. Reactor design issues. *J. Appl. Polym. Sci.* **1998**, *69*, 1775–1788. [CrossRef]

73. Ma, Y.; Agarwal, U.S.; Sikkema, D.J.; Lemstra, P.J. Solid-state polymerization of PET: Influence of nitrogen sweep and high vacuum. *Polymer* **2003**, *44*, 4085–4096. [CrossRef]

74. Ma, Y.; Agarwal, U.S. Solvent assisted post-polymerization of PET. *Polymer* **2005**, *46*, 5447–5455. [CrossRef]

75. Bikiaris, D.N.; Achilias, D.S.; Giliopoulos, D.J.; Karayannidis, G.P. Effect of activated carbon black nanoparticles on solid state polymerization of poly(ethylene terephthalate). *Eur. Polym. J.* **2006**, *42*, 3190–3201. [CrossRef]

76. Achilias, D.S.; Karandrea, E.; Triantafyllidis, K.S.; Ladavos, A.; Bikiaris, D.N. Effect of organoclays type on solid-state polymerization (SSP) of poly(ethylene terephthalate): Experimental and modelling. *Eur. Polym. J.* **2015**, *63*, 156–167. [CrossRef]

77. Achilias, D.S.; Gerakis, K.; Giliopoulos, D.J.; Triantafyllidis, K.S.; Bikiaris, D.N. Effect of high surface area mesoporous silica fillers (MCF and SBA-15) on solid state polymerization of PET. *Eur. Polym. J.* **2016**, *81*, 347–364. [CrossRef]

78. Kaushik, A.; Gupta, S.K. A molecular model for solid-state polymerization of nylon 6. *J. Appl. Polym. Sci.* **1992**, *45*, 507–520. [CrossRef]

79. Kulkarni, M.R.; Gupta, S.K. Molecular Model for Solid-state Polymerization of Nylon 6. II. An Improved Model. *J. Appl. Polym. Sci.* **1994**, *53*, 85–103. [CrossRef]

80. Gaymans, R.J.; Amirtharaj, J.; Kamp, H. Nylon 6 Polymerization in the Solid State. *J. Appl. Polym. Sci.* **1982**, *27*, 2513–2526. [CrossRef]

81. Li, L.F.; Huang, N.X.; Tang, Z.L.; Hagen, R. Reaction kinetics and simulation for the solid-state polycondensation of nylon 6. *Macromol. Theory Simul.* **2001**, *10*, 507–517. [CrossRef]

82. Duh, B. Semiempirical rate equation for solid state polymerization of poly(ethylene terephthalate). *J. Appl. Polym. Sci.* **2002**, *84*, 857–870. [CrossRef]

83. Wu, D.; Chen, F.; Li, R.; Shi, Y. Reaction kinetics and simulations for solid-state polymerization of poly(ethylene terephthalate). *Macromolecules* **1997**, *30*, 6737–6742. [CrossRef]

© 2018 by the authors. Licensee MDPI, Basel, Switzerland. This article is an open access article distributed under the terms and conditions of the Creative Commons Attribution (CC BY) license (http://creativecommons.org/licenses/by/4.0/).

![polymers logo] *polymers*

MDPI

Article

Structural Investigation of Poly(ethylene furanoate) Polymorphs

Lucia Maini [1], Matteo Gigli [2,3], Massimo Gazzano [4,*], Nadia Lotti [3], Dimitrios N. Bikiaris [5] and George Z. Papageorgiou [6]

[1] Department of Chemistry "G. Ciamician", Via Selmi 2, University of Bologna, 40126 Bologna, Italy; l.maini@unibo.it
[2] Department of Chemical Science and Technologies, University of Roma Tor Vergata, Via della Ricerca Scientifica 1, 00133 Roma, Italy; matteo.gigli@uniroma2.it
[3] Civil, Chemical, Environmental and Materials Engineering Department, University of Bologna, Via Terracini 28, 40131 Bologna, Italy; nadia.lotti@unibo.it
[4] Organic Synthesis and Photoreactivity Institute, ISOF-CNR, Via Gobetti 101, 40129 Bologna, Italy
[5] Laboratory of Polymer Chemistry and Technology, Department of Chemistry, Aristotle University of Thessaloniki, Thessaloniki 54124, Greece; dbic@chem.auth.gr
[6] Chemistry Department, University of Ioannina, P.O. Box 1186, Ioannina 45110, Greece; gzpap@cc.uoi.gr
*   Correspondence: massimo.gazzano@cnr.it; Tel.: +39-051-2099552

Received: 26 January 2018; Accepted: 7 March 2018; Published: 9 March 2018

Abstract: α and β crystalline phases of poly(ethylene furanoate) (PEF) were determined using X-ray powder diffraction by structure resolution in direct space and Rietveld refinement. Moreover, the α' structure of a PEF sample was refined from data previously reported for PEF fiber. Triclinic α-PEF a = 5.729 Å, b = 7.89 Å, c = 9.62 Å, α = 98.1°, β = 65.1°, γ = 101.3°; monoclinic α'-PEF a = 5.912 Å, b = 6.91 Å, c = 19.73 Å, α = 90°, β = 90°, γ = 104.41°; and monoclinic β-PEF a = 5.953 Å, b = 6.60 Å, c = 10.52 Å, α = 90°, β = 107.0°, γ = 90° were determined as the best fitting of X-ray diffraction (XRD) powder patterns. Final atomic coordinates are reported for all polymorphs. In all cases PEF chains adopted an almost planar configuration.

Keywords: poly(ethylene furanoate); PEF; 2,5-furan dicarboxylate; crystal structure; polymorphism

1. Introduction

In recent years, in view of a greener and more sustainable economy, many efforts from both academic and industrial sectors have been devoted to the development of bio-based alternatives to fossil-based plastics. Bioplastic production is expected to increase from the actual 4.2 million tons to over 6.1 million tons in 2021, thus highlighting a very fast growing rate [1].

Among different renewable starting materials that have been used for the preparation of bioplastics, furan-based monomers have attracted considerable attention. In particular, 2,5-furandicarboxylic acid (FDCA), whose initial diffusion was hampered by the difficulty to produce large amounts with high purity, has been lately the object of much research [2,3]. Currently, FDCA can be readily obtained from the oxidation of hydroxymethylfurfural, in turn derived from the dehydration of (poly)saccharides or by exploiting new synthetic routes, such as through 2-furoic acid and CO_2 [4–6]. From this framework, in October 2016, the Dutch company Avantium announced the establishment of a new joint venture with BASF corporation, named Synvina, for the large-scale production and marketing of FDCA [7].

The reason behind the success of FDCA mostly lies in its use for the synthesis of poly(ethylene-2,5-furanoate) (PEF), considered the most credible bio-based alternative to poly(ethylene terephthalate) (PET), thanks to its very interesting physical/mechanical and barrier properties.

PEF can be processed with outstanding results into films, fibers, and, above all, bottles for beverage packaging. From a comparison between the barrier properties of PEF and PET, the following outcomes emerged: PEF exhibited 11× reduction in oxygen permeability [8], 19× reduction in carbon dioxide permeability [9], and 5× lower water diffusion coefficient [10] compared to PET. In addition, PEF displays more attractive thermal and mechanical properties than PET: higher T_g (85 °C vs. 76 °C), lower T_m (211 °C vs. 247 °C) [11], and 1.6× higher Young's modulus [2]. Lastly, the production of PEF would decrease non-renewable energy use by about 40–50% and greenhouse gas emissions by ca. 45–55% with respect to PET [12].

Several papers in the literature have been devoted to PEF characterization [10,13,14]. The relationships between the thermal and structural properties of PEF have been investigated to explain the complex behavior exhibited during isothermal crystallization. Two phases, called α' and α, were identified by Stoclet et al. [15]. The influence of different experimental conditions on the crystallization, stability, and transformation of PEF polymorphs were reported by Tsanatkis et al.; they also described a new β-phase obtained after solvent crystallization [16,17].

Nevertheless, the crystal structures of PEF polymorphs have yet to be defined, and further investigation is therefore necessary. Indeed, the triclinic cell proposed in an early study [18] with dimensions of a = 5.75 Å, b = 5.35 Å, c = 20.1 Å, α = 113.5°, β = 90°, and γ = 112° is not compatible with the data reported by all other authors whatever PEF phase is considered, as it would cause a wrong positioning of the main peaks in the XRD pattern. Although in a recent work by Mao et al. [19] the high-resolution structure of a PEF fiber is documented, a comparison between the powder and fiber data is needed. During the review process of this paper, the structural evolution of PEF upon uniaxial stretching was deeply investigated. Evidence of a mesomorphic phase at low stretching temperatures and of a crystalline phase, different from the thermally induced ones, was reported [20]. Therefore, the main aim of the present work is to shed light on the crystal structures of α-, α'-, and β-PEF obtained during usual PEF preparations, with no need for fibers or stretching conditions, from powder pattern X-ray diffraction. The study of structure-property relationships is indeed of crucial importance for the optimization of the end-use behavior of a polymeric material.

2. Materials and Methods

2.1. Materials and Sample Preparation

2,5-furan dicarboxylic acid (purum 97%), ethylene glycol, and tetrabutyl titanate catalyst of analytical grade were purchased from Sigma-Aldrich Co. (St. Louis, MO, USA) All other materials and solvents used were of analytical grade. Poly(ethylene furandicarboxylate) polyester was synthesized as previously reported by the two-stage melt polycondensation method (esterification and polycondensation) in a glass batch reactor [21,22].

The PEF sample displays an intrinsic viscosity value 0.45 dL/g. Its weight- and number-average molecular weight (M_w, M_n), measured by gel permeation chromatography (GPC) apparatus equipped with differential refractometer as detector (Waters Inc., Milford, MA, USA), are respectively M_w = 24,640 g/mol and M_n = 11,200 g/mol (Polydispersity Index, PDI = 2.2). To attain a high degree of crystallinity and to favor the selective formation of pure crystalline phases, α-PEF was obtained by annealing an "as-synthesized" sample at 195 °C for 24 h, α'-PEF was obtained by isothermal crystallization at 150 °C of a melt-quenched sample, and β-PEF was obtained by slow solvent evaporation at room temperature after dissolution in a trifluoroacetic acid/chlorofom 1/5 *v/v* mixture and precipitation in cold methanol. An overall sample preparation procedure is reported in Scheme 1.

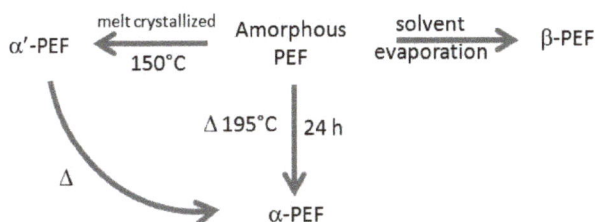

Scheme 1. Preparation of different crystal forms of poly(ethylene-2,5-furanoate) PEF.

2.2. Diffraction and Infrared Measurements

X-ray diffraction (XRD) patterns were collected in Bragg-Brentano geometry with a flat sample holder, over the 2θ range 3°–80° (40 kW–40 mA; Cu-Kα radiation λ = 1.5418 Å, step size 0.05°, 1500 s/step) on a X'Pert PRO automated diffractometer (Panalytical, Almelo, The Netherlands) equipped with a fast X'Celerator detector. Crystallinity was determined as $X_c = A_c/A_{tot}$, where A_c represents the integrated crystalline scattering and A_{tot} the integrated total scattering, both crystalline and amorphous; air and incoherent scattering were taken in due consideration. Attenuated Total Reflectance Fourier Transform Infrared (ATR-FTIR) spectra were recorded on an Alpha FT IR spectrometer (Bruker Optik GmbH, Ettlingen, Germany) with a platinum ATR single reflection diamond module. The background spectrum of air was collected before the acquisition of each sample spectrum. Spectra were recorded with a resolution of 2 cm^{-1}, and 64 scans were averaged for each spectrum (scan range 4000–450 cm^{-1}).

2.3. Structure Determinations

Topas 5 software package (Coelho Software, Brisbane, Australia) was used for indexing and structure determination in direct space and Rietveld refinement. A Pawely refinement was performed with the best cells, and the best one for each phase was chosen for structure solution. It was run in direct space by a simulated annealing algorithm. During the solution, the two torsional angles of the monomer were allowed to move freely. The background was described by the Chebyshev function with four parameters, while the amorphous content was modeled with two peaks centered at 2θ = 22.1° and 45.8° based on a fully amorphous PEF profile, and suitably scaled on each pattern of α-, α'-, β-PEF, respectively. The peak profile was described by a combination of Gaussian and Lorentzian functions. Microstructural parameters are reported in Table S1 (Supplementary Materials).

3. Results and Discussion

3.1. Sample Characterization

To the best of our knowledge, PEF displays three different XRD patterns [16,17]. Various experimental conditions were screened to obtain highly crystalline samples showing pure crystal phases. The diffraction patterns of the samples, used for structure resolution and refinements, are reported in Figure 1 together with the profile of an amorphous 'as-synthesized' sample. The degrees of crystallinity, as measured by XRD, are 47, 45, and 27% for α, α', and β samples, respectively. α and α' profiles are very similar, although α-PEF shows sharper reflections and an extra peak at 19.3° (2θ). On the other hand, β-PEF exhibits five main reflections roughly positioned at the same angular values as α'-PEF. However, they are broader and display completely different relative intensities. Lastly, an additional low-intensity reflection is detectable at 9.5°.

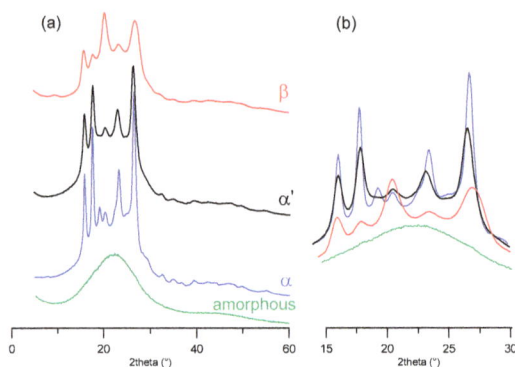

Figure 1. X-ray diffraction (XRD) patterns of PEF: (**a**) from top to bottom: β-, α'-, α-phase and amorphous sample; (**b**) superposition of the patterns in the range 15°–30°.

The Temperature Modulated Differential Scanning Calorimetry (TMDSC) scans reported in Figure S1 show the presence of narrow melting peaks for the solution crystallized sample (β crystals) and for the sample crystallized at 195 °C (α crystals). In particular, in the non-reversing signal curves, only a large non-reversing melting appeared, indicating the high perfection and thermal stability of the crystals, which do not suffer any recrystallization/reorganization upon heating. This result is consistent with the highly stable unique crystalline phases. For the sample crystallized at 170 °C a small recrystallization peak appeared, but only after non-reversing melting and at a very high temperature, just before the end of melting. On the other hand, PEF crystallized at 150 °C, showed a broad melting. A recrystallization exothermic peak was also observed above 170 °C, in the non-reversing signal curve, as well as in the case of the melt-quenched sample. This behavior indicates that crystal perfection occurred upon heating, due to the poor nature of the original crystals. This behavior is consistent with the transition from the less perfect α' to the α crystal phase at 170 °C, as already reported in the literature [15–17], Scheme 1. The ATR-FTIR spectra analysis, reported in Figure S2, confirm the presence of the right functional groups of the polymer (comparisons in Figure S3), but do not provide any specific insight about the solid phase, since no particular differences were detected in the bands position. A careful analysis of XRD data can give further confidence about the phase purity. As can be seen in Figure 1b, the peak at 19.3° (2θ) can be taken as a α-phase marker because at this angular value both of the other two samples display no peaks, thus indicating a complete absence of the α-phase. Similarly, a verification of the base line intensity around 9.5° in the α'-, α-samples excludes the presence of the β-phase because of the detection of a flat background.

3.2. Crystal Solutions and Refinements

Structural analysis was carried out on powder patterns and each dataset was treated as completely new. Patterns were indexed using 16 peaks and the cells were chosen based on cell volume and calculated density. Cells with a density higher than 1.60 (g/cm^3) or with a volume corresponding to a non-integer number of molecules were discarded.

The values found, although unusually high for organic polymers, were not surprising since comparably high densities for PEF have been reported in the literature [11,19]. A pretty good agreement between the observed and calculated diffraction profiles was obtained for all of the polymorphs (Figure 2). Cell parameters and discrepancy factors are reported in Table 1. The R$_p$ factors achieved, lower than 13%, are satisfactory for polymer structures determined by powder diffraction. The fractional atomic coordinates of the asymmetric unit for each of the phases are reported in Tables S2–S4, and the atomic distances and angles are reported in Tables S5–S7.

Table 1. Structural data; estimated standard deviations (e.s.d.) in parentheses.

Cell Parameters	α-PEF	α'-PEF	β-PEF
a (Å)	5.729(7)	5.912(3)	5.953(3)
b (Å)	7.89(3)	6.913(4)	6.600(3)
c (Å)	9.62(6)	19.73(2)	10.52(1)
α (°)	98.1(3)	90.0	90.0
β (°)	65.1(4)	90.0	107.0(1)
γ (°)	101.3(4)	104.41(3)	90.0
S.G.	P-1	P1	P1
Vol (Å^3)	385.85	780.84	394.30
ρ_{calc} (g/cm^3)	1.567	1.549	1.482
R_p (%)	12.5	11.4	13.2
R_{wp} (%)	13.2	12.1	13.5

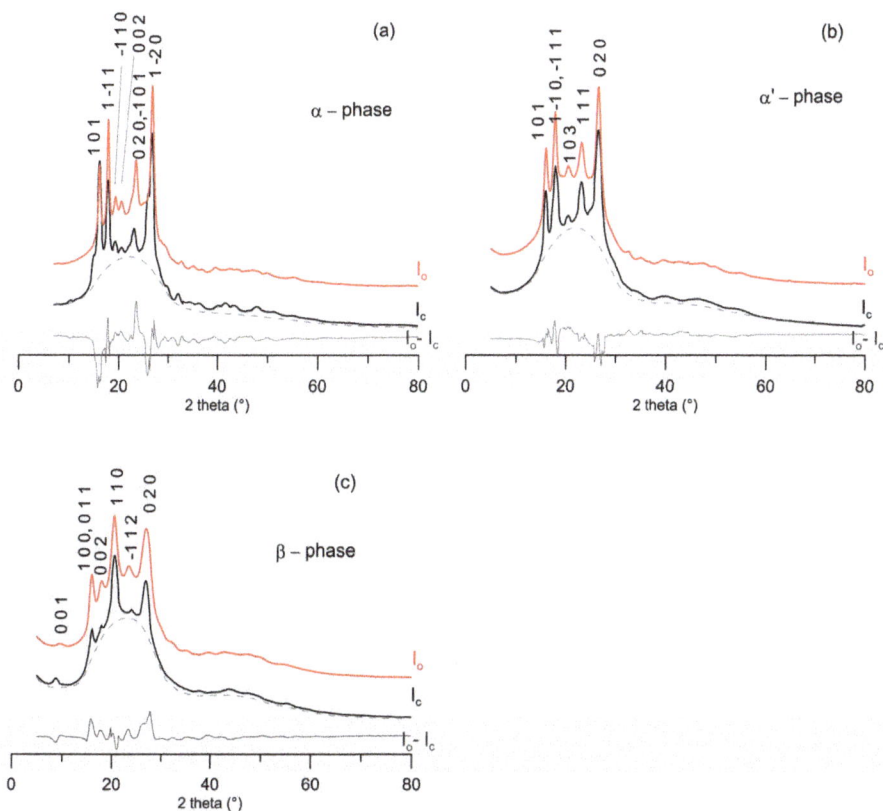

Figure 2. Comparison between observed (red), calculated (black), and Io-Ic difference (grey) powder-diffraction patterns: (**a**) α-PEF; (**b**) α'-PEF; (**c**) β-PEF. Dashed line (grey) represents the sum of background and amorphous phase contributions. The Miller indices of the reflections that mainly contribute to the peak intensities are reported.

3.3. α-PEF

The structure of α-PEF is displayed in Figure 3. The unit cell contains two monomers. The polymer chain expands in opposite directions with respect to the furanic ring due to the orientation of the two ester groups. The chains are aligned in the *c*-axis direction and lie roughly parallel to the 2 -5 0 plane (see Figure 3b,c).

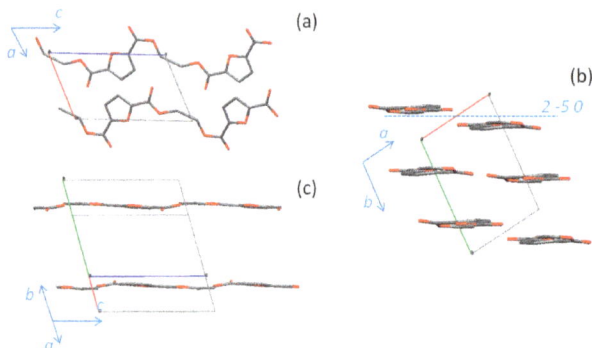

Figure 3. Views of the chain arrangements in the unit cell for the α-PEF structure: (**a**) along the *b*-axis; (**b**) projection down the *c*-axis; (**c**) a view to highlight chain section. Oxygen atoms are in red; dotted line represents the intersection of the 2 -5 0 plane with the *a, b* plane.

3.4. α′-PEF

The less stable α′-phase shows a different structure. During the course of this study the work of Mao et al. appeared in the literature, reporting the PEF structure of a fiber sample investigated with synchrotron light [19]. We found that one of the proposed structures showed a calculated powder pattern very similar to that of α′-PEF except for the amorphous "bump", see Figure S4. For this reason, we refined the α′-PEF phase starting from the data of the 3/12 structure, as proposed by Mao et al. With respect to the cell determined from a fiber sample [19], we found longer *a* and *b* axes and a shorter *c* axis. The pseudo monoclinic unit cell displays a double volume as compared to α-PEF and hosts four molecular units organized in two chains. Each of these has two monomers in a "trans" planar conformation with a 2_1 axis between adjacent monomers, as shown in Figure 4. The two chains in the unit cell are parallel to the *a, c* plane, staggered by 3/12 of the *c*-axis in the chain direction, and slightly bent.

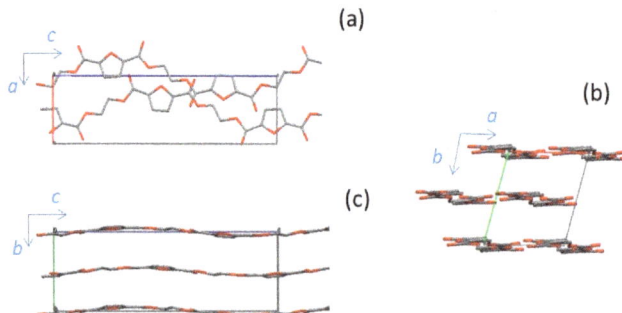

Figure 4. Views of the chain arrangement for the α′-PEF structure: (**a**) projection down the *b*-axis; (**b**) along the *c*-axis; (**c**) projection down the *a*-axis. Oxygen atoms are in red.

3.5. β-PEF

The crystal structure of the β-phase, generated by solvent-induced crystallization, is displayed in Figure 5. Two planar molecules are contained in the monoclinic cell with a volume slightly bigger than that in the α-phase. Two flat chains lie parallel to the *a*, *c* plane, shifted 0.4 units in the *c*-axis direction.

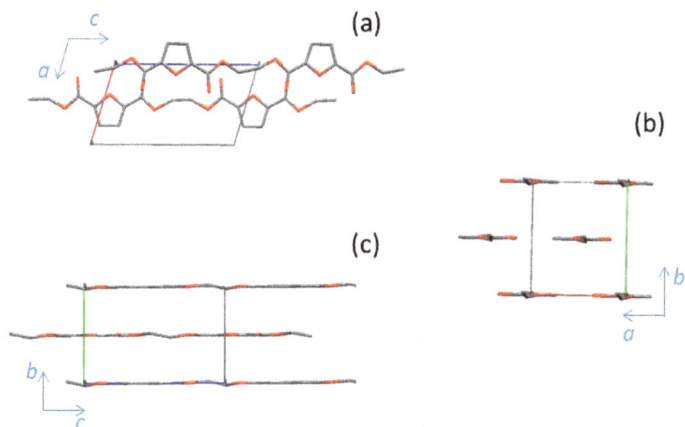

Figure 5. Views of the chain packing for the β-PEF structure: (**a**) along the *b*-axis; (**b**) projection down the *c*-axis; (**c**) projection down the *a*-axis. Oxygen atoms are in red.

Taking in the due consideration the thermal expansion of the unit cell, β-PEF may also be a good candidate for comparison with the structure of the strain-induced phase followed by in situ heat treatment, as reported by Stoclet et al. [20].

4. Conclusions

Suitable experimental conditions to obtain PEF samples with a unique crystal phase and very good crystallinity have been found. By structure determination in direct space (α-, β-phases) or by structure refinement (α′-phase), the solutions that gave the best fit with the experimental X-ray powder patterns have been determined. The unusual high density values of PEF have been confirmed and justified. The structure previously reported for a PEF fiber corresponds to the α′-phase and shows a double cell volume as compared to the other forms. None of the PEF phases is isomorphous to PET crystal lattice. PEF structures show unit cell dimensions different from PET [23]; nevertheless, the chain packing in α-PEF is similar to that of the polymer derived from terephthalic acid (see Figure S5). The presence of several polymorphs for PEF, as compared to the unique phase so far known for PET, could be associated with different factors, such as the intrinsic lower symmetry of furane with respect to benzene, which causes much more conformations, and to the lower chain mobility and higher rigidity of furanoate with respect to terephthalate [11].

Supplementary Materials: The following are available online at http://www.mdpi.com/2073-4360/10/3/296/s1, Figure S1: TMDSC scans, Figure S2: ATR-FTIR scans, Figure S3: comparisons of IR spectra, Figure S4: calculated pattern of 3/12 structure reported by Mao, Figure S5: overlap of the crystal structures of PET and α-PEF, Table S1: microstructural parameters, Table S2: crystal data of α-PEF, Table S3: crystal data of α′-PEF, Table S4: crystal data of β-PEF, Table S5: bond distances and angles of α-PEF, Table S6: bond distances and angles of α′-PEF, Table S7: bond distances and angles of β-PEF.

Author Contributions: George Z. Papageorgiou, Dimitrios N. Bikiaris, Massimo Gazzano, and Nadia Lotti conceived and designed the paper. George Z. Papageorgiou, Dimitrios N. Bikiaris, and Matteo Gigli performed PEF syntheses and sample preparations. Massimo Gazzano and Lucia Maini performed structural determinations, data refinement, and ATR-FTIR measurements. Matteo Gigli, Massimo Gazzano, and Nadia Lotti wrote the paper, which was revised by all of the authors.

Conflicts of Interest: The authors declare no conflict of interest.

References

1. European Bioplastics, Bioplastics Facts and Figures. Available online: http://www.european-bioplastics. org/news/publications/ (accessed on 31 October 2017).
2. Papageorgiou, G.Z.; Papageorgiou, D.G.; Terzopoulou, Z.; Bikiaris, D.N. Production of bio-based 2,5-furan dicarboxylate polyesters: Recent progress and critical aspects in their synthesis and thermal properties. *Eur. Polym. J.* **2016**, *83*, 202–229. [CrossRef]
3. Sousa, A.F.; Vilela, C.; Fonseca, A.C.; Matos, M.; Freire, C.S.R.; Gruter, G.J.M.; Coelho, J.F.J.; Silvestre, A.J.D. Biobased polyesters and other polymers from 2,5-furandicarboxylic acid: A tribute to furan excellency. *Polym. Chem.* **2015**, *6*, 5961–5983. [CrossRef]
4. Van Putten, R.J.; van der Waal, J.C.; de Jong, E.; Rasrendra, C.B.; Heeres, H.J.; de Vries, J.G. Hydroxymethylfurfural, A Versatile Platform Chemical Made from Renewable Resources. *Chem. Rev.* **2013**, *113*, 1499–1597. [CrossRef] [PubMed]
5. Gandini, A.; Lacerda, T.M.; Carvalho, A.J.F.; Trovatti, E. Progress of Polymers from Renewable Resources: Furans, Vegetable Oils, and Polysaccharides. *Chem. Rev.* **2016**, *116*, 1637–1669. [CrossRef] [PubMed]
6. Dick, G.R.; Frankhouser, A.D.; Banerjee, A.; Kanan, M.W. A scalable carboxylation route to furan-2,5-dicarboxylic acid. *Green Chem.* **2017**, *19*, 2966–2972. [CrossRef]
7. "Who we are page" of Synvina web site. Available online: https://www.synvina.com/about-us/who-we-are/ (accessed on 8 March 2018).
8. Burgess, S.K.; Karvan, O.; Johnson, J.R.; Kriegel, R.M.; Koros, W.J. Oxygen sorption and transport in amorphous poly(ethylene furanoate). *Polymer* **2014**, *55*, 4748–4756. [CrossRef]
9. Burgess, S.K.; Kriegel, R.M.; Koros, W.J. Carbon Dioxide Sorption and Transport in Amorphous Poly(ethylene furanoate). *Macromolecules* **2015**, *48*, 2184–2193. [CrossRef]
10. Burgess, S.K.; Mikkilineni, D.S.; Yu, D.B.; Kim, D.J.; Mubarak, C.R.; Kriegel, R.M.; Koros, W.J. Water sorption in poly(ethylene furanoate) compared to poly(ethylene terephthalate). Part 2: Kinetic sorption. *Polymer* **2014**, *55*, 6870–6882. [CrossRef]
11. Burgess, S.K.; Leisen, J.E.; Kraftschik, B.E.; Mubarak, C.R.; Kriegel, R.M.; Koros, W.J. Chain Mobility, Thermal, and Mechanical Properties of Poly(ethylene furanoate) Compared to Poly(ethylene terephthalate). *Macromolecules* **2014**, *47*, 1383–1391. [CrossRef]
12. Eerhart, A.J.J.E.; Faaija, A.P.C.; Patela, M.K. Replacing fossil based PET with biobased PEF; process analysis, energy and GHG balance Energy. *Environ. Sci.* **2012**, *5*, 6407–6422. [CrossRef]
13. Guigo, N.; van Berkel, J.; de Jong, E.; Sbirrazzuoli, N. Modelling the non-isothermal crystallization of polymers: Application to poly(ethylene 2,5-furandicarboxylate). *Thermochim. Acta* **2017**, *650*, 66–75. [CrossRef]
14. Dimitriadis, T.; Bikiaris, D.N.; Papageorgiou, G.Z.; Floudas, G. Molecular Dynamics of Poly(ethylene-2,5-furanoate) (PEF) as a Function of the Degree of Crystallinity by Dielectric Spectroscopy and Calorimetry. *Macromol. Chem. Phys.* **2016**, *217*, 2056–2062. [CrossRef]
15. Stoclet, G.; Gobius du Sart, G.; Yeniad, B.; de Vos, S.; Lefebvre, J.M. Isothermal crystallization and structural characterization of poly(ethylene-2,5-furanoate). *Polymer* **2015**, *72*, 165–176. [CrossRef]
16. Tsanaktsis, V.; Papageorgiou, D.G.; Exarhopoulos, S.; Bikiaris, D.N.; Papageorgiou, G.Z. Crystallization and Polymorphism of Poly(ethylene furanoate). *Cryst. Growth Des.* **2015**, *15*, 5505–5512. [CrossRef]
17. Nadia Lotti, N.; Munari, A.; Gigli, M.; Gazzano, M.; Tsanaktsis, V.; Bikiaris, D.N.; Papageorgiou, G.Z. Thermal and structural response of in situ prepared biobased poly(ethylene 2,5-furan dicarboxylate) nanocomposites. *Polymer* **2016**, *103*, 288–298. [CrossRef]
18. Kazaryan, L.G.; Medvedeva, F.M. X-ray study of poly(ethylene furan-2,5-dicarboxylate) structure. *Vysokomol. Soedin. Ser. B Kratk. Soobshcheniya* **1968**, *10*, 305–306.
19. Mao, Y.; Kriegel, R.M.; Bucknall, D.G. The crystal structure of poly(ethylene furanoate). *Polymer* **2016**, *102*, 308–314. [CrossRef]
20. Stoclet, G.; Lefebvre, J.M.; Yeniad, B.; Gobius du Sart, G.; de Vos, S. On the strain-induced structural evolution of Poly(ethylene-2,5-furanoate) upon uniaxial stretching: An in-situ SAXS-WAXS study. *Polymer* **2018**, *134*, 227–241. [CrossRef]

21. Papageorgiou, G.Z.; Tsanaktsis, V.; Bikiaris, D.N. Synthesis of poly(ethylene furandicarboxylate) polyester using monomers derived from renewable resources: Thermal behavior comparison with PET and PEN. *Phys. Chem. Chem. Phys.* **2014**, *16*, 7946–7958. [CrossRef] [PubMed]

22. Konstantopoulou, M.; Terzopoulou, Z.; Nerantzaki, M.; Tsagkalias, J.; Achilias, D.S.; Bikiaris, D.N.; Exarhopoulos, S.; Papageorgiou, D.G.; Papageorgiou, G.Z. Poly(ethylene furanoate-co-ethylene terephthalate) biobased copolymers: Synthesis, thermal properties and cocrystallization behavior. *Eur. Polym. J.* **2017**, *89*, 349–366. [CrossRef]

23. Tse, J.S.; Mak, T.C.W. Refinement of the crystal structure of polyethylene terephthalate. *J. Cryst. Mol. Struct.* **1975**, *5*, 75–80. [CrossRef]

© 2018 by the authors. Licensee MDPI, Basel, Switzerland. This article is an open access article distributed under the terms and conditions of the Creative Commons Attribution (CC BY) license (http://creativecommons.org/licenses/by/4.0/).

![polymers logo] *polymers*

MDPI

Article

Renewable Resources and a Recycled Polymer as Raw Materials: Mats from Electrospinning of Lignocellulosic Biomass and PET Solutions

Rachel Passos de Oliveira Santos, Patrícia Fernanda Rossi, Luiz Antônio Ramos and Elisabete Frollini *

Macromolecular Materials and Lignocellulosic Fibers Group, Center of Research on Science and Technology of BioResources, Institute of Chemistry of São Carlos, University of São Paulo, CP 780, 13560-970 São Carlos, SP, Brazil; rachelpassos@gmail.com (R.P.O.S.); patricia.rossi@usp.br (P.F.R.); ramos@iqsc.usp.br (L.A.R.)
* Correspondence: elisabete@iqsc.usp.br; Tel.: +55-16-3373-9923

Received: 6 April 2018; Accepted: 14 May 2018; Published: 17 May 2018

Abstract: Interest in the use of renewable raw materials in the preparation of materials has been growing uninterruptedly in recent decades. The aim of this strategy is to offer alternatives to the use of fossil fuel-based raw materials and to meet the demand for materials that are less detrimental to the environment after disposal. In this context, several studies have been carried out on the use of lignocellulosic biomass and its main components (cellulose, hemicelluloses, and lignin) as raw materials for polymeric materials. Lignocellulosic fibers have a high content of cellulose, but there has been a notable lack of investigations on application of the electrospinning technique for solutions prepared from raw lignocellulosic biomass, even though the presence of cellulose favors the alignment of the fiber chains during electrospinning. In this investigation, ultrathin (submicrometric) and nanoscale aligned fibers were successfully prepared via electrospinning (room temperature) of solutions prepared with different contents of lignocellulosic sisal fibers combined with recycled poly(ethylene terephthalate) (PET) using trifluoroacetic acid (TFA) as solvent. The "macro" fibers were deconstructed by the action of TFA, resulting in solutions containing their constituents, i.e., cellulose, hemicelluloses, and lignin, in addition to PET. The "macro" sisal fibers were reconstructed at the nanometer and submicrometric scale from these solutions. The SEM micrographs of the mats containing the components of sisal showed distinct fiber networks, likely due to differences in the solubility of these components in TFA and in their dielectric constants. The mechanical properties of the mats (dynamic mechanical analysis, DMA, and tensile properties) were evaluated with the samples positioned both in the direction (*dir*) of and in opposition (*op*) to the alignment of the nano and ultrathin fibers, which can be considered a novelty in the analysis of this type of material. DMA showed superior values of storage modulus (E' at 30 °C) for the mats characterized in the preferential direction of fiber alignment. For example, for mats obtained from solutions prepared from a 0.4 ratio of sisal fibers/PET, Sisal/PET$_{0.40}$*dir* presented a high E' value of 765 MPa compared to Sisal/PET$_{0.40}$*op* that presented an E' value of 88.4 MPa. The fiber alignment did not influence the T_g values (from tan δ peak) of electrospun mats with the same compositions, as they presented similar values for this property. The tensile properties of the electrospun mats were significantly impacted by the alignment of the fibers: e.g., Sisal/PET$_{0.40}$*dir* presented a high tensile strength value of 15.72 MPa, and Sisal/PET$_{0.40}$*op* presented a value of approximately 2.5 MPa. An opposite trend was observed regarding the values of elongation at break for these materials. Other properties of the mats are also discussed; such as the index of fiber alignment, average porosity, and surface contact angle. To our knowledge, this is the first time that the influence of fiber alignment on the properties of electrospun mats based on untreated lignocellulosic biomass combined with a recycled polymer, such as PET, has been evaluated. The mats obtained in this study have potential for diversified applications, such as reinforcement for polymeric matrices in nanocomposites, membranes for filtration, and support for enzymes, wherein the fiber alignment, together with other evaluated properties, can impact their effectiveness in these applications.

Polymers **2018**, *10*, 538

Keywords: lignocellulosic sisal fibers; recycled PET; electrospinning; mechanical properties

---

1. Introduction

Renewable resources are viable, biodegradable, and low-energy consumption raw materials. They are a current subject of interest due to the benefits of increasingly replacing fossil-based raw materials in the preparation of polymeric materials [1–8].

Use of lignocellulosic fibers as reinforcing agents in polymeric composites is an example of the successful application of renewable resources [9–14]. In a previous study, composites based on recycled poly(ethylene terephthalate) (PET) reinforced with sisal fibers were prepared via a thermopressing process [9]. In this study, plasticizers derived from renewable raw materials were used to decrease the melting point of the recycled PET, which is sufficiently high to initiate thermal decomposition of the lignocellulosic fibers. Therefore, composites were successfully prepared from a recycled polymer that is widely available globally and from other components derived from renewable sources, i.e., sisal fibers and plasticizers [9]. Given our continued interest in recycled PET and lignocellulosic sisal fibers, the study progressed with the aim of continuously adding more value to these two raw materials, and this paper is focused on the electrospinning of solutions obtained from both of them.

The electrospinning technique has gained much attention because it is a versatile method that enables the production of submicro- and nanoscale fibers, including from biopolymer solutions [15–18]. However, there is a lack of investigations on the application of this technique to produce fibers based on raw lignocellulosic biomass [19,20], probably due to the shortage of solvents able to deconstruct the fibers to generate solutions of their components (cellulose, hemicelluloses, and lignin) that simultaneously have properties suitable for electrospinning. The linear structure of cellulose, the most abundant biopolymer in the world, favors the alignment of its chains during the electrospinning process, as opposed to hemicelluloses, which are heteropolysaccharides composed of branched chains. Lignin is a large-scale renewable source of aromatic functionalities, which gives it unique properties. However, the complex nonlinear structure of lignin prevents the electrospinning of lignin solutions [21].

Lignocellulosic sisal fibers have a high content of cellulose, which favors the electrospinning of hemicelluloses and lignin when the three components are together. Brazil is the world's largest producer of sisal with a production in 2016 of 84.551 tons, and approximately 70% of the fibers produced were exported [21,22].

Rodrigues et al. [19] first reported the electrospinning of untreated lignocellulosic biomass using trifluoroacetic acid (TFA) as a solvent. The sisal fibers were deconstructed by dissolution and reconstructed as a homogeneous network of ultrathin and nanofibers.

PET is one of the major polymers recycled and reused for a wide range of applications [9]. This polymer, both recycled and new, has also been widely used as a raw material for the electrospinning process to produce ultrathin fibers [21,23–25]. These fibers can be used, e.g., in combination with graphitic carbon nitride as an easily recycled photocatalyst for the degradation of antibiotics under solar irradiation [26]

Santos et al. [20] pioneered the combination of lignocellulosic sisal fibers and recycled PET to produce non-aligned ultrathin fibers and nanofibers. Hybrid mats composed of non-aligned fibers were successfully prepared with different final properties than those presented by PET electrospun mats.

The electrospinning technique can also be successfully used to produce aligned fibers with a high degree of orientation [27–29]. Those fibers can present remarkable properties [30–33] and can be used in applications such as reinforcing agents in nanocomposites [33–36], as well as substituents of extracellular matrices (ECMs) in areas such as tissue engineering [37].

In the present study, mats composed of aligned fibers and based on lignocellulosic sisal fibers/recycled PET were produced, and their mechanical and surface properties were evaluated.

2. Materials and Methods

2.1. Materials

The recycled PET (MFI of 36.4 g·(10 min)$^{-1}$ [9]) was a donation from Gruppo Mossi & Ghisolfi (M&G, São Paulo, Brazil).

The sisal fibers (chemical composition of 64.9% cellulose; 11.7% lignin; 25.4% hemicellulose; 7.1% moisture; 0.4% ash content; crystallinity index of 58% [9]) were acquired from Sisal Sul Indústria e Comércio LTDA (São Paulo, Brazil). Prior to their use, the lignocellulosic fibers were submitted to an extraction process using a mixture of ethanol/cyclohexane (experimental conditions: 1:1, *v/v*, 10 min). This pretreatment aimed to remove waxes (composed mostly of apolar molecules) and inorganic compounds. The extracted fibers were milled in a MARCONI MA048 (São Paulo, Brazil) vertical rotor mill (30 mesh stainless steel) and dried in a vacuum oven at 60 °C until they reached a constant weight. Trifluoroacetic acid (TFA) was purchased from Mallinckrodt Chemicals (Dublin, Ireland), and it was used as received.

2.2. Electrospinning Process

The electrospun solutions were prepared based on the study by Santos et al. [20]. Briefly, a reference solution of recycled PET (15 g·dL^{-1}) and four solutions containing mixtures with different ratios of this polymer and sisal fibers were prepared, as summarized in Table 1. In all cases, the solutions were kept under vigorous magnetic stirring at room temperature until complete dissolution of the components (one-phase solution) was reached.

Table 1. Composition and respective reference codes of the electrospun mats.

Sisal fiber (g)	Recycled PET (g)	Sisal/PET ratio	Sample code
-	0.45	Reference sample	PET$_{ref}$
0.06	0.60	0.10	Sisal/PET$_{0.10}$
0.06	0.45	0.13	Sisal/PET$_{0.13}$
0.06	0.30	0.20	Sisal/PET$_{0.20}$
0.06	0.15	0.40	Sisal/PET$_{0.40}$

The concentration of sisal fibers was chosen based on a previous study on non-aligned electrospun mats [19], where it was found that 2 g·dL^{-1} corresponded to the threshold to avoid gelation after dissolution. The sisal fiber-based solutions presented high viscosity, which prevented determine the viscosities of the electrospun solutions using the falling-ball viscometer from GILMONT Instruments (Cole-Parmer scientific experts, Vernon Hills, IL, USA). The high volatility of the TFA represents a major drawback to the use of other methods to determine the viscosity of the solutions, such as the capillary viscometry.

The electrospinning process was performed using an EC-DIG electrospinning apparatus from IME Technologies (Geldrop, The Netherlands) and a metallic rotator drum from Instor (Porto Alegre, RS, Brazil). All of the electrospinning conditions adopted, such as the needle-collector distance, voltage, solution flow rate, rotational speed of the collector and process time, were established after previous tests in which the viscosity of the solutions, lack of formation of beads, and diameter of the fibers were also taken into account. The one-phase solutions were placed into capillary tubes (ID: 1.00 mm and OD: 1.6 mm) and these tubes were connected to a mechanical pump (IME Technologies, Geldrop, The Netherlands). Figure 1 shows a scheme of the electrospinning parameters adopted in all experiments.

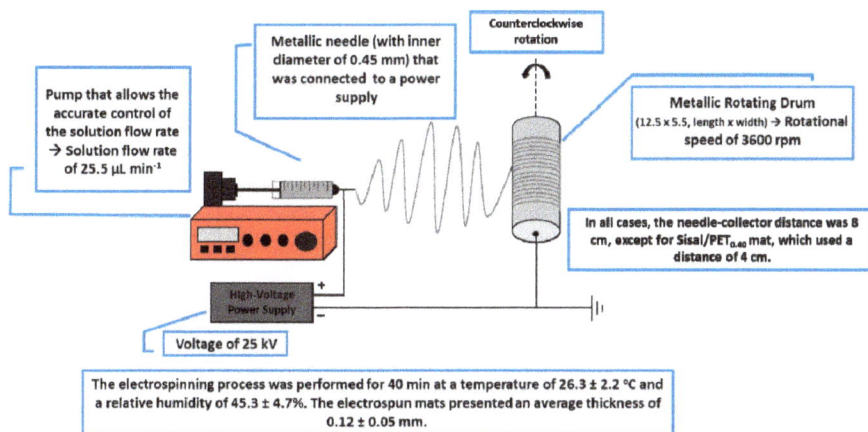

Figure 1. Experimental scheme with the electrospinning parameters adopted.

2.3. Characterization of the Materials

Scanning electron microscopy (SEM) was performed on a 440 Zeiss DSM 940 instrument (Oberkochen, Germany) using an accelerating voltage of 20 kV. Prior to this analysis, to improve the image acquisition, the samples were coated with an ultrathin gold layer using a sputter-coating system. ImageJ 1.45 image processing software (National Institutes of Health, Bethesda, MD, USA) was used to process and analyze the SEM images, including determination of the average fiber diameter, the average pore area and porosity, and the alignment index and average preferred orientation of the fibers.

The contact angle (25 °C) between droplets of deionized water and the surface of the mats were obtained from a CAM 200 goniometer (KSV Instruments Ltd., Helsinki, Finland) at Bernhard Gross Physics Institute of São Carlos, University of São Paulo, equipped with image analysis software (CAM 2008). Drops of water (with a volume of approximately 5 µL) were deposited on the surface of the mats (1 cm × 1 cm), and the contact angles (left and right) were calculated from the digitalized image. For this analysis, 500 measurements were individually collected at one second intervals, which allowed calculation of the resulting angle as a function of time. All measurements were carried out at least three times for each mat.

Dynamic mechanical analysis (DMA) was carried out on a thermal analyzer model Q800 from TA Instruments (New Castle, DE, USA) equipped with a tension clamp for films in multifrequency mode. The samples (6.3 mm in width and 5 mm in gauge length) were analyzed using the following parameters: amplitude of 4 µm, frequency of 1 Hz, static force of 0.25 N and heating rate of $3\ ^\circ C \cdot min^{-1}$ (from 0 to 200 °C). At least three specimens were tested from each sample group.

Tensile tests were performed at 25 °C using a TA Instruments model Q800 in tension mode. The samples (6.4 mm wide with a gauge length of 5 mm) were strained to 18 N or failure at a constant rate of $1\ N \cdot min^{-1}$. These tests were carried out with at least three specimens from each sample group.

The mechanical properties of the mats (DMA and tensile properties) were evaluated with the samples positioned both in the direction (*dir*) of and in opposition (*op*) to the alignment of the nano and ultrathin fibers to evaluate the influence of fiber alignment on these properties.

3. Results and Discussion

Mats composed of aligned fibers were produced via an electrospinning technique from solutions of a renewable resource—namely, sisal fibers—and a recycled polymer, i.e., recycled PET. These mats were characterized regarding their morphological and surface properties, as well as with respect to their mechanical properties, in order to evaluate the influence of fiber alignment on these properties.

TFA, the solvent used to dissolve the recycled polymer and to deconstruct the lignocellulosic fibers into their components, can esterify the hydroxyl groups present in the chemical structures of cellulose, hemicelluloses, and lignin, as shown in Figure 2. In a previous study, mats composed of sisal fiber solutions were analyzed via Fourier transform infrared spectroscopy (FTIR). The initial FTIR spectrum showed, in addition to the characteristic bands of a typical lignocellulosic material, a small shoulder at 1790 cm^{-1}, which could be attributed to the trifluoroacetyl groups resulting from esterification [17], as shown in Figure 2.

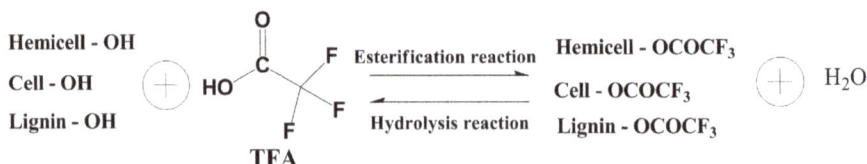

Figure 2. Reactions of esterification and hydrolysis of the lignocellulosic components.

However, exposure to air led to hydrolysis of the trifluoroacetyl groups resulting from esterification (hydrolysis reaction, Figure 2) [17], regenerating the constituents of the lignocellulosic biomass (Cell-OH, Hemicell-OH, Lignin-OH, Figure 2).

3.1. Scanning Electron Microscopy (SEM)

Figure 3 displays the SEM micrographs, respective color-coded images and data related to the alignment index (A.I.) and average preferred orientation (A.P.O.) of the electrospun mats PET$_{ref}$, Sisal/PET$_{0.10}$, Sisal/PET$_{0.13}$, Sisal/PET$_{0.20}$, and Sisal/PET$_{0.40}$.

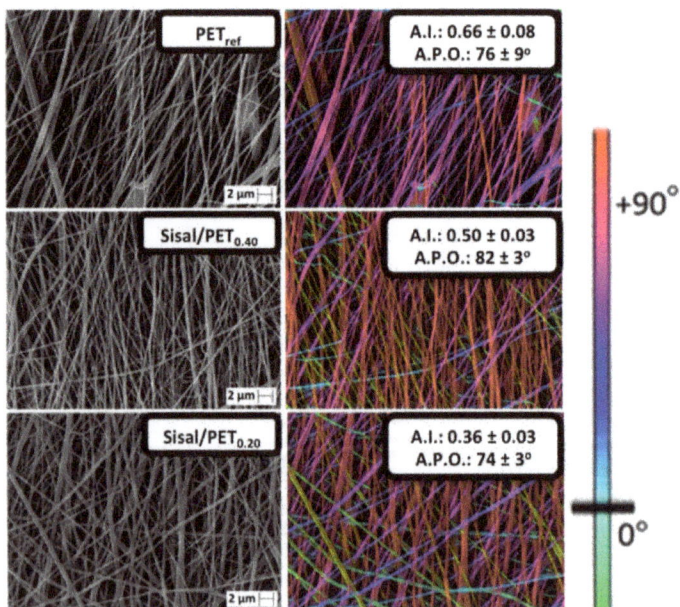

Figure 3. *Cont.*

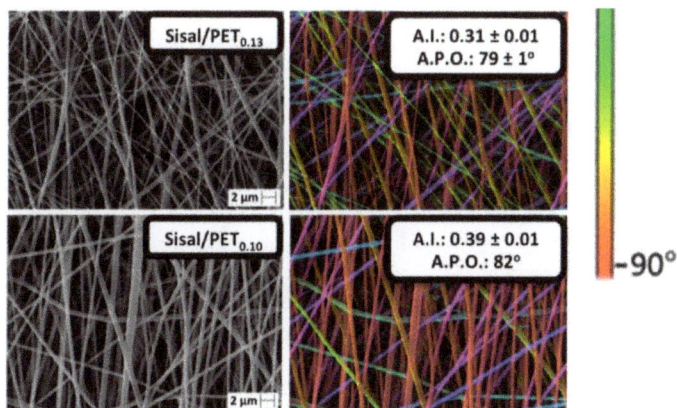

Figure 3. SEM micrographs of the electrospun mats and their respective color-coded images, alignment index (**AI**) and average preferred orientation (**APO**) of the fibers.

Figure 4 shows the histograms related to the fiber diameter frequency of the electrospun mats.

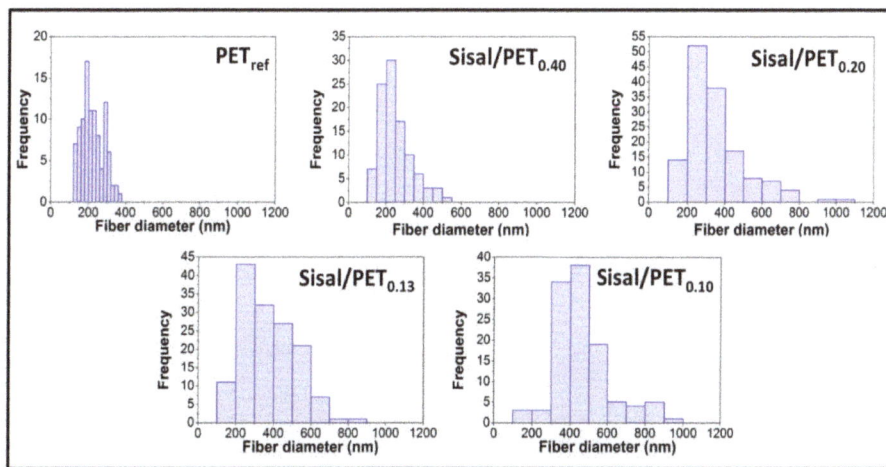

Figure 4. Histograms of fiber diameter frequency for the electrospun mats.

Few beads were observed in the Sisal/$PET_{0.13}$ fiber network. The presence of beads along the fibers can be attributed to instability of the jet of polymer solution during the electrospinning process, as indicated by Yarin [38]. The formation of beads was also observed in the PET/sisal-based mats produced using the stationary fiber collector [20].

The SEM micrographs of the mats containing sisal in their compositions (mainly for Sisal/$PET_{0.10}$, Figure 3) showed two distinct fiber networks. The fiber network with thicker fibers corresponded to fibers from recycled PET (as observed in the SEM micrograph of this polymer (Figure 3, $PET_{ref.}$)), and the other network composed of thinner fibers probably corresponded to the components of the sisal fibers. The phenomenon of separate electrospinning of recycled PET and sisal fibers observed in the PET/sisal fiber-based mats (Figure 3) can be attributed to the differences in the solubility of these components in TFA and in their dielectric constants (parameters that can affect the electrospinning process of hybrid solutions). Additionally, PET and components of the sisal fibers may have partially electrospun together, generating hybrid fibers with an intermediate diameter between those mentioned above

Regarding the histograms of fiber diameter frequency in the electrospun mats, it was observed that the materials containing sisal in their compositions presented a larger distribution of fiber diameters than PET_{ref}, as shown in Figure 4. This was the result of the possibility that at least part of the components of the lignocellulosic sisal fibers were electrospun in combination with the PET chains (probably mostly lignin due to its aromatic structure and affinity for the aromatic structure of PET), in addition to having electrospun separately from the PET chains, as mentioned above. Thus, mats containing sisal in their compositions presented a network formed by thicker fibers (251.9 ± 87.5 nm (Sisal/$PET_{0.40}$), 346.7 ± 160.7 nm (Sisal/$PET_{0.20}$), 375.8 ± 143.2 nm (Sisal/$PET_{0.13}$) and 462.9 ± 145.4 nm (Sisal/$PET_{0.10}$) (Figure 5a)) and another one composed of fibers so thin that their diameters were below the threshold for determination via the ImageJ 1.45 image processing software. It should be highlighted that the electrospun raw lignocellulosic sisal fibers and sisal pulp fibers presented, respectively, a web of ultrathin fibers with diameters ranging from 120–510 nm, and nano/ultrathin fibers with diameters ranging from 65–200 nm, depending on the different values of solution flow rate adopted [17]. Further evidence of the large distribution of fiber diameters (Figure 4) presented by the sisal-based mats came from comparison between the standard deviations of the average diameter of PET_{ref} (± 59.10 nm) and of the mats containing sisal fibers (± 87.5, ± 160.7, ± 143.2, ± 145.4 nm for Sisal/$PET_{0.40}$, Sisal/$PET_{0.20}$, Sisal/$PET_{0.13}$ and Sisal/$PET_{0.10}$, respectively).

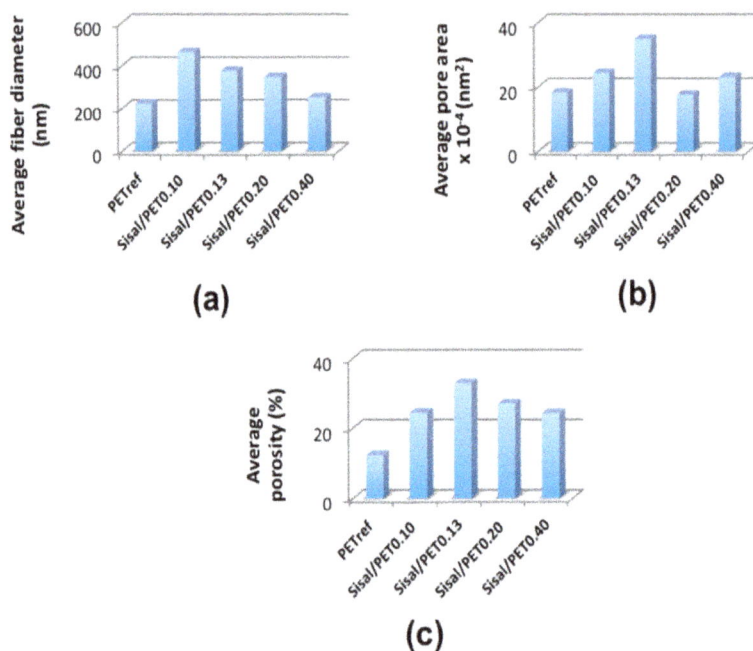

Figure 5. Average values of (**a**) fiber diameter (errors from ± 59.10 (PET_{ref}) to ± 160.70 (Sisal/$PET_{0.20}$)); (**b**) pore area (errors from ± 1.00 (Sisal/$PET_{0.10}$) to ± 17.80 (Sisal/$PET_{0.40}$)); and (**c**) porosity (errors from ± 0.20 (PET_{ref}) to ± 7.10 (Sisal/$PET_{0.40}$)) of the electrospun mats.

The alignment index, A.I., ranged from 0.31 ± 0.01 (Sisal/$PET_{0.13}$) to 0.66 ± 0.08 (PET_{ref}), Figure 3. The differences observed between PET_{ref} and the Sisal/PET mats were probably due to the higher viscosity of the Sisal/PET solutions compared to PET$_{ref}$, which led to an increase in the viscoelastic forces and hence to a decrease in the stretching and alignment of the fibers. Regarding the average preferred orientation (A.P.O.), there was no significant difference between any of the electrospun mats, as seen in Figure 3. Figure 5 presents the results of the average fiber diameter, average pore area, and average porosity of the electrospun mats.

The average pore areas ranged from $35.0 \pm 4.2 \times 10^4$ nm^2 (Sisal/PET$_{0.13}$) to 17.6×10^4 nm^2 (Sisal/PET$_{0.20}$), which included PET$_{ref}$ ($18.2 \pm 1.7 \times 10^4$ nm^2) (Figure 5b), showing that most of the Sisal/PET mats had higher average pore areas than PET$_{ref}$. The same trend was observed when comparing the values of average porosity of the electrospun mats, which ranged from $12.3 \pm 0.2\%$ (PET$_{ref}$) to $32.8 \pm 6.9\%$ (Sisal/PET$_{0.13}$), as shown in Figure 5c. These results can be attributed to the network formed by the lignocellulosic components in the electrospun Sisal/PET-mats. The fibers present in this network presented smaller diameters, as mentioned above, which impacted both the average porosity and average pore area [39]. This high porosity combined with the high specific surface area and the low basic weight presented by this type of material, makes it suitable for use as a high-performance air filter, for example [40–42].

3.2. Contact Angle (CA) Measurements

Evaluation of the hydrophobic and hydrophilic characteristics of the electrospun mats is important for predicting the potential applications of these materials. For filtration purposes, for example, depending on the object to be intercepted, the ability to control the hydrophobic/hydrophilic characteristics of the mat can be very useful.

The advancing and receding contact angles between the water droplets and the surfaces of the electrospun mats are displayed in Figure 6. Typical curves can be found in Supplementary Material.

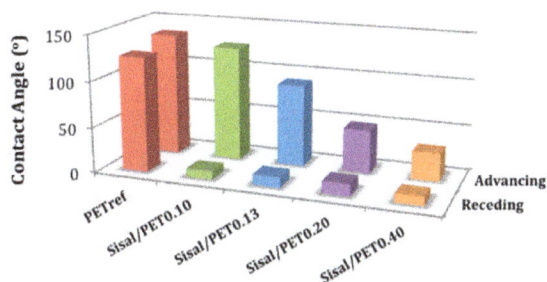

Figure 6. Advancing (errors from ± 0.52 (PET$_{ref}$) to ± 16.31 (Sisal/PET$_{0.10}$)) and receding (errors from ± 2.21 (Sisal/PET$_{0.10}$) to ± 7.53 (Sisal/PET$_{0.13}$)) contact angles between the water droplet and the surface of the electrospun mats.

Figure 6 depicts the decrease in the advancing contact angles (ACA) as the ratio of sisal/PET in the composition of these materials increased. Therefore, the addition of sisal fiber led to formation of hydrophilic materials with surface properties considerably different from those shown by PET$_{ref}$.

The ACA ranged from $134.6 \pm 2.4°$ (PET$_{ref}$), which indicated a highly hydrophobic surface, to $32.5 \pm 2.3°$ (highly hydrophilic surface, Sisal/PET$_{0.40}$). The significant decrease in ACA values and the small values of receding contact angle (RCA) compared to those presented by PET$_{ref}$ (Figure 6) indicated that in the Sisal/PET-mats, many polar groups—mostly those from the cellulose structure (major sisal fiber component)—were oriented towards the surface.

The increase in the average porosity of the sisal-based mats compared to that of PET$_{ref}$ (Figure 5c) may have also contributed to the increase in the hydrophilicity of these materials. Therefore, the addition of different ratios of sisal fiber/recycled PET in the composition of the electrospun mats allows control and modulation of the surface properties of the materials for specific applications.

3.3. Dynamic Mechanical Analysis (DMA)

Figure 7a presents the storage modulus values (at 30 °C) of the electrospun mats. The results of the glass transition temperature for the recycled PET (T_g obtained from the maximum of the tan δ curve) are displayed in Figure 7b. As mentioned, the mats were evaluated with the samples positioned both in the direction (*dir*) of and in opposition (*op*) to the alignment of the ultrathin and nanofibers. Typical curves can be found in Supplementary Material.

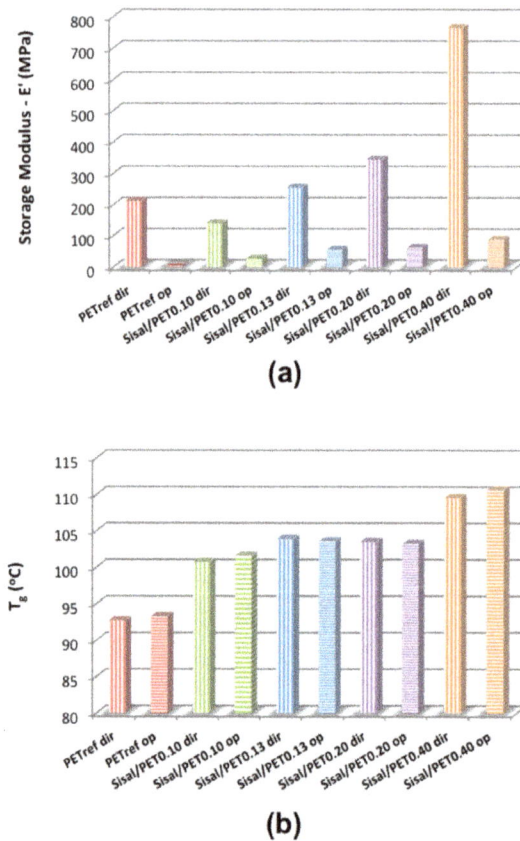

(a)

(b)

Figure 7. (a) Storage Modulus (at 30 °C) (errors from ±0.90 (Sisal/PET$_{0.13}$*op*) to ±50.10 (Sisal/PET$_{0.13}$*dir*)) and (b) T_g values (errors from ±0.04 (PET$_{ref}$*op*) to ±0.90 (Sisal/PET$_{0.20}$*op*)) of the electrospun mats. All materials were characterized in the preferential direction of fiber alignment (acronyms containing "*dir*"), and in the opposite direction of fiber alignment (acronyms containing "*op*").

An increase in the E' values (at 30 °C) with the increase in the sisal/PET ratio in the composition of the electrospun mats can be observed from Figure 7a. In the materials characterized in the preferential direction of fiber alignment, PET$_{ref}$*dir* presented a lower value of E' (212 ± 10.2 MPa) than Sisal/PET$_{0.40}$*dir* (E' = 765 MPa), indicating that the sisal fiber (mainly the cellulose component) was responsible for the increase in E'. The storage modulus is related to the stiffness during load. These results indicated that the presence of sisal components in the composition of the electrospun mats led to strong interactions between the recycled polymer and sisal components at the molecular level, increasing the stiffness of the mats [43].

The values of E' (at 30 °C) in Figure 7a for the materials characterized in the preferred direction of fiber alignment were higher than those presented by the materials of corresponding compositions characterized in the opposite direction. These results indicated that the alignment of the fibers relative to the force applied positively impacted the resistance of the material against the applied load, as well as the capacity of the material to recover its shape.

Thus, it can be concluded that there was a significant effect of the fiber orientation on the property of E' and hence on the stiffness of the material.

It can be observed in Figure 7b that the incorporation of sisal fibers into the mat composition led to an increase in the PET T_g values; e.g., the PET$_{ref}$*dir* presented a T_g of 92.7 ± 0.3 °C and Sisal/PET$_{0.40}$*dir* presented a T_g value of 109.5 ± 0.8 °C. The increase in the sisal fiber/PET ratio in the mat composition favored the establishment of strong intermolecular interactions—i.e., hydrogen bond interactions and hydrophobic interactions—between these components, which progressively hampered the rotational movements of the covalent bonds present in the segments of the polymer chain [44]. The variation observed in the T_g values of the mats reinforced was similar to that observed in the storage modulus results in Figure 7a, which indicated that the interactions between the segments of the PET chain and the sisal components occurred at the molecular level. These results indicated that hybrid fibers (in which intermolecular distances favored strong interactions between the polar and apolar groups of PET segments and sisal components) were produced via electrospinning and that the fraction of hybrid fibers produced impacted the T_g of PET.

According to Figure 7b, there was no significant difference in T_g values between the mats with the same composition but characterized in the preferential direction of fiber alignment and in the opposite direction. The glass transition is a consequence of the movement of segments of the PET chains at the molecular level, and the alignment of the fibers in the direction of or in opposition to the applied load did not influence the movement of the segments.

3.4. Tensile Tests

Figure 8 presents the results of tensile tests regarding the rupture strength, elastic modulus, and elongation at break of the electrospun mats. All materials were characterized in the perpendicular direction relative to the collector axis (preferential direction of fiber alignment) and in the parallel direction (opposite direction of fiber alignment). Typical curves can be found in Supplementary Material.

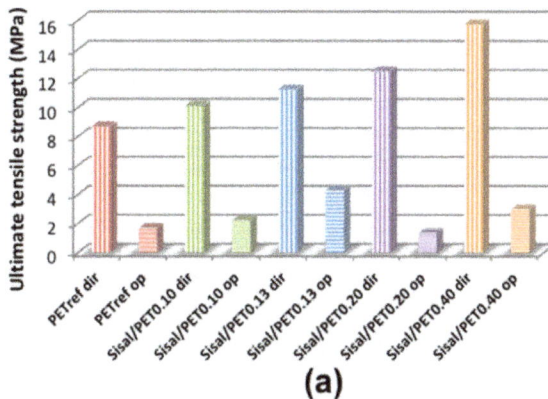

Figure 8. *Cont.*

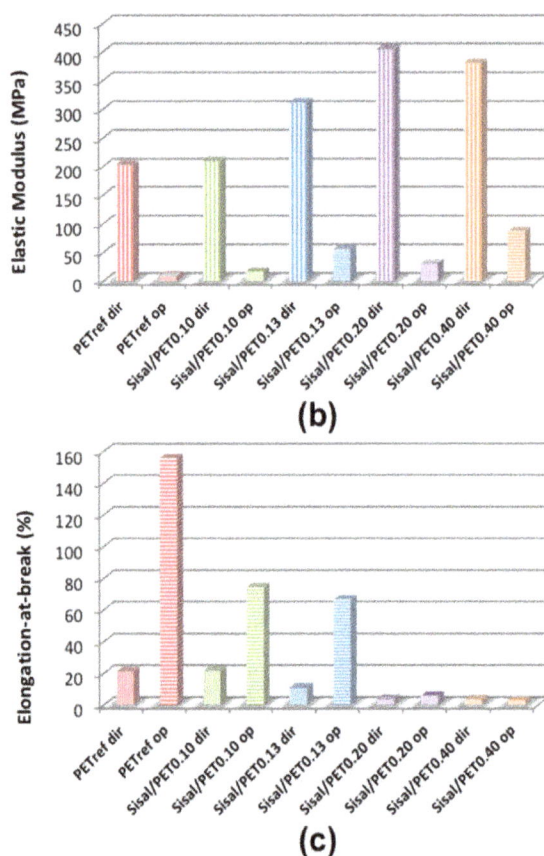

Figure 8. (**a**) Ultimate tensile strength (errors from ±0.10 (Sisal/PET$_{0.10}$op) to ±1.20 (Sisal/PET$_{0.10}$dir)); (**b**) elastic modulus (errors from ±0.10 (Sisal/PET$_{0.13}$op) to ±15.63 (Sisal/PET$_{0.40}$dir)); and (**c**) elongation at break (errors from ±0.01 (Sisal/PET$_{0.20}$dir) to ±17.53 (Sisal/PET$_{0.13}$op)) of the electrospun mats. All materials were characterized in the preferential direction of fiber alignment (acronyms containing "*dir*"), and in the opposite direction of fiber alignment (acronyms containing "*op*").

It can be observed in Figure 8a that an increase in the tensile strength values occurred with the increase in the sisal/PET ratio in the composition of the electrospun mats, especially for those characterized in the preferential direction of fiber alignment. Regarding these mats, PET$_{ref}$*dir* presented the lowest value of tensile strength (8.7 ± 0.5 MPa), and Sisal/PET$_{0.40}$*dir*, with the highest ratio of sisal/PET in its composition, presented the highest value for this property (15.72 ± 0.2 MPa). These results indicated that sisal fiber (mainly the cellulose component) was responsible for the increase in the tensile strength values [45]. Thus, it can be concluded that the presence of cellulose, hemicelluloses, and lignin in the composition of the materials favored the establishment of strong intermolecular interactions between this polysaccharide and PET chains, making the orientation of the polymer chains with the traction axis during the final stage of deformation difficult, causing an increase in mechanical resistance of the electrospun mats.

The tensile strength values of the materials characterized in the preferential direction of fiber alignment (Figure 8a) were superior and more divergent compared to the values of this property presented by the mats of the corresponding composition characterized in the opposite direction.

These results are probably a consequence of the fact that when a mechanical force is applied in the preferential direction of fiber alignment, the fibers only need a minimum rotation to align with the direction of the applied force. Consequently, a high percentage of fibers readily resists the force applied, endowing the material with a high tensile strength. Conversely, when the force is applied in the direction opposite of fiber alignment, a greater rotation and reorientation of the fibers are necessary to align with the axis of the applied force, which leads to low tensile strength.

In general, the elastic modulus of the electrospun mats (Figure 8b) exhibited the same trend as that observed for E' (Figure 6b) and tensile strength (Figure 8a), evidencing the influence of the increasing sisal/PET ratio on the composition of the materials and the influence of fiber orientation on the mechanical properties. According to Figure 8b, $PET_{ref}op$ presented the lowest value of elastic modulus (9.5 ± 2.5 MPa), and $Sisal/PET_{0.40}dir$, with the highest ratio of sisal/PET in its composition, presented one of the highest values for this property (380.55 ± 15.63 MPa).

An opposite trend to that presented for E' (at 30 °C) (Figure 7a), tensile strength (Figure 8a) and elastic modulus (Figure 8b) were observed regarding the values of elongation at break for the electrospun mats (Figure 8c). The mats with lower sisal/PET ratios in their compositions characterized in the direction opposite to fiber alignment presented the highest values of elongation at break (Figure 8c). When a mechanical force is applied in the direction opposite to fiber alignment, a greater rotation and reorientation of the fibers are required to align them with the axis of the applied tension, as mentioned above, which leads to higher elongation of these materials compared to that in situations in which the mechanical force is applied in the fiber alignment direction [46,47]. According to Figure 7c, $PET_{ref}op$ presented the highest value of elongation at break (156 ± 2%), and $Sisal/PET_{0.40}dir$, with the highest ratio of sisal/PET in its composition, presented one of the lowest values for this property (3.6 ± 0.0%).

4. Conclusions

Ultrathin and nanoscale aligned fibers were successfully prepared via electrospinning of solutions based on lignocellulosic sisal fibers and recycled PET with different ratios of sisal/PET in their compositions. The SEM images, as well as the average diameters and average porosities, indicated that mats composed of networks consisting of PET fibers, of the components of the lignocellulosic fibers, and of hybrid fibers (PET/cellulose-hemicelluloses-lignin) were obtained. These results, together with those from contact angle measurements, showed that it was possible to control the fiber diameters and porosities and the hydrophilic/hydrophobic character of the mats using different ratios of sisal fiber/PET to prepare solutions subsequently submitted to electrospinning. The possibility of adjusting these parameters is of importance for several applications, for example, for the use of mats as enzyme supports.

The results of DMA and tensile tests revealed that fiber alignment in the direction of the applied force had a strong influence of and that the increase in the sisal fiber/PET ratio positively impacted the storage and tensile moduli, as well as the tensile strength of the mats. These results are of particular importance for applications in which mechanical properties are important, such as in filter systems and as reinforcements in nanocomposites.

This set of results discloses the potential of obtaining mats with diversified properties, opening an opportunity for a wide range of applications. The approach of this study, to our knowledge, is unprecedented.

Supplementary Materials: The supplementary materials are available online at http://www.mdpi.com/2073-43 60/10/5/538/s1.

Author Contributions: Conceptualization, E.F. and R.P.O.S.; Methodology, R.P.O.S., L.A.R. and P.F.R.; Software, R.P.O.S.; Validation, R.P.O.S., L.A.R. and P.F.R; Formal Analysis, R.P.O.S and L.A.R.; Investigation, R.P.O.S. and E.F.; Resources, E.F.; Data Curation, R.P.O.S and E.F.; Writing-Original Draft Preparation, R.P.O.S; Writing-Review & Editing, E.F.; Visualization, R.P.O.S. and E.F.; Supervision, E.F.; Project Administration, E.F.; Funding Acquisition, E.F.

Acknowledgments: The authors would like to thank CAPES (Coordination for the Improvement of Higher Level or Education Personnel) for providing the fellowship for Rachel Passos de Oliveira Santos, as well as CNPq (National Council of Scientific Research) for the fellowship for Patrícia Fernanda Rossi, for a research productivity fellowship to Elisabete Frollini and for financial support (process 426847/2016-4), as well as FAPESP (State of São Paulo Research Foundation, Brazil) for financial support (process 2012/00116-6).

Conflicts of Interest: The authors declare no conflict of interest.

References

1. Li, M.-C.; Wu, Q.; Song, K.; Cheng, H.N.; Suzuki, S.; Lei, T. Chitin nanofibers as reinforcing and antimicrobial agents in carboxymethyl cellulose films: Influence of partial deacetylation. *ACS Sustain. Chem. Eng.* **2016**, *4*, 4385–4395. [CrossRef]

2. Li, M.-C.; Mei, C.; Xu, X.; Lee, S.; Wu, Q. Cationic surface modification of cellulose nanocrystals: Toward tailoring dispersion and interface in carboxymethyl cellulose films. *Polymer* **2016**, *107*, 200–210. [CrossRef]

3. Amarasekara, A.S.; Ha, U.; Okorie, N.C. Renewable polymers: Synthesis and characterization of poly(levulinic acid-pentaerythritol). *J. Polym. Sci. Part A* **2018**, 1–4. [CrossRef]

4. Barbosa, V.; Ramires, E.C.; Razera, I.A.T.; Frollini, E. Biobased composites from tannin-phenolic polymers reinforced with coir fibers. *Ind. Crops Prod.* **2010**, *32*, 305–312. [CrossRef]

5. Jiang, Y.; Loos, K. Enzymatic synthesis of biobased polyesters and polyamides. *Polymers* **2016**, *8*. [CrossRef]

6. Ramires, E.C.; Megiatto, J.D.; Gardrat, C.; Castellan, A.; Frollini, E. Valorization of an industrial organosolv-sugarcane bagasse lignin: Characterization and use as a matrix in biobased composites reinforced with sisal fibers. *Biotechnol. Bioeng.* **2010**, *107*, 612–621. [CrossRef] [PubMed]

7. Schoon, I.; Kluge, M.; Eschig, S.; Robert, T. Catalyst influence on undesired side reactions in the polycondensation of fully bio-based polyester itaconates. *Polymers* **2017**, *9*, 693. [CrossRef]

8. Vijjamarri, S.; Streed, S.; Serum, E.M.; Sibi, M.P.; Du, G. Polymers from bioderived resources: Synthesis of poly(silylether)s from furan derivatives catalyzed by a salen–Mn(V) complex. *ACS Sustain. Chem. Eng.* **2018**, *6*, 2491–2497. [CrossRef]

9. Santos, R.P.O.; Castro, D.O.; Ruvolo-Filho, A.C.; Frollini, E. Processing and thermal properties of composites based on recycled PET, sisal fibers, and renewable plasticizers. *J. Appl. Polym. Sci.* **2014**, *131*, 1–13. [CrossRef]

10. Frollini, E.; Bartolucci, N.; Sisti, L.; Celli, A. Biocomposites based on poly(butylene succinate) and curaua: Mechanical and morphological properties. *Polym. Test.* **2015**, *45*, 168–173. [CrossRef]

11. Castro, D.O.; Passador, F.; Ruvolo-Filho, A.; Frollini, E. Use of castor and canola oils in "biopolyethylene" curauá fiber composites. *Compos. Part A Appl. Sci. Manuf.* **2017**, *95*, 22–30. [CrossRef]

12. De Oliveira, F.; da Silva, C.G.; Ramos, L.A.; Frollini, E. Phenolic and lignosulfonate-based matrices reinforced with untreated and lignosulfonate-treated sisal fibers. *Ind. Crops Prod.* **2017**, *96*, 30–41. [CrossRef]

13. Hung, K.; Yeh, H.; Yang, T.; Wu, T.; Xu, J.; Wu, J. Characterization of wood-plastic composites made with different lignocellulosic materials that vary in their morphology, chemical composition and thermal stability. *Polymers* **2017**, *9*, 726. [CrossRef]

14. Oliver-Ortega, H.; Méndez, J.A.; Mutjé, P.; Tarrés, Q.; Espinach, F.X.; Ardanuy, M. Evaluation of thermal and thermomechanical behaviour of bio-based polyamide 11 based composites reinforced with lignocellulosic fibres. *Polymers* **2017**, *9*. [CrossRef]

15. De Dicastillo, C.L.; Roa, K.; Garrido, L.; Pereira, A.; Galotto, M.J. Novel polyvinyl alcohol/starch electrospun fibers as a strategy to disperse cellulose nanocrystals into poly(lactic acid). *Polymers* **2017**, *9*. [CrossRef]

16. Li, R.; Tomasula, P.; de Sousa, A.M.M.; Liu, S.C.; Tunick, M.; Liu, K.; Liu, L. Electrospinning pullulan fibers from salt solutions. *Polymers* **2017**, *9*, 32. [CrossRef]

17. Zander, N.E. Hierarchically structured electrospun fibers. *Polymers* **2013**, *5*, 19–44. [CrossRef]

18. Fahami, A.; Fathi, M. Fabrication and characterization of novel nanofibers from cress seed mucilage for food applications. *J. Appl. Polym. Sci.* **2018**, *135*, 4–9. [CrossRef]

19. Rodrigues, B.V.M.; Ramires, E.C.; Santos, R.P.O.; Frollini, E. Ultrathin and nanofibers via room temperature electrospinning from trifluoroacetic acid solutions of untreated lignocellulosic sisal fiber or sisal pulp. *J. Appl. Polym. Sci.* **2015**, *132*, 1–8. [CrossRef]

20. Santos, R.P.O.; Rodrigues, B.V.M.; Ramires, E.C.; Ruvolo-Filho, A.C.; Frollini, E. Bio-based materials from the electrospinning of lignocellulosic sisal fibers and recycled PET. *Ind. Crops Prod.* **2015**, *72*, 69–76. [CrossRef]

21. De Oliveira Santos, R.P.; Rodrigues, B.V.M.; dos Santos, D.M.; Campana-Filho, S.P.; Ruvolo-Filho, A.C.; Frollini, E. Electrospun recycled PET-based mats: Tuning the properties by addition of cellulose and/or lignin. *Polym. Test.* **2017**, *60*, 422–431. [CrossRef]

22. Brazil Sisal Fibre Production. Available online: http://www.londonsisalassociation.org/brazilian-sisal-fibre.php (accessed on 19 March 2018).

23. Shi, H.H.; Naguib, H.E. Highly flexible binder-free core-shell nanofibrous electrode for lightweight electrochemical energy storage using recycled water bottles. *Nanotechnology* **2016**, *27*, 325402. [CrossRef] [PubMed]

24. Zander, N.E.; Gillan, M.; Sweetser, D. Recycled PET nanofibers for water filtration applications. *Materials* **2016**, *9*, 247. [CrossRef] [PubMed]

25. Pezzoli, D.; Cauli, E.; Chevallier, P.; Farè, S.; Mantovani, D. Biomimetic coating of cross-linked gelatin to improve mechanical and biological properties of electrospun PET: A promising approach for small caliber vascular graft applications. *J. Biomed. Mater. Res. Part A* **2017**, *105*, 2405–2415. [CrossRef] [PubMed]

26. Qin, D.; Lu, W.; Wang, X.; Li, N.; Chen, X.; Zhu, Z.; Chen, W. Graphitic carbon nitride from burial to re-emergence on polyethylene terephthalate nanofibers as an easily recycled photocatalyst for degrading antibiotics under solar irradiation. *ACS Appl. Mater. Interfaces* **2016**, *8*, 25962–25970. [CrossRef] [PubMed]

27. Liao, C.C.; Wang, C.C.; Chen, C.Y. Stretching-induced crystallinity and orientation of polylactic acid nanofibers with improved mechanical properties using an electrically charged rotating viscoelastic jet. *Polymer* **2011**, *52*, 4303–4318. [CrossRef]

28. Lui, Y.S.; Lewis, M.P.; Chye, S.; Loo, J. Sustained-release of naproxen sodium from electrospun-aligned PLLA—PCL scaffolds. *J. Tissue Eng. Regen. M.* **2017**, *11*, 1011–1021. [CrossRef] [PubMed]

29. Fee, T.; Surianarayanan, S.; Downs, C.; Zhou, Y.; Berry, J. Nanofiber alignment regulates NIH3T3 cell orientation and cytoskeletal gene expression on electrospun PCL+gelatin nanofibers. *PLoS ONE* **2016**, *11*, 1–12. [CrossRef] [PubMed]

30. Zhu, Y.; Pyda, M.; Cebe, P. Electrospun fibers of poly(L-lactic acid) containing lovastatin with potential applications in drug delivery. *J. Appl. Polym. Sci.* **2017**, *134*, 45287. [CrossRef]

31. Zhou, X.; Yang, A.; Huang, Z.; Yin, G.; Pu, X.; Jin, J. Enhancement of neurite adhesion, alignment and elongation on conductive polypyrrole-poly(lactide acid) fibers with cell-derived extracellular matrix. *Colloids Surfaces B Biointerfaces* **2017**, *149*, 217–225. [CrossRef] [PubMed]

32. Song, Y.; Sun, Z.; Xu, L.; Shao, Z. Preparation and characterization of highly aligned carbon nanotubes/polyacrylonitrile composite nanofibers. *Polymers* **2017**, *9*, 1. [CrossRef]

33. Wang, G.; Yu, D.; Kelkar, A.D.; Zhang, L. Electrospun nanofiber: Emerging reinforcing filler in polymer matrix composite materials. *Prog. Polym. Sci.* **2017**. [CrossRef]

34. Wu, M.; Wu, Y.; Liu, Z.; Liu, H. Optically transparent poly(methyl methacrylate) composite films reinforced with electrospun polyacrylonitrile nanofibers. *J. Compos. Mater.* **2012**, *46*, 2731–2738. [CrossRef]

35. Zhang, J.; Song, M.; Wang, X.; Wu, J.; Yang, Z.; Cao, J.; Chen, Y.; Wei, Q. Preparation of a cellulose acetate/organic montmorillonite composite porous ultrafine fiber membrane for enzyme immobilization. *J. Appl. Polym. Sci.* **2016**, *133*. [CrossRef]

36. Spackman, C.C.; Frank, C.R.; Picha, K.C.; Samuel, J. 3D printing of fiber-reinforced soft composites: Process study and material characterization. *J. Manuf. Process.* **2016**, *23*, 296–305. [CrossRef]

37. He, X.; Xiao, Q.; Lu, C.; Wang, Y.; Zhang, X.; Zhao, J.; Zhang, W.; Zhang, X.; Deng, Y. Uniaxially aligned electrospun all-cellulose nanocomposite nanofibers reinforced with cellulose nanocrystals: Scaffold for tissue engineering. *Biomacromolecules* **2014**, *15*, 618–627. [CrossRef] [PubMed]

38. Yarin, A.L. *Free Liquid Jets and Films: Hydrodynamics and Rheology*; Longman Scientific & Technical: New York, NY, USA, 1993.

39. Ngadiman, N.H.A.; Noordin, M.Y.; Idris, A.; Kurniawan, D. Effect of electrospinning parameters setting towards fiber diameter. *Adv. Mater. Res.* **2013**, *845*, 985–988. [CrossRef]

40. Ahn, Y.C.; Park, S.K.; Kim, G.T.; Hwang, Y.J.; Lee, C.G.; Shin, H.S.; Lee, J.K. Development of high efficiency nanofilters made of nanofibers. *Curr. Appl. Phys.* **2006**, *6*, 1030–1035. [CrossRef]

41. Amin, A.; Merati, A.A.; Bahrami, S.H.; Bagherzadeh, R. Effects of porosity gradient of multilayered electrospun nanofibre mats on air filtration efficiency. *J. Text. Inst.* **2016**, 1–9. [CrossRef]

42. Balgis, R.; Murata, H.; Goi, Y.; Ogi, T.; Okuyama, K.; Bao, L. Synthesis of dual-size cellulose—Polyvinylpyrrolidone nanofiber composites via one-step electrospinning method for high-performance air filter. *Langmuir* **2017**, *33*, 6127–6134. [CrossRef] [PubMed]

43. Fan, Y.; Li, X.; Jang, S.H.; Lee, D.H.; Li, Q.; Cho, U.R. Reinforcement of solution styrene-butadiene rubber by incorporating hybrids of rice bran carbon and surface modified fumed silica. *J. Vinyl Addit. Technol.* **2018**, 1–7. [CrossRef]

44. Zhang, Y.; Choi, J.R.; Park, S.-J. Thermal conductivity and thermo-physical properties of nanodiamond-attached exfoliated hexagonal boron nitride/epoxy nanocomposites for microelectronics. *Compos. Part A* **2017**, *101*, 227–236. [CrossRef]

45. Deng, F.; Zhang, Y.; Ge, X.; Li, M.; Li, X.; Cho, U.R. Graft copolymers of microcrystalline cellulose as reinforcing agent for elastomers based on natural rubber. *J. Appl. Polym. Sci.* **2016**, *133*, 1–11. [CrossRef]

46. Mubyana, K.; Koppes, R.A.; Lee, K.L.; Cooper, J.A.; Corr, D.T. The influence of specimen thickness and alignment on the material and failure properties of electrospun polycaprolactone nanofiber mats. *J. Biomed. Mater. Res. Part A* **2016**, *104*, 2794–2800. [CrossRef] [PubMed]

47. Kumar, P.; Vasita, R. Understanding the relation between structural and mechanical properties of electrospun fiber mesh through uniaxial tensile testing. *J. Appl. Polym. Sci.* **2017**, *134*, 1–11. [CrossRef]

© 2018 by the authors. Licensee MDPI, Basel, Switzerland. This article is an open access article distributed under the terms and conditions of the Creative Commons Attribution (CC BY) license (http://creativecommons.org/licenses/by/4.0/).

![polymers logo] *polymers*

MDPI

Article

Impact of Nanoclays on the Biodegradation of Poly(Lactic Acid) Nanocomposites

Edgar Castro-Aguirre [1], Rafael Auras [1,*], Susan Selke [1], Maria Rubino [1] and Terence Marsh [2]

[1] School of Packaging, Michigan State University, East Lansing, MI 48824, USA; castroag@msu.edu (E.C.-A.); sselke@msu.edu (S.S.); mariar@msu.edu (M.R.)

[2] Department of Microbiology and Molecular Genetics, Michigan State University, East Lansing, MI 48824, USA; marsht@msu.edu

* Correspondence: aurasraf@msu.edu; Tel.: +1-517-432-3254

Received: 9 January 2018; Accepted: 12 February 2018; Published: 17 February 2018

Abstract: Poly(lactic acid) (PLA), a well-known biodegradable and compostable polymer, was used in this study as a model system to determine if the addition of nanoclays affects its biodegradation in simulated composting conditions and whether the nanoclays impact the microbial population in a compost environment. Three different nanoclays were studied due to their different surface characteristics but similar chemistry: organo-modified montmorillonite (OMMT), Halloysite nanotubes (HNT), and Laponite® RD (LRD). Additionally, the organo-modifier of MMT, methyl, tallow, bis-2-hydroxyethyl, quaternary ammonium (QAC), was studied. PLA and PLA bio-nanocomposite (BNC) films were produced, characterized, and used for biodegradation evaluation with an in-house built direct measurement respirometer (DMR) following the analysis of evolved CO_2 approach. A biofilm formation essay and scanning electron microscopy were used to evaluate microbial attachment on the surface of PLA and BNCs. The results obtained from four different biodegradation tests with PLA and its BNCs showed a significantly higher mineralization of the films containing nanoclay in comparison to the pristine PLA during the first three to four weeks of testing, mainly attributed to the reduction in the PLA lag time. The effect of the nanoclays on the initial molecular weight during processing played a crucial role in the evolution of CO_2. PLA-LRD5 had the greatest microbial attachment on the surface as confirmed by the biofilm test and the SEM micrographs, while PLA-QAC0.4 had the lowest biofilm formation that may be attributed to the inhibitory effect also found during the biodegradation test when the QAC was tested by itself.

Keywords: montmorillonite; halloysite; Laponite®; composting; biofilm; degradation; bio-based

1. Introduction

Biodegradable polymers like poly(lactic acid) (PLA), poly(butylene adipate-*co*-terephthalate) (PBAT), and thermoplastic starch (TPS), have great potential to replace fossil-based polymers, avoid landfill disposal of most non-recyclable polymers, and help reduce environmental impacts. However, these materials have some properties and processing shortcomings that have limited their use in many applications, for example, brittleness, water sensitivity, low heat distortion temperature, medium to high gas permeability, and low melt viscosity [1,2]. Therefore, the creation of bio-nanocomposites (BNCs) in which the reinforcements have at least one dimension in the nanoscale dimension and the matrix is a biodegradable polymer, preferably a bio-based polymer, has garnered attention [1,3,4]. Ideally, BNCs could be recycled or treated together with other organic wastes in composting facilities and produce compost, a valuable soil conditioner and fertilizer [5].

One particularly useful class of nanofillers used to produce BNCs is inorganic layered silicate minerals, or nanoclays, due to their commercial availability, low cost, significant property enhancement and relatively simple processability [3]. Natural nanoclays, such as montmorillonite

(MMT) with chemical structure $[Na_{0.38}K_{0.01}][Si_{3.92}Al_{0.07}O_8][Al_{1.45}Mg_{0.55}O_2(OH)_2]\cdot7H_2O$, and synthetic nanoclays, such as Laponite® RD (LRD) with chemical structure $Na_{0.7}[(Si_8Mg_{5.5}Li_{0.3})O_{20}(OH)_4]_{0.7}$, and halloysite nanotubes (HNT) with chemical structure $Al_2(OH)_4Si_2O_5(2H_2O)$, offer a unique route for enhancing the mechanical, physical and barrier properties of biodegradable polymers at low levels of loading (<5 wt %), especially when the nanoclay particles are well dispersed in the polymer matrix [2,6]. However, the dispersion of hydrophilic nanofillers in a polymer matrix is challenging. Organophilization, or organic modification, is a technique that improves clay compatibility with organic polymers by reducing the surface energy between the clay layers. Increasing the clay inter-gallery spacing facilitates the intercalation and exfoliation of the clay in the polymer matrix [2,3]. The exfoliation into individual layers depends on the clay's ability for surface modification in which the interlayer inorganic ions are exchanged with organic cations [4,7].

The most broadly studied organo-modifiers are ammonium alkyls. When the clay inorganic ions are exchanged with these organic cations, the inter-gallery spacing increases due to the bulkiness of the alkyl–ammonium ions [7]. For example, organo-modified montmorillonite (OMMT), in which its inorganic ions (e.g., Na^+, K^+, Ca^{2+}, and Mg^{2+}) have been replaced by organic alkyl-ammonium ions, improving the wetting with the polymer chains [1,3]. Several researchers have reported improvement in the properties and performance of PLA with addition of OMMT. For example, Ray et al., through a series of papers, demonstrated that the addition of montmorillonite has a significant effect in the improvement of PLA properties (in both solid and melt states), crystalline behavior, and biodegradability in comparison with pristine PLA. Among the different mechanical properties that have been improved are storage modulus, flexural modulus, flexural strength, tensile modulus and elongation at break [8–10]. Additional benefits in performance have been reported such as increased glass transition and thermal degradation temperatures [3,11]. Another reported advantage, other than enhancement of the mechanical and thermal properties, is improvement in the barrier properties due to the enhanced tortuous path provided by the silicate layers to gases like oxygen [9,12,13]. The decreased transparency is a minor disadvantage of these BNCs [3]. Other researchers have found significant improvement in thermo-mechanical and barrier properties of BNCs based on PLA and OMMT [14,15].

Halloysite is another type of nanoclay that has received great attention as filler for polymer/clay nanocomposites due to its biocompatibility, natural abundance, and relatively low cost. HNT has almost no surface charge and does not require organic modification for adequate dispersion [16,17]. However, functionalized HNT has shown improved dispersion during processing and enhanced mechanical and thermal properties [18,19]. HNT has been used as filler for several polymers like poly(propylene) (PP), vinyl ester, polyamide (PA), poly(vinyl chloride) (PVC), epoxy, and natural rubber for enhancing properties such as mechanical, thermal, crystallinity, and fire resistance [18,19]. Researchers have found that PLA-HNT nanocomposites exhibited improvement in properties like tensile strength, Young's modulus, impact properties, flexural properties, and storage modulus, but no significant modification in the thermal properties in comparison with pure PLA [16,20–22]. The addition of HNT promotes crystallization and formation of different crystalline phases [21,22]. HNT was also found to slightly increase water absorption [23]. Other researchers found increased thermal and flame retardant properties besides improvement in mechanical properties [19]. Esma et al. also found enhanced thermal properties, but in their case mechanical properties were not significantly improved [24]. Similarly, Kim et al. found decreased tensile strength with clay loading higher than 5 wt % but enhanced rheological properties [17].

Laponite® (LRD), another clay that might lead to novel properties, has not been widely investigated for the development of PLA-based nanocomposites. LRD is an entirely synthetic hectorite clay that belongs to the group of smectite phyllosilicate minerals, and it has great capacity for swelling and exfoliation [25,26]. The advantage of using synthetic clays like LRD is the high structural regularity, single layer dispersions of nanoparticles, and low level of impurities (e.g., silica, iron oxides, and carbonates). Due to its gelation properties, LRD has been used for different pharmaceutical and cosmetic applications; for example, toothpastes, creams, and glazes [27–30]. Zhou et al. studied PLA-LRD composite films and found improvement in the thermal stability, tensile strength and hydrophilicity of PLA, especially when the LRD content is below 0.2 wt %. [31,32]. Similarly, Tang et al. studied nanocomposites based on starch, poly vinyl alcohol (PVOH), and LRD and found that an increase in LRD content (0–20%) enhanced tensile strength and decreased water vapor permeability [26].

Besides performance limitations, one of the drawbacks of some biodegradable polymers, like PLA, is that they do not biodegrade as fast as other organic wastes during composting, which in turn affects their general acceptance in industrial composting facilities [33]. Therefore, increasing their biodegradation rate in the composting environment should facilitate and encourage their disposal through these facilities by degrading in a time frame comparable with other organic materials.

Several researchers studied the effect of OMMT on the biodegradation of biodegradable polymers like polycaprolactone (PCL) [34], poly(3-hydroxybutyrate-*co*-3-hydroxyvalerate) (PHBV) [35], TPS [36], and PLA [10,33,37–45]. Their results indicated that, in general, these BNCs biodegraded faster than their respective pristine polymer. Therefore, the incorporation of nanoclays into a biodegradable polymer matrix represents a promising approach not only for enhancing the polymer performance but also for increasing its biodegradation rate in composting conditions. However, the effect of different nanoclays and organo-modifiers on the abiotic and biotic degradation of PLA is still unclear and needs further investigation. Even though it is well known that the biodegradation mechanism of PLA involves chemical hydrolysis, the role of microorganisms and how they are affected by the presence of nanoparticles is still not well understood [44].

Thus, this study aimed to understand the biodegradation mechanisms of BNCs made of PLA and compounded with OMMT, HNT, and LRD, and to identify the main factors contributing to their biodegradation rate such as those related to the polymer structure and also those related to the soil/compost environments or to the microbial populations that could be impacted by the presence of nanoparticles.

2. Materials and Methods

2.1. Materials

Poly(lactic acid) resin (Ingeo™ 2003D) was obtained from NatureWorks LLC. (Minnetonka, MN, USA) with 3.8–4.2% D-LA, number average molecular weight (Mn) of 121.1 ± 7.5 kDa, polydispesity index (PDI) of 1.9 ± 0.1, and melt flow index (MFI) of 6 g/10 min (210 °C, 2.16 kg). Cellulose powder (particle size ~20 μm) and Halloysite nanotubes (HNT) were purchased from Sigma-Aldrich (St. Louis, MO, USA). Organo-modified montmorillonite (OMMT) (Cloisite® 30B) and Laponite® RD (LRD) were acquired from BYK Additives Inc. (Gonzales, TX, USA). Additionally, Tomamine™ Q-T-2 (QAC) with 60–70% purity of a methyl, tallow, bis-2-hydroxyethyl, quaternary ammonium, the organo-modifier of Cloisite® 30B, was obtained from Air Products and Chemicals Inc. (Butler, IN, USA). Tetrahydrofuran (THF) was obtained from Pharmco-AAPER (North East, CA, USA). The composition per liter of the R2 broth (R2B) used was 0.5 g yeast extract, 0.5 g proteose peptone #3, 0.5 g casamino acids, 0.5 g dextrose, 0.5 g soluble starch, 0.3 g sodium pyruvate, 0.3 g dipotassium phosphate, and 0.05 g magnesium sulfate. The composition per liter of the M9 minimal medium was 12.8 g $Na_2HPO_4 \cdot 7H_2O$, 3 g KH_2PO_4, 0.5 g NaCl, 1 g NH_4Cl, and 1 g of 1 mM $MgSO_4$, 1 mM $CaCl_2$, 3×10^{-9} M $(NH_4)_6Mo_7O_{24} \cdot 4H_2O$, 4×10^{-7} M H_3BO_3, 3×10^{-8} M $CoCl_2 \cdot 6H_2O$, 1×10^{-8} M $CuSO_4 \cdot 5H_2O$, 8×10^{-8} M $MnCl_2 \cdot 4H_2O$, 1×10^{-8} M $ZnSO_4 \cdot 7H_2O$, 1×10^{-6} M $FeSO_4 \cdot 7H_2O$. All the chemicals and reagents were commercial products of the highest available grade.

2.2. Processing of the PLA Bio-Nanocomposites

PLA-BNCs (PLA-OMMT, PLA-LRD, and PLA-HNT) were produced in a two-step process. First, masterbatches were prepared in a ZSK 30 twin-screw extruder (Werner Pfleiderer, NJ, USA) and pelletized. Second, PLA-BNC films (1 and 5 wt % nanoclay) were produced in a cast film microextruder model RCP-0625 (Randcastle Extrusion Systems, Inc., Cedar Grove, NJ, USA). Two PLA-QAC films (0.4 and 1.5 wt % organo-modifier) were produced in a similar fashion. Three PLA films (PLA1, PLA2, and PLA3) with different molecular weight were obtained by varying the processing conditions, and used as control films. In all cases, the materials were dried at 60 °C for 8 h under vacuum (85 kPa) prior to processing. The thickness of the films was measured using a digital micrometer (Testing Machines Inc., New Castle, DE, USA). More details regarding the film processing are provided in Table S1.

2.3. Characterization of the PLA Bio-Nanocomposites

To evaluate the presence and dispersion of the nanoclays in the PLA matrix, X-ray diffraction (XRD) and transmission electron microscopy (TEM) were performed. PLA and BNC films were embedded in paraffin blocks and microtomed in 100-nm sections for bright field imaging using an Ultramicrotome MYX (RMC Boeckeler Instruments, Tucson, AZ, USA). TEM micrographs were obtained using a JEOL 2200FS transmission electron microscope (JEOL USA, Inc., Peabody, MA, USA) operating at an acceleration voltage of 200 kV. XRD analysis was conducted in a Rigaku Rotaflex Ru-200BH X-ray diffractometer equipped with a Ni-filtered Cu Kα radiation source setting at 45 kV and 100 mA. The interlayer spacing was calculated according to Bragg's Law [46]. The carbon, hydrogen and nitrogen content, as well as the amount of nanoclay present in each BNC film was determined by elemental analysis (CHN) and are reported in Table S2. Additional methodologies, such as differential scanning calorimetry (DSC), thermal gravimetric analysis (TGA), moisture isotherm, electrical conductivity, and contact angle, used for characterization of the BNCs are provided in Section S2.

2.4. Biodegradation Evaluation

The aerobic biodegradation of PLA and BNCs was evaluated through a series of experiments (Table 1) by analysis of evolved CO_2 under controlled composting conditions (at 58 °C), using an in-house built direct measurement respirometer (DMR) with a CO_2 non-dispersive infrared gas analyzer (NDIR). Manure compost from the MSU Composting Facility (East Lansing, MI, USA) was used. The compost was sieved on a 10 mm screen and preconditioned at 58 °C for three days prior to use. Deionized water was incorporated to adjust the moisture content to about 50%. Saturated vermiculite premium grade (Sun Gro Horticulture Distribution Inc., Bellevue, WA, USA) was mixed with the compost (1:4 parts, dry wt. compost) for better aeration. Compost samples were sent to the Soil and Plant Nutrient Laboratory at Michigan State University (East Lansing, MI, USA) for determination of the physicochemical parameters (dry solids, volatile solids, C/N ratio, and pH) and are reported in Table S6. Detailed information about the methods used for compost characterization can be found elsewhere [47].

Table 1. Key for biodegradation test and labels of the samples.

Test ID	Samples Tested
I	Blank, Cellulose, OMMT, HNT, LRD, PLA1, PLA-OMMT5
II	Blank, Cellulose, OMMT, OMMT5, QAC, QAC5, PLA1, PLA-OMMT1, PLA-OMMT5, PLA-OMMT7.5
III	Blank, Cellulose, PLA2, PLA-OMMT1, PLA-OMMT5, PLA-HNT1, PLA-HNT5, PLA-LRD1, PLA-LRD5, PLA-QAC1.5, PLA-QAC0.4
IV	Blank, Cellulose, PLA1, PLA2, PLA3, PLA-OMMT5, PLA-QAC0.4

The bioreactors were loaded with 400 g of compost (or vermiculite) and mixed thoroughly with 8 g of polymer sample (unless otherwise specified). Film samples were cut to 1 cm^2 pieces and triplicates of each test material were analyzed. Additionally, triplicates of blank bioreactors (with compost or vermiculite only) were evaluated. To simulate composting conditions, the bioreactors were placed in an environmental chamber set at a constant temperature of 58 ± 2 °C. Water-saturated CO_2-free air was provided to each bioreactor with a flow rate of 40 ± 2 sccm (cm^3/min at standard temperature and pressure). The bioreactors were incubated in the dark for at least 45 d or until the evolved CO_2 reached a plateau. For all the biodegradation studies, the results are presented as average ($n = 3$) and standard deviation.

2.5. Size Exclusion Chromatography (SEC)

The number average molecular weight (*Mn*), weight average molecular weight (*Mw*), and polydispersity index (*PDI*) of PLA and BNCs before and during composting were determined by SEC with a system from Waters Inc. (Milford, MA, USA) as previously described [47]. Shortly, 20 mg of films were dissolved in 10 cm^3 of THF and filtered with a hydrophobic polytetrafluoroethylene (0.45 µm pore size) filter. Then, 100 µL of each sample solution were injected. A third-order polynomial calibration curve was obtained from polystyrene (PS) standards ranging 0.5–2,480 kDa, and the Mark–Houwink constants, $K = 0.000164$ dL/g and $\alpha = 0.704$, for PS were used.

2.6. Microbial Attachment

Biofilm Assay: The biofilm forming ability of microorganisms on the surface of PLA and BNCs was assessed with a biofilm assay in 24-well polystyrene plates as described elsewhere [48,49]. For this test, sterilized PLA films and BNC films were added to the wells of a microtiter plate (24 wells). The films were sterilized by rinsing with 70% ethanol, followed by irradiation with ultraviolet light for 5 min prior to testing. Four replicates of each sample were tested. Each well contained 600 µL of R2B and 200 µL of compost extract (CE), which was prepared by vigorously mixing dry compost with deionized water (1:2 wt./vol.) on vortex for 2 min. The mix was allowed to settle for 20 min and then the supernatant was passed through a sieve with 1 mm mesh. A sterile compost extract (SCE) was prepared for a control by passing the CE twice through a 0.22 µm filter. The inoculated plates were incubated for 48 h at 58 °C gently shaking at 100 rpm. *Pseudomonas aeruginosa* (PA) strain PAO1, a biofilm producing bacterium, was used as a positive control at 23 °C, and uninoculated wells were considered as a negative control. To determine the level of biofilm formed on the surface of PLA and BNCs after incubation, the films were transferred to clean Eppendorf tubes and treated in parallel with the microtiter plates. The broth was removed from the plates and the wells and films were gently washed with water three times. The biofilm was stained with 800 µL of 0.5% crystal violet for 15 min followed by washing three times with water. After the plates and films had air-dried, 800 µL of 30% acetic acid were added, followed by incubation for 15 min. Measurements were done using an Epoch™ Microplate Spectrophotometer (BioTek Instruments, Inc., Winooski, VT, USA) at 600 nm directly on the wells and following decantation of the films. Decanted acetic acid from films was transferred into clean microtiter plates for absorbance measurement at 600 nm. The biofilm formation was quantified by subtracting the average absorbance of the cognate controls from the average absorbance of the inoculated samples.

Scanning Electron Microscopy (SEM): Similar to the biofilm test, sterilized PLA films and PLA-LRD5 films were added to an Erlenmeyer flask containing R2B ($2\times$) and an overnight culture of the compost extract (CE) on R2B at 58 °C (3:1 vol.). The samples were incubated for 48 h at 58 °C. The films were removed from the flasks, gently washed with water three times, and air-dried. The samples were mounted on aluminum stubs using high vacuum carbon tabs (SPI Supplies, West Chester, PA, USA), and coated with osmium. SEM micrographs were obtained at various magnifications using a JEOL 6610LV (tungsten hairpin emitter) scanning electron microscope (JEOL Ltd., Tokyo, Japan) operating at a voltage of 10 kV to observe the biofilm formation.

2.7. Statistical Analysis

All statistical analyses were performed using Minitab18 software (Minitab Inc., State College, PA, USA) by analysis of variance (one way ANOVA), and Tukey test with a *p*-value threshold of 0.05 as for level of significance. Data are reported as mean and standard deviations.

3. Results and Discussion

3.1. Characterization of the PLA Bio-Nanocomposites

Figures 1 and 2 show the XRD spectra and TEM micrographs of the BNCs, respectively. These methods were used to evaluate the presence and dispersion of the nanoclays in the PLA matrix. Depending on the degree of dispersion, a layered silicate nanocomposite can be either intercalated or exfoliated. Intercalation occurs when the polymer chains penetrate into the interlayer regions of the clay, while exfoliation is observed when the clay layers are delaminated and randomly dispersed in the polymer matrix [3]. As observed in Figure 1a, in the case of PLA-OMMT5 film, OMMT is not fully exfoliated but intercalated in the PLA matrix, which is represented by the shift of the peak to the left, i.e., the increase in the interlayer distance from 1.85 nm, for the pristine OMMT, to 3.42 nm, for the OMMT present in the film. The organic modification of the MMT through exchange of cationic ions allows for better dispersion and exfoliation of the silicate layers into the PLA matrix [1,3,7]. However, in the case of PLA-OMMT5 it was not enough to obtain a fully exfoliated BNC. This was confirmed by the TEM micrograph (Figure 2a), which shows some small agglomerations. However, it seems that the OMMT is evenly distributed in the PLA matrix. PLA-OMMT1 showed a better dispersion of the OMMT in the polymer matrix, but in general, full exfoliation is difficult to achieve, and most nanocomposites are a mixture of both structures, which is usually referred to as disordered morphology or orderly exfoliated morphology [4].

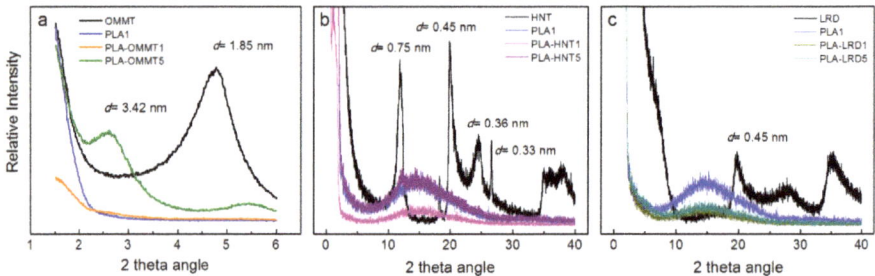

Figure 1. XRD spectra of the different nanoclays, PLA1, and (**a**) OMMT, (**b**) HNT, and (**c**) LRD bio-nanocomposite films.

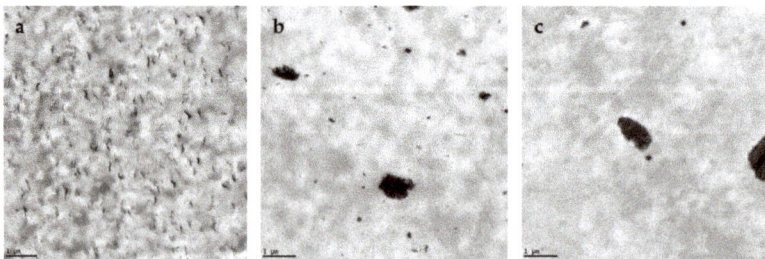

Figure 2. TEM micrographs of (**a**) PLA-OMMT5, (**b**) PLA-HNT5, and (**c**) PLA-LRD5 bio-nanocomposites at 10k×. The bar in the left bottom represents 1 μm.

Similarly, Figure 1b,c show the XRD spectra of HNT and LRD nanocomposites, respectively. In both cases, the profiles showed broad peaks around a 2θ angle of 16, which are representative of amorphous PLA samples [50,51]. HNT is an alumina-silicate clay with an elongated hollow tubular structure consisting of an external surface composed of siloxane (Si-O-Si) groups and an inner side and edges consisting of (Al-OH) groups [16,24,52]. In the XRD spectrum of the HNT nanoclay (Figure 1b), the presence of three main peaks at 2θ angles of 12.02, 19.99, and 24.54 can be observed, corresponding to the basal *d*-spacing of 0.75, 0.45, and 0.36 nm, respectively. Similar diffraction patterns are reported elsewhere [24,53–57]. In the case of PLA-HNT5, the presence of a peak at 2θ angle of 12.25 was observed. The small shift to the right, from the 12.02 of the pristine HNT, indicates a reduction in the *d*-spacing. This behavior has been observed by other researchers, and was attributed to the formation of a micro-filled composite [24,54]. The disappearance of the other peaks, such in the case of PLA-HNT5 and PLA-HNT1, has been explained as due to the interaction of the polymer chains with the nanotubes, and also due to the preferential orientation of nanotubes during processing of the film [19,24]. It was also observed that the intensity of the characteristic peaks depends on the level of loading of nanoclay [53,54].

LRD particles have a disk-like shape with two external tetrahedral silica sheets that present continuous corner-shared tetrahedral SiO_4 units arranged in hexagonal rings, and a central octahedral magnesia sheet that is composed of bivalent or trivalent cations sharing the edges coordinated to hydroxyl groups. The excess negative charge is compensated by the presence of Na ions between the silicate layers [25,27–29]. In the XRD spectrum of the LRD nanoclay (Figure 1c), the presence of the characteristic LRD peak at 2θ angle of 19.8 can be observed, corresponding to the basal *d*-spacing of 0.45 nm. Similar diffraction patterns are reported for LRD elsewhere [25,26]. In the XRD spectra of the PLA-LRD, no diffraction peaks were observed. This behavior has been attributed, in the literature, to separated LRD platelets dispersed individually in the polymer matrix [25]. The nanoclay dispersion was also confirmed by TEM.

Figure 2b,c show the TEM micrographs of HNT and LRD nanocomposites, respectively. In the case of PLA-HNT5, Figure 2b shows the presence of big agglomerations indicating that HNT was not evenly distributed in the PLA matrix. Similar observations have been reported in the literature for PLA-HNT nanocomposites [20,53]. A similar distribution was also found for the PLA-LRD5 film (Figure 2c).

Other factors influencing the nanoclay dispersion in the PLA matrix are the level of loading and the size of the nanoparticles [26]. For example, HNT and LRD are bigger particles than MMT. While MMT has layers with 1 nm thickness and tangential dimensions from 300 Å to a few microns [1,3,7], HNT has inner and outer diameters of the tube ranging from 10 nm to 40 nm and 40 nm to 70 nm, respectively, while the length ranges from 0.2 μm to 3 μm [16,24,52]. LRD usually has dimensions around 25–30 nm in diameter and 1 nm in thickness [26,27,29].

3.2. Biodegradation Evaluation

The DMR system was used to perform four different biodegradation tests in which the CO_2 evolved from each bioreactor was measured with controlled temperature, RH, and air flow rate. For the data analysis, the average cumulative CO_2 and % mineralization of each test material was calculated and plotted as a function of time. Detailed information about the concepts and calculations is provided elsewhere [47,58–60]. The blank bioreactors contain the solid media only (i.e., compost or vermiculite). In all cases, cellulose powder was used as a positive reference material since it is a well-known easily biodegradable material. The cumulative CO_2 and % mineralization curves obtained from the different biodegradation tests for the evaluation of PLA and PLA-BNCs, as well as the different nanoclays and surfactant, are presented in Figures 3–11.

To evaluate the effect of the nanoclays on the compost microbial population, the three different nanoclays were tested in the powder form as received. Figure 3 shows the CO_2 evolved from the bioreactors containing the three different nanoclays. A significant difference between the CO_2 evolved from cellulose and the one from the nanoclays was observed. During the first 40 days of the test, OMMT and LRD bioreactors produced a significantly higher amount of CO_2 than the blank indicating that there was no inhibition. On the contrary, the HNT bioreactors produced equal or less CO_2 than the blank, especially after 35 days, indicating some kind of inhibition in which HNT may limit the availability and/or the distribution of carbon and other nutrients for basic microorganism functions.

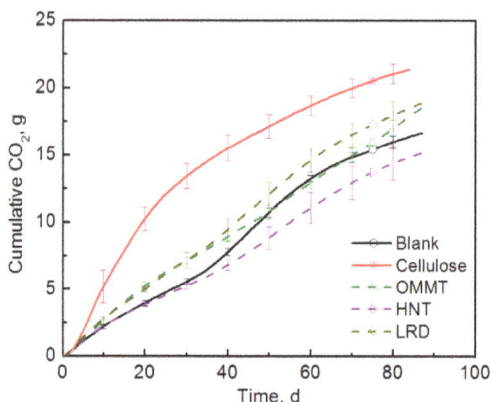

Figure 3. CO_2 evolution of the three different nanoclays (Test I in compost).

Figure 4 shows the CO_2 and % mineralization of the pristine PLA film and PLA-OMMT5. The typical PLA biodegradation behavior with the presence of a lag time of around 25 days was observed [47,61]. The lag time observed in the biodegradation of PLA has been explained by the low diffusion rate of the byproducts formed during the hydrolytic degradation and present inside the sample [62]. Cellulose reached a maximum mineralization of 65.7% after 34 days while PLA and PLA-OMMT5 reached 53.2 and 59.6% after 87 days, respectively. The decrease in the mineralization curve of cellulose indicates that these bioreactors were no longer producing more CO_2 than the blank bioreactors. This behavior may be explained by a rapid and large increase of the microbial population at the beginning of the test when there are plenty of resources easily available for microbial assimilation. Then, a decrease in the mineralization curve is observed when these resources are depleted and/or limited [47]. Even though by the end of the test, the mineralization of PLA and PLA-OMMT5 was not significantly different, it was clearly observed that the lag phase of the pristine PLA was longer than the PLA-OMMT5. The mineralization of PLA-OMMT5 was significantly higher before day 60. Among the different explanations for this accelerated biodegradation due to OMMT found in the literature is the relatively high hydrophilicity of the nanoclay, which improves the diffusion of water into the PLA polymeric matrix and in turn promotes hydrolytic degradation [33,37,38,44,62]. Another reason is that the presence of terminal hydroxyl groups in the silicate layers and in some organo-modifiers promotes the hydrolytic degradation of PLA [10,44,63]. However, the molecular weight of the PLA-OMMT5 films and the thickness can also play a crucial role and influence the observed results [47].

To evaluate the effect of clay loading on the biodegradation of PLA, three films with different loadings of OMMT (1, 5, and 7.5 wt %) were tested. Figure 5 shows the CO_2 evolution and % mineralization of PLA and PLA-OMMT films. Cellulose reached a maximum mineralization of 61.7% after 45 days of testing. The biodegradation behavior of the pristine PLA and PLA-OMMT1 was similar, again with a typical lag time at the beginning of the biodegradation test. The negative mineralization values observed in Figure 5b are generated as an artifact when the blank bioreactors produce more CO_2 than the sample material bioreactors. This effect might be caused because of the physical barrier offered by the polymer film at this early stage of the test, contrary to the PLA-OMMT5 and PLA-OMMT7.5 in which their biodegradation phase started much earlier. The observed shorter lag time of PLA-OMMT5 is in agreement with the previous test results, but in this case the average mineralization was significantly higher than the PLA control. It seems that PLA-OMMT7.5 has the highest average mineralization and the fastest biodegradation rate in which the lag time was only around five days. However, mineralization values above 100% indicate the presence of a priming effect, in which the additional carbon converted to CO_2, is not coming from the sample material but from the over-degradation of the indigenous organic carbon present in the compost [47,64]. Again, the initial molecular weight of the films should influence the observed results. It is important to mention that during the processing of the films, with different nanoclay loading, the resulting molecular weight was affected even though, in this case, the same processing conditions were maintained, with the higher clay loading corresponding to the lower molecular weight. Furthermore, Roy et al. analyzed the water-soluble degradation products by electrospray ionization-mass spectrometry (ESIMS), and their results indicated a catalytic effect of MMT in hydrolysis of PLA since shorter lactic acid oligomers were formed in the case of PLA/MMT composites [41]. Some researchers have attributed a plasticizing effect to the degradation byproducts (i.e., lactic acid oligomers and monomers), represented by a decrease in the T_g of PLA and BNCs. In this context, faster biodegradation of the PLA and BNC could also be induced by the increased segmental mobility of backbone chains and the expanded amorphous regions of the polymeric matrix [44,62,65]. Another factor influencing the biodegradation rate of the BNCs is the crystallinity of the material. The presence of nanoclays could affect the degree of crystallization of PLA (Table S3), with the amorphous parts preferentially biodegrading [47].

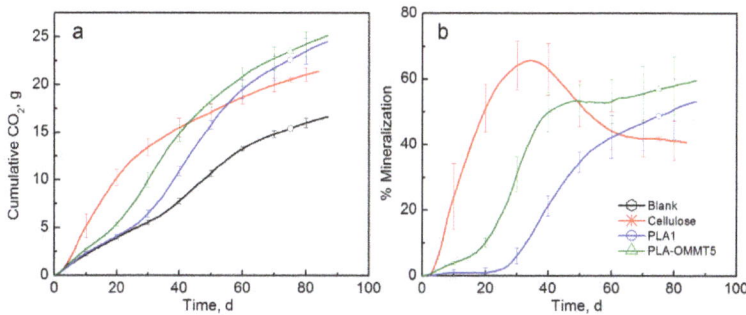

Figure 4. (**a**) CO_2 evolution and (**b**) % Mineralization of PLA and PLA-OMMT5 films (Test I in compost).

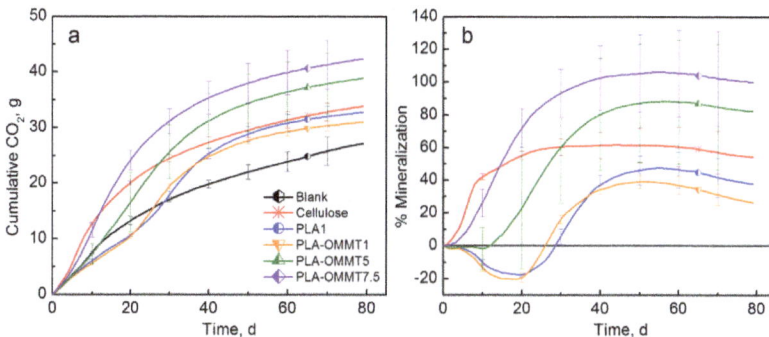

Figure 5. (**a**) CO_2 evolution and (**b**) % Mineralization of PLA and PLA-OMMT films with three different levels of loading (1, 5, and 7.5%) (Test II in compost).

The effect of the amount/concentration of clay and surfactant on the compost microbial populations was evaluated and the results are shown in Figure 6. In this case, OMMT refers to 8 g of the tested sample material, while OMMT5 refers to the theoretical amount of nanoclay contained in 8 g of PLA-OMMT5 film. Similarly, QAC refers to 8 g of the tested sample material and QAC5 to the theoretical amount of surfactant contained in 8 g of PLA-OMMT5 film. Regardless of the concentration of either OMMT or QAC, the CO_2 evolution was always significantly lower than the blank, indicating that there was clear inhibition of the microbial activity when these materials were present by themselves.

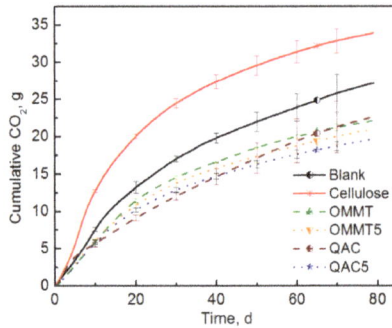

Figure 6. CO_2 evolution of OMMT nanoclay and QAC surfactant (Test II in compost).

Figure 7 shows the results of a different biodegradation test in which the PLA-OMMT and the PLA-QAC films were evaluated. Cellulose reached a mineralization of 85.5% after 38 days of testing, while the PLA control reached 74.2% after 69 days. As in the previous test, there was no significant difference between the pristine PLA and the PLA-OMMT1 films (Figure 7b). However, PLA-OMMT5 had significantly higher mineralization and a shorter lag time than the PLA control. A priming effect was observed with mineralization values over 100%. The PLA films containing the surfactant (QAC) also showed reduced lag time and a significantly higher amount of evolved CO_2 than the PLA control, and in both cases a priming effect was observed (Figure 7d). This may be due to the lower initial molecular weight of these films. In our previous work [47], it was demonstrated that the PLA film with the lowest M_n presented a priming effect when tested in compost, but it was not observed in inoculated vermiculite, having mineralization values closer to the other two tested PLA films with higher M_n. PLA-OMMT5 and PLA-QAC0.4 were also tested in inoculated and uninoculated vermiculite, and the results are later shown in Figure 11. Similarly, the priming effect was not observed in this case.

Figure 8 shows that the mineralization of PLA-HNT films was not significantly different from the PLA control by the end of the test (90 days). However, it can be clearly observed that with both levels of loading the lag time was reduced and the mineralization was significantly different before day 45. A higher variability and also a priming effect were observed in the biodegradation of PLA-HNT1 film. PLA-HNT films reached their maximum mineralization after 50 days of testing with an average of 86.9 and 74.6% for PLA-HNT1 and PLA-HNT5, respectively.

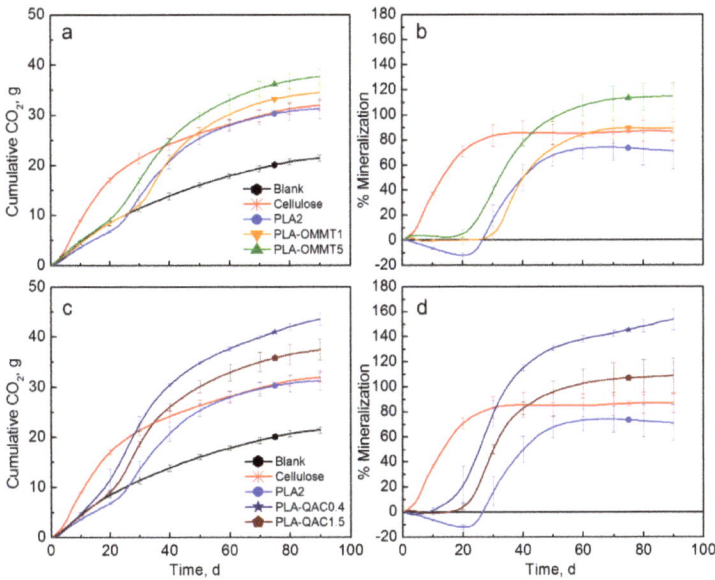

Figure 7. CO_2 evolution and % Mineralization of PLA-OMMT films (**a,b**) and PLA-QAC films (**c,d**) (Test III in compost).

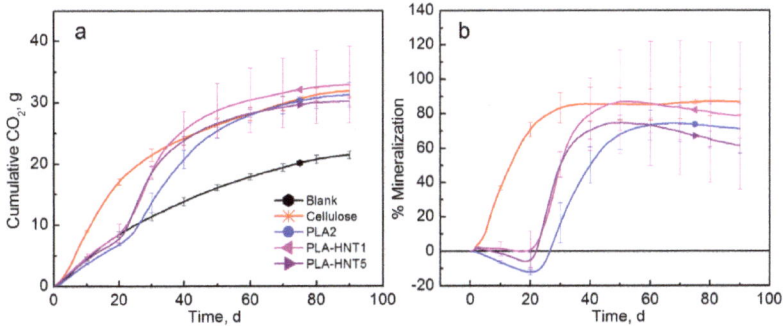

Figure 8. (**a**) CO_2 evolution and (**b**) % Mineralization of PLA-HNT films (Test III in compost).

As observed in Figure 9, PLA-LRD5 showed significantly higher mineralization than the pristine PLA and the PLA-LRD1 films. In this case, the lag time was not reduced but the PLA-LRD5 showed a priming effect. PLA-LRD films reached their maximum mineralization by the end of the test with an average of 82.5 and 112.5% for PLA-LRD1 and PLA-LRD5, respectively.

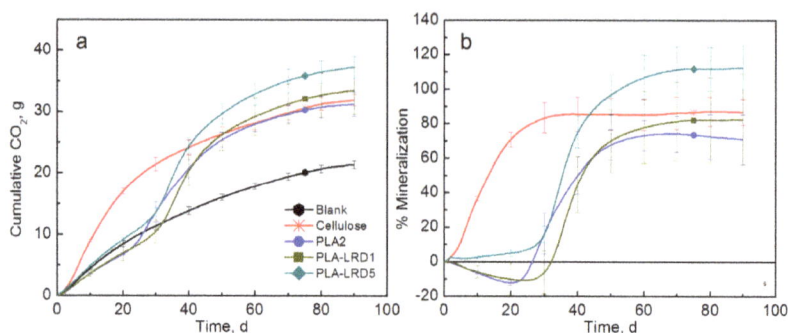

Figure 9. (a) CO_2 evolution and (b) % Mineralization of PLA-LRD films (Test III in compost).

To avoid the priming effect observed in the previous tests, a specific new biodegradation test was performed in three different solid environments (compost, inoculated vermiculite, vermiculite) as described elsewhere [47]. When tested in compost (Figure 10), there was no significant difference in the mineralization of these materials by the end of the test (132 days). However, it seems that the mineralization of PLA-OMMT5 was significantly higher than the PLA during the first 45 days of testing. Similarly to the previous tests, PLA-OMMT5 showed a reduced lag time and a priming effect could be occurring due to the low molecular weight of both films. The maximum average mineralization for PLA and PLA-OMMT5 was 110.4 and 100.2%, respectively.

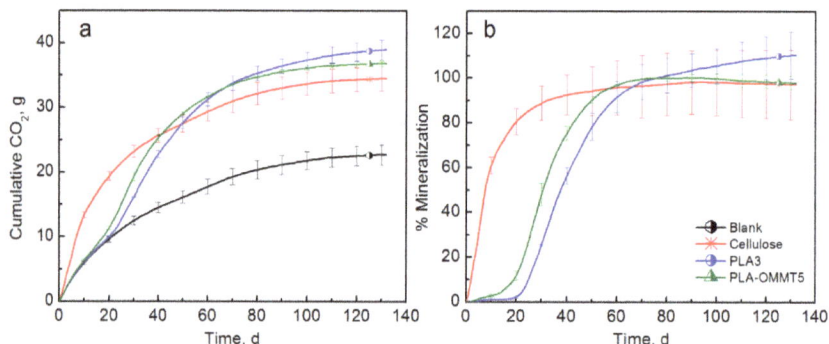

Figure 10. (a) CO_2 evolution and (b) % Mineralization of PLA and PLA-OMMT5 films (Test IV in compost).

The biodegradation test with inoculated vermiculite should avoid the priming effect as previously demonstrated [47,64,66]. Figure 11 shows that there was no significant difference in the mineralization of the tested materials at the end of the test (132 days). However, both PLA-OMMT5 and PLA-QAC0.4 showed significantly higher mineralization than the PLA control before 70 days of testing, and a much shorter lag time. The PLA control reached 77.7% mineralization after 132 days while PLA-OMMT5 reached the same mineralization after 83 days of testing and a maximum average mineralization of 83.3%. PLA-QAC reached a mineralization of 77.3%. It is important to mention that longer testing times were expected in this case since the biodegradation in inoculated vermiculite occurs at a slower rate than in compost. Even though the initial molecular weight of the films has a strong effect on their mineralization and priming effect, it seems that the addition of OMMT also accelerated the initial degradation of the samples. As previously mentioned, this behavior may be explained by the improved diffusion of water into PLA due to the high hydrophilicity of the nanoclay, which in turn promotes hydrolytic degradation [33,37,38,44,62].

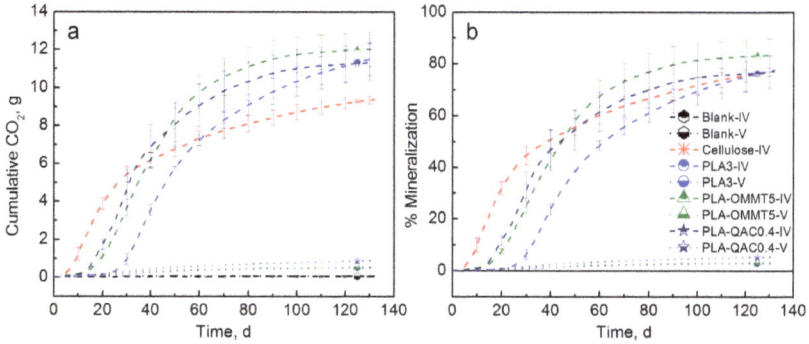

Figure 11. (**a**) CO_2 evolution and (**b**) % Mineralization of PLA, PLA-OMMT5, and PLA-QAC0.4 (Test IV in inoculated vermiculite (dashed lines) and uninoculated vermiculite (dotted lines)).

Figure 11 also shows the results when testing with uninoculated vermiculate. As expected, there was no significant evolution of CO_2 in the abiotic degradation test, and there was no significant difference in the mineralization values. For the biodegradation test III, film samples were taken at different periods of time in order to track the changes in the molecular weight and the results are explained in Section 3.3.

3.3. Molecular Weight

Figure 12 shows the initial molecular weight distribution (MWD) of the PLA film and BNCs. As previously mentioned, the addition of nanoclay resulted on a reduction of the M_n during processing. This reduction in M_n was more pronounced in the case of PLA-OMMT5, PLA-QAC1.5, and PLA-QAC0.4. More detailed information about the initial M_n, M_w, and *PDI*, of PLA and BNCs films is provided in Table S7.

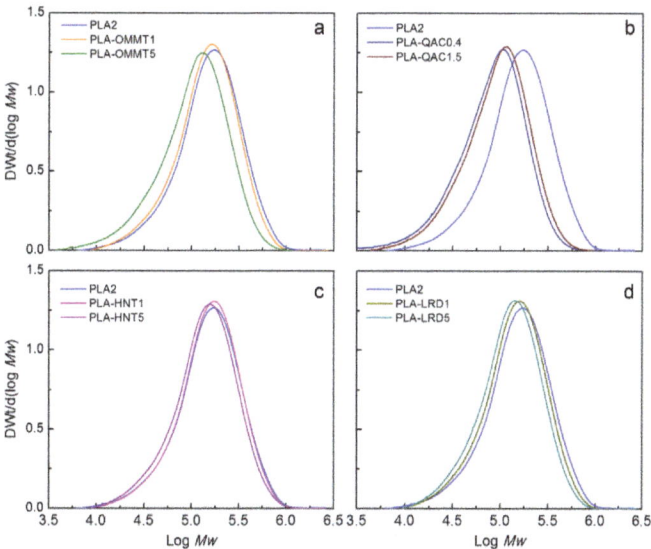

Figure 12. Initial molecular weight of PLA and BNCs.

Figure 13 shows the decrease of molecular weight of the PLA control film as a function of time during the biodegradation test III, represented by the shift of the peak to the left. This behavior was previously reported in the literature during the hydrolytic degradation of PLA, and was attributed to the chain scission preferentially occurring in the bulk of the polymer matrix rather than the surface [67]. The broadening of the peaks over time indicates an increase in the *PDI* due to the fragmentation of the PLA chains. The change in the MWD from monomodal to multimodal after day 14 has also been previously observed during hydrolytic degradation of PLA and was attributed to the formation of crystalline residues due to the rearrangement of the new shorter polymer chains into a more stable configuration (i.e., crystalline structures) [51,67]. The formation of more defined and higher peaks, as observed at days 42 and 56, has been attributed to the predominant degradation of the amorphous regions [68]. During the biodegradation tests a whitening effect in PLA and BNC was observed. It has been reported that this effect indicates increased crystallinity and opacity due to the beginning of the hydrolytic degradation phase of the biodegradation process [44,45,62]. The whitening effect occurs because a change in the refraction index of the polymer is induced by the absorbed water and/or the byproducts, e.g., carboxylic end-groups that are able to catalyze ester hydrolysis [45,62].

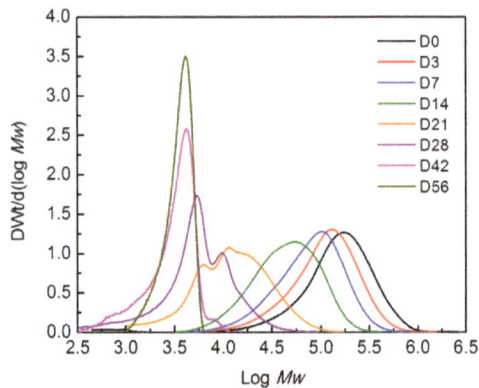

Figure 13. Change in molecular weight of PLA2 film (Test III in compost).

Figure 14 shows the changes in the MWD of the BNCs as function of time until day 28 since it was not possible to collect samples for SEC analysis after that period of time (except for PLA control as shown in Figure 13). Similarly to the PLA control, the BNCs showed multimodal peaks after day 14, although more evidently after day 21. In general, this behavior was less pronounced for PLA-OMMT1, PLA-LRD1, and PLA-LRD5, and it may be attributed to a slower formation of crystalline residuals. From Figure 14, it can be observed that the reduction of molecular weight was slower for PLA-OMMT1 and PLA-LRD1, in comparison with the pristine PLA. Similarly, the MWD of PLA-OMMT5 and PLA-QAC15 have a similar trend with an evident multimodal peak at day 21, while the reduction of molecular weight of PLA-HNT5 and PLA-LRD5 films seems to be slower than PLA control.

Deconvolution of the peaks was performed due to the multimodal MWD observed in the previous results, followed by kinetics analysis (Section S4). The M_n reduction rate (k) constant was calculated for PLA and the BNCs, fitting of a first order reaction of the form $M_n/M_{n0} = \exp(-kt)$, where M_{n0} is the initial M_n, k is the rate constant, and t is the time. The results (Figure S6 and Table S8) show that the BNCs, especially PLA-LRD films, have a lower M_n reduction rate than the PLA control ($k = 0.1008 \pm 0.0037$) until day 28. Ray and Okamoto analyzed the molecular weight of PLA and PLA nanocomposites and found that the reduction was almost the same for all the samples [10]. In contrast, Paul et al. found that the M_n of PLA decreased ~40% with respect to its initial value while for the PLA nanocomposites M_n decreased 70–80% [38]. In this case, even though the M_n reduction rate of the BNC was the same or lower than the PLA control, a higher evolution of CO_2 from the bioreactors supplemented with the BNC was generally observed during the biodegradation tests. Therefore, it is also relevant to understand the role of the microorganisms and how they are affected by the presence of these nanoclays. For example, Annamalai et al. suggested that the clay nanoparticles improve the absorption of UV light and promote polymer photo-oxidation due to the catalytic effect of metal ion impurities. That increased oxidation at the surface of the nanocomposites could favor the adhesion, accumulation and growth of the microorganisms [69].

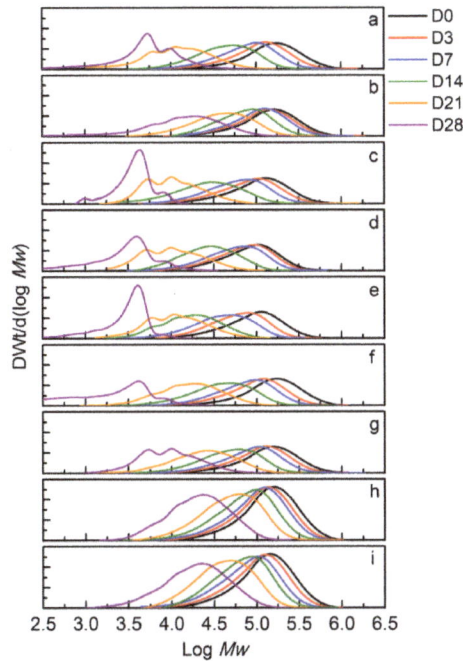

Figure 14. Change in molecular weight of (**a**) PLA2, (**b**) PLA-OMMT1, (**c**) PLA-OMMT5, (**d**) PLA-QAC0.4, (**e**) PLA-QAC1.5, (**f**) PLA-HNT1, (**g**) PLA-HNT5, (**h**) PLA-LRD1, and (**i**) PLA-LRD5 films (Test III in compost).

3.4. Microbial Attachment

Biofilm assays were performed to evaluate the ability of the microorganisms present in the compost to attach to the surface of PLA film and BNCs (i.e., PLA-OMMT5, PLA-QAC0.4, PLA-HNT5, and PLA-LRD5). Even though biofilm formation does not necessarily mean that the material is biodegraded by the attached populations [70], it is an important aspect of microbial performance and survival [71]. When biofilm-forming microorganisms release exopolymeric substances (EPS) (e.g., carbohydrates, nucleic acids, and proteins) such resources become available for other microorganisms, including secreted enzymes that degrade PLA and derivatives. Secreting extracellular digestive enzymes after forming a biofilm would localize the effect of extracellular digestion and increase the benefit to biofilm-forming strains. Biofilm production is a common trait among microorganisms living in soil, which are usually exposed to low moisture conditions. Biofilms can contribute to water retention in the soil matrix, prevent microorganisms from being washed out, and confer tolerance to other environmental stressors [71].

An initial test of the biofilm assay is shown in the Supplementary Materials (Section S5). Figure 15 and Tables S11 and S12 show the results of the biofilm test. A positive control was performed using

Pseudomonas aeruginosa (PA) strain PAO1, a high biofilm forming strain, at 23 °C [72,73]. Looking at the control with PA at 23 °C (Figure 15a), it was observed that the positive control wells (PA + R2B) had an absorbance (600 nm) of 1.226–1.332, with uninoculated control wells ranging from 0.060 to 0.065, which is in agreement with the values reported by Satti et al. [49]. The wells containing PLA, PLA-QAC0.4, PLA-HNT5, and PLA-LRD5 were approximately the same as the control lacking any film (R2B only). However, the wells containing PLA-OMMT5 showed significantly more biofilm formation (average 2.042), suggesting that the OMMT had an indirect stimulation on biofilm formation by PA. For the biofilm formed on the surface of the films by PA at 23 °C, PLA ranged from 0.501 to 0.752, which is also in agreement with the values previously observed [49]. In this case, the values of PLA-OMMT5 and PLA-HNT5 were significantly different from PLA-QAC0.4. PLA-HNT5 had one of the highest average values with 1.254. Looking at the total biofilm formation, PLA-OMMT5 and PLA-QAC0.4 were significantly different from pristine PLA and the rest of the BNCs with the highest (2.917) and lowest (1.107) values, respectively. The total average biofilm values (wells + film) for PA at 23 °C in descending order are as follows PLA-OMMT5 > PLA-HNT5 > PLA > PLA-LRD5 > PLA-QAC0.4.

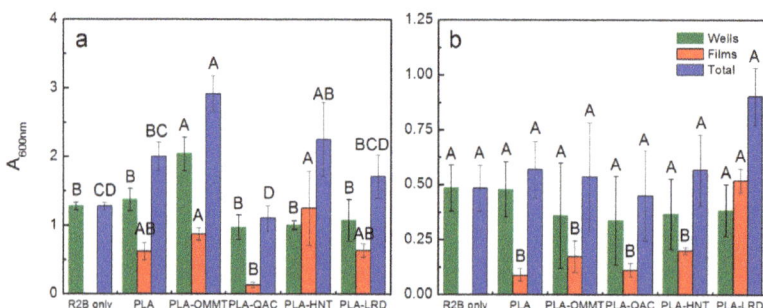

Figure 15. Absorbance (600 nm) of (**a**) PA at 23 °C, and (**b**) CE at 58 °C for second biofilm test. Columns with the same letter within a group (i.e., wells, films, or total) are not significantly different at $p \leq 0.05$ (Tukey test).

Regarding the biofilm estimates with CE at 58 °C (Figure 15b), the sterile controls (SCE) have values that are between 0.101 and 0.124, which are slightly greater than what was seen with low nutrient media at 23 °C. This is probably due to significant amounts of humic material in the CE. The control wells (CE only) have values of 0.381–0.588. These values are less than the ones for PA at 23 °C, which is expected since PA is a well-known biofilm former and because microbial growth and survival is generally more challenging at 58 °C and CE contains a diverse collection of microbial populations, many of which do not form biofilm under these conditions. The wells supplemented with PLA and BNCs ranged from 0.122–0.603 with no statistically significant difference among them. Biofilm formation was observed on the surface of PLA and BNCs with CE at 58 °C. PLA-LRD5 has significantly higher value (0.519) than the rest of the BNCs. The lowest average values were observed for PLA-QAC0.4 and PLA with 0.113 and 0.090, respectively. In this case, the total biofilm was also not significantly different among the sample materials.

In general, the PLA-LRD5 biofilm was the largest among the different samples, indicating that population in CE have a preference for PLA-LRD5 at 58 °C. In contrast, a pure culture, *Pseudomonas aeruginosa*, clearly preferred PLA-OMMT5 at 23 °C. Overall the biofilms at 58 °C were smaller than the biofilm at 23 °C. At both temperatures, PLA-QAC0.4 was the film producing the lowest average amount of biofilm, which may be attributed to inhibition due to the surfactant. This is supported by the biodegradation test where the surfactant was tested alone. Further investigation is recommended to understand which specific microbial strains present in the compost bind to and preferentially degrade PLA and the BNCs.

Due to the significant differences between pristine PLA and PLA-LRD5 found in the biofilm formed on the surface of the films during the test at 58 °C with CE, several SEM micrographs were taken from samples coated with osmium. Figure 16 shows the difference in microbial attachment between pristine PLA and PLA-LRD5 at a magnification of 1000×. It can be clearly observed that the surface of PLA-LRD5 is much more heavily populated by microorganisms, in agreement with the biofilm test results (Figure 15b).

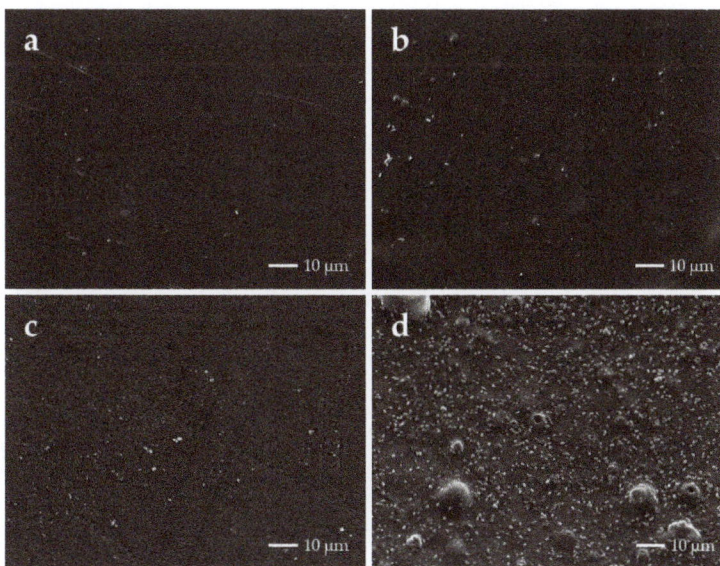

Figure 16. SEM micrographs of (**a**) PLA and (**b**) PLA-LRD at 1000× before incubation, (**c**) PLA and (**d**) PLA-LRD5 after incubation for 48 h at 58 °C with compost extract in R2B.

4. Conclusions

The effect of three different nanoclays, OMMT, HNT, and LRD, as well as the OMMT organo-modifier (QAC) on the biodegradation of PLA was evaluated with an in-house built DMR system following the analysis of evolved CO_2 approach. The results obtained from four different biodegradation tests along with the study of microbial attachment on the surface of PLA and its BNCs show that the biodegradation phase of the films containing nanoclay started earlier than that for pristine PLA. This behavior was confirmed by the results obtained from different tests for PLA-OMMT5, even when tested in inoculated vermiculite. The tests performed in vermiculite allowed untangling the observed priming effect even though longer testing times were required. The effect of the nanoclays on the initial molecular weight during processing played a crucial role in the biodegradation studies, also since a lower M_{n0} ($\leq$60 kDa) seems to be correlated to the priming effect in compost. Further investigation is recommended using PLA and BNCs with the same initial molecular weight and thickness, a task not easy to achieve in lab settings. When the different nanoclays and surfactant were tested alone, it was observed that HNT, OMMT, and QAC showed some inhibition regardless of the amount introduced in the bioreactors. PLA-LRD5 showed a priming effect with mineralization values exceeding 100%. This behavior may be explained by the lower initial molecular weight and by the results observed during the microbial attachment tests, in which PLA-LRD5 showed the greatest biofilm formation on the surface as confirmed by the SEM micrographs. PLA-QAC0.4 had the lowest biofilm

Polymers **2018**, *10*, 202

formation, which may be attributed to the inhibitory effect also found during the CO_2 evolution test when QAC was tested alone. Under the experimental conditions used to investigate biofilm formation, it was noted that significant biofilm was established in only 48 h; however, the timing may be different in composting conditions. Further investigation is required on the specific microbial strains that are capable of biodegrading PLA and its BNCs and how they can affect the biodegradation rate. Disposable products like packaging would greatly benefit from the biodegradable features of PLA since it would allow its disposal along with other organic wastes in composting facilities.

Supplementary Materials: Information about material processing and characterization, physicochemical characteristics of the compost, molecular weight determination, and biofilm test formation are available online at www.mdpi.com/2073-4360/10/2/202/s1.

Acknowledgments: Edgar Castro-Aguirre thanks the Mexican National Council for Science and Technology (CONACYT) and the Mexican Secretariat of Public Education (SEP) for providing financial support through a fellowship, and the Center for Advanced Microscopy at Michigan State University for assistance with the TEM and SEM analyses. The authors also thank the undergraduate and graduate students and visiting scholars that helped during the four different tests reported in this manuscript, the School of Packaging (SoP), and the Center for Packaging Innovation and Sustainability (CPIS) for partially funding this project. Rafael Auras acknowledges the partial support of the USDA and the MI AgBioResearch, Hatch.

Author Contributions: Edgar Castro-Aguirre, Rafael Auras, Susan Selke, Maria Rubino, and Terence Marsh conceived and designed the experiments; Edgar Castro-Aguirre performed the experiments and initially analyzed the data; Edgar Castro-Aguirre drafted the manuscript; all the authors contributed in the writing process and approved the final version of the paper.

Conflicts of Interest: The authors declare no conflict of interest.

References

1. Kumar, A.P.; Depan, D.; Tomer, N.S.; Singh, R.P. Nanoscale particles for polymer degradation and stabilization—Trends and future perspectives. *Prog. Polym. Sci.* **2009**, *34*, 479–515. [CrossRef]
2. Lagaron, J.M. Nanotechnology for bioplastics: Opportunities, challenges and strategies. *Trends Food Sci. Technol.* **2011**, *22*, 611–617. [CrossRef]
3. Azeredo, H.M.C. De Nanocomposites for food packaging applications. *Food Res. Int.* **2009**, *42*, 1240–1253. [CrossRef]
4. Raquez, J.-M.; Habibi, Y.; Murariu, M.; Dubois, P. Polylactide (PLA)-based nanocomposites. *Prog. Polym. Sci.* **2013**, *38*, 1504–1542. [CrossRef]
5. Kijchavengkul, T.; Auras, R. Compostability of polymers. *Polym. Int.* **2008**, *57*, 793–804. [CrossRef]
6. De Abreu, D.A.P.; Losada, P.P.; Angulo, I.; Cruz, J.M. Development of new polyolefin films with nanoclays for application in food packaging. *Eur. Polym. J.* **2007**, *43*, 2229–2243. [CrossRef]
7. Reddy, M.M.; Vivekanandhan, S.; Misra, M.; Bhatia, S.K.; Mohanty, A.K. Biobased plastics and bionanocomposites: Current status and future opportunities. *Prog. Polym. Sci.* **2013**, *38*, 1653–1689. [CrossRef]
8. Ray, S.S.; Maiti, P.; Okamoto, M.; Yamada, K.; Ueda, K. New Polylactide/Layered Silicate Nanocomposites. 1. Preparation, Characterization, and Properties. *Macromolecules* **2002**, *35*, 3104–3110.
9. Ray, S.S.; Yamada, K.; Okamoto, M.; Ueda, K. New polylactide-layered silicate nanocomposites. 2. Concurrent improvements of material properties, biodegradability and melt rheology. *Polymer (Guildf)* **2003**, *44*, 857–866.
10. Ray, S.S.; Yamada, K.; Okamoto, M.; Ogami, A.; Ueda, K. New polylactide/layered silicate nanocomposites, 4. Structure, properties and biodegradability. *Compos. Interfaces* **2003**, *10*, 435–450. [CrossRef]
11. Bourbigot, S.; Fontaine, G.; Duquesne, S.; Delobel, R. PLA nanocomposites: Quantification of clay nanodispersion and reaction to fire. *Int. J. Nanotechnol.* **2008**, *5*, 683–692. [CrossRef]
12. Ray, S.S.; Yamada, K.; Okamoto, M.; Fujimoto, Y.; Ogami, A.; Ueda, K. New polylactide/layered silicate nanocomposites. 5. Designing of materials with desired properties. *Polymer (Guildf)* **2003**, *44*, 6633–6646.
13. Picard, E.; Espuche, E.; Fulchiron, R. Effect of an organo-modified montmorillonite on PLA crystallization and gas barrier properties. *Appl. Clay Sci.* **2011**, *53*, 58–65. [CrossRef]
14. Re, G.L.; Benali, S.; Habibi, Y.; Raquez, J.; Dubois, P. Stereocomplexed PLA nanocomposites: From in situ polymerization to materials properties. *Eur. Polym. J.* **2014**, *54*, 138–150. [CrossRef]

15. Ligot, S.; Benali, S.; Ramy-Ratiarison, R.; Murariu, M.; Snyders, R.; Dubois, P. Mechanical, Optical and Barrier Properties of PLA-layered silicate nanocomposites coated with Organic Plasma Polymer Thin Films. *Mater. Sci. Eng. Adv. Res.* **2015**, *1*, 1–11.

16. Chen, Y.; Geever, L.M.; Killion, J.A.; Lyons, J.G.; Higginbotham, C.L.; Devine, D.M. Halloysite Nanotube Reinforced Polylactic Acid Composite. *Polym. Compos.* **2015**. [CrossRef]

17. Kim, Y.H.; Kwon, S.H.; Choi, H.J.; Choi, K.; Kao, N.; Bhattacharya, S.N.; Gupta, R.K. Thermal, Mechanical, and Rheological Characterization of Polylactic Acid/Halloysite Nanotube Nanocomposites. *J. Macromol. Sci. Part B* **2016**, *55*, 680–692. [CrossRef]

18. Kausar, A. Review on Polymer/Halloysite Nanotube Nanocomposite. *Polym. Plast. Technol. Eng.* **2017**. [CrossRef]

19. Liu, M.; Jia, Z.; Jia, D.; Zhou, C. Recent advance in research on halloysite nanotubes-polymer nanocomposite. *Prog. Polym. Sci.* **2014**, *39*, 1498–1525. [CrossRef]

20. Murariu, M.; Dechief, A.-L.; Paint, Y.; Peeterbroeck, S.; Bonnaud, L.; Dubois, P. Polylactide (PLA)—Halloysite Nanocomposites: Production, Morphology and Key-Properties. *J. Polym. Environ.* **2012**, *20*, 932–943. [CrossRef]

21. Prashantha, K.; Lecouvet, B.; Sclavons, M.; Lacrampe, M.F.; Krawczak, P. Poly(lactic acid)/Halloysite Nanotubes Nanocomposites: Structure, Thermal, and Mechanical Properties as a Function of Halloysite Treatment. *J. Appl. Polym. Sci.* **2013**, *128*, 1895–1903. [CrossRef]

22. Wu, W.; Cao, X.; Zhang, Y.; He, G. Polylactide/Halloysite Nanotube Nanocomposites: Thermal, Mechanical Properties, and Foam Processing. *J. Appl. Polym. Sci.* **2013**, *130*, 443–452. [CrossRef]

23. Russo, P.; Cammarano, S.; Bilotti, E.; Peijs, T.; Cerruti, P.; Acierno, D. Physical Properties of Poly Lactic Acid/Clay Nanocomposite Films: Effect of Filler Content and Annealing Treatment. *J. Appl. Polym. Sci.* **2014**, *131*. [CrossRef]

24. Esma, C.; Erpek, Y.; Ozkoc, G.; Yilmazer, U. Effects of Halloysite Nanotubes on the Performance of Plasticized Poly (lactic acid)-Based Composites. *Polym. Compos.* **2015**. [CrossRef]

25. Aouada, F.A.; Mattoso, L.H.C.; Longo, E. A simple procedure for the preparation of lapo- nite and thermoplastic starch nanocomposites: Structural, mechanical, and thermal characterizations. *J. Thermoplast. Compos. Mater.* **2011**. [CrossRef]

26. Tang, X.; Alavi, S. Structure and Physical Properties of Starch/Poly Vinyl Alcohol/Laponite® RD Nanocomposite Films. *J. Agric. Food Chem.* **2012**, *60*, 1954–1962. [CrossRef] [PubMed]

27. Loyens, W.; Jannasch, P.; Maurer, F.H.J. Poly(ethylene oxide)/Laponite® nanocomposites via melt-compounding: Effect of clay modification and matrix molar mass. *Polymer (Guildf)* **2005**, *46*, 915–928. [CrossRef]

28. Utracki, L.A.; Sepehr, M.; Boccaleri, E. Synthetic, layered nanoparticles for polymeric nanocomposites (PNCs). *Polym. Adv. Technol.* **2007**, *18*, 1–37. [CrossRef]

29. Perotti, G.F.; Tronto, J.; Bizeto, M.A.; Izumi, C.M.S.; Temperini, M.L.A.; Lugao, A.B.; Parra, D.F.; Constantino, V.R.L. Biopolymer-Clay Nanocomposites: Cassava Starch and Synthetic Clay Cast Films. *J. Brazilian Chem.* **2014**, *25*, 320–330. [CrossRef]

30. Wu, C.-J.; Gaharwar, A.K.; Schexnailder, P.J.; Schmidt, G. Development of Biomedical Polymer-Silicate Nanocomposites: A Materials Science Perspective. *Materials (Basel)* **2010**, *3*, 2986–3005. [CrossRef]

31. Zhou, G.X.; Yuan, M.W.; Jiang, L.; Yuan, M.L.; Li, H.L. The Preparation and Property Research on Laponite®-Poly (L-Lactide) Composite Film. *Adv. Mater. Res.* **2013**, *750*, 1919–1923. [CrossRef]

32. Li, H.L.; Zhou, G.X.; Shan, Y.K.; Yuan, M.L. The Mechanical Properties and Hydrophilicity of Poly (L-Lactide)/Laponite® Composite Film. *Adv. Mater. Res.* **2013**, *706*, 340–343. [CrossRef]

33. Stloukal, P.; Pekařová, S.; Kalendova, A.; Mattausch, H.; Laske, S.; Holzer, C.; Chitu, L.; Bodner, S.; Maier, G.; Slouf, M.; et al. Kinetics and mechanism of the biodegradation of PLA/clay nanocomposites during thermophilic phase of composting process. *Waste Manag.* **2015**, *42*, 31–40. [CrossRef] [PubMed]

34. Wu, T.; Xie, T.; Yang, G. Preparation and characterization of poly (ε-caprolactone)/Na + -MMT nanocomposites. *Appl. Clay Sci.* **2009**, *45*, 105–110. [CrossRef]

35. Correa, M.C.S.; Branciforti, M.C.; Pollet, E.; Agnelli, J.A.M.; Nascente, P.A.P.; Averous, L. Elaboration and Characterization of Nano-Biocomposites Based on Plasticized Poly(Hydroxybutyrate-Co-Hydroxyvalerate) with Organo-Modified Montmorillonite. *J. Polym. Environ.* **2012**, *20*, 283–290. [CrossRef]

36. Magalhães, N.F.; Andrade, C.T. Thermoplastic corn starch/clay hybrids: Effect of clay type and content on physical properties. *Carbohydr. Polym.* **2009**, *75*, 712–718. [CrossRef]

37. Lee, S.-R.; Park, H.; Lim, H.; Kang, T.; Li, X.; Cho, W.-J.; Ha, C.-S. Microstructure, tensile properties, and biodegradability of aliphatic polyester/clay nanocomposites. *Polymer (Guildf)* **2002**, *43*, 2495–2500. [CrossRef]
38. Paul, M.A.; Delcourt, C.; Alexandre, M.; Degee, P.; Monteverde, F.; Dubois, P. Polylactide/montmorillonite nanocomposites: Study of the hydrolytic degradation. *Polym. Degrad. Stab.* **2005**, *87*, 535–542. [CrossRef]
39. Lee, Y.H.; Lee, J.H.; An, I.; Kim, C.; Lee, D.S.; Lee, Y.K.; Nam, J. Electrospun dual-porosity structure and biodegradation morphology of Montmorillonite reinforced PLLA nanocomposite scaffolds. *Biomaterials* **2005**, *26*, 3165–3172. [CrossRef] [PubMed]
40. Fukushima, K.; Abbate, C.; Tabuani, D.; Gennari, M.; Camino, G. Biodegradation of poly (lactic acid) and its nanocomposites. *Polym. Degrad. Stab.* **2009**, *94*, 1646–1655. [CrossRef]
41. Roy, P.K.; Hakkarainen, M.; Albertsson, A. Nanoclay effects on the degradation process and product patterns of polylactide. *Polym. Degrad. Stab.* **2012**, *97*, 1254–1260. [CrossRef]
42. Molinaro, S.; Romero, M.C.; Boaro, M.; Sensidoni, A.; Lagazio, C.; Morris, M.; Kerry, J. Effect of nanoclay-type and PLA optical purity on the characteristics of PLA-based nanocomposite films. *J. Food Eng.* **2013**, *117*, 113–123. [CrossRef]
43. Souza, P.M.S.; Morales, A.R.; Marin-Morales, M.A.; Mei, L.H.I. PLA and Montmorilonite Nanocomposites: Properties, Biodegradation and Potential Toxicity. *J. Polym. Environ.* **2013**, *21*, 738–759. [CrossRef]
44. Machado, A.V.; Araújo, A.; Oliveira, M. Assessment of Polymer-Based Nanocomposites Biodegradability. In *Biodegradable Polymers. Volume 1: Advancement in Biodegradation Study and Applications*; Nova Publishers: Hauppauge, NY, USA, 2015.
45. Balaguer, M.P.; Aliaga, C.; Fito, C.; Hortal, M. Compostability assessment of nano-reinforced poly(lactic acid) films. *Waste Manag.* **2016**, *48*, 143–155. [CrossRef] [PubMed]
46. Cowley, J.M. *Diffraction Physics*, 3rd ed.; Elsevier B.V.: Amsterdam, The Netherlands, 1995; ISBN 0-444-82218-6.
47. Castro-Aguirre, E.; Auras, R.; Selke, S.; Rubino, M.; Marsh, T. Insights on the aerobic biodegradation of polymers by analysis of evolved carbon dioxide in simulated composting conditions. *Polym. Degrad. Stab.* **2017**, *137*, 251–271. [CrossRef]
48. Merritt, J.H.; Kadouri, D.E.; O'Toole, G.A. Growing and analyzing static biofilms. *Curr. Protoc. Microbiol.* **2011**, 1–18. [CrossRef]
49. Satti, S.M.; Shah, A.A.; Auras, R.; Marsh, T.L. Isolation and characterization of bacteria capable of degrading poly(lactic acid) at ambient temperature. *Polym. Degrad. Stab.* **2017**, *144*, 392–400. [CrossRef]
50. Gorrasi, G.; Pantani, R. Effect of PLA grades and morphologies on hydrolytic degradation at composting temperature: Assessment of structural modification and kinetic parameters. *Polym. Degrad. Stab.* **2013**, *98*, 1006–1014. [CrossRef]
51. Iñiguez-Franco, F.; Auras, R.; Burgess, G.; Holmes, D.; Fang, X.; Rubino, M.; Soto-Valdez, H. Concurrent solvent induced crystallization and hydrolytic degradation of PLA by water-ethanol solutions. *Polymer* **2016**, *99*, 315–323. [CrossRef]
52. Tham, W.L.; Poh, T.; Arifin, Z.; Ishak, M.; Chow, W.S. Characterisation of Water Absorption of Biodegradable Poly(lactic Acid)/Halloysite Nanotube Nanocomposites at Different Temperatures. *J. Eng. Sci.* **2016**, *12*, 13–25.
53. De Silva, R.T.; Soheilmoghaddam, M.; Goh, K.L.; Wahit, M.U.; Bee, S.; Hamid, A.; Chai, S.; Pasbakhsh, P. Influence of the Processing Methods on the Properties of Poly(lactic acid)/Halloysite Nanocomposites. *Polym. Compos.* **2014**, 1–9. [CrossRef]
54. Touny, A.H.; Jones, A.D. Effect of electrospinning parameters on the characterization of PLA/HNT nanocomposite fibers. *J. Mater. Res.* **2010**, *25*, 857–865. [CrossRef]
55. Cai, N.; Dai, Q.; Wang, Z. Toughening of electrospun poly(L-lactic acid) nanofiber scaffolds with unidirectionally aligned halloysite nanotubes. *J. Mater. Sci.* **2015**, *50*, 1435–1445. [CrossRef]
56. Dong, Y.; Marshall, J.; Haroosh, H.J.; Mohammadzadehmoghadam, S.; Liu, D.; Qi, X.; Lau, K. Composites: Part A Polylactic acid (PLA)/halloysite nanotube (HNT) composite mats: Influence of HNT content and modification. *Compos. Part A* **2015**, *76*, 28–36. [CrossRef]
57. Gorrasi, G.; Pantani, R.; Murariu, M.; Dubois, P. PLA/Halloysite Nanocomposite Films: Water Vapor Barrier Properties and Specific Key Characteristics. *Macromol. Mater. Eng.* **2014**, *299*, 104–115. [CrossRef]
58. ASTM Standard D5338-15. *Standard Test Method for Determining Aerobic Biodegradation of Plastic Materials Under Controlled Composting Conditions, 2015*; West Conshohocken, PA, USA, 2015.

59. International Organization for Standardization. *Determination of the Ultimate Aerobic Biodegradability of Plastic Materials under Controlled Composting Conditions—Method by Analysis of Evolved Carbon Dioxide—Part 1: General Method*; ISO 14855-1:2012; ISO: Geneva, Switzerland, 2012; p. 20.

60. Kijchavengkul, T.; Auras, R.; Rubino, M.; Ngouajio, M.; Thomas Fernandez, R. Development of an automatic laboratory-scale respirometric system to measure polymer biodegradability. *Polym. Test.* **2006**, *25*, 1006–1016. [CrossRef]

61. Castro-Aguirre, E.; Iñiguez-Franco, F.; Samsudin, H.; Fang, X.; Auras, R. Poly(lactic acid)—Mass production, processing, industrial applications, and end of life. *Adv. Drug Deliv. Rev.* **2016**, *107*, 333–366. [CrossRef] [PubMed]

62. Fukushima, K.; Giménez, E.; Cabedo, L.; Lagarón, J.M.; Feijoo, J.L. Biotic degradation of poly (DL-lactide) based nanocomposites. *Polym. Degrad. Stab.* **2012**, *97*, 1278–1284. [CrossRef]

63. Fukushima, K.; Tabuani, D.; Dottori, M.; Armentano, I.; Kenny, J.M.; Camino, G. Effect of temperature and nanoparticle type on hydrolytic degradation of poly (lactic acid) nanocomposites. *Polym. Degrad. Stab.* **2011**, *96*, 2120–2129. [CrossRef]

64. Bellia, G.; Tosin, M.; Degli-Innocenti, F. Test method of composting in vermiculite is unaffected by the priming effect. *Polym. Degrad. Stab.* **2000**, *69*, 113–120. [CrossRef]

65. Bikiaris, D.N. Nanocomposites of aliphatic polyesters: An overview of the effect of different nanofillers on enzymatic hydrolysis and biodegradation of polyesters. *Polym. Degrad. Stab.* **2013**, *98*, 1908–1928. [CrossRef]

66. Bellia, G.; Tosin, M.; Floridi, G.; Degli-Innocenti, F. Activated vermiculite, a solid bed for testing biodegradability under composting conditions. *Polym. Degrad. Stab.* **1999**, *66*, 65–79. [CrossRef]

67. Tsuji, H.; Ikada, Y. Blends of crystalline and amorphous poly(lactide). 3. Hydrolysis of solution-cast blend films. *J. Appl. Polym. Sci.* **1997**, *63*, 855–863. [CrossRef]

68. Tsuji, H.; Saeki, T.; Tsukegi, T.; Daimon, H.; Fujie, K. Comparative study on hydrolytic degradation and monomer recovery of poly(L-lactic acid) in the solid and in the melt. *Polym. Degrad. Stab.* **2008**, *93*, 1956–1963. [CrossRef]

69. Annamalai, P.K.; Martin, D.J.; Annamalai, P.K.; Martin, D.J. Can clay nanoparticles accelerate environmental biodegradation of polyolefins? *Mater. Sci. Technol.* **2014**, *30*, 593–602. [CrossRef]

70. Eubeler, J.P.; Bernhard, M.; Knepper, T.P. Environmental biodegradation of synthetic polymers II. Biodegradation of different polymer groups. *Trensds Anal. Chem.* **2010**, *29*, 84–100. [CrossRef]

71. Lennon, J.T.; Lehmkuhl, B.K. A trait-based approach to bacterial biofilms in soil. *Environ. Microbiol.* **2016**, *18*, 2732–2742. [CrossRef] [PubMed]

72. Ali, S.G.; Ansari, M.A.; Khan, H.M.; Jalal, M.; Mahdi, A.A.; Cameotra, S.S. Crataeva nurvala nanoparticles inhibit virulence factors and biofilm formation in clinical isolates of Pseudomonas aeruginosa. *J. Basic Microbiol.* **2017**, *57*, 193–203. [CrossRef] [PubMed]

73. Overhage, J.; Lewenza, S.; Marr, A.K.; Hancock, R.E.W. Identification of genes involved in swarming motility using a Pseudomonas aeruginosa PAO1 mini-Tn5-lux mutant library. *J. Bacteriol.* **2007**, *189*, 2164–2169. [CrossRef] [PubMed]

© 2018 by the authors. Licensee MDPI, Basel, Switzerland. This article is an open access article distributed under the terms and conditions of the Creative Commons Attribution (CC BY) license (http://creativecommons.org/licenses/by/4.0/).

![polymers logo] *polymers*

MDPI

Article

Crystallization and Stereocomplexation of PLA-*mb*-PBS Multi-Block Copolymers

Rosa M. D'Ambrosio [1], Rose Mary Michell [1], Rosica Mincheva [2], Rebeca Hernández [3], Carmen Mijangos [3], Philippe Dubois [2] and Alejandro J. Müller [1,4,5,*]

[1] Grupo de Polímeros USB, Departamento de Ciencia de los Materiales, Universidad Simón Bolívar, Apartado 89000, 1080-A Caracas, Venezuela; rdambrosio@usb.ve (R.M.D.); rmichell@usb.ve (R.M.M.)
[2] Laboratory of Polymeric and Composite Materials, Center of Innovation and Research in Materials & Polymers (CIRMAP), University of Mons-Hainaut, Place du Parc 20, B-7000 Mons, Belgium; rosica.mincheva@umons.ac.be (R.M.); philippe.dubois@umons.ac.be (P.D.)
[3] Instituto de Ciencia y Tecnología de Polímeros, CSIC, c/Juan de la Cierva 3, 28006 Madrid, Spain; rhernandez@ictp.csic.es (R.H.); cmijangos@ictp.csic.es (C.M.)
[4] Polymat and Polymer Science and Technology Department, Faculty of Chemistry, University of the Basque Country UPV/EHU, Paseo Manuel de Lardizabal 3, 20018 Donostia-San Sebastián, Spain
[5] Ikerbasque, Basque Foundation for Science, 48013 Bilbao, Spain
*   Correspondence: alejandrojesus.muller@ehu.es; Tel.: +34-943-018-191

Received: 29 November 2017; Accepted: 20 December 2017; Published: 22 December 2017

Abstract: The crystallization and morphology of PLA-*mb*-PBS copolymers and their corresponding stereocomplexes were studied. The effect of flexible blocks (i.e., polybutylene succinate, PBS) on the crystallization of the copolymers and stereocomplex formation were investigated using polarized light optical microscopy (PLOM), differential scanning calorimetry (DSC), infrared spectroscopy (FTIR), and carbon-13 nuclear magnetic resonance spectroscopy (^{13}C-NMR). The PLA and PBS multiple blocks were miscible in the melt and in the glassy state. When the PLA-*mb*-PBS copolymers are cooled from the melt, the PLA component crystallizes first creating superstructures, such as spherulites or axialites, which constitute a template within which the PBS component has to crystallize when the sample is further cooled down. The Avrami theory was able to fit the overall crystallization kinetics of both semi-crystalline components, and the *n* values for both blocks in all the samples had a correspondence with the superstructural morphology observed by PLOM. Solution mixtures of PLLA-*mb*-PBS and PLDA-*mb*-PBS copolymers were prepared, as well as copolymer/homopolymer blends with the aim to study the stereocomplexation of PLLA and PDLA chain segments. A lower amount of stereocomplex formation was observed in copolymer mixtures as compared to neat L_{100}/D_{100} stereocomplexes. The results show that PBS chain segments perturb the formation of stereocomplexes and this perturbation increases with the amount of PBS in the samples. However, when relatively low amounts of PBS in the copolymer blends are present, the rate of stereocomplex formation is enhanced. This effect dissappears when higher amounts of PBS are present. The stereocomplexation was confirmed by FTIR and solid state ^{13}C-NMR analyses.

Keywords: poly(lactic acid) (PLA); crystallization kinetics; stereocomplexes; crystallization in multi-block copolymers; polybutylene succinate (PBS)

---

1. Introduction

Poly(lactic acid) (PLA) is a biodegradable and biocompatible polymer that can be obtained from renewable resources and represents an interesting alternative for replacing petroleum-based polymers [1]. PLA is usually obtained in the amorphous state after being processed, in view of its slow crystallization rate in comparison to the cooling rates employed during processing. The resulting physical properties limit PLA practical applications, as the amorphous material has a low distortion temperature (related to its low T_g value of around 55 °C) and is fragile at room temperature.

PLA has three different stereoisomers, the poly(L-lactide) (PLLA), poly(D-lactide) (PDLA), and the poly(L,D-lactide). The first two can crystallize, while the last one is always amorphous as it has no chain regularity for crystallization. Another interesting fact about PLA is the possibility to form stereocomplexes from the equimolecular mixture of PLLA and PDLA. The stereocomplex is a crystal formed by the helical configuration of the PLLA and PDLA chains assembled together in the same unit cells (i.e., co-crystallization) and has a higher stability than the PLA homocrystals, with melting temperatures that are approximately 50 °C higher than those of PLA homocrystals [2–11].

Tsuji et al. have systematically studied the stereocomplexation ability of PLA isomers [2–9]. They found that the optimal condition for stereocomplexation from solution is the use of the same molar amount of PLLA and PDLA with similar molecular weights. On the other hand, the stereocomplexation behavior of PLA chains within a copolymer has been studied by Michell et al. [12]. The poly(D-lactide)-b-poly(N,N-dimethylamino-2-ethyl methacrylate) (PDLA-b-PDMAEMA) and poly(L-lactide)-b-poly(N,N-dimethylamino-2-ethyl methacrylate) (PLLA-b-PDMAEMA) copolymers were able to form stereocomplex structures, however, the crystallization temperature was lower than that of the stereocomplex formed by the homopolymers, this indicated that the crystallization of the stereocomplex is impaired by the amorphous block of PDMAEMA in the copolymers [12].

To enhance the physical properties of PLA, various chemical modifications and blending methods have been employed [13–16]. PBS was selected in this work due to its biocompatibility and reported application in blending with PLA to improve its mechanical performance [17–28]. The copolymerization of PBS and PLA as multi-block copolymers may lead to a material with improved mechanical properties as compared with PLAs [20,26,29]. Additionally, the crystallization process of such multi-block copolymers has not been studied in detail, and the stereocomplexation ability of such copolymers has not been reported in the literature so far. Hence, in this publication, we study the morphology and isothermal crystallization of PLA-*mb*-PBS multi-block copolymer and the stereocomplexes that can be formed by mixtures of copolymers and by PLA-*mb*-PBS copolymer/PLA homopolymer blends.

2. Materials and Methods

2.1. Materials

The molecular weights and polydispersities of the employed materials are shown in Table 1, the nomenclature used is as follows: (L or D)L$_{xx}{}^{zz}$-*mb*-BS$_{yy}{}^{zz}$, where *xx* and *yy* represent the content (in weight percent) of the PLA and PBS blocks, respectively. The superscript *zz* represents the number-average molecular weight in kg/mol. Some of us have reported the complete synthesis procedure of all studied block copolymers and homopolymers previously [21].

Table 1. Molecular characteristics of the PLA and PBS homopolymers and copolymers.

Sample	M_n (g/mol)	M_w/M_n [c]
LL$_{100}{}^{5.4}$	5400 [a]	1.14
DL$_{100}{}^{6.1}$	6050 [a]	1.19
BS$_{100}{}^{7.4}$	7400 [b]	1.77
LL$_{70}{}^{5.4}$-*mb*-BS$_{30}{}^{7.4}$	14,800	1.91
DL$_{70}{}^{6.1}$-*mb*-BS$_{30}{}^{7.4}$	12,000	1.90
LL$_{20}{}^{5.4}$-*mb*-BS$_{80}{}^{7.4}$	7000	1.87
DL$_{20}{}^{6.1}$-*mb*-BS$_{80}{}^{7.4}$	6700	2.86

[a] The calculated number average molecular weights using reported Mark-Houwink parameters; [b] the apparent number average molecular weights determined by size exclusion chromatography in THF at 30 °C with reference to polystyrene standards; [c] the molecular weight dispersities.

2.2. Stereocomplex Preparation

PLLA and PDLA homopolymers were dissolved in dichloromethane, separately, at a fixed concentration of 1 g/dL. Then PLLA and PDLA solutions with similar molecular weights (see Table 2) were mixed together under vigorous stirring for 3 h, the molar ratio of PLLA/PDLA in all the blends was fixed at 1:1. Finally, the mixed solutions were casted onto Petri dishes and the solvent was evaporated at room temperature for 24 h. The product was dried under vacuum at 25 °C for 72 h.

Table 2. Identification of PLA stereocomplexes from PLA homopolymers and copolymers blends.

Sample	Components		Global Composition	
	PLLA	PDLA	%PLA	%PBS
L_{100}/D_{100}	$LL_{100}^{5.4}$	$DL_{100}^{6.1}$	100	0
$L_{70}B_{30}/D_{70}B_{30}$	$LL_{70}^{5.4}\text{-}mb\text{-}BS_{30}^{7.4}$	$DL_{70}^{6.1}\text{-}mb\text{-}BS_{30}^{7.4}$	70	30
$L_{20}B_{80}/D_{20}B_{80}$	$LL_{20}^{5.4}\text{-}mb\text{-}BS_{80}^{7.4}$	$DL_{20}^{6.1}\text{-}mb\text{-}BS_{80}^{7.4}$	20	80
$L_{70}B_{30}/D_{100}$	$LL_{70}^{5.4}\text{-}mb\text{-}BS_{30}^{7.4}$	$DL_{100}^{6.1}$	84	16
$D_{70}B_{30}/L_{100}$	$LL_{100}^{5.4}$	$DL_{70}^{6.1}\text{-}mb\text{-}BS_{30}^{7.4}$	81	19
$L_{20}B_{80}/D_{100}$	$LL_{20}^{5.4}\text{-}co\text{-}BS_{80}^{7.4}$	$DL_{100}^{6.1}$	35	65
$D_{20}B_{80}/L_{100}$	$LL_{100}^{5.4}$	$DL_{20}^{6.1}\text{-}co\text{-}BS_{80}^{7.4}$	32	68

The same solution method was employed to prepare PLA-*mb*-PBS copolymers and copolymer/homopolymer blends. Table 2 summarizes all stereocomplex samples prepared with homopolymers, copolymers, and their blends.

2.3. Polarized Light Optical Microscopy (PLOM)

The superstructural morphology was observed employing a ZEISS Mc 80 polarized light optical microscope (Carl Zeiss Microscopy, Jena, Germany). The homopolymers and copolymers were melted at 180 °C and the stereocomplexes at 230 °C in order to erase the thermal history. Next, the temperature was quickly lowered until the desired isothermal crystallization temperature was reached. The morphology was recorded with a digital camera and the temperature was controlled using a Linkam TP 91 hot plate (Linkam Scientific Instruments, Tadworth, UK).

2.4. Differential Scanning Calorimetry (DSC)

Homopolymer/homopolymer equimolar blends were characterized by differential scanning calorimetry with a Perkin-Elmer DSC-7. For copolymer/copolymer and homopolymer/copolymer blends, DSC characterization was performed with a Perkin-Elmer DSC Diamond™ (Perkin Elmer, Waltham, MA, USA). The samples were studied under inert atmosphere employing ultra-high purity nitrogen and the instruments were calibrated with indium and tin standards. The non-isothermal crystallization studies were performed at 20 °C/min. For isothermal crystallization, the samples were quenched from the melt at 60 °C/min to each isothermal crystallization temperature investigated and then measurements as a function of time were performed until the crystallization was complete, following the procedure reported by Lorenzo et al. [30].

2.5. Infrared Spectroscopy (FTIR)

For FTIR experiments of PLLA/PDLA stereocomplexes, the samples were prepared by compression moulding of discs of KBr/sample blends. The discs were placed in the sample compartment of a Nicolet Magna 760 spectrometer (Thermo Fisher Scientific, Waltham, MA, USA) at room temperature. Thirty-two scans were co-added in order to achieve an acceptable signal-to-noise ratio.

2.6. Carbon-13 Nuclear Magnetic Resonance Spectroscopy (¹³C-NMR)

Solid-state NMR spectra were obtained using a Bruker Avance TM 400 WB spectrometer (Bruker, Billerica, MA, USA) operating at 400 MHz for ^{13}C using cross-polarization (CP), magic-angle spinning (MAS), and high-power ^{1}H decoupling. The contact time was set to 3 ms, and recycle time between subsequent acquisitions was set to 3 s. A total of 2092 data points were acquired using a spectral width of 30 kHz. A total of 2000 and 1536 transients were averaged for each spectrum. MAS speed was between 5 kHz and 7.5 kHz.

3. Results and Discussion

3.1. Homopolymers and Copolymers

3.1.1. Non-Isothermal Crystallization

The Fox equation (see Equation (1)) was applied to estimate the T_g values for a miscible blend or copolymer, depending on the composition and on the T_g values of the original homopolymers [31]. Table 3 shows the values estimated using the Fox equation and the experimental ones.

$$\frac{1}{T_g} = \frac{X_A}{T_{gA}} + \frac{X_B}{T_{gB}}$$ (1)

Table 3. Values of T_g estimated employing the Fox equation ($T_{g\,FOX}$) and the experimental values ($T_{g\,exp}$) from the second heating scan at 20 °C/min for the indicated samples.

Sample	$T_{g\,FOX}$ (°C)	$T_{g\,exp}$ (°C)
LL$_{100}$$^{5.4}$	-	51.0
DL$_{100}$$^{6.1}$	-	51.8
LL$_{70}$$^{5.4}$-*mb*-BS$_{30}$$^{7.4}$	17	3.8
DL$_{70}$$^{6.1}$-*mb*-BS$_{30}$$^{7.4}$	17	25.4
LL$_{20}$$^{5.4}$-*mb*-BS$_{80}$$^{7.4}$	−26	−16.0
DL$_{20}$$^{6.1}$-*mb*-BS$_{80}$$^{7.4}$	−26	−17.7

A single glass transition temperature was found for all copolymers employed here. The T_g values experimentally determined are between those of PBS and PLA and they decrease with the increase of PBS content within the copolymer. The experimental values differ from those predicted by the Fox equation; however, their decrease with the increase in PBS content suggests that the copolymers form a single phase in the melt (i.e., homogeneous melt), where both PBS and PLA chains are mixed.

The influence of the copolymerization process on the crystallization and melting of both PBS and PLA blocks is shown in Figure 1. The crystallization process can be observed in Figure 1a for the PBS homopolymer and for the PBS block within the copolymers with 80% of PBS (LL$_{20}$$^{5.4}$-*mb*-BS$_{80}$$^{7.4}$ and DL$_{20}$$^{6.1}$-*mb*-BS$_{80}$$^{7.4}$). In the case of those copolymers with 30% PBS, the crystallization of the PBS blocks cannot be detected during cooling from the melt.

During the cooling scans from the melt, it is not possible to detect any crystallization exotherm corresponding to PLA homopolymers or PLA phases within the copolymers. However, during the subsequent heating scans, it is possible to observe cold crystallization exotherms that can be attributed to PLA homopolymers or to the PLA blocks within the copolymers. It is well known that PLA homopolymers have a slow crystallization kinetics when cooled from the melt, but during cooling and vitrification their nucleation density can be greatly enhanced. Therefore, when heated from the glassy state they can undergo cold-crystallization [32–35].

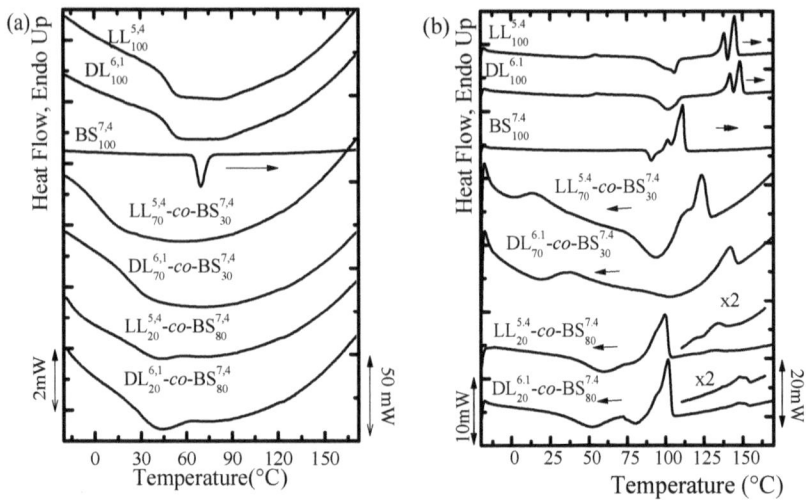

Figure 1. DSC scans at 20 °C/min for the homopolymers and copolymers: (a) cooling; and (b) second heating.

During the second heating scan, PLLA and PDLA homopolymers melt at around 140 °C (see Table 4) while PBS melts at 110 °C. This large difference in melting temperatures is useful to distinguish between the melting of PBS block crystals from those of PLA block crystals within the copolymers. The multi-block copolymers with the lowest PBS content ($LL_{70}{}^{5.4}$-*co*-$BS_{30}{}^{7.4}$ and $DL_{70}{}^{6.1}$-*co*-$BS_{30}{}^{7.4}$) show only one melting peak that corresponds to the fusion of the PLA block crystals. Therefore, in these copolymers the PBS blocks are unable to crystallize during cooling or during the subsequent heating and remain amorphous.

Table 4. Enthalpy of fusion, glass transition, and melting temperature for the homopolymer and copolymers indicated, during the first heating, cooling, and second heating at 20 °C/min.

Sample	Crystal	Cooling		2nd Heating					X_c
		T_c	ΔH_c	T_g	T_{cc}	ΔH_{cc}	T_m	ΔH	
		(°C)	(J/g)	(°C)	(°C)	(J/g)	(°C)	(J/g)	(%)
$LL_{100}{}^{5.4}$	PLA	-	-	51.0	106.1	31.6	144.5 137.8	10.2	-
$DL_{100}{}^{6.1}$	PLA	-	-	51.8	102.3	33.4	148.6 142.2	11.6	-
$BS_{100}{}^{7.4}$	PBS	69.2	74.4	-	91.0	14.6	111.0	67.6	48
$LL_{70}{}^{5.4}$-*co*-$BS_{30}{}^{7.4}$	PLA	-	-	3.8	93.9	13.6	125.1	11.7	-
$DL_{70}{}^{6.1}$-*co*-$BS_{30}{}^{7.4}$	PLA	-	-	25.4	102.3	13.6	141.5	6.3	-
$LL_{20}{}^{5.4}$-*co*-$BS_{80}{}^{7.4}$	PLA PBS	- 40.3	- 3.0	-16.0	- 55.2	- 26.3	131.5 97.9	9.5 42.8	10 39
$DL_{20}{}^{6.1}$-*co*-$BS_{80}{}^{7.4}$	PLA PBS	- 42.7	- 6.1	-17.7	79.5 50.0	19.0 30.1	147.2 101.7	12.0 47.9	- 22

On the other hand, the copolymers with 80% PBS are double crystalline, as they exhibit two clear melting endotherms that can be ascribed to the sequential melting of PBS block crystals (at around 100 °C) and PLA block crystals (at temperatures between 125–150 °C).

The melting process in both PLLA and PDLA is characterized by two endothermic peaks. PLAs have a tendency to reorganize during heating and this is probably the origin of such double endotherms. Another possibility, would be the presence of two polymorphs, i.e., the α' and α phases, but such a possibility would have to be assessed by wide angle X-ray diffraction. The difference between the melting temperatures of PDLA and PLLA homopolymers may originate from their small difference in molecular weight, as in this molecular weight range (i.e., very low molecular weights), the melting point is a strong function of the average molecular weight [36].

In the copolymers case, the incorporation of PBS segments covalently bonded to PLA chains decreases the melting point values corresponding to the PLA blocks; this could be attributed to a dilution effect, since these di-block copolymers are miscible in the melt [32–34]. Therefore, the PBS blocks act like a macromolecular solvent surrounding the PLA crystals in the melt. In addition, the covalent link between the PLA and PBS will affect the crystallization process of both blocks and, consequently, their corresponding T_c and T_m values [37].

The PBS endotherm has a complex behavior as well, as it is possible to observe a cold crystallization peak before the double melting endotherm. In this case, cold crystallization is followed by the melting of the crystals formed during previous cooling (first endothermic peak) while the second peak is a consequence of the melting of the reorganized crystals during the scan.

3.1.2. Polarized Light Optical Microscope: Morphology and Superstructural Growth Kinetics

The samples were isothermally crystallized from the melt on a hot plate and were observed employing PLOM. The superstructural morphology of the homopolymers and copolymers are shown in Figure 2.

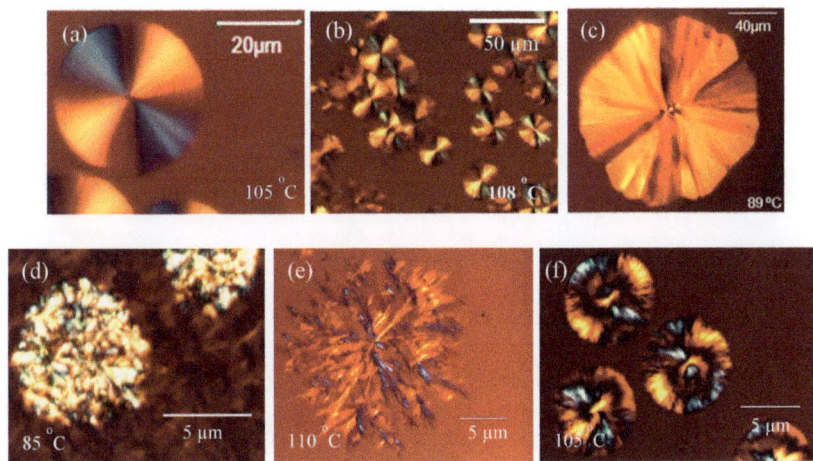

Figure 2. PLOM micrographs for (**a**) $DL_{100}^{6.1}$; (**b**) $LL_{100}^{5.4}$; (**c**) $BS_{100}^{7.4}$; (**d**) $LL_{70}^{5.4}$-*mb*-$BS_{30}^{7.4}$; (**e**) $LL_{20}^{5.4}$-*mb*-$BS_{80}^{7.4}$; and (**f**) $DL_{20}^{6.1}$-*mb*-$BS_{80}^{7.4}$, during isothermal crystallization at the indicated T_c.

Both PLA homopolymers ($DL_{100}^{6.1}$ and $LL_{100}^{5.4}$) display negative spherulites at low crystallization temperatures (with no banding) which are similar to those shown in Figure 2a,b. At higher temperatures, the spherulites show banding extinction patterns.

In the case of neat PBS, the superstructure was spherulite-like at lower crystallization temperatures, although the spherulites did not appear to have a perfectly circular cross-section. When the crystallization temperature was increased, hedrites were formed, instead of spherulites, as revealed by the example shown in Figure 2c.

For the copolymers with high PLA content, the superstructures observed at high T_c values (higher than the crystallization temperature of the PBS block) were distorted spherulites with ill-defined banding, as shown in Figure 2d. The most interesting fact is the crystallization of the PBS component within the interlamellar regions of the PLA template (formed at higher temperatures) when the crystallization temperature is reduced to values where PBS is able to crystallize (see Figure 3). In Figure 2d it is possible to observe the birefringence due to PBS interspherulitic lamellae (brighter regions of the sample) inside the PLA component template superstructure. In the case of the copolymers with a lower content of PLA, the superstructures observed are more similar to dendrites (see Figure 2e).

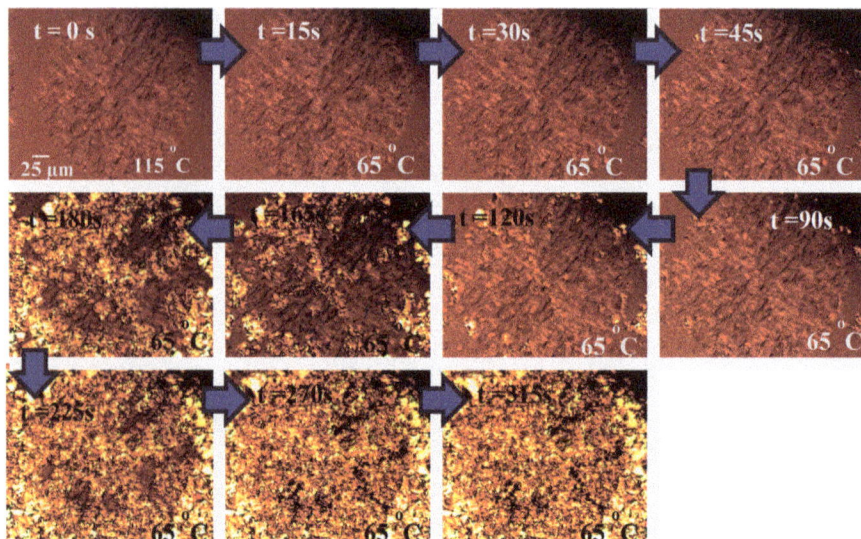

Figure 3. PBS crystallization at 65 °C from a PLLA spherulite (T_c = 115 °C) within a $LL_{20}{}^{5.4}$-*mb*-$BS_{80}{}^{7.4}$ copolymer.

Figure 3 shows a sequence of polarized light micrographs taken at different times and temperatures for copolymer $LL_{20}{}^{5.4}$-*mb*-$BS_{80}{}^{7.4}$. The sample was first isothermally crystallized at 115 °C, a temperature at which the PBS blocks are molten. A super-structural template is formed by the part of the PLA component that can crystallize, as can be observed in the first micrograph (top left-hand corner of Figure 3). As this copolymer sample contains only 30% PLA, the amount of crystalline superstructural phase observed in the micrograph could be at most 15% of the total sample by weight. Then, the sample was quenched to 65 °C, since at this temperature the isothermal crystallization of the PBS component can be followed as a function of time.

Figure 3 shows how the PBS component is nucleated by the PLA block crystals, forming PBS crystalline aggregates inside and on top of the PLA previously formed a loose template superstructure. As time increases, the entire PLA template is filled with PBS crystals. PBS and PLA blocks are melt mixed, but will phase segregate as PLA and PBS crystallize. Their amorphous regions still remain mixed, forming a single phase characterized by a single T_g value. As the amount of initial PLA crystals is in minority (i.e., equal or less than 15% of the sample), confinement of the PBS blocks during crystallization is not expected, as PBS blocks constitute the more abundant component in the multi-block component.

We tried to determine the growth rate of the different crystalline structures but their growth was very difficult to follow, especially when the morphology was not spherulitic. Therefore, we decided to determine, instead, the overall crystallization kinetics of selected samples by DSC.

3.1.3. Overall Isothermal Crystallization Kinetics

The study of the isothermal crystallization from the melt was performed for the samples that contained PLLA, instead of PDLA, as the measurements were facilitated by their generally faster crystallization from the melt. The overall crystallization rate, expressed as the inverse of the peak-crystallization time ($1/\tau_{50\%}$), as a function of the isothermal crystallization temperature, is shown in Figure 4.

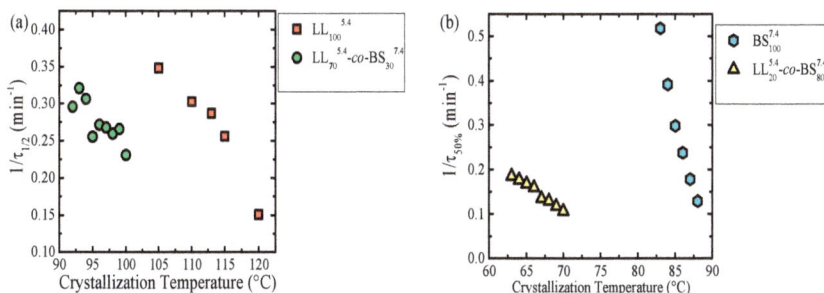

Figure 4. Overall crystallization rate ($1/\tau_{50\% \text{ exp}}$) versus crystallization temperature (T_c) for (**a**) PLLA homopolymer and PLLA blocks within the $LL_{70}{}^{5.4}\text{-}mb\text{-}BS_{30}{}^{7.4}$ copolymer; and (**b**) PBS homopolymer and PBS blocks within $LL_{20}{}^{5.4}\text{-}co\text{-}BS_{80}{}^{7.4}$ copolymer.

The overall crystallization kinetics comprises both primary nucleation and growth. The balance between these two processes will determine how the overall crystallization rate will be affected by the crystallization temperature and other factors, like molecular weight or composition. In Figure 4, it is possible to observe that the crystallization of the PLLA and PBS blocks need a higher undercooling than the respective homopolymers. This is a consequence of the miscibility between the blocks which changes the relative supercooling, as the equilibrium melting temperature must be reduced in the copolymers, as compared to the homopolymers. For the copolymer with a low content of PLLA it was impossible to measure the isothermal crystallization, as in this case the amount of crystallizable material was too small, and the calorimeter could not detect any signal in the isothermal mode.

Before the PBS isothermal crystallization, the PLLA block within the $LL_{20}\text{-}mb\text{-}BS_{80}$ copolymer was isothermally crystallized until saturation. Then the sample was quickly cooled (at 60 °C/min) to the PBS crystallization temperatures. It is worth pointing out that only the copolymer sample with 80% PBS content was examined, as in the samples with minor amounts of PBS, this component was unable to crystallize. The presence of rigid PLA semi-crystalline blocks retarded the isothermal crystallization kinetics of the PBS blocks within the $LL_{20}{}^{5.4}\text{-}co\text{-}BS_{80}{}^{7.4}$ copolymer, in comparison to the analog neat PBS.

The experimental data obtained during the overall isothermal crystallization experiments were analyzed using the Avrami equation, which can be expressed as follows:

$$1 - V_c(t - t_0) = e^{(-k(t-t_0)^n)} \tag{2}$$

where t is the experimental time, t_0 is the induction time, V_c is the relative volumetric transformed fraction, n is the Avrami index, and k is the overall crystallization rate constant [30,38].

The results are shown in Table S1 for PLLA and Table S2 for PBS homopolymers and corresponding blocks. The Avrami indices obtained for the PLLA homopolymer can be approximated to values between 3 and 4. These values are in agreement with instantaneous or sporadic three-dimensional spherulitic structures, as observed by PLOM (see above). For the PLLA block within $LL_{70}{}^{5.4}\text{-}mb\text{-}BS_{30}{}^{7.4}$, the most frequently obtained values were around 3 (with a few exceptions), indicating instantaneously nucleated three-dimensional superstructures in correspondence with those observed in Figure 2.

Polymers 2018, 10, 8

According to PLOM observation, the low molecular weight PBS employed here displayed hedrites, which are 2D lamellar aggregates that characterize the transition from axialites to spherulites. The Avrami index obtained for neat PBS was between 2.5 and 2.7, suggesting 2D aggregates with a sporadic nucleation, in agreement with PLOM observations. In the case of the PBS blocks crystallized within $LL_{20}^{5.4}$-co-$BS_{80}^{7.4}$ copolymer, similar n values (between 3.2 and 2.8) were found.

4. Stereocomplexes

4.1. Non-Isothermal Crystallization

As previously observed for the copolymers (see Table 3), all the blends exhibited a single glass transition, denoting that the samples have a single phase in the melt and in the amorphous state. The T_g results are summarized in Figure 5. In the copolymer/copolymer and copolymer/homopolymer blends, T_g decreases as the PBS content increases. This behavior corresponds to the plasticizing ability of the PBS segments. However, for the blends with DL homopolymer the glass transition is lower than the samples with LL homopolymer (with the same composition) (see Figure 5 and Table 5).

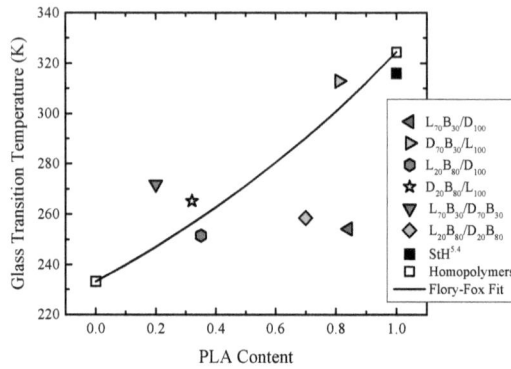

Figure 5. Glass transition temperature (T_g) versus PLA content (XPLA) for different homopolymers, copolymers and copolymer/homopolymer blends. The solid line represents the fitting to the Fox equation.

Table 5. Enthalpy of fusion, glass transition, and melting temperature of PLA stereocomplexes during the first heating, cooling, and second heating at 20 °C/min.

Sample	Crystal	1st Heating		Cooling		T_g	2nd Heating				X_c
		T_m	ΔH_m	T_c	ΔH_c		T_{cc}	ΔH_{cc}	T_m	ΔH_m	
		(°C)	(J/g)	(°C)	(J/g)	(°C)	(°C)	(J/g)	(°C)	(J/g)	(%)
L_{100}/D_{100}	SC$_{PLA}$	210.9	88.9	143.9	−86.2	42.9	-	-	206.8	89.1	63
$L_{70}B_{30}/D_{70}B_{30}$	PBS	91.9	3.3	-		−14.7	63.7	−6.0	92.6	1.7	-
	SC$_{PLA}$	191.2	37.4	105.2	−52.9		-		190.2	41.1	29
$L_{20}B_{80}/D_{20}B_{80}$	PBS	101.6	56.4	44.1	−45.4	−8.7	78.8	−6.9	101.3	47.1	36
	SC$_{PLA}$	198.6	11.0	107.3	−35.0		-		196.2	19.0	13
$L_{70}B_{30}/D_{100}$	PBS	87.6	3.1			−19.0	-		-	-	-
	SC$_{PLA}$	196.6	63.0	162.8	−67.1		-		192.9	59.5	42
$D_{70}B_{30}/L_{100}$	SC$_{PLA}$	203.9	59.5	128.5	−62.1	39.8	-		201.2	57.3	40
$L_{20}B_{80}/D_{100}$	PBS	100.3	33.8			−21.6	74.0	−5.4	99.5	28.0	21
	PLA	147.9	38.9	65.3	−90.3		123.7	−12.0	144.8	24.3	13
	SC$_{PLA}$	202.6	52.9				-		193.9	48.0	34
$D_{20}B_{80}/L_{100}$	PBS	103.6	37.8			−7.9	77.0	−18.7	101.2	29.9	10
	PLA	145.5	45.3				123.3	−2.8	140.4	24.1	23
	SC$_{PLA}$	206.3	62.2	108.5	−92.2		-		199.2	54.7	39

The stereocomplex crystals were obtained by precipitation from solutions of equimolecular mixtures. The first DSC scans of the samples show the characteristics of the crystals formed from solution. Figure 6a shows DSC first heating scans for all the blends prepared. All the samples show melting peaks at around 190–210 °C, these T_m values are higher than those of PLA homocrystals (crystals formed by homopolymers), indicating the presence of stereocrystals. Stereocomplexation of PLLA and PDLA chains is well known. PLLA and PDLA display a strong interaction with each other and can share a new crystal lattice. The stereocomplex unit cell is formed by segments of PLLA and PDLA packed in parallel, and the complex forms a 3_1 helical conformation, which is derived from the 10_3 helical structure characteristic of PLA homopolymers α phase. The temperature difference obtained for T_m stereocomplex crystals and T_m PLA homopolymers crystals (of approximately 50 °C) implies a significant change between the two crystal structures [4–6,9]. As expected, the L_{100}/D_{100} equimolar blend only forms stereocomplexes when crystallized from solution (i.e., no homocrystals are formed) and its T_m value is the highest shown in Figure 6a.

Figure 6a also shows the first heating scan of the copolymer/copolymer and copolymer/homopolymer blends, and it is possible to observe multiple endothermic peaks in almost all the samples. The endotherms at around 100 °C correspond to the melting of PBS crystals and they are present in most samples, with the exception of the sample $D_{70}B_{30}/L_{100}$ where the PBS component does not crystallize.

The samples $L_{20}B_{80}/D_{100}$ and $D_{20}B_{80}/L_{100}$ in Figure 6a are the only ones that show endothermic peaks associated with the melting of the PLA homocrystals (at temperatures around 140 °C). The presence of homocrystals in these samples could be due to their difficulty in forming stereocomplexes. For example, in the $L_{20}B_{80}/D_{100}$ blend, PDLA has a higher probability of finding another PDLA segment to form homocrystals than the possibility of finding a PLLA chain to form a stereocrystal. These two samples are also the only ones where the PBS blocks exhibit cold crystallization exotherms just before melting in double-peaked endotherms, which are due to reorganization during the scan.

The highest melting temperature (T_m) observed in Figure 6a corresponds to homopolymer/homopolymer blends. Since it is higher than those of stereocomplexes formed by copolymer blends, this is an indication that the presence of PBS in the multi-block copolymers disturbs the formation of the stereocrystals. The latent enthalpy of fusion of the PLLA/PDLA homopolymer/homopolymer stereocomplex is the highest of all stereocomplexes reported in Table 5. The enthalpy of fusion for the stereocomplexes prepared from copolymer/copolymer blends decreases as the content of PLA decreases, as stereocomplexation becomes more difficult. A similar effect is observed for copolymer/homopolymer blends in Table 5.

Figure 6b shows the DSC cooling scans from the melt, while Figure 6c shows the subsequent heating scans (i.e., second heating scans). Upon cooling from the melt, it is expected that the L_{100}/D_{100} mixture exhibits the highest crystallization temperature as stereocomplexation would be easier for this sample in comparison with all others. This is, in fact, observed, as almost all samples crystallize at lower temperatures, with the exception of the $L_{70}B_{30}/D_{100}$ sample. In the case of samples with small amounts of PLA, the crystallization exotherms are broad and multimodal, as at least three different types of crystals are formed: PLA stereocomplex crystals, PLA homocrystals and, finally, PBS crystals.

Figure 6c shows the heating scan after melt crystallization. When Figure 6c is compared to Figure 6a (for solution crystallized samples), it can be seen that the behavior in both cases is very similar. The stereocomplexes in the copolymer/copolymer and copolymer/homopolymer blends were formed from the melt without any problems. The only exception observed was that of sample $D_{70}B_{30}/L_{100}$, which was able to form a very small amount of PLLA homocrystals when crystallized from the melt, and none when crystallized from solution.

The PBS blocks act as a plasticizer for the PLA blocks in the copolymers and in their blends, as was demonstrated for studies in PBS-*b*-PLA-*b*-PBS triblock copolymer [20] and as was discussed in the previous sections. Figure 7 was drawn in order to understand the role of PBS content in the crystallization of copolymer/copolymer and copolymer/homopolymer stereocomplexes. In general, increasing the amount of PBS in the blends depresses the crystallization temperature, as expected, since PBS causes a dilution effect on PLA stereocomplexes.

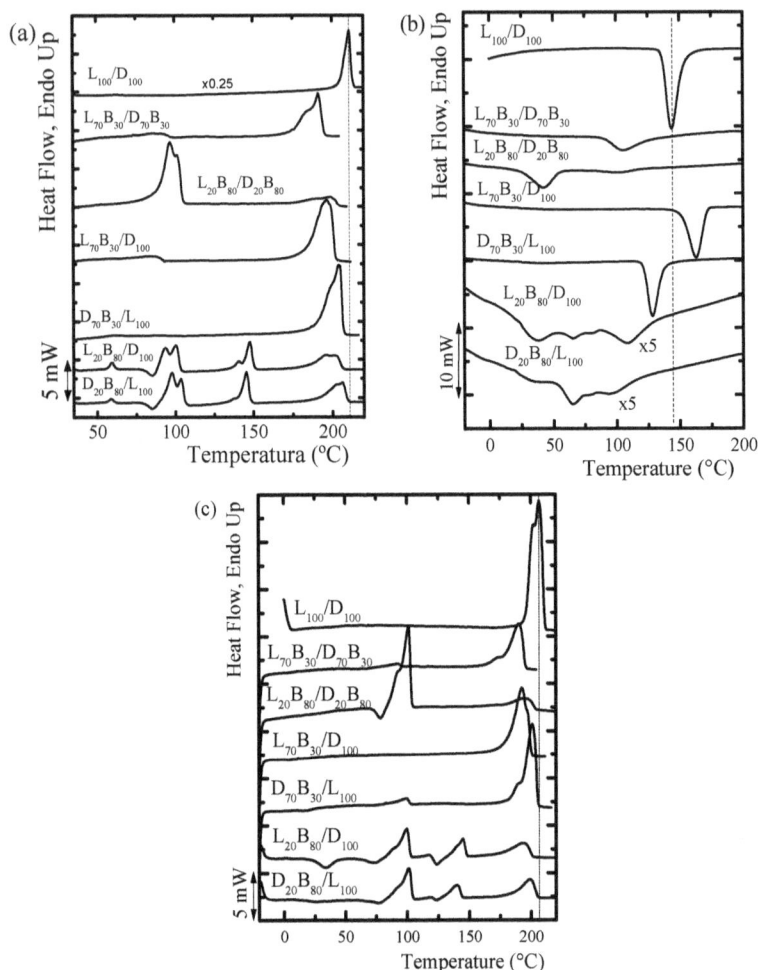

Figure 6. Non-isothermal DSC (**a**) first heating, (**b**) cooling, and (**c**) second heating scans at 20 °C/min for the indicated stereocomplexes. The dashed line represents the peak crystallization and melting temperature for the L_{100}/D_{100} stereocomplex.

The degree of crystallinity was calculated from ΔH_m values (melting enthalpies determined from the second heating scans) for stereocrystals and homocrystals, employing the following values for the enthalpies of fusion of 100% crystalline materials: $\Delta H_m°$: PBS (110.3 J/g) (66), PLA (93.6 J/g) (67), and PLA stereocomplex (142 J/g) (68). In general, crystallinity degree of PLA stereocomplexes decreases with increasing PBS content (see Figure 7). Consequently, fewer PLA stereocomplex crystals were formed for the $L_{20}B_{80}/D_{20}B_{80}$ blend and a higher proportion of PBS crystals. However, PBS does not prevent PLA stereocomplex formation even if PBS is the major component in this blend. The amount of stereocomplexes decreases when PLA content decreases, as too much PBS can generate a "macromolecular solvent" matrix where dispersed PLA molecules will have large mobility, but are highly diluted.

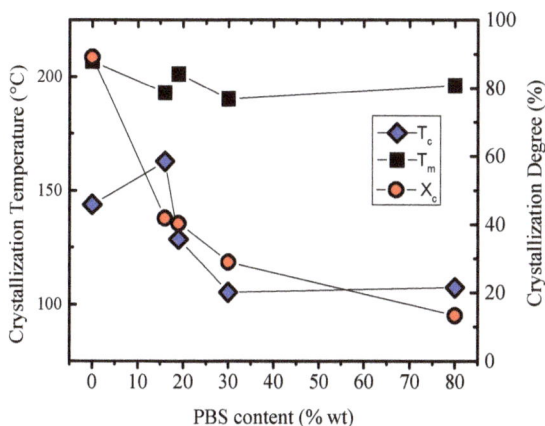

Figure 7. Crystallization degree, crystallization, and melting temperature versus the PBS content for the copolymer/copolymer and copolymer/homopolymer stereocomplexes.

4.2. Polarized Light Optical Microscope: Stereocomplex Superstructural Morphology

Figure 8 shows the stereocomplex superstructures observed in a polarized light optical microscope during isothermal crystallization at high T_c values, where only stereocomplex can crystallize (i.e., the T_c values employed are higher than the melting points of PLA homocrystals and PBS crystals).

For PLA homopolymer stereocomplexes the structures observed correspond to negative spherulites with Maltese cross extinction patterns. On the other hand, the copolymer/copolymer and copolymer/homopolymer blends exhibit a variety of superstructural morphologies. The stereocomplexes formed by $LL_{70}^{5.4}$-*mb*-$BS_{30}^{7.4}$/$DL_{70}^{6.1}$-*mb*-$BS_{30}^{7.4}$, $LL_{70}^{5.4}$-*mb*-$BS_{30}^{7.4}$/$DL_{100}^{6.1}$, and $DL_{70}^{6.1}$-*mb*-$BS_{30}^{7.4}$ also display negative spherulites; however, the other two stereocomplexes (i.e., $L_{20}B_{80}$/D_{100} and $D_{20}B_{80}$/L_{100}) exhibit a dendritic structure in view of their low PLA content [39]. The obtained morphologies are a function of composition. When the majority of the blends are constituted by PLA isomers, the morphology corresponds to 3D spherulites, and when the PLA content is small, then 2D aggregates (or axialites) developed. The results parallel those obtained for PLA homocrystals above (see Figure 2).

The morphology for the blends with high amounts of PBS segments ($L_{20}B_{80}$/$D_{20}B_{80}$, $L_{20}B_{80}$/D_{100} and $D_{20}B_{80}$/L_{100}) was always dendritic, as the concentration of PLA chains is rather low. Figure 8c,f,g represent mixtures with a higher PBS proportion. In this case, dendrites with secondary branches, called dizzi dendrites, can be observed. These dizzi dendrites are polycrystalline structures formed by sequential deformation of the tips of dendrites. Their formation mechanism has not been identified [39].

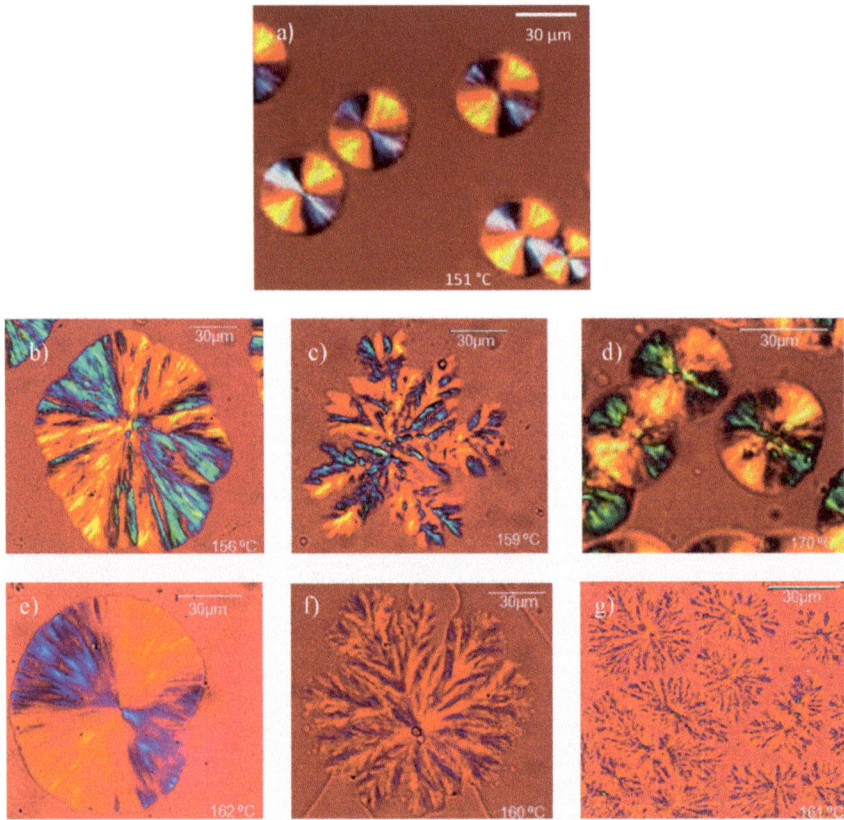

Figure 8. PLOM micrographs for (**a**) L_{100}/D_{100}; (**b**) $L_{70}B_{30}/D_{70}B_{30}$; (**c**) $L_{20}B_{80}/D_{20}B_{80}$; (**d**) $L_{70}B_{30}/D_{100}$; (**e**) $D_{70}B_{30}/L_{100}$; (**f**) $L_{20}B_{80}/D_{100}$; and (**g**) $D_{20}B_{80}/L_{100}$, during isothermal crystallization at the indicated T_c.

4.3. Overall Isothermal Crystallization Kinetics of Stereocomplexes

The overall crystallization rate (estimated as the inverse of the half-peak crystallization time) versus crystallization temperature for selected stereocomplexes is plotted in Figure 9. The results obtained are a function of PLA content in the samples.

In the case of copolymer/copolymer or copolymer/homopolymer blends, where the copolymer contains 70% PLLA, the stereocomplexes were formed at lower supercoolings than L_{100}/D_{100} stereocomplexes (see Figure 9). This is a remarkable result, especially when a comparison is made with the plasticizing effect of the PBS block chains have on PLLA homocrystals, in the case of neat copolymers (see Figure 4). It would seem that for stereocomplex formation, molten PBS segments plasticize the PLA chains, helping them find the necessary PDLA/PLLA pairs for stereocomplexation. As a result, the stereocomplex formation occurs faster when PBS chains are present in the copolymer/homopolymer blends. The situation changes when the number of PLA chains in the copolymer is reduced. In that case, it seems that too much PBS causes a dilution effect on PLLA and PDLA chains, causing difficulties for the formation of stereocomplexes and, therefore, a higher supercooling is needed for stereocomplexation, as compared to the simpler L_{100}/D_{100} case.

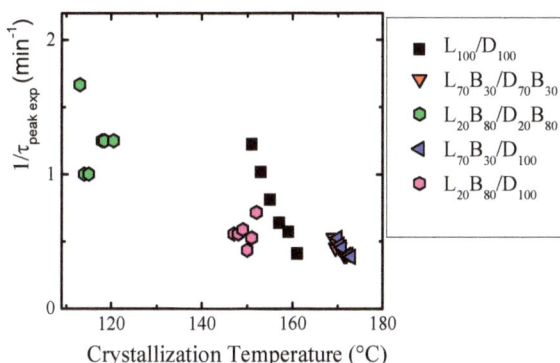

Figure 9. Overall crystallization rate ($1/\tau_{\text{peak exp}}$) versus crystallization temperature (T_c) for PLA stereocomplexes.

According to the results presented in Figures 6 and 9, the effect of the PBS blocks on the formation of PLA stereocomplexation can be ascertained. According to their melting points and enthalpies after non-isothermal crystallization, the PBS segments induce a lower amount of stereocomplex formation as compared to neat L_{100}/D_{100} stereocomplexes. It seems that PBS perturbs the formation of stereocomplexes and this perturbation increases with the amount of PBS in the samples. However, when relatively low amounts of PBS in the copolymer blends is present, the rate of stereocomplex formation is enhanced (if not the crystallinity degree achieved). This effect is lost when higher amounts of PBS are present.

4.4. Infrared Spectroscopy for PLA Stereocomplexes (FTIR)

According to the literature, PLA stereocomplex formation involves the formation of double helices. The molecules adopt a 3_1 helical conformation known as the β-form and a segment of PLLA molecule is paired with a segment of PDLA molecule, resulting in a triclinic unit cell [40]. In a permanently polar bond, the partial charge of this dipole will attract the partial opposite charge in the other molecule. This is a result of the asymmetric distribution of regions with positive and negative charges. This distribution can induce dipoles in nonpolar molecules. A positively-charged molecule is capable of generating an attractive force with the negative end of an adjacent molecule, which justifies the formation of C-H···O bonds presented by PLA stereocomplexes.

The IR band at 909 cm^{-1} is characteristic of the 3_1 helical conformation of stereocomplex PLA crystals. The presence of this band depends on the crystallinity of the polymer and the helix type [41]. The intensity of the stereocomplex band is higher in the samples with lower molecular weight (see Figure 10) and higher PLA content, this is a consequence of a larger content of PLA stereocomplex in these samples, as demonstrated by DSC. The FTIR results presented in Figure 10 are able to prove the existence of the PLA stereocrystals, however, it is necessary that a high amount of stereocrystals are present to produce a signal. For some of the samples in Figure 10, the 909 cm^{-1} signal is reduced because of the presence of PBS chain segments and PLA homocrystals.

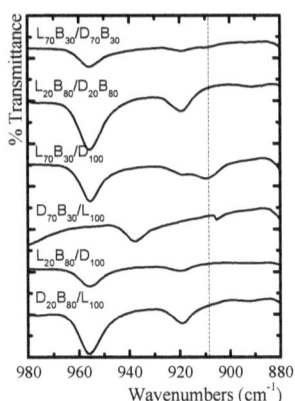

Figure 10. Infrared spectrum of the PLA stereocomplexes. Sensitive bands to the formation of PLA stereocomplexes. The dashed line indicates the position of the 909 cm^{-1} band.

4.5. Carbon-13 Nuclear Magnetic Resonance Spectroscopy (^{13}C-NMR) for PLA Stereocomplexes Prepared from Copolymer-Homopolymer Blends

In order to further characterize the presence of stereocrystals, ^{13}C-NMR experiments were performed in the solid state. The carbon-13 nuclear magnetic resonance spectra of selected samples are shown in Figure 11. The spectra corresponding to the homopolymers PDLA and PLLA present two broad peaks fixed at 170.4 ppm assigned to the non-crystalline component and 172.0 ppm assigned to the crystalline component. The spectrum corresponding to neat PBS presents a resonance peak at 174.5 ppm that corresponds to the carbonyl carbon. The stereocomplex formation in the sample L_{100}/D_{100} is ascertained by the presence of a resonance peak at 173.7 ppm which has been previously assigned to the rigid racemic crystalline component overlapped with disordered racemic crystalline component. The spectra corresponding to the copolymers do not show the presence of the peak assigned to the stereocomplex formation. However, in the spectrum corresponding to the sample $L_{70}B_{30}/D_{100}$ this peak is shifted to 173.3 ppm, whereas for sample $L_{20}B_{80}/D_{100}$ it appears as a shoulder at 173.7 ppm. These results confirm the formation of the stereocomplex in the case of the copolymer-homopolymer blends.

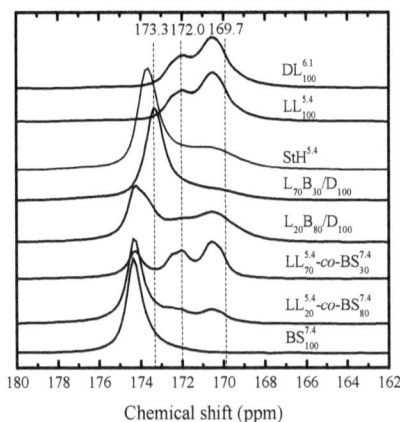

Figure 11. Proton-decoupled ^{13}C spectrum of homopolymers, copolymers, and mixtures thereof to form PLA stereocomplexes (400 MHz).

5. Conclusions

The thermal properties and morphologies of homopolymers, copolymers and stereocomplexes of PLA-*mb*-PBS multi-block copolymers were investigated. The PLA and PBS multi-blocks within the PLA-*mb*-PBS copolymers are miscible according to the observation of a single T_g value. The PBS has a dilution effect on PLA blocks, increasing the PLA chain mobility, thus lowering the crystallization and melting temperature and increasing the undercooling during isothermal crystallization. The Avrami theory was able to fit properly the data of all analyzed homopolymers and copolymers, and the Avrami indices obtained are in line with PLOM observations. Both homopolymers and copolymers with a higher PLA content exhibit negative spherulites, however, the copolymers with a lower PLA content had a dendritic morphology.

The PLA chains stereocomplexes were formed from solution and from the melt and exhibited melting and crystallization temperatures higher than the original homocrystals, as expected. The decrease of both the melting temperature and the crystallinity degree with PBS content, indicates that the PBS blocks perturb the PLA blocks stereocomplexation. However, the PBS blocks could retard or increase the crystallization kinetics, depending on the PBS amount in the blend. Lower contents of PBS induce a faster crystallization rate of the stereocomplexes, however, a higher content retards it.

The superstructural morphology of the homopolymer/homopolymer stereocomplexes is composed of three-dimensional spherulites, in a similar manner, the copolymer/copolymer and copolymer/homopolymer with the lowest PBS content also displayed a spherulitic texture. However, the superstructures change when the PBS content increases, and in the samples with the largest PBS content the morphology of the stereocrystals was dentritic.

The formation of stereocrystals was verified employing solid state ^{13}C-NMR and FTIR. The characteristic infrared absorption band at 909 cm^{-1}, associated with stereocomplex formation, was observed only in copolymer-homopolymer blends with a higher PLA content.

Supplementary Materials: The following are available online at www.mdpi.com/2073-4360/10/1/8/s1, Table S1: Avrami fit parameters for the PLA homopolymers and copolymers, Table S2: Avrami fit parameters for the PBS homopolymer and copolymers.

Acknowledgments: The USB team acknowledges support from DID-USB for the funding of USB Polymer Group 1 (GPUSB1-DIDG02). The two research groups from Spain acknowledge funding received through the coordinated project: "Mineco MAT2014-53437". Rosica Mincheva and Philippe Dubois gratefully acknowledge funding from Wallonia and Europe in the frame of the (FEDER) POLYTISS-POLYEST and BIOMAT projects.

Author Contributions: Rosa M. D'Ambrosio, performed the MOLP, DSC, and FTIR experiments, Rosica Mincheva synthetized the homopolymers and copolymers. Rebeca Hernández and Carmen Mijangos, design, performed and analyzed the NMR experiments. Philippe Dubois design the synthesis of the homopolymers and copolymers. Alejandro J. Müller conceived and designed the experiments, analyzed the data and edited the paper. Rose Mary Michell analyzed the data and wrote the paper.

Conflicts of Interest: The authors declare no conflict of interest.

References

1. Hirata, M.; Kimura, Y. Structure and properties of stereocomplex-type poly(lactic acid). In *Poly(lactic acid)*; John Wiley & Sons, Inc.: Hoboken, NJ, USA, 2010.
2. Bouapao, L.; Tsuji, H.; Tashiro, K.; Zhang, J.; Hanesaka, M. Crystallization, spherulite growth, and structure of blends of crystalline and amorphous poly(lactide)s. *Polymer* **2009**, *50*, 4007–4017. [CrossRef]
3. Ikada, Y.; Jamshidi, K.; Tsuji, H.; Hyon, S.H. Stereocomplex formation between enantiomeric poly(lactides). *Macromolecules* **1987**, *20*, 904–906. [CrossRef]
4. Okihara, T.; Tsuji, M.; Kawaguchi, A.; Katayama, K.-I.; Tsuji, H.; Hyon, S.-H.; Ikada, Y. Crystal structure of stereocomplex of poly(L-lactide) and poly(D-lactide). *J. Macromol. Sci. B* **1991**, *30*, 119–140. [CrossRef]
5. Tsuji, H. Poly(lactide) stereocomplexes: Formation, structure, properties, degradation, and applications. *Macromol. Biosci.* **2005**, *5*, 569–597. [CrossRef] [PubMed]

6. Tsuji, H.; Hyon, S.H.; Ikada, Y. Stereocomplex formation between enantiomeric poly(lactic acid)s. 4. Differential scanning calorimetric studies on precipitates from mixed solutions of poly(D-lactic acid) and poly(L-lactic acid). *Macromolecules* **1991**, *24*, 5657–5662. [CrossRef]

7. Tsuji, H.; Hyon, S.H.; Ikada, Y. Stereocomplex formation between enantiomeric poly(lactic acids). 5. Calorimetric and morphological studies on the stereocomplex formed in acetonitrile solution. *Macromolecules* **1992**, *25*, 2940–2946. [CrossRef]

8. Tsuji, H.; Tezuka, Y. Stereocomplex formation between enantiomeric poly(lactic acid)s. 12. Spherulite growth of low-molecular-weight poly(lactic acid)s from the melt. *Biomacromolecules* **2004**, *5*, 1181–1186. [CrossRef] [PubMed]

9. Tsuji, H.; Wada, T.; Sakamoto, Y.; Sugiura, Y. Stereocomplex crystallization and spherulite growth behavior of poly(L-lactide)-*b*-poly(D-lactide) stereodiblock copolymers. *Polymer* **2010**, *51*, 4937–4947. [CrossRef]

10. Miyata, T.; Masuko, T. Crystallization behaviour of poly(L-lactide). *Polymer* **1998**, *39*, 5515–5521. [CrossRef]

11. Kakuta, M.; Hirata, M.; Kimura, Y. Stereoblock polylactides as high-performance bio-based polymers. *Polym. Rev.* **2009**, *49*, 107–140. [CrossRef]

12. Michell, R.M.; Müller, A.J.; Spasova, M.; Dubois, P.; Burattini, S.; Greenland, B.W.; Hamley, I.W.; Hermida-Merino, D.; Cheval, N.; Fahmi, A. Crystallization and stereocomplexation behavior of poly(D- and L-lactide)-*b*-poly(*N*,*N*-dimethylamino-2-ethyl methacrylate) block copolymers. *J. Polym. Sci. Part B* **2011**, *49*, 1397–1409. [CrossRef]

13. Castillo, R.V.; Müller, A.J.; Lin, M.-C.; Chen, H.-L.; Jeng, U.S.; Hillmyer, M.A. Confined crystallization and morphology of melt segregated PLLA-*b*-PE and PLDA-*b*-PE diblock copolymers. *Macromolecules* **2008**, *41*, 6154–6164. [CrossRef]

14. Averous, L. Polylactic acid: Synthesis properties and applications. In *Monomers Oligomers, Polymers and Composites from Renewable Resourses*; Belgacem, N.G., Gandini, A., Eds.; Elsevier: Oxford, UK, 2008.

15. Zeng, J.-B.; Liu, C.; Liu, F.-Y.; Li, Y.-D.; Wang, Y.-Z. Miscibility and crystallization behaviors of poly(butylene succinate) and poly(L-lactic acid) segments in their multiblock copoly(ester urethane). *Ind. Eng. Chem. Res.* **2010**, *49*, 9870–9876. [CrossRef]

16. Jia, L.; Yin, L.; Li, Y.; Li, Q.; Yang, J.; Yu, J.; Shi, Z.; Fang, Q.; Cao, A. New enantiomeric polylactide-block-poly(butylene succinate)-block-polylactides: Syntheses, characterization and in situ self-assembly. *Macromol. Biosci.* **2005**, *5*, 526–538. [CrossRef] [PubMed]

17. Harada, M.; Ohya, T.; Iida, K.; Hayashi, H.; Hirano, K.; Fukuda, H. Increased impact strength of biodegradable poly(lactic acid)/poly(butylene succinate) blend composites by using isocyanate as a reactive processing agent. *J. Appl. Polym. Sci.* **2007**, *106*, 1813–1820. [CrossRef]

18. Homklin, R.; Hongsriphan, N. Mechanical and thermal properties of PLA/PBS co-continuous blends adding nucleating agent. *Energy Procedia* **2013**, *34*, 871–879. [CrossRef]

19. Lan, X.; Li, X.; Liu, Z.; He, Z.; Yang, W.; Yang, M. Composition, morphology and properties of poly(lactic acid) and poly(butylene succinate) copolymer system via coupling reaction. *J. Macromol. Sci. A* **2013**, *50*, 861–870. [CrossRef]

20. Ba, C.; Yang, J.; Hao, Q.; Liu, X.; Cao, A. Syntheses and physical characterization of new aliphatic triblock poly(L-lactide-*b*-butylene succinate-*b*-L-lactide)s bearing soft and hard biodegradable building blocks. *Biomacromolecules* **2003**, *4*, 1827–1834. [CrossRef] [PubMed]

21. Mincheva, R.; Raquez, J.-M.; Lison, V.; Duquesne, E.; Talon, O.; Dubois, P. Stereocomplexes from biosourced lactide/butylene succinate-based copolymers and their role as crystallization accelerating agent. *Macromol. Chem. Phys.* **2012**, *213*, 643–653. [CrossRef]

22. Park, J.W.; Im, S.S. Phase behavior and morphology in blends of poly(l-lactic acid) and poly(butylene succinate). *J. Appl. Polym. Sci.* **2002**, *86*, 647–655. [CrossRef]

23. Yokohara, T.; Yamaguchi, M. Structure and properties for biomass-based polyester blends of PLA and PBS. *Eur. Polym. J.* **2008**, *44*, 677–685. [CrossRef]

24. Stoyanova, N.; Paneva, D.; Mincheva, R.; Toncheva, A.; Manolova, N.; Dubois, P.; Rashkov, I. Poly(L-lactide) and poly(butylene succinate) immiscible blends: From electrospinning to biologically active materials. *Mater. Sci. Eng.* **2014**, *41*, 119–126. [CrossRef] [PubMed]

25. Jompang, L.; Thumsorn, S.; On, J.W.; Surin, P.; Apawet, C.; Chaichalermwong, T.; Kaabbuathong, N.; O-Charoen, N.; Srisawat, N. Poly(lactic acid) and poly(butylene succinate) blend fibers prepared by melt spinning technique. *Energy Procedia* **2013**, *34*, 493–499. [CrossRef]

26. Supthanyakul, R.; Kaabbuathong, N.; Chirachanchai, S. Random poly(butylene succinate-*co*-lactic acid) as a multi-functional additive for miscibility, toughness, and clarity of pla/pbs blends. *Polymer* **2016**, *105*, 1–9. [CrossRef]

27. Supthanyakul, R.; Kaabbuathong, N.; Chirachanchai, S. Poly(L-lactide-*b*-butylene succinate-*b*-L-lactide) triblock copolymer: A multi-functional additive for pla/pbs blend with a key performance on film clarity. *Polym. Degrad. Stabil.* **2017**, *142*, 160–168. [CrossRef]

28. Papageorgiou, G.Z.; Achilias, D.S.; Bikiaris, D.N. Crystallization kinetics of biodegradable poly(butylene succinate) under isothermal and non-isothermal conditions. *Macromol. Chem. Phys.* **2007**, *208*, 1250–1264. [CrossRef]

29. Chen, H.-B.; Wang, X.-L.; Zeng, J.-B.; Li, L.-L.; Dong, F.-X.; Wang, Y.-Z. A novel multiblock poly(ester urethane) based on poly(butylene succinate) and poly(ethylene succinate-*co*-ethylene terephthalate). *Ind. Eng. Chem. Res.* **2011**, *50*, 2065–2072. [CrossRef]

30. Lorenzo, A.T.; Arnal, M.L.; Albuerne, J.; Müller, A.J. DSC isothermal polymer crystallization kinetics measurements and the use of the avrami equation to fit the data: Guidelines to avoid common problems. *Polym. Test.* **2007**, *26*, 222–231. [CrossRef]

31. Gedde, U.W. *Polymer Physics*; Chapman & Hall: London, UK, 1995.

32. Mathot, V.B. *Calorimetry and Thermal Analysis of Polymers*; Hanser Publishers: New York, NY, USA, 1994.

33. Wunderlich, B. *Macromolecular Physics, Vol. 2 Crystal Nucleation, Growth, Annealing*; Academic Press: New York, NY, USA, 1976.

34. Mandelkern, L. *Crystallization of Polymers: Volume 2: Kinetics and Mechanisms*, 2nd ed.; Cambridge University Press: Cambridge, UK, 2004.

35. Rizzuto, M.; Marinetti, L.; Caretti, D.; Mugica, A.; Zubitur, M.; Muller, A.J. Can poly(ε-caprolactone) crystals nucleate glassy polylactide? *CrystEngComm* **2017**, *19*, 3178–3191. [CrossRef]

36. Muller, A.J.; Avila, M.; Saenz, G.; Salazar, J. Chapter 3 crystallization of pla-based materials. In *Poly(lactic acid) Science and Technology: Processing, Properties, Additives and Applications*; The Royal Society of Chemistry: London, UK, 2015.

37. Strobl, G. Crystallization and melting of bulk polymers: New observations, conclusions and a thermodynamic scheme. *Prog. Polym. Sci.* **2006**, *31*, 398–442. [CrossRef]

38. Avrami, M. Granulation, phase change, and microstructure kinetics of phase change. III. *J. Chem. Phys.* **1941**, *9*, 177–184. [CrossRef]

39. Granasy, L.; Pusztai, T.; Warren, J.A.; Douglas, J.F.; Borzsonyi, T.; Ferreiro, V. Growth of 'dizzy dendrites' in a random field of foreign particles. *Nat. Mater.* **2003**, *2*, 92–96. [CrossRef] [PubMed]

40. Bourque, H.; Laurin, I.; Pézolet, M.; Klass, J.M.; Lennox, R.B.; Brown, G.R. Investigation of the poly(L-lactide)/poly(D-lactide) stereocomplex at the air–water interface by polarization modulation infrared reflection absorption spectroscopy. *Langmuir* **2001**, *17*, 5842–5849. [CrossRef]

41. Rathi, S.R.; Coughlin, E.B.; Hsu, S.L.; Golub, C.S.; Ling, G.H.; Tzivanis, M.J. Effect of midblock on the morphology and properties of blends of aba triblock copolymers of PDLA-mid-block-PDLA with PLLA. *Polymer* **2012**, *53*, 3008–3016. [CrossRef]

© 2017 by the authors. Licensee MDPI, Basel, Switzerland. This article is an open access article distributed under the terms and conditions of the Creative Commons Attribution (CC BY) license (http://creativecommons.org/licenses/by/4.0/).

![polymers logo] *polymers*

MDPI

Article

Bleached Kraft Eucalyptus Fibers as Reinforcement of Poly(Lactic Acid) for the Development of High-Performance Biocomposites

Marc Delgado-Aguilar [1], Rafel Reixach [2], Quim Tarrés [1], Francesc X. Espinach [3,*], Pere Mutjé [1] and José A. Méndez [1]

[1] LEPAMAP Research Group, Department of Chemical Engineering, University of Girona, 17004 Girona, Spain; m.delgado@udg.edu (M.D.-A); joaquimagusti.tarres@udg.edu (Q.T.); pere.mutje@udg.edu (P.M.); jalberto.mendez@udg.edu (J.A.M.)

[2] Department of Architecture and Construction Engineering, University of Girona, 17004 Girona, Spain; rafel.reixach@udg.edu

[3] Design, Development and Product Innovation, Department of Organization, Business Management and Product Design, University of Girona, 17004 Girona, Spain

* Correspondence: francisco.espinach@udg.edu; Tel.: +34-650-20-6884

Received: 13 April 2018; Accepted: 21 June 2018; Published: 24 June 2018

Abstract: Poly(lactic acid) (PLA) is one of the most well-known biopolymers. PLA is bio-based, biocompatible, biodegradable, and easy to produce. This polymer has been used to create natural fiber reinforced composites. However, to produce high-performance and presumably biodegradable composites, the interphase between PLA and natural fibers still requires further study. As such, we aimed to produce PLA-based composites reinforced with a commercial bleached kraft eucalyptus pulp. To become a real alternative, fully biodegradable composites must have similar properties to commercial materials. The results found in this research support the competence of wood fiber reinforced PLA composites to replace other glass fiber reinforced polypropylene composites from a tensile property point of view. Furthermore, the micromechanics analysis showed that obtaining strong interphases between the PLA and the reinforcement is possible without using any coupling agent. This work shows the ability of totally bio-based composites that fulfill the principles of green chemistry to replace composites based on polyolefin and high contents of glass fiber. To the best knowledge of the authors, previous studies obtaining such properties or lower ones involved the use of reagents or the modification of the fiber surfaces.

Keywords: natural fibers; green composites; micro-mechanics; Kelly-Tyson; interphase

---

1. Introduction

A wide variety of biodegradable and bio-based polymers are available, including thermoplastic starch (TPS), polyhydroxybutyrate (PHB), thermoplastic lignin, and polycaprolactones [1–5]. However, biopolymers still have some drawbacks in terms of costs and properties. Although producers have bargaining power due to low demand, the introduction of such biopolymers into the technosphere will be a challenging task. Fortunately, environmental awareness is increasing the demand and more products based on these biopolymers are being produced and introduced.

Poly(lactic acid) (PLA) is becoming one of the most promising biopolymers in several fields, mainly due to its good mechanical properties, biocompatibility, and biodegradability. Although the potential applications of PLA are different than those of other commodities such as polypropylene (PP), polyethylene (PE), or even poly(vinyl chloride) (PVC) and poly(ethylene terephthalate) (PET). Previous works demonstrated that the use of properly modified, natural fibers increased the mechanical properties of PLA-based composites up to levels that made these materials competitive with the abovementioned materials [6]. The environmental advantage of PLA compared to other commodities is clear and has been extensively discussed [7–9]. Figure 1 shows the main difference, in terms of lifecycle, between PLA and PP, which was selected as the most representative oil-based commodity.

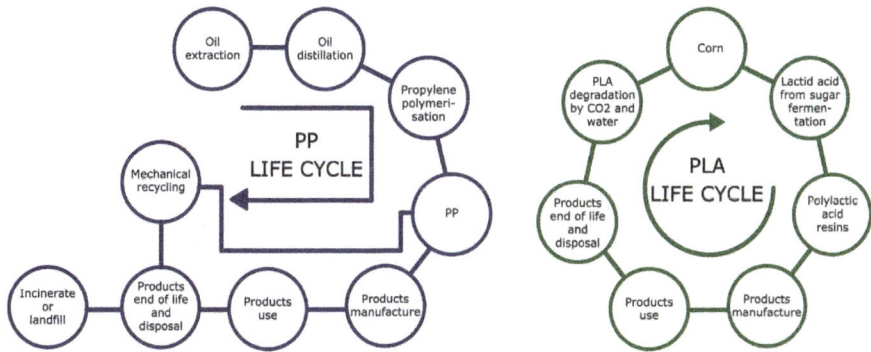

Figure 1. Simplified life cycle diagrams of polypropylene (PP) and poly(lactic acid) (PLA) resins used to manufacture consumer products.

Thermoplastic polymers are usually combined with fibers to further improve their properties. In fact, PP-based composites are mostly reinforced with glass fibers (GF), preventing their reuse and recycling, thus limiting their end of life to landfilling or incineration to recover some part of the material in the form of energy [10]. The substitution of these mineral fibers has become an interesting research topic, mainly due to increasing environmental awareness, safety issues, and governmental regulations [11]. The use of glass fibers has been reported to be harmful for thermoplastic processing equipment, such as injection molding machines or extruders, mainly due to their rigidity, hardness, and abrasiveness. Potential sustainable candidates as substitutes for these fibers have included natural fibers, either coming from wood [12], annual plants [13], or even recovered paper [14].

PLA has also been studied as a matrix in thermoplastic composites. However, as discussed in a previous study, the interphase between PLA and natural fibers requires further research. Although some evidence supports the presence of OH groups on fibers' surface promoting the interaction between the polymer and reinforcement. Nonetheless, the resulting mechanical properties are still far from the market requirements, especially when compared to PP/GF composites [12]. Furthermore, like polyolefin, where removing lignin form the fibers surfaces has shown beneficial in order to enhance the mechanical properties of natural fiber reinforced composites, some authors found that lignin increased the interactions between PLA and the natural fiber surfaces [7,15]. This is of interest for the present research as supports the use of untreated natural fibers as PLA reinforcement. There are also other authors that propose mechanisms to increase the interactions between the phases of the composite, but involve the use of expensive coupling agents like montmorillonite or chemical modification, difficultly scalable to industry level, like acetylation [16].

Granda et al. (2016) used bleached kraft pine fibers as a PLA reinforcement, obtaining significant improvements when 30 wt % of fiber was used [6]. The tensile strength of PLA increased from 49.85 to 68.80 MPa, a value in the same magnitude of PP/GF composites with a 20 wt % content of reinforcement. Nonetheless, when the amount of GF increased to 30 wt %, a tensile strength of about 80 MPa was achieved. To the date, this value is not attainable with injection molding PLA grades and natural fibers.

Given the above, the present work aimed to produce PLA composites reinforced with bleached kraft eucalyptus fibers, without any chemical modification or the use of any cross-linking agent,

Polymers **2018**, *10*, 699

with comparable properties to PP/GF composites with the purpose of creating competitive and presumably biodegradable high-performance materials. The macromechanic and micromechanic properties, and interphase were studied to determine the influence of fiber morphology and composition. The results proved that it is possible to formulate and obtain totally bio-based composite materials with mechanical properties comparable to glass fiber reinforces polypropylene composites. The strength of the obtained materials was only 94% lower than a PP-based composite with 30% glass fiber content. To the best knowledge of the authors, this is the first time that these values have been obtained for a totally bio-based composite material, without the use of further reagents of coupling agents.

2. Materials and Methods

2.1. Materials

A PLA Ingeos Biopolymer 3251D, PLA-based polymer, by Nature Works (Blair, NE, USA) was used as biodegradable matrix for the composites. This thermoplastic has a volumetric index of 30 cc/10 min at 190 °C/2.1 kg. The bleached kraft hardwood fibers (BKHF) from eucalyptus were provided by LECTA, SA (Madrid, Spain). Diethylene glycol dimethyl ether (Diglyme), with a molecular weight of 134.17 g/mol, and a boiling point of 162 °C, provided by Clariant (Tarragona, Spain), was used as a dispersing agent during the composite compounding.

All the chemical reagents used for fibers characterization, extraction, and bleaching were supplied by Scharlau, S.L. (Sentmenat, Spain).

2.2. Composite and Sample Preparation

The BKHF were pulped using an aqueous solution of 2/3 of Diglyme to avoid the formation of hydrogen bonds between cellulose fibers. These hydrogen bonds usually form during pulping and compounding. After pulping, fibers were dried at 105 °C until constant weight and then shred in a knife mill.

The composites were mixed in a Gelimat kinetic mixer (model G5S, Draiswerke, Mahwah, NJ, USA). The matrix and the reinforcement were introduced in the mixer at 300 rpm, then the speed was increased to 2500 rpm for 2 min, reaching a discharge temperature of 195 °C. Composites with 10, 20, and 30% *w/w* of BKHF were produced.

Composites were pelletized in a knife miller. The pellets were stored in an oven at 80 °C until needed in order to prevent moisture absorption.

Composite materials were mold injected using a Meteor 40 injection-machine by Mateu & Solé (Barcelona, Spain) to obtain standard ISO 527-1:2000 specimens. The pressure during mold filling was 120 kgf/cm^2; afterward, a 37.5 kgf/cm^2 holding pressure was applied. The machine has three heating areas that were tuned to 175, 175, and 190 °C. To ensure at least five specimens were available for the tensile tests, a minimum of 10 specimens for each composite were created.

2.3. Mechanical Characterization

The methodology to tensile test the specimens agreed with the ISO 527-1:200 standard specifications. The specimens were stored in a Dycometal conditioning chamber at 23 °C and 50% relative humidity for 48 h prior to the mechanical testing. Then, at least 10 samples were tested to obtain the tensile strength, strain at break, and Young's modulus of the composites. The tests were performed in a dynamometer DTC-10 supplied by IDMtest (San Sebastian, Spain), fitted with a 5 kN load cell and operating at a rate of 2 mm/min. For the evaluation of the Young's modulus, a MFA 2 extensometer, by MF Mess and Feienwerktechnik GMBH (Velbert, Germany), was used to more precisely measure the deformation.

2.4. Fiber Extraction from the Composites

The reinforcing fibers were extracted from the composite by using a Soxhlet apparatus to dissolve the matrix. Small pieces of composite obtained from the mold-injected specimens were placed into the Soxhelt apparatus. Decalin vapors were refluxed for 24 h until the PLA matrix was completely dissolved. Then, the fibers were extracted from the apparatus and washed with acetone and water to remove all the remaining residues. Finally, the fibers were kept in an oven at 105 °C for 24 h to obtain dried fibers.

2.5. Morphologic Analysis of the Fibers

The length and width distributions of the reinforcement were measured with a MorFi Compact (morphological fiber analyzer) from Techpap SAS (Gières, France), following the ISO/FDIS 160652 standard. The equipment measured between 25,000 and 30,000 fibers. Four samples of each type of fiber were analyzed. The equipment returns the arithmetic (l_a) and weighted (l_l) mean lengths, computed as

$$l_a = \frac{\sum_i n_i \cdot l_i}{\sum_i n_i}; \ l_l = \frac{\sum_i n_i \cdot l_i^2}{\sum_i n_i \cdot l_i} \tag{1}$$

In some cases, if the amount of long fibers is scarce, a double weighted mean length (l_w) is used with the micromechanics models. This double weighted length solves the statistic misrepresentation of the impact of the long fibers on the mechanical properties of the composite materials.

3. Micromechanics

The micromechanics study encompassed the impact of the phase's contents on the properties of a composite material. In the present work, the micromechanics of the Young's modulus and the tensile strength were examined and calculated. The intrinsic properties of the reinforcements can be obtained experimentally, but this is difficult in some cases due to the morphology of the fibers. However, some authors defend that the intrinsic properties of a fiber inside and outside a composite can vary considerably [17,18]. In such cases, the literature recommends the use of micromechanics models, mainly in situations with good interphases between the reinforcement and the matrix [19–22].

3.1. Hirsch's Model

Hirsch's model combines the parallel and perpendicular models to equalize the orientation effects of the fibers on the Young's modulus of the composite [23,24]

$$E_t^C = \beta \cdot \left(E_t^F \cdot V^F + E_t^M \cdot \left(1 - V^F \right) \right) + (1 - \beta) \frac{E_t^F \cdot E_t^M}{E_t^M \cdot V^F + E_t^F \cdot (1 - V^F)} \tag{2}$$

where E_t^C, E_t^F, and E_t^M are the Young's modulus of the composite, the reinforcement, and the matrix, respectively. V^F is the reinforcement volume fraction, and the factor β equalizes the parallel and perpendicular models. For natural fiber composites, a value of $\beta = 0.4$ has been reported to adequately reproduce the results obtained experimentally [25,26]. Once the Young's modulus of the composite and matrix are experimentally obtained, it is possible to solve the equation to obtain the intrinsic Young's modulus.

3.2. Modified Rule of Mixtures for the Young's Modulus

Although some formulations have been introduced for a rule of mixtures for the Young's modulus of short fiber semi-aligned reinforced composites, one of the most used is

$$E_t^C = \eta_l \cdot \eta_o \cdot E_t^F \cdot V^F + \left(1 - V^F \right) \cdot E_t^M \tag{3}$$

where $E_t{}^C$, $E_t{}^M$, and $E_t{}^F$ are the Young's modulus of the composite, the matrix, and the reinforcement, respectively; and η_l and η_o are the modulus length and orientation efficiency factors, respectively, used to equalize the contribution of the semi-aligned short reinforcement fibers. The modulus efficiency factor η_e is obtained by multiplying the abovementioned efficiency factors ($\eta_e = \eta_l \cdot \eta_o$). In the equation, the term $\eta_e \cdot E_t{}^F$ is referred to as the neat contribution of the fibers to the Young's modulus of the composite [24,27,28].

Once the intrinsic Young's modulus of the reinforcement is obtained the Hirsch model, Equation (3) can be used to compute the value of η_e. The modulus efficiency factor is a measure of the stiffening yield of the reinforcement.

3.3. Cox and Krenchel's Model

The main factors impacting the efficiency of the reinforcement contribution are the morphology and the orientation of the fibers. To evaluate the impact of the morphology, the Cox and Krenchel's model [29,30] was used

$$\eta_l = 1 - \frac{tanh\left(\beta \cdot l^F /2\right)}{\left(\beta \cdot l^F /2\right)} \tag{4}$$

with

$$\beta = \frac{1}{r^F} \sqrt{\frac{E_t^M}{E_t^F (1-v) \cdot Ln\sqrt{\pi/4 \cdot V^F}}} \tag{5}$$

where β is a coefficient for the stress concentration rate at the ends of the fibers, r^F is the mean fiber radius, l^F is the fiber's weighted length, and v is the Poisson's ratio of the matrix (0.42 for PLA) [31]. The efficiency factor η_e can be expressed as $\eta_e = \eta_o \cdot \eta_l$ and the identity was used to calculate η_o.

3.4. Modified Rule of Mixtures for the Tensile Strength

Like the Young's modulus, some formulations exist for the rule of mixtures for the tensile strength of a composite. With short semi-aligned fiber reinforced composites, the literature has introduced a set of modifiers to equalize the contribution of the reinforcement to the tensile strength of the composite. The main parameters impacting this contribution are the morphology of the reinforcement, its orientation relative to the loads, its grade of dispersion, the chemical nature of the phases, its mechanical properties, and its content [27,32,33]. An accepted formulation of this modified rule of mixtures (mRoM) for the tensile strength is

$$\sigma_t^C = \chi_1 \cdot \chi_2 \cdot \sigma_t^F \cdot V^F + \left(1 - V^F\right) \cdot \sigma_t^{M*} \tag{6}$$

where $\sigma_t{}^X$ refers to the tensile strength of X. X can be the composite (C), the reinforcement (F), or the matrix (M). The asterisk after M indicates that the tensile strength of the matrix is not its ultimate value but the contribution of the matrix to the strength of the composite or the corresponding stress of the matrix at the strain at break of the composite. The composite is reinforced with semi-aligned short fibers and the model adds the parameters χ_1 and χ_2 as the orientation and the length factors, respectively. A more general model presents both parameters as a coupling factor f_c ($f_c = \chi_1 \cdot \chi_2$). The literature shows that composites with strong interphases have coupling factors in the range of 0.18 to 0.2 [12,34]. The term $f_c \cdot \sigma_t{}^F$ has been defined by some authors as the neat contribution of the reinforcement to composite's strength [18,32].

Similarly to the Young's modulus, the tensile strength of the composites and the matrix, but not the intrinsic strength of the fibers, can be obtained experimentally. As mentioned above, some authors found noticeable differences between the intrinsic values obtained experimentally and the back-calculated values [17]. We propose the use of micromechanics models, validated by the literature, to obtain the intrinsic tensile strength of the reinforcement.

3.5. Modified Kelly and Tyson Equation

Kelly and Tyson presented an evolution of the modified rule of mixtures separating the contributions of the subcritical and supercritical fibers [35]. The concept of a critical length was provided by the shear-lag model used to analyze the stress distribution in reinforcements inside a composite. This model states that the matrix transmits its load to the reinforcement in the interface via shear loads. Thus, the fibers will show a null load in its ends and a full load at its center. Depending on the length of the fiber, the load at the center will be higher or lower than its intrinsic tensile strength. The fibers with enough length to be fully loaded will be called supercritical, and shorter fibers will be called subcritical (Figure 2).

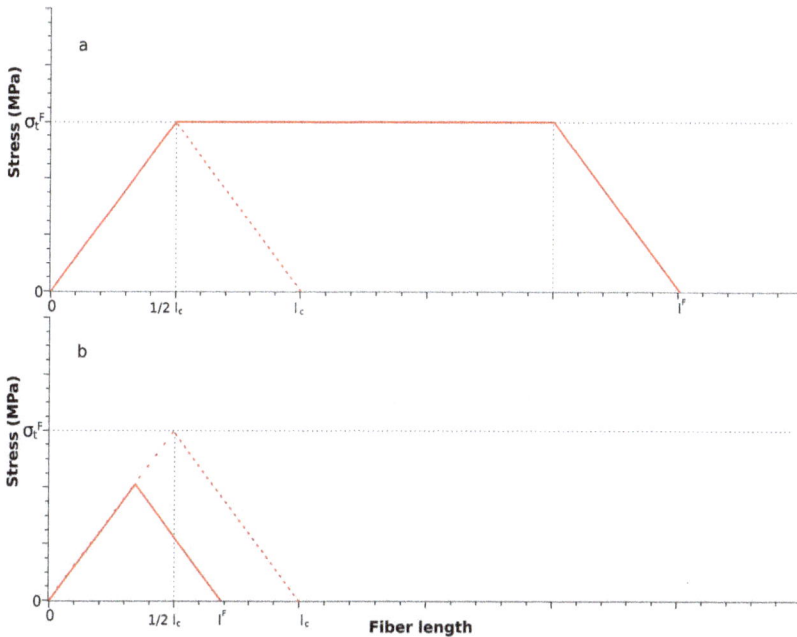

Figure 2. Axial load diagrams for (**a**) supercritical and (**b**) subcritical length fibers.

The length required to be fully loaded is impacted by the exterior area of the fibers and the quality of the interphase, or its ability to pass the shear loads from the matrix to the reinforcement. The critical length (l_c) can be computed as

$$l_c = \frac{r^F \cdot \sigma_t^F}{\tau} \tag{7}$$

where τ is the interfacial shear strength that limits the shear load transfer between the reinforcement and the matrix.

The modified Kelly and Tyson model has the formulation of

$$\sigma_t^C = \chi_1 \left(\sum_{l=0}^{l=lc} \left[\frac{\tau \cdot l \cdot V_l^F}{2r^F} \right] + \sum_{l=lc}^{\infty} \left[\sigma_t^F \cdot V_l^F \cdot \left(1 - \frac{\sigma_t^F \cdot 2r^F}{4 \cdot \tau \cdot l} \right) \right] \right) + \left(1 - V^F \right) \cdot \sigma_t^{M*} \tag{8}$$

The reinforcements do not have a regular size and show a distribution of lengths and diameters, as shown by the morphological analysis. Thus, for any of the length ranges, obtaining a volume fraction (V_l^F) is possible. Equation (8) adds an orientation factor (χ_1) to the original Kelly and Tyson equation, formulated for aligned reinforcements [13,36,37].

In its present form, Equation (8) has three unknowns, σ_t^F, τ, and χ_1. A method was developed by Bowyer and Bader to solve the equation [38].

3.6. Bowyer and Bader Method

The solution proposed by Bowyer and Bader suggested changing the intrinsic tensile strength of the fiber by $\varepsilon_t^C E_t^F$, assuming that the matrix, the composite, and the fibers will show the same strain under the same loads. This will be only true for low deformations and in the elastic zone of the composite. Bowyer and Bader proposed a solution based on collecting experimental data from two strain-stress load states between 0 and the ultimate stress of the composite (Figure 3).

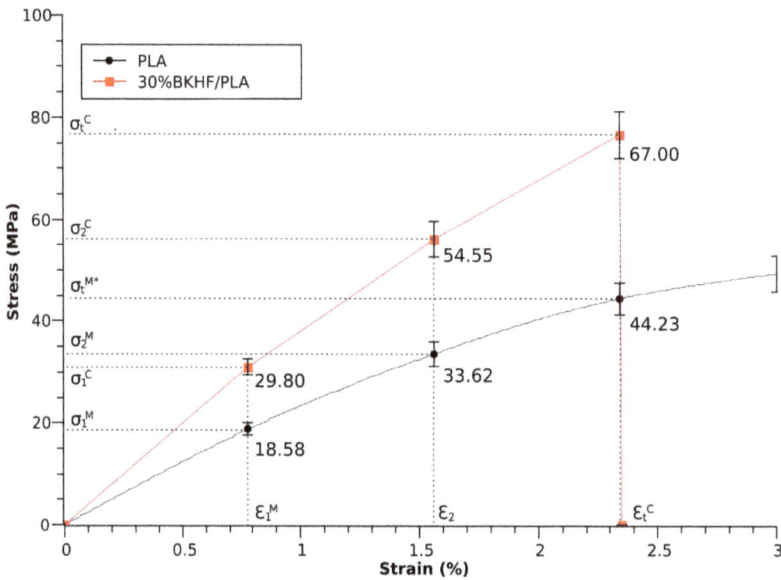

Figure 3. Stress-strain curves of the PLA matrix and the composite reinforced with a 30% bleached kraft hardwood fibers (BKHF). The intermediate strain points used to solve the Kelly and Tyson modified equation are indicated with the subscripts 1 and 2, respectively.

Then, a numerical method was used to find the values of the interfacial shear strength and the orientation factor that solve the equation. Then, Equation (8) was used to obtain the intrinsic tensile strength of the reinforcement.

4. Results and Discussion

4.1. Bleached Kraft Hardwood Fiber Morphology

The morphology of the fibers, including length and width, changes during the preparation of composite materials [39]. The length of the fibers decreases under attrition during the mixing and injection processes. The decrease in the reinforcement mean lengths are higher when the fiber content is increased, due to the higher viscosity of the composite materials, implying that more energy is required to perform a correct mixing [37]. However, the width of the fibers can change drastically due to the collapse of their lumen [40,41]. Therefore, the real impact of the morphology of the fibers on the properties of a composite must be evaluated from the fibers extracted from these composites. Figure 4 shows the distribution of the fiber's lengths and diameters.

The recorded mean arithmetic length and diameter of the BKSF fibers were 191.0 and 18.7 μm, respectively. Thus, the aspect ratio of the reinforcement (l_a/d^f) was 10.21. It is accepted that reinforcements with aspect ratios higher than 10 have good strengthening and stiffening capabilities. Figure 4 shows a high content of short fibers, whereas fewer longer fibers are present. Thus, the distribution is far from normal. Consequently, the weight and double weight lengths were computed as 336.5 and 453.7 μm, respectively. These weighted lengths were used later during the micromechanics analysis.

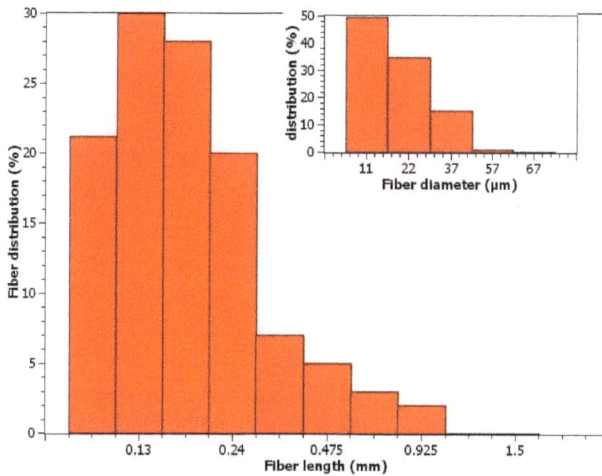

Figure 4. Distribution of the lengths and diameters of the BKHF, extracted from a composite with a 30% w/w reinforcement content.

4.2. Tensile Properties of the Composites

Table 1 shows the means of the tensile strength $(\sigma_t{}^C)$, strain at break $(\varepsilon_t{}^C)$, and Young's moduli $(E_t{}^C)$ of the BKHF-based composites. The table also shows the contribution of the matrix to the tensile strength of the composite $(\sigma_t{}^{M*})$, evaluated as the tensile strength of the matrix at the ultimate strain of the composite.

Table 1. Tensile strength, strain at break, and Young's modulus of the composite. Contribution of the matrix to the tensile strength of the composite.

BKHF (%)	V^F	$\sigma_t{}^C$ (MPa)	$\varepsilon_t{}^C$ (MPa)	$E_t{}^C$ (GPa)	$\sigma_t{}^{M*}$ (MPa)
0	-	49.6 ± 0.23	3.3 [+] ± 0.18	3.4 ± 0.11	-
10	0.085	57.3 ± 0.48	2.9 ± 0.12	4.4 ± 0.18	48.4
20	0.172	68.7 ± 1.08	2.6 ± 0.15	5.7 ± 0.22	46.2
30	0.263	76.5 ± 1.31	2.3 ± 0.07	6.8 ± 0.26	44.2

[+] The strain of the PLA matrix was measured at its maximum strength.

The tensile strength of the composites evolved linearly against the reinforcement content with a correlation coefficient (R^2) of 0.9934. Usually a linear behavior implies a good dispersion of the reinforcement and the presence of a medium or strong interphase between the reinforcement and the matrix. The tensile strengths of the composites with a 10% to 30% BKHF contents were 10.5%, 38.6%, and 54.3% higher than the PLA matrix, respectively. These percentages were lower than those obtained when a polypropylene matrix was used for natural fiber-based composites [13,42]. Nonetheless, the nominal values of the PLA-based composites were higher, mainly due to the initial tensile strength of. PLA with a tensile strength of 49.6 ± 1.54 MPa. PP matrixes previously used by the researchers showed tensile strengths around 28 MPa [42]. Thus, PLA's tensile strength is 43% higher than that of PP. However, the interphase of the PP-based composites was optimized with the use of coupling agents. Compared to commodity composites like glass fiber-reinforced PP, the tensile strength of such composites is similar to those that are PLA-based [12,43]. Thus, from a tensile strength point of view, BKHF-reinforced PLA composites can replace glass fiber-reinforced PP composites.

The Young's modulus of the composites behaved similarly to the tensile strength, evolving and increasing linearly with the reinforcement content, with a R^2 of 0.9981. It is accepted that the Young's modulus is barely impacted by the quality of the interphase. Thus, a linear behavior against the reinforcement content indicates a good dispersion of the reinforcement. The Young's modulus of the composites with 10%, 20%, and 30% BKHF contents were 29.4%, 67.6%, and 100.0% higher than those of PLA, respectively. Like the tensile strength, the percentage values obtained with PP-based composites were higher [44,45]. Regardless, the nominal values of the PLA-based composites were still higher than those of the PP-based composites at the same reinforcement ratio. The reason for this result is that the initial Young's modulus of the PLA is more than double that of PP (1.5 GPa) [45]. Compared with glass fiber-reinforced PP composites, the Young's moduli of the PLA-based composites were higher at the same reinforcement content [24,45]. Thus, from a stiffness point of view, the PLA-based composites are also an alternative to glass fiber-reinforced PP composites.

PLA-based composites also showed comparatively good strains at break. PP-based composites usually show high decreases [32,42].

The study of the tensile properties of BKHF-reinforced PLA composites revealed materials with similar or better properties than those of commodity materials that are actually being commercialized and used in areas like automotive, product design, or as building materials.

PP-based composites use coupling agents to obtain good interphases due to its hydrophobic nature in front of hydrophilic reinforcements [46–48]. The literature shows that uncoupled composites tend to maintain or decrease in tensile strength when the reinforcement content increases. Unlike these composites, uncoupled PLA-based composites showed linear increases in their tensile strength against the reinforcement contents. Thus, a compatibility exists between PLA and natural fibers. The interphase is probably composed of hydrogen bonds or van der Waals interactions between the fiber surface cellulose and holocelluloses and the PLA [43].

However, the interphase between poly(lactic acid) and natural fibers is still controversial. On the one hand, several authors demonstrated that lignin could hinder the interaction between natural fibers and poly(lactic acid) [6]. However, proper reinforcement dispersion within the matrix becomes a challenge if the hydrophilic characteristic of the fibers is too high, directly affecting the final mechanical properties of the composites. On the other hand, other studies demonstrated that moderate amounts of lignin could promote the interaction between both phases [49]. In this work, bleached kraft eucalyptus fibers were used; the surface is mainly composed of hydroxyl groups. For this reason, mimicking the strategy adopted by Granda et al., diglyme was used as the dispersant [11].

4.3. Net Contribution of the Fibers to the Tensile Properties of the Composite

A possible measure of the quality of the interphase is the evaluation of the net contribution of the reinforcements to the tensile strength and the Young's modulus of the composite. The literature shows that this contribution is considerably impacted by the nature of the matrix [50,51]. Similar contributions were obtained for the same reinforcement used with different polyolefin like PP or high density polyethylene (HDPE), but when this reinforcement was used with another polymer chemical family, the contributions changed drastically [43,51].

The contributions of BKHF to the tensile strength and the Young's modulus of the matrix were evaluated by means of a fiber tensile strength factor (FTSF) and a fiber tensile modulus factor (FTMF), respectively. Both factors were obtained by rearranging the corresponding modified rule of mixtures and isolating the contribution of the fibers [37,43]. Then, the factors were the slope of the linear regression curve passing through the origin (Figure 5).

When used as reinforcement for PLA, the BKHF had a FTSF value of 176.75 MPa. This value is clearly higher than softwood as a reinforcement for the same matrix (123.98 MPa), indicating that BKHF were able to create a better interphase than the softwood [43]. The FTSF of BKHF was also higher than that of wood fiber as a PP reinforcement, with a value of 109.4 MPa. Regardless, the value was also clearly inferior to that of glass fiber as a PP reinforcement with 273.85 and 427 MPa for the uncoupled and coupled composites, respectively. The neat contribution of the reinforcement indicates the presence of a strong interface or a high intrinsic tensile strength.

The FTMF revealed similar conclusions, at 16.375 MPa, it was higher than that of a natural fibers as PP reinforcement (10.87 MPa), but lower than glass fiber as a PP reinforcement (32.6 MPa) [41]. As mentioned above, the quality of the interphase has little impact on the Young's modulus of a short fiber reinforced composite. Thus, the relatively high FTMF indicates a high value for the efficiency factor or BKHF's high intrinsic Young's modulus.

With the objective of finding the intrinsic tensile properties of BKHF and the quality of the interphase, a micromechanics analysis was proposed.

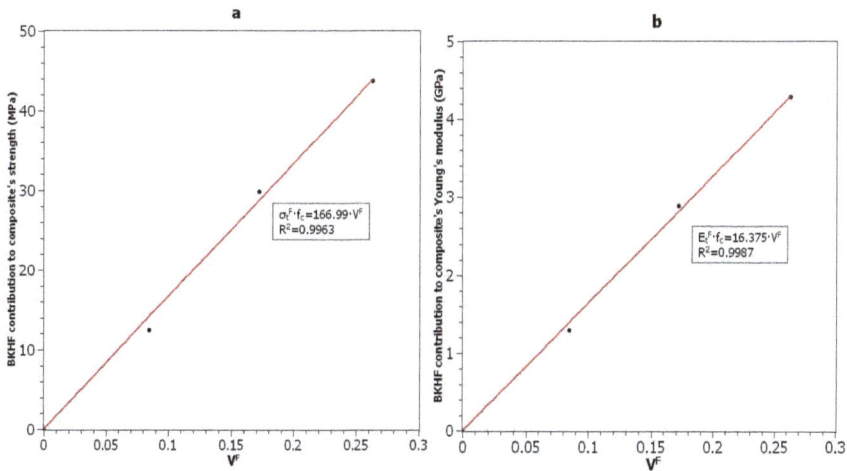

Figure 5. Net contributions of the reinforcement to the (**a**) tensile strength of the composite and (**b**) Young's modulus of the composite.

4.4. Micromechanics

4.4.1. Micromechanics of the Young's Modulus

The first step involved computing the intrinsic tensile strength of the fibers using Hirsch's model (Equation (2)). The results are shown in Table 2.

Table 2. Intrinsic Young's modulus of the bleached kraft hardwood fibers (BKHF) and micromechanics parameters for a modified rule of mixtures of the Young's modulus of the composites

BKHF (%)	E_t^F (Gpa)	η_e	η_l	η_o	α
10	28.09	0.541	0.905	0.598	46.6°
20	31.41	0.533	0.917	0.581	47.9°
30	29.84	0.547	0.931	0.587	47.5°
Mean	28.78	0.541	0.918	0.589	47.3
S.D.	1.661	0.007	0.013	0.008	0.66

The fibers showed a mean intrinsic Young's modulus of 28.8 ± 1.661 GPa. The value is higher than that obtained for a bleached kraft softwood fiber, with a value of 21.2 GPa, showing the advantage in terms of the stiffening capabilities of the hardwood over softwood. Other reinforcements provided by softwood, like stone groundwood (SGW), also showed lesser intrinsic tensile strengths (18.2 GPa) [12]. Conversely, other lignocellulosic fibers, like hemp strands, were reported to have similar intrinsic Young's modulus, with a 26.8 GPa [45]. As hemp strands can be considered high quality lignocellulosic reinforcements, the values obtained with BKHF are notable. Regardless, compared with commodity reinforcements like glass fibers that have a 71.6 GPa modulus, BKHF is clearly inferior [44]. Notwithstanding, the relative Young's moduli, which is the ratio between the reinforcement modulus and its density, reduces the differences.

The mean efficiency factor of 0.541 was in line with those obtained for other lignocellulosic reinforcements. However, SGW-based composites were reported to have slightly higher values and hemp strand-based composites to have lower values [44].

The length efficiency factors showed a mean value of 0.92, very similar to those obtained for other lignocellulosic reinforced composites [44,45]. The obtained length efficiency factors were noticeably higher than the orientation factors, revealing the greater impact of the morphology of the reinforcements than its relative orientation on the Young's modulus of the composites.

Two scientific studies completed by Fukuda and Kawada [52] and Sanomura and Kawamura [53] linked the orientation factor with a theoretical mean orientation angle of the fibers (α_o). The rectangular distribution (square packing) has already been reported as adequate for short fiber semi-aligned reinforced composites [28,41,43,54]; its equation is

$$\eta_l = \frac{\sin(\alpha_o)}{\alpha_o} \cdot \left(\frac{3-v}{4} \cdot \frac{\sin(\alpha_o)}{\alpha_o} + \frac{1+v}{4} \cdot \frac{\sin(3\alpha_o)}{3\alpha_o} \right) \tag{9}$$

The obtained mean orientation angle was 47.3°, very similar to other lignocellulosic-reinforced composites, despite the matrix.

The good Young's modulus of the composites can be explained by the notable intrinsic Young's modulus of BKHF and the high Young's modulus of PLA. The other parameters were very similar to those of other composites.

4.4.2. Micromechanics of the Tensile Strength

A micromechanics analysis of the tensile strength was performed to support the conclusions of the Young's modulus study, focusing on the important role of the intrinsic properties of the reinforcement instead of the other parameters (morphology and orientation). The analysis of the tensile strength allowed obtaining a measurement of the quality of the interphase.

The modified Kelly and Tyson equation (Equation (7)), with the solution provided by Bowyer and Bader, were used to obtain the intrinsic tensile strength of BKHF. The study was performed for the composites with a 30% BKHF content, as the fiber extraction and morphological analysis was completed on these composites (Figure 4). The literature reports that the morphology of the fibers changes with the reinforcement content, and the composites with higher reinforcement content had shorter mean fiber lengths [36,55]. The intermediate points used to solve the equation were one-third and two-thirds of the ultimate strain. Figure 3 provides the numerical values.

The intrinsic tensile strength of BKHF was calculated as 768.4 MPa. The value was notably higher than other lignocellulosic or wood fibers, like hemp strands or stone groundwood [12,34]. Thus, as expected, the intrinsic property of the reinforcement had an important role in the tensile strength of the composite. Notably, this value was also inferior to that of glass fiber, as expected when comparing the FTSF of both reinforcements. We also found that the orientation factor was inside the expected parameters, with a value of 0.284. It is accepted that the orientation of the fibers in a mold-injected specimen is mainly impacted by the geometry of the mold and the equipment used to mold the specimen [37]. Past analyses showed that with the equipment at our labs, the orientation factor ranges from 0.25 to 0.35 [19,56]. Finally, the interfacial shear strength was evaluated at a value of 27.8 MPa. This value is very near to 28.6 MPa, the von Mises prediction for a strong interphase ($\sigma_t^M / 3^{1/2}$). Thus, the interphase between PLA and BKHF was considered as strong, in agreement with the findings of the macro-mechanical properties analysis.

The Kelly and Tyson equation allowed the comparison of the contribution of the fibers and the matrix. The reinforcements contributed 57.4% to the tensile strength of the composite, showing its high strengthening capabilities, already demonstrated during the FTSF analysis.

The modified rule of mixtures (Equation (6)) was used to compute the value of the coupling factor (Table 3).

Table 3. Intrinsic tensile strength of the BKHF and micromechanics parameters for a modified rule of mixtures of the tensile strength of the composites.

BKHF (%)	σ_t^F (Mpa)	f_c	χ_1	χ_2	α
10	768	0.20	0.284	0.704	61°
20	768	0.23	0.284	0.810	61°
30	768	0.22	0.284	0.774	61°
Mean	768	0.22	0.284	0.763	61°
S.D.	-	0.015	-	0.054	-

Composites with strong to optimal interphases have coupling factors around 0.2 [13]. All the composites showed this or slightly higher values. Thus, the hypothesis about a strong interphase between PLA and BKHF was doubly confirmed by the values of the interfacial shear strength and the coupling factor. The morphology and the interphase quality impacting the composite more than the orientation of the fibers was supported by the higher values of the length and interphase factor compared to the orientation factor.

Finally, the literature supports a relation between the orientation factor and a theoretical mean fiber orientation [13]

$$\chi_1 = cos^4(\alpha) \tag{10}$$

The resulting theoretical orientation angle was 61°. The discrepancies between the theoretical orientation angles preview by the Young's modulus (Table 2) and the tensile strength analysis (Table 3) are worth nothing. These discrepancies have been already reported in the literature [13]. The establishment of the mean orientation angle in three dimensions is a difficult and subjective task. The mathematical function that predicts the fiber orientation factor in the modified rule of mixtures appears to be different from the mathematical function that predicts the orientation factor (η_o).

5. Conclusions

Bleached kraft hardwood fibers from eucalyptus were used to formulate, prepare, and test PLA-based composites. These composites, with reinforcement contents ranging from 10% to 30% *w/w* showed noticeably good tensile strengths, in line with glass fiber-reinforced polypropylene composites. Compared with other lignocellulosic or wood fiber-reinforced polypropylene composites, the BKHF/PLA composites always demonstrated better tensile strength and Young's modulus at the same reinforcement content. Thus, obtaining presumably biodegradable composites with mechanical properties comparable to commercially available non-biodegradable composites is possible.

The BKHF-reinforced PLA composites had a good interphase without using any further coupling agent or chemical treatment. This places these composites under the umbrella of green chemistry. However, further research is still needed to assess the environmental impact of such composites and the advantages compared to recycling.

The micromechanics analysis revealed the high intrinsic tensile strength and Young's modulus of BKHF. These properties were higher than other wood fibers and comparable to high quality lignocellulosic reinforcements such as hemp strands. The analysis also revealed the higher importance of the morphology of BKHF and the quality of the interphase compared to its orientation to obtain good composite tensile properties.

Author Contributions: R.R. and Q.T. proposed the research and analyzed the macro mechanical data. J.A.M. and M.D.-A. conceived and designed the experiments. Q.T. performed the experiments. F.X.E. performed the micromechanics analysis. P.M. reviewed the analysis.

Acknowledgments: Authors wish to acknowledge the financial support of the Spanish Economy and Competitiveness Ministry to the project GREENCOMP, reference: MAT2017-83347-R. We would also like to thank to the University of Girona for the financial support to publish on Open Access Journals.

Conflicts of Interest: The authors declare no conflict of interest.

References

1. La Mantia, F.P.; Morreale, M. Green composites: A brief review. *Compos. Part A* **2011**, *42*, 579–588. [CrossRef]
2. Vilaseca, F.; Mendez, J.A.; Pèlach, A.; Llop, M.; Cañigueral, N.; Gironès, J.; Turon, X.; Mutjé, P. Composite materials derived from biodegradable starch polymer and jute strands. *Process Biochem.* **2007**, *42*, 329–334. [CrossRef]
3. Saad, G.R.; Seliger, H. Biodegradable copolymers based on bacterial poly ((R)-3-hydroxybutyrate): Thermal and mechanical properties and biodegradation behaviour. *Polym. Degrad. Stab.* **2004**, *83*, 101–110. [CrossRef]
4. Dufresne, A.; Vignon, M.R. Improvement of starch film performances using cellulose microfibrils. *Macromolecules* **1998**, *31*, 2693–2696. [CrossRef]
5. Scott, G. 'Green'polymers. *Polym. Degrad. Stab.* **2000**, *68*, 1–7. [CrossRef]
6. Granda, L.A.; Espinach, F.X.; Tarrés, Q.; Méndez, J.A.; Delgado-Aguilar, M.; Mutjé, P. Towards a good interphase between bleached kraft softwood fibers and poly(lactic) acid. *Compos. Part B* **2016**, *99*, 514–520. [CrossRef]
7. Yusoff, R.B.; Takagi, H.; Nakagaito, A.N. Tensile and flexural properties of polylactic acid-based hybrid green composites reinforced by kenaf, bamboo and coir fibers. *Ind. Crops Prod.* **2016**, *94*, 562–573. [CrossRef]
8. Bledzki, A.K.; Jaszkiewicz, A.; Scherzer, D. Mechanical properties of PLA composites with man-made cellulose and abaca fibres. *Compos. Part A* **2009**, *40*, 404–412. [CrossRef]
9. Oksman, K.; Skrifvars, M.; Selin, J.F. Natural fibres as reinforcement in polylactic acid (PLA) composites. *Compos. Sci. Technol.* **2003**, *63*, 1317–1324. [CrossRef]
10. Netravali, A.N.; Chabba, S. Composites get greener. *Mater. Today* **2003**, *6*, 22–29. [CrossRef]
11. Granda, L.; Tarres, Q.; Espinach, F.X.; Julian, F.; Mendes, A.; Delgado-Aguilar, M.; Mutje, P. Fully biodegradable polylactic composites reinforced with bleached softwood fibers. *Cellul. Chem. Technol.* **2016**, in press.
12. Lopez, J.P.; Mendez, J.A.; El Mansouri, N.E.; Mutje, P.; Vilaseca, F. Mean intrinsic tensile properties of stone groundwood fibers from softwood. *BioResources* **2011**, *6*, 5037–5049.
13. Vallejos, M.E.; Espinach, F.X.; Julian, F.; Torres, L.; Vilaseca, F.; Mutje, P. Micromechanics of hemp strands in polypropylene composites. *Compos. Sci. Technol.* **2012**, *72*, 1209–1213. [CrossRef]

14. Serrano, A.; Espinach, F.X.; Tresserras, J.; Pellicer, N.; Alcala, M.; Mutje, P. Study on the technical feasibility of replacing glass fibers by old newspaper recycled fibers as polypropylene reinforcement. *J. Clean. Prod.* **2014**, *65*, 489–496. [CrossRef]

15. Bax, B.; Müssig, J. Impact and tensile properties of PLA/cordenka and PLA/flax composites. *Compos. Sci. Technol.* **2008**, *68*, 1601–1607. [CrossRef]

16. Kovacevic, Z.; Bischof, S.; Fan, M. The influence of spartium junceum l. Fibres modified with montmorrilonite nanoclay on the thermal properties of PLA biocomposites. *Compos. Part B* **2015**, *78*, 122–130. [CrossRef]

17. Shah, D.U.; Nag, R.K.; Clifford, M.J. Why do we observe significant differences between measured and 'back-calculated' properties of natural fibres? *Cellulose* **2016**, *23*, 1481–1490. [CrossRef]

18. Serra, A.; Tarrés, Q.; Claramunt, J.; Mutjé, P.; Ardanuy, M.; Espinach, F. Behavior of the interphase of dyed cotton residue flocks reinforced polypropylene composites. *Compos. Part B* **2017**, *128*, 200–207. [CrossRef]

19. Oliver-Ortega, H.; Granda, L.A.; Espinach, F.X.; Mendez, J.A.; Julian, F.; Mutjé, P. Tensile properties and micromechanical analysis of stone groundwood from softwood reinforced bio-based polyamide11 composites. *Compos. Sci. Technol.* **2016**, *132*, 123–130. [CrossRef]

20. Serrano, A.; Espinach, F.X.; Tresserras, J.; del Rey, R.; Pellicer, N.; Mutje, P. Macro and micromechanics analysis of short fiber composites stiffness: The case of old newspaper fibers-polypropylene composites. *Mater. Design* **2014**, *55*, 319–324. [CrossRef]

21. Wang, F.; Chen, Z.Q.; Wei, Y.Q.; Zeng, X.G. Numerical modeling of tensile behavior of fiber-reinforced polymer composites. *J. Compos. Mater.* **2010**, *44*, 2325–2340. [CrossRef]

22. Zuccarello, B.; Scaffaro, R. Experimental analysis and micromechanical models of high performance renewable agave reinforced biocomposites. *Compos. Part B* **2017**, *119*, 141–152. [CrossRef]

23. Hirsch, T. Modulus of elasticity of concrete affected by elastic moduli of cement paste matrix and aggregate. *J. Am. Concr. Inst.* **1962**, *59*, 427–451.

24. Reixach, R.; Espinach, F.X.; Franco-Marquès, E.; Ramirez de Cartagena, F.; Pellicer, N.; Tresserras, J.; Mutjé, P. Modeling of the tensile moduli of mechanical, thermomechanical, and chemi-thermomechanical pulps from orange tree pruning. *Polym. Compos.* **2013**, *34*, 1840–1846. [CrossRef]

25. Kalaprasad, G.; Joseph, K.; Thomas, S.; Pavithran, C. Theoretical modelling of tensile properties of short sisal fibre-reinforced low-density polyethylene composites. *J. Mater. Sci.* **1997**, *32*, 4261–4267. [CrossRef]

26. Vilaseca, F.; Valadez-Gonzalez, A.; Herrera-Franco, P.J.; Pelach, M.A.; Lopez, J.P.; Mutje, P. Biocomposites from abaca strands and polypropylene. Part i: Evaluation of the tensile properties. *Bioresour. Technol.* **2010**, *101*, 387–395. [CrossRef] [PubMed]

27. Thomason, J.L. Interfacial strength in thermoplastic composites—At last an industry friendly measurement method? *Compos. Part A* **2002**, *33*, 1283–1288. [CrossRef]

28. Jiménez, A.M.; Delgado-Aguilar, M.; Tarrés, Q.; Quintana, G.; Fullana-i-Palmer, P.; Mutjé, P.; Espinach, F.X. Sugarcane bagasse reinforced composites: Studies on the young's modulus and macro and micro-mechanics. *BioResources* **2017**, *12*, 3618–3629. [CrossRef]

29. Cox, H.L. The elasticity and strength of paper and other fibrous materials. *Br. J. Appl. Phys.* **1952**, *3*, 72–79. [CrossRef]

30. Krenchel, H. *Fibre Reinforcement*; Akademisk Forlag: Copenhagen, Denmark, 1964.

31. Rezgui, F.; Swistek, M.; Hiver, J.; G'sell, C.; Sadoun, T. Deformation and damage upon stretching of degradable polymers (PLA and PCL). *Polymer* **2005**, *46*, 7370–7385. [CrossRef]

32. Reixach, R.; Franco-Marquès, E.; El Mansouri, N.-E.; de Cartagena, F.R.; Arbat, G.; Espinach, F.X.; Mutjé, P. Micromechanics of mechanical, thermomechanical, and chemi-thermomechanical pulp from orange tree pruning as polypropylene reinforcement: A comparative study. *BioResources* **2013**, *8*, 3231–3246. [CrossRef]

33. Thomason, J.L. The influence of fibre length and concentration on the properties of glass fibre reinforced polypropylene: 5. Injection moulded long and short fibre PP. *Compos. Part A* **2002**, *33*, 1641–1652. [CrossRef]

34. Lopez, J.P.; Mendez, J.A.; Espinach, F.X.; Julian, F.; Mutje, P.; Vilaseca, F. Tensile strength characteristics of polypropylene composites reinforced with stone groundwood fibers from softwood. *BioResources* **2012**, *7*, 3188–3200. [CrossRef]

35. Kelly, A.; Tyson, W. Tensile porperties of fibre-reinforced metals: Copper/tungsten and copper/molybdenum. *J. Mech. Phys. Solids* **1965**, *13*, 329–338. [CrossRef]

36. Li, Y.; Pickering, K.L.; Farrell, R.L. Determination of interfacial shear strength of white rot fungi treated hemp fibre reinforced polypropylene. *Compos. Sci. Technol.* **2009**, *69*, 1165–1171. [CrossRef]

37. Granda, L.A.; Espinach, F.X.; Lopez, F.; Garcia, J.C.; Delgado-Aguilar, M.; Mutje, P. Semichemical fibres of leucaena collinsii reinforced polypropylene: Macromechanical and micromechanical analysis. *Compos. Part B* **2016**, *91*, 384–391. [CrossRef]

38. Bowyer, W.H.; Bader, H.G. On the reinforcement of thermoplastics by imperfectly aligned discontinuous fibres. *J. Mater. Sci.* **1972**, *7*, 1315–1312. [CrossRef]

39. Karmaker, A.C.; Youngquist, J.A. Injection molding of polypropylene reinforced with short jute fibers. *J. Appl. Polym. Sci.* **1996**, *62*, 1147–1151. [CrossRef]

40. Valente Nabais, J.M.; Laginhas, C.; Ribeiro Carrott, M.M.L.; Carrott, P.J.M.; Crespo Amoros, J.E.; Nadal Gisbert, A.V. Surface and porous characterisation of activated carbons made from a novel biomass precursor, the esparto grass. *Appl. Surf. Sci.* **2013**, *265*, 919–924. [CrossRef]

41. Granda, L.A.; Espinach, F.X.; Mendez, J.A.; Tresserras, J.; Delgado-Aguilar, M.; Mutje, P. Semichemical fibres of leucaena collinsii reinforced polypropylene composites: Young's modulus analysis and fibre diameter effect on the stiffness. *Compos. Part B* **2016**, *92*, 332–337. [CrossRef]

42. Reixach, R.; Espinach, F.X.; Arbat, G.; Julián, F.; Delgado-Aguilar, M.; Puig, J.; Mutjé, P. Tensile properties of polypropylene composites reinforced with mechanical, thermomechanical, and chemi-thermomechanical pulps from orange pruning. *BioResources* **2015**, *10*, 4544–4556. [CrossRef]

43. Delgado-Aguilar, M.; Julián, F.; Tarrés, Q.; Méndez, J.A.; Mutjé, P.; Espinach, F.X. Bio composite from bleached pine fibers reinforced polylactic acid as a replacement of glass fiber reinforced polypropylene, macro and micro-mechanics of the young's modulus. *Compos. Part B* **2017**, *125*, 203–210. [CrossRef]

44. Lopez, J.P.; Mutje, P.; Pelach, M.A.; El Mansouri, N.E.; Boufi, S.; Vilaseca, F. Analysis of the tensile modulus of PP composites reinforced with stone grounwood fibers from softwood. *BioResources* **2012**, *7*, 1310–1323.

45. Espinach, F.X.; Julian, F.; Verdaguer, N.; Torres, L.; Pelach, M.A.; Vilaseca, F.; Mutje, P. Analysis of tensile and flexural modulus in hemp strands/polypropylene composites. *Compos. Part B* **2013**, *47*, 339–343. [CrossRef]

46. Sullins, T.; Pillay, S.; Komus, A.; Ning, H. Hemp fiber reinforced polypropylene composites: The effects of material treatments. *Compos. Part B* **2017**, *114*, 15–22. [CrossRef]

47. Fuqua, M.A.; Chevali, V.S.; Ulven, C.A. Lignocellulosic byproducts as filler in polypropylene: Comprehensive study on the effects of compatibilization and loading. *J. Appl. Polym. Sci.* **2013**, *127*, 862–868. [CrossRef]

48. Zabhizadeh, S.M.; Ebrahimi, G.; Enayati, A.A. Effect of compatibilizer on mechanical, morphological, and thermal properties of chemimechanical pulp-reinforced PP composites. *J. Thermoplast. Compos. Mater.* **2011**, *24*, 221–231. [CrossRef]

49. Saha, P.; Manna, S.; Chowdhury, S.R.; Sen, R.; Roy, D.; Adhikari, B. Enhancement of tensile strength of lignocellulosic jute fibers by alkali-steam treatment. *Bioresour. Technol.* **2010**, *101*, 3182–3187. [CrossRef] [PubMed]

50. Jiménez, A.M.; Espinach, F.X.; Granda, L.; Delgado-Aguilar, M.; Quintana, G.; Fullana-i-Palmer, P.; Mutje, P. Tensile strength assessment of injection-molded high yield sugarcane bagasse-reinforced polypropyene. *BioResources* **2016**, *11*, 6346–6361. [CrossRef]

51. Jiménez, A.M.; Espinach, F.X.; Delgado-Aguilar, M.; Reixach, R.; Quintana, G.; Fullana-i-Palmer, P.; Mutjé, P. Starch-based biopolymer reinforced with high yield fibers from sugarcane bagasse as a technical and environmentally friendly alternative to high density polyethylene. *BioResources* **2016**, *11*, 9856–9868. [CrossRef]

52. Fukuda, H.; Kawata, K. On young's modulus of short fibre composites. *Fibre Sci. Technol.* **1974**, *7*, 207–222. [CrossRef]

53. Sanomura, Y.; Kawamura, M. Fiber orientation control of short-fiber reinforced thermoplastics by ram extrusion. *Polym. Compos.* **2003**, *24*, 587–596. [CrossRef]

54. Oliver-Ortega, H.; Granda, L.A.; Espinach, F.X.; Delgado-Aguilar, M.; Duran, J.; Mutjé, P. Stiffness of bio-based polyamide 11 reinforced with softwood stone ground-wood fibres as an alternative to polypropylene-glass fibre composites. *Eur. Polym. J.* **2016**, *84*, 481–489. [CrossRef]

55. Vallejos, M.E.; Canigueral, N.; Mendez, J.A.; Vilaseca, F.; Corrales, F.; Lopez, A.; Mutje, P. Benefit from hemp straw as filler/reinforcement for composite materials. *Afinidad* **2006**, *63*, 354–361.

56. Espinach, F.X.; Granda, L.A.; Tarrés, Q.; Duran, J.; Fullana-i-Palmer, P.; Mutjé, P. Mechanical and micromechanical tensile strength of eucalyptus bleached fibers reinforced polyoxymethylene composites. *Compos. Part B* **2017**, *116*, 333–339. [CrossRef]

© 2018 by the authors. Licensee MDPI, Basel, Switzerland. This article is an open access article distributed under the terms and conditions of the Creative Commons Attribution (CC BY) license (http://creativecommons.org/licenses/by/4.0/).

polymers

MDPI

Article

Preparation and Characterization of Thermoplastic Potato Starch/Halloysite Nano-Biocomposites: Effect of Plasticizer Nature and Nanoclay Content

Jiawei Ren [1,2], Khanh Minh Dang [2,3], Eric Pollet [2,*] and Luc Avérous [2,*]

[1] Polymer Processing Laboratory, Key Laboratory for Preparation and Application of Ultrafine Materials of Ministry of Education, School of Material Science and Engineering, East China University of Science and Technology, Shanghai 200237, China; jiaweiren1991@gmail.com

[2] BioTeam/ICPEES-ECPM, UMR CNRS 7515, Université de Strasbourg, 25 rue Becquerel, 67087 Strasbourg CEDEX 2, France; minhkhanh238@gmail.com

[3] Department of Packaging and Materials Technology, Faculty of Agro-Industry, Kasetsart University, Bangkok 10900, Thailand

* Correspondence: eric.pollet@unistra.fr (E.P.); luc.averous@unistra.fr (L.A.); Tel.: +33-368-852-786 (E.P.)

Received: 25 June 2018; Accepted: 22 July 2018; Published: 24 July 2018

Abstract: Nano-biocomposites based on halloysite nanoclay and potato starch were elaborated by melt blending with different polyol plasticizers such as glycerol, sorbitol or a mixture of both. The effects of the type of plasticizer and clay content on potato starch/halloysite nano-biocomposites were studied. SEM analyses combined with ATR-FTIR results showed that a high content of sorbitol had a negative effect on the dispersion of the halloysite nanoclay in the starchy matrix. XRD results demonstrated that incorporation of halloysite nanoclay into glycerol-plasticized starch systems clearly led to the formation of a new crystalline structure. The addition of halloysite nanoclay improved the thermal stability and decreased the moisture absorption of the nano-biocomposites, whatever the type of plasticizer used. Halloysite addition led to more pronounced improvement in mechanical properties for glycerol plasticized system compared to nanocomposites based on sorbitol and glycerol/sorbitol systems with a 47% increase in tensile strength for glycerol-plasticized starch compared to 10.5% and 11% for sorbitol and glycerol/sorbitol systems, respectively. The use of a mixture of polyols was found to be a promising way to optimize the mechanical properties of these starch-based nanocomposites.

Keywords: halloysite nanoclay; plasticized starch; glycerol; sorbitol; microstructure; mechanical properties

1. Introduction

In recent decades, biopolymers (e.g., polymers directly extracted from the biomass) such as polysaccharides, and more generally bio-based polymers from renewable resources have attracted a great deal of attention from researchers and industry because of the increasing awareness of environment protection and the lack of certain specific fractions extracted from fossil reserves [1,2]. Among the biopolymers, starch is commonly considered as a promising alternative to traditional non-renewable, non-biodegradable and fossil-based polymers due to its availability, renewability, biodegradability, and biocompatibility [3,4]. Native starch is a complex polysaccharide sourced from plants, composed of two types of α-glucans, linear amylose and highly-branched amylopectin. These biomacromolecules are organized into a highly complex semi-crystalline structure that results from the biosynthesis of the starch granules by the plant [5,6]. Neat starch exhibits high brittleness and poor mechanical properties. However, the addition of plasticizers can improve the material flexibility and processability. It is now very well documented that with plasticizers and under thermomechanical input, the highly organized native starch can be disrupted and destructured into a molten continuous amorphous phase to obtain thermoplastic starch (TPS). Two types of plasticizers are usually used and combined with starch: a volatile plasticizer, mainly water, which also acts as a destructuring agent, and a non-volatile plasticizer such as polyols (sorbitol, glycerol) [7]. TPS shows great potential for short term applications e.g., agricultural mulch films and packaging [7–10]. Nevertheless, TPS still exhibits multiple shortcomings that limit its usage, such as weak mechanical properties compared to conventional thermoplastics, long post-processing aging before stabilizing, and high water sensitivity [11,12]. Some of these issues can be addressed by developing multiphasic systems (blends, composites, etc.). Recently, the use of nanofillers and the elaboration of starch-based nano-biocomposites have been recognized as a powerful solution to overcome these weaknesses [9,13–15]. Due to their high surface areas, nanofillers bring increased mechanical properties, improved thermal resistance, and reduced gas permeability [16] while preserving the biodegradability and biocompatibility of the starchy matrix [17].

Up to now, researchers have focused on layered silicates, especially montmorillonites (MMT) [16]. However, improving the material properties of these silicates requires a great deal of exfoliation of the MMT platelets in the starch-based nano-biocomposites, which leads to increased fabrication cost [18]. Alternatively, halloysite nanoclay with its large aspect ratio, easy availability, high functionality, good biocompatibility, and high mechanical strength, has the potential to elaborate nanocomposites with promising performances [19–24]. This nanofiller is a multi-wall kaolinite nanotube with a theoretical unit cell formulation $Al_2Si_2O_5(OH)_4 \cdot nH_2O$, with n from 0 to 2, for hydrated and dehydrated halloysite, respectively. The tubular structure of halloysite results from the wrapping of the constitutive 1:1 clay layers under specific geological conditions. Typically, halloysite ranges in length from 300 nm to 1500 nm, with an external diameter of 40–120 nm and internal diameter of 15–100 nm [25]. In contrast to other silicates such as kaolinite and MMT, most of the hydroxyl groups are located in the interior of halloysite tubes while siloxane groups are located on the external surface. Some silanols/aluminols are also present on the outer surface, mainly at the edges of the platelets. This unique feature gives halloysite a low surface energy, which in turn reduces the extent of filler-filler aggregation in the matrix compared to other nanofillers [26]. Consequently, halloysite nanoclay has been successfully used by some authors to enhance the properties of glycerol-plasticized starch matrix [27,28]. However, the behavior and efficiency of other types of polyol-based plasticizers on the thermal and mechanical properties of starch/halloysite nanocomposites prepared by the melt-blending method have not been systematically evaluated so far.

Thus, the aim of this work is to study the effect of the addition of halloysite into a starch-based matrix which is plasticized by glycerol, sorbitol or by a mixture of them, so as to reduce the water sensitivity and improve the thermal stability and mechanical strength. In particular, the effects of the type of plasticizer and nanoclay loading on the structure and properties of the obtained starch nano-biocomposites will be investigated using different characterization techniques such as scanning electron microscopy (SEM), attenuated total reflectance-Fourier transform infrared (ATR-FTIR), X-ray diffraction (XRD), thermogravimetric analysis (TGA), moisture content measurement and uniaxial tensile testing methods.

2. Materials and Methods

2.1. Materials

Potato starch was supplied by Roquette (Lestrem, France) (80% starch content, 19.5% moisture content, 0.05% proteins and 0.2% ash). The amylose and amylopectin contents were 20% and 80%, respectively. Glycerol was a 99.5% purity product (Thermo Fisher Scientific, Illkirch-Graffenstaden, France). Sorbitol was kindly supplied by Tereos (Origny-Sainte-Benoite, France) with 98% purity. Polysorb®, a glycerol/sorbitol mixture (59/41 by weight) was kindly supplied by Roquette (Lestrem,

France). The halloysite nanoclay with diameter of 30–70 nm and length of 1–3 μm (CAS 1332-58-7) was purchased from Sigma-Aldrich (Lyon, France).

2.2. Preparation of Nanocomposites

The nanocomposites were obtained from a two-step process with the first being the elaboration of a dry-blend (powder) by the mixing of starch, water and polyol. Such dry-blend protocol allows the subsequent preparation of plasticized starch with high plasticizer content without the exudation phenomenon, mainly thanks to the strong interactions established between the polysaccharide chains and the polyols [29]. Then, the nano-biocomposites resulting from the dry-blend powder were elaborated by thermo-mechanical input with the addition of nanoclay.

2.2.1. Preparation of Plasticized Starch Dry-Blends

First, the starch/plasticizer dry-blends were prepared. In this process, native potato starch was dried overnight at 70 °C in a ventilated oven to remove the free water (ca. 10 wt.% of the materials depending on the atmosphere relative humidity and temperature). Then, the dried starch powder was introduced into a Papenmeier turbo-mixer, and the plasticizer (glycerol, sorbitol, or the mixture) was slowly added under mixing until a homogenous mixture was obtained. The mixture was then dried at 170 °C in a ventilated oven for 40 min and occasionally stirred, and then the dry-blend was recovered. To obtain adequate moisture content, a pre-determined quantity of water was added to the dry-blend after cooling and mixed in the turbo-mixer, resulting in a formulation containing 54 wt.% potato starch, 23 wt.% plasticizer and 23 wt.% water. Finally, the powder was stored in a polyethylene bag in a refrigerator at 6 °C overnight prior to processing.

2.2.2. Nano-Biocomposites Elaboration

To obtain nano-biocomposites, the plasticized starch powder was processed, with the addition of 3 wt.%, 5 wt.% and 7 wt.% of halloysite nanotubes compared to the weight of the dry-blend powder, in a counter-rotating internal batch mixer, Rheomix OS (Haake Thermo Fisher Scientific, Illkirch-Graffenstaden, France), at 70 °C for 20 min with rotor speed of 150 rpm. After melt mixing, the mixtures were then compression molded (Labtech Engineering Company, Muang, Thailand) at 110 °C, applying 18 MPa for 15 min. The obtained films (thickness approx. 1 mm) were then stored at 57% relative humidity at 23 °C for at least three weeks before characterization to obtain stabilized properties. The samples are designated as XY where X stands for the type of plasticizer (G for glycerol, S for sorbitol or P for the mixture of glycerol and sorbitol—Polysorb®) and Y for the weight percentage of halloysite nanoclay.

2.3. Characterization Techniques

2.3.1. Scanning Electron Microscopy (SEM)

The fracture surfaces of the samples were observed with a VEGA3 LM scanning electron microscope (TESCAN, Brno, Czech Republic). The film samples were mounted on a stub using double-sided adhesive tape and coated with a thin layer of gold (10–20 nm). The images were obtained at an operating voltage of 5 kV and ×10,000 magnification.

2.3.2. Attenuated Total Reflectance-Fourier Transform Infrared (ATR-FTIR)

IR spectra were recorded using a Thermo Scientific Nicolet iS10 FTIR spectrometer (Thermo Fisher Scientific, Illkirch-Graffenstaden, France) in attenuated total reflectance (ATR) mode. IR spectra of films were obtained for wavenumbers ranging from 4000 to 650 cm^{-1} by accumulation of 32 scans at a resolution of 4 cm^{-1}.

2.3.3. X-ray Diffraction (XRD)

X-ray diffraction analysis was carried out on a Bruker AXS D8 ADVANCE (Bruker, Wissembourg, France) using Cu-Kα radiation ($\lambda = 0.1542$ nm) operating at 50 kV and 40 mA. The scanning region of the diffraction angle (2θ) was from $10°$ to $50°$ with a step size of $0.02°$.

2.3.4. Thermogravimetric Analysis (TGA)

The thermogravimetric analysis was performed on a TGA Q5000 machine (TA Instruments, New Castle, DE, USA). Sample masses ranging from 2 mg to 3 mg were heated from 35 °C to 600 °C at the rate of 10 °C/min under air atmosphere (flow rate of 25 mL/min).

2.3.5. Density and Moisture Measurements

Samples films were cut into square shapes. Then, the weight, length, width and thickness of the film were measured, allowing the calculation of the material density. Experiments were done in triplicate. The water contents of samples were estimated using a MB45 moisture analyzer (Ohaus, Parsippany, NJ, USA) considering the weight loss measured after 1 h of drying at 105 °C.

2.3.6. Mechanical Properties

The uniaxial tensile strength, elongation at break and Young's modulus were determined using a Universal Testing Machine MTS 2M (MTS, Eden Prairie, MN, USA). The samples were cut into a dumbbell shape and the tests were performed at room temperature with a constant deformation rate of 10 mm/min and a distance grip separation of 4 cm. Five specimens of each formulation were tested and the average values were calculated. Samples were conditioned at ambient temperature with relative humidity of $52 \pm 2\%$ for at least 3 weeks prior to testing.

3. Results and Discussion

3.1. Morphological Characterization

SEM images of plasticized starch and starch nano-biocomposites are displayed in Figure 1. It can be observed that the control glycerol plasticized starch film (Figure 1A) exhibited uniform morphology without any obvious remaining starch granules, indicating that native starch structure was completely disrupted during the thermos-mechanical process. However, some voids were observed, resulting in discontinuous phases. By contrast, the addition of sorbitol or glycerol/sorbitol as non-volatile plasticizers (Figure 1E,I) resulted in more uniform morphologies and also no remaining starch granules. These observations could be explained by the greater shear force induced by sorbitol addition, leading to better plasticizer dispersion and continuous phases.

Adding halloysite into glycerol plasticized starch resulted in a more uniform and continuous morphology of the starchy matrix. No aggregates were visible for nano-biocomposites based on glycerol plasticized starch (Figure 1B–D). Even at the highest loading (7 wt.%) of halloysite (Figure 1D), homogenous clay dispersion with predominantly individual nanotubes was observed. The halloysite nanotubes were embedded into the matrix and no interfacial voiding was visible for glycerol plasticized samples, which implies a good interfacial adhesion between halloysite and the glycerol plasticized starch matrix (Figure 1C,D). However, large aggregates were clearly visible in the nanocomposites based on sorbitol plasticized starch at high loading (5 wt.% and 7 wt.%) of halloysite (red circles on Figure 1K,L).

Besides, the presence of voids between the clay aggregates and the starchy matrix suggested a poor interface quality (Figure 1K,L). For samples plasticized by the mixture of both polyols, halloysite nanotubes were also found to be uniformly dispersed in the matrix at the lowest content (3 wt.%) (Figure 1F). Limited amounts of very small aggregates and a large proportion of almost individually and randomly dispersed nanotubes were observed (Figure 1G,H) when the nanofiller content increased.

The behavior of the nanocomposites based on starch plasticized by the mixture of glycerol and sorbitol was in-between those of the nanocomposites based on glycerol and sorbitol as plasticizers.

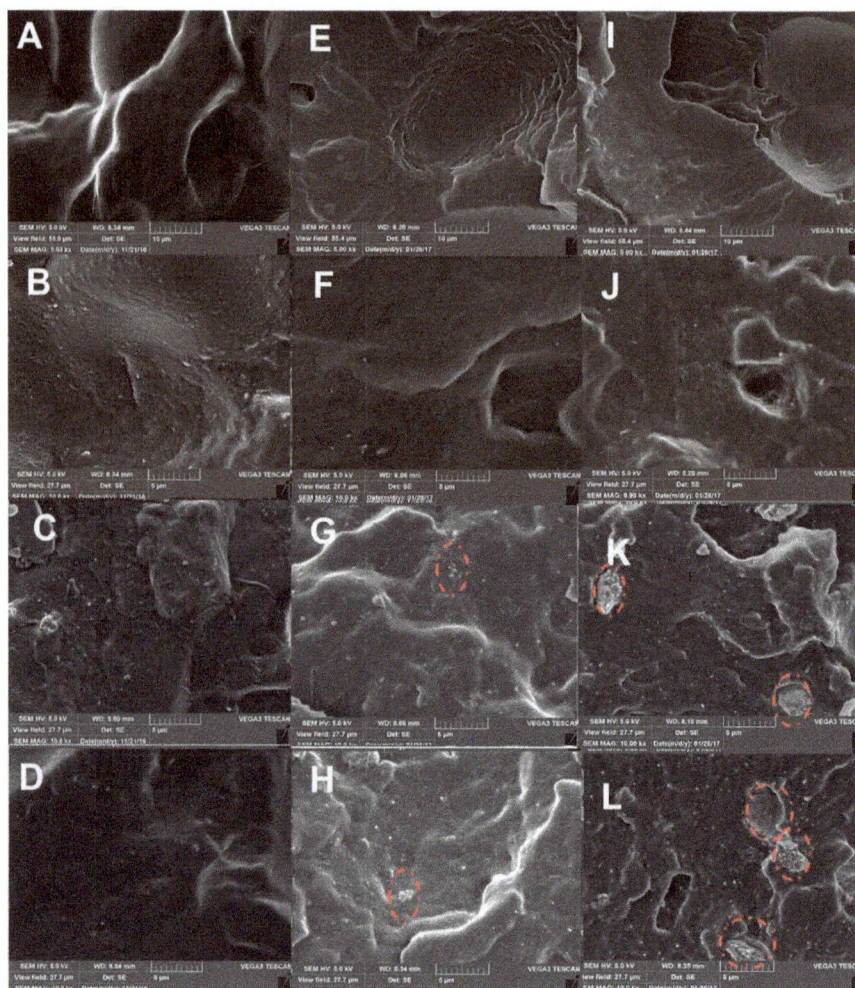

Figure 1. SEM images of the various plasticized starch systems and corresponding plasticized starch nano-biocomposites: (**A**) G0; (**B**) G3; (**C**) G5; (**D**) G7; (**E**) P0; (**F**) P3; (**G**) P5; (**H**) P7; (**I**) S0; (**J**) S3; (**K**) S5 and (**L**) S7. Scale bars are 5 or 10 microns.

3.2. Chemical Interactions

ATR-FTIR was used to investigate the interactions between plasticized starch and halloysite nanoclay. The spectra of the native potato starch, halloysite nanoclay and plasticized starch nano-biocomposites containing 7 wt.% halloysite nanoclay are shown in Figure 2.

The broad band between 3000 and 3600 cm^{-1} in the spectrum of native potato starch was attributed to the complex vibrational stretches associated with the free, inter- and intramolecular bound hydroxyl groups [30]. In the case of starch plasticized with glycerol, this band shifted to lower wavenumbers indicating strong and stable hydrogen bonds formation between the plasticizer and the starch macromolecules [31]. Tang, et al. reported similar behavior when incorporating nano-SiO$_2$ into corn starch film [32] with the absorption band of O–H stretching shifted to lower wavenumbers, indicating an increase in intermolecular hydrogen bonds between nano-SiO$_2$ and starch. The same trend was observed for the nanocomposites plasticized with sorbitol, alone or mixed with glycerol.

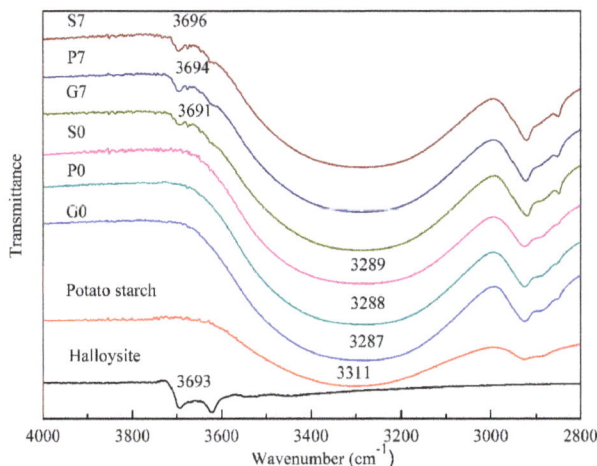

Figure 2. ATR-FTIR spectra of the native potato starch, halloysite nanoclay, unfilled plasticized starch matrices (G0, S0, P0) and plasticized starch nano-biocomposites containing 7 wt.% halloysite nanoclay (G7, S7, P7).

In the spectrum of halloysite nanoclay, the peaks at 3693 and 3623 cm^{-1} were related to the O–H stretching of inner-surface hydroxyl groups and inner hydroxyl groups, respectively. After addition into the glycerol plasticized matrix, the O–H stretching peak of inner-surface hydroxyl groups slightly shifted to a lower wavenumber, 3691 cm^{-1}. Similar findings were previously observed by Schmitt et al. [27]. This shift could be attributed to the formation of interactions between the inner-surface hydroxyl groups of halloysite and the C–O–C groups of starch and/or glycerol. On the contrary, when halloysite nanotubes were added to sorbitol-plasticized matrices, the O–H stretching peak of inner-surface hydroxyl groups shifted to a higher wavenumber, 3694 and 3696 cm^{-1}, respectively. This suggested a decrease in the intermolecular interactions between the inner-surface hydroxyl groups of halloysite and the C–O–C groups of starch and/or plasticizers. Since the main difference between these nano-biocomposites was the nature of the plasticizer, it could be concluded that glycerol formed stronger and more stable hydrogen bonds with the inner-surface hydroxyl groups of halloysite compared with sorbitol. This explains why a better dispersion of nanofillers could be obtained in the case of the glycerol plasticized starch matrix.

3.3. Microstructure and Crystallinity Studies

Figure 3 shows the XRD patterns recorded for native potato starch, the halloysite nanoclay and the plasticized starch/halloysite nano-biocomposites.

Halloysite displayed the typical pattern of the dehydrated form with characteristic peaks at 2θ of 12.1°, 18.6°, 20.3°, 25.0° and 27.0° [33]. The pattern of native potato starch showed diffraction peaks at 2θ of 15.4°, 17.4°, 19.9°, 22.5° and 24.2° which are characteristic of B-type starch crystalline structures [34]. After processing with glycerol, the B-type crystalline structure was transformed into E$_H$-type and V$_H$-type structures with characteristic peaks at 2θ of 17°, 19.9°, and 22.0°, respectively [35,36]. The E$_H$-type structure is linked to amylopectin recrystallization while the

V$_H$-type structure is attributed to amylose crystallization into single helical structure [36]. Except for these three peaks, sorbitol plasticized starch showed weak and small diffraction peaks at around 12.0° and 18.8° which were assigned to conventional sorbitol crystallization during storage [37]. Importantly, matrices based on the glycerol/sorbitol mixture showed similar patterns to glycerol-based matrices indicating an absence of sorbitol recrystallization in such systems.

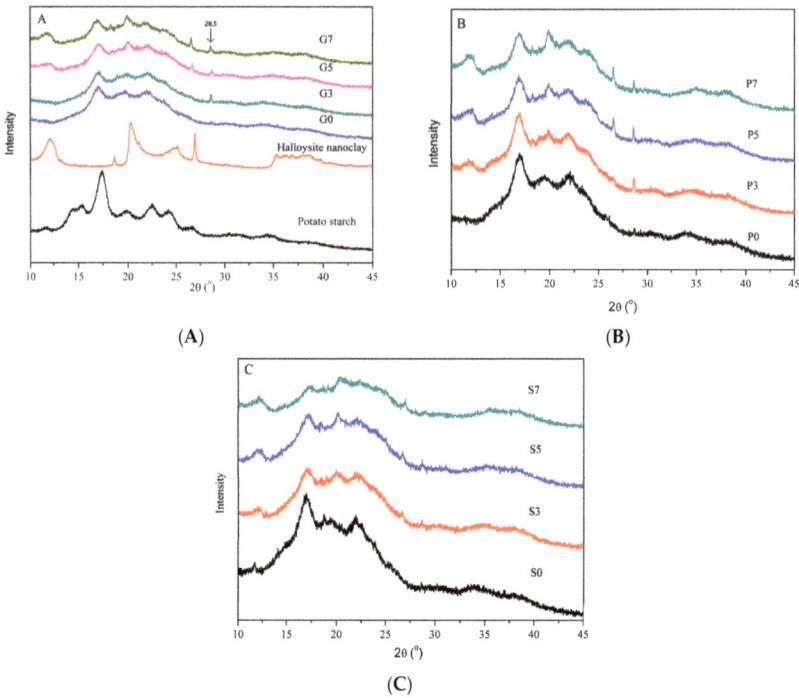

Figure 3. XRD patterns of the raw materials and the different nano-biocomposites with various nanofiller contents. From top to bottom, (**A**) native potato starch alone, halloysite nanoclay alone, and nano-biocomposites based on glycerol-plasticized starch; (**B**) nano-biocomposites based on glycerol/sorbitol mixture, and (**C**) nano-biocomposites based on sorbitol.

After dispersion in the starch matrix, the halloysite characteristic peaks at 2θ of 12.1° and 27.0° remained visible in the patterns of the nano-biocomposites. The peak intensity increased with the halloysite content while the peak position remained almost unchanged with only a slight broadening of the peak at 2θ of 12.1°. In addition, the characteristic peaks of plasticized starch were also observed in the diffractogram of the nano-biocomposites, attesting that halloysite nanotubes were well dispersed in the starch matrix [38]. No significant changes in the E$_H$-type and V$_H$-type crystallization peaks were observed for the starch/halloysite nano-biocomposites. The XRD patterns of nanocomposites based on glycerol displayed a new peak at 28.5°, compared with the pattern of unloaded plasticized matrix. Similar findings have been reported for starch/tunicin whiskers [39] and starch/sepiolite nanocomposites [40]. This was attributed to the formation of strong hydrogen bonds between the filler and the macromolecular system, facilitating amylopectin crystallization at the nanofiller interface. Thus, according to the ATR-FTIR analyses, it can be inferred that the interactions between the inner-surface hydroxyl groups of halloysite and the C–O–C groups of starch resulted in the appearance of a new crystalline structure. This phenomenon has been reported in the case of PP/halloysite [41], PA6/halloysite [42] and PVDF/halloysite nanocomposites [43]. However, the peak of the new crystalline structure was weak in the XRD patterns of sorbitol-based nanocomposites at high loading (5 wt.% and 7 wt.%) of halloysite. On the basis of SEM observation, this could be explained by the formation of halloysite aggregates in the sorbitol-plasticized matrix, decreasing their nucleating ability due to the smaller specific surface area, which hampered the amylopectin crystallization at the interface. Consequently, the new peak was more obvious in XRD patterns of nanocomposites based on the mixed sorbitol/glycerol compared to sorbitol–based nanocomposites due to the better dispersion state of halloysite nanoclays in the matrix.

3.4. Thermal Stability

Generally, the addition of nanoclays can generate two opposite effects on the thermal stability of polymer/clay nanocomposites: (i) a barrier effect which improves the thermal stability and (ii) a promoter effect which increases the thermal degradation process [44]. An enhancement in the material thermal stability is commonly observed in nanocomposite systems and is linked to the clay aspect ratio and dispersion state. The clay dispersion into the matrix increases the tortuosity of the combustion gas diffusion pathway and favors the formation of char at the material surface. Higher thermal stability can also result from strong interactions between the clay nanoparticles and the polymer matrix [45]. Also, the thermal degradation promoter effect is mainly due to the presence of hydroxyl groups on the edges of the clay which can catalyze the polymer degradation with a thermal dependence [46,47]. The barrier effect is predominant when the nanoclays are well dispersed in the matrix [47]. Thus, thermogravimetric analyses were performed on starch nano-biocomposites to analyze the impact of halloysite nanoclay on their thermal properties. Figure 4 presents the typical curves for glycerol plasticized starch and its nanocomposite with 7 wt.% halloysite. The curve of pristine plasticized starch shows three different degradation steps. The first step corresponds to the volatilization of water and glycerol, and the two other weight losses correspond to the degradation of amylose and amylopectin [48]. After incorporation of the nanoclay, the nanocomposite exhibits a further degradation at each step together with a slightly higher maximum degradation temperature, demonstrating an enhancement of thermal stability. Similar findings were also reported for starch/sepiolite nanocomposites [40].

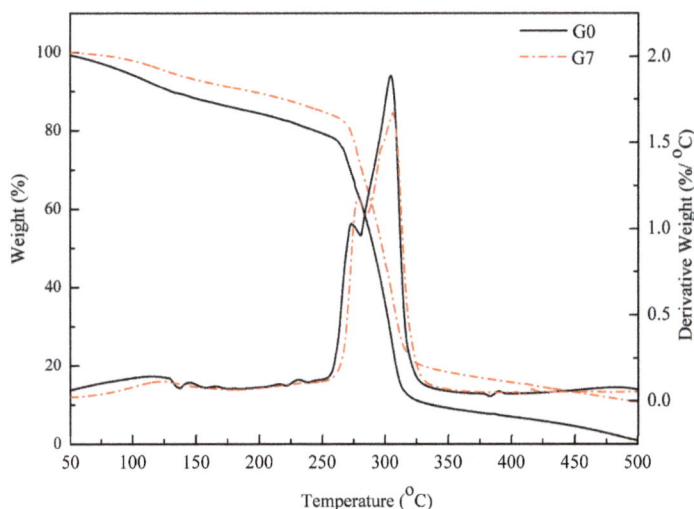

Figure 4. Weight loss (TG) and corresponding derivative (DTG) curves recorded for the unfilled glycerol-plasticized starch (G0) and the glycerol-plasticized starch nano-biocomposite containing 7 wt.% halloysite (G7).

The detailed TGA data for starch-based nanocomposites, and raw compounds are summarized in Table 1. The major parameters $T_{10\%}$, $T_{50\%}$ and $T_{90\%}$ refer to temperatures at which weight loss is 10%, 50% and 90%, respectively, and T_{max} corresponds to maximum degradation rate. The char residue (%) is the unburnt residue at 600 °C.

An increasing trend was observed for $T_{10\%}$ and correlated to the sorbitol content. This behavior was attributed to the higher thermal stability of sorbitol compared to glycerol. For all starch-based nanocomposites containing 3 wt.% halloysite nanoclay, $T_{50\%}$ was enhanced in comparison to the glycerol plasticized matrix, indicating that nanoclay had a barrier effect with an impact on degradation. This was attributed to the good dispersion of nanoclay in the matrix, even at low content. Considering the tube-like structure of halloysite nanoclay, an increase in the tortuosity was less obvious, thus the enhancement of thermal stability was more likely related to the strong interactions between the halloysite nanoclays and the matrix. Similar findings have been reported for polyethylene/halloysite [40], silicone rubber/halloysite [26] and polyurethane/halloysite nanocomposites [49].

Table 1. TGA data for the starch-based nanocomposites and neat glycerol, sorbitol and halloysite.

Sample Code	$T_{10\%}$ (°C)	$T_{50\%}$ (°C)	$T_{90\%}$ (°C)	T_{max} (°C)	Char Residue (%)
G0	131	291	337	287	0
G3	161	293	373	289	2.1
G5	195	295	463	291	4.9
G7	194	298	511	306	7.0
P0	189	297	361	288	0
P3	230	301	401	288	3.1
P5	226	300	429	290	4.6
P7	232	300	496	304	7.1
S0	223	297	349	297	0
S3	239	300	401	298	2.2
S5	248	300	448	299	4.7
S7	245	300	467	306	6.3
Glycerol	127	154	162		-
Sorbitol	218	251	268		-
Halloysite	405	>600	>600		-

$T_{50\%}$ gradually increased with halloysite nanoclay content in glycerol plasticized nanocomposites. However, there was no significant difference in $T_{50\%}$ as the nanoclay content increased from 3 wt.%, to 5 wt.% to 7 wt.% in nanocomposites of starch plasticized with sorbitol, alone or combined with glycerol. This finding suggests that some degradation promoter effect occurred and impacted the thermal stability with the addition of high levels of halloysite nanoclay. According to the SEM images, this could be related to the formation of aggregates in nanocomposites plasticized with sorbitol, alone or blended with glycerol, at high halloysite content (5 wt.% and 7 wt.%) which limited the barrier effect. Since the nanocomposites show different moisture content it is difficult to make a direct comparison of $T_{50\%}$ values, especially for such multiple-step degradation. Then T_{max}, which corresponds to the temperature at maximum degradation rate was determined from the DTG curves maximum. In this case, the overall gradual increase in thermal stability with the clay content is more obvious and seems particularly marked for 7 wt.%. For $T_{90\%}$ and the char residue, whatever the type of plasticizer, both increased with the halloysite content due to the higher thermal stability of the halloysite nanoclays.

3.5. Density and Moisture Content

The densities and moisture content of the different samples were determined after three weeks of equilibration, at room temperature and 52 ± 2 %RH. Results are summarized in Table 2.

It can be seen that the nanocomposites showed a slightly increased density when nanofiller content increased. This is due to the higher density of halloysite nanoclay. Since the density of glycerol is lower than that of sorbitol, nanocomposites based on glycerol exhibited lower density than nanocomposites based on sorbitol. For the same reasons, the densities of the nanocomposites with the mixed polyols system were in-between those of glycerol and sorbitol-based plasticized starch nanocomposites.

Table 2. Mechanical properties, moisture content and densities of the different samples.

Sample Code	Stress at Break (MPa)	Strain at Break (%)	Young's Modulus (MPa)	Moisture Content (%)	Density (g/cm³)
G0	2.28 ± 0.13	26.1 ± 2.3	22.0 ± 2.2	16.7 ± 0.2	1.45 ± 0.02
G3	2.64 ± 0.11	27.3 ± 2.6	28.3 ± 1.9	15.0 ± 0.2	1.47 ± 0.01
G5	2.94 ± 0.05	29.3 ± 2.3	34.6 ± 2.2	14.2 ± 0.2	1.48 ± 0.01
G7	3.36 ± 0.24	34.5 ± 3.6	37.0 ± 2.7	13.8 ± 0.2	1.50 ± 0.01
P0	7.13 ± 0.06	42.4 + 1.7	119.2 ± 4.9	11.1 ± 0.1	1.48 ± 0.01
P3	7.36 ± 0.22	39.6 ± 1.9	144.7 ± 12.0	10.6 ± 0.1	1.49 ± 0.01
P5	7.77 ± 0.12	37.0 ± 0.4	194.1 ± 15.3	10.4 ± 0.1	1.50 ± 0.01
P7	7.88 ± 0.22	33.2 ± 3.0	174.5 ± 8.1	10.0 ± 0.1	1.52 ± 0.01
S0	9.70 ± 0.79	43.3 ± 3.0	256.0 ± 19.0	9.6 ± 0.1	1.49 ± 0.03
S3	10.24 ± 0.48	39.1 ± 2.7	276.5 ± 28.8	8.4 ± 0.1	1.50 ± 0.02
S5	10.52 ± 0.51	36.5 ± 5.4	281.8 ± 19.0	8.7 ± 0.1	1.52 ± 0.01
S7	10.78 ± 0.37	35.0 ± 5.6	292.1 ± 34.9	7.5 ± 0.1	1.54 ± 0.02

Whatever the halloysite clay loading, the moisture content for nano-biocomposites with glycerol were higher than those recorded for nano-biocomposites with sorbitol alone or blended with glycerol. This is attributed to the higher hydrophilic character of glycerol compared to sorbitol [50]. Besides, the moisture content of nano-biocomposites showed an overall decrease with the increase in halloysite content. This was likely due to the relatively hydrophobic nature of the halloysite [25]. This last behavior could also be linked to the formation of interaction between the inner-surface hydroxyl groups of halloysite and the different groups of starch and plasticizers, thus decreasing the interaction between water molecules and polysaccharide chains and/or plasticizers molecules. The theoretical water contents of nanocomposites with the mixture of polyols were also calculated according to Equation (1):

$$W_{mixture} = W_{glycerol} \, V_{glycerol} + W_{sorbitol} \, V_{sorbitol} \tag{1}$$

where $W_{glycerol}$ is the water content of the nanocomposite with neat glycerol, $W_{sorbitol}$ is the water content of the nanocomposite with neat sorbitol, $V_{glycerol}$ and $V_{sorbitol}$ are the respective nanocomposite volume fractions in the composites. The calculated values for sample P0, P3, P5 and P7 were 13.9%, 12.4%, 12.0% and 11.3% respectively. Interestingly, the experimental values were slightly lower but still in good agreement with the theoretical ones.

3.6. Uniaxial Tensile Properties

Young's modulus, tensile strength and elongation at break of the different unfilled plasticized starch matrices and corresponding nano-biocomposites were measured. The main results are summarized in Table 2. It can be seen that the plasticized starch stiffness was greatly affected by the type of plasticizer. The sorbitol plasticized starch had the highest Young's modulus while the glycerol plasticized starch showed the lowest value. This phenomenon was due to the difference in the water content of the materials which directly affects the Young's modulus since water acts as a plasticizer [50]. For all starch nano-biocomposites, whatever the type of plasticizer, a reinforcement effect with an increase in the matrix stiffness was clearly observed with addition of halloysite. Such enhancement in modulus had been widely reported in the literature for other starch-based nanocomposites systems [40,50]. This has been attributed to (i) the addition of the halloysite nanoclay which has a high elastic modulus of about 140 GPa [25], and (ii) the interactions between the nanofillers and the plasticized starch matrix. Besides, the new crystalline structure induced from halloysite addition led to strong interactions which contributed to the stiffness increase through improved load transfer between the matrix and the nanofiller. The theoretical Young's moduli of nanocomposites with polyols mixture were calculated according to Equation (2):

$$Y_{mixture} = Y_{glycerol} V_{glycerol} + Y_{sorbitol} V_{sorbitol} \tag{2}$$

where $Y_{glycerol}$ is the Young's modulus of the glycerol-plasticized starch nanocomposite, $Y_{sorbitol}$ is the Young's modulus of the sorbitol-plasticized starch nanocomposite, $V_{glycerol}$ and $V_{sorbitol}$ are the corresponding nanocomposite volume fractions in the systems. The ratio of $V_{glycerol}/V_{sorbitol}$ is 0.6/0.4. Since the Young's moduli of nanocomposites with sorbitol as plasticizer were much higher than those of nanocomposites with glycerol, the theoretical values were mainly dependent on sorbitol nanocomposites. The calculated values for P0, P3, P5 and P7 were 115.6, 127.6, 133.5 and 139.0 MPa, respectively. These values were lower than the experimental ones. This was probably due to (i) the much better dispersion state of nanofillers in the matrix based on the polyols mixture compared to that in the sorbitol-based matrix, and (ii) the lower experimental water content compared to the theoretical values [50].

Similar to Young's modulus evolutions, the tensile strength of plasticized starch was significantly impacted by the nature of the plasticizer. The tensile strength of starch plasticized with the polyols mixture was higher than the glycerol plasticized one. The highest tensile strength value was obtained with sorbitol. Such a trend was also recently observed in plasticized alginate obtained by thermo-mechanical mixing [51]. This was due to the higher plasticizing efficiency of glycerol compared to sorbitol and to higher water uptake after equilibration with glycerol. Zhang and Han [52] studied the plasticizing effects of polyols on pea starch films and concluded that glycerol had higher plasticization efficiency due to the small size of the molecule, which can easily locate between the starch chains and disrupt intermolecular polymer interactions while sorbitol had larger sized molecules which can reduce its plasticizing effect. Compared to the neat matrix, whatever the type of plasticizer, an increase in the nanocomposites tensile strength was obtained and was correlated to the clay content. This enhancement seemed to be more pronounced for the glycerol plasticized nano-biocomposites, with an increase of up to 47% for the highest clay loading compared to a roughly 11% increase for the sorbitol and polyols mixture plasticized systems. This was also attributed to better nanofiller dispersion and stronger halloysite/matrix interactions leading to better load transfer between the main components during the uniaxial test. On the other hand, it should be noted that the nanocomposites based on both sorbitol and polyols mixture as plasticizers exhibited an insignificant increase in tensile strength, probably due to aggregation of halloysite at 5 wt.% and 7 wt.% loadings. The theoretical tensile strength of nanocomposites with polyols mixture were also calculated according to Equation (3):

$$T_{mixture} = T_{glycerol} V_{glycerol} + T_{sorbitol} V_{sorbitol}, \tag{3}$$

where $T_{glycerol}$ is the tensile strength of the nanocomposite with glycerol-plasticized starch, $T_{sorbitol}$ is the tensile strength of the nanocomposite with sorbitol-plasticized starch, $V_{glycerol}$ and $V_{sorbitol}$ are the corresponding nanocomposite volume fractions in the systems. The calculated values for P0, P3, P5 and P7 were 5.25, 5.68, 5.97 and 6.33 MPa, respectively. As expected, the experimental tensile strength values were higher than the theoretical ones.

The nature of the plasticizer also impacted the elongation at break values of neat plasticized matrices. The elongation at break of the glycerol plasticized starch was lower than the sorbitol ones (alone or in mixture) probably due to the nanofiller dispersion state in the matrix. According to the morphological analyses, nanocomposites based on glycerol displayed good halloysite dispersion while nanocomposites based on sorbitol and the polyols mixture as plasticizers showed some aggregation of nanofiller at a high degree of loading. These results were in agreement with some previous observations off plasticized starch stored at high RH [37,53]. Due to the higher hydrophilic character of glycerol compared to sorbitol, the overall content of plasticizer (water included) for materials based on glycerol was higher than that in materials based on sorbitol or in the polyols mixture (sorbitol and glycerol). In the presence of a high amount of water, the relatively strong hydrogen bonds between starch-polyol and starch-starch molecules are partially replaced by the weaker hydrogen bonds between starch-water and polyol-water, resulting in decreased deformation [37]. Besides, according to XRD analysis and compared to glycerol, sorbitol crystallized during storage, which decreased the amount of efficient plasticizing polyol in the starch matrix and could also have enhanced "crosslink" formation in the starch network and then impacted the elongation at break properties [37]. This behavior has also been observed in plasticized alginate with high plasticizer content [51,54]. The elongation properties of the nanocomposites showed clear and different trends depending on the plasticizer's nature. The incorporation of halloysite nanoclay into the glycerol plasticized starch induced slightly increased elongation at break values related to the homogenous dispersion of nanofiller in the matrix and decreased water content. On the contrary, the values for polyols mixture and sorbitol-based systems decreased with increases in nanoclay content. Such behaviors were attributed to aggregates of halloysite nanoclay in the matrix, as observed by SEM, which embrittled the materials. Chivrac et al. [50] also confirmed that glycerol plasticized starch/montmorillonite (MMT) nanocomposites showed no variation in the strain at break with increasing clay contents because of high exfoliation and the good dispersion state of MMT in the starch matrix, while sorbitol plasticized nano-biocomposites showed slightly decreased extensibility due to remaining small clay tactoids which embrittled the plasticized starch matrices.

4. Conclusions

Plasticized starch/halloysite nano-biocomposites were successfully prepared. The influence of plasticizer type and filler loading on the microstructure and properties of the resulting materials were studied. Glycerol contributed to a more homogenous dispersion of halloysite nanotubes in the matrix compared to sorbitol, used alone or in combination, due to stronger and more stable hydrogen bonds between glycerol plasticized starch and halloysite, as revealed by ATR-FTIR analyses. The XRD results showed that incorporation of halloysite nanoclay led to the formation of a new crystalline structure attributed to amylopectin crystallization at the nanofiller interface which was mainly observed for nanocomposites based on glycerol or the mixture of glycerol and sorbitol as plasticizers. Whatever the type of plasticizer, the incorporation of halloysite increased the thermal stability and reduced the moisture content of the starch nano-biocomposites. The mechanical properties were also greatly impacted by the nature of the plasticizer. Glycerol-plasticized starch exhibited lower Young's modulus, tensile strength and elongation at break than the sorbitol one. After addition of halloysite nanoclay, the improvement in tensile properties was more pronounced for glycerol plasticized systems due to the better dispersion state of halloysite compared to the systems based on sorbitol-plasticized starch. Finally, the use of mixtures of these polyols was found to be a good and promising way to optimize the mechanical properties of these starch-based nanocomposites, since the values of Young's modulus and tensile strength were in-between those of glycerol- and sorbitol-based nanocomposites and higher than the theoretically expected ones.

This work clearly showed that halloysite nanoclay is an effective and promising clay for enhancing the properties of plasticized starch. Nano-biocomposites based on starch/halloysite nanoclay represent an interesting alternative to replace some non-biodegradable materials in different fields such as packaging, agriculture or biomedical applications. For instance, in active packaging or in specific biomedical applications, halloysite nanoclay could be a good carrier for some active drugs due to its specific structure and surface properties. In future works, we will focus on developing different functional biomaterials based on starch/halloysite nanocomposites in order to broaden the properties and applications of starch-based materials.

Author Contributions: J.R., E.P. and L.A. conceived the experiments; J.R. performed the experiments; J.R., K.M.D. and E.P. analyzed the data; J.R., K.M.D., E.P. and L.A. wrote the paper.

Funding: This research received no external funding

Acknowledgments: J.R. would like to thank the China Scholarship Council (CSC) for its financial support. Thanks are also extended to Roquette and Tereos for their kind supply of the potato starch and plasticizers.

Conflicts of Interest: The authors declare no conflict of interest.

References

1. Yu, L.; Dean, K.; Li, L. Polymer blends and composites from renewable resources. *Prog. Polym. Sci.* **2006**, *31*, 576–602. [CrossRef]
2. Mohanty, A.K.; Misra, M.; Drzal, L.T. Sustainable bio-composites from renewable resources: Opportunities and challenges in the green materials world. *J. Polym. Environ.* **2002**, *10*, 19–26. [CrossRef]
3. Liu, H.; Xie, F.; Yu, L.; Chen, L.; Li, L. Thermal processing of starch-based polymers. *Prog. Polym. Sci.* **2009**, *34*, 1348–1368. [CrossRef]
4. Xie, F.; Halley, P.J.; Avérous, L. Rheology to understand and optimize processibility, structures and properties of starch polymeric materials. *Prog. Polym. Sci.* **2012**, *37*, 595–623. [CrossRef]
5. Ren, J.; Zhang, W.; Yu, Y.; Zhang, G.; Guo, W. Preparation and structure characterization of linear long-chain dextrin obtained from pullulanase debranching of cassava starch. *Starch-Stärke* **2015**, *67*, 884–891. [CrossRef]
6. Tester, R.F.; Karkalas, J.; Qi, X. Starch—Composition, fine structure and architecture. *J. Cereal Sci.* **2004**, *39*, 151–165. [CrossRef]
7. Ren, J.; Zhang, W.; Lou, F.; Wang, Y.; Guo, W. Characteristics of starch-based films produced using glycerol and 1-butyl-3-methylimidazolium chloride as combined plasticizers. *Starch-Stärke* **2017**, *69*, 160161. [CrossRef]
8. Averous, L. Biodegradable multiphase systems based on plasticized starch: A review. *J. Macromol. Sci. Polym. Rev.* **2004**, *44*, 231–274. [CrossRef]
9. Xie, F.; Pollet, E.; Halley, P.J.; Avérous, L. Starch-based nano-biocomposites. *Prog. Polym. Sci.* **2013**, *38*, 1590–1628. [CrossRef]
10. Glenn, G.M.; Orts, W.; Imam, S.; Chiou, B.-S.; Wood, D.F. Starch plastic packaging and agriculture applications. In *Starch Polymers*; Halley, P., Avérous, L., Eds.; Elsevier: Amsterdam, The Netherland, 2014; pp. 421–452, ISBN 978-0-444-53730-0.
11. Dean, K.M.; Do, M.D.; Petinakis, E.; Yu, L. Key interactions in biodegradable thermoplastic starch/poly(vinyl alcohol)/montmorillonite micro- and nanocomposites. *Compos. Sci. Technol.* **2008**, *68*, 1453–1462. [CrossRef]
12. Sarazin, P.; Li, G.; Orts, W.J.; Favis, B.D. Binary and ternary blends of polylactide, polycaprolactone and thermoplastic starch. *Polymer* **2008**, *49*, 599–609. [CrossRef]
13. Chang, P.R.; Wu, D.; Anderson, D.P.; Ma, X. Nanocomposites based on plasticized starch and rectorite clay: Structure and properties. *Carbohydr. Polym.* **2012**, *89*, 687–693. [CrossRef] [PubMed]
14. Miculescu, F.; Maidaniuc, A.; Voicu, S.I.; Thakur, V.K.; Stan, G.E.; Ciocan, L.T. Progress in hydroxyapatite–starch based sustainable biomaterials for biomedical bone substitution applications. *ACS Sustain. Chem. Eng.* **2017**, *5*, 8491–8512. [CrossRef]
15. Madhumitha, G.; Fowsiya, J.; Mohana Roopan, S.; Thakur, V.K. Recent advances in starch–clay nanocomposites. *Polymers* **2018**, *10*, 467. [CrossRef]
16. Avérous, L.; Pollet, E. Green nano-biocomposites. In *Environmental Silicate Nano-Biocomposites*; Avérous, L., Pollet, E., Eds.; Springer: London, UK, 2012; pp. 1–11, ISBN 978-1-4471-4101-3.
17. Chivrac, F.; Pollet, E.; Schmutz, M.; Avérous, L. New approach to elaborate exfoliated starch-based nanobiocomposites. *Biomacromolecules* **2008**, *9*, 896–900. [CrossRef] [PubMed]
18. Chen, B.; Evans, J.R.G. Thermoplastic starch–clay nanocomposites and their characteristics. *Carbohydr. Polym.* **2005**, *61*, 455–463. [CrossRef]
19. Ismail, H.; Pasbakhsh, P.; Fauzi, M.N.A.; Abu Bakar, A. Morphological, thermal and tensile properties of halloysite nanotubes filled ethylene propylene diene monomer (EPDM) nanocomposites. *Polym. Test.* **2008**, *27*, 841–850. [CrossRef]
20. Liu, M.; Guo, B.; Du, M.; Lei, Y.; Jia, D. Natural inorganic nanotubes reinforced epoxy resin nanocomposites. *J. Polym. Res.* **2008**, *15*, 205–212. [CrossRef]
21. Makaremi, M.; Pasbakhsh, P.; Cavallaro, G.; Lazzara, G.; Aw, Y.K.; Lee, S.M.; Milioto, S. Effect of morphology and size of halloysite nanotubes on functional pectin bionanocomposites for food packaging applications. *ACS Appl. Mater. Interfaces* **2017**, *9*, 17476–17488. [CrossRef] [PubMed]

22. Bertolino, V.; Cavallaro, G.; Lazzara, G.; Milioto, S.; Parisi, F. Biopolymer-targeted adsorption onto halloysite nanotubes in aqueous media. *Langmuir* **2017**, *33*, 3317–3323. [CrossRef] [PubMed]

23. Yang, Y.; Chen, Y.; Leng, F.; Huang, L.; Wang, Z.; Tian, W. Recent advances on surface modification of halloysite nanotubes for multifunctional applications. *Appl. Sci.* **2017**, *7*, 1215. [CrossRef]

24. Lazzara, G.; Cavallaro, G.; Panchal, A.; Fakhrullin, R.; Stavitskaya, A.; Vinokurov, V.; Lvov, Y. An assembly of organic-inorganic composites using halloysite clay nanotubes. *Curr. Opin. Colloid Interface Sci.* **2018**, *35*, 42–50. [CrossRef]

25. Liu, M.; Jia, Z.; Jia, D.; Zhou, C. Recent advance in research on halloysite nanotubes-polymer nanocomposite. *Prog. Polym. Sci.* **2014**, *39*, 1498–1525. [CrossRef]

26. Berahman, R.; Raiati, M.; Mehrabi Mazidi, M.; Paran, S.M.R. Preparation and characterization of vulcanized silicone rubber/halloysite nanotube nanocomposites: Effect of matrix hardness and HNT content. *Mater. Des.* **2016**, *104*, 333–345. [CrossRef]

27. Schmitt, H.; Prashantha, K.; Soulestin, J.; Lacrampe, M.F.; Krawczak, P. Preparation and properties of novel melt-blended halloysite nanotubes/wheat starch nanocomposites. *Carbohydr. Polym.* **2012**, *89*, 920–927. [CrossRef] [PubMed]

28. Xie, Y.; Chang, P.R.; Wang, S.; Yu, J.; Ma, X. Preparation and properties of halloysite nanotubes/plasticized *Dioscorea opposita* Thunb. starch composites. *Carbohydr. Polym.* **2011**, *83*, 186–191. [CrossRef]

29. Avérous, L.; Fringant, C.; Moro, L. Plasticized starch–cellulose interactions in polysaccharide composites. *Polymer* **2001**, *42*, 6565–6572. [CrossRef]

30. Muscat, D.; Adhikari, B.; Adhikari, R.; Chaudhary, D.S. Comparative study of film forming behaviour of low and high amylose starches using glycerol and xylitol as plasticizers. *J. Food Eng.* **2012**, *109*, 189–201. [CrossRef]

31. Yang, J.; Yu, J.; Ma, X. Study on the properties of ethylenebisformamide and sorbitol plasticized corn starch (ESPTPS). *Carbohydr. Polym.* **2006**, *66*, 110–116. [CrossRef]

32. Tang, S.; Zou, P.; Xiong, H.; Tang, H. Effect of nano-SiO$_2$ on the performance of starch/polyvinyl alcohol blend films. *Carbohydr. Polym.* **2008**, *72*, 521–526. [CrossRef]

33. Levis, S.R.; Deasy, P.B. Characterisation of halloysite for use as a microtubular drug delivery system. *Int. J. Pharm.* **2002**, *243*, 125–134. [CrossRef]

34. Zhu, J.; Li, L.; Chen, L.; Li, X. Study on supramolecular structural changes of ultrasonic treated potato starch granules. *Food Hydrocoll.* **2012**, *29*, 116–122. [CrossRef]

35. Nuessli, J.; Putaux, J.L.; Bail, P.L.; Buléon, A. Crystal structure of amylose complexes with small ligands. *Int. J. Biol. Macromol.* **2003**, *33*, 227–234. [CrossRef] [PubMed]

36. Van Soest, J.J.G.; Hulleman, S.H.D.; de Wit, D.; Vliegenthart, J.F.G. Changes in the mechanical properties of thermoplastic potato starch in relation with changes in B-type crystallinity. *Carbohydr. Polym.* **1996**, *29*, 225–232. [CrossRef]

37. Talja, R.A.; Helén, H.; Roos, Y.H.; Jouppila, K. Effect of various polyols and polyol contents on physical and mechanical properties of potato starch-based films. *Carbohydr. Polym.* **2007**, *67*, 288–295. [CrossRef]

38. He, Y.; Kong, W.; Wang, W.; Liu, T.; Liu, Y.; Gong, Q.; Gao, J. Modified natural halloysite/potato starch composite films. *Carbohydr. Polym.* **2012**, *87*, 2706–2711. [CrossRef]

39. Anglès, M.N.; Dufresne, A. Plasticized starch/tunicin whiskers nanocomposites. 1. Structural analysis. *Macromolecules* **2000**, *33*, 8344–8353. [CrossRef]

40. Chivrac, F.; Pollet, E.; Schmutz, M.; Avérous, L. Starch nano-biocomposites based on needle-like sepiolite clays. *Carbohydr. Polym.* **2010**, *80*, 145–153. [CrossRef]

41. Liu, M.; Guo, B.; Du, M.; Chen, F.; Jia, D. Halloysite nanotubes as a novel β-nucleating agent for isotactic polypropylene. *Polymer* **2009**, *50*, 3022–3030. [CrossRef]

42. Guo, B.; Zou, Q.; Lei, Y.; Du, M.; Liu, M.; Jia, D. Crystallization behavior of polyamide 6/halloysite nanotubes nanocomposites. *Thermochim. Acta* **2009**, *484*, 48–56. [CrossRef]

43. Tang, X.-G.; Hou, M.; Zou, J.; Truss, R. Poly(vinylidene fluoride)/halloysite nanotubes nanocomposites: The structures, properties, and tensile fracture behaviors. *J. Appl. Polym. Sci.* **2013**, *128*, 869–878. [CrossRef]

44. Bordes, P.; Pollet, E.; Avérous, L. Nano-biocomposites: Biodegradable polyester/nanoclay systems. *Prog. Polym. Sci.* **2009**, *34*, 125–155. [CrossRef]

45. Alexandre, M.; Dubois, P. Polymer-layered silicate nanocomposites: Preparation, properties and uses of a new class of materials. *Mater. Sci. Eng. R Rep.* **2000**, *28*, 1–63. [CrossRef]

46. Nikkhah, S.J.; Ramazani, S.A.A.; Baniasadi, H.; Tavakolzadeh, F. Investigation of properties of polyethylene/clay nanocomposites prepared by new in situ Ziegler–Natta catalyst. *Mater. Des.* **2009**, *30*, 2309–2315. [CrossRef]

47. Chrissafis, K.; Bikiaris, D. Can nanoparticles really enhance thermal stability of polymers? Part I: An overview on thermal decomposition of addition polymers. *Thermochim. Acta* **2011**, *523*, 1–24. [CrossRef]

48. Olivato, J.B.; Marini, J.; Pollet, E.; Yamashita, F.; Grossmann, M.V.E.; Avérous, L. Elaboration, morphology and properties of starch/polyester nano-biocomposites based on sepiolite clay. *Carbohydr. Polym.* **2015**, *118*, 250–256. [CrossRef] [PubMed]

49. Fu, H.; Wang, Y.; Li, X.; Chen, W. Synthesis of vegetable oil-based waterborne polyurethane/silver-halloysite antibacterial nanocomposites. *Compos. Sci. Technol.* **2016**, *126*, 86–93. [CrossRef]

50. Chivrac, F.; Pollet, E.; Dole, P.; Avérous, L. Starch-based nano-biocomposites: Plasticizer impact on the montmorillonite exfoliation process. *Carbohydr. Polym.* **2010**, *79*, 941–947. [CrossRef]

51. Gao, C.; Pollet, E.; Averous, L. Innovative plasticized alginate obtained by thermo-mechanical mixing: Effect of different biobased polyols systems. *Carbohydr. Polym.* **2017**, *157*, 669–676. [CrossRef] [PubMed]

52. Zhang, Y.; Han, J.H. Plasticization of pea starch films with monosaccharides and polyols. *J. Food Sci.* **2006**, *71*, E253–E261. [CrossRef]

53. Averous, L.; Moro, L.; Dole, P.; Fringant, C. Properties of thermoplastic blends: Starch–polycaprolactone. *Polymer* **2000**, *41*, 4157–4167. [CrossRef]

54. Jost, V.; Kobsik, K.; Schmid, M.; Noller, K. Influence of plasticiser on the barrier, mechanical and grease resistance properties of alginate cast films. *Carbohydr. Polym.* **2014**, *110*, 309–319. [CrossRef] [PubMed]

© 2018 by the authors. Licensee MDPI, Basel, Switzerland. This article is an open access article distributed under the terms and conditions of the Creative Commons Attribution (CC BY) license (http://creativecommons.org/licenses/by/4.0/).

MDPI

Article

Effects of Surface Functionalization of Lignin on Synthesis and Properties of Rigid Bio-Based Polyurethanes Foams

Xuefeng Zhang *, Dragica Jeremic, Yunsang Kim, Jason Street and Rubin Shmulsky *

Department of Sustainable Bioproducts, Mississippi State University, Mississippi State, MS 39762, USA; dragica.jn@gmail.com (D.J.); ysk13@msstate.edu (Y.K.); jason.street@msstate.edu (J.S.)
* Correspondence: njfuxf@gmail.com (X.Z.); rs26@msstate.edu (R.S.);
  Tel.: +1-662-325-2119 (X.Z.); +1-662-325-2243 (R.S.)

Received: 30 May 2018; Accepted: 25 June 2018; Published: 26 June 2018

Abstract: We report the preparation of lignin-based rigid polyurethane (RPU) foams from surface functionalized kraft lignin via a simple and environmentally benign process. Lignin was functionalized with polyisocyanate at 80 °C for 1 h, the resulting lignin-polyisocyanate prepolymer was confirmed by increased viscosity and Fourier-transform infrared spectroscopy (FTIR). The RPU foams containing up to 30% surface functionalized lignin as a substitute for petroleum-based polyols exhibited comparable thermal and mechanical properties to conventional RPU foams. The lignin-based RPU foams prepared from surface functionalization outperformed RPU foams without the surface functionalization, showing up to 47% and 45% higher specific compressive strength and modulus, respectively, with a 40% lignin substitution ratio. Thermal insulation and temperature-stability of the two types of the foams were comparable. The results indicate that the surface functionalization of lignin increases reactivity and homogeneity of the lignin as a building block in RPU foams. The life cycle assessment for the lignin-based RPU foams shows that the surface functionalization process would have overall lesser environmental impacts when compared with the traditional manufacturing of RPU foams with synthetic polyols. These findings suggest the potential use of surface functionalized lignin as a sustainable core material replacement for synthetic polyols in building materials.

Keywords: lignin; surface functionalization; rigid polyurethane foam; compressive strength and modulus; thermal insulation; building materials

---

1. Introduction

Rigid polyurethane (RPU) foams are one of the most commonly used polymeric materials in building applications such as wall panels, flooring, and structural insulated panels (SIPs) because of their low density, high dimensional stability, good adhesion, and excellent thermal insulation and mechanical properties [1]. Despite their superior properties, the use of petroleum-based polyols for the production of the RPU foams can be problematic due to the depletion of fossil fuel resources and negative environmental impacts of the industrial production of synthetic polyols [2]. Therefore, the development of renewable bio-based polyols derived from natural resources has garnered a strong global interest both in academia and industry.

Lignin is the most abundant aromatic biopolymer on Earth and is available in large quantities from wood-pulping and bio-ethanol industries [3–5]. Extensive efforts have been made to convert lignin to polyurethane compounds because lignin contains a large amount of aliphatic and aromatic hydroxyl groups [6–8]. The use of lignin as a polyol substitute in the preparation of RPU foams commonly follows two general approaches. In the first approach, lignin undergoes chemical modification processes, i.e., oxypropylation and liquefaction, to increase the number of hydroxyl groups with enhanced accessibility for reactions with polyisocyanates [9–15]. Although this approach increases the reactivity of lignin, it is difficult to prepare RPUs containing relatively high quantities of lignin [16]. In addition, the chemical modification processes increase the production cost due to the extensive use of energy and larger space requirements for accommodation of equipment [17], such as high pressure or microwave reactors [1,18], thereby making this approach less attractive to the polyurethane industries [17]. Increased environmental footprint from the chemical modification of lignin due to the use of strong acid or toxic substances is an additional concern. Alternatively, RPU foams can also be prepared by a relatively simple approach, in which unmodified lignin is mixed with petroleum-based polyols and directly combined with polyisocyanates. Various unmodified types of lignin i.e., lignosulfonates, kraft, hydrolysis, and organosolv lignin, have been used to replace up to 40 wt. % of petroleum-based polyols in RPU foams [7,19–22]. Although this simple and fast approach is appealing because of its potential to reduce the process cost, the use of unmodified lignin as a polyol substitute often negatively impacts the mechanical properties of the RPU foams [7,20,23]. The reduced mechanical properties of the lignin-containing RPU foams are generally attributed to low reactivity of lignin with polyisocyanates and to poor dispersion of lignin in the cellular structure of the foams [7,20].

In the present study, in order to overcome cost-associated issues related to lignin oxypropylation and liquefaction as well as the poor mechanical properties of unmodified lignin-based RPU foams, methyl diphenyl diisocyanate (MDI)-based polyisocyanate (pMDI) was used to functionalize kraft lignin. The highly reactive isocyanate group in pMDI can interact with lignin surface hydroxyl group to form lignin urethanes moiety, thereby improving phase compatibility when it would mix with polyol and polyisocyanate during RPU foam synthesis (Scheme 1). This may improve dispersion of lignin in the crosslinking RPU system, which is expected to improve the cellular structure uniformity and the mechanical properties of RPU foams.

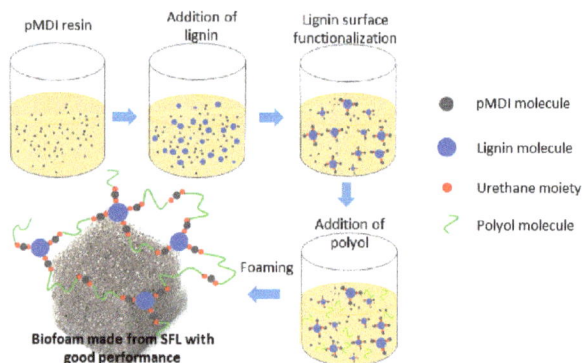

Scheme 1. Illustration of the process and formation mechanism of surface functionalized lignin-based rigid polyurethane (SFL-RPU) foam.

In this article, surface functionalized lignin was used to substitute 0–40% of conventional petroleum-based polyol for the synthesis of bio-based RPU foams. As-synthesized surface functionalized lignin-based RPU (SFL-RPU) foam was compared to conventional petroleum-based RPU foams and to non-functionalized lignin-based RPU (L-RPU) foams in regard to its morphological, physicochemical, mechanical, and thermal properties. The SFL-RPU foams with up to 30% lignin-to polyol substitution ratio exhibited comparable thermal and mechanical properties when compared to conventional RPU foams. The specific compressive strength and modulus of the SFL-RPU foam surpassed the performance of L-RPU foams up to 47% and 45%, respectively. These findings along with decreased environmental cost supported by life cycle assessment suggest potential use of surface functionalized lignin as a sustainable core material replacing synthetic polyols in building applications.

2. Materials and Methods

2.1. Materials

A commercial USA softwood kraft lignin was used as lignin source for RPU foam preparation. The lignin sample was oven-dried at 105 °C for 24 h before it was used for the preparation of RPU foams. Commercial polyester polyol (R-23-015) and pMDI (A-23-015) were purchased from NCFI Polyurethanes (Mount Airy, NC, USA). Based on the specification provided by NCFI Polyurethanes, the polyol R-23-015 contains tertiary amine catalyst (<1%) and a physical blowing agent containing 1,1,1,3,3-Pentafluoropropane (<3%) and 1,1,1,3,3-Pentafluorobutane (<10%). The total amount of blowing agents in polyol R-23-015 measured by a gravimetric method in our laboratory was 9%. For lignin-based RPU foams, Dabco® 33-LV (33 wt. % solution of triethylenediamine in dipropylene glycol, Sigma Aldrich, St. Louis, MO, USA) was used as a catalyst. 1,1,1,3,3-Pentafluorobutane (Fisher Scientific, Hampton, NH, USA) was used as a physical blowing agent for the preparation of lignin-based RPU foams. Other chemicals were purchased from Sigma-Aldrich as analytical grade and used as received.

2.2. Surface Functionalization of Lignin

Surface functionalization of lignin was carried out by pMDI under heat. Lignin, silicone oil, and excess pMDI were mixed into a 300-mL glass jar and stirred in a vacuum desiccator subjected to two cycles of vacuum (5 min at 250 Torr), followed by nitrogen gas (>99.99%) purging, to prepare a lignin-pMDI premix (ratios of the components listed in Table 1). The lignin-pMDI premix was then heated in a water bath at 80 °C for 1 h under stirring (800 rpm) to initiate surface functionalization. At the end of reaction, the mixture of surface functionalized lignin (SFL) and pMDI were termed as lignin-pMDI prepolymer, which was used immediately for SFL-RPU foam preparation.

Table 1. Amounts of components used for the preparation of RPU foams.

Lignin Substitution Ratio (%)	Lignin-pMDI Premix (g)				Polyol Premix (g)		NCO Index
	Lignin	pMDI	Silicone Oil *	Polyol R-23-015	Dabco 33-LV *	1,1,1,3,3-Pentafluorobutane *	
0	0	51.5	0	50	0	0	157
10	5	51.5	0.25	45	0.025	0.45	156
20	10	51.5	0.50	40	0.050	0.90	154
30	15	51.5	0.75	35	0.075	1.35	153
40	20	51.5	1.00	30	0.100	1.80	152

Note: The amounts of silicone oil, Dabco® 33-LV, and 1,1,1,3,3-Pentafluorobutane are 5%, 0.5%, and 9% in terms of lignin weight, respectively.

2.3. Preparation of RPU Foams

Prior to the preparation of lignin-based RPU foams, 1,1,1,3,3-Pentafluorobutane and Dabco® 33-LV were added into polyol R-23-015 in appropriate amounts (Table 1) in order to keep their ratios in pure R-23-015 and polyol/lignin mixtures constant. The polyol-premix was then sealed in a glass jar and shaken vigorously for 10 min, before conditioning for at least 24 h at 10 °C.

The RPU foams were prepared by adding polyol-premix into either lignin-pMDI premix or preprolymer, and stirring the combination at 2500 rpm for 10 s. Two premix mixtures were prepared for each formulation. The NCO index (isocyanate group equivalents/hydroxyl group equivalents × 100) between 150 and 160, as recommended by NCFI Polyurethanes, was used for all foams (Table 1). Each of the resultant mixtures were then immediately transferred into a $15 \times 15 \times 15$ cm^3 open mold and allowed to freely rise at room temperature.

All RPU foams were conditioned for at least 48 h under ambient conditions. The RPU foams prepared using SFL with the lignin-to-polyol substitution ratio of 10, 20, 30, and 40% were labeled as SFL-RPU10, SFL-RPU20, SFL-RPU30, and SFL-RPU40, respectively. The RPU foams prepared using non-functionalized lignin were labeled as L-RPU10, L-RPU20, L-RPU30, and L-RPU40, for the equivalent amount of lignin. The reference foam without lignin was labeled as RPU0. The foams were cut into samples of different sizes, depending on the type of analysis.

2.4. Lignin Structure Characterization

The ash content of kraft lignin was measured following American Society for Testing Materials (ASTM) D1102. In a typical example, three replicates of 1 g of lignin sample was calcined in a muffle furnace at 525 °C for at least 4 h.

Size exclusion chromatography (SEC) of acetylated lignin was used to determine molecular weight and polydispersity index of lignin. Acetylation was performed by suspending approximately 500 mg of oven-dry lignin sample in 20 mL of pyridine/acetic acid (v/v = 1:1) solution and stirring at room temperature. After 48 h, 12.5 mL of hydrochloric acid (0.1 mol/L) was gradually added to the cooled solution and kept at 0 °C in an ice bath in order to precipitate acetylated lignin. The precipitates were washed and neutralized with excessive deionized water, and then vacuum-dried at 40 °C overnight. SEC analysis of three lignin replicate samples was carried out by injecting 40 μL of the freshly filtered (0.45 μm PTFE syringe filter) acetylated lignin solution (1 mg/mL in tetrahydrofuran) into a PLgel 10 μm Mixed-D column (300 × 7.5 mm, Agilent Technologies, Inc., Santa Clara, CA, USA) equilibrated at 30 °C, and using tetrahydrofuran as an eluent at a flow rate of 1 mL/min. The analysis was performed on Thermo Finnigan Surveyor system coupled with Refractomax 520 refractive index (RI) detector. The calibration of the system was performed with polystyrene standards (Polymer Standards Servic-USA, Inc., Amherst, MA, USA) under the same conditions.

Lignin phosphitylation and phosphorus nuclear magnetic resonance (^{31}P-NMR) analysis were performed to quantify hydroxyl functional groups in lignin. For phosphitylation, 20 mg of lignin sample replicates was dissolved in 0.5 mL of pyridine/CDCl$_3$ solution (1.6:1 by v/v) containing 1 mg/mL of chromium acetylacetonate (relaxation agent) and 2 mg/mL of endo-N-hydroxy-5-norbornene-2,3-dicarboximide (internal standard), followed by adding 0.1 mL of 2-Chloro-1,3,2-dioxaphopholane (phosphitylation reagent) into the solvent. Quantitative ^{31}P-NMR spectra were acquired after in situ phosphitylation of the lignin sample. ^{31}P-NMR analysis of the lignin replicates was carried out using a Bruker Avance 400 MHz NMR spectrometer (Bruker, Inc., Billerica, MA, USA) housed in The Department of Chemistry, Mississippi State University. The NMR spectra were acquired at 25 °C using an inverse gated decoupling pulse sequence with a 90° pulse angle, 5 s pulse delay, and 128 scans. The same method was applied for determination of commercial polyol R-23-015 hydroxyl content.

2.5. Characterization of SFL

The rheological properties of 2 replicates of lignin-pMDI premix and prepolymer were determined using an AR 1500EX (TA Instruments, New Castle, DE, USA) rheometer equipped with a DIN concentric at an angular frequency of 0.1 rad/s were recoded for comparison.

The SFL was extracted from lignin-pMDI prepolymer (20 wt. % lignin content) by washing the lignin-pMDI prepolymer with benzene to remove the unreacted pMDI. The Fourier-transform infrared spectroscopy (FTIR) spectra of five sub-samples of the SFL were recorded with Spectrum Two attenuated total reflection (ATR) spectrometer (PerkinElmer, Waltham, MA, USA) over the range of 720 to 4000 cm^{-1} (at a resolution of 2 cm^{-1}) at the average of 10 scans. The spectra were baseline-corrected manually using Spectrum® Quant software (PerkinElmer, Waltham, MA, USA). The baseline-corrected spectra were normalized over the total spectral area and analyzed by principal component analysis (PCA) using Unscrambler 10.3 X (CAMO Software Inc., Magnolia, TX, USA).

2.6. Characterization of RPU Foams

The apparent density of RPU foams was determined according to ASTM D1622-08. The cell morphology of the foams was analyzed by a scanning electron microscopy (SEM, JSM-6110 LV, JOEL, Akishima, Tokyo). A cross-section parallel to the foam rise direction was examined under the SEM.

The samples ($10 \times 10 \times 2$ mm^3) were sputter coated with 25 nm-thick platinum prior to SEM imaging. The average cell diameter was estimated from more than 200 measurements of each foam by using ImageJ software.

Thermal conductivity of foams was determined using a KD2 Pro (Decagon Devices Inc., Pullman, WA, USA) thermal conductivity meter. Ten measurements were conducted for one sample ($50 \times 50 \times 100$ mm^3; 100 mm in rise direction) cut out of one foam of each sample group. The measurements were performed over 10 min, as suggested in previous reports [18].

To verify if the surface functionalization could increase the crosslinking density between SFL and pMDI, the RPU foams were extracted with 1,4-dioxane-water binary solutions (dioxane/water = 8/2, *w/w*), a good solvent of lignin. In the experiment, the foam was cut into three small pieces of approximately $8 \times 8 \times 8$ mm^3 and extracted with dioxane in a Soxhlet extractor for 24 h. The weight loss of RPU foams (of its original weight) after extraction was calculated for comparison. The mechanical properties of the foams were measured on eight replicates ($50 \times 50 \times 25$ mm^3; 25 mm in rise direction), cut from two foam samples of each sample group. The measurements were carried out by Instron 3382 universal testing machine (Instron Corp., Norwood, MA, USA). Compressive strength (σ) and compressive modulus (E) in direction parallel to foam rise were measured and calculated according to ASTM D 1621-10 (crosshead movement rate at 2.5 mm/min). The specific compressive strength (σ_ρ, compressive strength per unit density) and specific modulus (E_ρ, compressive modulus per unit density) were calculated using the following equations:

$$\sigma_\rho (kPa\ kg^{-1}\ m^3) = \frac{\sigma}{\rho} \tag{1}$$

$$E_\rho (kPa\ kg^{-1}\ m^3) = \frac{E}{\rho} \tag{2}$$

where ρ is the density of foams.

In order to quantify the effects of lignin surface functionalization on the foam's specific compressive properties, the following equations were used to calculate the changes of σ_ρ and E_ρ ($\Delta\sigma_\rho$ and ΔE_ρ) at a given lignin substitution ratio:

$$\Delta\sigma_\rho\ (\%) = \frac{\sigma_\rho^{SFL} - \sigma_\rho^{L}}{\sigma_\rho^{L}} \times 100 \tag{3}$$

$$\Delta E_\rho\ (\%) = \frac{E_\rho^{SFL} - E_\rho^{L}}{E_\rho^{L}} \times 100 \tag{4}$$

where σ_ρ^{SFL} is specific compressive strength of SFL-RPU foam (L represents lignin-to-polyol substitution ratio, 10 to 40%); σ_ρ^{L} is specific compressive strength of L-RPU foam; E_ρ^{SFL} is specific compressive modulus of SFL-RPU foam (L represents lignin substitute ratio); E_ρ^{L} is specific compressive modulus of L-RPU foam.

FTIR spectra of three sub-samples of the foams were recorded with Spectrum Two ATR spectrometer (PerkinElmer, Waltham, MA, USA) in the range of 450 to 4000 cm^{-1} (at a resolution of 2 cm^{-1}) as an average of 10 scans. The spectra were baseline-corrected by "Data Tune-up" transformation option, and normalized with respect to aromatic rings band (at 1510 cm^{-1}) using Spectrum® Quant software. The intensity ratio of N=C=O band to C=O band (I_{NCO}/I_{CO}) was calculated based on the normalized spectra.

Thermogravimetric analyses (TGA) of the foams were performed by 50 H thermo-gravimetric analyzer (TA Instruments, New Castle, DE, USA). Two replicates (~5 mg each) of each sample type were heated from 50 °C to 750 °C at a ramping rate of 10 °C/min in flowing nitrogen (99.99%, 100 mL/min) atmosphere under ambient pressure. The differences between the replicates were negligible in comparison to the differences between the samples.

The foam density, cell size, thermal conductivity, mechanical properties, and I_{NCO}/I_{CO} were analyzed and compared by analysis of variance (ANOVA) using SAS software (SAS Institute, Cary, NC, USA). A one-factor ANOVA was performed to analyze significance of differences among the sample groups. All statistical analyses were performed at a significance level of 0.05.

2.7. Life Cycle Assessment

The life cycle assessment (LCA) was performed with openLCA 1.6.3 (GreenDelta, Berlin, Germany). The ecoinvent database 2.2 was used to model the life cycle inventory concerning the process flows. The kraft lignin lifecycle data was downloaded from the USDA LCA Commons public domain from previous research [24]. Tables S1–S4 shows the inventory data from the ecoinvent database, which was slightly modified so that equivalent comparisons of the processes could be made. Table S5 shows the inventory data from the kraft lignin lifecycle dataset [25]. Electricity needed to produce the polyurethane and waste heat generation were ignored involving the polyurethane formation. The tool for reduction and assessment of chemicals and other environmental impacts (TRACI—from the EPA) life cycle impact assessment (LCIA) protocol used to determine the effects of the flows used in this study.

3. Results and Discussion

3.1. Lignin Properties

Ash content, molecular weight, and hydroxyl value of lignin are important parameters as they influence overall properties of the resultant RPU foams. The hydroxyl content and molecular weight of lignin were estimated from ^{31}P-NMR spectrum (shown in Figure S1 in Supporting Information) and SEC analysis, and the results summarized in Table 2. The comparable hydroxyl contents of lignin (5.24 mmol/g) and the commercial polyol (5.34 mmol/g) suggest potential of lignin to be a good alternative for polyols in the case of preparation of RPU foams. However, the higher content of phenolic hydroxyl groups in lignin (58% of total OH) in comparison to commercial polyol (2.1% of total OH) indicates lower reactivity of lignin.

Table 2. Composition of lignin and polyol R-23-015.

Samples	Ash (wt. %)	Mn (g/mol)	Mw (g/mol)	Mw/Mn	Al-OH (mmol/g)	Ph-OH (mmol/g)	COOH (mmol/g)	Total OH (mmol/g)
Lignin	1.65 ± 0.04	816	3374	4.13	1.91 ± 0.12	3.04 ± 0.21	0.29 ± 0.02	5.24 ± 0.35
Polyol	0	-	-	-	5.21 ± 0.25	0.11 ± 0.01	0.02 ± 0.00	5.34 ± 0.25

3.2. Rheological Behavior of Lignin-pMDI Premix and Prepolymer

Figure 1a shows the viscosity of the lignin-pMDI premix and the prepolymer with respect to shear rate. The viscosity is strongly dependent on the lignin content in the mixture. The viscosity of mixture at a shear rate of 10 s^{-1} increased from 0.2 to 1.2 Pa·s upon addition of lignin in the amount of 40%, probably due to the filler effects of lignin [26]. At the same lignin content, the lignin-pMDI prepolymer exhibited apparently higher viscosity than lignin-pMDI premix, which suggests lignin was successfully functionalized by pMDI and formed lignin-pMDI macromolecules. Figure 1b shows the storage (G′) and loss (G″) modulus of various lignin-pMDI premixes and the prepolymers at an angular frequency of 0.1 rad/s. Both G′ and G″ were found increased with the increase of lignin content. The increase of G′ and G″ at high lignin content could be related to the filler effects of lignin particles and the interaction of lignin to pMDI [27]. At the same lignin content, the lignin-pMDI prepolymer exhibited higher G′ and G″ than lignin-pMDI premix, which also suggests the surface functionalization of lignin and the formation of cross-linked lignin-pMDI macromolecules in lignin-pMDI prepolymers.

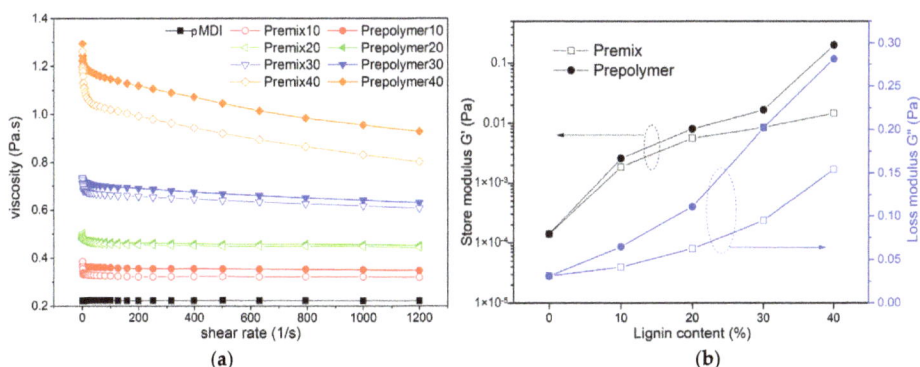

Figure 1. (a) Viscosity vs. shear rate curves of raw polyisocyanate methyl diphenyl diisocyanate (pMDI) resin (black), lignin-pMDI premix (hollow symbols), and lignin-pMDI prepolymer (solid symbols) containing 10–40% lignin; (b) storage (G′) and loss (G″) modulus of various lignin-pMDI premixes and the prepolymers at an angular frequency of 0.1 rad/s.

Our hypothesis was that the isocyanate groups in pMDI react with lignin surface hydroxyl groups to form urethane bonds on lignin surface. To verify this hypothesis, the extracted SFL from lignin-pMDI prepolymer was investigated by FTIR and the data was analyzed by PCA. Figure 2 shows the normalized FTIR spectra of pristine lignin and the SFL. The SFL shows higher vibration intensity of the isocyanate (–N=C=O, 2278 cm^{-1}), carbonyl (–C=O, 1630 cm^{-1}) and amine –NH (1530 cm^{-1}) bands [28]. The score plot (Supporting Figure S2a) of FTIR spectra shows clusters of the two sample groups according to PC1, where pristine lignin has negative loading on PC1 and the SFL have positive loading on PC1. The loading plot (Supporting Figure S2b) shows positive values at absorption bands of –C=O (1630 cm^{-1}), –N=C=O (2278 cm^{-1}), and –NH (1530 cm^{-1}), confirming presence of the isocyanate groups in SFL. The shift of N=C=O and NH from 2242 and 1522 cm^{-1} in MDI (Supporting Figure S3) to 2278 and 1530 cm^{-1} in SFL suggests and the formation of urethane bonds between lignin and pMDI.

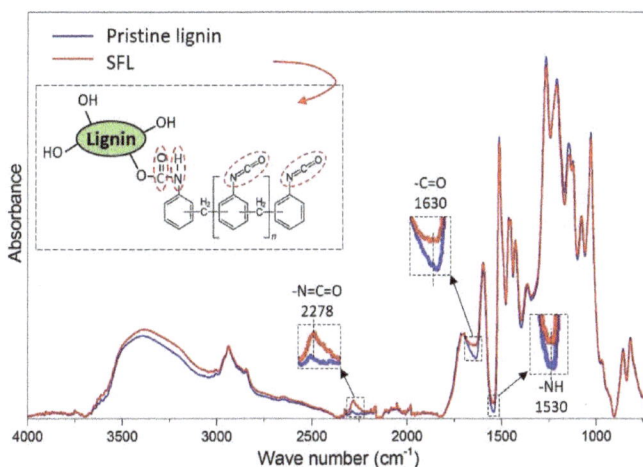

Figure 2. Baseline corrected and normalized Fourier-transform infrared spectroscopy (FTIR) spectra of pristine lignin and SFL.

3.3. Physical Properties of RPU Foams

The photographs and the SEM images of RPU foams are shown in Figure 3 and Supporting Figure S4. RPU foams became darker with increase in lignin substitution ratio, which can be attributed to the light-absorbing property of lignin. The dark color of lignin-based RPU foams is acceptable because the foams are used as a core material in building applications. The number of small particles (dark spots in the images) and agglomerates observed in the lignin-based RPU foams, increased with the rise of lignin content. RPU foams were also scanned by SEM to determine homogeneity of the cellular structure. RPU foams with up to 20 wt. % lignin replacement ratio revealed homogeneous cell structure (Figure 3). With the increase in lignin content, the cell shape became inhomogeneous and less regular, with higher number of defective and distorted cells. The SFL-RPU foams displayed more defective cells and lignin agglomerates (indicated by red arrows) than L-RPU foams, which indicates that the lignin surface functionalization can increase the isotropy of lignin-based foam cellular structure.

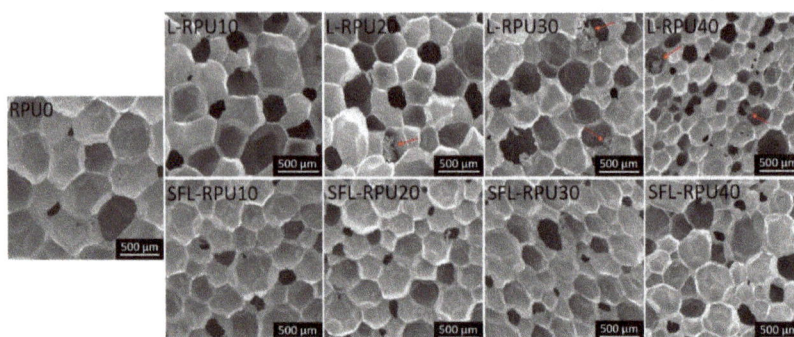

Figure 3. Scanning electron microscopy (SEM) images of L-RPU (**top row**) and SFL-RPU (**bottom row**) foams made with different amounts (0–40%) of lignin.

Table 3 lists the properties of the lignin-based (L-RPU and SFL-RPU) and the reference (RPU0) foams including cell diameter, apparent density, and thermal conductivity. The average cell diameter decreased significantly with the increase of lignin content. The apparent density of RPU foams increased significantly with the increase of lignin substitution ratio.

Table 3. Cell diameter, density, and thermal conductivity of RPU foams containing 0–40% lignin.

Samples	Cell Diameter (μm)	Density (kg/m^3)	Thermal Conductivity (mW·m^{-1}·K^{-1})
RPU0	588 ± 85	34.6 ± 1.8	24.1 ± 0.5
L-RPU10	548 ± 154	33.9 ± 2.3	23.8 ± 0.4
L-RPU20	448 ± 135	36.2 ± 1.9	24.0 ± 0.4
L-RPU30	393 ± 93	41.7 ± 1.7	23.9 ± 0.3
L-RPU40	255 ± 67	53.3 ± 3.3	25.0 ± 0.7
SFL-RPU10	419 ± 98	35.5 ± 1.3	24.0 ± 0.6
SFL-RPU20	347 ± 132	38.5 ± 1.8	23.6 ± 0.5
SFL-RPU30	305 ± 135	41.6 ± 1.8	23.3 ± 0.5
SFL-RPU40	304 ± 99	42.2 ± 1.6	24.2 ± 0.4

At the 40% of lignin substitution ratio, the density increased by 58% and 25% in L-RPU40 and SFL-RPU40 foams, respectively. For the same foaming parameters (i.e., NCO index and amount of blowing agent), the higher foam density indicates the poor foaming process, which can be attributed to lignin particle agglomeration. Thus, the lower density of SFL-RPU40 than L-RPU40 echoes with the

SEM observations and confirms our hypothesis that surface functionalization improves dispersion of lignin in the crosslinking RPU system. The increase in lignin substitution ratio resulted in higher density and smaller cell diameter of foams was also reported by Luo et al. and Camila et al. [17,29]. This can be attributed to three reasons. Firstly, the addition of lignin increases the mixture viscosity (Figure 1) which restrains the expansion of the pores with smaller diameters, and reduces the free rise volume of foams [17,29,30]. Secondly, lignin has lower reactivity than a commercial polyol and the gelation reaction rate is lower, which allows more gases to escape from the foam structure, consequently decreasing the pore size and free rise volume [22]. Lastly, lignin powder could act as a nucleation site to facilitate the nucleation of bubbles and lead to a smaller cell size [18].

Thermal conductivity is an important factor for the application of RPU foams as building materials and it is closely related to the foam density and cell diameter. Generally, the thermal conductivity value is directly proportional to the foam apparent density and inversely proportional to the cell size. The thermal conductivity values of lignin-based RPU foams meet the thermal conductivity requirements for structural insulation materials specified in ASTM E1730. Compared with RPU0, the lignin-based RPU foams with up to 30% lignin substitution ratio showed slight reduction in thermal conductivity, which can be attributed to the cell size reduction. However, the thermal conductivity of L-RPU40 was higher than RPU0, which may be due to high density and defective cells. Therefore, the use of lignin in the substitution ratio up to 30% seems beneficial for increase of thermal insulation properties of as-synthesized RPU foams.

To verify if the surface functionalization could increase the crosslinking density between SFL and pMDI, the RPU foams were extracted with a dioxane-water solution. The weight loss of RPU foams after extraction was calculated and shown in Figure 4. The reference foam (RPU0) is found to have the lowest weight loss of 1.10% after the extraction. With the increase of lignin substitution ratio from 10% to 40%, the weight loss increased from 3.57 to 6.89 and 3.36 to 5.34% for L-RPU and SFL-RPU foams, respectively. At the same lignin substitution ratio, the SFL-RPU foams exhibit lower weight loss than that of L-RPU, which confirms the surface functionalization increased the crosslinking density between SFL and pMDI in the foams.

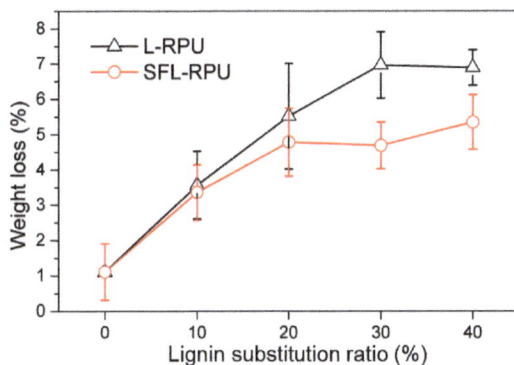

Figure 4. Weight loss of PU0, SFL-RPU and L-RPU foams after extraction.

3.4. Mechanical Properties of RPU Foams

The RPU foams obtained in this study showed average compressive strength of 136 to 205 kPa (Table S6), which meets the compressive properties requirements for insulation materials for wall panel applications specified in ANSI/APA PRS 610.1-2013. It is worthwhile to mention that the compressive strength (σ) and modulus (E) of the RPU foam prepared with 30% functionalized lignin (SFL-RPU30) are above 200 kPa, which is higher than what has been reported by Camila et al. (<100 kPa) and Xue et al. (140 kPa) [17,20].

It is well known that the mechanical strength of RPU foams is proportional to foam apparent density [17,31]. As described before, the foam apparent density increased with the increase of lignin substitution ratio. Thus, to exclude the effect of density on the σ and E of the foams, the specific compressive strength (σ_ρ, compressive strength per unit density) and specific modulus (E_ρ, compressive modulus per unit density) were calculated (Table S6). All RPU foams prepared without surface functionalization (L-RPU10 to L-RPU40) showed lower σ_ρ and E_ρ (Table S6 and Figure 5a) than the reference (RPU0), and in the case of L-RPU30 and L-RPU40 foams, the differences were significant. This is attributed to the lower reactivity and poor dispersion of lignin than substituted polyols that inevitably decreased the density of cross-linked polyurethane chains in the foams. Additionally, the introduction of lignin aggregates would increase the number of defective foam cells and decrease the uniformity of the foam cellular structure, as shown in SEM images in Figure 3. By contrast, the σ_ρ and E_ρ of the SFL-RPU foams with up to the 30% substitution ratio showed slightly higher values than that of the reference RPU0 (Table S6 and Figure 5b), although the differences were not significant. As the lignin ratio increased from 0 to 20%, the σ_ρ and E_ρ of RPU foams showed a trend of increasing values. Further increase in lignin content leads a reduction in both values.

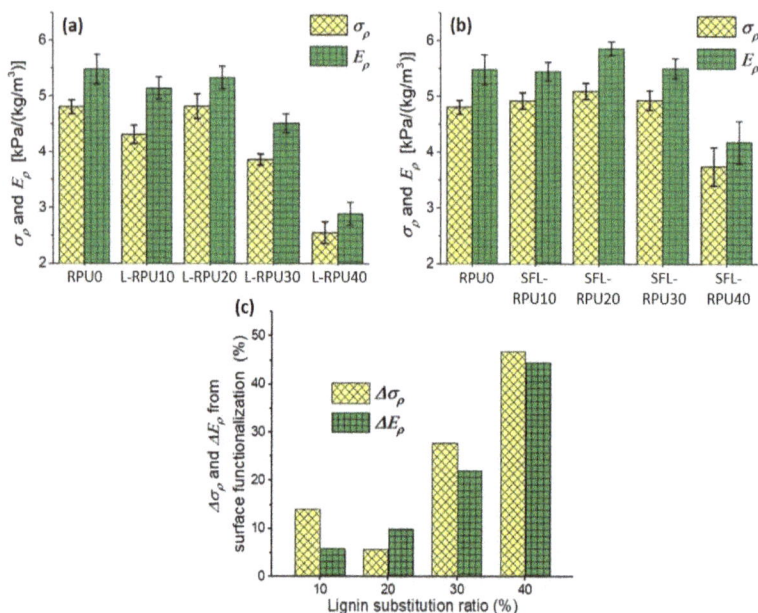

Figure 5. Specific compressive strength (σ_ρ) and modulus (E_ρ) of L-RPU (**a**) and SFL-RPU (**b**) containing different lignin substitution ratios, and the change of σ_ρ and E_ρ ($\Delta\sigma_\rho$ and ΔE_ρ) of lignin-containing RPU foams at given lignin substitution ratios (**c**).

For the RPU foams containing lignin substitute of up to 20%, there was no significant difference between L-RPU and SFL-RPU foams. However, the foams containing lignin substitute greater than 30%, the SFL-RPU foams showed significantly higher σ_ρ and E_ρ (Table S6) than L-RPU foams. As shown in Figure 5c, the change in specific compressive strength, $\Delta\sigma_\rho$ was 28% and 47%, and the change in specific modulus, ΔE_ρ, was 22% and 45% for the foams with lignin substitute of 30% and 40%, respectively. We believe the improvement in σ_ρ and E_ρ of the RPU foams prepared using SFL is attributed to the increase in crosslinking density of RPU foams and enhanced compatibility of lignin and polyisocyanate, as well as the good dispersion of lignin in foam cellular structure, as indicated by the SEM images in Figure 3.

3.5. FTIR Analysis of RPU Foams

Figure 6 shows the baseline corrected and normalized FTIR spectra of the all RPU foams. The urethane-related linkages such as N–H (3200–3400 cm^{-1}), C=O (1714 cm^{-1}), C–O (1069 cm^{-1}), and C–N (1218 cm^{-1}) peaks can be observed in all spectra [10,18]. The peaks at 1510 and 1596 cm^{-1} can be attributed to the vibration bands of aromatic rings originating from pMDI, polyester polyol, and lignin raw materials [18]. Additionally, the isocyanate peak at 2274 cm^{-1}, observed in all the RPU foams, is attributed to the residual isocyanate groups. This residual isocyanate group is expected to exist in the foams since the amount of MDI available for the reaction is greater than stoichiometric amount for the polyols added (initial NCO index was ~155).

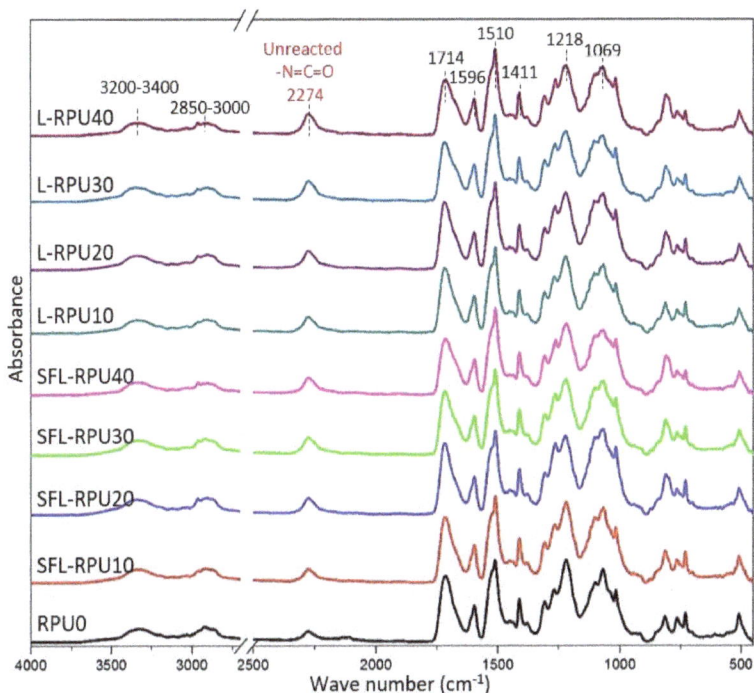

Figure 6. FTIR spectra RPU foams containing 0–40% lignin.

During the formation of RPU foams, isocyanate groups in pMDI undergo urethane formation with polyol and lignin hydroxyls (Scheme 1), and as a result, the intensity of N=C=O (I_{NCO}) decreases and intensity of C=O (I_{CO}) increases [32]. Thus, the amount of residual, i.e., unreacted pMDI in the foams prepared from different lignin amounts can be estimated through the peak intensity ratios (I_{NCO}/I_{CO}) shown in Table S7. At the given lignin substitution ratio, the I_{NCO}/I_{CO} of SFL-RPU foams is always lower than that of L-RPU, which indicates improved reactivity of SFL toward urethane formation in the foams and echoes with the overall improved mechanical properties of the SFL-RPU foams.

3.6. Thermal Stability of RPU Foams

Thermogravimetric (TG) curves of the RPU foams are displayed in Figure 7. All the foam samples degraded in one broad temperature range of 150–650 °C (with the maximum derivative thermogravimetric (DTG) peak at ~310 °C). The thermal degradation temperatures and char yield are presented in Table S8. The onset of thermal degradation temperature ($T_{5\%}$, defined as the temperature at 5% mass loss) shifted to higher temperature as the lignin content increased. This improved thermal stability is possibly a result of the highly cross-linked thermostable segments produced from the interactions between lignin and the foam matrix [12,31]. The maximum mass loss rate (DTG-max) also showed a decreasing trend as lignin content in the foams increased. This is mainly due to the higher aromatic density in the lignin-based RPU foam network [12]. The char yield increased with the increase of lignin content in foams, which is also due to the high aromatic density and thermostable nature of lignin. These properties confirmed that use of lignin is beneficial for production of thermal insulating foams.

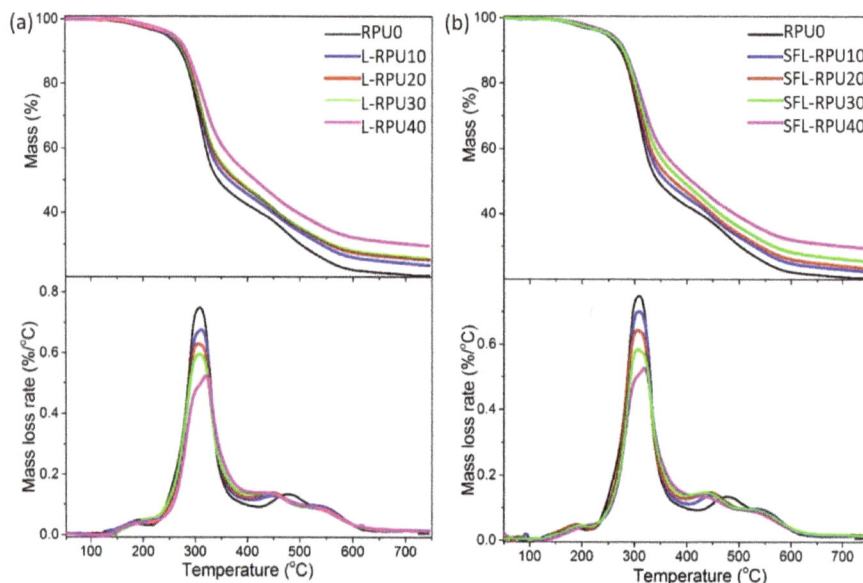

Figure 7. Thermogravimetry (TG) and the maximum derivative thermogravimetric (DTG) curves of L-RPU (**a**) and SFL-RPU (**b**) foams containing 0–40% lignin.

3.7. Life Cycle Assessment

Although the use of lignin as a substitute for synthetic polyols in the production of RPU foams is not new, chemical modification of lignin to polyol such as liquefaction and oxypropylation usually involves harmful chemicals including H_2SO_4 and KOH, which leaves a substantial environmental footprint. In contrast, the proposed lignin surface functionalization with isocyanates is free of those toxic substances. A cradle-to-grave analysis was performed to quantitatively assess environmental impacts associated with the production of RPU foams with using kraft lignin. As summarized in Figure 8, the LCA results showed that the lignin surface functionalization could surpass traditional manufacturing of polyurethane in both environmental and human health areas of concern such as acidification, eutrophication, global warming, photochemical oxidation, carcinogenics, and respiratory effects. The lignin surface functionalization had a lesser effect on the environment concerning ecotoxicity when compared to both the liquefaction and oxypropylation processes. The lignin surface functionalization also had a less of an effect on human health concerning carcinogen creation when compared with the liquefaction and oxyproplyation processes and less of an effect on non-carcinogens when compared with the oxypropylation process. Details outlining the measurements of the units concerning the impact and the model output stipulations were shown in Tables S9 and S10, respectively.

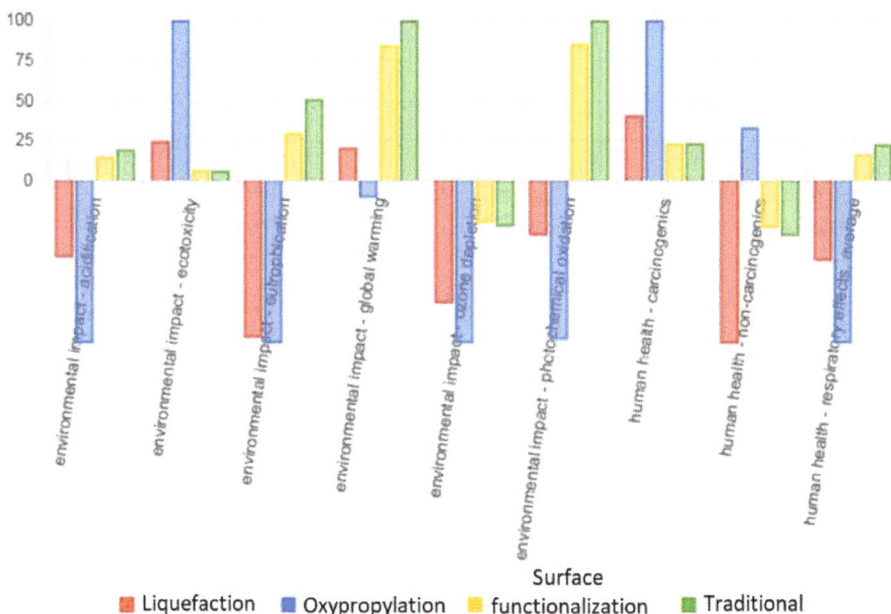

Figure 8. Life Cycle Assessment comparison of liquefaction, oxypropylation, surface functionalization, and traditional processing to manufacture polyurethane. For each indicator, the maximum result is set to 100% and the results of the other variants are displayed in relation to this result.

4. Conclusions

In summary, a simple lignin surface functionalization with polyisocyanate poses an attractive strategy for the development of high-lignin-content RPU foams for construction and structural applications. Lignin-based RPU foams with as high as 30% of lignin substitution of petroleum-based polyol exhibited comparable thermal and mechanical properties to conventional RPU foams. The mechanical performance of the SFL-RPU foams outperformed the L-RPU foams by up to an ~50% increase in compressive strength. Lignin surface characteristics had changed during the surface functionalization due to the conversion of lignin hydroxyl groups into lignin urethanes moiety, thereby enabling good lignin dispersion and reactivity, which is beneficial for the enhancement of RPU foam mechanical properties.

Supplementary Materials: The supplementary materials are available online at www.mdpi.com/2073-4360/10/7/706/s1.

Author Contributions: X.Z. conducted the overall process of the experimental design, characterization, data analysis, and the manuscript drafting. D.J. and Y.K. assisted with the experiments, reviewed the manuscript, and made comments. J.S. conducted the life cycle assessment for this research. R.S. supervised the whole project, reviewed the manuscript, and made comments.

Funding: This research received no external funding.

Acknowledgments: This work was supported by the U.S. Forest Service Wood Innovations Grant No. 16-DG-11083150-052. This material is supported by the National Institute of Food and Agriculture, U.S. Department of Agriculture, and McIntire Stennis under accession number 1009735. This manuscript is publication #SB932 of the Department of Sustainable Bioproducts, Mississippi State University.

Conflicts of Interest: The authors declare no conflicts of interest.

References

1. Li, Y.; Ragauskas, A.J. Kraft lignin-based rigid polyurethane foam. *J. Wood Chem. Technol.* **2012**, *32*, 210–224. [CrossRef]
2. Li, H.; Mahmood, N.; Ma, Z.; Zhu, M.; Wang, J.; Zheng, J.; Yuan, Z.; Wei, Q.; Xu, C. (Chunbao) Preparation and characterization of bio-polyol and bio-based flexible polyurethane foams from fast pyrolysis of wheat straw. *Ind. Crops Prod.* **2017**, *103*, 64–72. [CrossRef]
3. Zhang, X.; Yan, Q.; Leng, W.; Li, J.; Zhang, J.; Cai, Z.; Hassan, E.B. Carbon nanostructure of kraft lignin thermally treated at 500 to 1000 °C. *Materials* **2017**, *10*, 975. [CrossRef] [PubMed]
4. Zhang, X.; Yan, Q.; Hassan, E.B.; Li, J.; Cai, Z.; Zhang, J. Temperature effects on formation of carbon-based nanomaterials from kraft lignin. *Mater. Lett.* **2017**, *203*, 42–45. [CrossRef]
5. Zhang, X.; Yan, Q.; Li, J.; Chu, I.-W.; Toghiani, H.; Cai, Z.; Zhang, J. Carbon-based nanomaterials from biopolymer lignin via catalytic thermal treatment at 700 to 1000 °C. *Polymers* **2018**, *10*, 183. [CrossRef]
6. Griffini, G.; Passoni, V.; Suriano, R.; Levi, M.; Turri, S. Polyurethane coatings based on chemically unmodified fractionated lignin. *ACS Sustain. Chem. Eng.* **2015**, *3*, 1145–1154. [CrossRef]
7. Pan, X.; Saddler, J.N. Effect of replacing polyol by organosolv and kraft lignin on the property and structure of rigid polyurethane foam. *Biotechnol. Biofuels* **2013**, *6*, 12. [CrossRef] [PubMed]
8. Lora, J.H.; Glasser, W.G. Recent industrial applications of lignin: A sustainable alternative to nonrenewable materials. *J. Polym. Environ.* **2002**, *10*, 39–48. [CrossRef]
9. Xue, B.-L.; Wen, J.-L.; Sun, R.-C. Producing lignin-based polyols through microwave-assisted liquefaction for rigid polyurethane foam production. *Materials* **2015**, *8*, 586–599. [CrossRef] [PubMed]
10. Mahmood, N.; Yuan, Z.; Schmidt, J.; Xu, C. Valorization of hydrolysis lignin for polyols/polyurethane foam. *J. Sci. Technol. For. Prod. Process.* **2014**, *3*, 26–31.
11. Li, H.-Q.; Shao, Q.; Luo, H.; Xu, J. Polyurethane foams from alkaline lignin-based polyether polyol. *J. Appl. Polym. Sci.* **2016**, *133*. [CrossRef]
12. Huang, X.; De Hoop, C.F.; Xie, J.; Hse, C.-Y.; Qi, J.; Hu, T. Characterization of biobased polyurethane foams employing lignin fractionated from microwave liquefied switchgrass. *Int. J. Polym. Sci.* **2017**. [CrossRef]
13. Cinelli, P.; Anguillesi, I.; Lazzeri, A. Green synthesis of flexible polyurethane foams from liquefied lignin. *Eur. Polym. J.* **2013**, *49*, 1174–1184. [CrossRef]
14. Bernardini, J.; Cinelli, P.; Anguillesi, I.; Coltelli, M.-B.; Lazzeri, A. Flexible polyurethane foams green production employing lignin or oxypropylated lignin. *Eur. Polym. J.* **2015**, *64*, 147–156. [CrossRef]
15. Cateto, C.A.; Barreiro, M.F.; Rodrigues, A.E.; Belgacem, M.N. Optimization study of lignin oxypropylation in view of the preparation of polyurethane rigid foams. *Ind. Eng. Chem. Res.* **2009**, *48*, 2583–2589. [CrossRef]
16. Langlois, A.; Drouin, M. Process for the Preparation of Lignin Based Polyurethane Products. U.S. Patent 9,598,529 B2, 21 March 2017.
17. Carriço, C.S.; Fraga, T.; Pasa, V.M.D. Production and characterization of polyurethane foams from a simple mixture of castor oil, crude glycerol and untreated lignin as bio-based polyols. *Eur. Polym. J.* **2016**, *85*, 53–61. [CrossRef]
18. Huang, X.; De Hoop, C.F.; Xie, J.; Wu, Q.; Boldor, D.; Qi, J. High bio-content polyurethane (PU) foam made from bio-polyol and cellulose nanocrystals (CNCs) via microwave liquefaction. *Mater. Des.* **2018**, *138*, 11–20. [CrossRef]
19. Hatakeyama, H.; Hatakeyama, T. Environmentally compatible hybrid-type polyurethan foams containing saccharide and lignin components. *Macromol. Symp.* **2005**, *224*, 219–226. [CrossRef]
20. Xue, B.-L.; Wen, J.-L.; Sun, R.-C. Lignin-based rigid polyurethane foam reinforced with pulp fiber: Synthesis and characterization. *ACS Sustain. Chem. Eng.* **2014**, *2*, 1474–1480. [CrossRef]
21. Hatakeyama, H.; Kosugi, R.; Hatakeyama, T. Thermal properties of lignin-and molasses-based polyurethane foams. *J. Therm. Anal. Calorim.* **2008**, *92*, 419. [CrossRef]
22. Mahmood, N.; Yuan, Z.; Schmidt, J.; Xu, C. (Charles) Preparation of bio-based rigid polyurethane foam using hydrolytically depolymerized Kraft lignin via direct replacement or oxypropylation. *Eur. Polym. J.* **2015**, *68*, 1–9. [CrossRef]
23. Faruk, O.; Sain, M. *Lignin in Polymer Composites*; Elsevier: Kidlington, Oxford, UK; Waltham, MA, USA, 2016; ISBN 978-0-323-35565-0.

24. Culbertson, C.; Treasure, T.; Venditti, R.; Jameel, H.; Gonzalez, R. Life Cycle Assessment of lignin extraction in a softwood kraft pulp mill. *Nord. Pulp Pap. Res. J.* **2016**, *31*, 30–40. [CrossRef]

25. Bernier, E.; Lavigne, C.; Robidoux, P.Y. Life cycle assessment of kraft lignin for polymer applications. *Int. J. Life Cycle Assess.* **2013**, *18*, 520–528. [CrossRef]

26. Chauhan, M.; Gupta, M.; Singh, B.; Singh, A.K.; Gupta, V.K. Effect of functionalized lignin on the properties of lignin–isocyanate prepolymer blends and composites. *Eur. Polym. J.* **2014**, *52*, 32–43. [CrossRef]

27. Kourki, H.; Famili, M.H.N.; Mortezaei, M.; Malekipirbazari, M.; Disfani, M.N. Highly nanofilled polystyrene composite: Thermal and dynamic behavior. *J. Elastomers Plast.* **2016**, *48*, 404–425. [CrossRef]

28. Ciobanu, C.; Ungureanu, M.; Ignat, L.; Ungureanu, D.; Popa, V.I. Properties of lignin–polyurethane films prepared by casting method. *Ind. Crops Prod.* **2004**, *20*, 231–241. [CrossRef]

29. Luo, X.; Mohanty, A.; Misra, M. Lignin as a reactive reinforcing filler for water-blown rigid biofoam composites from soy oil-based polyurethane. *Ind. Crops Prod.* **2013**, *47*, 13–19. [CrossRef]

30. Cateto, C.A.; Barreiro, M.F.; Ottati, C.; Lopretti, M.; Rodrigues, A.E.; Belgacem, M.N. Lignin-based rigid polyurethane foams with improved biodegradation. *J. Cell. Plast.* **2014**, *50*, 81–95. [CrossRef]

31. Paruzel, A.; Michałowski, S.; Hodan, J.; Horák, P.; Prociak, A.; Beneš, H. Rigid polyurethane foam fabrication using medium chain glycerides of coconut oil and plastics from end-of-life vehicles. *ACS Sustain. Chem. Eng.* **2017**, *5*, 6237–6246. [CrossRef]

32. Cateto, C.A.; Barreiro, M.F.; Rodrigues, A.E. Monitoring of lignin-based polyurethane synthesis by FTIR-ATR. *Ind. Crops Prod.* **2008**, *27*, 168–174. [CrossRef]

© 2018 by the authors. Licensee MDPI, Basel, Switzerland. This article is an open access article distributed under the terms and conditions of the Creative Commons Attribution (CC BY) license (http://creativecommons.org/licenses/by/4.0/).

![polymers logo] *polymers*

MDPI

Article

Synthesis and Characterization of Cellulose Nanofibril-Reinforced Polyurethane Foam

Weiqi Leng [1], Jinghao Li [2,*] and Zhiyong Cai [1,*]

[1] U.S. Department of Agriculture, Forest Service, Forest Products Laboratory, Madison, WI 53726, USA; wleng@fs.fed.us

[2] Department of Biomaterials, International Center for Bamboo and Rattan, Beijing 10000, China

* Correspondence: jli@ fs.fed.us (J.L.); zcai@fs.fed.us (Z.C.); Tel.: +1-608-628-6602 (J.L.); +1-608-231-9446 (Z.C.)

Received: 11 October 2017; Accepted: 8 November 2017; Published: 10 November 2017

Abstract: In this study, traditional polyol was partially replaced with green, environmentally friendly cellulose nanofibrils (CNF). The effects of CNF on the performance of CNF-reinforced polyurethane foam nanocomposites were investigated using scanning electron microscopy, Fourier transform infrared spectroscopy (FT-IR), X-ray diffraction (XRD) analysis, thermogravimetric analysis (TGA), differential scanning calorimetry (DSC), dynamic mechanical analysis (DMA), and a compression test. The results showed that the introduction of CNF into the polyurethane matrix not only created stronger urethane bonding between the hydroxyl groups in the cellulose chain and isocyanate groups in polymethylene polyphenylisocyanate, but also developed an additional filler–matrix interaction between CNF and polyurethane. With the increase of the CNF replacement ratio, a higher glass transition temperature was obtained, and a higher amount of char residue was generated. In addition, an increase of up to 18-fold in compressive strength was achieved for CNF-PUF (polyurethane foam) nanocomposites with a 40% CNF replacement ratio. CNF has proved to be a promising substitute for traditional polyols in the preparation of polyurethane foams. This study provides an interesting method to synthesize highly green bio-oriented polyurethane foams.

Keywords: cellulose nanofibrils; polyurethane foam; reinforced nanocomposite

1. Introduction

Polyurethane foam, first produced and then commercialized in the 1950s, have attracted much attention due to their low density, high mechanical properties, and use in a wide variety of applications including the construction and automotive industries, depending on the type of foam [1–3]. PUF has been used extensively since their commercialization. The global PUF market value was about 49 billion US dollars in 2015, and there is expected growth up to 92 billion US dollars by 2024 [4]. PU foams are usually prepared by the reaction of petroleum-based polyols (either polyol polyether or polyol polyester) with isocyanate, forming urethane linkages. In addition, catalysts, surfactant and blowing agents are needed to regulate their properties and cell morphology [5,6]. There are wide varieties of polyols and poly-isocyanates that can be used to synthesize PUF. The foam properties vary significantly depending on the selected raw materials. The PUF can be flexible, semi-rigid, and rigid [5].

Although PUF has many merits, there is a significant drawback that causes much environmental concern. The raw materials for PUF are petroleum-based and non-renewable, which are difficult to degrade in nature [6]. There is an urgent need to find raw materials that are environmentally friendly and competitive with petroleum-based counterparts, in terms of price and properties [7,8]. Extensive research has been conducted to modify the polyols and make them biodegradable [9]. However, the chemically modified PUF had inferior mechanical properties compared to its petroleum-based counterpart [10]. Hence, additives were introduced to improve the mechanical properties of PUF. The idea was that natural materials containing hydroxyl groups could play the same role as polyols did [11]. Many natural materials including starch, soy flour and cellulose were extensively investigated, with focus on cellulose and its derivatives, due to their extraordinary properties [6].

Cellulose is the most abundant natural polymer in the world with global reserves of up to 75 billion tons, which accounts for approximately 40% of plant biomass [12–17]. Cellulose can be obtained from a variety of sources including wood, non-woody plants, agricultural residues, algae and bacteria [18]. Cellulose consists of several hundred to over ten thousand β-1,4-D-linked glucose chains, in which the glucose units are joined by single oxygen atoms (acetal linkages) between the C-1 of one glucose unit and the C-4 of the next unit. There are a large number of hydroxyl groups on the glucose unit that can easily form hydrogen bonds with each other to hold the chain together [19,20]. Cellulose I and II are the first and second most extensively studied allomorphs. Cellulose I has the native crystalline structure, and is insoluble in water. It can be converted to cellulose II, with a crystalline structure via a modification or regeneration treatment, during which the native crystalline structure is altered, and the mechanical properties decrease [12,21–24].

Cellulose nanofibrils (CNFs), an aggregation of 10–50 cellulose elementary fibrils with a complex web-like network structure, have gained increasing interest due to their excellent properties including high Young's modulus (estimated at ~140 GPa in the crystal region along the longitudinal direction) and specific strength, making it an ideal building block for products with desirable mechanical properties [25,26]. CNFs have a diameter in the range of 5 to 50 nm and length of several μm, respectively [27]. They also exhibit a hierarchical order in the supramolecular structure and organization, high aspect ratio, high surface area, and reactive surfaces containing –OH groups. All these unique characteristics make CNF a promising candidate as a reinforcing material and may at least partially replace polyols [28–35].

CNF is usually manufactured from wood pulp via mechanical defibrillation with chemical, enzymatic or physical pre-treatments. Various methods have been employed to defibrillate cellulose, e.g., blending, high-pressure homogenization, steam explosion, and grinding [14,36–38]. However, all means of mechanical disintegration of cellulose require high energy inputs. Consequently, chemical pretreatment methods have been developed to reduce energy use, including 2,2,6,6-tetramethylpiperidine-1-oxyl radical (TEMPO)-mediated oxidation, carboxymethylaion, periodate-chlorite oxidation, enzymatic reactions, acidified chlorite, ultra-sonication, or a combination of two or more of these methods [27,39–43].

CNF is usually dispersed in water and stored in a cold room. However, in PUF synthesis, water is a popular blowing agent. Special forms of CNF should be used to replace traditional polyols in the production of PUF. Research has been conducted using up to 10% CNF as the additive to improve the mechanical properties of PUF. However, higher CNF weight ratios have rarely been investigated [44–46]. In this study, a CNF weight ratio (based on total weight) of up to 20% was added to replace petroleum based polyols, i.e., polyethylene glycol (PEG-400). The compressive properties were significantly improved. In addition, the morphology, thermal stability, and spectroscopic characterization of the foam composites were analyzed in this study.

2. Materials and Methods

2.1. Materials

Polymethylene polyphenylisocyanate (PAPI™ 27 Polymeric MDI) was donated by Dow Chemical, Midland, MI, USA, Polyethylene Glycol (PEG-400) was procured from Sigma Aldrich, St. Louis, MO, USA. Spray-dried CNF was supplied by the process development center at University of Maine. DABCO T12 catalyst and DABCO DC5357 surfactant were donated by Air Products, Allentown, PA, USA. Deionized water was used as the blowing agent.

2.2. Fabrication of Pure PUF and CNF-PUF

A detailed foaming formulation is listed in Table 1. For the PUF control, PEG-400, DABCO T12, DABCO DC5357, and deionized water were first mixed for about 5 min under mechanical stirring in a plastic beaker until homogeneous mixture was obtained. For the CNF-PUF, a specific amount of spray-dried CNF was first added to PEG-400 and mixed for 5 min under mechanical stirring in a plastic beaker until a homogeneous mixture was obtained. DABCO T12, DABCO DC5357, and deionized water were then added into the mixture and further stirred for another 5 min. Finally, PAPI 27 was added into the mixture and vigorously stirred for 30 s, after which the foaming process started. The final products were cured overnight in a vacuum oven at 50 °C The cured foams were then cut into different sizes and conditioned at 20 °C and 50% relative humidity conditioning room before characterization. In this study, the –NCO/–OH index was set at 1.1 to ensure complete reaction of all –OH groups.

Table 1. Foaming formulation.

Chemicals	Parts by weight					Role
	Control	A	B	C	D	
PEG-400	100	90	80	70	60	Polyol
Spray-dried CNF	0	10	20	30	40	Polyol, reinforcing agent
DABCO T12	3	3	3	3	3	Catalyst
DABCO DC5357	1	1	1	1	1	Surfactant
Deionized water	0.8	0.8	0.8	0.8	0.8	Blowing agent
PAPI™ 27	88	89	89	90	90	Reactive prepolymer

2.3. Characterization

2.3.1. Scanning Electron Microscopy (SEM)

A scanning electron microscope (Zeiss LEO 1530 Gemini, Oberkochen, Germany) with an acceleration voltage of 5 kV was used to evaluate the morphology of PUF and CNF-PUF samples. Thin slices of 10 mm × 10 mm × 3 mm were cut from the foams using a stainless steel blade and then mounted onto carbon tapes on the aluminum stubs. Samples were then sputter coated with gold (Denton High Vacuum Coating System, Moorestown, NJ, USA) for 1 min under vacuum. The working distance was set at 5 mm.

2.3.2. Fourier Transform Infrared Spectroscopy (FT-IR)

Both pure PUF and CNF-PUF samples were characterized by a Thermo Nicolet iZ10 FTIR spectrometer, attenuated total reflection (ATR) probe (Thermo Scientific, Verona, WI, USA) using a smart iTR™ Basic accessory. A diamond crystal with 45° incident angle was used. The absorbance spectra were taken for an average of 64 scans in the range of 4000–400 cm^{-1} with resolution of 4 cm^{-1}. The spectra were baseline corrected, averaged, and normalized using Omnic v9.0 software (Thermo Scientific, Verona, WI, USA).

2.3.3. X-ray Diffraction (XRD)

XRD patterns of PUF and CNF-PUF samples were obtained with a Bruker Discover 8 diffractometer (The Woodlands, TX, USA) using a Cu Kα rotation tube at 50 kV and 1000 μA with a scanning range from 5° to 50°. The scanning speed was 10°/min. The crystallinity index (CI) was expressed by measuring the peak height of the crystalline area (I_{002}) and amorphous area (I_{am}), as shown in Equation (1) [47,48]:

$$\text{CI}(\%) = \frac{I_{002} - I_{am}}{I_{002}} * 100\% \tag{1}$$

2.3.4. Thermogravimetric Analysis (TGA) and Differential Scanning Calorimetry (DSC)

The foam's thermal stability was tested using a Pyris 1 TGA (PerkinElmer, Shelton, CT, Waltham, MA, USA). Samples were heated from 25 to 800 °C at a heating rate of 5 °C/min under nitrogen gas environment with a flow rate of 20 mL/min. Approximately 5 mg of samples were used for each test. The loss of weight was recorded and normalized against the initial weight. Another group of samples were analyzed with a DSC Q2000 (TA Instruments, New Castle, DE, USA). Samples of 4–6 mg were placed in hermetic aluminum sample pans and first cooled down from room temperature to −30 °C and then heated to 400 °C at a rate of 5 °C/min.

2.3.5. Dynamic Mechanical Analysis (DMA)

The viscoelastic properties of polyurethane and its nanocomposites were measured using DMA (DMA-Q800, TA instruments, New Castle, DE, USA) in penetration mode. The square samples (10 mm × 10 mm × 4 mm) were placed onto a compression clamp, and heated from 30 to 150 °C at a rate of 2 °C/ min with a dynamic strain of 0.1% and a preload force of 0.01 N at a single frequency of 1 Hz.

2.4. Compression Test

The compressive properties of the foam samples were tested using a universal testing machine (Instron 5544, Norwood, MA, USA) equipped with a 1 kN loading cell. The speed of the crosshead was set at 1.27 mm/min. The dimension of the samples was 12.7 mm × 12.7 mm × 12.7 mm. The compression direction was parallel to the foam rising direction. The compressive strength was calculated at 30% compression ratio. Fifteen samples were tested for each group.

3. Results and Discussions

3.1. Foam Structure

Figure 1 shows the microstructure of the PUF control and CNF-PUFs. SEM images confirmed open-cell structure of the foams. Figure 1a shows that the microstructure of the PUF control was a simple alignment of open cells. With the introduction of CNF, the open-cell structure was disrupted, and as the CNF replacement ratio increased, there were large amounts of CNF deposited on and dispersed among the open cells, providing a chance to interact with the PUF matrix. The CNF played an important role as fillers interacting with the PUF cells, which potentially resulted in improved mechanical properties. It was reported in another study that the CNF can strongly interact through hydrogen bonding and improve the nanocomposite behavior [49].

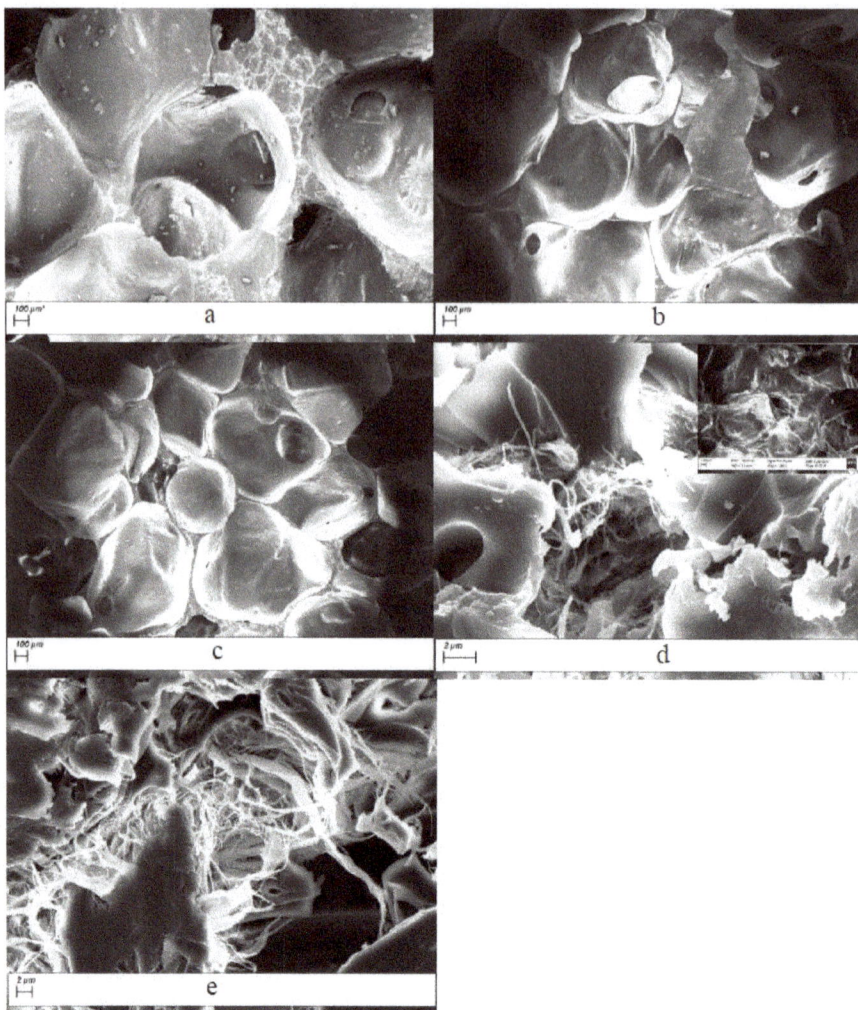

Figure 1. SEM images of (**a**) PUF control and CNF-PUF with (**b**) 10%, (**c**) 20%, (**d**) 30%, (**e**) 40% CNF replacement ratios.

3.2. Fourier Transform Infrared Spectroscopy (FT-IR)

The FTIR spectra of CNF, PUF control, and CNF-PUF with different replacement ratios are shown in Figure 2. The –OH groups in CNF are obvious at 3300 cm^{-1}, while the –OH peak shifted slightly to the right in the PUF control and CNF-PUF samples (insert in Figure 2), and the intensity dropped down sharply, compared with that in CNF. This was direct evidence showing that the –OH groups in CNF reacted with isocyanate groups in PAPI27. However, not all the –OH groups in CNF reacted with isocyanate groups; even though the isocyanate groups were overdosed in the original formulation (–OH/NCO = 1:1.1), it was possible that due to molecular Steric hindrance, some –OH groups were not accessible by the isocyanate groups. Both PUF control and CNF-PUF show peaks at 3200 and 1720 cm^{-1}, which corresponded to urethane –NH stretching, and urethane carbonyl groups, respectively [50]. It was unexpected that no isocyanate peak was observed at 2270 cm^{-1}, since isocyanate groups were overdosed in the original formulation. One possible reason was that excess isocyanate reacted with moisture in the air during the final cure step; since isocyanate is very reactive with –OH groups, it could capture –OH groups in the moisture and form amine and carbon dioxide [51]. The two peaks at 2850 cm^{-1} and 900 cm^{-1} were ascribed to –CH stretching and –CH bending vibrations [52]. A sharp C–O–C peak appeared at 1100 cm^{-1}, which corresponded to the polyether (PEG-400) used as the polyol (not polyester) in this study [52]. In general, there was no significant difference between the PUF control and CNF-PUF in terms of functional groups, since CNF did not introduce any new functional groups into the polyurethane.

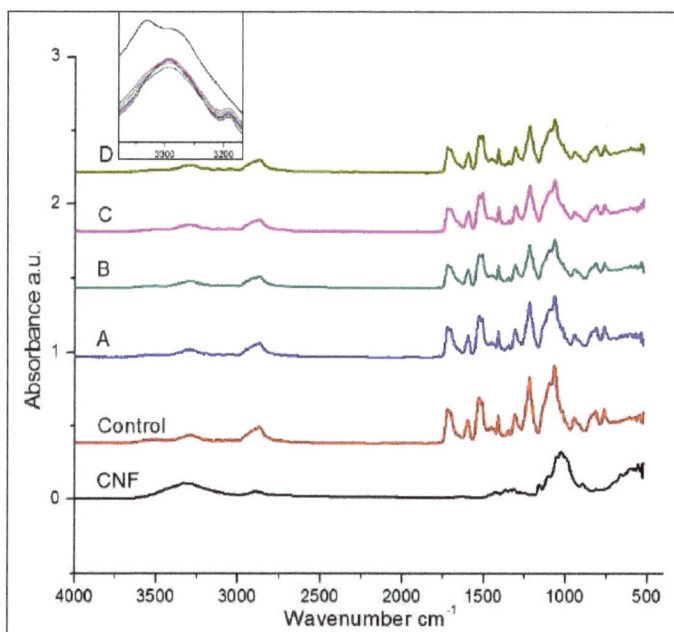

Figure 2. FTIR spectra of CNF, PUF control, and CNF-PUF with 10–40% CNF replacement ratios.

3.3. X-ray Diffraction (XRD) Analysis

Wide angle XRD was used to determine the effect of CNF on the macro- and microstructure changes of PUF. Figure 3 shows the XRD patterns of pure CNF, PUF control, and CNF-PUF from 2θ = 5–40°, because all the characteristic peaks were in this range. CNF had three peaks at 2θ = 16.3°, 18°, and 22.4°, corresponding to the diffractions of amorphous cellulose II, amorphous cellulose I, and crystalline cellulose I respectively [53,54]. The crystallinity of CNF was 41.7%. For the PUF control, a wide diffraction from 15–25° with a maximum peak appeared at approximately 21° [51]. It was reported that there were a few sharp peaks between 15–25°, making it easier to accurately calculate the crystallinity index [55–57]. All CNF-PUF composites showed similar diffraction patterns to the PUF control in Figure 3. Introducing CNF into the PUF matrix resulted in the two characteristic diffraction peaks of CNF overlapping with the PUF peak in the nanocomposites, and decreased intensity after the initial incorporation of CNF, followed by greater intensity with the increase of the CNF replacement ratio. At first, the intensity decreased because the introduction of CNF disrupted the originally uniform PUF structure and made the nanocomposite more amorphous [51,58]. Increasing the replacement ratio of CNF gradually made it more compatible with the PUF matrix and formed more uniform composites. Hence, the intensity increased again.

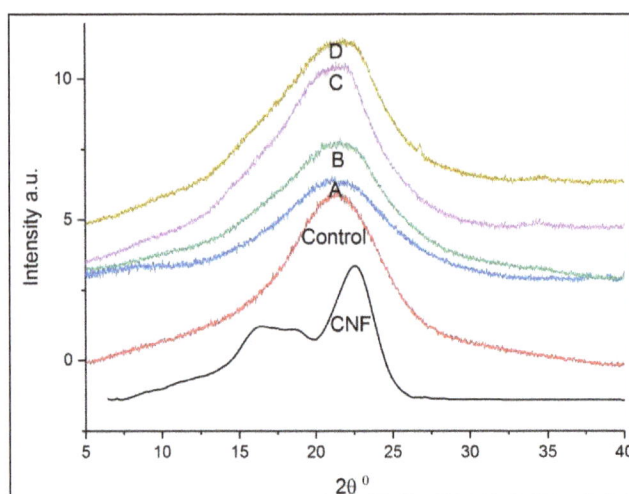

Figure 3. XRD spectra of CNF, PUF control, and CNF-PUF with 10–40% CNF replacement ratios.

3.4. Thermal Properties of PUR and CNF-PUR

Thermal stability is an important characteristic for structural materials, and polyol plays a vital role in determining thermal stability [59]. The thermogravimetry (TG) and differential thermogravimetry (DTG) (derivative weight change) results for PUF control and CNF-PUF with various CNF replacement ratios are shown in Figure 4. The TG curve (Figure 4a) shows that all CNF-PUF started to degrade at 240 °C, which was lower than the PUF control (at 270 °C). It was reported that the thermal degradation of pure CNF aerogel occurred at a temperature around 215 °C in nitrogen atmosphere, which was lower than that of PUF [60]. Hence, the initial thermal degradation temperature for CNF-PUF was lower than the pure PUF control. In addition, the DTG curve (Figure 4b) shows that the greatest weight loss occurred at around 360 °C for the PUF control and 340 °C for CNF-PUF. The incorporation of CNF also lowered the temperature for the greatest weight loss. For both the PUF control and CNF-PUF, there was only one major weight loss between 240 and 650 °C.

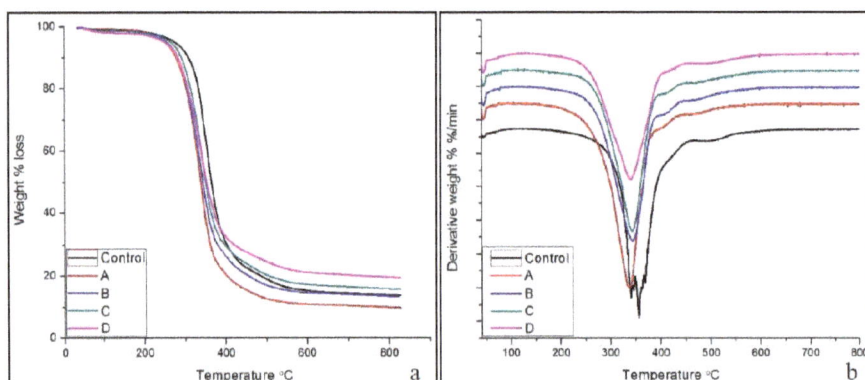

Figure 4. (**a**) TGA and (**b**) DTG of PUF control and CNF-PUF with 10–40% CNF replacement ratios.

The amount of char residue for the PUF control was 14% at 650 °C. For the CNF-PUF, the amount of char residue increased concurrently with the CNF replacement ratio, up to 20% gain for samples with 40% CNF replacement ratio. CNF is known as a radical scavenger during thermal degradation, resulting in higher char residue [61]. DSC analysis was further used to characterize the thermal properties of foam samples. The DSC curve (Figure 5) shows that there were three significant endothermic peaks for both PUF control and CNF-PUF. The first endothermic peak appeared at 30 °C for PUF control, and that shifted up to 70 °C for CNF-PUF with a 40% CNF replacement ratio. This first endothermic peak was related to the glass–rubber transition of the foams. The reaction between CNF and the isocyanate group resulted in a stronger cross-linking matrix than the pure PUF [49], since there are three available –OH groups in CNF skeleton. Hence, more energy was required to mobilize the foam structure, and the glass transition temperature (T_g) increased after partially replacing the PEG-400 with CNF. The second endothermic peak appeared at 270 °C for both the PUF control and CNF-PUF with a CNF replacement ratio of 10%, 30%, and 40%, while that for CNF-PUF, with a CNF replacement ratio of 20%, appeared at 250 °C. The second endothermic peak was caused by the decomposition of the urea bond which was formed by the reaction of isocyanate with water, as well as to the decomposition of urethane group [45,59]. The DSC results showed that the replacement of CNF did not affect the decomposition of the urea bond. The last endothermic peak was ascribed to the decomposition of isocyanurate bonds, which occurred at temperatures above 300 °C [59].

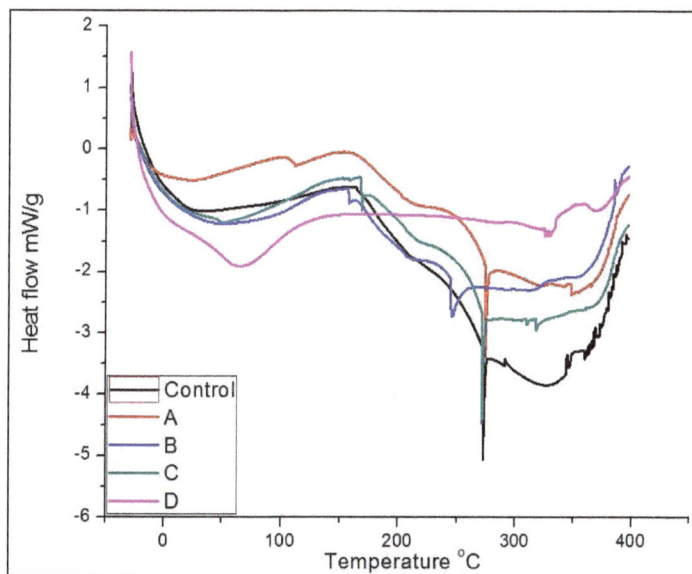

Figure 5. DSC curves of PUF control and CNF-PUF with 10–40% CNF replacement ratios.

3.5. Dynamic Mechanical Analysis (DMA)

Although T_g can be determined using the DSC curve, it was not as sensitive as that measured by the DMA curve [62]. In this study, the T_g was all determined from the temperature position of the maximum in tan δ in the DMA curve. Figure 6 shows that the T_g increased from 40 °C for the PUF control up to 100 °C for the CNF-PUF with 40% CNF replacement ratio. The increasing trend of T_g with the increase of CNF replacement ratio agreed with the results from the DSC curve, except that the T_g values were slightly different. Possible reasons for the increase of T_g were that the reaction between CNF and isocyanate resulted in a higher crosslinking density than that between PEG-400 and isocyanate [49]. In addition, as shown in the SEM, the replacement of CNF generated many interlocks between the cells due to the entanglement of long CNF fibers. This more complicated web-like structure limited the mobility of the polyurethane matrix, requiring more energy to reach the glass-to-rubber transition.

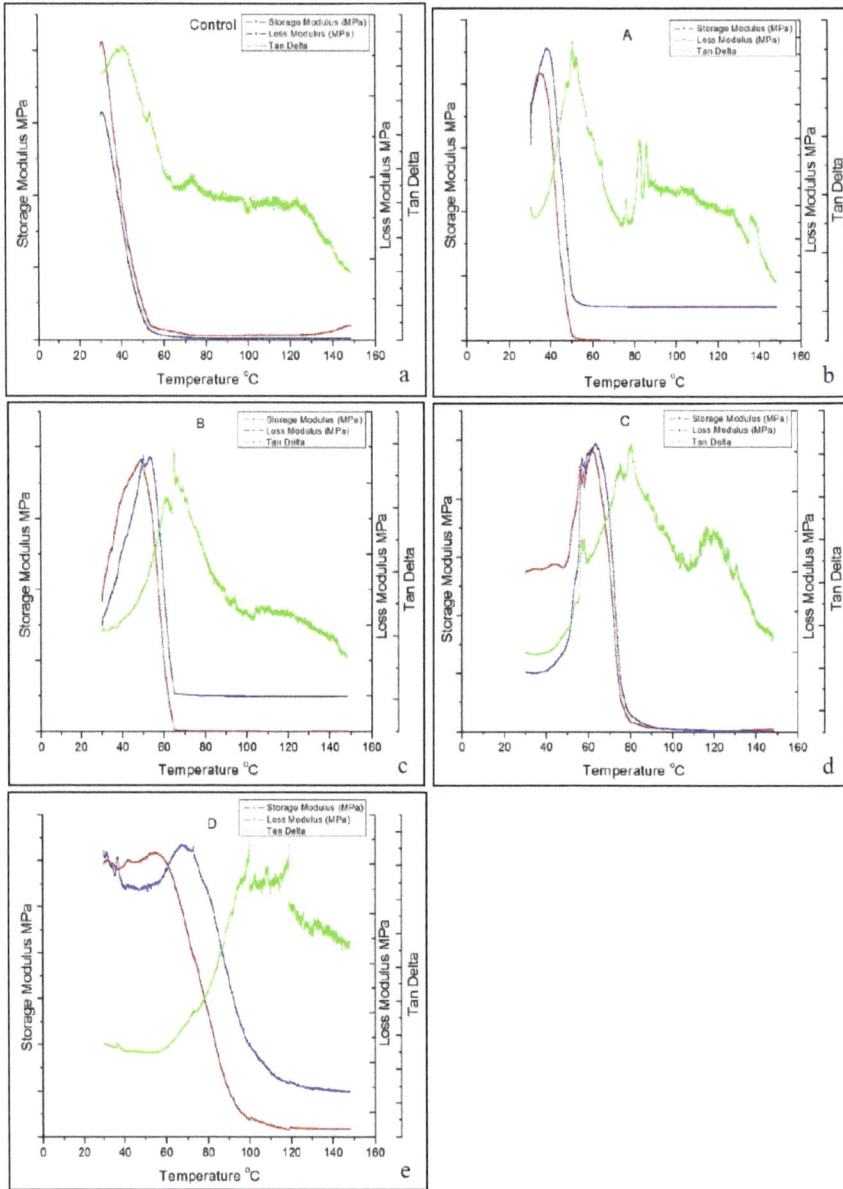

Figure 6. DMA curves of (**a**) PUF control and (**b–e**) CNF-PUF with 10–40% CNF replacement ratios.

3.6. Mechanical Properties of PUF and CNF-PUF

In this study, the compressive properties of PUF control and CNF-PUF were evaluated. Although the target density (approximately 90 kg/m^3) was set to be the same during the experiment design, the actual density was slightly varied between different groups of samples. Hence, normalized compressive strength was used to compare the effect of CNF on the mechanical properties. Figure 7 shows that with the increasing CNF replacement ratio, the normalized compressive strength increased up to 18 times that of the PUF control [45]. As discussed in the thermal stability section, the introduction of CNF into the PUF matrix created new urethane bonding between the –OH groups in CNF and the isocyanate, resulting in a higher crosslinking density than in the PUF control [49]. Additionally, the high mechanical properties and web-like entanglement of CNF itself, acting as a filler to the PUF matrix, contributed to the improvement of compressive strength for the CNF-PUF nanocomposites. In this study, a maximum of 40% CNF replacement ratio was reported. Formulations of higher than 40% CNF replacement ratios resulted in a failure of uniform mixing during foam preparation. As shown in Figure 7, the CNF replacement ratio of 30% and 40% did not generate much difference in the normalized compressive strength, indicating that a 30% CNF replacement ratio might be optimal in terms of compressive strength improvement. However, higher amounts of CNF resulted in much greener and more competitive environmentally-friendly foam composites.

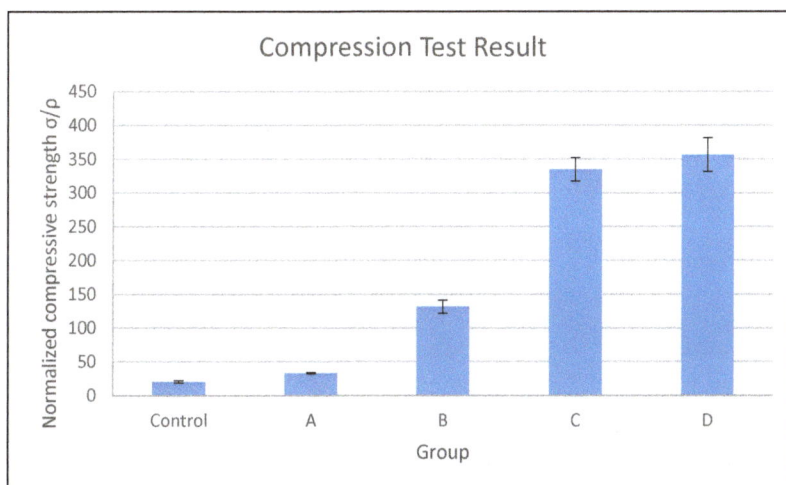

Figure 7. Compression results of PUF control and CNF-PUF with 10–40% CNF replacement ratios.

4. Conclusions

In this study, up to 40% of PEG-400 polyol was replaced with CNF to synthesize PUF. The introduction of CNF disrupted the original open-cell structure of PUF and made the nanocomposite more amorphous. As the CNF replacement ratios increased, large amounts of CNFs deposited on the open-cells and also dispersed among the open-cells, acting as a bridge connecting the cells. The incorporation of CNF also resulted in stronger crosslinking between the CNF-PUF matrix. Evidently, the T_g increased from 40 °C for the PUF control up to 100 °C for the CNF-PUF with a 40% CNF replacement ratio. Additionally, the introduction of CNF rendered a significant increase in the normalized compressive strength up to 18 times the original value for the PUF control. This study provides an interesting way to synthesize a much greener bio-oriented PUF.

Acknowledgments: The authors would like to acknowledge the EMRSL group at the USDA Forest Products Lab for the mechanical testing, Jane O'Dell for providing DMA testing equipment, and Neil Gribbins for the manuscript review.

Author Contributions: Weiqi Leng and Jinghao Li evenly contributed to the overall process of the experiment design, characterization, data analysis, and the manuscript drafting. Zhiyong Cai supervised the whole project, reviewed the draft, and made comments.

Conflicts of Interest: The authors declare no conflict of interest.

References

1. Efstathiou, K. *Synthesis and Characterization of a Polyurethane Prepolymer for the Development of a Novel Acrylate-Based Polymer Foam*; Budapest University of Technology and Economics: Budapest, Hungary, 2011; pp. 1–57.

2. Szycher, M. *Szycher's Handbook of Polyurethanes*, 2nd ed.; CRC Press: Boca Raton, FL, USA, 2012; ISBN 9780439839584.

3. Seydibeyoglu, M.O.; Misra, M.; Mohanty, A.; Blaker, J.J.; Lee, K.-Y.; Bismarck, A.; Kazemizadeh, M. Green polyurethane nanocomposites from soy polyol and bacterial cellulose. *J. Mater. Sci.* **2013**, *48*, 2167–2175. [CrossRef]

4. Grand View Research, Inc. Polyurethane (PU) Foam Market Analysis by Product (Rigid Foam, Flexible Foam), by Application (Bedding & Furniture, Transportation, Packaging, Electronics, Footwear) and Segment Forecasts to 2024. Available online: https://www.prnewswire.com/news-releases/polyurethane-pu-foam-market-analysis-by-product-rigid-foam-flexible-foam-by-application-bedding--furniture-transportation-packaging-electronics-footwear-and-segment---forecasts-to-2024---research-and-markets-300278945.html (accessed on 2 October 2017).

5. Rivera-Armenta, J.L.; Heinze, T.; Mendoza-Martínez, A.M. New polyurethane foams modified with cellulose derivatives. *Eur. Polym. J.* **2004**, *40*, 2803–2812. [CrossRef]

6. Zhou, X.; Sain, M.M.; Oksman, K. Semi-rigid biopolyurethane foams based on palm-oil polyol and reinforced with cellulose nanocrystals. *Compos. A* **2016**, *83*, 56–62. [CrossRef]

7. Karlsson, K.; Schuster, E.; Stading, M.; Rigdahl, M. Foaming behavior of water-soluble cellulose derivatives: Hydroxypropyl methylcellulose and ethyl hydroxyethyl cellulose. *Cellulose* **2015**, *22*, 2651–2664. [CrossRef]

8. Kumari, S.; Chauhan, G.S.; Ahn, J. Novel cellulose nanowhiskers-based polyurethane foam for rapid and persistent removal of methylene blue from its aqueous solutions. *Chem. Eng. J.* **2016**, *304*, 728–736. [CrossRef]

9. D'Souza, J.; Camargo, R.; Yan, N. Polyurethane foams made from liquefied bark-based polyols. *J. Appl. Polym. Sci.* **2014**, *131*. [CrossRef]

10. Gu, R.; Konar, S.; Sain, M. Preparation and Characterization of Sustainable Polyurethane Foams from Soybean Oils. *J. Am. Oil Chem. Soc.* **2012**, *89*, 2103–2111. [CrossRef]

11. Hinrichsen, G. *Polyurethane Handbook*, 2nd ed.; Hanser Publishers: Munich, Germany, 1993; 770p; ISBN 3-446-17198-3.

12. Chen, W.; Yu, H.; Li, Q.; Liu, Y.; Li, J. Ultralight and highly flexible aerogels with long cellulose I nanofibers. *Soft Matter* **2011**, *7*, 10360. [CrossRef]

13. Klemm, D.; Heublein, B.; Fink, H.-P.; Bohn, A. Cellulose: Fascinating biopolymer and sustainable raw material. *Angew. Chem. Int. Ed.* **2005**, *44*, 3358–3393. [CrossRef] [PubMed]

14. Liu, Q.; Jing, S.; Wang, S.; Zhuo, H.; Zhong, L.; Peng, X.; Sun, R. Flexible nanocomposites with ultrahigh specific areal capacitance and tunable properties based on a cellulose derived nanofiber-carbon sheet framework coated with polyaniline. *J. Mater. Chem. A* **2016**, *4*, 13352–13362. [CrossRef]

15. Silva, T.C.F.; Habibi, Y.; Colodette, J.L.; Elder, T.; Lucia, L.A. A fundamental investigation of the microarchitecture and mechanical properties of tempo-oxidized nanofibrillated cellulose (NFC)-based aerogels. *Cellulose* **2012**, *19*, 1945–1956. [CrossRef]

16. Zanini, M.; Lavoratti, A.; Lazzari, L.K.; Galiotto, D.; Pagnocelli, M.; Baldasso, C.; Zattera, A.J. Producing aerogels from silanized cellulose nanofiber suspension. *Cellulose* **2017**, *24*, 769–779. [CrossRef]

17. Zhao, J.; Zhang, X.; He, X.; Xiao, M.; Zhang, W.; Lu, C. A super biosorbent from dendrimer poly(amidoamine)-grafted cellulose nanofibril aerogels for effective removal of Cr(VI). *J. Mater. Chem. A* **2015**, *3*, 14703–14711. [CrossRef]

18. Jack, A.A.; Nordli, H.R.; Powell, L.C.; Powell, K.A.; Kishnani, H.; Johnsen, P.O.; Pukstad, B.; Thomas, D.W.; Chinga-Carrasco, G.; Hill, K.E. The interaction of wood nanocellulose dressings and the wound pathogen P. aeruginosa. *Carbohydr. Polym.* **2017**, *157*, 1955–1962. [CrossRef] [PubMed]

19. Xiao, S.; Gao, R.; Lu, Y.; Li, J.; Sun, Q. Fabrication and characterization of nanofibrillated cellulose and its aerogels from natural pine needles. *Carbohydr. Polym.* **2015**, *119*, 202–209. [CrossRef] [PubMed]

20. Zheng, Q.; Zhang, H.; Mi, H.; Cai, Z.; Ma, Z.; Gong, S. High-performance flexible piezoelectric nanogenerators consisting of porous cellulose nanofibril (CNF)/poly(dimethylsiloxane) (PDMS) aerogel films. *Nano Energy* **2016**, *26*, 504–512. [CrossRef]

21. Duchemin, B.J.C.; Staiger, M.P.; Tucker, N.; Newman, R.H. Aerocellulose based on all-cellulose composites. *J. Appl. Polym. Sci.* **2010**, *115*, 216–221. [CrossRef]

22. Fu, J.; He, C.; Huang, J.; Chen, Z.; Wang, S. Cellulose nanofibril reinforced silica aerogels: Optimization of the preparation process evaluated by a response surface methodology. *RSC Adv.* **2016**, *6*, 100326–100333. [CrossRef]

23. Heath, L.; Thielemans, W. Cellulose nanowhisker aerogels. *Green Chem.* **2010**, *12*, 1448. [CrossRef]

24. Kettunen, M.; Silvennoinen, R.J.; Houbenov, N.; Nykänen, A.; Ruokolainen, J.; Sainio, J.; Pore, V.; Kemell, M.; Ankerfors, M.; Lindström, T.; et al. Photoswitchable Superabsorbency Based on Nanocellulose Aerogels. *Adv. Funct. Mater.* **2011**, *21*, 510–517. [CrossRef]

25. Donius, A.E.; Liu, A.; Berglund, L.A.; Wegst, U.G.K. Superior mechanical performance of highly porous, anisotropic nanocellulose–montmorillonite aerogels prepared by freeze casting. *J. Mech. Behav. Biomed. Mater.* **2014**, *37*, 88–99. [CrossRef] [PubMed]

26. Gao, K.; Shao, Z.; Li, J.; Wang, X.; Peng, X.; Wang, W.; Wang, F. Cellulose nanofiber–graphene all solid-state flexible supercapacitors. *J. Mater. Chem. A* **2013**, *1*, 63–67. [CrossRef]

27. Kim, C.H.; Youn, H.J.; Lee, H.L. Preparation of cross-linked cellulose nanofibril aerogel with water absorbency and shape recovery. *Cellulose* **2015**, *22*, 3715–3724. [CrossRef]

28. Barari, B.; Ellingham, T.K.; Ghamhia, I.I.; Pillai, K.M.; El-Hajjar, R.; Turng, L.-S.; Sabo, R. Mechanical characterization of scalable cellulose nano-fiber based composites made using liquid composite molding process. *Compos. B* **2016**, *84*, 277–284. [CrossRef]

29. Chen, B.; Zheng, Q.; Zhu, J.; Li, J.; Cai, Z.; Chen, L.; Gong, S. Mechanically strong fully biobased anisotropic cellulose aerogels. *RSC Adv.* **2016**, *6*, 96518–96526. [CrossRef]

30. Chen, W.; Li, Q.; Wang, Y.; Yi, X.; Zeng, J.; Yu, H.; Liu, Y.; Li, J. Comparative study of aerogels obtained from differently prepared nanocellulose fibers. *ChemSusChem* **2014**, *7*, 154–161. [CrossRef] [PubMed]

31. Meng, Y.; Young, T.M.; Liu, P.; Contescu, C.I.; Huang, B.; Wang, S. Ultralight carbon aerogel from nanocellulose as a highly selective oil absorption material. *Cellulose* **2015**, *22*, 435–447. [CrossRef]

32. Meng, Y.; Wang, X.; Wu, Z.; Wang, S.; Young, T.M. Optimization of cellulose nanofibrils carbon aerogel fabrication using response surface methodology. *Eur. Polym. J.* **2015**, *73*, 137–148. [CrossRef]

33. Olsson, R.T.; Azizi Samir, M.A.S.; Salazar-Alvarez, G.; Belova, L.; Ström, V.; Berglund, L.A.; Ikkala, O.; Nogués, J.; Gedde, U.W. Making flexible magnetic aerogels and stiff magnetic nanopaper using cellulose nanofibrils as templates. *Nat. Nanotechnol.* **2010**, *5*, 584–588. [CrossRef] [PubMed]

34. Wang, M.; Anoshkin, I.V.; Nasibulin, A.G.; Ras, R.H.A.; Nonappa, N.; Laine, J.; Kauppinen, E.I.; Ikkala, O. Electrical behaviour of native cellulose nanofibril/carbon nanotube hybrid aerogels under cyclic compression. *RSC Adv.* **2016**, *6*, 89051–89056. [CrossRef] [PubMed]

35. Zhai, T.; Zheng, Q.; Cai, Z.; Turng, L.-S.; Xia, H.; Gong, S. Poly(vinyl alcohol)/Cellulose Nanofibril Hybrid Aerogels with an Aligned Microtubular Porous Structure and their Composites with Polydimethylsiloxane. *ACS Appl. Mater. Interfaces* **2015**, *7*, 7436–7444. [CrossRef] [PubMed]

36. Srithep, Y.; Turng, L.-S.; Sabo, R.; Clemons, C. Nanofibrillated cellulose (NFC) reinforced polyvinyl alcohol (PVOH) nanocomposites: Properties, solubility of carbon dioxide, and foaming. *Cellulose* **2012**, *19*, 1209–1223. [CrossRef]

37. Yildirim, N.; Shaler, S.M.; Gardner, D.J.; Rice, R.; Bousfield, D.W. Cellulose nanofibril (CNF) reinforced starch insulating foams. *Cellulose* **2014**, *21*, 4337–4347. [CrossRef]

38. Sajab, M.S.; Chia, C.H.; Chan, C.H.; Zakaria, S.; Kaco, H.; Chook, S.W.; Chin, S.X.; Noor, A.M. Bifunctional graphene oxide–cellulose nanofibril aerogel loaded with Fe(III) for the removal of cationic dye via simultaneous adsorption and Fenton oxidation. *RSC Adv.* **2016**, *6*, 19819–19825. [CrossRef]

39. Chook, S.W.; Chia, C.H.; Chan, C.H.; Chin, S.X.; Zakaria, S.; Sajab, M.S.; Huang, N.M. A porous aerogel nanocomposite of silver nanoparticles-functionalized cellulose nanofibrils for SERS detection and catalytic degradation of rhodamine B. *RSC Adv.* **2015**, *5*, 88915–88920. [CrossRef]

40. Javadi, A.; Zheng, Q.; Payen, F.; Javadi, A.; Altin, Y.; Cai, Z.; Sabo, R.; Gong, S. Polyvinyl Alcohol-Cellulose Nanofibrils-Graphene Oxide Hybrid Organic Aerogels. *ACS Appl. Mater. Interfaces* **2013**, *5*, 5969–5975. [CrossRef] [PubMed]

41. Melone, L.; Altomare, L.; Alfieri, I.; Lorenzi, A.; De Nardo, L.; Punta, C. Ceramic aerogels from TEMPO-oxidized cellulose nanofibre templates: Synthesis, characterization, and photocatalytic properties. *J. Photochem. Photobiol. Chem.* **2013**, *261*, 53–60. [CrossRef]

42. Zheng, Q.; Cai, Z.; Gong, S. Green synthesis of polyvinyl alcohol (PVA)–cellulose nanofibril (CNF) hybrid aerogels and their use as superabsorbents. *J. Mater. Chem. A* **2014**, *2*, 3110. [CrossRef]

43. Zheng, Q.; Javadi, A.; Sabo, R.; Cai, Z.; Gong, S. Polyvinyl alcohol (PVA)–cellulose nanofibril (CNF)–multiwalled carbon nanotube (MWCNT) hybrid organic aerogels with superior mechanical properties. *RSC Adv.* **2013**, *3*, 20816. [CrossRef]

44. Li, Y.; Ren, H.; Ragauskas, A.J. Rigid polyurethane foam reinforced with cellulose whiskers: Synthesis and characterization. *Nano-Micro Lett.* **2010**, *2*, 89–94. [CrossRef]

45. Li, Y.; Ragauskas, A.J. Ethanol organosolv lignin-based rigid polyurethane foam reinforced with cellulose nanowhiskers. *RSC Adv.* **2012**, *2*, 3347–3351. [CrossRef]

46. Zhu, M.; Bandyopadhyay-Ghosh, S.; Khazabi, M.; Cai, H.; Correa, C.; Sain, M. Reinforcement of soy polyol-based rigid polyurethane foams by cellulose microfibers and nanoclays. *J. Appl. Polym. Sci.* **2012**, *124*, 4702–4710. [CrossRef]

47. Zhang, X.; Yan, Q.; Leng, W.; Li, J.; Zhang, J.; Cai, Z.; Hassan, E. Carbon Nanostructure of Kraft Lignin Thermally Treated at 500 to 1000 °C. *Materials* **2017**, *10*, 975. [CrossRef] [PubMed]

48. Zhang, X.; Yan, Q.; Hassan, E.B.; Li, J.; Cai, Z.; Zhang, J. Temperature effects on formation of carbon-based nanomaterials from kraft lignin. *Mater. Lett.* **2017**, *203*, 42–45. [CrossRef]

49. Marcovich, N.E.; Auad, M.L.; Bellesi, N.E.; Nutt, S.R.; Aranguren, M.I. Cellulose micro/nanocrystals reinforced polyurethane. *J. Mater. Res.* **2006**, *21*, 870–881. [CrossRef]

50. Gunashekar, S.; Abu-Zahra, N. Characterization of Functionalized Polyurethane Foam for Lead Ion Removal from Water. *Int. J. Polym. Sci.* **2014**, *2014*, 1–7. [CrossRef]

51. Strankowski, M.; Włodarczyk, D.; Piszczyk, Ł.; Strankowska, J. Polyurethane nanocomposites containing reduced graphene oxide, FTIR, Raman, and XRD Studies. *J. Spectrosc.* **2016**, *2016*, 1–6. [CrossRef]

52. Silverstein, R.M.; Bassler, G.; Morrill, T. *Spectrometric Identification of Organic Compounds*, 4th ed.; John Wiley and Sons: New York, NY, USA, 1981; ISBN 978-0-471-02990-8.

53. Xu, X.; Liu, F.; Jiang, L.; Zhu, J.Y.; Haagenson, D.; Wiesenborn, D.P. Cellulose Nanocrystals vs. Cellulose Nanofibrils: A Comparative Study on their Microstructures and Effects as Polymer Reinforcing Agents. *ACS Appl. Mater. Interfaces* **2013**, *5*, 2999–3009. [CrossRef] [PubMed]

54. Wulandari, W.T.; Rochliadi, A.; Arcana, I.M. Nanocellulose prepared by acid hydrolysis of isolated cellulose from sugarcane bagasse. *IOP Conf. Ser. Mater. Sci. Eng.* **2016**, *107*, 012045. [CrossRef]

55. Zhou, C. Bulk Preparation of Radiation Crosslinking Poly (Urethane-Imide). In *New Polymers for Special Applications*, 1st ed.; De Souza Gomes, A., Ed.; InTech: London, UK, 2012; ISBN 978-953-51-0744-6.

56. Li, F.; Hou, J.; Zhu, W.; Zhang, X.; Xu, M.; Luo, X.; Ma, D.; Kim, B.K. Crystallinity and morphology of segmented polyurethanes with different soft-segment length. *J. Appl. Polym. Sci.* **1996**, *62*, 631–638. [CrossRef]

57. Pistor, V.; de Conto, D.; Ornaghi, F.G.; Zattera, A.J. Microstructure and crystallization kinetics of polyurethane thermoplastics containing trisilanol isobutyl POSS. *J. Nanomater.* **2012**, *2012*, 1–8. [CrossRef]

58. Liu, H.; Dong, M.; Huang, W.; Gao, J.; Dai, K.; Guo, J.; Zheng, G.; Liu, C.; Shen, C.; Guo, Z. Lightweight conductive graphene/thermoplastic polyurethane foams with ultrahigh compressibility for piezoresistive sensing. *J. Mater. Chem. C* **2017**, *5*, 73–83. [CrossRef]

59. Liszkowska, J.; Czupryński, B.; Sadowska, J.P. Thermal properties of polyurethane-polyisocyanurate (PUR-PIR) foams modified with tris(5-Hydroxypenthyl) gitrate. *J. Adv. Chem. Eng.* **2016**, *6*. [CrossRef]

60. Li, J.; Wei, L.; Leng, W.; Hunt, J.F.; Cai, Z. Fabrication and characterization of CNF/Epoxy nanocomposite foam. *J. Mater. Sci.* **2017**, accepted.

61. Lvov, Y.; Guo, B.; Fakhrullin, R.F. *Functional Polymer Composites with Nanoclays*, 1st ed.; Royal Society of Chemistry: London, UK, 2016; 433p; ISBN 978-1-78262-672-5.
62. Tien, Y.I.; Wei, K.H. The effect of nano-sized silicate layers from montmorillonite on glass transition, dynamic mechanical, and thermal degradation properties of segmented polyurethane. *J. Appl. Polym. Sci.* **2002**, *86*, 1741–1748. [CrossRef]

© 2017 by the authors. Licensee MDPI, Basel, Switzerland. This article is an open access article distributed under the terms and conditions of the Creative Commons Attribution (CC BY) license (http://creativecommons.org/licenses/by/4.0/).

polymers

MDPI

Article

Gas Dissolution Foaming as a Novel Approach for the Production of Lightweight Biocomposites of PHB/Natural Fibre Fabrics

Heura Ventura [1,*], Luigi Sorrentino [2], Ester Laguna-Gutierrez [3], Miguel Angel Rodriguez-Perez [3] and Monica Ardanuy [1]

[1]   Departament de Ciència dels Materials i Enginyeria Metal·lúrgica (CMEM), Universitat Politècnica de Catalunya (UPC): C/Colom 11, TR4, 08222 Terrassa, Spain; monica.ardanuy@upc.edu

[2]   Istituto per i Polimeri, Compositi e Biomateriali (IPCB), Consiglio Nazionale delle Ricerche (CNR): P/Enrico Fermi 1, Loc. Granatello, 80055 Portici, Italy; luigi.sorrentino@cnr.it

[3]   Cellular Materials Laboratory (CellMat), Condensed Matter Physics Department, Facultad de Ciencias, Universidad de Valladolid (UVa): P° de Belén 7, 47011 Valladolid, Spain; ester.laguna@fmc.uva.es (E.L.-G.); marrod@fmc.uva.es (M.A.R.-P.)

*   Correspondence: heura.ventura@upc.edu; Tel.: +34-93-739-8182

Received: 12 January 2018; Accepted: 27 February 2018; Published: 28 February 2018

Abstract: The aim of this study is to propose and explore a novel approach for the production of cellular lightweight natural fibre, nonwoven, fabric-reinforced biocomposites by means of gas dissolution foaming from composite precursors of polyhydroxybutyrate-based matrix and flax fabric reinforcement. The main challenge is the development of a regular cellular structure in the polymeric matrix to reach a weight reduction while keeping a good fibre-matrix stress transfer and adhesion. The viability of the process is evaluated through the analysis of the cellular structure and morphology of the composites. The effect of matrix modification, nonwoven treatment, expansion temperature, and expansion pressure on the density and cellular structure of the cellular composites is evaluated. It was found that the nonwoven fabric plays a key role in the formation of a uniform cellular morphology, although limiting the maximum expansion ratio of the composites. Cellular composites with a significant reduction of weight (relative densities in the range 0.4–0.5) were successfully obtained.

Keywords: biopolymer; biocomposite; fabric reinforcement; natural fibres; foaming

---

1. Introduction

Over the last three decades, concern about the depletion of fossil resources, waste accumulation, and other environmental problems has prompted researchers to investigate biocomposite materials (of bio-based polymeric matrix and natural fibre reinforcement, for instance), which can offer sustainable and biodegradable replacements for conventional materials. In general, composites are appropriate for applications requiring light weight and good mechanical performance. Therefore, the development of cellular composites presents an interesting topic, since the combination of composite reinforcement with a cellular structure in the matrix can lead to a class of materials with high specific mechanical properties [1]. However, to our knowledge, the production of cellular biocomposites, with the combination of a biopolymer and a fabric reinforcement of natural fibres, has not been yet reported in the literature.

Among other biopolymers, polyhydroxyalkanoates (PHAs) offer a bio-based and biodegradable alternative. PHAs are bio-polyesters generated by bacteria under specific feeding conditions, and can be produced from sources that do not compete with food crops [2]. Although their properties are of great interest, their high price nowadays limits their competitiveness against the synthetic commodity polymers. Nonetheless, their use as matrix in cellular biocomposites presents an opportunity to increase their competitiveness, owing to a partial substitution of the matrix by natural fibres and gas, which reduces the amount of polymer required, and hence the cost.

Reinforcements based on natural fibres (such as cellulosic fibres) have been shown to increase the stiffness and strength of PHAs [3–5], but a good fibre-matrix interaction (chemical, mechanical, or both) is required. Natural fibre reinforcements are often randomly dispersed short fibres, although they can also be used in the form of nonwoven fabrics [6,7]. This use of fibre mats instead of random fibres leads to a good distribution of fibres throughout the final structure while keeping a low production cost and allowing easy handling. Moreover, the mechanical reinforcement capacity of these fabrics allows the development of composites for structural applications.

To achieve a cellular structure in the composite's matrix, a foaming process is required, where a blowing agent (such as carbon dioxide or nitrogen) is used to generate porosity, resulting in a lightweight cellular material. Nonetheless, the foaming process presents some challenges in this case. On one hand, PHAs present an intrinsic difficulty to be foamed owing to their low molecular weight, low melt strength, high crystallinity, and narrow processing window. Despite this, they have been successfully foamed by injection foaming and extrusion foaming [8–12], and chain extender additives (CE) have been reported to enhance foamability in biopolyesters such as polylactic acid [13–15] and PHAs [9,16]. On the other hand, techniques such as injection foaming, extrusion foaming, and supercritical CO_2-assisted extrusion have been used for the production of cellular biocomposites [17–21], but the use of a nonwoven structure is a limiting factor for those techniques (i.e., the entanglement of fibres obtained in such a structure does not allow the required flow for processing). In this respect, a gas dissolution foaming technique can overcome these technical limitations. Gas dissolution foaming is a discontinuous (batch) foaming process, which consists of the saturation of a polymer with a physical blowing agent (gas) at a given temperature and pressure inside a high-pressure vessel, forming a single-phase polymer-gas solution. The expansion is achieved by applying a thermodynamic instability that abruptly reduces the gas solubility limit in the polymer, leading to the phase separation of the dissolved gas and promoting the nucleation and growth of cells. This thermodynamic instability can be induced by a temperature increase (temperature soak method) or by a quick pressure release (pressure quench method).

The aim of this study is to propose and explore a novel approach for the production of cellular lightweight fibre-reinforced green composites of polyhydroxybutyrate-based matrix and flax nonwoven fabric reinforcement. The goal is to achieve a regular expansion of the precursor with a homogeneously distributed porosity in the polymeric matrix, to reach a weight reduction of the composites while keeping a good fibre-matrix stress transfer and adhesion. The approach consists of foaming—by a gas dissolution process—the composite precursors in which the reinforcement is a nonwoven structure (fabric). For the foaming process, after CO_2 dissolution at high temperature, the phase separation in the matrix is induced by a pressure quench. The viability of the system is evaluated through the analysis of the cellular structure and morphology of the composites. Preliminary tests are performed to determine the conditions in which the lowest density values are reached, where the influence of the nonwoven fabrics on the matrix's foaming behaviour is observed. Further tests, considering variations of four factors (matrix type, nonwoven fabric's treatment, expansion temperature, and expansion pressure) are accomplished to determine the effects of these factors on the density and cellular structure of the cellular composite samples.

2. Materials and Methods

2.1. Foamed Composites Production

Figure 1 summarizes the procedure followed to prepare the foamed composites. As shown, the preparation of composite precursors required firstly the production and treatment of nonwoven fabrics (NW), and secondly the production of films made of modified or non-modified polymeric matrices, further processed by film-stacking. The solid composite precursors were then foamed after the CO_2 dissolution by a quick pressure release. A detailed description of this procedure is presented hereunder.

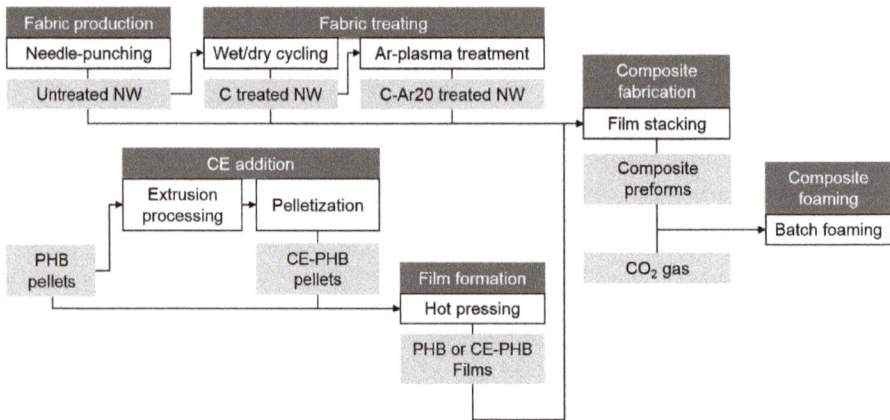

Figure 1. Scheme for the samples preparation.

2.1.1. Production and Treatment of the Natural Fibre Nonwoven Fabrics

The production and treatment of the flax NW fabrics (summarized in Figure 1) was addressed in a previous study [22,23]. The NW flax reinforcements with a density of 284 ± 23 g/m^2 and a thickness of 1.9 ± 0.1 mm were used in three different conditions: as fabricated (untreated), after wet/dry cycling (C), and after wet/dry cycling and treatment with argon plasma (C-Ar20).

2.1.2. CE Addition to the Matrix

The biopolymer used was Mirel P3001, a proprietary polyhydroxybutyrate (PHB)-based resin of thermoforming grade provided by Metabolix (Cambridge, MA, USA), designated hereafter as PHB. For its use in the matrix, it was used as received, as well as modified with a chain extender (CE) to improve its foamability (referred to as CE-PHB). For the production of this CE-PHB matrix, the PHB pellets were dried at 50 °C overnight, and mixed with 1 wt % of the CE Joncryl ADR-4368-C, kindly provided by BASF Española (Barcelona, Spain). The mixture was then processed in a COLLIN ZK 25 T co-rotating twin-screw extruder (Dr. Collin GmbH, Ebersberg, Germany), using a reverse temperature profile linearly decreasing from 170 °C to 150 °C at a screw-speed of 70 rpm and at a feeding rate of ~70 g/min. The extrudate was cooled in a water bath and pelletized.

2.1.3. Fabrication of Solid Composite Precursors

Both PHB and CE-PHB matrices were used to produce films of 0.6 mm thickness, with the help of a frame, in a COLLIN P 300P Hot Plate Press (Dr. Collin GmbH, Ebersberg, Germany). The process consisted of heating at 175 °C for 5 min to melt the polymer and a further 5 min under a 50 MPa pressure, followed by cooling to room temperature under a 50 MPa pressure. Composites (1 mm thick) were produced in a sandwich-like sequence consisting of film/NW/film by means of the film-stacking method. The temperature was set to 175 °C, and the pressure profile consisted of 6 min under no pressure for melting the polymer, followed by 4 min under 50 MPa for the fibre impregnation, and a cooling step under 50 MPa. Six solid precursors (with an average fibre fraction of 19.8 ± 1.5 wt %) were prepared. In order to identify the composite precursors as a function of the type of matrix and reinforcement, the samples will be referred to hereafter as: untreated/PHB, untreated/CE-PHB, C/PHB, C/CE-PHB, C-Ar20/PHB and C-Ar20/CE-PHB. Five specimens of 40 mm × 60 mm were cut from each of these solid precursors.

2.1.4. Foaming Process of the Solid Composite Precursors

The pressure quench foaming method was used to foam the former solid precursors, which were dried at 80 °C overnight under vacuum. The samples were placed inside the high-pressure vessel. The dissolution of the blowing agent, CO_2 (99.9% purity) supplied by Rivoira SpA (Milan, Italy), was performed at high temperature, above the melting temperature of the polymer, to achieve the highest gas solubility available in the amorphous state. Foaming was achieved by a quick pressure release after cooling to the desired expansion temperature. For better understanding, a scheme of the pressure and temperature evolution is shown in Figure 2.

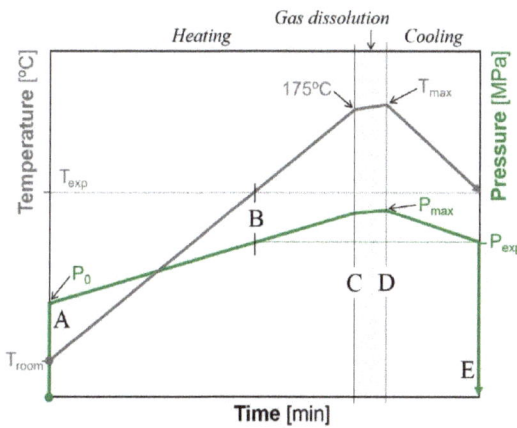

Figure 2. Definition of the temperature and pressure curves in the batch-foaming process.

It is worth noting that the foaming times were dependent on the heating/cooling capabilities of the experimental setup. The steps from A to D were performed as follows. (A) the pressure vessel was filled with CO_2 at an initial pressure (P_0) dependent on the required expansion pressure (P_{exp}), where P_0 was 3.75 ± 0.25 MPa, 6.25 ± 0.25 MPa, and 7.25 ± 0.25 MPa for P_{exp} of 5, 10, and 20 MPa, respectively. (A–C) the temperature was raised (heating), with the consequent pressure increase, up to 175 °C. The heating rates, which were affected by the CO_2 pressure in the vessel, were determined as 5.8, 5.1, and 4.6 °C/min for expansion pressures of 5, 10, and 20 MPa, respectively. (B) During heating, the pressure was controlled and regulated, if required, to be coincident with the final targeted conditions. (C,D) After reaching 175 °C, a 3-min step was set for dissolution of the CO_2 in the polymer. The short dissolution time is due to the sensitivity of PHAs to thermal degradation at temperatures just above their melting point. (D) Owing to the thermal inertia of the experimental apparatus, the maximum temperature (T_{max}) and pressure (P_{max}) were recorded at the end of the gas dissolution step. (D,E) the vessel was cooled to the targeted foaming temperature T_{exp} (120 or 140 °C) at a rate of −4.9 ± 0.3 °C/min (cooling). (E) Once the desired conditions were reached, the expansion was achieved by a fast pressure release at a rate of 40 MPa/s conducted by means of a gas evacuation system. The total time spent on the process was longer for the samples foamed at 120 °C than for the samples foamed at 140 °C due to the need for longer cooling from 175 °C to a lower temperature. Finally, the vessel was quickly opened and the samples extracted to avoid the modification/collapse of the cellular structure.

2.2. Samples Characterisation

2.2.1. Rheology Measurements

For the evaluation of the effect of the nonwoven fabric on the rheology of the composites, the dependence of viscosity on the shear rate of the two matrices (PHB and CE-PHB) and the untreated/PHB solid composite precursor was measured. The steady state flow curves at 175 °C were obtained in a TA Instruments Rheometer AR 2000 EX (TA Instruments, New Castle, DE, USA), equipped with electrically heated parallel plates of 25 mm diameter, with a gap set to 1 mm. The viscosity was measured, for comparative purposes, under shear rates between 0.001 and 10 s^{-1} in nitrogen atmosphere.

2.2.2. Density

The geometric density (ρ) of the samples was determined by dividing the mass of the specimen by its external volume, obtained from its dimensions, before (ρ_{solid}) and after expansion (ρ_{foam}). The relative density (ρ_{rel}) was calculated as $\rho_{rel} = \rho_{foam}/\rho_{solid}$.

2.2.3. Cell Size and Cell Density

The mean cell size (ϕ) of the foams was determined with a user-interactive image analysis adaptation of the ASTM D3576-04 method, based on [24], from scanning electron microscopy (SEM) images of fragile fractures taken in a JEOL Scanning Electron Microscope JSM 820 (JEOL Ltd, Tokyo, Japan).

The cell density per solid volume (N_0) was determined according to Equation (1):

$$N_0 = \frac{\frac{6}{\pi\phi^3}V_f}{1 - V_f} \tag{1}$$

where the porous fraction V_f was calculated as $V_f = 1 - (\rho_{foam}/\rho_{solid})$.

2.2.4. Mechanical Properties

The flexural modulus was measured at room temperature in a three-point bending mode with a span of 20 mm by means of a Perkin-Elmer DMTA 7 equipment (Perkin-Elmer Inc., Waltham, MA, USA). The forces were adjusted to obtain an indenter displacement amplitude of 20 ± 2 μm (oscillation frequency of 1 Hz), and the static stress was 120% the dynamic stress. The values, taken 3 min after applying the loads, were an average of six measurements (three specimens measured on both sides). The specimens were of 22 mm × 6 mm, with variable thickness according to the expansion of the samples. From the tests, the specific stiffness (defined as flexural modulus divided by density, E/ρ) was determined.

3. Results and Discussion

3.1. Preliminary Tests

Preliminary tests were performed with the unreinforced polymer, with the aim of evaluating the foamability of the neat matrix in the selected experimental conditions. Some samples from this preliminary evaluation are presented in Figure 3. The expansion of unreinforced specimens, performed under a quick pressure release at 40 MPa/s, was unsuccessful. The samples were irregular and presented uneven surfaces, with some large bubbles or huge air-traps and a cellular structure that was highly heterogeneous throughout the sample thickness, as shown in Figures 3a and 4 (left images). On the contrary, the composites reached very homogeneous expansions regardless of the conditions used (Figure 3b). For preliminary tests on composites, T_{exp} ranging from 100 °C to 170 °C, and P_{exp} of 5, 7.5, 10, and 20 MPa were tested.

(a)

(b)

Figure 3. Examples of the specimens from the preliminary tests. (**a**) PHB and CE-PHB without fibres expanded and sections showing the irregularity of the cellular structure throughout the thickness; (**b**) Comparison of the untreated/PHB composites before and after expansion.

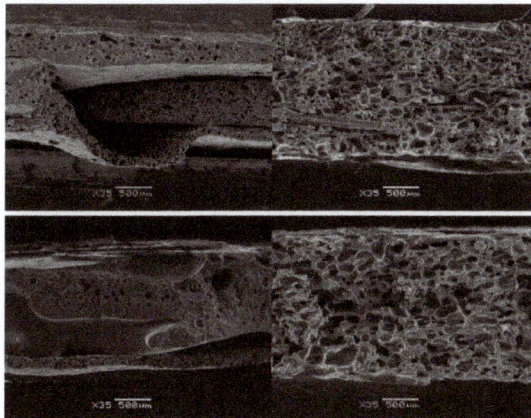

Figure 4. Comparison of unreinforced (**left**) and reinforced (**right**) foam sections showing large differences in the homogeneity of their cellular structure.

The exploratory results are presented in Figure 5. In Figure 5a, the relative densities achieved are plotted against the expansion temperature. The effect of T_{exp} on the relative density was investigated at a fixed expansion pressure of 10 MPa to restrict the temperatures at which to perform the foaming experiments. The relative density decreased with the increase of the temperature, reaching the minimum values between 120 and 140 °C, and then it started to rise again. Such temperatures were selected for the foaming experiments. In Figure 5b, the relative density results for the tests performed at 120, 130, and 140 °C at different P_{exp} are presented. The lowest densities were achieved at the highest expansion pressure of 20 MPa, regardless of the T_{exp} used. Densities ranging between 0.45 and 0.6 were obtained under 10 or 7.5 MPa at each foaming temperature, while under 5 MPa a high relative density was shown at T_{exp} = 120 °C. No clear differences could be observed between 7.5 and 10 MPa, and hence the 7.5 MPa expansion pressure was not further considered for the following tests.

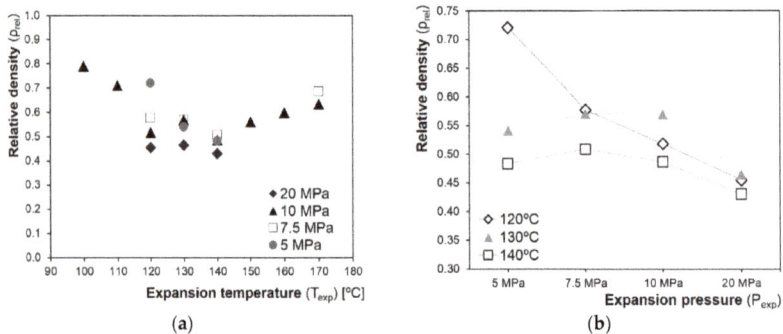

Figure 5. Relative density values of the untreated/PHB composite samples foamed on the exploratory tests against T_{exp} (a) and P_{exp} (b). In (b), the purpose of the lines is to guide the eye, and the P_{exp} axis is categorical (not in scale).

The previous results were used to design the following experiments. The parameters considered for evaluating the foamability of the composites were the polymer matrix (PHB and CE-PHB), the nonwoven fabric type (untreated, C, and C-Ar20), and the expansion conditions. According to the foamability range observed in Figure 5, two temperatures and three pressures were considered to define the foaming conditions of the cellular composites. In particular, expansions were made under 10 and 20 MPa at 120 °C, and under 5, 10, and 20 MPa at 140 °C. All those parameters combined gave rise to 30 different experimental conditions. The effects of all these factors on the properties of the cellular composites are discussed in Section 3.3.

3.2. Influence of Fibres on the Foaming Process

Neat matrices revealed some difficulties in foaming under the experimental conditions used. As aforementioned, very irregular shapes with large voids and inhomogeneous cellular structures were observed. The low quality of the cellular morphology in the matrices was mainly attributed to the low viscosity, owing to the high temperatures required for the foaming process [25] and the T_m reduction due to the CO_2 sorption [26], and to the reduced capability of stabilizing the developed cellular structure. In particular, the developed cellular structure, fostered by the phase separation of the gas after the quick pressure release, could not be uniformly stabilized after cell nucleation and growth by the increase of the polymer viscosity alone, and cell coalescence, collapse, or both, took place.

Nevertheless, under the same conditions, very good foaming of the composites was easily achieved, showing a uniform expansion, homogeneous cellular morphology, and relative densities ranging between 0.45 and 0.6 for most of the specimens expanded at T_{exp} equal to 120 and 140 °C and P_{exp} range between 5 and 20 MPa. The comparison of the cellular structures of some unreinforced foams and their cellular composite counterparts (examples in Figure 4) showed smaller cells (on average) in the unreinforced foams, but these did not develop a regular structure. Large voids were present and a largely irregular morphology was detected. The lower apparent mean cell size can be related to the migration of gas into the large voids and, consequently, most of the cells collapsed and reduced their cell size. On the contrary, the fibres had a clear effect on the cell growth, playing a key role in the stabilization of the morphology and in hindering cell collapse. It must be pointed out that the homogeneous distribution of fibres in the precursor was a key parameter, since areas with low fibre content experienced the same issues as neat matrices. According to these results, the improved foamability in the presence of fibres can be attributed to an increase of the overall viscosity, due to the polymer flow being hindered, and to a stabilizing effect of the fibre network after cell growth.

A rheological characterisation was performed in the PHB, CE-PHB and C-PHB (precursor) samples in order to determine possible changes in the viscosity. The steady state flow curves are shown in Figure 6.

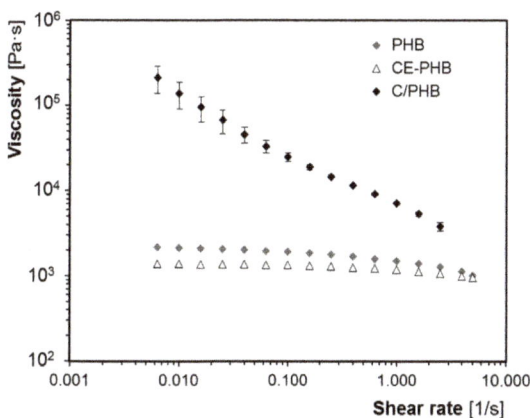

Figure 6. Steady state flow curves for the solid materials with and without nonwoven fabric reinforcement.

As can be observed, the viscosity curve for the fibre-reinforced material is around two orders of magnitude higher compared to the unreinforced matrices. Therefore, the addition of fibres clearly limited the polymer flow of the material in the melt state, and the contribution of the viscosity increase led to a higher foamability in terms of the regularity of the cellular morphology throughout the thickness for the composite samples.

As aforementioned, all cellular composite samples showed similar relative densities. By doubling the expansion pressure from 10 MPa to 20 MPa, the density values were barely lowered. This could point to a capping effect of the flax fabric reinforcement, which could be helping to achieve a regular foaming although limiting the maximum expansion at the same time. This degree of constriction could be a consequence of the mechanical entanglements of long fibres in the NW structure, produced by its own fabrication method where a 3D fibre network is obtained. Moreover, since flax fibres are stable at the T_m of the matrix (~174 °C) and cannot flow, the structural integrity of NWs remains throughout almost all of the batch-process, this guaranteeing the shape stability of the specimens.

3.3. Influence of the Processing Conditions on the Cellular Composites' Properties

3.3.1. Density

The results of relative density achieved against the expansion pressure are plotted in Figure 7. Specimens foamed at P_{exp} equal to 10 and 20 MPa presented relative densities ranging from 0.45 to 0.5 and from 0.4 to 0.45, respectively.

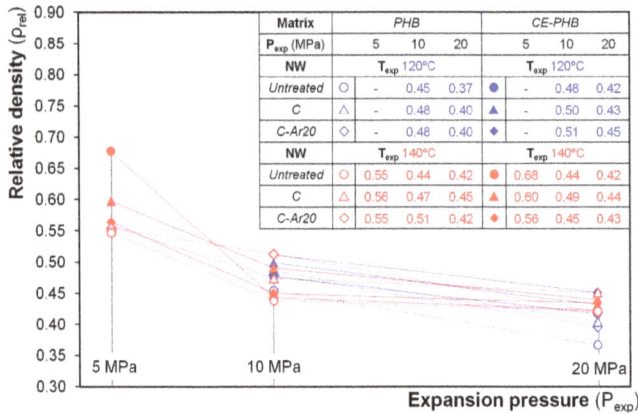

Matrix	PHB			CE-PHB		
P_{exp} (MPa)	5	10	20	5	10	20
NW	T_{exp} 120°C			T_{exp} 120°C		
Untreated ○	-	0.45	0.37	● -	0.48	0.42
C △	-	0.48	0.40	▲ -	0.50	0.43
C-Ar20 ◇	-	0.48	0.40	◆ -	0.51	0.45
NW	T_{exp} 140°C			T_{exp} 140°C		
Untreated ○	0.55	0.44	0.42	● 0.66	0.44	0.42
C △	0.56	0.47	0.45	▲ 0.60	0.49	0.44
C-Ar20 ◇	0.55	0.51	0.42	◆ 0.56	0.45	0.43

Figure 7. Relative densities of the cellular composites obtained according to the variables of study. The lines only serve to guide the eye.

Significantly higher densities (i.e., a lower expansion) and variability were obtained at P_{exp} = 5 MPa. This can be related to the lower solubilisation pressure during the gas dissolution process. The lower amount of blowing agent available has multiple effects: (a) reduces the maximum expansion ratio achievable; (b) reduces the thermodynamic instability during the pressure quench (thus reducing the nucleation of cells); (c) results in a weaker viscosity reduction (due to the lower amount of plasticisation) and hence in higher forces to be overcome by the growing bubble. The higher amount of absorbed CO_2 in the polymers at P_{exp} = 10 MPa lowered the density, thus levelling the expansion ratio, and an even higher solubilisation pressure of 20 MPa (albeit lowering the viscoelastic properties of the polymer) did not result in a larger density reduction. This limited density reduction under P_{exp} = 20 MPa with respect to P_{exp} = 10 MPa could be attributed to the effect of the reinforcement structure, owing to the degree of constriction explained before.

3.3.2. Cellular Structure

To evaluate the cellular structure, images of fractures (observed by SEM) of the 30 specimens were analysed. Some examples are presented in Figure 8, where the cellular structure throughout the thickness can be observed.

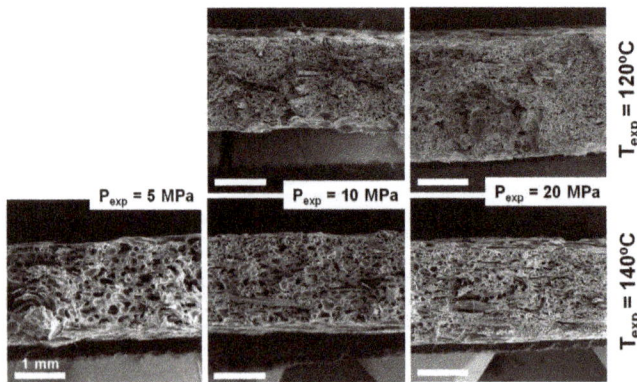

Figure 8. SEM images of the C-Ar20/PHB composite series. Scale bars correspond to 1 mm.

In general terms, a good distribution of the fibres through all the specimens was observed. This homogeneity was attributed to the NW nature of the reinforcement fabric and a good impregnation achieved in the precursors during the film-stacking. On the other hand, a good overall adhesion was observed (and hence a good load transfer is expected), although all the specimens presented some fibre pull-out, which pointed to certain heterogeneities in the fibre-matrix adhesion.

The morphology of the cellular structures at a higher magnification is presented in Figure 9, in order to show the clear differences observed in the cell size and the cell density. In general, the increase of the expansion pressure led to a reduction of cell size and a higher cell density for all the systems. The higher amount of gas dissolved was translated not only into a higher cell nucleation, but also into an increasing cell coalescence, given the low capability of the polymer to withstand extensional stresses and the higher available gas after the higher solubilisation pressure. The overall effect on cell size and cell density of such competing parameters seemed to be primarily governed by temperature and pressure, and to a lesser extent by cell coalescence.

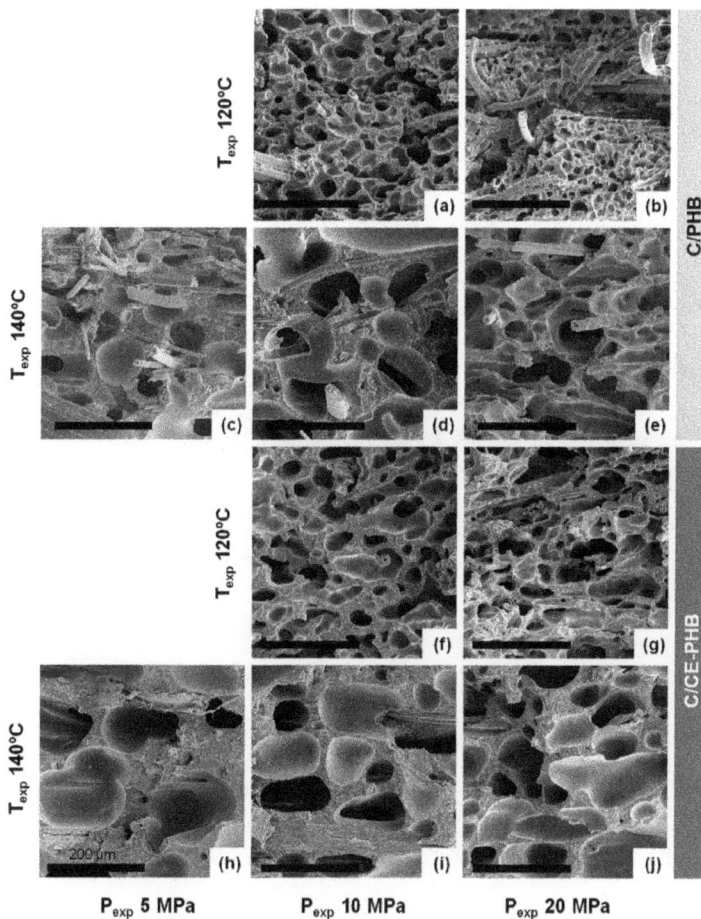

Figure 9. SEM images of the (a–e) C/PHB series and (f–j) C/CE-PHB series, as an example of the different cellular structures observed regarding the different expansion conditions. Scale bars correspond to 200 μm.

Results on the cell density for T_{exp} = 120 °C and T_{exp} = 140 °C have been plotted against P_{exp} in Figure 10, showing a clear influence of both the temperature and pressure on the amount of nucleate cells. The cell density decreased with the increase of temperature (in the range of 10^5–10^7 cells/cm^3 at 140 °C and 10^6–10^8 cells/cm^3 for 120 °C) but increased with the increase of P_{exp}, reaching maximum values for the samples expanded at 20 MPa. This increase of nucleated cells with the solubilisation pressure (related to the expansion pressure) is in accordance with the classical theory of bubble nucleation, which predicts a larger amount of bubbles with the increase of the blowing agent content at constant temperature and the thermodynamic instability (pressure decrease to induce foaming). On the other hand, the increase of temperature at the same time lowered the viscosity, which favours the cell growth but also the coalescence of bubbles due to the low capability of the investigated polymer to bear extensional stresses, leading to a reduction of cell density.

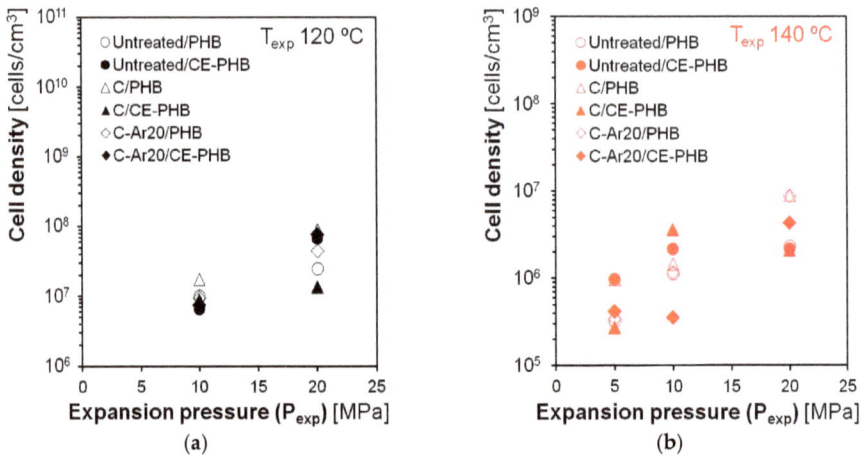

Figure 10. Cell densities of the cellular composites obtained according to the variables of study for the composites expanded at 120 °C (**a**) and at 140 °C (**b**).

While the cellular structure was highly influenced by the processing conditions for foaming, the fibre treatments and the presence of CE did not seem to affect the cellular structure in the processing conditions used in the present work. The modifications applied to the fibres did not foster the formation of the cellular structure. On the other hand, as shown in Figure 6, the CE addition led to a lower viscosity of the CE-PHB matrix (possibly due to the degradation derived from the extra thermal processing required), which would explain the lack of improvement in the cellular morphology.

3.3.3. Mechanical Behaviour

As can be observed in Figures 8 and 9, the specimens presented a good distribution of the fibres, which were well embedded in the matrix (mainly inside the cell walls). The good adhesion promotes high load transfer capability. In Table 1, the specific stiffness values are presented.

The addition of CE increased the stiffness of the matrix (the specific stiffness was ~30% higher). For that reason, the specific stiffness of all the CE-PHB-based composite precursors was higher than for those with PHB matrix. Additionally, the presence of fibres increased the specific stiffness of the composite precursors by around 60% with respect to the unreinforced matrices. The reinforcement type revealed some differences in the specific stiffness of the solid precursors, showing the best overall results for those with the C treatment.

Regarding the cellular composites, the specific stiffness for the foams was equal to or higher than for the unreinforced matrix, although lower than the specific stiffness of the solid precursors, due to the effect of the cellular structure (the properties of cellular materials decrease with density). The C/PHB and C-Ar20/CE-PHB cellular composites presented the best overall results among foams.

Table 1. Flexural modulus and relative density of some samples.

Reference	Sample Type	T_{exp} (°C)	P_{exp} (Mpa)	Specific Stiffness, E/ρ (GPa/g·cm^{-3})
PHB	Matrix	-	-	1.11
Untreated/PHB	Precursor	-	-	1.78
	Foam	120	10	1.40
		140	10	1.47
		140	20	1.43
C/PHB	Precursor	-	-	1.75
	Foam	120	10	1.81
		140	10	1.74
		140	20	1.44
C-Ar20/PHB	Precursor	-	-	1.66
	Foam	120	10	1.23
		140	10	1.50
		140	20	1.43
CE-PHB	Matrix	-	-	1.41
Untreated/CE-PHB	Precursor	-	-	2.21
	Foam	120	10	1.48
		140	10	1.58
		140	20	1.37
C/CE-PHB	Precursor	-	-	2.42
	Foam	120	10	1.62
		140	10	1.64
		140	20	1.45
C-Ar20/CE-PHB	Precursor	-	-	2.04
	Foam	120	10	1.70
		140	10	1.65
		140	20	1.65

The properties of the cellular materials obtained and the good regularity of the foaming method proposed point to a possible industrial use, with potential applications in the automotive industry (e.g., for interior door panels), biodegradable trays for packaging or as sustainable alternatives for lightweight structural panels, for instance.

4. Conclusions

The gas dissolution/pressure quench foaming of PHB-based matrix with flax nonwoven fabric reinforcement showed a high regularity in the density reduction, with homogeneous cellular morphology and a uniform distribution of the fibres, which were found to be embedded in the cellular matrix. Furthermore, an overall good fibre–matrix adhesion was achieved, all thus resulting in a good stress transfer as evidenced by the high specific stiffness measured.

The presence of the fibres increased the viscosity of the composites in the melt state. The presence of the nonwoven structure played a key role in stabilizing the cellular morphology regardless of the treatment applied. Moreover, the use of the chain extender increased the stiffness of the matrix, leading to higher specific stiffness of both solid precursors and cellular composites.

The foaming parameters (foaming temperature and expansion pressure) influenced the developed cellular structure and the density reduction. The cell size increased with the temperature but decreased with the expansion pressure. Relative densities around 0.4–0.5 were achieved, with the lowest values obtained when expanding the samples at 140 °C and 20 MPa. It is worth noting that the use of a higher pressure had a marginal effect with respect to 10 MPa.

The cellular composites produced presented a porosity of ~50% and a fibre content of ~20 wt %, both of which can contribute to the reduction of the cost of PHA-matrix-based materials, and offered higher specific properties with respect to the neat polymer, thus enhancing its competitiveness.

Polymers **2018**, *10*, 249

Acknowledgments: This work was supported by the Ministerio de Educación Cultura y Deporte, Government of Spain (grant numbers FPU12/05869, EST14/00273); Ministerio de Economia, Industria y Competitividad, FEDER, UE (grant numbers BIA2014-59399-R, MAT2015-69234-R); and the Junta of Castile and Leon (grant number VA011U16).

Author Contributions: The study was conceived and designed by M.A.R.-P., L.S., M.A., and H.V. Data acquisition and interpretation of data were mainly performed by H.V. and L.S., with the substantial contribution of E.L.-G. in the rheological characterisation of the materials. The manuscript was prepared by H.V. and supported by L.S., M.A., and M.A.R.-P.

Conflicts of Interest: The authors declare no conflict of interest.

References

1. Sorrentino, L.; Cafiero, L.; D'Auria, M.; Iannace, S. Cellular thermoplastic fibre reinforced composite (CellFRC): A new class of lightweight material with high impact properties. *Compos. Part A Appl. Sci. Manuf.* **2014**, *64*, 223–227. [CrossRef]
2. Chanprateep, S. Current trends in biodegradable polyhydroxyalkanoates. *J. Biosci. Bioeng.* **2010**, *110*, 621–632. [CrossRef] [PubMed]
3. Luo, S.; Netravali, A.N. Interfacial and mechanical properties of environment-friendly "green" composites made from pineapple fibers and poly(hydroxybutyrate-*co*-valerate) resin. *J. Mater. Sci.* **1999**, *34*, 3709–3719. [CrossRef]
4. Graupner, N.; Müssig, J. A comparison of the mechanical characteristics of kenaf and lyocell fibre reinforced poly(lactic acid) (PLA) and poly(3-hydroxybutyrate) (PHB) composites. *Compos. Part A Appl. Sci. Manuf.* **2011**, *42*, 2010–2019. [CrossRef]
5. Peterson, S.; Jayaraman, K.; Bhattacharyya, D. Forming performance and biodegradability of woodfibre–Biopol^{TM} composites. *Compos. Part A Appl. Sci. Manuf.* **2002**, *33*, 1123–1134. [CrossRef]
6. Barkoula, N.M.; Garkhail, S.K.; Peijs, T. Biodegradable composites based on flax/polyhydroxybutyrate and its copolymer with hydroxyvalerate. *Ind. Crops Prod.* **2010**, *31*, 34–42. [CrossRef]
7. Bodros, E.; Pillin, I.; Montrelay, N.; Baley, C. Could biopolymers reinforced by randomly scattered flax fibre be used in structural applications? *Compos. Sci. Technol.* **2007**, *67*, 462–470. [CrossRef]
8. Jeon, B.; Kim, H.K.; Cha, S.W.; Lee, S.J.; Han, M.-S.; Lee, K.S. Microcellular foam processing of biodegradable polymers—Review. *Int. J. Precis. Eng. Manuf.* **2013**, *14*, 679–690. [CrossRef]
9. Ventura, H.; Laguna-Gutiérrez, E.; Rodriguez-Perez, M.A.; Ardanuy, M. Effect of chain extender and water-quenching on the properties of poly(3-hydroxybutyrate-*co*-4-hydroxybutyrate) foams for its production by extrusion foaming. *Eur. Polym. J.* **2016**, *85*, 14–25. [CrossRef]
10. Liao, Q.; Tsui, A.; Billington, S.; Frank, C.W. Extruded foams from microbial poly(3-hydroxybutyrate-*co*-3-hydroxyvalerate) and its blends with cellulose acetate butyrate. *Polym. Eng. Sci.* **2012**, *52*, 1495–1508. [CrossRef]
11. Szegda, D.; Daungphet, S.; Song, J.; Tarverdi, K. Extrusion foaming and rheology of PHBV. In Proceedings of the 8th International Conference on Foam Materials & Technology (Foams), Seattle, WA, USA, 28 September–1 October 2010.
12. Wright, Z.C.; Frank, C.W. Increasing cell homogeneity of semicrystalline, biodegradable polymer foams with a narrow processing window via rapid quenching. *Polym. Eng. Sci.* **2014**, *54*, 2877–2886. [CrossRef]
13. Pilla, S.; Kim, S.G.; Auer, G.K.; Gong, S.; Park, C.B. Microcellular extrusion-foaming of polylactide with chain-extender. *Polym. Eng. Sci.* **2009**, *49*, 1653–1660. [CrossRef]
14. Ludwiczak, J.; Kozlowski, M. Foaming of polylactide in the presence of chain extender. *J. Polym. Environ.* **2015**, *23*, 137–142. [CrossRef]
15. Wang, J.; Zhu, W.; Zhang, H.; Park, C.B. Continuous processing of low-density, microcellular poly(lactic acid) foams with controlled cell morphology and crystallinity. *Chem. Eng. Sci.* **2012**, *75*, 390–399. [CrossRef]
16. Duangphet, S.; Szegda, D.; Song, J.; Tarverdi, K. The effect of chain extender on poly(3-hydroxybutyrate-*co*-3-hydroxyvalerate): Thermal degradation, crystallization, and rheological behaviours. *J. Polym. Environ.* **2013**, *22*, 1–8. [CrossRef]
17. Javadi, A.; Srithep, Y.; Lee, J.; Pilla, S.; Clemons, C.; Gong, S.; Turng, L.-S. Processing and characterization of solid and microcellular PHBV/PBAT blend and its RWF/nanoclay composites. *Compos. Part A Appl. Sci. Manuf.* **2010**, *41*, 982–990. [CrossRef]

18. Javadi, A.; Srithep, Y.; Pilla, S.; Lee, J.; Gong, S.; Turng, L.-S. Processing and characterization of solid and microcellular PHBV/coir fiber composites. *Mater. Sci. Eng. C* **2010**, *30*, 749–757. [CrossRef]

19. Boissard, C.I.R.; Bourban, P.-E.; Tingaut, P.; Zimmermann, T.; Manson, J.-A.E. Water of functionalized microfibrillated cellulose as foaming agent for the elaboration of poly(lactic acid) biocomposites. *J. Reinf. Plast. Compos.* **2011**, *30*, 709–719. [CrossRef]

20. Le Moigne, N.; Sauceau, M.; Benyakhlef, M.; Jemai, R.; Benezet, J.-C.; Rodier, E.; Lopez-Cuesta, J.-M.; Fages, J. Foaming of poly(3-hydroxybutyrate-*co*-3-hydroxyvalerate)/organo-clays nano-biocomposites by a continuous supercritical CO_2 assisted extrusion process. *Eur. Polym. J.* **2014**, *61*, 157–171. [CrossRef]

21. Chauvet, M.; Sauceau, M.; Fages, J. Extrusion assisted by supercritical CO_2: A review on its application to biopolymers. *J. Supercrit. Fluids* **2017**, *120*, 408–420. [CrossRef]

22. Ventura, H.; Ardanuy, M.; Capdevila, X.; Cano, F.; Tornero, J.A. Effects of needling parameters on some structural and physico-mechanical properties of needle-punched nonwovens. *J. Text. Inst.* **2014**, *105*, 1065–1075. [CrossRef]

23. Ventura, H.; Claramunt, J.; Navarro, A.; Rodriguez-Perez, M.; Ardanuy, M. Effects of wet/dry-cycling and plasma treatments on the properties of flax nonwovens intended for composite reinforcing. *Materials* **2016**, *9*, 93. [CrossRef] [PubMed]

24. Pinto, J.; Solorzano, E.; Rodriguez-Perez, M.A.; De Saja, J.A. Characterization of the cellular structure based on user-interactive image analysis procedures. *J. Cell. Plast.* **2013**, *49*, 555–575. [CrossRef]

25. Colton, J.S. The nucleation of microcellular foams in semi-crystalline thermoplastics. *Mater. Manuf. Process.* **1989**, *4*, 253–262. [CrossRef]

26. Takahashi, S.; Hassler, J.C.; Kiran, E. Melting behavior of biodegradable polyesters in carbon dioxide at high pressures. *J. Supercrit. Fluids* **2012**, *72*, 278–287. [CrossRef]

© 2018 by the authors. Licensee MDPI, Basel, Switzerland. This article is an open access article distributed under the terms and conditions of the Creative Commons Attribution (CC BY) license (http://creativecommons.org/licenses/by/4.0/).

polymers

MDPI

Article

Fully Biodegradable Biocomposites with High Chicken Feather Content

Ibon Aranberri *, Sarah Montes, Itxaso Azcune, Alaitz Rekondo and Hans-Jürgen Grande

CIDETEC Research Centre, Paseo de Miramón, 196, 20014 Donostia-San Sebastián (Gipuzkoa), Spain;
smontes@cidetec.es (S.M.); iazcune@cidetec.es (I.A.); arekondo@cidetec.es (A.R.); hgrande@cidetec.es (H.-J.G.)
*   Correspondence: iaranberri@cidetec.es; Tel.: +34-943-30-90-22

Received: 16 October 2017; Accepted: 8 November 2017; Published: 9 November 2017

Abstract: The aim of this work was to develop new biodegradable polymeric materials with high loadings of chicken feather (CF). In this study, the effect of CF concentration and the type of biodegradable matrix on the physical, mechanical and thermal properties of the biocomposites was investigated. The selected biopolymers were polylactic acid (PLA), polybutyrate adipate terephthalate (PBAT) and a PLA/thermoplastic copolyester blend. The studied biocomposites were manufactured with a torque rheometer having a CF content of 50 and 60 wt %. Due to the low tensile strength of CFs, the resulting materials were penalized in terms of mechanical properties. However, high-loading CF biocomposites resulted in lightweight and thermal-insulating materials when compared with neat bioplastics. Additionally, the adhesion between CFs and the PLA matrix was also investigated and a significant improvement of the wettability of the feathers was obtained with the alkali treatment of the CFs and the addition of a plasticizer like polyethylene glycol (PEG). Considering all the properties, these 100% fully biodegradable biocomposites could be adequate for panel components, flooring or building materials as an alternative to wood–plastic composites, contributing to the valorisation of chicken feather waste as a renewable material.

Keywords: biodegradable biocomposites; thermoplastics; fibres; chicken feathers

---

1. Introduction

There is a growing interest in the research and development of new materials obtained from natural resources as new alternatives to decrease or even replace petroleum-derived plastics. The incorporation of bio-based fillers into polymeric matrices results in a biocomposite [1]. Natural fibres such as flax, hemp and sisal have been widely studied as reinforcements for thermoplastics [2]. Animal fibres such as wool [3], silk [4] and horns [5], which are made of proteins, can also be used as alternative sources of reinforcements and fillers for polymers. All of them represent an opportunity to make the most of renewable and under-utilized raw materials in new application areas.

The poultry industry (including duck, turkey, goose and chicken breeding) generates a huge amount of waste each year. Despite figures regarding feather waste generation varying considerably depending on the source, it is estimated that over 65 million tons of poultry feathers are produced worldwide [6]. According to the European Commission, 13.1 million tons of poultry meat was produced in the European Union (EU-28) only in 2014 [7], with an estimated generation of 3.1 million tons of feather waste. At present, the majority of poultry feathers are disposed of in landfills or incinerated, and a minor part converted into low-nutritional-value animal food. The current solutions do not exploit the opportunity that this proteinaceous material represents, and more importantly, the management of environmental and health concerns as overall waste rises.

CFs, which are composed of 90% keratin, are abundant, cheap, biodegradable [8] and provide an opportunity to replace non-environmentally-friendly raw materials in many applications [9–11].

Serine is the most abundant amino acid in the keratin structure and contains a hydroxyl side group, which gives CFs the ability to attract moisture from the air [12]. This particular property makes keratin an effective material for insulation applications [13]. Despite the low mechanical properties of the feathers compared to other natural fibres [2], they have been used, for example, as reinforcement in cement-bonded composites [14], polyurethanes [15], polypropylene (PP)-based composites [16,17] and recycled PP composites [18,19]. Furthermore, a few studies have been performed reinforcing PLA with CFs, obtaining fully biodegradable materials [20–22]. Reddy et al. [23] also obtained 100% biodegradable composites using CFs as matrix and jute fibres as reinforcement. When feathers are used as fillers it is important to pay attention to the compatibility with the polymer matrix in order to improve the adhesion with the polymer. According to its amino acid sequence, keratin can be considered to have both hydrophilic and hydrophobic properties, since 39 out of the 95 amino acids in the keratin structure are hydrophilic [24]. Several studies have modified the CF [25] and the polymer matrix [26] for enhancing their compatibility.

Among the aliphatic polyesters, PLA is the most well-known biodegradable polymer and has the greatest commercial potential due to its good aesthetics, mechanical strength, thermal plasticity, biocompatibility and ability to be easily processed with natural fibres [27]. PLA is produced from renewable resources such as corn starch, which is fermented to lactic acid followed by ring-opening or gradual polycondensation polymerization into PLA [28]. The combination of properties of PLA makes it an ideal thermoplastic for a wide array of applications, but it still has some drawbacks when compared to other traditional materials, which are related to its production limitations and brittleness. The latter has a relatively easy solution, since the processability of the biocomposites can be addressed by modifying the PLA with plasticizers such as polyethylene glycol (PEG), citrate ester and glycerol [29]. This is not the case for the former drawback. A huge amount of starting raw material (e.g., corn) is required for PLA production, and having in mind the food category of corn, economic and sustainability issues arise. One approach to reduce the required PLA amount while keeping its benefits is the combination of PLA with natural fibres to create biocomposites.

Among other biodegradable polymers, PBAT is synthesized from fossil fuel-based monomers and shows similar properties to nonbiodegradable polymers like polyethylene. PBAT is flexible and has a higher elongation at break than most biodegradable polyesters, such as PLA, being therefore more suitable for packaging [30].

Bio-Flex® is a commercial product based on polylactic acid and a thermoplastic copolyester, and only a very few articles have published on this system [31].

Even though several studies regarding biocomposites containing CFs up to 20 wt % and PLA have been published, to the best of our knowledge, this is the first example of biocomposites obtained from biodegradable thermoplastics with high loadings of CFs. CF concentrations were selected based on the highest amount of CFs that was possible to incorporate into the polymer matrices, keeping an optimum processability of the blends. Hence, the aim of this study was to develop sustainable biocomposites comprised of 50 and 60 wt % CFs and biodegradable thermoplastic matrices (Figure 1) contributing to the responsible use of materials and zero-waste policies. The selected polymer matrices were PLA, PBAT and a PLA/copolyester blend, which are commercially available, biodegradable and compostable according to EN 13432. Additionally, the effect of the addition of plasticisers to the composite and alkali pre-treatments of the feathers were investigated. The mechanical properties, thermal stability and diffusivity of the resulting biocomposites were studied in order to identify potential applications of these new materials.

Figure 1. Chicken feather/PLA biocomposite containing 50 wt % of feather.

2. Materials and Methods

2.1. Materials

Sanitized CFs from Grupo SADA (Madrid, Spain) were ground in a universal cutting mill Pulverisette 19 (Fritsch, Germany) at a rotor rotational speed of 2800 rpm and at a sieve insert size of 1 mm. According to the optical microscope images (Figure 2), ground feathers showed a wide size distribution ranging from 100 μm to a few mm. The different polymers used in this work are detailed in Table 1. According to the EN 13432, all the matrices are biodegradable and compostable. Sodium hydroxide (NaOH, 97%) and polyethylene glycol with molecular weight of 400 g/mol (PEG 400) were purchased from Aldrich (St. Louis, MO, USA).

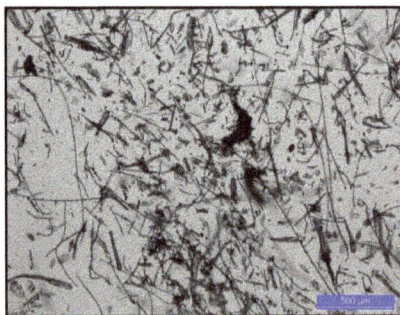

Figure 2. Optical image of ground chicken feathers.

Table 1. Properties of the biopolymers obtained from their technical datasheets.

Polymer (Sample Code)	Type of Matrix	Melting Temperature (°C)	Density (g/cm^3)	Supplier
Ingeo 2003D (ING)	PLA	145–160	1.24	NatureWorks (Minnetonka, MN, USA)
Ecoflex C1200 (ECO)	PBAT	120	1.25–1.27	BASF (Ludwigshafen, Germany)
Bio-Flex 6611 (BIO)	PLA/copolyester blend	150–170	1.29	FKuR (Willich, Germany)

2.2. Alkali Treatment of the CFs

CFs were immersed in NaOH solution (5% w/v) for 2 h at room temperature. The feathers were then washed with distilled water until the rinse water no longer indicated any alkalinity [32]. After washing, the feathers were kept in an oven at 80 °C for 6 h.

2.3. Biocomposite Preparation

The biocomposites were manufactured using a torque rheometer HAAKE PolyLab QC (Karlsruhe, Germany). Prior to blending, ground CFs and the polymers were dried at 80 °C for 6 h. The contents of CF in the biocomposites were 50 and 60 wt %. ING, ECO, BIO biopolymers were first melted at 170, 140 and 170 °C, respectively, and CFs were added later while mixing for 3 min. For the ING/PEG400 blends, 10 wt % of PEG400 was mixed before the CFs were incorporated. The blends were compression-molded in a Vogt 600T laboratory hot press machine (Maschinen + Technik Vogt GmbH, Möhnesee, Germany) into sheets of 90 × 90 × 2.1 mm^3. Consolidation was carried out at 200 °C for 3 min and finally, the square plates were cooled with air. TGA analyses of the biocomposites were performed previously in order to set the consolidation conditions.

2.4. Characterization of the Biocomposites

2.4.1. Density of the Biocomposites

The density of the biocomposites was determined experimentally according to the ISO 9427 [33]. Three rectangular samples of each composite with a known volume were weighed and the density was determined as the ratio of the mass to volume. The average and standard deviation were reported.

2.4.2. Water Absorption of Biocomposites

Water absorption of the biocomposites was determined by immersion of the specimens vertically in distilled water at 25 °C for 24 h (ASTM D570-98) [34]. First, rectangular specimens (24 mm × 12 mm × 2.1 mm) were cut from tensile testing fracture specimens and air-dried at 60 °C for 24 h, cooled in a desiccator and weighed (conditioned weight). Then, samples of the biocomposites were soaked in water for 48 h and wiped with paper to remove the excess of water on the surface of the specimens before weighing (wet weight) at fixed time intervals. Three specimens were tested with an analytical balance of 0.1 mg precision and the average and standard deviation were reported. The percentage of water absorption (*WA* in %) was calculated using Equation (1):

$$WA\ (\%) = \frac{\text{wet weight} - \text{conditioned weight}}{\text{conditioned weight}} \times 100 \tag{1}$$

2.4.3. Thermogravimetric Analysis (TGA)

The thermal stability was measured by thermogravimetric analysis using a TGA Q500 (TA Instruments, New Castle, DE, USA). Dynamic measurements were performed from 25 to 600 °C at a heating rate of 10 °C/min by using constant nitrogen flow of 60 mL/min to prevent thermal oxidation processes of the polymer sample. The temperatures at 5%, 10%, 25% and 50% of weight loss were calculated. Sample weight was approximately 10 mg.

2.4.4. Dynamic Scanning Calorimetry (DSC)

Thermal properties of CFs, ING, BIO, ECO and the biocomposites containing 50 and 60 wt % CFs were determined with a Discovery DSC 25 auto (TA Instruments, New Castle, DE, USA) at a scan rate of 10 °C·min^{-1} over the temperature range 25–200 °C for CF and ING series and −70–200 °C for ECO and BIO series due to their differences in thermal properties. The measurements were carried out using 5.00 ± 0.50 mg samples under a nitrogen atmosphere (50 mL·min^{-1}). All of the thermal properties were obtained from the second heating curves in order to evaluate the effect of the filler, as nucleating agent, during the processing. The degree of crystallinity (χ_c) was calculated from the following Equation (2):

$$\chi_c(\%) = \frac{100}{w} \times \frac{|\Delta H_m + \Delta H_{cc}|}{\Delta H_m^0},$$ (2)

where ΔH_m and ΔH_{cc} are the enthalpy of fusion and cold-crystallization at melting and crystallization temperatures, respectively, w is the weight fraction of neat polymer in the sample and ΔH_m^0 is the melting enthalpy of the 100% crystalline polymer. ΔH_m^0 was taken as 93.7 and 114 J/g for PLA and PBAT, respectively. For the BIO series, ΔH_m^0 was also taken as 93.7 J/g after the assumption that Bio-Flex 6611 is mainly based on PLA, and no other melting peak was observed in the analysed range.

2.4.5. Mechanical Testing

The tensile properties of the biocomposites (Young's modulus, tensile strength, and elongation at break) were evaluated using a tensile test according to the ISO 527 [35] standard with a universal testing machine model 3365 (Instron, Norwood, MA, USA) and controlled by Bluehill Lite software developed by Instron (Norwood, MA, USA). The initial length of the test specimens was 25.4 mm and a cross-head speed of 10 mm/min was used. The number of tested specimens for the mechanical properties was 5 for average calculations.

2.4.6. Thermal Diffusivity

The thermal properties were measured utilizing a light flash analyzer (LFA, Nanoflash 447 by Netzsch, Ahlden, Germany) following the ASTM E1461 [36]. This analyser operated a xenon flash light that induced a pulse of energy on one side of the sample. Such pulse increased the sample temperature and an indium antimonide (InSb) infrared detector measured the temperature response time to the pulse of energy on the other side of the sample. The response time was used to calculate the thermal diffusivity.

2.4.7. FE–SEM

The CF/polymer interface was analysed by scanning electron microscopy of the fractured surface of the composites. The microphotographs were taken with a Carl Zeiss Ultra Plus field-emission–scanning electron microscope (FE–SEM, Oberkochen, Germany) equipped with an energy dispersive X-ray spectrometer (EDXS). Prior to FE–SEM analysis, samples were Au-coated.

3. Results and Discussions

3.1. Density

One of the main reasons of using biocomposites containing a high concentration of CF is the high-density reduction that is expected due to the low density of these feathers. Compared to other natural fibres like wool 1.31 g/cm³, jute 1.3 g/cm³ and coir 1.2 g/cm³, the density of the chicken feather fibres ranges between 0.8 g/cm³ [37] and 0.9 g/cm³ [38]. This variation may be due to the different source of feathers. Hence, chicken feather inclusion in a thermoplastic matrix could potentially lower biocomposite density more than any other reinforcing natural fibre.

In Table 2 are summarised density values and the percent reduction of the biocomposites compared to the neat polymers. The density values determined for the processed ING, ECO and BIO matrices are 1.16, 1.22 and 1.28 g/cm³, respectively. As expected, the addition of CFs decreased the density values of all the biocomposites, as has been reported by other researchers [39]. The higher density of the neat matrix (ECO and BIO), the more pronounced was the reduction of the composite material. The addition of 50 wt % of CF led to 13.11% and 15.63% density reductions, whereas the addition of 60 wt % of CF lead to 17.21% and 20.31%, respectively.

Table 2. Density values and percentage of density reduction of the biocomposites compared to neat polymer.

Matrix	Polymer/CF Ratio	Density (g/cm³)	% of Density Reduction Compared to Neat Polymer
ING	100/0	1.16	-
	50/50	1.10	5.17
	40/60	1.00	13.79
ECO	100/0	1.22	-
	50/50	1.06	13.11
	40/60	1.01	17.21
BIO	100/0	1.28	-
	50/50	1.08	15.63
	4960	1.02	20.31

The reduction is attributed to the high void volume of the biocomposites due to the presence of the hollow structure of the CFs [40] which were preserved after the compression molding conditions.

3.2. Water Absorption

In order to study the dimensional stability of the biocomposites, all samples were immersed in water at 25 °C for 48 h. Figure 3 shows the water gain after 24 and 48 h of immersion of water. The water absorption of the composite samples was calculated with (1). The water gain in the neat polymers after 24 h was 0.46%, 1.1% and 0.32% for ING, BIO and ECO, respectively. These low values were expected since the polymeric matrices are hydrophobic. After 48 h of immersion, all the composite samples gained water in the range between 12.5–14.5% for 50 wt % and 16.6–17.6% for 60 wt % of CF. As can been seen, the mass uptake of the biocomposites increased as the CF loading and the immersion time increased. The water gain of the composites is mainly attributed to the hydrophilic polymer backbone of the CF [41]. ING-based composites showed the highest water absorption. This effect is probably due to neat PLA samples containing more polar groups in the polymer chain than the other polymer matrices. These results are in agreement with the results published by Carrillo et al., in which the addition of CFs promotes water absorption in biocomposites based on PLA [20,39] and polyolefins [39].

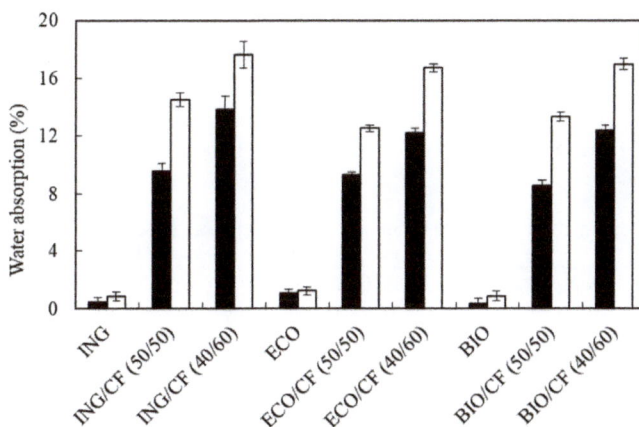

Figure 3. Water absorption of the polymers and biocomposites after 24 h (black) and 48 h (white) of immersion in water at 25 °C.

3.3. Thermogravimetric Analysis (TGA)

The thermal stability of neat polymers and polymer/CF biocomposites was investigated with TGA. The TGA curves presented in Figure 4a–c show thermal decomposition of CFs, neat polymers (ING, ECO and BIO) and the CF-containing biocomposites. The weight loss of the CFs is represented in the three figures as reference. Three weight-loss steps can be seen in the case of CFs. The weight loss in the first stage, from 50 to 250 °C, is due to the evaporation of absorbed water from the hydrophilic groups of the CFs; the second step, with a higher rate, between 250 and 400 °C, undergoes degradation associated with the destruction of disulphide bonds and the elimination of H_2S originating from amino acid cysteine [42]; and from 400 °C onwards, the keratin partially decomposes. According to Figure 4a–c, the biocomposites also degrade through three stages and the degradation starting temperature depends on the type of matrix. The degradation starting temperature of the ING and the corresponding biocomposites is slightly lower than that of ECO and BIO and their biocomposites. Moreover, the degradation rate is faster for ING-based biocomposites. All biocomposites are thermally stable until 230–240 °C; at this temperature, composites begin thermal degradation due to presence of CFs also in three stages. For temperatures above 400 °C, all the polymer matrices were completely decomposed and an inorganic residue coming from the CFs was left behind after the decompositions of the CF-containing composites. Composites containing 60 wt % CF started degrading at lower temperatures than composites containing 50 wt % CF due to the higher keratin content. These results are in agreement with previous studies of PLA/CF biocomposites containing up to 10 wt % [21] and 30 wt % of CFs [25].

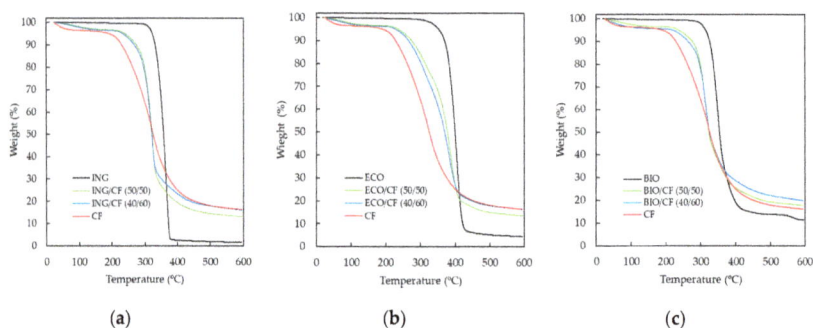

Figure 4. TGA curves of (**a**) ING-based biocomposites; (**b**) ECO-based biocomposites and (**c**) BIO-based biocomposites.

In Table 3, the 5%, 25% and 50% weight-loss temperatures are listed for all the specimens shown in Figure 4a–c.

Table 3. Thermal characterization of the neat polymers and biocomposites.

Samples	T (5%) (°C)	T (25%) (°C)	T (50%) (°C)
CF	191	276	325
ING	323	345	358
ING/CF (50/50)	238	305	322
ING/CF (40/60)	231	302	321
ECO	352	386	398
ECO/CF (50/50)	236	330	377
ECO/CF (40/60)	227	316	369
BIO	314	339	353
BIO/CF (50/50)	239	304	323
BIO/CF (40/60)	221	303	325

3.4. DSC Analysis

In order to study the effect of the CFs on the thermal properties of the neat polymers, DSC analysis of CFs, neat polymers and the corresponding biocomposites was performed. For CFs (shown in Figure 5a–c as reference), a large low-temperature endothermic peak was observed at 77 °C. This peak ranges from 30 to 130 °C approximately and shows the amount of bound water in the keratin structure, and on occasion is referred as the "denaturation" temperature. Figure 5a shows the DSC curves of CF/ING biocomposites. The glass transition temperature decreased from 53.6 to 47.4 °C and increased to 55.6 °C with addition of 50 and 60 wt % of CF, respectively. This lack of trend of the T_g of PLA may be due to: (i) heterogeneity of the material. The high concentration of CF may contribute to a poor mixture of both phases; (ii) the DSC thermogram of the CFs may affect the output of the T_g values of the biocomposites; and (iii) high disparity in the thermal properties of the different part of the feathers [43].

For the ECO series shown in Figure 5b, the T_g increased and the ΔH_m, T_m and the degree of the crystallinity of the biocomposites decreased in the presence of CFs. In this series, since the melting temperatures of the samples were between 116.8 and 120.4 °C, the DSC curve of the CFs in this range overlaps with the DSC curves of the ECO series, affecting the measurements of the melting enthalpy.

In Figure 5c the DSC curves of the BIO series are shown. All the samples based on a copolymer that contains PLA showed two T_g, both increasing significantly after addition of 50 and 60 wt % of CFs. The melting temperatures T_m increased slightly and ΔH_m decreased more than 50%. The crystallinity of PLA in the BIO series increased from 30.5% to 36.0% when the biocomposite was reinforced with 60 wt % of CFs, observing the same effect as observed in the ING series. According to [44,45], there are two main factors affecting the crystallinity of polymer composites: (i) additives having a nucleating effect that results in an increased crystallinity, and (ii) additives hindering the migration and diffusion of polymer chains to the surface of the growing polymer crystal, lowering crystallinity.

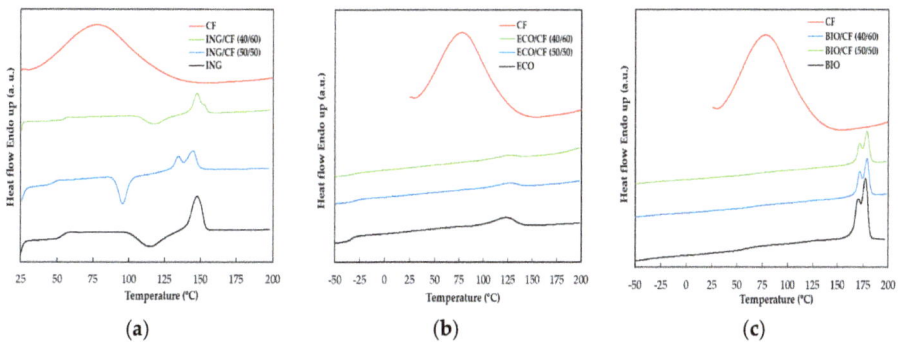

Figure 5. DSC curves of (**a**) ING-based biocomposites; (**b**) ECO-based biocomposites and (**c**) BIO-based biocomposites.

The crystallinity was found to increase in the ING and BIO series, both containing PLA, in agreement with Cheng [20], who reported that chicken feather fibres play the role having a nucleating effect in PLA. In the same study, it was observed that the crystallization peaks became narrower with the addition of up to 10 wt % of CFs, indicating an increase of the crystallization rate. A similar effect was observed for the ING/CF (50/50) biocomposite. However, a broader crystallization peak was found for the ING/CF (40/60), probably due to a more heterogeneous crystallization process, which can be attributed to a higher CF loading. Regarding the ECO-based biocomposites, the crystallinity decreased. In this case, the addition of CFs may disrupt the crystallite formation of PBAT due to the increasing of viscosity, hence leading to less-ordered and smaller crystals. Detailed results of the thermal properties obtained from DSC are summarized in Table 4.

Table 4. Thermal properties of ING, ECO, BIO and their corresponding biocomposites.

Sample	T_{g1} (°C)	T_{g2} (°C)	ΔHcc (J/g)	T_m (°C)	ΔH_m (J/g)	χ (%)
ING		53.6	19.5	147.6	22.7	3.48
ING/CF (50/50)		47.4	15.2	144.6	17.9	5.78
ING/CF (40/60)		55.6	9.00	147.4	11.8	7.50
ECO	−34.1			120.4	9.8	8.60
ECO/CF (50/50)	−30.4			116.8	4.3	7.54
ECO/CF (40/60)	−30.7			116.6	2.7	5.92
BIO	−37.8	58.7		176.9	28.6	30.5
BIO/CF (50/50)	−33.1	65.6		178.7	14.4	30.7
BIO/CF (40/60)	−32.1	64.8		178.5	13.5	36.0

3.5. Mechanical Properties of the Biocomposites

Previous work by Cheung [46] proved that the mechanical properties of chicken feather fibre/PLA biocomposites were determined by the feather loading. As a matter of fact, the tensile strength of PLA decreases with increasing the content of CF, which is ascribed to the fact that the strength of CF is insufficient. Cheng et al. [21] investigated 2, 5, 8 and 10 wt % feather concentration and showed that even though the tensile properties of the composites were improved with 2 and 5 wt % loading compared to the neat PLA, increasing feather amount resulted in a significant worsening of the mechanical properties. Therefore, it was expected that 50 and 60 wt % of feather content could decrease the mechanical properties considerably. The three polymer matrices that have been investigated in this work have different initial mechanical properties and thus have shown different performance upon the addition of CFs.

Under some conditions, alkali treatment could significantly improve adhesion at the interface and contribute to penetrate the fibre into the polymer matrix. Alkali treatment's main purpose is to remove some wax and oils, thereby increasing surface roughness and reducing its hydrophilic nature [47].

Figures 6–8 show how three main variables affect the mechanical properties (Young's modulus, tensile strength and elongation at break) of the biocomposites: (1) the type of the polymer matrix, (2) the CF content and (3) the alkali treatment of the feathers and blending of ING with PEG.

In Figure 6, the Young's moduli of the biocomposites are presented. This parameter measures the stiffness of the materials. ING and BIO showed a similar trend upon the addition of CFs.

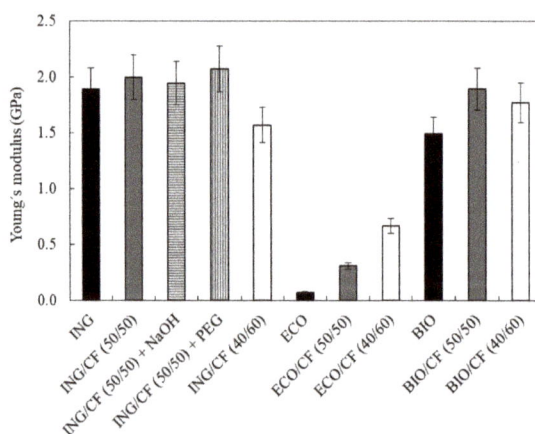

Figure 6. Young's moduli values of neat polymers and the different biocomposites containing 50 and 60 wt % of CFs.

For biocomposites based on ING, the addition of 50 wt % of CFs has no major effect on the Young's modulus compared to the neat polymer. Even when alkali-treated CFs and PEG as plasticiser were added, only a minor increase was observed, reaching a maximum value of 2.1 GPa. However, increasing the feather loading, that is, the ING/CF (40/60) biocomposite, showed a pronounced decrease of Young's modulus of 20%.

For ECO/CF biocomposites, the reference value of the neat polymer is lower than that of ING and BIO and hence, the effect of reinforcing ECO with CF is substantial. The Young's moduli are 70, 300 and 670 MPa for 0, 50 and 60 wt % of CFs, respectively, with an increase up to 1000%. For BIO/CF biocomposites, adding feathers has a positive effect, but as observed in the first case, the increase of CF loading to 60 wt % is detrimental. The differences observed on the trends of the mechanical properties of the biocomposites are probably due to the different degrees of compatibility between the matrices and the feathers.

In Figure 7, the tensile strengths of the biocomposites are shown. This property is a measurement of the force required to break the material per unit area. For all the series, the effect on the tensile strength was similar, decreasing when 50 and 60 wt % of CF was added. This effect is probably a combination of the high CF content and hence poor mixing between the components, and low adhesion between the matrices and the fibres [43]. For ING/CF specimens, the alkali treatment of the CFs increased the tensile strength, with respect to the composite with untreated CFs, from 13.7 to 24 MPa.

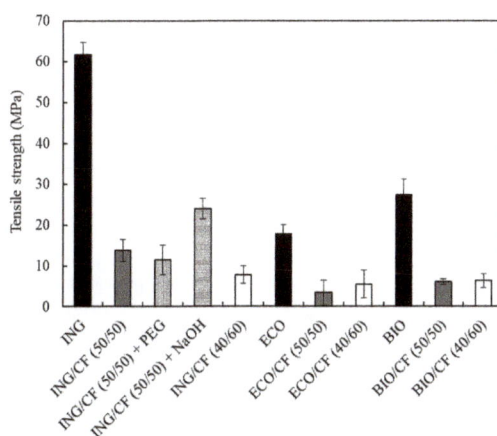

Figure 7. Tensile strength values of neat polymers and the different biocomposites containing 50 and 60 wt % of CFs.

Figure 8 shows the elongation at break of the different samples. The elongation at break, which is the ratio between changed length and initial length after breakage, showed a dramatic decrease with CFs obtaining less ductile materials. The most obvious case is the ECO/CF series, varying the elongation at break from 570% for pure ECO to 2.5% for ECO/CF (50/50). When the percentage of the CF loading was increased in this series, the elongation at break further decreased. For all the series, when the percentage of CF was 50 wt %, the ductility of the biocomposites decreased, which indicates that the CFs had hardened the biocomposites and reduced their ductility. For the ING series, no clear effect of the alkali treatment or PEG was observed. When the CF loading was increased to 60 wt %, the elongation at break of ING/CF (40/60) and ECO/CF (40/60) was reduced. The elongation at break of ING/CF (40/60) was not further reduced.

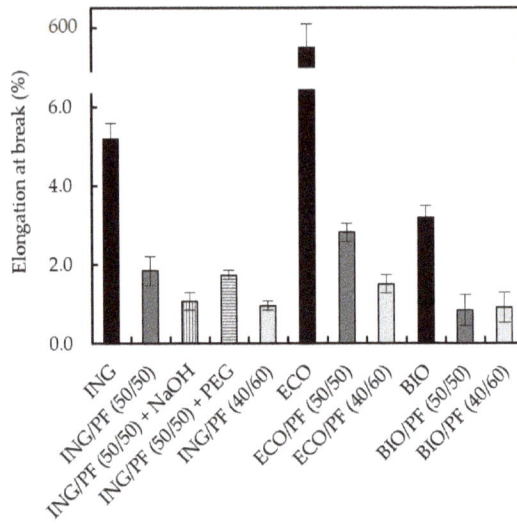

Figure 8. Elongation-at-break values of neat polymers and the different biocomposites containing 50 and 60 wt % of CFs.

3.6. Morphology of the Biocomposites

The effects of the two different treatments on the adhesion between the PLA matrix and CFs were investigated by FE–SEM. Micrographs of raw CFs and NaOH-treated CFs are shown in Figure 9a,b, respectively. As can be observed, alkali treatment increases the roughness of the CFs, increasing the surface area of the fibres available for contact to the matrix. Bismarck et al. [48] also observed this effect in natural fibres. Additionally, alkali treatment also removes hydrophilic components located on the fibre surface [47]. The micrographs of the fractured surface of the untreated CF/ING biocomposites can be seen in Figure 9c. This figure shows individual CFs, which indicates that the fibres have been not mixed well with the matrix due to a low adhesion between the matrix and the fibres. As can be seen in Figure 9d, the ING/NaOH–CF biocomposites show a good dispersion of the NaOH–CFs in the ING matrix compared to the appearance of the ING/CF biocomposites (Figure 9c), enhancing the CF wettability in the polymer matrix and improving the tensile strength of the biocomposites. In the case of the biocomposites containing the PEG, see Figure 9e; chicken fibres were completely covered by the polymer, however, no improvement in the tensile strength and modulus was observed.

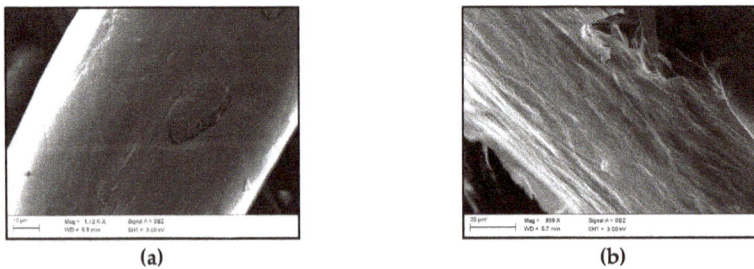

(a)

(b)

Figure 9. *Cont.*

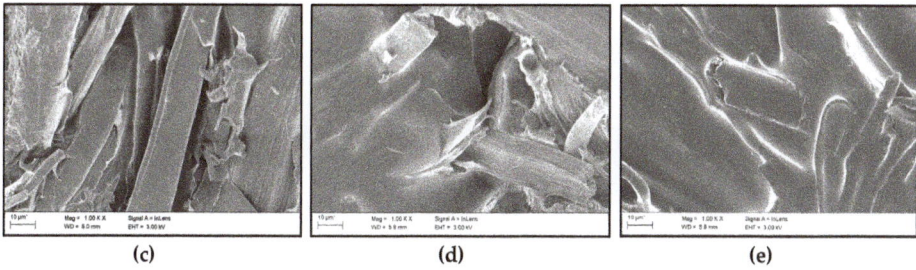

Figure 9. FE–SEM images of (**a**) untreated CFs; (**b**) NaOH-treated CFs; (**c**) fractured ING/CF (50/50); (**d**) fractured ING/NaOH–CF (50/50) and (**e**) fractured PEG–ING/CF (50/50).

3.7. Thermal Diffusivity

Thermal diffusivity measures the rate at which heat flows through a material. It is a measure of the rate at which a body with a non-uniform temperature reaches a state of thermal equilibrium. The measured thermal diffusivity of the biocomposites at 25 °C and containing 50 wt % of CFs are shown in Figure 10. It can be seen that when the CFs were added, the thermal diffusivity of the biocomposites was reduced. The thermal diffusivity reduction varied from 5% for ING and 9.5% for BIO to 18.9% for ECO. The CFs showed a hollow honeycomb-shaped structure, which acted as air and heat insulators causing a decrease of thermal diffusivity of the CF-containing biocomposites [40]. These results mean that biocomposites containing CFs will require longer time to be heated or cooled than the unfilled polymer, indicating that the CF-reinforced biocomposites show improved thermal insulation properties. Qin, X. [13] reported a similar trend when PLA sheets were reinforced with CF mats.

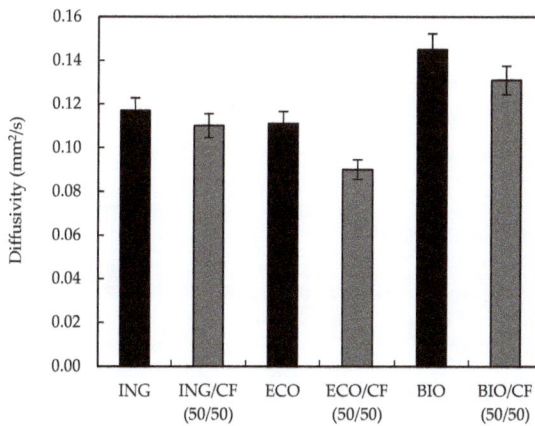

Figure 10. Diffusivity values of neat polymers and polymers reinforced with 50 wt % CFs.

4. Conclusions

In this work, fully biodegradable biocomposites with 50 and 60 wt % of CFs were successfully developed and their density, water absorption, thermal stability, mechanical properties and thermal diffusivity were investigated.

CF/polymer biocomposites with high content of CF showed a decrease in density values unlike any other natural or synthetic fibres, forming lightweight fully biodegradable materials. Regarding

Polymers **2017**, *9*, 593

water absorption, the biocomposites gained from 12% to 17% of weight after 48 h of immersion, due to the hydrophilic amino acid groups found in the structure of the feathers.

The thermogravimetric curves revealed that the thermal stability of the biocomposites is lowered by the addition of feathers, due to their high content of thermally unstable keratin.

Mechanical properties of the CF-containing biocomposites showed lower tensile strength and lower elongation-at-break values compared to those of neat polymers, whereas the elastic moduli depended on the matrix type. The elastic moduli of the biocomposites were increased for ECO- and BIO-based biocomposites due to the low values of the neat polymers. In these cases, the biocomposites became stiffer. In the ING series, the moduli were not affected by the addition of the CFs for the ING/CF (50/50) biocomposite.

The presence of honeycomb structures in CFs provides air- and heat-insulating capabilities to the biocomposites, lowering the thermal diffusivity of the biocomposites. The results showed that the CFs could be considered as isolator components in buildings to reduce heat transfer and hence decrease energy consumption.

Taking into account the overall properties observed, these materials could be used as an alternative to wood–plastic composites in medium-density fibreboards (MDF) composites [49], decking materials and many other applications [50], in which low density, low thermal diffusivity and low water absorption are also required.

This approach would contribute to a more efficient use of natural resources and take advantage of this material that is produced in huge amounts and is currently underutilized by the poultry industry. It is intended that the use of 100% biodegradable materials will contribute to sustainability and a reduction in the environmental impact associated with the disposal of non-biodegradable polymers.

Acknowledgments: This work was supported by KaRMA2020 project. This project has received funding from the European Union's Horizon 2020 Research and Innovation program under Grant Agreement n° 723268.

Author Contributions: Ibon Aranberri and Sarah Montes performed the experiments and wrote the paper; Itxaso Azcune conceived and designed the experiments and contributed on the interpretation of the results and Alaitz Rekondo and Hans-Jürgen Grande contributed on the final discussion of the results.

Conflicts of Interest: The authors declare no conflict of interest.

References

1. Mohanty, A.; Misra, M.; Drzal, L. *Natural Fibers, Biopolymers and Biocomposites*; CRC Press, Taylor & Francis: Boca Ratón, FL, USA, 2005; pp. 1–36. ISBN 9780849317415.
2. Pickering, K.L.; Efendy, M.G.A.; Le, T.M. A review of recent developments in natural fibre composite and their mechanical performance. *Compos. Part A Appl. Sci. Manuf.* **2016**, *83*, 98–112. [CrossRef]
3. Conzatti, L.; Giunco, F.; Stagnaro, P.; Patrucco, A.; Marano, C.; Rink, M.; Marsano, E. Composites based on polypropylene and short wool fibres. *Compos. Part A Appl. Sci. Manuf.* **2013**, *47*, 165–171. [CrossRef]
4. Alam, A.K.M.M.; Shubhra, Q.T.H.; Al-Imran, G.; Barai, S.; Islam, M.R.; Rahman, M.M. Preparation and characterization of natural silk fiber-reinforced polypropylene and synthetic E-glass fiber-reinforced polypropylene composites: A comparative study. *J. Compos. Mater.* **2011**, *45*, 2301–2308. [CrossRef]
5. Kumar, D.; Boopathy, S.R.; Sangeetha, D.; Bharathiraja, G. Investigation of Mechanical Properties of Horn Powder-Filled Epoxy Composite. *J. Mech. Eng.* **2017**, *63*, 138–147. [CrossRef]
6. Zhao, W.; Yang, R.; Zhang, Y.; Wu, L. Sustainable and practical utilization of feather keratin by a novel physicochemical pretreatment: High density steam flash-explosion. *Green Chem.* **2012**, *14*, 3352–3360. [CrossRef]
7. European Commission. Available online: http://ec.europa.eu/agriculture/poultry_en (accessed on 9 November 2017).
8. Korniłłowicz-Kowalska, T.; Bohacz, J. Biodegradation of keratin waste: Theory and practical aspects. *Waste Manag.* **2011**, *31*, 1689–1701. [CrossRef] [PubMed]
9. Reddy, N. Non-food industrial applications of poultry feathers. *Waste Manag.* **2015**, *45*, 91–107. [CrossRef] [PubMed]

10. Yin, J.; Rastogi, S.; Terry, A.E.; Popescu, C. Self-organization of polypeptides obtained on dissolution of feather keratins in superheated water. *Biomacromolecules* **2007**, *8*, 800–806. [CrossRef] [PubMed]

11. Yin, X.C.; Li, F.Y.; He, Y.F.; Wang, Y.; Wang, R.M. Study on effective extraction of chicken feather keratin and their films for controlling drug release. *Biomater. Sci.* **2013**, *1*, 528–536. [CrossRef]

12. Schmidt, W.F.; Line, M.J. Physical and chemical structures of poultry feather fiber fractions in fiber process development. In Proceedings of the 1996 TAPPI Nonwovens Conference, Charlotte, NC, USA, 11–13 March 1996; pp. 135–141.

13. Qin, X. Chicken Feather Fibre Mat/PLA Composites for Thermal Insulation. Master's Thesis, University of Waikato, Hamilton, New Zealand, 2015.

14. Acda, M.N. Waste chicken feather as reinforcement in cement-bonded composites. *Philipp. J. Sci.* **2010**, *139*, 161–166.

15. Saucedo-Rivalcoba, V.; Martínez-Hernández, A.L.; Martínez-Barrera, G.; Velasco-Santos, C.; Castaño, V.M. (Chicken feathers keratin)/polyurethane membranes. *Appl. Phys. A* **2011**, *104*, 219–228. [CrossRef]

16. Reddy, N.; Yang, Y. Light-weight polypropylene composites reinforced with whole chicken feathers. *J. Appl. Polym. Sci.* **2010**, *116*, 3668–3675. [CrossRef]

17. Huda, S.; Yang, Y. Composites from ground quill and polypropylene. *Compos. Sci. Technol.* **2008**, *68*, 790–798. [CrossRef]

18. Jiménez-Cervantes, E.; Velasco-Santos, C.; Martínez-Hernández, A.L.; Rivera-Armenta, J.L.; Mendoza-Martínez, A.M.; Castaño, V.M. Composites from chicken feathers quill and recycled polypropylene. *J. Compos. Mater.* **2015**, *49*, 275–283.

19. Yang, Y.; Reddy, N. Utilizing discarded plastic bags as matrix material for composites reinforced with chicken feathers. *J. Appl. Polym. Sci.* **2013**, *130*, 307–312. [CrossRef]

20. Cañavate, J.; Aymerich, J.; Garrido, N.; Colom, X.; Macanás, J.; Molins, G.; Álvarez, M.D.; Carrillo, F. Properties and optimal manufacturing conditions of chicken feather/poly(lactic acid) biocomposites. *J. Compos. Mater.* **2016**, *50*, 1671–1683. [CrossRef]

21. Cheng, S.; Lau, K.T.; Liu, T.; Zhao, Y.; Lam, P.M.; Yin, Y. Mechanical and thermal properties of chicken feather fiber/PLA Green composites. *Compos. Part B Eng.* **2009**, *40*, 650–654. [CrossRef]

22. Baba, B.O.; Özmen, U. Preparation and mechanical characterization of chicken feather/PLA composites. *Polym. Compos.* **2015**, *38*, 837–845. [CrossRef]

23. Reddy, N.; Jiang, J.; Yang, Y. Biodegradable composites containing chicken feathers as matrix and jute fibers as reinforcement. *J. Polym. Environ.* **2014**, *22*, 310–317. [CrossRef]

24. Arai, K.M.; Takahashi, R.; Yokote, Y.; Akahane, K. Amino-acid sequence of feather keratin from fowl. *Eur. J. Biochem.* **1983**, *132*, 501–507. [CrossRef] [PubMed]

25. Huda, M.S.; Schmidt, W.F.; Misra, M.; Drzal, L.T. Effect of fiber surface treatment of poultry feather fibers on the properties of their polymer matrix composites. *J. Appl. Polym. Sci.* **2013**, *128*, 1117–1124. [CrossRef]

26. Ghani, S.A.; Tan, S.J.; Yeng, T.S. Properties of chicken feather fiber-filled low density polyethylene composites: The effect of polyethylene grafted maleic anhydride. *Polym. Plast. Technol. Eng.* **2013**, *52*, 495–500. [CrossRef]

27. Kuma, P.; Bajpai, P.K.; Singh, I.; Madaan, J. Development and characterization of PLA-based green composites: A review. *J. Thermoplast. Compos. Mater.* **2012**, *10*, 1–30.

28. Masutani, K.; Kimura, Y. PLA Synthesis. From the Monomer to the Polymer. In *Poly(lactic acid) Science and Technology: Processing, Properties, Additives and Applications*; RSC: London, UK, 2014.

29. Martin, O.; Avérous, L. Poly(lactic acid): Plasticization and properties of biodegradable multiphase system. *Polymer* **2001**, *45*, 6209–6219. [CrossRef]

30. Bordes, P.; Pollet, E.; Avérous, L. Nano-biocomposites: Biodegradable polyester/nanoclay systems. *Prog. Polym. Sci.* **2009**, *34*, 125–155. [CrossRef]

31. Briassoulis, D.; Babou, E.; Hiskakis, M. Degradation Behaviour and Field Performance of Experimental Biodegradable Drip Irrigation Systems. *J. Polym. Environ.* **2011**, *19*, 341–361. [CrossRef]

32. Islam, M.S.; Pickering, K.L.; Foreman, N.J. Influence of Alkali Treatment on the Interfacial and Physico-Mechanical Properties of Industrial Hemp Fibre Reinforced Polylactic Acid Composites. *Compos. Part A Appl. Sci. Manuf.* **2010**, *41*, 596–603. [CrossRef]

33. *ISO 9427:2003. Wood-Based Panels. Determination of Density*; International Organization for Standardization: Geneva, Switzerland, 2003.

34. *ASTM D570-98. Standard Test Method for Water Absorption of Plastics*; ASTM International: West Conshohocken, PA, USA, 2010.

35. *ISO 527-1:2012. Plastics—Determination of Tensile Properties*; International Organization for Standardization: Geneva, Switzerland, 2012.
36. *ASTM E1461-13. Standard Test Method for Thermal Diffusivity by the Flash Method*; ASTM International: West Conshohocken, PA, USA, 2013.
37. Barone, J.R.; Schmidt, W.F. Polyethylene reinforced with keratin fibers obtained from chicken feathers. *Compos. Sci. Technol.* **2005**, *65*, 173–181. [CrossRef]
38. Carrillo, F.; Rahalli, A.; Cañavate, J.; Colom, X. Composites from keratin biofibers. Study of compatibility using polyolephinic matrices. In Proceedings of the 15th European Conference on Composite Materials (ECCM-15) Conference, Venice, Italy, 24–28 June 2012; pp. 1–15.
39. Carrillo, F.; Rahhali, A.; Cañavate, J.; Colom, X. Biocomposites using waste whole chicken feathers and thermoplastic matrices. *J. Reinf. Plast. Compos.* **2013**, *32*, 1419–1429. [CrossRef]
40. Reddy, N.; Yang, Y. Structure and Properties of Chicken Feather Barbs as Natural Protein Fibers. *J. Polym. Environ.* **2007**, *15*, 81–87. [CrossRef]
41. Barone, J.R.; Gregoire, N.T. Characterization of fibre-polymer interactions and transcrystallity in short keratin fiber-polypropylene composites. *Plast. Rubber Compos.* **2006**, *35*, 287–293. [CrossRef]
42. Martínez-Hernández, A.L.; Velasco-Santos, C.; de Icaza, M.; Castaño, V.M. Microstructural characterization of keratin fibres from chicken feathers. *Int. J. Environ. Pollut.* **2005**, *23*, 162–178. [CrossRef]
43. Takahashi, K.; Yakamoto, H.; Yokote, Y.; Hattori, M. Thermal Behavior of Fowl Feather Keratin. *Biosci. Biotechnol. Biochem.* **2004**, *68*, 1875–1881. [CrossRef] [PubMed]
44. Huda, M.S.; Drzal, L.T.; Mohanty, A.K.; Misra, M. Chopped glass and recycled newspaper as reinforcement fibers in injection molded poly(lactic acid) (PLA) composites: A comparative study. *Compos. Sci. Technol.* **2006**, *66*, 1813–1824. [CrossRef]
45. Houshyar, S.; Shanks, R.A.; Hodzic, A. The effect of fiber concentration on mechanical and thermal properties of fiber-reinforced polypropylene composites. *J. Appl. Polym. Sci.* **2005**, *96*, 2260–2272. [CrossRef]
46. Cheung, H.Y.; Lau, K.T.; Tao, X.M.; Hui, D. A potential material for tissue engineering: Silkworm silk/PLA biocomposite. *Compos. Part B Eng.* **2008**, *39*, 1026–1033. [CrossRef]
47. Ghani, S.A.; Jahari, M.H.; Soo-Jin, T. Effects of sodium hydroxide treatment on the properties of low-density polyethylene composites filled with chicken feather fiber. *J. Vinyl Addit. Technol.* **2014**, *20*, 36–41. [CrossRef]
48. Bismarck, A.; Mohanty, A.K.; Aranberri-Askargorta, I.; Czapla, S.; Misra, M.; Hinrichsen, G.; Springer, J. Surface characterization of natural fibers; surface properties and the water up-take behavior of modified sisal and coir fibers. *Green Chem.* **2001**, *3*, 100–107. [CrossRef]
49. Winandy, J.E.; Muehl, J.H.; Micales, J.A.; Raina, A.; Schmidt, W. Potential of chicken feather fibre in wood MDF composites. In Proceedings of the EcoComp, London, UK, 1–2 September 2003; p. 20.
50. Peter, S. *Material Revolution: Sustainable and Multi-Purpose Materials for Design and Architecture*; Birkhäuser: Basel, Switzerland, 2011; ISBN 13:978-3034606639.

© 2017 by the authors. Licensee MDPI, Basel, Switzerland. This article is an open access article distributed under the terms and conditions of the Creative Commons Attribution (CC BY) license (http://creativecommons.org/licenses/by/4.0/).

polymers

MDPI

Article

3D Printable Filaments Made of Biobased Polyethylene Biocomposites

Daniel Filgueira [1], Solveig Holmen [2], Johnny K. Melbø [3], Diego Moldes [1],
Andreas T. Echtermeyer [2] and Gary Chinga-Carrasco [3,*]

[1] Department of Chemical Engineering, Edificio Isaac Newton, Lagoas-Marcosende s/n, University of Vigo,
36310 Vigo, Spain; danmartinez@uvigo.es (D.F.); diego@uvigo.es (D.M.)

[2] Department of Mechanical and Industrial Engineering, NTNU, 7491 Trondheim, Norway;
solveig.holmen91@gmail.com (S.H.); andreas.echtermeyer@ntnu.no (A.T.E.)

[3] RISE PFI, Høgskoleringen 6b, 7491 Trondheim, Norway; johnny.melbø@rise-pfi.no

* Correspondence: gary.chinga.carrasco@rise-pfi.no; Tel.: +47-908-36-045

Received: 21 February 2018; Accepted: 10 March 2018; Published: 13 March 2018

Abstract: Two different series of biobased polyethylene (BioPE) were used for the manufacturing of biocomposites, complemented with thermomechanical pulp (TMP) fibers. The intrinsic hydrophilic character of the TMP fibers was previously modified by grafting hydrophobic compounds (octyl gallate and lauryl gallate) by means of an enzymatic-assisted treatment. BioPE with low melt flow index (MFI) yielded filaments with low void fraction and relatively low thickness variation. The water absorption of the biocomposites was remarkably improved when the enzymatically-hydrophobized TMP fibers were used. Importantly, the 3D printing of BioPE was improved by adding 10% and 20% TMP fibers to the composition. Thus, 3D printable biocomposites with low water uptake can be manufactured by using fully biobased materials and environmentally-friendly processes.

Keywords: laccase; grafting; TMP; BioPE; biocomposites; lauryl gallate; octyl gallate; 3D printing

1. Introduction

Biocomposites are expected to contribute to the production of environmentally sound products [1]. Materials are classified as biocomposites if at least one of the constituents is derived from biological material [2]. Most biocomposites used today are made of a synthetic polymer reinforced with lignocellulosic fibers. Such fibers are low density biodegradable materials with low cost, high availability worldwide, and acceptable specific strength properties [3,4].

Lignocellulosic fibers need to be processed at low temperatures since their degradation is initiated at about 200 °C [5–7]. This characteristic limits the use of several polymers, such as polyethylene teraphtalate (PET), as matrix phase in the manufacturing of biocomposites, since their melting temperature exceeds the fiber´s degradation temperature. A suitable polymer is polyethylene (PE), a thermoplastic polymer with a processing temperature low enough to avoid the degradation of the lignocellulosic fibers [8]. Hence, PE is one of the most used polymers in the manufacturing of biocomposites [3]. Traditionally, the PE used is an oil-derived polymer with a carbon footprint higher than biopolymers like polylactic acid [9]—not the best choice from an environmental perspective. However, BioPE is available and industrially manufactured from materials, such as sugarcane, sugar beet, or wheat grains. It is worth mentioning that plants are a renewable feedstock, which consume CO_2 in each annual growth cycle. Hence, the manufacturing of 1 Ton of biobased PE (BioPE) from sugarcane could capture 2.5 Tons of CO_2 from the atmosphere, whenever solar energy is used for energy production [10]. Therefore, BioPE clearly contributes to reducing the carbon footprint, compared to fossil PE. On the other hand, BioPE exhibits the same physical and chemical characteristics as fossil-based PE which allows its direct implementation in industrial manufacturing processes [10].

Regarding the properties of biocomposites, the physical characteristics of the fibers such as fiber length and aspect ratio have a dramatic impact on the mechanical properties of the biocomposites [11–13]. This effect is already well known from traditional composites [14]. Thermomechanical pulp (TMP) fibers are a relatively cheap raw material with a higher aspect ratio than other biobased materials (e.g., wood flour), leading to the manufacturing of biocomposites with good mechanical properties [15,16]. One challenge of using TMP fibers in a PE matrix is that the fibers are highly hydrophilic [17] (like all natural fibers) and the matrix is hydrophobic. The fibers and matrix are chemically incompatible resulting in poor interfacial adhesion. The incompatibility also leads to clustering of the fibers, preventing good dispersion and possibly formation of voids. A further problem is swelling of the TMP-reinforced biocomposites due to absorption of water by the fibers from the environment, which can affect the dimensional stability of the biocomposite products.

Surface modification of lignocellulosic fibers is a promising strategy to improve the interfacial adhesion of the fiber-matrix system and dispersion of the fibers. Different compatibilizers, but also chemical and physical treatments have been proposed for the surface modification of the lignocellulosic fibers with interesting results [3,18]. However, the problem with the existing solutions is their environmental footprint. Treatments may be based on chemicals that are not ideal from an environmental point of view. Environmentally friendly alternatives are enzymatic treatments such as laccase-mediated reactions. Laccase is a phenoloxidase enzyme that has the capability to oxidize phenolic compounds leading to the formation of their corresponding phenoxy radicals which could be grafted onto the surface of the fibers, modifying their bio-chemical properties [19]. Although laccase cannot oxidize high-redox potential compounds, the addition of a mediator to the laccase-assisted treatment enables the oxidation of other chemical structures beyond phenolic compounds. Therefore, in the so-called laccase-mediator system (LMS), the oxidized mediators may oxidize chemical compounds such as sterols or fatty alcohols, which could not be oxidized by laccase itself [20,21]. In addition, the mediators could penetrate in regions of the lignocellulosic fibers which laccase cannot access due to steric hindrance. Previous studies have shown that TMP fibers could be modified by means of laccase-assisted reactions for the manufacturing of biocomposites [22] and particleboards [23,24], or for removing lipophilic extractives [25]. Additionally, biocomposites with low water uptake are expected to be an interesting material, for example, for injection molding of automotive parts, where the dimensional stability is a critical characteristic.

Most of the applications of the biocomposites reinforced with lignocellulosic fibers are related to the construction and automotive industries [26]. Nonetheless, new manufacturing processes are required in order to increase the applications of biocomposites and provide added-value to the lignocellulosic fibers. 3D printing offers a new perspective in the manufacturing of biocomposite products, which could be used in very specific applications such as medical prosthesis, regenerative medicine, or drug delivery [27–30]. Its possibilities of rapid prototyping and direct part fabrication make 3D printing one of the most promising and time/cost-efficient production processes for the industry. Moreover, 3D printing is not only cost efficient, but also facilitates production of parts with complex geometry, repairs, and assembly. In fact, important industrial sectors (e.g., medical, aerospace) are already applying 3D printing technology [31–33].

In the present study, BioPE and TMP fibers are used for biocomposite manufacturing. In order to reduce the water uptake of TMP fibers and of the corresponding biocomposites, the grafting of hydrophobic compounds was carried out by means of an eco-friendly enzymatic process. Hence, the new biocomposites are expected to have improved compatibility between their components, but also lower water absorption and improved suitability for 3D printing.

2. Materials and Methods

2.1. Materials

Two series of biobased polyethylene (BioPE) were kindly provided by Braskem (Sao Paulo, Brazil). They were both high-density polyethylene (HDPE) with different melt flow index (MFI), 20 g/10 min (BioPE1) and 4.5 g/10 min for (BioPE2). The density of BioPE was practically the same, 0.955 and 0.954 g/cm^3 respectively for BioPE1 and BioPE2.

Spruce TMP fibers were kindly provided by Norske Skog Saugbrugs (Halden, Norway). The chemical composition of TMP was 48.2% cellulose, 25.6% hemicellulose, 26% lignin, and 0.2% extractives. The average fiber length of the collected TMP fibers was 1.5 mm and the diameter was 33 μm [22].

Laccase from *Myceliophthora termophila* (NS51003) was supplied by Novozymes (Bagsværd, Denmark). The activity of the enzyme was calculated by the 2,2′-azino-bis(3-ethylbenzothiazoline-6-sulphonic acid) (ABTS) oxidation assay. One unit of activity was defined as the amount of enzyme that oxidized 1 μmol of ABTS per minute at 25 °C and pH 7 (0.1 M phosphate buffer).

Compatibilizer Licocene maleic anhydride polyethylene (MAPE) 4351 Fine Grain was provided by Clariant (Basel, Switzerland). The compatibilizer has an acid value of 42–49 mg KOH/g, a density (23 °C) of 0.98–1.00 g/cm^3, a drop point of 120–126 °C, and a viscosity between 200–500 mPa·s.

The remaining chemical reagents were purchased from Sigma-Aldrich (St. Louis, MO, USA) at reagent grade and used without further purification.

2.2. TMP Fibers Modification

TMP fibers (18 g of oven dried pulp (odp)) were suspended in a 2 L reactor with 1.8 L of phosphate buffer (0.1 M, pH 7) at 50 °C for 30 min. Laccase enzyme (175 U/g odp) and guaiacol (G) (10 mM) were added to the solution 30 min before adding 80 mL of an acetone solution containing Octyl Gallate (OG) or Lauryl Gallate (LG) (0.15 M). After 2 h under agitation, the fibers were dried at room temperature and then washed with distilled water/acetone solution (60:40%, v/v) for 1 h at 50 °C. Finally, the fibers were repeatedly washed with distilled water and dried at 50 °C for 12 h.

2.3. Extrusion of Biocomposite Filaments

An overview of the series that were prepared is given in Table 1. In order to obtain a homogeneous blend, BioPE pellets and TMP fibers were ground in a Thomas Wiley Mini-Mill Cutting mill to mesh 10 and 30, respectively. The average fiber length of the milled TMP fibers was 0.4 mm and the diameter was 38 μm [22]. The milled BioPE and the fibers were oven dried (105 °C during 1 h) and the blending was performed at two different TMP fiber loads, 10% and 20% weight fraction. MAPE compatibilizer was added to the blends depending on the TMP fiber load; 1% and 2% MAPE for weight fraction 10% and 20% of fibers loads, respectively. The blend was extruded in a Noztek Xcalibur (Shoreham, UK), which has a single screw and three different heating chambers for the total control of the extrusion temperature.

Different temperatures were tested during the extrusion process of BioPE1 and BioPE2 with the TMP fibers. For temperatures lower than 150 °C, filaments with a high roughness and porosity were obtained due to a low melt flow of both BioPE1 and BioPE2. At the same time, for temperatures above 170 °C, significant foaming and deterioration of the fibers cell structure was observed. Hence, the best temperature conditions were found in the range of 150 and 165 °C. Such results were in accordance with Guo et al., who found that the critical temperature for the extrusion of HDPE/wood fibers composites was 170 °C [34]. Therefore, BioPE1 was extruded at 150, 155, and 160 °C and BioPE2 at 155, 160, and 165 °C, respectively for chambers 1, 2, and 3. The differences in the extrusion temperatures between BioPE1 and BioPE2 were due to the different MFI (20 and 4.5 g/10 min, respectively, for BioPE1 and BioPE2).

Table 1. Composition of biocomposite filaments for 3D printing.

Code	BioPE	BioPE (Weight %)	TMP fiber (Weight %)	TMP fiber modification
B1	BioPE1	100	-	-
B1-10T	BioPE1	90	10	-
B1-20T	BioPE1	80	20	-
B1-10LGT	BioPE1	90	10	LG
B1-20LGT	BioPE1	80	20	LG
B1-10OGT	BioPE1	90	10	OG
B1-20OGT	BioPE1	80	20	OG
B2	BioPE2	100	-	-
B2-10T	BioPE2	90	10	-
B2-20T	BioPE2	80	20	-
B2-10LGT	BioPE2	90	10	LG
B2-20LGT	BioPE2	80	20	LG
B2-10OGT	BioPE2	90	10	OG
B2-20OGT	BioPE2	80	20	OG

The speed of the screw extruder was 12 mm/s and the fan was set at 65%. Filaments with a diameter of approx. 2 mm were obtained. All the filaments were spooled with a Filabot spooler at the output of the extruder.

The Biobased polyethylene (BioPE) filaments were extruded once and twice to assess the evolution of filament' thickness variation and porosity. Hence, the filaments obtained after the first extrusion were cut in small pellets (20 mm length), which were extruded again under the same conditions of the first extrusion.

2.4. Filament Morphology

Three random pieces of filaments, each 20 mm in length, were scanned in an Epson Perfection V750 (Long Beach, CA, USA) for quantification of thickness variation. The images were acquired in reflection and transmission modes with a resolution of 2400 dots per inch. The images acquired in transmission mode were automatically thresholded and binarized. The variation in thickness of the binarized filaments were quantified with a plugin for ImageJ developed for this purpose. The thickness variation is considered a measurement of the filament roughness and corresponds to the variation in thickness along each single filament piece (20 mm length, 3 replicates).

Pieces of filaments were used to estimate the void fraction of the filaments considering the weight of the filaments (W_i, in g), the cross-sectional area (A_i, in cm^2) and length (L_i, in cm) of the pieces, the estimated density of TMP fibers (1.56×10^{-6} g/cm^3) and BioPE (0.955 g/cm^3 for BioPE1 and 0.954 g/cm^3 for BioPE2), and the mass fraction of fibers (X_F) and BioPE (X_{BioPE}) in the filaments. The void fraction was calculated as follows:

$$\text{Void fraction (\%)} = ((\text{Theoretical density} - \text{Real density})/\text{Real density}) \times 100 \qquad (1)$$

$$\text{Real density} = W_i/(A_i \times L_i) \qquad (2)$$

$$\text{Theoretical density} = (X_F \times 1.56 \times 10^{-6}) + (X_{BioPE} \times d_{BioPE}) \qquad (3)$$

It should be noted that the cross-sectional area A_i is obtained by measuring the diameter of the filaments and assuming a circular cross-section. This is a reasonably simple approach, but it considers the surface roughness as voids.

2.5. SEM Analysis

The filaments were embedded in epoxy resin and prepared for scanning electron microscopy (SEM), in backscatter electron imaging (BEI) mode [35]. The prepared samples were coated with carbon before visualization in BEI mode. Additionally, and for exemplification purposes, the surfaces of some fracture areas were visualized in secondary electron imaging (SEI) mode after coating the surface with a thin layer of gold. A Hitachi SU3500 Scanning Electron Microscope (Tokyo, Japan) was used for the analyses. The applied acceleration voltage and magnification were 5 kV and 50×, respectively.

2.6. Water Uptake

Three test specimens (length = 60 ± 1 mm) of each filament were immersed in 40 mL of distilled water for 32 days. The samples were initially dried (50 °C for 24 h) and the dried weight (W_0, in g) was measured. The samples were weighted every 24 h (W_i). The water uptake of the filaments was measured by the following equation,

$$\text{Water uptake (weight \%)} = ((W_i - W_0)/W_0) \times 100 \tag{4}$$

2.7. 3D Printing

BioPE filaments complemented with hydrophobized-TMP fibers were used for printing 3D model figures (Ø = 20 mm) in an Ultimaker Original 3D printer (Geldermalsen, The Netherlands). The diameter of the 3D printer nozzle was 0.4 mm and the print speed and temperature were set at 15 mm/s and 210 °C, respectively. The design of the model figures was performed with the ImageJ program (version 1.50i, National Institutes of Health, Bethesda, MD, USA).

3. Results and Discussion

3.1. Filaments Morphology and Porosity

All the manufactured filaments showed high roughness after the first extrusion. SEM images evidenced a heterogeneous distribution of the TMP fibers and relatively big pores in the polymer matrix (Figure 1a). Hence, a second extrusion under the same conditions was carried out to enhance the dispersion of the fibers in the BioPE matrix. As exemplified in Figure 1b, the fiber dispersion of the sample B2-20OGT was remarkably improved and relatively big pores were not detected after a second extrusion process.

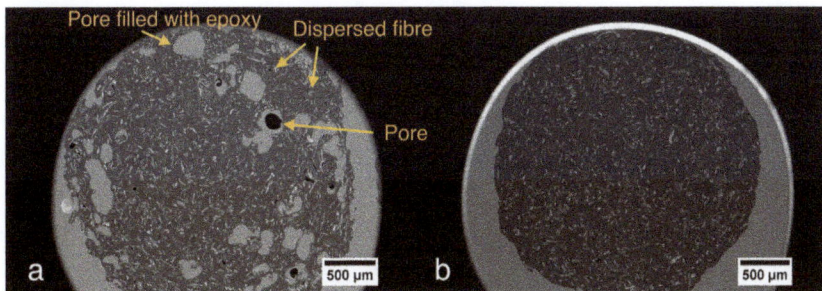

Figure 1. SEM cross-sectional image at 50× magnification. (**a**) B2-OGT20 after one extrusion; (**b**) B2-OGT20 after two extrusions.

One parameter that could influence the reduction of the filaments' porosity after the second extrusion could be the different shape of the feeding in the extruder. In the second extrusion, the feeding was added in the form of small pellets (20 mm in length; 2 mm in diameter) whereas in the first extrusion the feeding was added as powder of milled fibers and BioPE. Therefore, the pelletizing of the fibers and the BioPE before extruding could improve the blending in the extrusion chambers, leading to the manufacture of filaments with a lower porosity. Additionally, it is likely that the second extrusion process could reduce the fiber agglomerations. Therefore, the roughness and visual appearance of the filaments were significantly improved after the second extrusion (Figure 2).

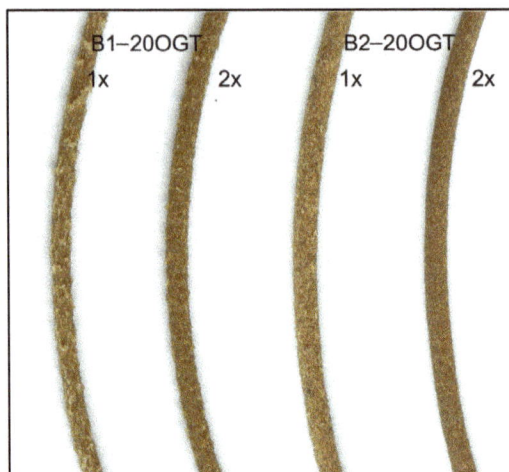

Figure 2. Representative filaments after first (1×) and second (2×) extrusion.

Regarding the different TMP fibers used to complement the BioPE matrix, it was observed that enzymatically LG-modified TMP fibers exhibited a more homogeneous distribution and apparently less fiber agglomeration in the BioPE2 matrix (Figure 3). A similar behaviour was observed with the OG-modified TMP fibers. Additionally, Figure 3a exemplifies some pores and cavities in the filament surface which may facilitate the water diffusion into the filament structure. The modified TMP fibers were hydrophobized by means of the enzymatic grafting of OG or LG. Such compounds possess an aliphatic chain, which apparently favored the chemical compatibility between the TMP fibers with the hydrophobic matrix.

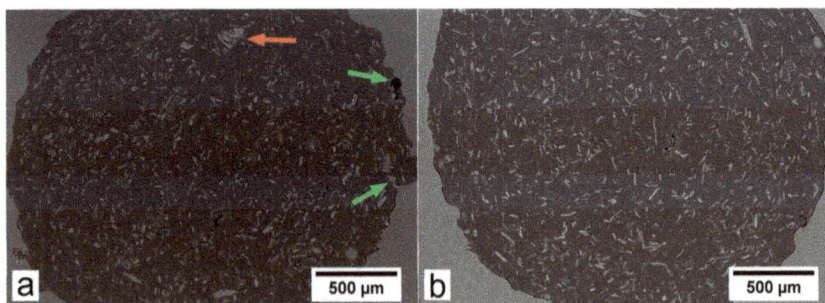

Figure 3. SEM cross-sectional image at 50× magnification. (**a**) B2-20T; (**b**) B2-20LGT. The red arrow indicates an agglomeration of fibers. The green arrows indicate surface pores probably caused by the relatively high surface roughness at this local area.

Nonetheless, the roughness of the filaments manufactured with BioPE1 was remarkably higher than those manufactured with BioPE2 (Figure 4), even after the second extrusion. The MFI of BioPE1 is 5-fold higher than the MFI of BioPE2, which means that BioPE1 has a lower viscosity and a higher capacity to flow than BioPE2. The difference in MFI is directly related to their average molecular weight. Although both biopolymers are HDPE, BioPE2 has a higher average molecular weight, which means that BioPE2 possesses an increased entanglement of chains and a less ordered structure than BioPE1 [36]. Hence, the higher proportion of entanglements of the BioPE2 could facilitate its blending with the TMP fibers during the extrusion process. On the other hand, due to the low viscosity, the speed at which the BioPE1 flows in the extruder could be remarkably higher than the flow of TMP fibers, probably leading to a heterogeneous distribution of the fibers along the filaments.

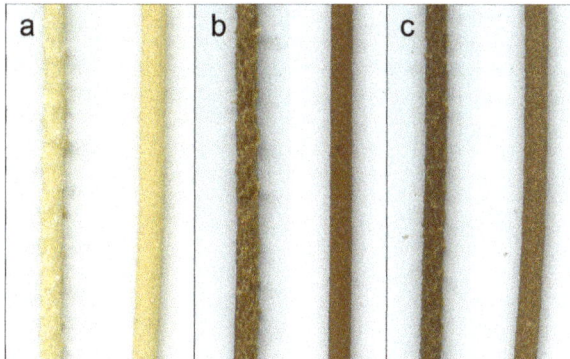

Figure 4. Comparison of the roughness in the filaments produced from BioPE1 (left) and BioPE2 (right) with fibers loads of 20%: (**a**) B1-T20 and B2-T20; (**b**) B1-LGT20 and B2-LGT20; (**c**) B1-OGT20 and B2-OGT20.

Therefore, the thickness variation of the filaments was assessed in order to measure the differences in the filaments' roughness. For all the manufactured filaments (diameter = 2 ± 0.1 mm), the thickness variation was lower for the BioPE2 series, compared to the corresponding BioPE1. Those differences were significantly larger for fiber loads of 20%, where the BioPE2-based filaments showed on average a 70% lower thickness variation respect to BioPE1-based filaments (Figure 5). Fewer differences between both matrices were found for fiber loads of 10% and especially with the use of enzymatically modified TMP fibers (10LGT and 10OGT). These results suggest that the lower MFI of BioPE2 enabled a better blending with the modified TMP fibers during the extrusion process, yielding relatively smooth and homogeneous filaments. On the contrary, the higher MFI of BioPE1 led to a poorer blending between the polymer matrix and the TMP fibers, thus yielding rough filaments with presumptively limited 3D printability.

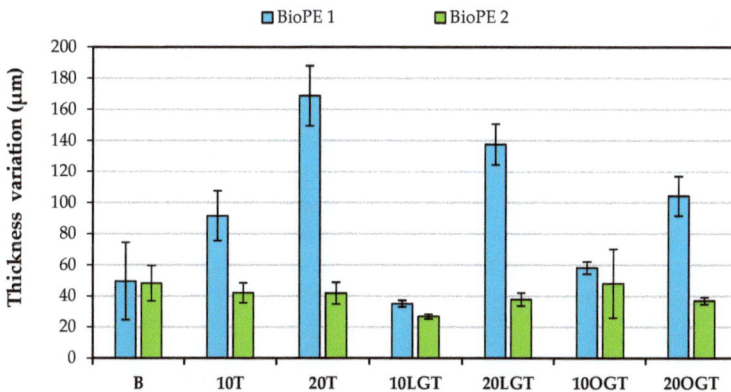

Figure 5. Thickness variation of the filaments as a function of the matrix (BioPE1 or BioPE2) and the thermomechanical pulp (TMP) fibers; BioPE (B), TMP fibers (T), LG-treated TMP fibers (LGT) and OG-treated TMP fibers (OGT).

When the filaments were manufactured with hydrophobic fibers, their roughness was reduced in most cases. The lowest thickness variation was observed in the B2-10LGT series. The differences between B2-10LGT and B2-10OGT could be caused by the longer aliphatic chain of LG (Figure 6), which could significantly improve the chemical compatibility with the hydrophobic polymer matrix. However, previous work with both fibers (LGT and OGT) and polylactic acid (PLA) as polymeric matrix evidenced that OGT fibers had a better chemical compatibility with the PLA matrix [22]. These results suggest that the interfacial adhesion of the fiber-matrix system depends on both the degree of hydrophobicity of the TMP fibers, but also the chemical structure of the polymeric matrix.

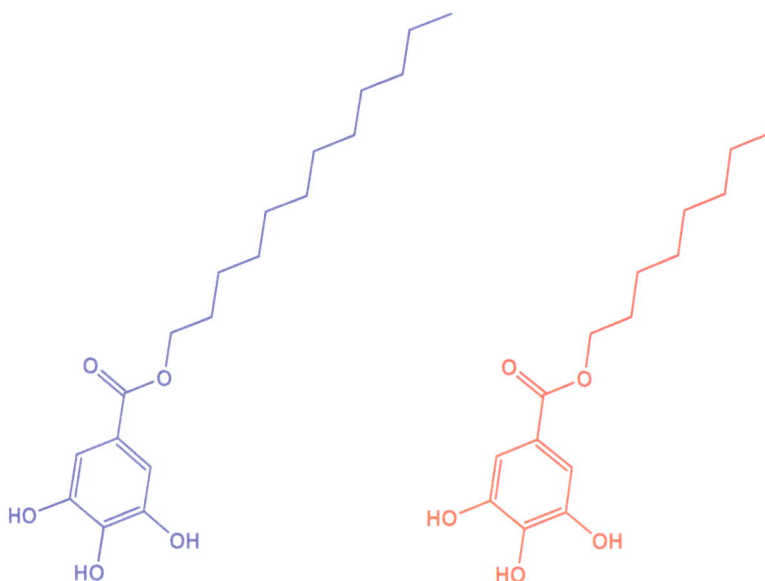

Figure 6. Chemical structure of Lauryl Gallate (LG) and Octyl Gallate (OG).

The void fraction of the filaments is a measure of the filaments' porosity. Hence, the density of the filaments was estimated and compared with the theoretical density of the filaments (Table 2). As expected, the use of TMP fibers produces a variable void volume depending on the matrix and fiber used, but mainly depending on the amount of fiber. Different void volumes, from 8% to 32% were obtained. When unmodified TMP was used for filaments manufacturing, BioPE1 produced filaments with higher void volume than BioPE2.

The use of hydrophobized TMP fibers led to a general reduction of void volume. However, the reduction of void volume does not seem to depend on the used biopolymer (BioPE1 and BioPE2). Thus, the chemical compatibility between matrix and fibers and the MFI of BioPE are probably not the only parameters to consider in the production of filaments.

The lowest void fraction for both BioPE1 and BioPE2 matrices was achieved in the filaments complemented with 10% of LGT fibers, 7.73% and 10.21% respectively. Additionally, among the filaments complemented with 20% of fibers, B2-20LGT showed the lowest void fraction. These results are in accordance with the results of the filaments' thickness variation. Therefore, the enzymatic grafting of LG onto TMP fibers' surface could leads to the manufacturing of BioPE-based biocomposite filaments with a relative low porosity and limited thickness variation. This could presumptively improve the 3D printability. On the other hand, the filaments' porosity was directly proportional to the fibers fraction in the filaments, since the higher the fibers load the higher the porosity.

3.2. SEM Analysis

The surface of filaments was analyzed by SEM in order to assess the chemical compatibility between the polymer and the TMP fibers, both unmodified and hydrophobized. Since the filaments complemented with 10% of fibers showed a relatively low void fraction, the main differences were found in the filaments with fiber loads of 20%. Such differences were notable for the BioPE1 series. As observed in Figure 7, the B1-20T filaments showed a high porosity, suggesting that the interfacial adhesion between the hydrophobic matrix and the hydrophilic TMP fibers was not satisfactory. However, the laccase-assisted grafting of LG onto TMP fibers enhanced the chemical compatibility with the BioPE1, since both B1-20LGT and B1-20OGT filaments showed a lower porosity than B1-20T. However, B1-20LGT filaments still exhibited a relatively high porosity, but lower than B1-20T. With respect to the BioPE2 series, the SEM pictures showed minor differences regarding the porosity of the filaments complemented with both unmodified and hydrophobized TMP fibers (Figure 7). Nevertheless, B2-20LGT filaments showed a homogeneous surface and a small fraction of micro-voids, suggesting that the laccase-assisted modification of TMP fibers with LG is a promising strategy for the manufacturing of low density filaments with a suitable roughness and porosity. Comparing both matrices, BioPE1-based filaments (Figure 7a–c) showed bigger voids and a notably higher porosity than BioPE2 series (Figure 7d–f). Therefore, the SEM images confirmed the trend observed in the measurement of the thickness variation and void fraction. It is worth mentioning that the low porosity of the BioPE2 series is expected to cause a lower water uptake and a better 3D printability, compared to the BioPE1 series.

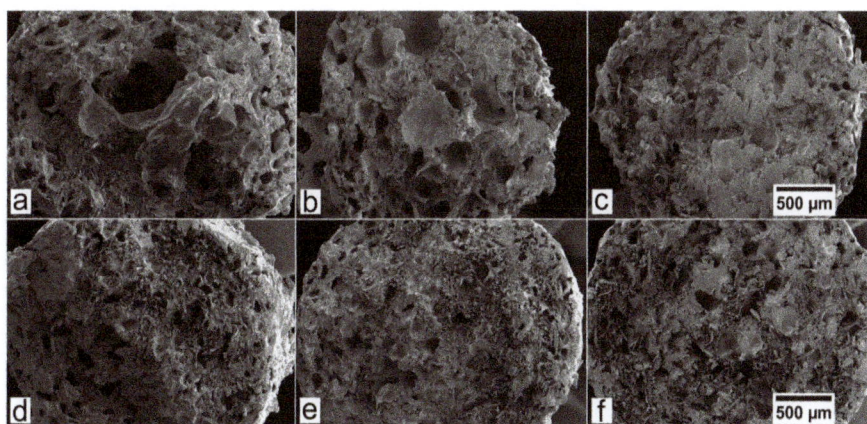

Figure 7. SEM images of the biocomposite filaments. BioPE1 containing 20% of untreated TMP fibers (**a**); 20% LG-treated TMP fibers (**b**); and 20% OG-treated TMP fibers (**c**). BioPE2 containing 20% of untreated TMP fibers (**d**); 20% LG-treated TMP fibers (**e**); and 20% OG-treated TMP fibers (**f**). All the images were acquired at 50× magnification in SEI mode.

Table 2. Measurement of the % of void fraction in the manufactured filaments.

Sample	Theoretical density (g/cm^3)	Measured density, Equations (1) and (2) (g/cm^3)	Void fraction (vol %)
B1	0.955	0.9538	0.1
B1-10T	0.8595	0.6761	27.1
B1-20T	0.764	0.5199	47.0
B1-10LGT	0.8595	0.7930	8.4
B1-20LGT	0.764	0.5922	29.0
B1-10OGT	0.8595	0.7746	11.0
B1-20OGT	0.764	0.6466	18.2

Table 2. *Cont.*

Sample	Theoretical density (g/cm^3)	Measured density, Equations (1) and (2) (g/cm^3)	Void fraction (vol %)
B2	0.954	0.953	0.1
B2-10T	0.8586	0.697	23.2
B2-20T	0.763	0.6143	24.2
B2-10LGT	0.8586	0.7709	11.4
B2-20LGT	0.763	0.6741	13.2
B2-10OGT	0.8586	0.739	16.2
B2-20OGT	0.763	0.5994	27.3

3.3. Water Uptake

The water uptake of the filaments was measured to compare the two BioPE polymer matrices and also to assess the hydrophobic behavior of the enzymatically treated-TMP fibers. The water absorption of the biocomposites depends on several factors, mainly on the hygroscopic behavior of the fibers and the chemical compatibility between fibers and matrix, which affects the void volume and roughness. A good chemical compatibility of the matrix-fiber system could improve their interfacial adhesion, reducing the void fraction in the biocomposite. In addition, a high roughness of the filaments leads to a relatively high specific surface area, which will increase the contact between the filament and the water molecules. However, the chemical characteristics of fibers is expected to have an effect on water uptake, since the fiber hydrophobization act as a water repellent. Importantly, we demonstrated recently that the enzymatic treatment hydrophobized the TMP fibers and thus reduced the water uptake of PLA-based biocomposite filaments [22].

As observed in Table 2, the laccase-assisted grafting of the hydrophobic compound onto the TMP fibers reduced the void fraction of the biocomposites, with the exception of B2-20OGT. In addition, SEM images (Figure 7) confirmed that the interfacial adhesion between the matrix and the fibers was improved after the enzymatic treatment. Such effects had a clear impact on the water uptake of the filaments complemented with hydrophobized-TMP fibers (Figures 8, A1 and A2). For the BioPE1-based series, the enzymatic hydrophobization of the TMP fibers resulted in a major reduction of the filaments' water uptake after 32 days. The water uptake has not completely flattened out after the 32 days (Figures A1 and A2), but it clearly shows the different speed of water uptake of filaments based on BioPE1 and BioPE2. Such hydrophobic effects were especially significant for the filaments containing 20% fibers. For BioPE1 series, filaments complemented with 20% of LG-treated fibers reduced the water uptake, but to a lesser extent than B1-20OGT. Nonetheless, B1-20LGT series showed a much higher thickness variation than the B1-20OGT series, confirming the importance of manufacturing smooth and homogeneous filaments. The BioPE2-based filaments complemented with unmodified TMP fibers showed, on average, 30% higher water absorption than those in which the TMP fibers were previously hydrophobized by means of the laccase-mediated treatment. Generally, BioPE2-based biocomposite filaments absorbed lower amount of water than the BioPE1 series, especially for TMP fibers loads of 20%. As observed in Table 2 and Figure 7 the BioPE1 series complemented with fiber loads of 20% showed a much higher porosity than the BioPE2-based filaments, which affected clearly their water uptake behavior. Such results suggest that the MFI of the BioPE conditioned the manufacturing of the biocomposite filaments as well as their water uptake behavior.

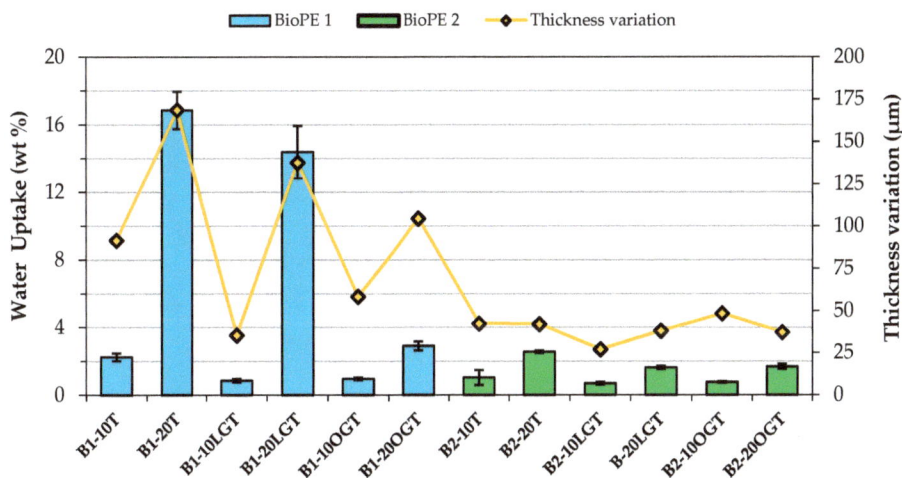

Figure 8. Water uptake (blue/green columns) of the biocomposite filaments after 32 days of water immersion (left axes). The thickness variation (yellow line) of the biocomposites is given in the right axis.

3.4. 3D Printing

Polyethylene (PE) and more concretely HDPE are hardly ever used for 3D printing. In fact, as far as we know there is only one commercially available fully HDPE-based filament for 3D printing [37]. However, the manufacturer advises on their website that they do not have a reliable way to print such filament. Moreover, we could not find any scientific article focused on the manufacturing of PE-based biocomposite filaments for 3D printing. Generally, PE tends to shrink, bend, and warp when its temperature cools down, hindering remarkably the 3D printing process. Thus, the 3D printing of the manufactured BioPE-based biocomposite filaments is demanding.

The biocomposite filaments made from BioPE1 polymer were not suitable for 3D printing. The high MFI (20 g/10 min), due to its low viscosity, promoted warping and shrinkage problems during the 3D printing. Additionally, the relative high thickness variation of the BioPE1-based filaments hindered their feeding into the 3D printer. Nonetheless, BioPE2 biocomposite filaments showed a good 3D printing performance. As it was mentioned BioPE2 has a higher viscosity and a lower MFI (4.5 g/10 min) than BioPE1, which probably improved its 3D printability. Moreover, BioPE2-TMP biocomposites showed a better printability than the neat BioPE2. It is likely that the addition of TMP fibers reduced the MFI of the neat BioPE2 polymer [38], restricting the swelling and the shrinkage of the printed layers, and facilitating the corresponding shape fidelity and layers adhesion [39,40].

It is known that in fusion deposition modeling (FDM) amorphous polymers works better than crystalline polymers. Amorphous polymers have a disordered structure and a viscosity high enough to facilitate the adhesion of the layers and also maintain the shape of the printed layers [41]. On the contrary, highly crystalline polymers like HDPE develop partially crystalline structures upon cooling, resulting in distortions and internal stress [36]. Thus, highly crystalline polymers tend to shrink, hindering the shape fidelity of the printed layers and, therefore, limiting the 3D printing process [40]. Such a drawback could be solved, in part, with a heated print bed, which could reduce the cooling rate of the printed layers. However, for the 3D printing of big objects this inconvenience will probably appear again. Some interesting results were previously obtained by the Washington Open Object Fabricators (WOOF) team, who were able to 3D print a boat from recycled HDPE from milk jugs [42]. Nevertheless, they had to attach a heater to the extruder of the 3D printer in order to avoid the cooling down of the printed layers. In addition, they created a PE-based print bed since PE does not stick to any material other than PE. However, in this study and due to limitations of the printing unit, the 3D printing of BioPE2-based biocomposites filaments was performed without a heated print bed.

Regarding the TMP fibers modification, objects printed with unmodified TMP showed a smoother surface for fibers loads of 10% than 20% fiber (Figure 9). Nonetheless, LG-TMP fibers showed a similar smoothness without warping or curling for both 10% and 20%, while OG-TMP fibers showed an improved quality in terms of smoothness at 20% fiber content. Thus, there was no major difference in quality between the 3D printed objects containing BioPE2 complemented with both modified and unmodified TMP fibers.

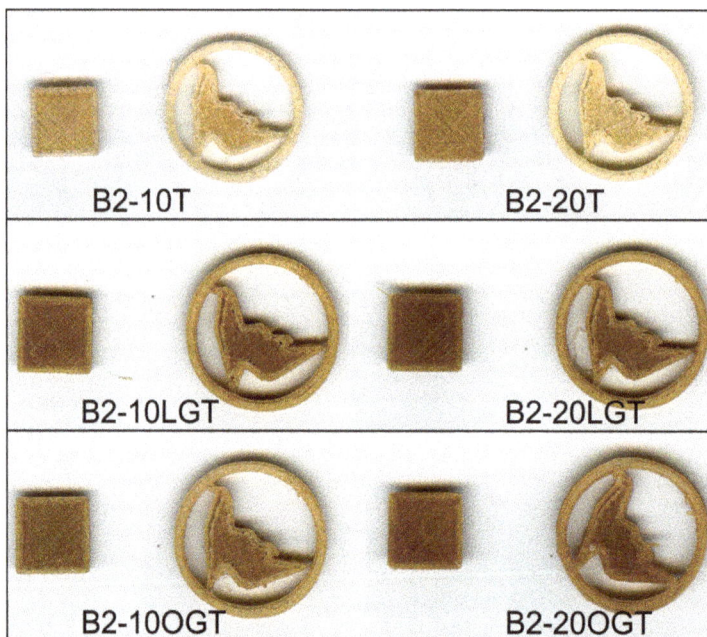

Figure 9. 3D printed objects of the BioPE2 series, containing unmodified, LG, and OG-modified TMP fibers (10% and 20%). The squares are 10 mm × 10 mm. The circle is 20 mm in diameter.

4. Conclusions

To the best of our knowledge, this is the first time that a scientific article focuses on the manufacturing of polyethylene-based biocomposite filaments for 3D printing. Two series of BioPE with different MFI were tested for the manufacturing of biocomposite filaments. BioPE1 with a relatively high MFI leads to the manufacturing of biocomposites with a high void fraction and high roughness. The high roughness had a notable impact on the water uptake behavior in some BioPE1 series. On the other hand, the relatively high MFI of the BioPE1 leads to warping and bending problems, and also to poor layer adhesion during 3D printing of BioPE1 biocomposite filaments. In addition, the relatively high porosity and thickness variation limited the 3D printing of BioPE1-based filaments.

The lower MFI of BioPE2 enabled the manufacturing of biocomposite filaments suitable for 3D printing. Moreover, the 3D printing of BioPE2 was improved with the addition of TMP fibers. The hydrophobicity of the fibers was tailored by means of laccase-assisted grafting of OG or LG compounds. Hence, filaments complemented with enzymatically treated TMP fibers showed a remarkably lower water uptake compared with those filled with unmodified TMP fibers. No major differences were observed with respect to the 3D print quality and water uptake behavior of the filaments containing OG and LG-treated TMP fibers. Finally, it is worth mentioning that the biocomposites developed in this study may be plausible materials for injection molding operations and products where low water uptake is required.

Acknowledgments: An important part of the present study was funded by the ValBio-3D project (Grant ELAC2015/T03-0715 Valorization of residual biomass for advanced 3D materials, Research Council of Norway, Grant no. 271054). The first author thanks the support of the COST action FP1306 for funding a STSM to RISE PFI, where these activities were performed. Funding from Xunta de Galicia (EM2014/041 Novos tratamentos biocatalíticos para a mellora da durabilidade da madeira. Valorización de extractivos da madeira e de lignina kraft) is also appreciated.

Author Contributions: Daniel Filgueira performed the experiments, analyzed the data and wrote the paper; Solveig Holmen performed the experiments and analyzed the data; Johnny K. Melbø performed the experiments; Diego Moldes contributed reagents/materials/analysis tools and analyzed the data; Andreas T. Echtermeyer analyzed the data and proof-read the paper; Gary Chinga-Carrasco conceived and designed the experiments, analyzed the data, contributed reagents/materials/analysis tools, and proof-read the paper.

Conflicts of Interest: The authors declare no conflict of interest.

Appendix A

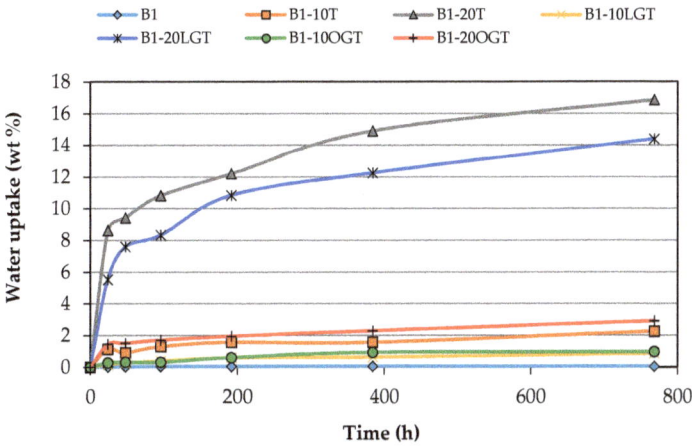

Figure A1. Water absorption (wt %) by the BioPE1-based filaments during 32 days of water immersion.

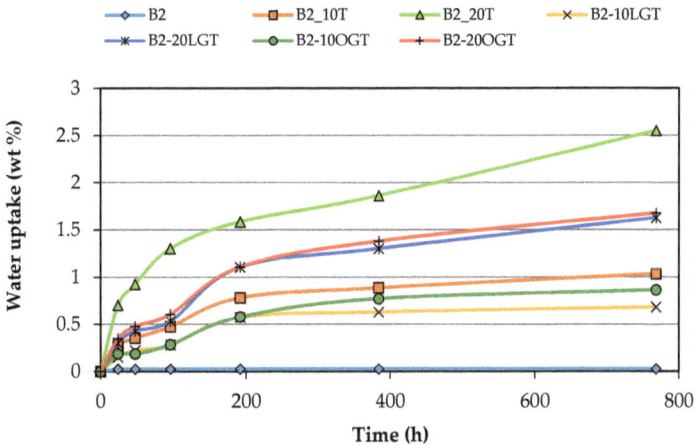

Figure A2. Water absorption (wt %) by the BioPE2-based filaments during 32 days of water immersion.

References

1. Mohanty, A.K.; Misra, M.; Drzal, L.T. Sustainable Bio-Composites from renewable resources: Opportunities and challenges in the green materials world. *J. Polym. Environ.* **2002**, *10*, 19–26. [CrossRef]
2. Abhilash, M.; Thomas, D. Biopolymers for Biocomposites and Chemical Sensor Applications. In *Biopolymer Composites in Electronics*; Sadasivuni, K.K., Ponnamma, D., Cabibihan, J.J., Al-Maadeed, M.A., Kim, J., Eds.; Elsevier Inc.: Oxford, UK, 2017; pp. 405–435. ISBN 9780128092613.
3. Gurunathan, T.; Mohanty, S.; Nayak, S.K. A review of the recent developments in biocomposites based on natural fibres and their application perspectives. *Compos. Part A* **2015**, *77*, 1–25. [CrossRef]
4. Faruk, O.; Bledzki, A.K.; Fink, H.; Sain, M. Biocomposites reinforced with natural fibers: 2000–2010. *Prog. Polym. Sci.* **2012**, *37*, 1552–1596. [CrossRef]
5. Ho, M.; Wang, H.; Lee, J.-H.; Ho, C.-K.; Lau, K.-T.; Leng, J.; Hui, D. Critical factors on manufacturing processes of natural fibre composites. *Compos. Part B* **2012**, *8*, 3549–3562. [CrossRef]
6. Saheb, D.N.; Jog, J.P. Natural Fiber Polymer Composites: A Review. *Adv. Polym. Technol.* **1999**, *18*, 351–363. [CrossRef]
7. Poletto, M.; Zattera, A.J.; Forte, M.M.C.; Santana, R.M.C. Thermal decomposition of wood: Influence of wood components and cellulose crystallite size. *Bioresour. Technol.* **2012**, *109*, 148–153. [CrossRef] [PubMed]
8. Kakroodi, A.R.; Kazemi, Y.; Cloutier, A.; Rodrigue, D. Mechanical performance of polyethylene (PE)-based biocomposites. In *Biocomposites: Design and Mechanical Performance*; Manjusri, M., Pandey, J.K., Mohanty, A.K., Eds.; Elsevier Inc.: Cambridge, UK, 2015; pp. 237–256. ISBN 9781782423942.
9. Narayan, R. Carbon footprint of bioplastics using biocarbon content analysis and life-cycle assessment. *MRS Bull.* **2011**, *36*, 716–721. [CrossRef]
10. Morschbacker, L.; Siqueira Campos, C.E.; Cassiano, L.C.; Roza, L.; Almada, F.; Werneck do Carmo, R. Bio-polyethylene. In *Handbook of Green Materials: Processing Technologies, Properties and Applications (in 4 Volumes): Materials Science*; Oksman, K., Mathew, A.P., Bismarck, A., Rojas, O., Sain, M., Eds.; World Scientific Publishing Co.: Singapore, 2014; Volume 4, pp. 89–104. ISBN 9789814566476.
11. Shinoj, S.; Panigrahi, S.; Visvanathan, R. Water absorption pattern and dimensional stability of oil palm fiber-linear low density polyethylene composites. *J. Appl. Polym. Sci.* **2010**, *117*, 1064–1075. [CrossRef]
12. Nygård, P.; Tanem, B.S.; Karlsen, T.; Brachet, P.; Leinsvang, B. Extrusion-based wood fibre-PP composites: Wood powder and pelletized wood fibres—A comparative study. *Compos. Sci. Technol.* **2008**, *68*, 3418–3424. [CrossRef]
13. Migneault, S.; Koubaa, A.; Erchiqui, F.; Chaala, A.; Englund, K.; Krause, C.; Wolcott, M. Effect of fiber length on processing and properties of extruded wood-fiber/HDPE composites. *J. Appl. Polym. Sci.* **2008**, *110*, 1085–1092. [CrossRef]
14. Agarwal, B.D.; Broutman, L.J. Short fiber composites. In *Analysis and Performance of Fiber Composites*; John Wiley: New York, NY, USA, 1990; pp. 193–196. ISBN 0-471-511528.
15. Mertens, O.; Gurr, J.; Krause, A. The utilization of thermomechanical pulp fibers in WPC: A review. *J. Appl. Polym. Sci.* **2017**, *134*, 45161. [CrossRef]
16. Peltola, H.; Pääkkönen, E.; Jetsu, P.; Heinemann, S. Wood based PLA and PP composites: Effect of fibre type and matrix polymer on fibre morphology, dispersion and composite properties. *Compos. Part A* **2014**, *61*, 13–22. [CrossRef]
17. Koljonen, K.; Österberg, M.; Johansson, L.S.; Stenius, P. Surface chemistry and morphology of different mechanical pulps determined by ESCA and AFM. *Colloids Surf. A Physicochem. Eng. Asp.* **2003**, *228*, 143–158. [CrossRef]
18. La Mantia, F.P.; Morreale, M. Composites: Part A Green composites: A brief review. *Compos. Part A* **2011**, *42*, 579–588. [CrossRef]
19. Grönqvist, S.; Rantanen, K.; Alén, R.; Mattinen, M.L.; Buchert, J.; Viikari, L. Laccase-catalysed functionalisation of TMP with tyramine. *Holzforschung* **2006**, *60*, 503–508. [CrossRef]
20. Morozova, O.V.; Shumakovich, G.P.; Shleev, S.V.; Yaropolov, Y.I. Laccase-mediator systems and their applications: A review. *Appl. Biochem. Microbiol.* **2007**, *43*, 523–535. [CrossRef]
21. Gutiérrez, A.; Rencoret, J.; Ibarra, D.; Molina, S.; Camarero, S.; Romero, J.; Del Río, J.C.; Martínez, Á.T. Removal of lipophilic extractives from paper pulp by laccase and lignin-derived phenols as natural mediators. *Environ. Sci. Technol.* **2007**, *41*, 4124–4129. [CrossRef] [PubMed]

22. Filgueira, D.; Holmen, S.; Melbø, J.K.; Moldes, D.; Echtermeyer, A.; Chinga-Carrasco, G. Enzymatic-assisted modification of Thermomechanical Pulp Fibres To Improve the Interfacial Adhesion with Poly(lactic acid) for 3D printing. *ACS Sustain. Chem. Eng.* **2017**, *5*, 9338–9346. [CrossRef]

23. Euring, M.; Rühl, M.; Ritter, N.; Kües, U.; Kharazipour, A. Laccase mediator systems for eco-friendly production of medium-density fiberboard (MDF) on a pilot scale: Physicochemical analysis of the reaction mechanism. *Biotechnol. J.* **2011**, *6*, 1253–1261. [CrossRef] [PubMed]

24. Schubert, M.; Ruedin, P.; Civardi, C.; Richter, M.; Hach, A.; Christen, H. Laccase-catalyzed surface modification of thermo-mechanical pulp (TMP) for the production of wood fiber insulation boards using industrial process water. *PLoS ONE* **2015**, *10*, e0128623. [CrossRef] [PubMed]

25. Gutiérrez, A.; Del Río, J.C.; Rencoret, J.; Ibarra, D.; Martínez, Á.T. Main lipophilic extractives in different paper pulp types can be removed using the laccase-mediator system. *Appl. Microbiol. Biotechnol.* **2006**, *72*, 845–851. [CrossRef] [PubMed]

26. Schwarzkopf, M.J.; Burnard, M.D. Wood-Plastic Composites—Performance and Environmental Impacts. In *Environmental Impacts of Traditional and Innovative Forest-Based Products*; Kutnar, A., Muthu, S.S., Eds.; Springer: Singapore, 2016; pp. 19–43. ISBN 9789811006555.

27. Chinga-Carrasco, G.; Ehman, N.V.; Pettersson, J.; Vallejos, M.; Brodin, M.; Felissia, F.E.; Hakansson, J.; Area, M.C. Pulping and pretreatment affect the characteristics of bagasse inks for 3D printing. *ACS Sustain. Chem. Eng.* **2018**, *6*, 4068–4075. [CrossRef]

28. Chiulan, I.; Frone, A.; Brandabur, C.; Panaitescu, D. Recent Advances in 3D Printing of Aliphatic Polyesters. *Bioengineering* **2018**, *5*, 2. [CrossRef] [PubMed]

29. Bandyopadhyay, A.; Bose, S.; Das, S. 3D printing of biomaterials. *MRS Bull.* **2015**, *40*, 108–114. [CrossRef]

30. Gbureck, U.; Vorndran, E.; Müller, F.A.; Barralet, J.E. Low temperature direct 3D printed bioceramics and biocomposites as drug release matrices. *J. Controll. Release* **2007**, *122*, 173–180. [CrossRef] [PubMed]

31. Ventola, C.L. Medical Applications for 3D Printing: Current and Projected Uses. *Pharm. Ther.* **2014**, *39*, 704–711. [CrossRef]

32. Shafiee, A.; Atala, A. Printing Technologies for Medical Applications. *Trends Mol. Med.* **2016**, *22*, 254–265. [CrossRef] [PubMed]

33. Liu, R.; Wang, Z.; Sparks, T.; Liou, F.; Newkirk, J.; States, U. Aerospace applications of laser additive manufacturing. In *Laser Additive Manufacturing: Materials, Design, Technologies, and Applications*; Elsevier: Amsterdam, The Netherlands, 2016; pp. 351–371, ISBN 9780081004333.

34. Guo, G.; Rizvi, G.M.; Park, C.B.; Lin, W.S. Critical Processing Temperature in the Manufacture of Fine-Celled Plastic/Wood-Fiber Composite Foams. *J. Appl. Polym. Sci.* **2004**, *91*, 621–629. [CrossRef]

35. Reme, P.A.; Helle, T. Assessment of transverse dimensions of wood tracheids using SEM and image analysis. *J. Pulp Pap. Sci.* **2002**, *28*, 122–128. [CrossRef]

36. Sperling, L.H. *Introduction to Physical Polymer Science*, 4th ed.; John Wiley and Sons Inc.: Chicester, UK, 2011; ISBN 9780471706069.

37. High-Density Polyethylene (HDPE) 3D Printer Filament. Available online: https://filaments.ca/products/hdpe-filament-natural-1-75mm?variant=42590589320 (accessed on 12 March 2018).

38. Tazi, M.; Erchiqui, F.; Godard, F.; Kaddami, H.; Ajji, A. Characterization of rheological and thermophysical properties of HDPE-wood composite. *J. Appl. Polym. Sci.* **2014**, *131*, 1–11. [CrossRef]

39. Shofner, M.L.; Lozano, K.; Rodríguez-Macías, F.J.; Barrera, E.V. Nanofiber-Reinforced Polymers Prepared by Fused Deposition Modeling. *J. Appl. Polym. Sci.* **2003**, *89*, 3081–3090. [CrossRef]

40. Chang, T.C.; Faison, E.I. Shrinkage Behavior and Optimization of Injection Molded Parts Studies by the Taguchi Method. *Polym. Eng. Sci.* **2001**, *41*, 703–710. [CrossRef]

41. Gibson, I.; Rosen, D.W.; Stucker, B. *Additive Manufacturing Technologies Rapid Prototyping to Direct Digital Manufacturing*; Springer: New York, NY, USA, 2010; ISBN 9781441911193.

42. D Printing a Functional Boat with Post-Consumer Milk Jugs. Available online: https://makezine.com/2013/05/30/large-format-3d-printing/ (accessed on 12 March 2018).

© 2018 by the authors. Licensee MDPI, Basel, Switzerland. This article is an open access article distributed under the terms and conditions of the Creative Commons Attribution (CC BY) license (http://creativecommons.org/licenses/by/4.0/).

![polymers logo]

Article

Two-Sided Surface Oxidized Cellulose Membranes Modified with PEI: Preparation, Characterization and Application for Dyes Removal

Wei Wang *, Qian Bai, Tao Liang, Huiyu Bai and Xiaoya Liu

Key Laboratory of Synthetic and Biological Colloids, Ministry of Education, School of Chemical and Material Engineering, Jiangnan University, Wuxi 214122, China; qianbaichem@163.com (Q.B.); taoliangchem@163.com (T.L.); bhy.chem@163.com (H.B.); lxy@jiangnan.edu.cn (X.L.)
* Correspondence: ww12230098@sina.com or weiwangpolymer@163.com

Received: 16 August 2017; Accepted: 13 September 2017; Published: 16 September 2017

Abstract: Porous regenerated cellulose (RC) membranes were prepared with cotton linter pulp as a raw material. These membranes were first oxidized on both sides by a modified (2,2,6,6-tetramethylpiperidin-1-yl)oxyl (TEMPO) oxidation system using a controlled oxidation reaction technique. Then, the oxidized RC membranes were functionalized with polyethylenimine (PEI) via the glutaraldehyde crosslinking method to obtain bifunctional (carboxyl and amino) porous RC membranes, as revealed by Fourier transform infrared spectroscopy (FT-IR), elemental analysis and zeta potential measurement. The scanning electron microscopy (SEM) and the tests of the mechanical properties and permeability characteristics of modified RC membranes demonstrated that the porous structure and certain mechanical properties could be retained. The adsorption performance of the modified membranes towards dyes was subsequently investigated. The modified membranes displayed good adsorption capacities, rapid adsorption equilibrium and removal efficiencies towards both anionic (xylenol orange (XO)) and cationic (methylene blue (MB)) dyes, making them suitable bioadsorbents for wastewater treatment.

Keywords: regenerated celluloses; bioadsorbents; dye removal; functional membranes

---

1. Introduction

Synthetic dyes are widely utilized in numerous industries (e.g., textile, paper, leather tanning, plastics, rubber, cosmetics and printing) owing to their high stability, relatively low costs and color uniformity characteristics [1,2]. Inevitably, a certain fraction of dyes end up in the effluent during the dyeing process. This colored waste water must be purified before being released into the environment since it is toxic, carcinogen and can damage the aquatic ecological balance [3]. Various methods (e.g., adsorption, membrane separation, chemical oxidation, coagulation-flocculation, photochemical degradation and bioremoval) have been used singly or in combination with other approaches for the removal of dyes [4–6]. Among them, adsorption is considered to be economically and environmentally superior to other conventional techniques owing to its easy operation, low cost and effectiveness [7]. However, mechanical agitation is often required to improve the adsorption efficiency by enhancing the contact between the dyes and the adsorbents. Moreover, numerous adsorbents usually require long times before reaching adsorption equilibrium [8].

One paradigmatic adsorbent for the removal of a target pollutant should exhibit a large surface area, good adsorption efficiency and an abundance of adsorption sites while being produced at low cost and in a sustainable and environmentally-friendly manner. Sharma et al. compiled a list of naturally available, low cost and eco-friendly adsorbents for the removal of hazardous dyes from aqueous waste streams via adsorption [4]. Various methods including carbonization, chemical activation and pyrolysis have been proposed to enhance the adsorption efficiency for dyes [9].

Biodegradable and biocompatible cellulose-based materials have been developed in recent decades taking advantage of the wide availability (i.e., the most abundant renewable biopolymer in nature) and low cost characteristics of this polymer. These materials have been used for water treatment purposes especially, as a new class of versatile adsorbents for the removal of dyes. Moreover, the high density of surface hydroxyl groups present on cellulose provides this material with excellent surface modification characteristics, thereby allowing a wide range of functionalization approaches for the adsorption of dyes from aqueous solutions [10]. Thus, different forms of cellulose-based materials have been used as dye adsorbents. Jin et al. prepared amino-functionalized nanocrystalline cellulose, and the sample was then applied as an adsorbent to remove anionic dyes in aqueous solutions [11]. A carboxylate-functionalized adsorbent based on CNCs was prepared, and adsorptive removal of multiple cationic dyes was investigated [12]. Cellulose recycled newspaper fibers were grafted with double quaternary ammonium groups, and the maximum adsorption capacity of this for RTB G-133 was 524 mg·g^{-1} [13]. Luo et al. developed millimeter-scale magnetic regenerated cellulose (RC) beads, and the adsorbent could efficiently adsorb the organic dyes from wastewater, as well as the used adsorbents could be recovered completely [14]. Cellulose nanosponges modified with methyltrioctylammonium chloride were prepared and used for pre-concentration, removal and determination of tartrazine dye, using UV–Vis spectrophotometry [15]. Cellulose powders functionalized with hyperbranched polyethylenimine were prepared, and used for selective dye adsorption and separation based on the unique selective adsorption [16].

Cellulose membranes can be also used as dye adsorbents, benefitting from the porous structure, the adsorption capacity for the removal of positively-charged dyes and good reusable performance, however in the form of multilayer [17], composites [18,19] and cellulose derivatives [20]. However, at present, many studies have focused on using membrane separation technology as the cellulose-based membranes are used for water treatment [21,22], and there are less reports on the pure cellulose functional membranes as adsorbents. Oshima et al., reported phosphorylated bacterial cellulose as an adsorbent for metal Ions [23]. As described above, when used as adsorbents, celluloses need to undergo chemical modification for achieving efficient adsorption. However, during the modification process, celluloses are often subjected to strong acid or oxidant, etc., treatments, so leading to the degradation of cellulose macromolecules [24], which further may damage the porous structures of the cellulose membranes and even destroy membranes themselves. These facts can significantly limit the application of cellulose membranes for adsorption purposes.

In this study, cotton linter pulp was used as a raw material to prepare porous RC membranes by using a phase-inversion method. Subsequently, the RC membranes underwent a two-sided surface oxidation treatment with a modified (2,2,6,6-tetramethylpiperidin-1-yl)oxyl (TEMPO) oxidation system to yield the corresponding C-6 carboxyl membrane materials. To prevent the cellulose membranes from being damaged via oxidation–degradation of the cellulose macromolecular chains, a control oxidation reaction was carried out by following the method described by Fitz-Binder et al. [25]. Amine groups were subsequently incorporated onto the oxidized cellulose membranes by grafting polyethylenimine (PEI) via the glutaraldehyde crosslinking method [26]. As a result, porous RC membranes containing both carboxyl and amino groups were obtained. Fourier transform infrared spectroscopy (FT-IR), elemental analysis and zeta potential measurement and scanning electron microscopy (SEM) were used to analyze the structure and morphology of the modified cellulose membranes. Furthermore, anionic and cationic dyes were both used as model pollutants to evaluate the dye removal efficiency of the modified cellulose membranes. The adsorption performance and kinetic behaviors during dye adsorption on the cellulose-based adsorbents were further investigated. With the aim to further evaluate the potential application of the modified RC membranes for water treatment purposes, the porosity, pure water permeability characteristics and the mechanical properties of the modified RC membranes were also tested.

2. Experimental Section

2.1. Materials

Cellulose (cotton linter pulp, α-cellulose ≥95%) was purchased from Hubei Chemical Fiber Group Ltd. (Xiangfan, China). PEI (molecular weight of 600 Da), TEMPO (AR), xylenol orange (XO) (AR) and methylene blue (MB) (AR) were supplied by Aladdin Chemical Reagent Corp., Shanghai, China. Other reagents used in this work were of analytical grade and purchased from Sinopharm Chemical Reagent Co., Shanghai, China.

2.2. Preparation of the RC Membranes

RC membranes were prepared according to a reported method [27]. Five grams of cotton linter pulp were added to a LiOH/urea/H_2O (8.76/12/79.24 in wt %, 100 g) solution, and the resulting aqueous solution was stored at 200 °C for 20 h. The frozen cellulose solution was then vigorously stirred for 5 min at ambient temperature to obtain a transparent cellulose dope. The cellulose dope was then subjected to centrifugation at 8000 rpm for 10 min at −4 °C, and the transparent supernatant fraction was immediately cast on a glass plate. Subsequently, the resulting gel sheets were coagulated with a sulfate aqueous solution to obtain a transparent membrane. The wet membrane was washed with deionized water and dried at ambient temperature to finally obtain the RC membrane.

2.3. Preparation of the TEMPO–Oxidized RC Membranes

Oxidized cellulose membranes were prepared according to a reported method [25]. First, the RC membrane was immersed in 1 L of a boric acid (0.1 M) buffer solution (pH = 10.5), and 2 g of NaBr and 0.3 g of TEMPO were subsequently added and mixed by magnetic stirring for 180 min. The pre-wetted membranes were then dried at 60 °C for 5 min. For the printing paste, 5 g of alginate were dissolved in 40 mL of the boric acid buffer solution to form a thickener, and 6 mL of NaOCl were added to the former paste. The mixtures were then stirred to obtain a homogenous paste. Subsequently, the as-prepared thickener paste was applied by printing it on the one-sided surface of the RC membranes containing the NaBr/TEMPO/buffer mixture, and the prints were rested for 60 min. Finally, the printing membrane was thoroughly washed with water. The other side of the cellulose membrane was impregnated in a similar way to generate the two-sided surface TEMPO-oxidized regenerated cellulose (TORC) membrane.

2.4. Preparation of Aminated TORC Membranes

The oxidized RC membranes were cut into approximately 80 mm × 70 mm × 0.1 mm pieces and saturated in 50 mL of an anhydrous methanol solution. Four grams of PEI were added, and the resulting solution was stirred at room temperature for 24 h. After that, the wet membranes were rinsed with water to remove the residual PEI and immediately immersed in 100 mL of an anhydrous methanol solution. Subsequently, 20 mL of a glutaraldehyde solution were added dropwise, and the resultant mixtures were stirred at 25 °C for 2 h. Finally, the modified cellulose membranes were repeatedly washed with deionized water to remove the unreacted material and denoted as PEI–TORC membranes.

2.5. Characterization of the Functionalized TORC Membranes

The RC and modified RC membranes were chemically characterized by attenuated total reflectance infrared (ATR-IR) spectroscopy (Nicolet 560, Nicolet Co., Ltd., Madison, WI, USA). The spectra were recorded from 600–4000 cm^{-1} with a resolution of 2 cm^{-1} and a minimum of 16 scans. The mass ratios of C, H, O and N of each sample were measured using Elemental Analyzer (Vario EL III, Elementar Co., Langenselbold, Germany). The surface and cross-section morphologies of the RC and modified RC membranes were assessed by using a scanning electron microscope (S-4800, Hitachi Corporation, Tokyo, Japan). Cross-sectional faces of membranes were prepared by being fractured in liquid nitrogen.

The samples were deposited on a glass plate and coated with a thin layer of gold/palladium using a sputter coater (K550X, Emitech Ltd., Kent, UK). The pore size analysis of the membrane surface was calculated by using the software of SEM image analysis (Nano Measurer System, Version 1.2.5, Fudan University, Shanghai, China). The surface charge of the PEI-TORC membrane was determined by zeta-potential measurement using an Electrokinetic Analyzer (SurPASS, Anton Paar, Graz, Austria), using a 2 mM KCl electrolyte solution. The tensile strength and strain at break of the membranes were measured on a universal testing machine (Instron 5967, Instron, Norwood, MA, USA), using a 250-N load cell at room temperature. The strain rate was set at 10 mm/min, and five measurements were taken for each sample.

The porosities of the modified membranes were calculated using a reported method [22]. The porosity (*P*) was calculated as Equation (1):

$$P = \frac{(M_1 - M_2)/q_1}{(M_1 - M_2)/q_1 + M_2/q_2} \times 100\% \tag{1}$$

The wet membranes were weighed as M_1 and then freeze dried overnight and weighed as M_2. The water content was calculated as $M_1 - M_2$. q_1 is water density, and q_2 is PEI–TORC density (calculated according to the density of bulk cellulose, 1.5 g cm^{-3}).

The water flux of the modified membranes was evaluated using a reported method [28]. The pure water permeability test of the wet membranes was carried out on miniature microfiltration equipment at ambient temperature under an operation pressure of 0.1 MPa, while the water flux was calculated according to Equation (2):

$$J_w = \frac{V}{S \times t \times P} \tag{2}$$

where *V* is the volume of solvent passing through the membrane; *t* is the measurement time; *S* is the effective membrane area; *P* is the pressure (0.1 MPa).

The mean pore radius, r_f, was calculated by employing Equation (3), derived based on the straight-through cylindrical pore model [29].

$$\eta = \sqrt{\frac{8 \times \eta \times I \times J}{P \times \Delta P}} \tag{3}$$

where η is the water viscosity (8.9×10^{-4} Pa s), *I* is the membrane thickness (m), *J* is the permeation flux (m^3·m^{-2}·s^{-1}) and ΔP is the load pressure (Pa).

2.6. Batch Adsorption Experiments

Batch adsorption experiments were conducted in some 100 mL glass conical flasks in a water bath shaker (25 °C, 200 rpm). Each flask contained 60 mL MB or XO solution and 70 mg adsorbent. The pH of the solution was adjusted by adding 0.1 M HCl or NaOH aqueous solutions. At predetermined time intervals, approximately 3 mL of dye solution were used for UV–Vis measurements and afterwards returned into the flask. This was repeated until equilibrium was reached. The dye removal efficiencies of the RC and PEI–TORC membranes towards XO and MB were investigated for 100 min (dye concentration, 30 mg·L^{-1}, pH = 6.8). The effect of pH on the removal efficiency was investigated in the range of 4–11 for 100 min (dye concentration: 30 mg·L^{-1}). The effect of initial dye concentration on the adsorption performance was investigated in the range from 30–1230 mg·L^{-1} (at pH values of 6.8 and 4.6) for 120 min. The effect of the contact time (10–100 min) was investigated (dye concentration, 30 mg·L^{-1}) at pH 6.8 and 4.6.

The removal efficiency and adsorption capacity of dyes were calculated according to Equations (4) and (5):

$$\text{adsorption capacity (mg·g}^{-1}) = \frac{(C_0 - C_t) \times V}{m} \times 100 \tag{4}$$

$$\text{Removal efficiency (\%)} = \frac{C_0 - C_t}{C_0} \times 100 \tag{5}$$

where C_0 (mg·L^{-1}) and C_t (mg·L^{-1}) are the initial concentration of the dye and the concentration of the dye at an adsorption time t, respectively, V is the volume of the dye solution and m is the weight of the adsorbent.

The evaluation of the reusability of the adsorbent was carried out at a 0.07 g·L^{-1} dosage of bioadsorbent added into 60 mL of 30 mg·L^{-1} dye solutions for 100 min, and then, the adsorbent was taken out from the solution. The desorption and regeneration of the used adsorbent was performed by immersing the adsorbent in 30 mL of a 0.1 M NaOH or HCl solution for 5 h at room temperature and subsequently washed using distilled water till neutral for the next adsorption. The generated adsorbent was used for another adsorption study in the subsequent cycles.

2.7. Adsorption Isotherm and Kinetic Model

The Langmuir and Freundlich models [30] can be used to describe the adsorption dynamic equilibrium process by Equations (6) and (7):

$$\frac{c_e}{q_e} = \frac{1}{k_1 q_{max}} + \frac{c_e}{q_{max}} \tag{6}$$

$$Inq_e = Ink_f + \frac{Inc_e}{n} \tag{7}$$

where q_e (mg·g^{-1}) is the equilibrium adsorption capacity, q_{max} (mg·g^{-1}) is the maximum adsorption capacity, C_e (mg·L^{-1}) is the equilibrium concentration of free dye molecules in the solution (mg·L^{-1}), K_1 is the Langmuir constant and K_f and $1/n$ are Freundlich constants.

In order to explore the adsorption mechanism of the rate limiting steps involved, the pseudo-first order and the pseudo-second order kinetic models [31] were used to study the type of adsorption and the adsorption mechanism. The first and second order rate equations can be expressed as Equations (8) and (9):

$$\log(q_e - q_t) = \log q_e - \frac{k_1 t}{2.303} \tag{8}$$

$$\frac{t}{q_t} = \frac{1}{k_2 q_e^2} + \frac{t}{q_e} \tag{9}$$

where q_e and q_t are the amounts of adsorbed dye per unit mass of adsorbent (mg·g^{-1}) at equilibrium and a time t, respectively, and k_1 and k_2 are the first order and the pseudo-second order adsorption rate constants, respectively.

3. Results and Discussion

3.1. Characterization of the Modified RC Membranes

The FTIR spectra of the RC, TORC and PEI-TORC membranes are shown in Figure 1a. Compared with the unmodified RC, the FTIR spectra of the TORC membranes showed a new peak at 1736 cm^{-1}, which was attributed to the C=O stretching frequency of the carboxyl group [32], thereby revealing a successful TEMPO-oxidation process. In the case of the PEI-TORC membranes, new absorption bands were observed at ca. 1653 cm^{-1} corresponding to the C=N stretching vibration, which was formed in the glutaraldehyde crosslinking process. Three new peaks appeared in the 1600–1800-cm^{-1} region and were ascribed to the C=O stretching vibration of carboxyl groups (1736 cm^{-1}), the N–H bending vibration of secondary (1615 cm^{-1}) and primary (1564 cm^{-1}) amines [33,34]. Furthermore, the large enhancement of the C–C skeleton vibration at 1157 cm^{-1} and the presence of the C–H characteristic peaks at 2924 and 2849 cm^{-1} for the PEI-TORC membranes further demonstrated the successful crosslinking reaction.

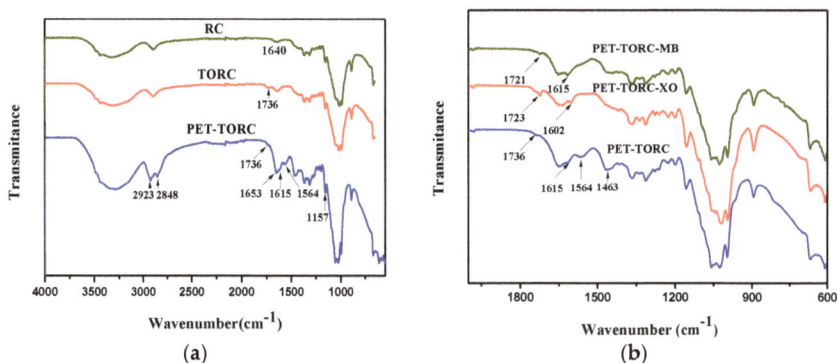

Figure 1. (a) FTIR spectra of RC, TORC, polyethylenimine(PEI)-(2,2,6,6-tetramethylpiperidin-1-yl)oxyl (TEMPO)-oxidized regenerated cellulose (TORC); (b) FTIR spectra of xylenol orange (XO)-adsorbed PEI–TORC and MB-adsorbed PEI–TORC.

In order to evaluate the change of the content of the elements in the whole material after modification, the contents of C, N, O and H in different cellulose-based membranes are listed in Table 1. After TEMPO oxidation, the oxygen content in cellulose increased from 49.11–51.29%. The nitrogen content of PEI-TORC increased significantly from 0–3.87% compared to that of TORC. The elemental analysis data and the FTIR results both confirmed the successful oxidation of cellulose and PEI grafting on TORC.

Table 1. Elements analysis results of unmodified and modified cellulose membranes.

Sample	C (%)	O (%)	H (%)	N (%)
RC	41.02	49.11	9.87	0
TORC	39.64	51.29	9.07	0
PEI-TORC	41.98	45.91	8.24	3.87

FT-IR is a useful tool to study the possible adsorbent-adsorbate interactions. The FT-IR spectra of the cellulose-based and the dye-loaded bioadsorbents are shown in Figure 1b. After dye adsorption, the adsorption peaks of the bioadsorbent shifted towards lower wavenumbers. For example, the adsorption peak attributed to the C=O stretching vibration shifted from 1736 to 1725 cm^{-1} (for XO) and 1724 cm^{-1} (for MB). The absorption bands assigned to the N–H stretching vibration shifted from 1615 to 1602 cm^{-1} (for XO) and 1613 cm^{-1} (for MB). After absorption, the disappearance of the N–H bending vibration peak of primary amines (1564 cm^{-1}) and the variation of the CH$_2$ bending peak (1463 cm^{-1}) also revealed the existence of electrostatic and hydrogen bonding interactions between the functional groups of the adsorbent and the dye molecules.

The SEM images for the RC, TORC and PEI-TORC membranes are shown in Figure 2. Surface morphology changes of cellulose-based membranes were observed; the porous structure of RC was maintained after TEMPO oxidation and crosslinking reaction. The average diameter of the pores of the pure cellulose membrane surface from SEM (78 nm, standard deviation of 18) was larger than that of PEI-TORC (43 nm, standard deviation of 10), revealing the reaction on the surface of the cellulose, introducing new groups and molecular chains on the membrane surface. As shown in the SEM images of the cross-section of cellulose-based membranes, compared with the cross-section morphology of RC membranes, the interior structures of TORC and PEI-TORC membranes had changed greatly. The cross-section of the RC membrane was relatively smooth. Interestingly, it can be observed that the cross-section image of the TORC membrane shows a non-homogeneous structure. This is because the control oxidation reaction leads to the fact that the pore structure and the microstructure of the material are different between the surface and the interior of the TORC membrane. When fractured in liquid nitrogen, there is a different morphology from the homogeneous RC membrane. Compared with the cross-section morphology of PET-TORC membranes, the cross-section of the PEI–TORC membrane became flat, suggesting that the uniformity of the membrane structure was improved due to the crosslinking reaction; however, the pore size and shape of the middle part of the membrane are different from those of the peripheral part, suggesting both oxidation and crosslinking reaction affected the morphology and structure of the PEI-TORC membranes. Moreover, this structural feature makes it unsuitable as a molecular sieve.

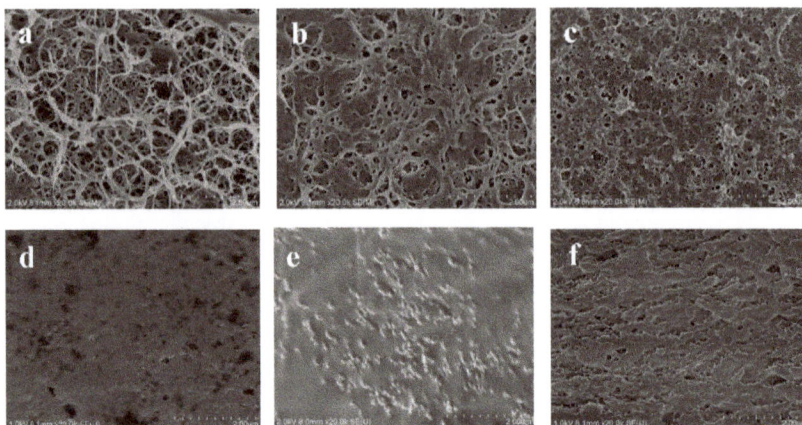

Figure 2. SEM images of the RC membrane: (**a**) surface, (**d**) cross-section; TORC membrane: (**b**) surface, (**e**) cross-section; PEI-TORC membrane: (**c**) surface, (**f**) cross-section.

As shown in Figure 3, the mechanical properties of the modified membranes were reduced. The tensile strength and elongation at break of RC membranes were 81.9 MPa, and 9.0%, respectively, while the tensile strength and elongation at break of PEI–TORC membranes were 31.0 MPa and 2.7%, respectively. The decrease of the mechanical properties of TORC membranes was due to the oxidative degradation of cellulose and the non-uniformity of the membrane materials caused by the control of the oxidation reaction. The decrease in elongation at break of the PEI-TORC membranes was due to the fact that the crosslinking reaction limits the movement of the molecular chain, and the descent in the tensile strength is likely due to the inhomogeneity of the material.

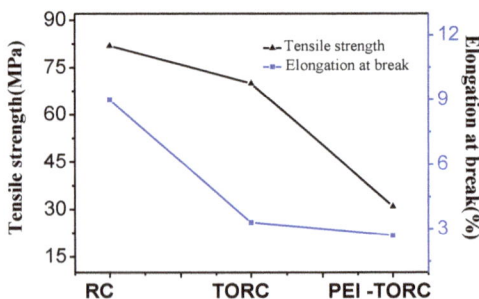

Figure 3. Tensile strength and elongation at break of RC, TORC and PEI-TORC membranes.

In order to explore the potential application value of membrane in water treatment, the structure and performance of the PEI–TORC membrane were further studied. The porosity, water flux and pore size of the PEI–TORC membrane were 72%, 6.14 $L \cdot m^{-2} \cdot h^{-1} \cdot bar^{-1}$ and 25.9 nm (the mean pore size), respectively. Importantly, it is possible to adjust the morphology, structure and properties of the PEI–TORC membrane by changing the various influencing factors in the preparation process, making it possible to further expand the application range of the modified cellulose membrane.

3.2. Adsorption Properties of the RC and PEI-TORC Membranes

The dye removal percentages of the RC and PEI–TORC membranes are also shown in Figure 4. RC membranes exhibited a poor adsorption capacity, meaning that raw cellulose did not have enough binding sites for dye adsorption. However, the adsorption capacity of hydroxyl for the removal of positively-charged dyes, the hydroxyl groups on the surface, is quite limited because most of them are included in the intra- and inter-molecule hydrogen-bond network. The removal efficiency of the PEI–TORC membrane was significantly higher than that of the RC material. The sorption of dyes onto adsorbents may well include chemical sorption, which could greatly improve the adsorption capacity. From the change in the FTIR of the functions in PEI–TORC upon the adsorption of MB or XO, it is apparent that the adsorption process is likely to involve chemical sorption.

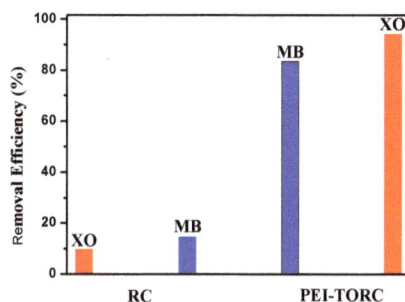

Figure 4. The removal efficiencies of RC and PEI-TORC (initial dye concentration: 30 $mg \cdot L^{-1}$, 60 mL MB or XO solution, 70 mg adsorbent, 100 min, pH = 6.8).

3.3. Effect of the pH on Adsorption

The solution pH significantly alters the level of electrostatic or molecular interaction between the adsorbent and the adsorbate due to the charge distribution on the material [11]. Thus, the pH of the solution is one of the determinants of the efficiency of the adsorbent for dye removal, as shown in Figure 5. The zeta potentials of PEI–TORC at various pH are also shown in Figure 5. The results showed that PEI–TORC exhibited a positively-charged surface at a pH lower than 5.7, while a negatively-charged surface at a pH higher than 5.7, which is shown to be electrically neutral (i.e., zero point charge). Xylenol orange is a negatively-charged species [35], while the methylene blue molecules are positively charged [36]. PEI–TORC exhibited positive surface charge at a pH lower than 5.7, leading to electrostatic attraction between the bioadsorbent and the anionic group of XO, resulting in maximum removal efficiency of 95.7% at pH 4.1. However, the PEI–TORC exhibited a negatively-charged surface, resulting in weaker electrostatic interactions between the bioadsorbent and XO and lower dye adsorption [37], with increasing the pH to higher than 5.7. The efficiency of the bioadsorbent for MB dye removal kept increasing in the pH range of 4.1–10.9, indicating the electronegativity of PEI–TORC continued to increase with increasing pH values. The results revealed the excellent dye removal efficiency of the bioadsorbent for both anionic and cationic dyes. The high adsorption of cationic or acidic dyes at higher pH may be due to the unique molecular structure of this bioadsorbent. However, considering the practical application and the zeta potential of the adsorbent, this paper mainly studied the adsorption performance of the bioadsorbent at pH 4.6 and 6.8.

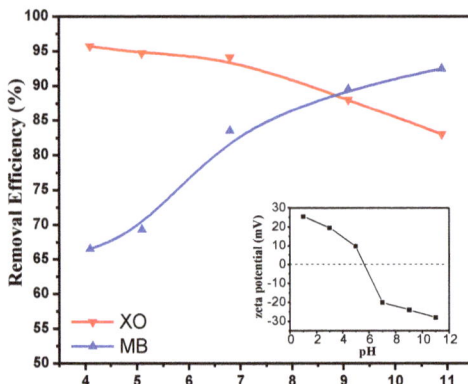

Figure 5. Effect of pH on the dye removal efficiencies of XO and MB (initial dye concentration: 200 mg·L^{-1}, 30 mL dye solution, 70 mg adsorbent).

3.4. Effect of the Initial Dye Concentration and the Adsorption Isotherm

As shown in Figure 6a, the dye adsorption capacity of PEI-TORC increased with the increase of the initial concentration and then tended to level off, resulting from the increasing driving force from the concentration gradient [38]. However, the equilibrium adsorption capacity and the adsorption behavior of PET–TORC varied greatly due to different types of dyes and different pH values. The adsorption capacity of XO was higher than that of MB, which was attributed to the limited number of carboxyl groups on the surface of the bioadsorbent, caused by controlled oxidation. The maximum adsorption capacities of XO and MB reached 403 and 74 mg·g^{-1}, at pH 4.6, respectively; while the maximum values of XO and MB reached 229 and 139 mg·g^{-1}, at pH 6.8, respectively.

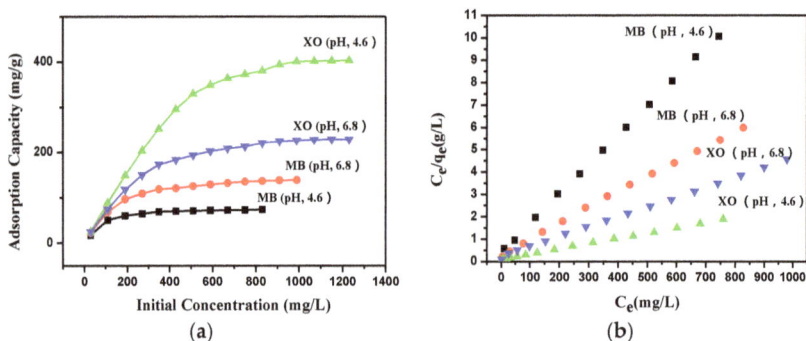

Figure 6. (a) Effect of initial concentration on dye adsorption by PEI-TORC membranes; (b) the plots of C_e/q_e against C_e for adsorption of XO and MB at different pH.

The adsorption isotherm study was carried out on two well-known isotherms, Langmuir and Freundlich. When C_e/q_e was plotted against C_e, a straight line with slope $1/q_{max}$ was obtained (Figure 6b), indicating that the adsorption of the both dyes onto PEI–TORC follows the Langmuir isotherm. The q_{max} values of the Langmuir model for adsorption of MB were 77 mg·g^{-1} (pH, 4.6) and 144 mg·g^{-1} (pH, 6.8), while the q_{max} values for adsorption of XO were 420 mg·g^{-1} (pH, 4.6) and

241 mg·g^{-1} (pH, 6.8), which was similar to the experimental q_{max} values obtained from Figure 6a. Thus, these results indicated that the PEI-TORC membrane exhibited relatively high effectiveness in removing the tested dyes, especially anionic dyes. Moreover, the isotherm parameters are summarized in Table 2. It is evident that the Langmuir isotherm model fits the experimental data better than the Freundlich model.

Table 2. Isotherm parameters for the adsorption of XO and MB onto PEI–TORC.

pH	Dyes	Langmuir Model			Freundlich Model		
		K_1 (L·mg^{-1})	q_{max} (mg·g^{-1})	R^2	n	K_f (mg·g^{-1})	R^2
6.8	XO	0.018	240.96	0.9981	2.89	25.28	0.9465
	MB	0.024	144.09	0.9987	3.12	18.64	0.8902
4.6	XO	0.030	420.17	0.9995	2.56	39.30	0.8922
	MB	0.031	76.57	0.9998	3.36	11.90	0.8170

3.5. Effect of the Contact Time and the Adsorption Kinetics

Figure 7 shows the dye removal efficiency and color change of XO and MB as a function of the adsorption time. The percentage of dye removal increased rapidly within the first 40 min, and the absorption equilibrium was reached at ca. 100 min. Furthermore, the color of the XO and MB solutions gradually became lighter with the adsorption time. The color of the XO solution completely disappeared after the adsorption process. Accordingly, the PEI–TORC membrane shifted from dark red to orange after adsorption, in line with the dye removal efficiency results. In the adsorption stage, the large number of hydrophilic hydroxyl, carboxyl and amino groups on the surface of the adsorbent resulted in electrostatic interactions that occurred upon rapid migration of the anionic XO or cationic MB dyes to the surface adsorption sites (i.e., –COOH and –NH$_2$) on the bioadsorbent, and the number of dye molecules adsorbed on the membrane increased with the adsorption time and levelled off at 50 min as a result of electrostatic repulsion forces between the adsorbed dye molecules. Therefore, the optimum contact time for the adsorption of XO and MB was ca. 50 min (i.e., 93% of XO and 83% of MB were removed). These results demonstrated that the PEI-TORC membrane is a good adsorbent for the rapid removal of XO and MB from waste waters.

As shown in Table 3, the R^2 values for MB and XO indicated that pseudo-second order (PSO) kinetic models could be used to predict the behavior over the whole range of the adsorption process, which indicating that the intraparticle diffusion was involved in the adsorption process. At a pH of 6.8, the q_e values for XO and MB calculated with a pseudo-second order model were 25 and 22 mg·g^{-1}, respectively, and these values changed to 25 and 18 mg·g^{-1} at a pH of 4.6. These results were in good agreement with the experimental data.

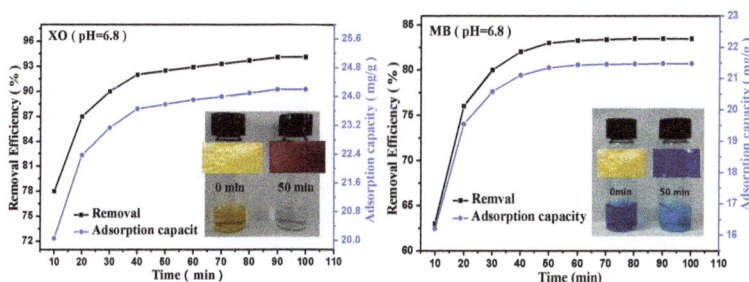

Figure 7. *Cont.*

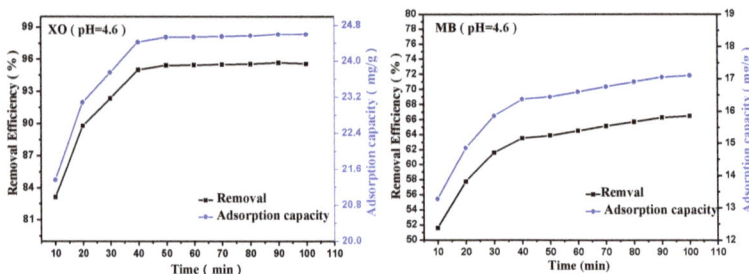

Figure 7. Effect of contact time on the dye removal efficiencies of XO and MB (initial dye concentration: 30 mg L^{-1}, 60 mL MB or XO solution, 70 mg adsorbent, 10–100 min, pH = 6.8 or 4.6).

Table 3. Kinetic parameters for the adsorption of XO and MB onto PEI–TORC.

pH	Dyes	Pseudo-First Order Model			Pseudo-Second Order Model		
		K_1 (g·mg^{-1} min^{-1})	q_e (mg·g^{-1})	R^2	K_2 (g·mg^{-1}·min^{-1})	q_e (mg·g^{-1})	R^2
6.8	XO	0.049	5.16	0.9753	0.0092	24.75	0.9999
	MB	0.092	13.01	0.9954	0.017	22.16	0.9996
4.6	XO	0.072	5.09	0.9144	0.028	25.03	0.9999
	MB	0.054	7.61	0.9631	0.014	17.84	0.9997

3.6. Reusability

Figure 8 shows the reusability of the PEI–TORC membrane. After three consecutive desorption-adsorption cycles, the dye removal rates towards XO and MB decreased by 6 and 11%, respectively. Therefore, the adsorbent showed high reusability for the removal of XO and MB.

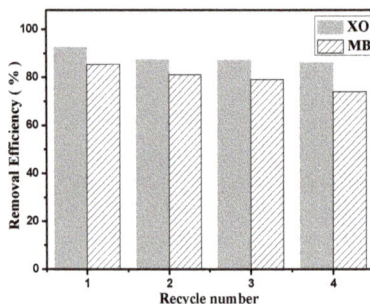

Figure 8. The dye removal efficiency of XO and MB after three desorption-adsorption cycles.

4. Conclusions

Functional membranes based on regenerated cellulose were prepared by grafting PEI onto controlled oxidized RC membranes. Modification conditions were screened to achieve the biofunctionalization of the regenerated cellulose membranes successfully and to retain the porous membrane structure. The maximum adsorption capacities of cationic and anionic dyes was observed, being the highest for xylenol orange (403 mg·g^{-1}), followed by methylene blue (139 mg·g^{-1}). Furthermore, the membranes showed low flux, supporting the usefulness of them as adsorbents. One may adjust the morphology, structure and properties of the PEI-TORC membrane by changing the various influencing factors in the preparation process, for exploring its potential application in the future.

Acknowledgments: Financial support from MOE&SAFEA for the 111 Project (B13025) is gratefully acknowledged.

Author Contributions: Wei Wang, Qian Bai, Huiyu Bai and Xiaoya Liu conceived of and designed the experiments. Wei Wang and Qian Bai performed the experiments and analyzed the data. Wei Wang, Qian Bai and Tao Liang wrote the paper. All authors discussed the results and improved the final text of the paper.

Conflicts of Interest: The authors declare no conflict of interest.

References

1. Sokolowska-Gajda, J.; Freeman, H.S.; Reife, A. Synthetic dyes based on environmental considerations. Part 2: Iron complexes formazan dyes. *Dyes Pigment.* **1996**, *30*, 1–20. [CrossRef]

2. Fleischmann, C.; Lievenbrück, M.; Ritter, H. Polymers and Dyes: Developments and Applications. *Polymers* **2015**, *7*, 717–746. [CrossRef]

3. Karadağ, E.; Saraydin, D.; Güven, O. Removal of some cationic dyes from aqueous solutions by acrylamide/itaconic acid hydrogels. *Water Air Soil Pollut.* **1998**, *106*, 369–378. [CrossRef]

4. Sharma, P.; Kaur, H.; Sharm, M.; Sahore, V. A review on applicability of naturally available adsorbents for the removal of hazardous dyes from aqueous waste. *Environ. Monit. Assess.* **2011**, *183*, 151–195. [CrossRef] [PubMed]

5. Liu, L.L.; Yu, C.X.; Zhou, W.; Zhang, Q.G.; Liu, S.M.; Shi, Y.F. Construction of Four Zn(II) Coordination Polymers Used as Catalysts for the Photodegradation of Organic Dyes in Water. *Polymers* **2016**, *8*, 3. [CrossRef]

6. Guendouz, S.; Khellaf, N.; Zerdaoui, M.; Ouchefoun, M. Biosorption of synthetic dyes (direct red 89 and reactive green 12) as an ecological refining step in textile effluent treatment. *Environ. Sci. Pollut. Res.* **2013**, *20*, 3822–3829. [CrossRef] [PubMed]

7. Ganesan, P.; Kamaraj, R.; Sozhan, G.; Vasudevan, S. Oxidized multiwalled carbon nanotubes as adsorbent for the removal of manganese from aqueous solution. *Environ. Sci. Pollut. Res.* **2013**, *20*, 987–996. [CrossRef] [PubMed]

8. Sanghi, R.; Bhattacharya, B. Review on decolorisation of aqueous dye solutions by low cost adsorbents. *Color Technol.* **2002**, *118*, 256–269. [CrossRef]

9. Kyzas, G.; Kostoglou, M. Green Adsorbents for Wastewaters: A Critical Review. *Materials* **2014**, *7*, 333–364. [CrossRef] [PubMed]

10. Zhou, Y.; Zhang, M.; Hu, X.; Wang, X.; Niu, J.; Ma, T. Adsorption of Cationic Dyes on a Cellulose-Based Multicarboxyl Adsorbent. *J. Chem. Eng. Data* **2013**, *58*, 413–421. [CrossRef]

11. Jin, L.; Li, W.; Xu, Q. Amino-functionalized nanocrystalline cellulose as an adsorbent for anionic dyes. *Cellulose* **2015**, *22*, 2443–2456. [CrossRef]

12. Qiao, H.; Zhou, Y.; Yu, F.; Wang, E.; Min, Y.; Huang, Q.; Pang, L.; Ma, T. Effective removal of cationic dyes using carboxylate-functionalized cellulose nanocrystals. *Chemosphere* **2015**, *141*, 297–303. [CrossRef] [PubMed]

13. Qi, Y.; Li, J.; Wang, L. Removal of Remazol Turquoise Blue G-133 from aqueous medium using functionalized cellulose from recycled newspaper fiber. *Ind. Crop. Prod.* **2013**, *50*, 15–22. [CrossRef]

14. Luo, X.G.; Zhang, L. High effective adsorption of organic dyes on magnetic cellulose beads entrapping activated carbon. *J. Hazard. Mater.* **2009**, *171*, 340–347. [CrossRef] [PubMed]

15. Shiralipour, R.; Larki, A. Pre-concentration and determination of tartrazine dye from aqueous solutions using modified cellulose nanosponges. *Ecotoxicol. Environ. Saf.* **2017**, *135*, 123–129. [CrossRef] [PubMed]

16. Zhu, W.; Liu, L.; Liao, Q.; Chen, X.; Qian, Z.; Shen, J.; Liang, J.; Yao, J. Functionalization of cellulose with hyperbranched polyethylenimine for selective dye adsorption and separation. *Cellulose* **2016**, *23*, 3785–3797. [CrossRef]

17. Ma, H.; Burger, C.; Hsiao, B.S.; Chu, B. Nanofibrous microfiltration membrane based on cellulose nanowhiskers. *Biomacromolecules* **2012**, *13*, 180–186. [CrossRef] [PubMed]

18. Karim, Z.; Mathew, A.P.; Grahn, M.; Mouzon, J.; Oksman, K. Nanoporous membranes with cellulose nanocrystals as functional entity in chitosan: removal of dyes from water. *Carbohydr. Polym.* **2014**, *112*, 668–676. [CrossRef] [PubMed]

19. Gopakumar, D.A.; Pasquini, D.; Henrique, M.A.; Morais, L.C.; Grohens, Y.; Thomas, S. Meldrum's Acid Modified Cellulose Nanofiber-Based Polyvinylidene Fluoride Microfiltration Membrane for Dye Water Treatment and Nanoparticle Removal. *ACS Sustain. Chem. Eng.* **2017**, *5*, 2026–2033. [CrossRef]

20. Wang, K.; Ma, Q.; Wang, S.D.; Liu, H.; Zhang, S.Z.; Bao, W.; Zhang, R.Q.; Ling, L.Z. Electrospinning of silver nanoparticles loaded highly porous cellulose acetate nanofibrous membrane for treatment of dye wastewater. *Appl. Phys. A* **2016**, *122*, 40. [CrossRef]

21. Carpenter, A.W.; Lannoy, C.F.; Wiesner, M.R. Cellulose nanomaterials in water treatment technologies. *Environ. Sci. Technol.* **2015**, *49*, 5277–5287. [CrossRef] [PubMed]

22. He, S.; Fang, H.; Xu, X. Filtering absorption and visual detection of methylene blue by nitrated cellulose acetate membrane. *Korean J. Chem. Eng.* **2016**, *33*, 1472–1479. [CrossRef]

23. Oshima, T.; Kondo, K.; Ohto, K.; Inoue, K.; Baba, Y. Preparation of phosphorylated bacterial cellulose as an adsorbent for metal ions. *React. Funct. Polym.* **2008**, *68*, 376–383. [CrossRef]

24. Ma, X.J.; Cao, S.L.; Lin, L. Hydrothermal pretreatment of bamboo and cellulose degradation. *Bioresour. Technol.* **2013**, *148*, 408–413. [CrossRef] [PubMed]

25. Fitz-Binder, C.; Bechtold, T. One-sided surface modification of cellulose fabric by printing a modified TEMPO-mediated oxidant. *Carbohydr. Polym.* **2014**, *106*, 142–147. [CrossRef] [PubMed]

26. Sun, X.F.; Wang, S.G.; Cheng, W.; Fan, M.; Tian, B.H.; Gao, B.Y.; Li, X.M. Enhancement of acidic dye biosorption capacity on poly(ethylenimine) grafted anaerobic granular sludge. *J. Hazard. Mater.* **2011**, *189*, 27–33. [CrossRef] [PubMed]

27. Zhang, L.; Mao, Y.; Zhou, J.; Cai, J. Effects of Coagulation Conditions on the Properties of Regenerated Cellulose Films Prepared in NaOH/Urea Aqueous Solution. *Ind. Eng. Chem. Res.* **2005**, *44*, 522–529. [CrossRef]

28. Cai, J.; Wang, L.X.; Zhang, L.N. Influence of coagulation temperature on pore size and properties of cellulose membranes prepared from NaOH–urea aqueous solution. *Cellulose* **2007**, *14*, 205–215. [CrossRef]

29. Mohamed, M.A.; Salleh, W.N.W.; Jaafar, J.; Ismail, A.F.; Mutalib, M.A.; Jamil, S.M. Feasibility of recycled newspaper as cellulose source for regenerated cellulose membrane fabrication. *J. Appl. Polym. Sci.* **2015**, *132*, 43. [CrossRef]

30. Lai, C.; Guo, X.; Xiong, Z. A comprehensive investigation on adsorption of Ca (II), Cr (III) and Mg (II) ions by 3D porous nickel films. *J. Colloid Interface Sci.* **2016**, *463*, 154–163. [CrossRef] [PubMed]

31. Yu, X.; Tong, S.; Ge, M. Synthesis and characterization of multi-amino-functionalized cellulose for arsenic adsorption. *Carbohydr. Polym.* **2013**, *92*, 380–387. [CrossRef] [PubMed]

32. Montanari, S.; Roumani, M.; Heux, L.; Vignon, M.R. Topochemistry of Carboxylated Cellulose Nanocrystals Resulting from TEMPO-Mediated Oxidation. *Macromolecules* **2005**, *38*, 1665–1671. [CrossRef]

33. Han, K.N.; Yu, B.Y.; Kwak, S.Y. Hyperbranched poly (amidoamine)/polysulfone composite membranes for Cd (II) removal from water. *J. Membr. Sci.* **2012**, *396*, 83–91. [CrossRef]

34. Sehaqui, H.; Larraya, U.P.D.; Liu, P.; Pfenninger, N.; Mathew, A.P. Enhancing adsorption of heavy metal ions onto biobased nanofibers from waste pulp residues for application in wastewater treatment. *Cellulose* **2014**, *21*, 2831–2844. [CrossRef]

35. Khan, M.N.; Bhutto, S. Kinetic study of the oxidatwe decolorization of xylenol orange by hydrogen peroxide in micellar medium. *JCCS* **2010**, *55*, 170–175.

36. Zhang, W.; Zhou, C.; Zhou, W.; Lei, A.; Zhang, Q.; Wan, Q.; Zou, B. Fast and Considerable Adsorption of Methylene Blue Dye onto Graphene Oxide. *Bull. Environ. Contam. Toxicol.* **2011**, *87*, 86–90. [CrossRef] [PubMed]

37. Chen, C.; Zhang, M.; Guan, Q. Kinetic and thermodynamic studies on the adsorption of xylenol orange onto MIL-101 (Cr). *Chem. Eng. J.* **2012**, *183*, 60–67. [CrossRef]

38. Min, L.U.; Zhang, Y.; Guan, X. Thermodynamics and kinetics of adsorption for heavy metal ions from aqueous solutions onto surface amino-bacterial cellulose. *Trans. Nonferrous Met. Soc.* **2014**, *24*, 1912–1917.

© 2017 by the authors. Licensee MDPI, Basel, Switzerland. This article is an open access article distributed under the terms and conditions of the Creative Commons Attribution (CC BY) license (http://creativecommons.org/licenses/by/4.0/).

MDPI

Article

Green Binder Based on Enzymatically Polymerized Eucalypt Kraft Lignin for Fiberboard Manufacturing: A Preliminary Study

Susana Gouveia [1], Luis Alberto Otero [2], Carmen Fernández-Costas [1], Daniel Filgueira [1], Ángeles Sanromán [1] and Diego Moldes [1,*]

[1]   Department of Chemical Engineering, University of Vigo, Lagoas Marcosende s/n., E-36310 Vigo, Spain; gouveia@uvigo.es (S.G.); mcarmenfc@uvigo.es (C.F.-C.); danmartinez@uvigo.es (D.F.); sanroman@vigo.es (Á.S.)

[2]   R & D Department of FORESA, Avda. Doña Urraca, 91, Caldas de Reis, 36650 Pontevedra, Spain; l.otero@foresa.com

*   Correspondence: diego@uvigo.es; Tel.: +34-986-818-723

Received: 10 April 2018; Accepted: 7 June 2018; Published: 9 June 2018

Abstract: The capability of laccase from *Myceliophthora thermophila* to drive oxidative polymerization of *Eucalyptus globulus* Kraft lignin (KL) was studied as a previous step before applying this biotechnological approach for the manufacturing of medium-density fiberboards (MDF) at a pilot scale. This method, which improves the self-bonding capacity of wood fibers by lignin enzymatic cross-linking, mimics the natural process of lignification in living plants and trees. An interesting pathway to promote these interactions could be the addition of lignin to the system. The characterization of *E. globulus* KL after enzymatic treatment showed a decrease of phenolic groups as well as the aromatic protons without loss of aromaticity. There was also an extensive oxidative polymerization of the biomolecule. In the manufacture of self-bonded MDF, the synergy generated by the added lignin and laccase provided promising results. Thus, whenever laccase was present in the treatment, MDF showed an increase in mechanical and dimensional stability for increasing amounts of lignin. In a pilot scale, this method produced MDF that meets the requirements of the European standards for both thickness swell (TS) and internal bonding (IB) for indoor applications.

Keywords: wood; *Eucalyptus globulus*; laccase; Kraft lignin; medium-density fiberboards; pilot scale

---

1. Introduction

Currently, the wood-based panels industry needs to find natural substitutes of synthetic resins to be a sustainable and eco-friendly industry. The bonding of wood components is achieved when a resinous matrix is formed. The wood boards produced with conventional gluing process show high mechanical strength and relative low swelling when exposed to water.

Up to the present, a great deal of research has been devoted to reducing the environmental impact of wood-based panels industries, more concretely the impact of oil-derived adhesives such as urea-formaldehyde or phenol-formaldehyde. Replacement of fossil phenol by biomass-derived phenols (e.g., lignin, tannins) seems to be a sustainable and environmentally friendly approach [1–3]. However, these authors substituted phenol with lignin, but the binders still included formaldehyde, a toxic compound with a significant impact in human health and environment. It is worth noting that the manufacturing of wood-based panels requires the bonding of wood elements, and such unions are due not only to the effect of synthetic resins but also by the intrinsic auto-adhesive properties of wood compounds [4,5]. The auto-adhesive ability suggests that a more environmentally friendly approach to reduce the use of resins is to boost up the inherent capacity of wood elements to bind among themselves.

Most of the studies related to the quest of a binderless fiberboard manufacture process are based on the oxidative modification of lignin [6,7]. The chemical structure of lignin possesses a broad range of chemical groups, i.e., hydroxyl (aliphatic and aromatic), methoxyl, carboxyl or carbonyl, which converts lignin in a raw material with several potential applications [8]. Moreover, such renewable biopolymer is industrially available because it is separated from the polysaccharides during the pulping process of lignocellulosic biomass. Nonetheless, technical lignins, mainly Kraft lignin (KL), are considered by-products and their use is normally limited to energy production in the pulp mills.

The use of phenol-oxidizing enzymes (e.g., laccases, peroxidases) for adhesive applications is based on the capability of enzymes to promote the polymerization and cross-linking of lignin [4,9,10] (Figure S1). The effects of laccases [11–14] and peroxidases [15,16] over the chemical structure of lignin has been examined by several studies. The authors concluded that both enzymes promote the oxidation of the phenolic groups to phenoxy radicals in a one-electron oxidation. Hence, the incubation of wood fibers with phenol oxidizing enzymes results in the cross-linking of lignin moieties by means of covalent bonds [4,17]. Such mechanism enables the wood fibers adhesion, leading to the manufacturing of wood-based panels with substantial improvements in both mechanical and physical properties. Among the wood composite panels, the manufacturing process for medium-density fiberboards (MDF) has some characteristics which make it particularly suitable for the use of this biotechnology.

The aim of this study is to test the laccase-assisted enzymatic treatment for the manufacturing of binderless *Eucalyptus globulus* fiberboard at pilot scale. Thus, a two-component system manufacturing process adding KL to wood fibers was tested and the effect of incubation time as well as the amount of lignin added to the board was studied. The effect on the physical and mechanical properties of the resulting wood boards was assessed. To a better understanding of the enzymatic reactions involving lignin during fiber enzymatic treatments, isolated KL was subjected to the same treatment and comparative studies of the lignins, before and after enzymatic action, were performed by several analytical techniques (FTIR, HPLC-SEC, 2D NMR).

The MDF manufacturing process tested in this study is presented as an interesting lignin valorization in the context of circular economy with the additional environmental benefits of removing toxic binding materials from the conventional manufacturing process.

2. Materials and Methods

2.1. Lignin

Black liquor from the Kraft cooking process of *Eucalyptus globulus* was supplied by ENCE (Pontevedra, Spain). Kraft lignin (KL) was obtained by acidic precipitation of the black liquor as follows. Black liquor was diluted in water (1:1) and the pH was lowered to 2.5 by addition of sulfuric acid 4 M. The solution was left under stirring for 30 min. The precipitate was centrifuged, and the isolated KL was washed twice with acidified water (pH 2.5). KL was oven-dried overnight at 60 °C, and then it was milled in an agate mortar and stored in amber glass bottle until use.

2.2. Laccase Enzyme

Laccase from *Myceliophthora thermophila* (NS51003) was kindly supplied by Novozymes (Bagsvaerd, Denmark). Commercial laccase was desalted using a PD-10 desalting column with Sephadex G-25 Medium (General Electric, Norwalk, CT, USA) following the supplier recommended protocol.

Laccase activity was determined spectrophotometrically by oxidation of 2,2'-azino-*bis* (3-ethylbenzothiazoline-6-sulphonic acid) (ABTS) at 436 nm ($\varepsilon = 29300$ $M^{-1}cm^{-1}$) in potassium phosphate buffer pH 7.3 at 25 °C. One activity unit (U) was defined as the amount of enzyme that oxidized 1 μmol of substrate per min.

2.3. Wood Fibers

Eucalyptus globulus fibers were supplied by the MDF board plant of FINSA industries (Padrón, Spain). The fibers were dried in a flash drier until a moisture content of 2%.

2.4. KL Enzymatic Polymerization

Lignin was solubilized in phosphate buffer pH 7.3 (100 mM) obtaining a solution of 1.5 $g \cdot L^{-1}$. The desalted laccase was added to 90 mL of the latter solution to reach a final activity of 2 $U \cdot mL^{-1}$. The reaction took place at 70 °C for 2 h in an orbital shaker. The reaction was stopped by lowering the pH to 2.0 causing enzyme deactivation and lignin precipitation. The reaction product was filtered and washed twice with acidified water (pH 2.5) and oven-dried at 60 °C overnight.

2.5. KL Characterization

2.5.1. Determination of Phenolic Content

Phenolic content was evaluated as described by [18]. Initially, KL solutions of 0.5 $g \cdot L^{-1}$ in NaOH 0.05 M were prepared. Then, 1 mL of KL solution, 3 mL of Folin and Ciocalteu reagent and 30 mL of distilled water were added into a volumetric flask and mixed thoroughly. After 5–8 min of stirring, 10 mL of 20% (w/w) sodium carbonate solution was added and the volume was adjusted to 50 mL with DI (deionized) water. The mixture was stirred for 2 h and finally, the absorbance at 760 nm was measured. The phenolic content test was run in duplicate. Vanillin standard solutions (0–5 mM) were used for calibration.

2.5.2. Fourier Transform Infrared Spectroscopy (FTIR)

Solid KL samples were placed to dry for 10 min under an infrared lamp. The FTIR spectra were recorded with a Jasco FT/IR-4100 (Easton, PA, USA), equipped with attenuated total reflectance (ATR) in absorbance mode using a frequency range of 650–4000 cm^{-1}. Each spectrum, which accumulated 32 scans at 4 cm^{-1} resolution, was then analyzed with the OMNIC 32 software.

The absorption bands were assigned as suggested by [19]. The spectrum was baselined and ATR corrected and the bands intensities were normalized referring aromatic skeletal vibration (around 1510 cm^{-1}).

2.5.3. Molecular Weight Distribution

Lignin samples were dissolved in NaOH 0.05 M (final concentration of 0.5 $g \cdot L^{-1}$), and left in agitation for 8 h at 300 rpm and 40 °C. Once totally dissolved, the samples were filtrated with PVDF 0.2 μm syringe filter.

Size-exclusion chromatography was performed in a Jasco HPLC system (Easton, PA, USA) (AS 1555 auto sampler; PU 2080 plus pump, UV 975 detector) equipped with two GPC columns (Phenomenex, Torrance, CA, USA) coupled in series (GPC P4000 and P5000, both 300 mm × 7.8 mm) and a safeguard column (35 mm × 7.8mm). The injection volume was 100 μL, and the isocratic flow (NaOH 0.05 M) was pumped at a rate of 1 $mL \cdot min^{-1}$ at 25 °C for 26 min. Detection was performed with a UV detector at 254 nm. Data was recorded and analyzed with ChromNAV GPC software. Calibration curve was obtained with polystyrene polymer standards (Phenomenex) with molecular weights of 891, 1670, 6430, 10,200, 33,500, 65,400, 158,000, 305,000, 976,000 and 2,350,000 Da. Concentrations and injection volumes were performed according to the manufacturer's specifications. The mobile phase was the same used as solvent for KL samples, i.e., 0.05 M NaOH.

2.5.4. Nuclear Magnetic Resonance (NMR)

NMR techniques, namely proton nuclear magnetic resonance spectroscopy (^1H), carbon nuclear magnetic resonance spectroscopy (^{13}C) and Heteronuclear Single Quantum Coherence (HSQC) were used for KL characterization. KL samples were previously acetylated to enhance their solubility in the NMR solvent. An adaptation of the method described by [20] was performed to acetylate the samples. Approximately 100 mg of KL was placed in a dry acetylation vial and 500 μL of pyridine were added. The vial was sealed and constantly stirred at 37 °C during 1 h or until the KL full sample dissolution. Afterwards, 1 mL of acetyl anhydride was added and left reacting with constant agitation at 37 °C for 48 h. Finally, once the acetylation process was over, 440 μL of methanol were added to remove the excess of pyridine and acetyl anhydride. After 2 h of stirring, the content of the vial was dried and milled with and agate mortar and pestle.

All spectra have been recorded at 25 °C on a Bruker Avance 400 MHz (Billerica, MA, USA), equipped with a z-gradient 5-mm QNP probe. Chemical shifts were referred to the solvent signals, $\delta(^1H) = 2.5$ ppm, $\delta(^{13}C) = 39.5$ ppm.

^1H NMR

50 mg of both acetylated and non-acetylated KL were dissolved in 750 μL of deuterated dimethyl sulfoxide (DMSO-*d6*). The relaxation delay was 1 s and a flip angle of 30° was used. The number of scans was 16 with an acquisition time of 6.3 s. Spectra were processed using an exponential weighting function of 0.3 Hz prior to Fourier transformation.

^{13}C NMR

50 mg of KL were dissolved in 750 μL of DMSO-*d6*. The relaxation delay was 5 s, and a flip angle of 30° was used. The number of scans was 50,000 with an acquisition time of 1.4 s.

HSQC

KL samples (150 mg) were dissolved in 750 μL of DMSO-*d6*. The number of complex points collected was 2048 for the ^1H dimension and a recycle delay of 5.28 s (5 s relaxation delay and 0.28 s acquisition time) was chosen. The number of transients was 128 and 256 time increments were recorded in the ^{13}C dimension.

2.6. Medium-Density Fiberboards Production

2.6.1. Enzymatic Fiber Pre-Treatments

E. globulus fibers were submitted to different treatments as listed in Table 1. Regarding to one-component system, laccase was air-pressurized sprayed to the dried fibers on a rotary drum blender until homogenization. However, in the case of two-component system (fiber and KL), KL was premixed with the fibers in the rotary blender prior to enzyme addition. Enzyme was added as commercially supplied and no pH adjustment was carried out. After homogenization, to allow sufficient time for the enzymatic reactions take place, fibers (or fibers and added KL) were left in an oven at 70 °C for 2 h, in a vessel without agitation. The vessel was covered with aluminum foil to avoid water evaporation, except when initial moisture was 20% or higher; in such cases the vessel was uncovered, and the moisture controlled every 30 min to ensure minimum moisture of 10–12%.

Table 1. Experimental conditions and fiberboards composition.

Treatment	One-Component System				Two-Component System				
	T1	T2	T3	T4	T5	T6	T7	T8	T9
Dry fiber (g)	520	520	520	520	520	421	421	325	145
KL (% w/w dry fiber)	0	0	0	15	15	42	42	78	280
Water added (% w/w dry fiber)	12			12			12		
Enzyme dosage (U·g^{-1} dry fiber)	0	29	29						
Enzyme dosage (U·g^{-1} KL)				0	190	0	190	190	190
Moisture content after blending (%)	12	12	12	11	11	11	20	26	36
Incubation time at 70°C (h)		2	24		2		2	2	2
Moisture content before pressing (%)	11	11	11	10	10	10	12	13	15

Fixed experimental conditions	
Wood fiber origin	*Eucalyptus globulus* (100%)
KL origin	*Eucalyptus globulus* (100%)
Target board density	650–700 kg·m^{-3}
Cold pressing time	120 s
Cold press plate position (H)	30 mm
Hot pressing temperature	200 °C
Hot press plate position (H)	22 mm 18.5 mm 16.7 mm
Hot pressing times	232 s 124 s 124 s
Final board size (W × L × H)	250 mm × 250 mm × 16.7 mm

2.6.2. MDF Preparation

After the enzymatic treatment, fibers were pressed in a mold, producing an approximately cubic mat (≈250 mm × 250 mm × 250 mm). The resulting mat was cold pressed for about 2 min to an approximate height of 40 mm. The mat was taken in hot press (press platens were maintained at 200°C) and reduced to its final thickness in a dynamic three step cycle with a total time of 8 min. In step one, the mat was pressed to 22 mm for 232 s and, in the second and third step, the board was pressed to a height of 18.5 mm and 16.7 mm respectively, both for 124 s.

The complete manufacturing process is described in Figure 1, and the process parameters are summarized in Table 1.

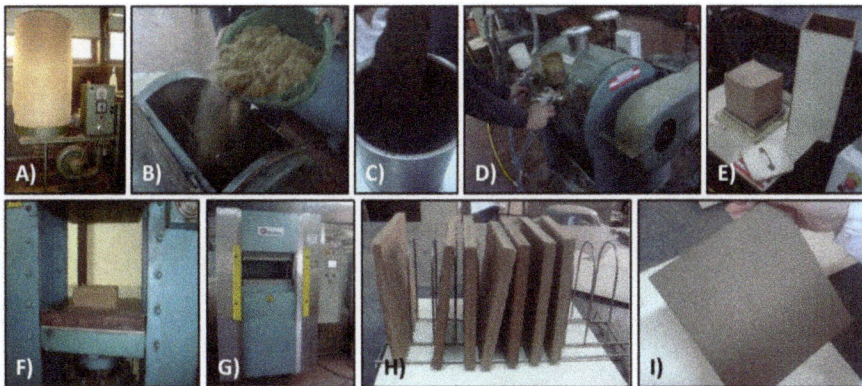

Figure 1. Binderless MDF pilot scale manufacturing procedure: (**A**) Drying of eucalyptus fibers; (**B**) Addition of fibers with 2% moisture to drum blender; (**C**) *E. globulus* KL addition; (**D**) Laccase spraying with air pressure; (**E**) Mat manually pressed; (**F**) Mat cold pressing; (**G**) 3 steps hot pressing cycle; (**H,I**) Final MDF boards.

2.6.3. Medium-Density Fiberboard Properties

The internal bond strength (IB) was determined according to EN 319 [21]. Thickness swell (TS) and water absorption (WA) were determined using specimens of (50 ± 1) mm × (50 ± 1) mm. After an aging cycle where the specimens were submersed for 24 h in an upright position in water at 20 °C, the excess of water was drained. Specimen's thickness and mass was measured prior and after the immersion.

3. Results and Discussion

3.1. Kraft Lignin Enzymatic Polymerization and Characterization

Laccase enzyme is one of the most inexpensive enzymes widely available in the market. Commercial laccase from *M. thermophila* showed the ability to oxidize and polymerize Kraft lignin (KL). The extension of this polymerization depended on several parameters such as lignin origin, pH, temperature, phenolic mediators or reaction time [11,12,22,23]. The aim of this study was to test the capability of the enzymatic treatment of KL in the manufacturing of MDF with no synthetic resins. The enzymatic treatment was tested to enhance the cross-linking among lignin molecules to produce stable and durable MDF from *E. globulus* fibers and KL. The experimental conditions used in these studies were those previously identified to favor the enzymatic polymerization of *E. globulus* KL [11].

The oxidative enzymatic treatment of KL with laccase induced clear changes in the structure of lignin as it was observed in the reaction vessel during the reaction. Hence, the resulting enzymatically-treated lignin showed a much darker brown color compared to the original lignin. Moreover, the polymerized KL required a very long stirring time (1 day) to be dissolved in NaOH 0.05 M whereas the original KL was easily dissolved in just a few seconds. Gel permeation chromatography, phenolic content, FTIR and NMR (^{1}H, ^{13}C and HSQC) analyses were used to characterize the enzymatically-polymerized KL. Detailed analytical procedures, results and discussion can be found in previous publications [11,12,24]. Gel permeation chromatography permits to compare the average molecular weight of the original and polymerized KL (Figure S2). As observed in Figure 2A, the enzymatically-treated KL showed a strong increase in the average molecular weight (17-fold) compared to the original KL. At the same time, the enzymatic polymerization was associated with a loss of the lignin's phenolic content (Figure 2B). Such results evidenced that laccase oxidized the KL's phenolic moieties to phenoxy radicals, which were coupled among themselves leading to a remarkable increase in the molecular weight and a lower phenolic content of the KL. The FTIR spectra analyses of both polymerized and original KL detected that polymerized KL showed a higher signal of the peaks associated with C=O bonds, for both non-conjugated (1718–1703 cm^{-1}) and conjugated C=O bonds (1655–1654 cm^{-1}) (Figure S3). It is important to stress that the conjugated C=O were not detected in the original KL (Figure 2C).

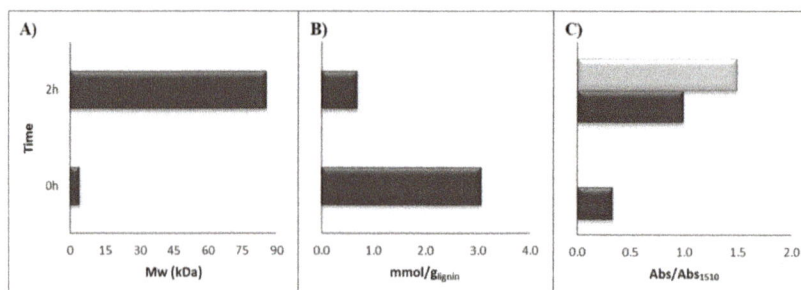

Figure 2. Comparison between untreated (0 h) and enzymatically treated (2 h) *E. globulus* KL: (**A**) Molecular weight; (**B**) Phenolic content; (**C**) Conjugated C=O FTIR absorbance referenced to aromatic skeletal vibration (light gray) and non-conjugated C=O FTIR absorbance referenced to aromatic skeletal vibration (dark gray).

NMR techniques, particularly 2D-NMR such as HSQC, are powerful tools in the identification and quantification of lignin main structures [25,26]. ^{1}H NMR analysis of non-acetylated KL (Figure 3) indicates a pronounced decrease of the aromatic protons in KL after laccase-mediated polymerization. The comparison of ^{1}H NMR spectra, before and after polymerization, shows a wide band associated to methoxyl groups (4.2–3.1 ppm) after enzymatic treatment whereas most of the aromatic protons disappeared (7.0–6.0 ppm). In addition, benzaldehyde protons, with a chemical shift of 9.78 ppm, also disappeared after the polymerized KL. The disappearance of aromatic proton signals was not totally unexpected as it has been previously reported with lignosulfonates by [27].

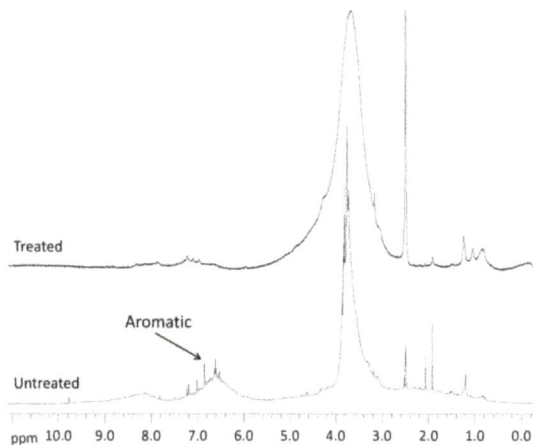

Figure 3. ^{1}H NMR spectra of non-acetylated *E. globulus* Kraft lignin samples.

Sample acetylation improved the KL solubilization and allows differentiating the aliphatic from the aromatic hydroxyl groups. However, the process of acetylation caused structural changes and the spectra obtained for the acetylated samples showed poorer signals. Nevertheless, ^{1}H NMR analysis of acetylated samples (Figure 4) evidenced that nearly all aromatic hydroxyls have been attacked by laccase while some aliphatic hydroxyls remained in the polymerized KL.

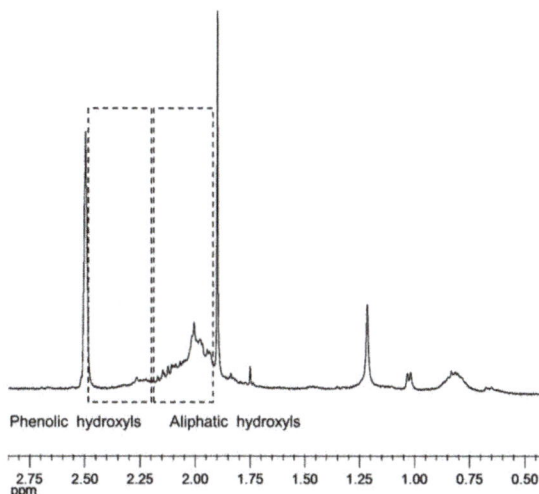

Figure 4. Expanded region of ^{1}H NMR spectrum of an enzymatically treated *E. globulus* KL subjected to acetylation.

This result was even clearer in the HSQC spectrum (Figure 5), where there was no correlation for aromatic hydroxyls and there was only a small area corresponding to the aliphatic hydroxyls.

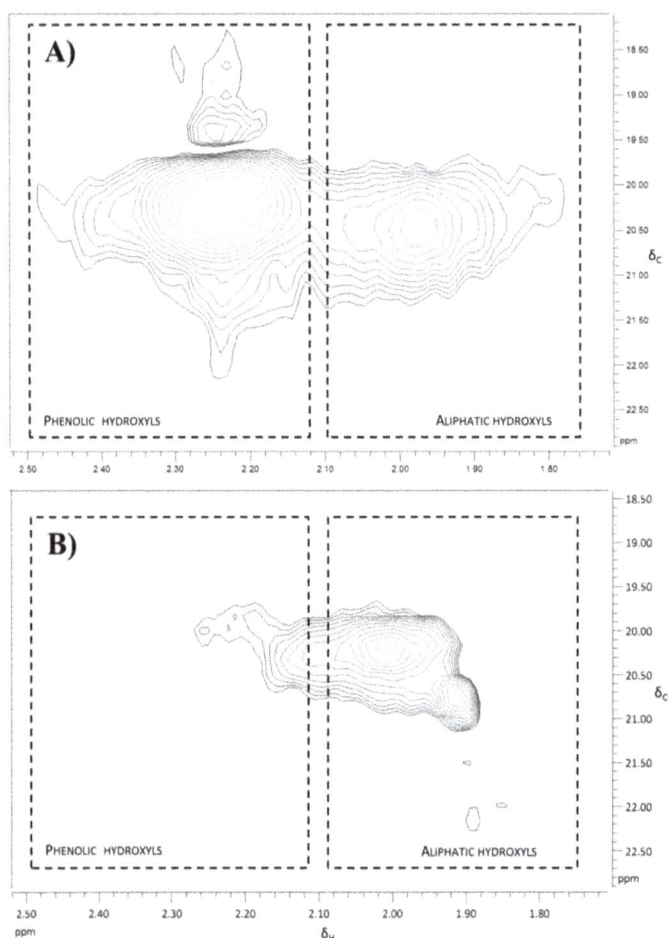

Figure 5. Hydroxyl expanded region of HSQC spectrum of acetylated *E. globulus* KL: (**A**) Untreated; (**B**) After enzymatic polymerization.

^{13}C NMR also revealed significant differences between KL samples (Figure 6). After the enzymatic treatment, a remarkable increase in the signal at 172.0 ppm (carboxyl groups) was accompanied by the disappearance of the signal corresponding to phenolic hydroxyls (168.1 ppm), whereas the weak signals of primary and secondary aliphatic hydroxyls remained almost unchanged (170.0 ppm and 169.4 ppm, respectively). On the other hand, after the enzymatic treatment, strong signals of aromatic carbons (160–100 ppm) remained in ^{13}C NMR spectrum proving that benzene rings were not affected in the polymerization. This suggests that the polymerization observed by size-exclusion chromatography could be explained by the laccase mode of action: the enzyme initiated the oxidation of phenolic groups into stabilized radicals that subsequently, undergone radical-radical coupling through phenyl ether-carbon and carbon-carbon links. These new bonds yield to the observed increase in molecular weight without destruction of the aromatic KL backbone. Similar behavior was observed by other authors in the polymerization of commercial lignosulphonates and lignin models by laccase [27,28].

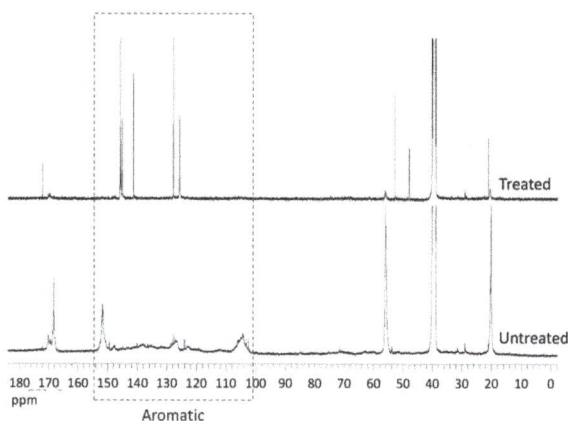

Figure 6. ^{13}C NMR spectra of acetylated *E. globulus* KL.

3.2. MDF Manufacture

Manufacturing of binderless MDF with an enzymatic pre-treatment by dry process were performed under the conditions specified in Table 1. The main objective of these tests was to assess the possible MDF manufacture with only an enzymatic treatment of the wood fibers and/or the addition of KL. The experimental conditions in the MDF manufacturing, such as the amount of enzyme and KL added were adjusted to improve the fiberboard characteristics, maximizing its internal bonding (IB) and minimizing the water absorption and thickness swelling (TS). The dynamic pressing cycle was not optimized in the present study. The temperature and press factor (s·mm^{-1} board thickness) used during hot pressing were the same used by the industrial partner when testing, in pilot plant, the production of MDF with synthetic resins. These conditions would have to be improved in an industrial scale, since the typical values in the industrial process are around 10 s pressing time per mm board thickness.

Nowadays, there are two main industrial processes to obtain high quality wood fibers for MDF manufacturing: thermal and thermo-mechanical pulping process. In both processes, high temperatures and pressure can efficiently defibrate wood, producing a raw material with suitable properties for fiberboard manufacturing [29]. Importantly, the heat associated with high pressures causes chemical changes in the fibers. On one hand, small fragments of lignin are formed in the defibration process, that finally settle in the surface of the fibers. These fragments show a high reactivity due to the presence of phenolic radicals in their structure [30,31]. On the other hand, a glassy layer of hardened lignin is formed on the surface of the fiber because defibration takes place at temperatures above lignin's glass transition temperature [32]. This glassy crust on the fiber surface is a barrier for the enzymatic action in wood fiber's lignin. However, when enzymes are used in the manufacturing process, the initially existing crust is loosened, and lignin is again available for laccase oxidative action. Furthermore, the results found in literature for MDF production, at laboratory and pilot-scale (Table 2), are revealing that laccases can oxidize and polymerize in an efficient way not only isolated lignins, but also lignins in a fiber-bound state. These results indicate that the glassy form of lignin, which covers the wood fibers after defibration, is not able to inhibit, at least completely, the oxidative action of laccase over lignin.

Table 2. Examples found in literature of binderless MDF glued by auto-adhesion of treated fibers.

Fiber		Laccase	Incubation Conditions			Enzyme Dosage (U·g⁻¹ Fiber)	Pressing Conditions		Scale	Board Properties						Reference
			T (°C)	Time (h)	pH		Press Factor (s/mm)	T (°C)		Density (kg·m⁻³)	MOE (GPa)	MOR (MPa)	IB (MPa)	WA (%)	TS (%)	
Fagus sylvatica	Control	Heat deactivated *Trametes versicolor*	20	1.0	5	0	100	200	Lab.	850	3.42	25.3	0.91	143	45	[9]
	Treated					3.5				895	4.02	41.7	1.57	72	19	
Fagus sylvatica	Control	Untreated *M. thermophila*	50	0.5	7	0	25	200	Pilot	820	(b)	(b)	0.33	224	146	[4]
	Treated					6				858	3.70	40.1	0.82	109	69	
						24				868	3.95	46.0	0.93	92	46	
Leaf sheath from commercial plants	Control	Untreated *Aspergillus oryzae*	30	1.0	6	0	160	200	Lab.	1100	1.3	13.3	-	268.3	218.8	[6]
	Treated					6				1100	13.3	17.5	-	82.4	67.5	
						12				1100	268.3	18.6	-	79.7	31.4	
						24				1100	218.8	18.7	-	80.5	30.1	
P. sylvestris (90%) *P. radiata* (10%)	Control	Heat deactivated *Trametes villosa*	≤120	≈0.5	6	0	22	200	Pilot	800	-	≈10	≈0.1	-	≈122	[33]
	Treated					100				800	-	≈20	≈0.38	-	≈62	
Harea brasiliensis	Control	*Trametes villosa*	25	1.0	5	-	40	200	Lab.	-	-	-	-	-	-	[34]
	Treated					9				750	3.6	9.3	0.67	-	-	
Spruce (80%) *Fir* (20%)	Control	Buffer + wax *Trametes villosa*	≤165	≈0.5	6	100	60	200	Pilot	850	-	19	<0.1	-	32	[35]
	Treated					3.5				850	-	37	0.32	-	20	
Spruce (80%) *Fir* (20%)	Control	Buffer *Trametes villosa*	-	-	6	100	12	190	Pilot	750	-	12	<0.1	-	100	[36]
	Treated					100				750	-	20	0.42	-	50	

(b) delaminated boards; MOE: modulus of elasticity; MOR: modulus of rupture; IB: internal bonding; WA: water absorption; TS: thickness swell.

Table 3 summarizes the properties of the binderless MDF manufactured under various experimental conditions. The enzyme dosage was kept constant throughout the tests (29 U/g of dry fiber). This value was selected considering our previous studies regarding lignin polymerization [11] and the information available in the literature (Table 2). A control experiment (T1) was performed by adding to the fiber nothing but the necessary water for proper heat conduction, i.e., 10–13% moisture content. These conditions were set to ensure that the temperature in the center of the board exceeds the lignin glass transition temperature because lignin plasticization have been considered partially responsible for board adhesion [37].

Table 3. Summary of MDF manufacturing conditions and MDF properties.

	Treatment								
	One-Component System				Two-Component System				
	T1	T2	T3	T4	T5	T6	T7	T8	T9
Treatment									
KL (% w/w dry fiber)	0	0	0	15	15	42	42	78	280
Enzyme dose (U·g^{-1}fiber)	0	29	29						
Enzyme dose (U·g^{-1} KL)				0	190	0	190	190	190
Incubation time (h)		2	24		2		2	2	2
Board properties									
Density (kg·m^{-3})	661	670	633	697	705	698	688	785	831
IB (MPa)	<0.01	<0.01	<0.01	<0.01	0.06	0.04	0.22	0.41	0.80
WA (%)	D	D	D	D	D	134.7	76.8	60.0	27.8
TS (%)	D	D	D	D	D	55.2	14.6	7.5	5.0

D—detached.

3.2.1. One-Component System

According to the results shown in Table 3, the fiber pre-treatment with laccase (T2) did not improve the MDF properties when compared to the untreated control (T1). The increase of the enzymatic treatment time, until 24 h (T3), did not show any improvement in the IB of the MDF. Furthermore, no dimensional stability was showed by the enzymatically-treated MDFs (both T2 and T3), which completely detached when exposed to water for 24 h. The poor results indicate an issue in the manufacturing conditions.

Among the few publications using dry incubation process for one-component binderless MDF production (Table 2), [9] were able to produce fiberboards with reasonably good characteristics. However, there are some noticeable differences between this study and [9]. One is the fiber source: lignins from different wood species have distinct chemical and physical characteristics, such as syringyl/guaiacyl ratio, phenolic content, molecular weight distribution, etc., that may considerably affect the enzymatic polymerization [11,38,39]. Also, the thickness and density of the MDF were very different in both studies, and this may have an important impact on the final MDF properties [40]. In a more recent work [36], Euring et al. were able to achieve MDF with good mechanical properties with an IB and modulus of rupture (MOR) of 0.42 and 20 Mpa respectively, yet, a TS of 50% was still an issue. To improve the dimensional stability of the fiberboards, the authors tested a laccase-mediator-system. The best results were obtained using 4-hydroxybenzoic acid as mediator, leading to a TS of 19% and a simultaneous improvement in the mechanical properties. Other authors [35] used a laccase-mediator-system to introduce a water-based wax as hydrophobic agent in the manufacturing of a binderless MDF. Until then, the combined use of hydrophobic agents and enzymes were supposed to be incompatible because the hydrophobic agents coated the fibers hindering the enzymatic access to lignin [4]. However, [35] obtained an MDF with IB as high as 0.9 MPa and TS of 17% using 1% of hydrowax in a laccase treatment with vanillic alcohol as mediator.

Other alternative to enhance the MDF properties could be the addition of technical lignin to the fibers before the enzymatic treatment. This is the so-called two-component system. Technical lignins are those obtained after the lignocellulosic pulping process (e.g., KL). These lignins have a molecular structure considerably different with respect to native lignin. Moreover, technical lignins will have distinct characteristics in composition and structure depending on the pulping process.

Among the several pulping processes to fragment lignocellulose into their major compounds, hemicellulose, cellulose and lignin, Kraft process is the predominant. During Kraft digestion, depolymerization of native lignin mainly occurs through the extensive cleavage of β-O-4′ ether bonds. The resulting KL has not only a higher amount of phenolic hydroxyl groups but also, biphenyl and other condensed structures are less formed than in other pulping processes [41]. As the main goal during the Kraft process is to achieve high quality cellulose, the large amount of KL obtained is considered a by-product.

In modern Kraft mills, on-site Kraft lignin burning for steam and energy production ensures not only the energy self-sufficiency of these industrial units but also an energetic surplus. However, the incineration (even with energy recovery) is considered the last option for any by-product valorization. As many of the major Kraft pulp producers are considering the conversion of their units in biorefineries, the valorization of Kraft lignin, which is produced in huge quantities, is crucial. Nowadays, Kraft powder lignin is already widely available in the market for various proposes other than incineration.

Therefore, the high availability and high phenolic content of KL, converts such by-product of the pulp industry in a suitable substrate to be used in laccase-assisted processes such as the manufacturing of eco-friendly binderless MDF.

3.2.2. Two-Component System

The two-component system, consisting on the addition of *E. globulus* KL and laccase to the fibers, was tested for the manufacturing of MDF at pilot-scale. It was expected that the enzymatic treatment enhanced the copolymerization between the added KL and the lignin on the wood fibers. Taking into account that solid *E. globulus* KL was added to the mixture of wood fibers and enzyme, the enzyme dosage was adjusted (T4–T9) as shown in Table 3. Hence, the enzymatic dosage used is no longer referred to fiber weight but to the amount of added KL (190 U of laccase per g of KL).

In the case of fiberboards composed only by wood fibers and KL (T4), with no enzyme addition, the properties obtained were similar to the control fiberboards (T1), suggesting the non-significant effect of single KL addition. When laccase was sprayed in the mixture of wood fibers jointly with KL (T5), the cohesion of the board was improved. Such small improvement (IB was 0.06 MPa) suggests that the laccase-KL system was able to form cross-linking between the wood fiber's lignin and the added KL. However, the water absorption and TS of the MDF were not measurable because the board detached before the end of the analysis.

Previous studies have combined the addition of lignin, or lignin-like phenolic compounds, in a two-component system binder for MDF production [36]. The phenolic groups offered by the added KL are more accessible to enzymatic action than those found within the wood fibers [42]. Thus, the advantage of a two-component system is not limited to provide a larger number of phenolic groups, but also ensure that those phenolic groups are readily available for enzymatic action, enabling the formation of a greater number of phenoxy radicals. In fact, [36], improved the IB of MDF with the addition of lignosulphonates to the enzymatic treatment of wood fibers. Nonetheless, the TS values obtained were still worse than those obtained with urea-formaldehyde resin. Probably, the hydrophilic nature of lignosulphonates led to less favorable results of MDF dimensional stability. Nonetheless, due to its higher hydrophobic character, KL could be an interesting alternative to lignosulfonates for the manufacturing of binderless MDF in a two components system.

The effect of the KL amount added to the manufacturing process of MDF was also assessed, but keeping constant the ratio of enzyme activity per g of added KL. The increase of KL amount added to the enzymatic treatment (T7) of the wood fibers had a remarkable effect. IB was clearly improved (5-fold), indicating that a higher interlinking between both the wood fiber's lignin and the KL was achieved. Moreover, the water absorption and the TS were significantly improved respect to previous treatments. Nevertheless, the amount of polymerized KL was not enough to give an optimal protection against water. It is worth noticing that T6 and T7 had the same amount of KL but, laccase was not present in T6. Therefore, there was no enzymatic induced cross-linking between the KL and the lignin moieties in the wood fibers. Furthermore, the decrease in the lignin phenolic groups (which are the hydrophilic groups of lignin) caused by laccase oxidation (Figures 2B, 4 and 6), could considerably contribute to the lower water absorption and TS observed in the MDF of T7.

Given these results, in T8 and T9, the amount of KL was increased even further. As a result, a clear improvement of board properties was obtained, enabling the manufacturing of MDF with both IB and dimensional stability (Figure 7). Moreover, the MDF properties from test T9 were even better than those reported by [36] in a lignin-laccase-mediator-system. It is noticeable that the T9 MDF (KL + laccase with no mediators) showed similar IB but much better TS than [36] results. As it was commented before, the different industrial lignin used in these studies could be the main reason for the different TS results.

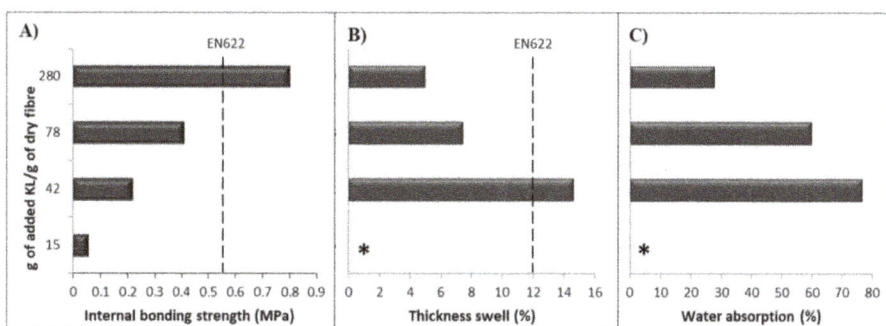

Figure 7. Effect of *E. globulus* KL addition in MDF properties. In all cases 190 U laccase·g^{-1} of KL was used. (**A**) Internal bond; (**B**) Thickness swell after 24 h immersion cycle; (**C**) Water absorption after 24 h immersion cycle. * Indicates MDF detachment when submersed.

Importantly, T8 and T9 binderless MDFs reached the standards required for indoor application. Such standards depend on the target market, for instance, the American norm for general interior uses, ANSI A.208.2-2002 [43], requires a minimum IB of 0.30 MPa and a TS no higher than 10% for an MDF Grade 110. The equivalent European standard EN 622-5:2010 [44] is more demanding for IB requirements (≥ 0.55 MPa) but more tolerant for the TS limit ($\leq 12\%$). Thus, T9 MDF manufactured in the present study met both the European and American standards for dry environments in both studied properties (IB and TS). This set of tests have proved that it is possible the manufacturing of fiberboards with no synthetic resins. In our tests, the binding among the wood fibers may be attained by the plasticizing effect of the lignin (added lignin and the lignin in the fibers) when heated, and the cross-linking catalyzed by the laccase enzyme. The plasticizing effect of lignin was observed in the tests T4 and T6 with no enzyme. The combination of enzyme and the plasticizing effect of lignin could result in a synergistic effect giving the fiberboard in test T9 the adequate properties to be marketed.

It is noteworthy that three quarters of T9 MDF weight is composed of *E. globulus* KL, so it could be considered a lignin-based matrix with wood fibers working as a reinforcing material. Thus, the proposed binderless process of MDF production, in addition to the environmental benefits resulting from the removal of synthetic resins, is a promising pathway of KL valorization, a waste stream of the pulp and paper industry.

4. Conclusions

The oxidation caused by laccase from *M. thermophila* led to a strong polymerization of an isolated industrial KL from *E. globulus*. Although a significant loss of hydroxyls and aromatic protons was detected by NMR, there was no disruption on the aromatic backbone of the polymer. The use of KL with laccase in a two-component system enabled the manufacturing of MDFs totally free of synthetic resins or additives, with *E. globulus* as the main raw material. Laccase catalyzed the cross-linking between different lignin molecules forming a 3D structure that confers to the MDF dimension stability, hydrophobicity, and mechanical resistance. The MDFs obtained with laccase + KL (two components system) showed remarkable high internal bonding and low thickness swelling. The boards met the European and American standards for indoor applications. These features and the high availability of KL can make the laccase-assisted polymerization of KL as a fully green strategy to substitute synthetic adhesives in the wood-panels industry.

Supplementary Materials: The following are available online at http://www.mdpi.com/2073-4360/10/6/642/s1, Figure S1: Representation of laccase redox cycle with: A) phenolic substrate; B) phenolic and non-phenolic substrate with a mediator title, Figure S2: Size exclusion chromatogram of untreated KL (blue line) and enzymatically treated KL (Black line), Figure S3: FTIR spectra of untreated KL (blue line) and enzymatically treated KL (Black line).

Author Contributions: S.G. designed and carried out the experiments, analyzed the data and wrote the paper, L.A.O. carried out the experiments, analyzed the data and contributed materials/analysis tools, C.F.-C. carried out the experiments and analyzed the data, D.F. carried out the experiments, and revised the paper, Á.S. contributed materials and revised the paper, D.M. conceived and designed the experiments, analyzed the data and revised the paper.

Funding: This research was funded by [ERDF Funds, Xunta de Galicia] grant number [09TMT012E], [EM2014/041] and [ED431C 2017/47]; [Ministry of Science and Innovation] grant number [CTQ2009-13651]; and by [Fundação para a Ciência e a Tecnologia—Portugal] grant number [POP-QREN BD 42684/2008].

Acknowledgments: Ence (Spain) is gratefully acknowledged for supplying the black liquor. Novozymes (Bagsvaerd, Denmark) is also kindly acknowledged for supplying the laccase from *M. thermophila*.

Conflicts of Interest: The authors declare no conflict of interest.

References

1. Kalami, S.; Arefmanesh, M.; Master, E.; Nejad, M. Replacing 100% of phenol in phenolic adhesive formulations with lignin. *J. Appl. Polym. Sci.* **2017**, *134*. [CrossRef]
2. Liu, C.; Zhang, Y.; Li, X.; Luo, J.; Gao, Q.; Li, J. "Green" bio-thermoset resins derived from soy protein isolate and condensed tannins. *Ind. Crop. Prod.* **2017**, *108*, 363–370. [CrossRef]
3. Zhang, W.; Ma, Y.; Wang, C.; Li, S.; Zhang, M.; Chu, F. Preparation and properties of lignin–phenol–formaldehyde resins based on different biorefinery residues of agricultural biomass. *Ind. Crop. Prod.* **2013**, *43*, 326–333. [CrossRef]
4. Felby, C.; Hassingboe, J.; Lund, M. Pilot-scale production of fiberboards made by laccase oxidized wood fibers: Board properties and evidence for cross-linking of lignin. *Enzyme Microb. Technol.* **2002**, *31*, 736–741. [CrossRef]
5. Zhang, D.; Zhang, A.; Xue, L. A review of preparation of binderless fiberboards and its self-bonding mechanism. *Wood Sci. Technol.* **2015**, *49*, 661–679. [CrossRef]
6. Álvarez, C.; Rojano, B.; Almaza, O.; Rojas, O.J.; Gañán, P. Self-Bonding Boards From Plantain Fiber Bundles After Enzymatic Treatment: Adhesion Improvement of Lignocellulosic Products by Enzymatic Pre-Treatment. *J. Polym. Environ.* **2011**, *19*, 182–188. [CrossRef]
7. Widsten, P.; Kandelbauer, A. Adhesion improvement of lignocellulosic products by enzymatic pre-treatment. *Biotechnol. Adv.* **2008**, *26*, 379–386. [CrossRef] [PubMed]
8. Pye, E. Kendall Industrial Lignin Production and Applications. In *Biorefineries-Industrial Processes and Products*; Wiley Online Books; WILEY-VCH Verlag GmbH & Co. KGaA: Weinheim, Germany, 2008; pp. 165–200, ISBN 978-3-527-31027-2.
9. Felby, C.; Pedersen, L.S.; Nielsen, B.R. Enhanced auto adhesion of wood fibers using phenol oxidases. *Holzforschung* **1997**, *51*, 281–286. [CrossRef]

10. Kharazipour, A.; Schindel, K.; Hüttermann, A. Enzymatic Activation of Wood Fibers for Wood Composite Production. In *Enzyme Applications in Fiber Processing*; ACS Symposium Series; American Chemical Society: Washington DC, USA, 1998; Volume 687, pp. 99–115, ISBN 0-8412-3547-3.

11. Gouveia, S.; Fernández-Costas, C.; Sanromán, M.A.; Moldes, D. Polymerisation of Kraft lignin from black liquors by laccase from Myceliophthora thermophila: Effect of operational conditions and black liquor origin. *Bioresour. Technol.* **2013**, *131*, 288–294. [CrossRef] [PubMed]

12. Gouveia, S.; Fernández-Costas, C.; Sanromán, M.A.; Moldes, D. Enzymatic polymerisation and effect of fractionation of dissolved lignin from Eucalyptus globulus Kraft liquor. *Bioresour. Technol.* **2012**, *121*, 131–138. [CrossRef] [PubMed]

13. Dashtban, M.; Schraft, H.; Syed, T.A.; Qin, W. Fungal biodegradation and enzymatic modification of lignin. *Int. J. Biochem. Mol. Biol.* **2010**, *1*, 36–50. [PubMed]

14. Leonowicz, A.; Cho, N.; Luterek, J.; Wilkolazka, A.; Wojtas-Wasilewska, M.; Matuszewska, A.; Hofrichter, M.; Wesenberg, D.; Rogalski, J. Fungal laccase: Properties and activity on lignin. *J. Basic Microbiol.* **2001**, *41*, 185–227. [CrossRef]

15. Hofrichter, M. Review: Lignin conversion by manganese peroxidase (MnP). *Enzyme Microb. Technol.* **2002**, *30*, 454–466. [CrossRef]

16. Wong, D.W.S. Structure and action mechanism of ligninolytic enzymes. *Appl. Biochem. Biotechnol.* **2009**, *157*, 174–209. [CrossRef] [PubMed]

17. Felby, C.; Thygesen, L.G.; Sanadi, A.; Barsberg, S. Native lignin for bonding of fiber boards–evaluation of bonding mechanisms in boards made from laccase-treated fibers of beech (*Fagus sylvatica*). *Ind. Crop. Prod.* **2004**, *20*, 181–189. [CrossRef]

18. Areskogh, D.; Li, J.; Gellerstedt, G.; Henriksson, G. Investigation of the molecular weight increase of commercial lignosulfonates by laccase catalysis. *Biomacromolecules* **2010**, *11*, 904–910. [CrossRef] [PubMed]

19. Faix, O. Fourier transform infrared spectroscopy. In *Methods in Lignin Chemistry*; Springer-Verlag: Berlin/Heidelberg, Germany, 1992; pp. 83–109, ISBN 3-540-50295-5.

20. El Mansouri, N.-E.; Salvadó, J. Analytical methods for determining functional groups in various technical lignins. *Ind. Crop. Prod.* **2007**, *26*, 116–124. [CrossRef]

21. European Standards EN 319:1993. *Particleboards and Fibreboards—Determination of Tensile Strength Perpendicular to the Plane of the Board*; European Commission: Brussels, Belgium, 1993.

22. Huber, D.; Pellis, A.; Daxbacher, A.; Nyanhongo, G.S.; Guebitz, G.M. Polymerization of Various Lignins via Immobilized Myceliophthora thermophila Laccase (MtL). *Polymers* **2016**, *8*, 280. [CrossRef]

23. Ortner, A.; Huber, D.; Haske-Cornelius, O.; Weber, H.K.; Hofer, K.; Bauer, W.; Nyanhongo, G.S.; Guebitz, G.M. Laccase mediated oxidation of industrial lignins: Is oxygen limiting? *Process Biochem.* **2015**, *50*, 1277–1283. [CrossRef]

24. Fernández-Costas, C.; Gouveia, S.; Sanromán, M.A.; Moldes, D. Structural characterization of Kraft lignins from different spent cooking liquors by 1D and 2D Nuclear Magnetic Resonance spectroscopy. *Biomass Bioenergy* **2014**, *63*, 156–166. [CrossRef]

25. Capanema, E.A.; Balakshin, M.Y.; Kadla, J.F. A comprehensive approach for quantitative lignin characterization by NMR spectroscopy. *J. Agric. Food Chem.* **2004**, *52*, 1850–1860. [CrossRef] [PubMed]

26. Maniet, G.; Schmetz, Q.; Jacquet, N.; Temmerman, M.; Gofflot, S.; Richel, A. Effect of steam explosion treatment on chemical composition and characteristic of organosolv fescue lignin. *Ind. Crop. Prod.* **2017**, *99*, 79–85. [CrossRef]

27. Nugroho Prasetyo, E.; Kudanga, T.; Østergaard, L.; Rencoret, J.; Gutiérrez, A.; del Río, J.C.; Ignacio Santos, J.; Nieto, L.; Jiménez-Barbero, J.; Martínez, A.T.; et al. Polymerization of lignosulfonates by the laccase-HBT (1-hydroxybenzotriazole) system improves dispersibility. *Bioresour. Technol.* **2010**, *101*, 5054–5062. [CrossRef] [PubMed]

28. Areskogh, D.; Li, J.; Nousiainen, P.; Gellerstedt, G.; Sipilä, J.; Henriksson, G. Oxidative polymerisation of models for phenolic lignin end-groups by laccase. *Holzforschung* **2010**, *64*, 21–34. [CrossRef]

29. Back, E.L.; Salmen, N.L. Glass transitions of wood components hold implications for molding and pulping processes. *Tappi* **1982**, *65*, 107–110.

30. Widsten, P.; Laine, J.E.; Qvintus-Leino, P.; Tuominen, S. Effect of high-temperature defibration on the chemical structure of hardwood. *Holzforschung* **2002**, *56*, 51–59. [CrossRef]

31. Widsten, P.; Laine, J.E.; Qvintus-Leino, P.; Tuominen, S. Effect of high-temperature fiberization on the chemical structure of softwood. *J. Wood Chem. Technol.* **2001**, *21*, 227–245. [CrossRef]

32. Müller, C.; Kües, U.; Schöpper, C.; Kharazipour, A. Natural Binders. In *Wood Production, Wood Technology, and Biotechnological Impacts*; Universitätsverlag Göttingen: Göttingen, Germany, 2007; pp. 433–467, ISBN 978-3-940344-11-3.

33. Euring, M.; Rühl, M.; Ritter, N.; Kües, U.; Kharazipour, A. Laccase mediator systems for eco-friendly production of medium-density fiberboard (MDF) on a pilot scale: Physicochemical analysis of the reaction mechanism. *Biotechnol. J.* **2011**, *6*, 1253–1261. [CrossRef] [PubMed]

34. Nasir, M.; Gupta, A.; Beg, M.D.H.; Chua, G.K.; Kumar, A. Fabrication of medium density fibreboard from enzyme treated rubber wood (Hevea brasiliensis) fibre and modified organosolv lignin. *Int. J. Adhes. Adhes.* **2013**, *44*, 99–104. [CrossRef]

35. Kirsch, A.; Ostendorf, K.; Kharazipour, A.; Euring, M. Phenolics as mediators to accelerate the enzymatically initialized oxidation of laccase-mediator-systems for the production of medium density fiberboards. *BioResources* **2016**, *11*, 7091–7101. [CrossRef]

36. Euring, M.; Kirsch, A.; Schneider, P.; Kharazipour, A. Lignin-Laccase-Mediator-Systems (LLMS) for the Production of Binderless Medium Density Fiberboards (MDF). *J. Mater. Sci. Res.* **2016**, *5*, 7. [CrossRef]

37. Bouajila, J.; Limare, A.; Joly, C.; Dole, P. Lignin plasticization to improve binderless fiberboard mechanical properties. *Polym. Eng. Sci.* **2005**, *45*, 809–816. [CrossRef]

38. Moya, R.; Saastamoinen, P.; Hernández, M.; Suurnäkki, A.; Arias, E.; Mattinen, M.-L. Reactivity of bacterial and fungal laccases with lignin under alkaline conditions. *Bioresour. Technol.* **2011**, *102*, 10006–10012. [CrossRef] [PubMed]

39. Van de Pas, D.; Hickson, A.; Donaldson, L.; Lloyd-Jones, G.; Tamminen, T.; Fernyhough, A.; Mattinen, M.-L. Characterization of fractionated lignins polymerized by fungal laccases. *BioResources* **2011**, *6*, 1105–1121.

40. Rathke, J.; Sinn, G.; Konnerth, J.; Müller, U. Strain measurements within fiber boards. Part i: Inhomogeneous strain distribution within medium density fiberboards (MDF) loaded perpendicularly to the plane of the board. *Materials* **2012**, *5*, 1115–1124. [CrossRef] [PubMed]

41. Lora, J. Industrial commercial lignins: Sources, properties and applications. In *Monomers, Polymers and Composites from Renewable Resources*; Elsevier: Amsterdam, The Netherlands, 2008; pp. 225–241, ISBN 978-0-08-045316-3.

42. Lund, M.; Eriksson, M.; Felby, C. Reactivity of a fungal laccase towards lignin in softwood kraft pulp. *Holzforschung* **2003**, *57*, 21–26. [CrossRef]

43. American National Standard Institute ANSI A.208.2-2002. *Medium Density Fiberboard (MDF) for Interior Application*; American National Standard Institute: Washington DC, USA, 2002.

44. European Standards EN 622-5:2010. *Fibreboards Specifications. Part 5: Requirements for Dry Process Boards (MDF)*; European Commission: Brussels, Belgium, 2010.

© 2018 by the authors. Licensee MDPI, Basel, Switzerland. This article is an open access article distributed under the terms and conditions of the Creative Commons Attribution (CC BY) license (http://creativecommons.org/licenses/by/4.0/).

polymers

MDPI

Article

Isothermal and Nonisothermal Crystallization Kinetics of Poly(ε-caprolactone) Blended with a Novel Ionic Liquid, 1-Ethyl-3-propylimidazolium Bis(trifluoromethanesulfonyl)imide

Chun-Ting Yang [1], Li-Ting Lee [1],* and Tzi-Yi Wu [2]

[1]   Department of Materials Science and Engineering, Feng Chia University, Taichung 40724, Taiwan;
      zipper0501@gmail.com
[2]   Department of Chemical and Materials Engineering, National Yunlin University of Science and Technology,
      Yunlin 64002, Taiwan; wuty@gemail.yuntech.edu.tw
*   Correspondence: ltlee@fcu.edu.tw; Tel.: +886-4-2451-7250 (ext. 5306)

Received: 29 March 2018; Accepted: 16 May 2018; Published: 18 May 2018

Abstract: Recently, ionic liquids (ILs) and biodegradable polymers have become crucial functional materials in green sustainable science and technology. In this study, we investigated the influence of a novel IL, 1-ethyl-3-propylimidazolium bis(trifluoromethanesulfonyl)imide ([EPrI][TFSI]), on the crystallization kinetics of a widely studied biodegradable polymer, poly(ε-caprolactone) (PCL). To obtain a comprehensive understanding, both the isothermal and nonisothermal crystallization kinetics of the PCL blends were studied. Incorporating [EPrI][TFSI] reduced the isothermal and nonisothermal crystallization rates of PCL. Regarding isothermal crystallization, the small k and $1/t_{0.5}$ values of the blend, estimated using the Avrami equation, indicated that [EPrI][TFSI] decreased the rate of isothermal crystallization of PCL. The Mo model adequately described the nonisothermal crystallization kinetics of the blends. Increasing the [EPrI][TFSI] content caused the rate-related parameter $F(T)$ to increase. This indicated that the crystallization rate of PCL decreased when [EPrI][TFSI] was incorporated. The spherulite appearance temperature of the blending sample was found to be lower than that of neat PCL under a constant cooling rate. The analysis of the effective activation energy proposed that the nonisothermal crystallization of PCL would not be favored when the [EPrI][TFSI] was incorporated into the blends. The addition of [EPrI][TFSI] would not change the crystal structures of PCL according to the results of wide angle X-ray diffraction. Fourier transform infrared spectroscopy suggested that interactions occurred between [EPrI][TFSI] and PCL. The crystallization kinetics of PCL were inhibited when [EPrI][TFSI] was incorporated.

Keywords: crystallization kinetics; ionic liquids; biodegradable polymer; poly(ε-caprolactone); specific interactions

---

1. Introduction

With increasing concern regarding environmental protection, biodegradable polymers are attracting substantially more attention [1,2]. Biodegradable polymers can be classified into three types according to their mode of formation: natural, biosynthetic, and chemosynthetic [3,4]. Chemically synthesized poly(ε-caprolactone) (PCL) is among the most attractive and commonly used polymers due to its biocompatibility, biodegradability, favorable miscibility with other polymers, and low-temperature adhesiveness [5,6]. Furthermore, PCL can be degraded by the hydrolysis of its ester linkages in its polymer chain under physiological conditions [5]. PCL is an environmentally friendly food packaging material and is used in different biomedical applications, such as scaffolding, tissue engineering, and the controlled release of drugs [6,7]. However, its poor thermal stability, mechanical properties, and barrier properties to water and gases have restricted its application [8]. Therefore, techniques such as blending, copolymerization, and adding inorganic fillers are used to obtain PCL composites with satisfactory performance [9–12]. Relevant studies have been conducted on the blending systems and copolymers of PCL. Numerous researchers have synthesized PCL copolymers or blends, such as PCL/PEG [9], PCL/polysaccharides [10], and PCL/PVC [11], and have incorporated various functional groups into PCL [12].

The physical properties of a polymer are closely related to the polymer's crystallization behavior [13,14]. In general, a polymer might be crystallized under isothermal and nonisothermal conditions in practical processing. Relevant kinetic characters are generally investigated to understand the mechanism and crystallization rate of polymeric materials. Isothermal crystallization kinetics are widely studied [15–19]. Insight regarding the crystallization of polymers can be obtained by theoretically analyzing the kinetic data. Nonisothermal crystallization of polymers is also an important area of study [20–22], and the behavior of polymeric samples is generally analyzed at a constant cooling rate. Studies and discussions focus on nonisothermal crystallization because this process closely represents the industrial processing conditions [23–25]. To control the rate of crystallization and obtain materials with superior physical properties, the kinetic studies should be performed with suitable models by using mathematical analysis [5].

Ionic liquids (ILs) are composed of organic cations and organic or inorganic anions [26,27]. At room temperature or temperatures approaching room temperature, ILs are usually liquid-like. ILs can be used as "green solvents" for dissolving polar and nonpolar organic compounds and inorganic chemicals. Generally, ILs possess properties such as nonflammability, negligible vapor pressure, high ionic conductivity, thermal stability, and a wide electrochemical window [28–32]. Among all types of ILs, the bis(trifluoromethanesulfonyl)imide-based ILs presents high thermal stability, a low melting point, and a wide liquid range. In addition, it has been found that intermolecular interactions occurred between the IL consisting of a imidazolium cation ring and a polymer with polar functional groups [33,34]. 1-Ethyl-3-propylimidazolium bis(trifluoromethanesulfonyl)imide ([EPrI][TFSI]) is a novel bis(trifluoromethanesulfonyl)imide-based IL containing an imidazolium cation [35]. [EPrI][TFSI] also presents a unique hydrophobic property and high thermal stability [35]. It can be expected that [EPrI][TFSI] can form intermolecular interactions with the polymer containing polar functional groups, and that their blends will be less sensitive to moisture.

To the best of our knowledge, blends of PCL with an IL have not yet been investigated. In this study, blends comprising a novel ionic liquid, [EPrI][TFSI], and PCL were analyzed. The aim of this study was to investigate the influence of an IL on the crystallization kinetics of PCL. The isothermal and nonisothermal kinetics of the [EPrI][TFSI]/PCL blends were thoroughly analyzed. We found that [EPrI][TFSI] influenced the crystallization kinetics of PCL in the blends. The isothermal and nonisothermal crystallization kinetics of PCL were inhibited when [EPrI][TFSI] was incorporated.

2. Materials and Methods

2.1. Materials and Blend Preparation

The PCL was obtained from Sigma-Aldrich (Sigma-Aldrich, St. Louis, MO, USA), and its molecular weight was 140,000 g/mol according to the manufacturer. [EPrI][TFSI] was synthesized from 1-ethyl-3-propylimidazolium bromide [35]. The detailed synthesis procedure is available in the literature [35]. [EPrI][TFSI]/PCL blends were prepared using the solution casting method. Tetrahydrofuran (THF) was used as the mutual solvent to prepare the blends. The film casting procedure was performed by evaporating the solvent at 45 °C, followed by vacuum drying at 60 °C for at least 48 h. Blend films with different compositions were obtained after evaporating the solvent completely.

2.2. Instruments and Experiments

We investigated the thermal behavior and crystallization of the [EPrI][TFSI]/PCL blends by using differential scanning calorimetry (DSC) (Perkin-Elmer DSC-8500, Perkin Elmer, Waltham, MA, USA). The thermal traces of the blends were obtained at a heating rate of 20 °C/min using a scan from −70 to 100 °C. The isothermal crystallization behavior of the blends was evaluated by first heating the samples to a temperature higher than the melting temperature of PCL (approximately 80 °C). The samples were then rapidly cooled to various crystallization temperatures (T_c) for isothermal crystallization. Exothermal heat flow curves as a function of time were recorded to analyze the isothermal crystallization kinetics. The samples used for the study of nonisothermal crystallization were first heated to a temperature higher than the melting temperature of PCL, after which they were cooled at different cooling rates, and cooling traces were recorded.

The phase morphology and spherulite appearance were observed by a polarizing optical microscope (Olympus CX41, Olympus, Tokyo, Japan). The polarizing optical microscope used in this study was equipped with a Linkam THMS-600 microscopic hot stage. The samples for the optical microscopic observations were cast as thin films on glass slides and then dried thoroughly in a vacuum before characterization. To discuss the phase morphology of the blends, the specimens were heated on a hot stage, gradually increasing the temperature to the molten state. On the other hand, the spherulite appearance was observed by cooling the samples on hot stage at 1 °C/min from the melt.

The phase morphology of the [EPrI][TFSI]/PCL blends was also determined by using the scanning electron microscopy (SEM) (Hitachi S3000, Hitachi, Tokyo, Japan). A blend was solution-casted to produce a film that was sufficiently thick for convenient examination of the fracture surface of the cross-section. Prior to observation, the fractured blend films were coated with gold under the vacuum deposition.

The interactions in the blends were studied using Fourier transform infrared spectroscopy (FTIR) (Perkin-Elmer Frontier™, Perkin Elmer, Waltham, MA, USA). Every spectrum was collected at a resolution of 4 cm⁻¹. The recorded spectra were all within the wavelength range of 400–4000 cm⁻¹. The films used for FTIR measurements were prepared by coating a blend solution onto KBr pellets. The samples were then vacuum dried at 60 °C for at least 48 h to remove the residual solvent.

The crystalline structures of the blends were analyzed by the wide-angle X-ray diffraction (WAXD) (Bruker D2 PHASER, Bruker, Billerica, MA, USA) with copper kα radiation (30 kV and 10 mA). The samples were scanned under scanning 2θ angles between 5° and 50° with a step speed of 5 °C/min.

3. Results and Discussion

3.1. Phase Morphology and Thermal Analysis of the [EPrI][TFSI]/PCL Blends

In this research, the phase morphology and thermal behavior of [EPrI][TFSI]/PCL blends with different compositions were analyzed. A preliminary study for the phase morphology was made by the optical microscopy (OM) observation. The OM results are presented in Figure S1 in the Supplementary Materials. The OM results in Figure S1 demonstrate that the [EPrI][TFSI]/PCL blends have the same morphology type, as they present optically clear images in the amorphous molten state. The study of OM could preliminary suggest the phase homogeneity for the [EPrI][TFSI]/PCL blends. The phase morphology of the blends was further confirmed by using high-resolution scanning electron microscopy (HRSEM). The HRSEM image of the [EPrI][TFSI]/PCL = 20/80 blend was selected as the typical image for the demonstration. It should note that all blends revealed similar morphology types to the [EPrI][TFSI]/PCL = 20/80 blend. Figure 1 displays the SEM morphology of the [EPrI][TFSI]/PCL = 20/80 blend. A typical feature showing the fracture surface was resolved. In addition, as indicated in Figure 1, the high-resolution SEM image displayed the homogeneous morphology of the blend rather than the morphology with the phase separation. We further performed the SEM mapping analysis to detect the distribution of the constituent atoms in the area of SEM observation. We found that the result as a uniform distribution could be presented, suggesting a homogeneous mixing state for the constituent atoms in the blends. The result of mapping analysis is presented in Figure S2 in the Supplementary Materials. In short, with the results of SEM, it can confirm the phase homogeneity for the blends of [EPrI][TFSI] and PCL.

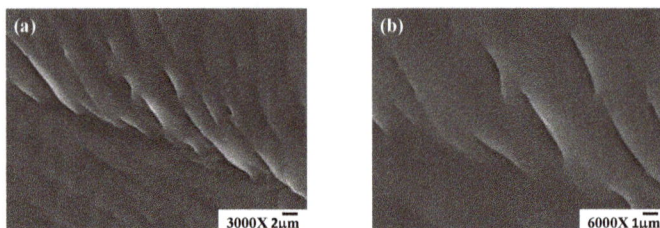

Figure 1. Images of scanning electron microscopy (SEM) for the 1-ethyl-3-propylimidazolium bis(trifluoromethanesulfonyl)imide/poly(ε-caprolactone) ([EPrI][TFSI]/PCL) blend at a composition of 20/80. The magnifications shown here are (**a**) 3000× and (**b**) 6000×.

We also performed DSC to characterize the thermal behavior of the [EPrI][TFSI]/PCL blends. The samples used for DSC measurements were first melted at a temperature higher than the T_m of PCL and then quenched for a sequential scan. The DSC heating scans for the blended samples after the melting/quenching treatment are displayed in Figure S3 of the Supplementary Materials. In Figure S3, the melting transitions of the blends are indicated. The glass transition temperatures of the blends could not be clearly determined by the DSC scans. This may be attributed to the resolution limit of the instrument used in the experiments. The important results of Figure S3 are summarized in Table 1. The melting enthalpy (ΔH_m) and melting temperature (T_m) of the [EPrI][TFSI]/PCL blends were estimated.

Table 1. ΔH_m and T_m values of the [EPrI][TFSI]/PCL blends.

[EPrI][TFSI]/PCL (wt %)	ΔH_m (J/g)	T_m (°C)
0/100	52.06	56.83
10/90	49.61	55.76
20/80	43.04	54.83
30/70	39.38	53.18

Table 1 provides the ΔH_m and T_m values of the [EPrI][TFSI]/PCL blends. ΔH_m decreased as the [EPrI][TFSI] content increased in the blend. The ΔH_m values decreased by 52.06–39.38 J/g with the addition of [EPrI][TFSI] to the blend. Moreover, T_m also decreased with an increase in the [EPrI][TFSI] content. Therefore, the [EPrI][TFSI] content in the blends influenced the thermal properties of PCL. This may further suggest that the [EPrI][TFSI] was a molecular diluent in the blends and influenced the physical state of the polymer chain and the polymer's physical properties such as its crystallization kinetics and melting point. Similar phenomena have been reported for other polymer–diluent pairs [36]. Generally, the presence of a diluent can reduce the equilibrium melting points (T_m^0) of a polymer. The value of T_m^0 represents the value of T_m in the thermodynamically stable state. We confirmed the decrease in the T_m^0 of the [EPrI][TFSI]/PCL blends by estimating the T_m^0 of PCL in the blends using the classical Hoffman–Weeks method [37]. In this procedure, the measured T_m of the specimens are plotted against the crystallization temperature T_c, and the line $T_m = T_c$ is extrapolated. The intercept of the line indicates the value of T_m^0. The specimens were crystallized at each T_c for 8 h. Figure S4 in the Supplementary Materials displays the DSC heating traces of the blends at different values of T_c, and Figure S5 in the Supplementary Materials illustrates Hoffman–Weeks plots of the [EPrI][TFSI]/PCL blends. The result of the Hoffman–Weeks plot is summarized in Table S1 of the Supplementary Materials. In the [EPrI][TFSI]/PCL blends, the T_m^0 of PCL decreased (from 70.34 to 66.49 °C) as the [EPrI][TFSI] content increased. Therefore, the [EPrI][TFSI] in the blends influenced the physical properties of PCL as a diluent. The isothermal and nonisothermal crystallization kinetics of the blends are discussed in the following sections to explore the influence of [EPrI][TFSI] on the crystallization kinetics and thermal behavior of PCL.

3.2. Isothermal Crystallization Kinetics of the [EPrI][TFSI]/PCL Blends

The isothermal crystallization thermograms for neat PCL and the [EPrI][TFSI]/PCL blends are presented in Figure 2. To perform isothermal crystallization, the blended samples were first heated to a temperature higher than the melting temperature of PCL and held at that temperature for 3 min to erase their thermal history. The samples were then rapidly cooled to various crystallization temperatures (T_c = 33, 35, 38, and 40 °C).

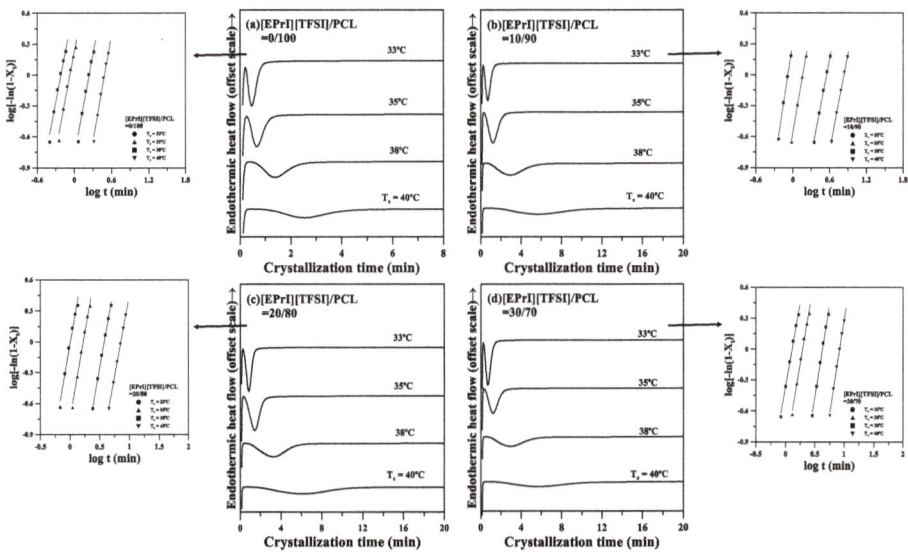

Figure 2. Differential scanning calorimetry (DSC) isothermal crystallization results of the [EPrI][TFSI]/PCL blends. The figures show the results for the blending compositions of (**a**) 0/100; (**b**) 10/90; (**c**) 20/80; and (**d**) 30/70. The inserts show the Avrami plots for the blends.

The Avrami equation [38,39] was applied to analyze the isothermal crystallization kinetics of the [EPrI][TFSI]/PCL blends. Theoretical background of the Avrami equation and its application to the analysis of the isothermal crystallization kinetics are summarized in the supplementary materials. The log–log representation of the Avrami equation was used to analyze the isothermal crystallization kinetics of the blends. The inserts of Figure 2 display the log–log representation of the Avrami plots for neat PCL and the [EPrI][TFSI]/PCL blends at different T_c. The isothermal crystallization kinetics of neat PCL and the [EPrI][TFSI]/PCL blends are suitably described using the Avrami equation. From the inserts of Figure 2, the Avrami exponent (n) and crystallization rate constant (k) were calculated. The parameters estimated from the Avrami equation as well as the crystallization half-time ($t_{0.5}$) and its reciprocal ($1/t_{0.5}$) are listed in Table 2. The n values of neat PCL at the measured T_c range are between 3 and 4, which are in agreement with the previous literature [40–42]. In addition, the average value of n for neat PCL and that of the blends are all close to 3.3. This result suggested that the addition of [EPrI][TFSI] did not considerably change the crystallization mechanism of PCL in the [EPrI][TFSI]/PCL blends. Based on the n values between 3 and 4, the crystallization mechanism for the blends may be considered as a three-dimensional truncated sphere growth [43]. It should also be noted that according to the Avrami theory, the n value for ideal three dimensional growth should be an integer without a fractional part. However, the n values of neat PCL and its blends do not satisfy the ideal state. The reason for this could be attributed to the slight difference between the actual process of growth and the ideal state. Since the samples for isothermal crystallization measurements were mostly confined to a thin film thickness in the DSC cells, the factor that the actual growth process may not be perfectly three-dimensional should be considered, and this factor might make a possible deviation from the ideal state. Similar results and statements have been also reported in recent literature [16]. In addition, it was also shown that k decreased as the [EPrI][TFSI] content increased. Furthermore, $1/t_{0.5}$ decreased as the [EPrI][TFSI] content increased. Generally, $1/t_{0.5}$ is a rate-dependent parameter. A larger $1/t_{0.5}$ indicates a higher crystallization rate. The values of k and $1/t_{0.5}$ decreased as the [EPrI][TFSI] content increased, which implied that the presence of [EPrI][TFSI] in the blends weakened the isothermal crystallization kinetics and decreased the crystallization rate of PCL. Figure 3a,b illustrated the variation in k and $1/t_{0.5}$, respectively, in the blends with different [EPrI][TFSI] content. Both k and $1/t_{0.5}$ gradually decreased as the [EPrI][TFSI] content increased. An increase in the [EPrI][TFSI] content of the blends reduced the isothermal crystallization rate of PCL.

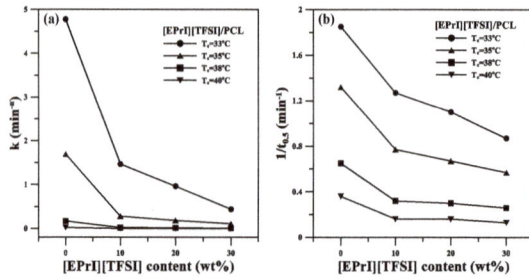

Figure 3. Plot of (a) k and (b) $1/t_{0.5}$ versus [EPrI][TFSI] content in the blends at different crystallization temperatures.

Table 2. Parameters estimated from the isothermal crystallization results of the Avrami equation.

[EPrI][TFSI]/PCL (wt %)	T_c (°C)	n	k (min^{-n})	$t_{0.5}$ (min)	$1/t_{0.5}$ (min^{-1})
0/100	33	3.14	4.775	0.54	1.85
	35	3.21	1.694	0.76	1.32
	38	3.26	0.167	1.55	0.65
	40	3.44	0.021	2.76	0.36
10/90	33	3.10	1.466	0.79	1.27
	35	3.45	0.28	1.29	0.77
	38	3.12	0.019	3.15	0.32
	40	3.08	2.6×10^{-3}	6.09	0.16
20/80	33	3.35	0.962	0.91	1.10
	35	3.32	0.185	1.49	0.67
	38	3.26	0.013	3.39	0.30
	40	3.15	2.046×10^{-3}	6.36	0.16
30/70	33	3.28	0.438	1.15	0.87
	35	3.36	0.106	1.75	0.57
	38	3.58	5.585×10^{-3}	3.85	0.26
	40	3.63	4.406×10^{-4}	7.60	0.13

3.3. Nonisothermal Crystallization Kinetics of the [EPrI][TFSI]/PCL Blends

The nonisothermal crystallization kinetics of the [EPrI][TFSI]/PCL blends were studied to further understand the influence of [EPrI][TFSI] on the crystallization kinetics of PCL. Figure 4 displays the

DSC results for neat PCL and the [EPrI][TFSI]/PCL blends at cooling rates of 2.5, 5, 7.5, and 10 °C/min. In Figure 5, we summarize the key information in Figure 4, which includes information related to the peak temperature (T_p) and heat of the nonisothermal crystallization ($\Delta H_{n,c}$). Figure 5a displays the plot of T_p versus [EPrI][TFSI] content at different cooling rates. According to Figure 5a, T_p gradually shifted to a lower temperature as the [EPrI][TFSI] content increased, regardless of the cooling rate. Figure 5b presents plots of $\Delta H_{n,c}$ versus [EPrI][TFSI] content at different cooling rates. $\Delta H_{n,c}$ gradually decreased as the [EPrI][TFSI] content increased. The two aforementioned phenomena suggest that [EPrI][TFSI] retarded the nonisothermal crystallization of PCL in the blends.

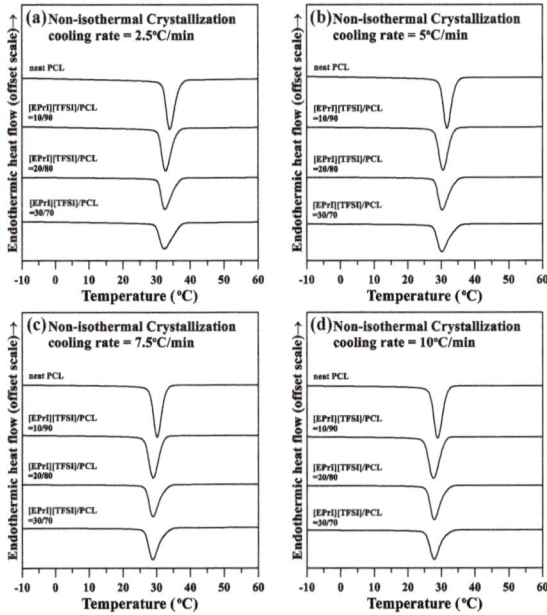

Figure 4. DSC results of nonisothermal crystallization at different cooling rates for different compositions of the [EPrI][TFSI]/PCL blends: (**a**) 0/100; (**b**) 10/90; (**c**) 20/80; and (**d**) 30/70.

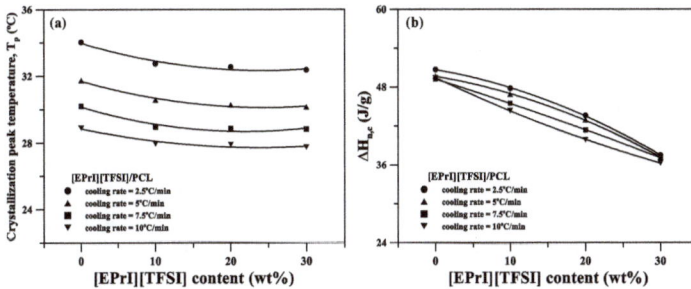

Figure 5. Plot of (**a**) T_p and (**b**) $\Delta H_{n,c}$ versus [EPrI][TFSI] content at different cooling rates.

The nonisothermal crystallization kinetics of the [EPrI][TFSI]/PCL blends were further analyzed using three different models: the modified Avrami equation [44], Ozawa analysis [45], and Mo model [46]. Theoretical backgrounds of the abovementioned equations and their applications on the analysis of nonisothermal crystallization kinetics are summarized in the supplementary materials.

Figure 6 presents logarithmic plots of the modified Avrami equation for the different [EPrI][TFSI]/PCL blends at various cooling rates. Each plot in Figure 6 exhibits a nonlinear relationship. Therefore, the modified Avrami equation was unsuitable for describing the nonisothermal crystallization kinetics of the [EPrI][TFSI]/PCL blends.

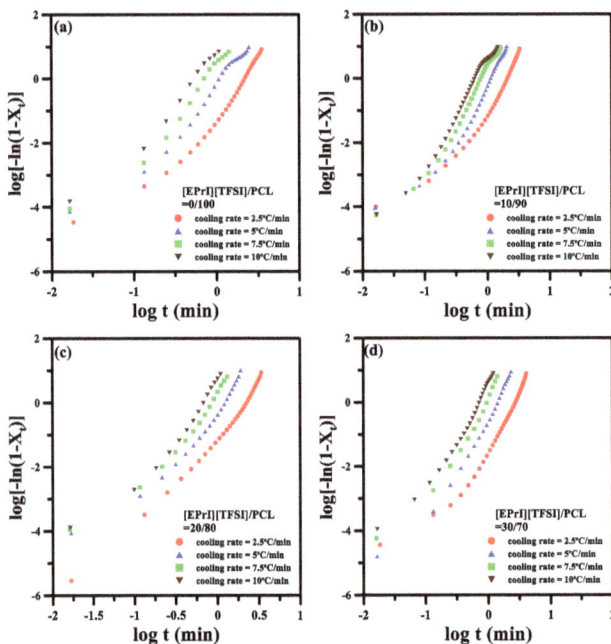

Figure 6. Avrami plots of $\log[-\ln(1 - X_t)]$ versus $\log(t)$ for the nonisothermal crystallization of the [EPrI][TFSI]/PCL blends. The figures show the results for the blending compositions of (a) 0/100; (b) 10/90; (c) 20/80; and (d) 30/70.

The Ozawa equation was also applied to study the nonisothermal crystallization kinetics of the [EPrI][TFSI]/PCL blends. Figure 7 presents logarithmic plots of the Ozawa equation for the different [EPrI][TFSI]/PCL blends. All the plots in Figure 7 are also nonlinear. Thus, the Ozawa equation failed to describe the nonisothermal crystallization kinetics of the [EPrI][TFSI]/PCL blends.

As described in the aforementioned text, the modified Avrami equation and Ozawa equation could not adequately describe the nonisothermal crystallization kinetics of the [EPrI][TFSI]/PCL blends, as has also been indicated in previous studies [47,48]. This failure of the modified Avrami equation and Ozawa equation could be due to secondary crystallization in the blends [47,48].

In this study, we also applied the Mo model for studying the nonisothermal crystallization kinetics of the [EPrI][TFSI]/PCL blends. Figure 8 displays $\log\Phi$ versus $\log t$ plots for the blends, and the plots are linear, which suggests that the Mo model is suitable for analyzing and describing the nonisothermal crystallization kinetics of the [EPrI][TFSI]/PCL blends. The nonisothermal crystallization parameters for the blends were estimated using the Mo model and are tabulated in Table 3. A high $F(T)$ is associated with a low crystallization rate [49]. At the same degree of crystallinity, $F(T)$ systematically increased with an increase in the [EPrI][TFSI] content of the blends, as indicated in Table 3. The Mo model analysis suggested that the presence of [EPrI][TFSI] in the blends decreased the nonisothermal crystallization rate of PCL. The isothermal and nonisothermal crystallization analyses thus indicated that [EPrI][TFSI] retarded both the isothermal and nonisothermal crystallization kinetics of PCL.

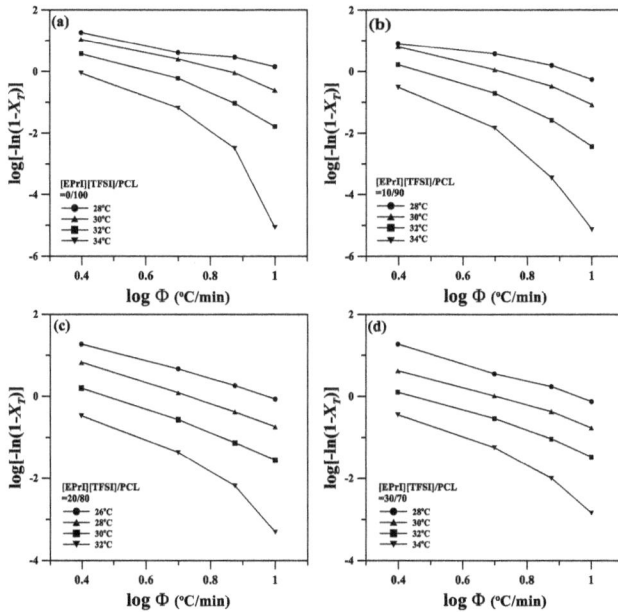

Figure 7. Ozawa plots of $\log[-\ln(1 - X_T)]$ versus $\log\Phi$ for the nonisothermal crystallization of the [EPrI][TFSI]/PCL blends. The figures show the results for the blending compositions of (**a**) 0/100; (**b**) 10/90; (**c**) 20/80; and (**d**) 30/70.

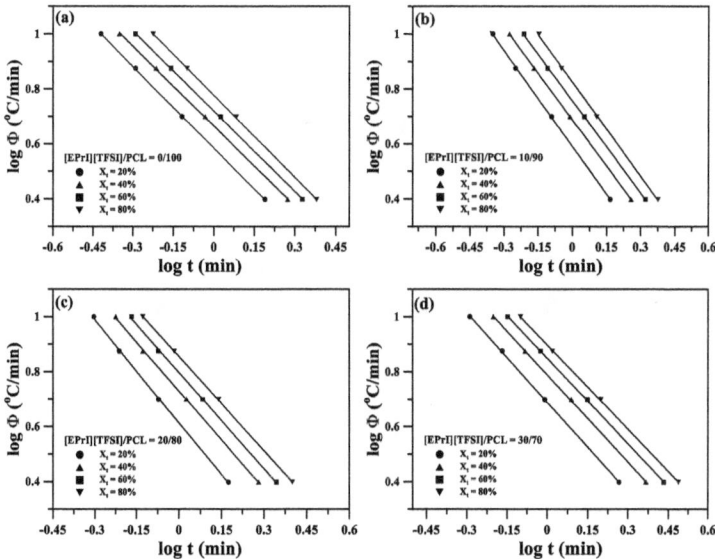

Figure 8. Mo model plots of $\log\Phi$ versus $\log(t)$ for the nonisothermal crystallization of the [EPrI][TFSI]/PCL blends. The figures show the results for the blending compositions of (**a**) 0/100; (**b**) 10/90; (**c**) 20/80; and (**d**) 30/70.

Table 3. Nonisothermal crystallization parameters of the [EPrI][TFSI]/PCL blends, estimated using the Mo model.

[EPrI][TFSI]/PCL (wt %)	X_t (%)	a	$F(T)$
0/100	20	0.99	3.80
	40	0.97	4.62
	60	0.97	5.25
	80	1.00	5.99
10/90	20	1.16	3.90
	40	1.13	4.87
	60	1.12	5.72
	80	1.15	6.69
20/80	20	1.25	4.11
	40	1.18	5.37
	60	1.17	6.27
	80	1.14	7.18
30/70	20	1.09	4.88
	40	1.06	6.15
	60	1.04	7.07
	80	1.02	7.90

3.4. Polarizing Optical Microscopy (POM) Observations for [EPrI][TFSI]/PCL Blends under Cooling

Polarizing optical microscopy (POM) observations were also performed for the blends of [EPrI][TFSI] and PCL. By cooling at a constant rate from the molten state of PCL, we found that the spherulite appearance temperature of neat PCL and that of the [EPrI][TFSI]/PCL blends were different. The spherulite appearance temperatures of the blends were found to be lower than that of neat PCL. A demonstration of this can be found in Figure 9. As shown in Figure 9, for the [EPrI][TFSI]/PCL = 20/80 blend under a cooling rate of 1 °C/min, the growth of spherulites appeared at about 37 °C. On the other hand, for neat PCL, the growth of spherulites was found to start at a higher temperature, approximately 42 °C. This phenomenon is consistent with the T_p results of the nonisothermal crystallization, as shown in the earlier section (Section 3.3). Both POM and DSC results of the nonisothermal crystallization demonstrated that the [EPrI][TFSI]/PCL blends crystallized at lower temperature compared to neat PCL, suggesting that the [EPrI][TFSI] retarded the nonisothermal crystallization of PCL in the blends.

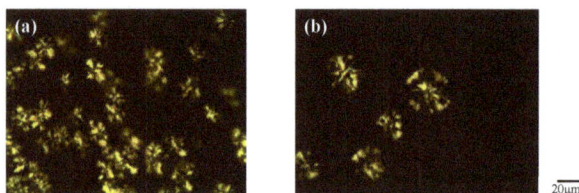

Figure 9. Polarizing optical microscopy (POM) pictures taken during the cooling process under the cooling rate of 1 °C/min for (**a**) neat PCL and (**b**) [EPrI][TFSI]/PCL = 20/80 blend.

3.5. Studies of Effective Ectivation Energy

The effective activation energy was also discussed for the nonisothermal crystallization of the blends; it intended to study the influence on effective activation energy when [EPrI][TFSI] was incorporated into PCL. The isoconversion method of Friedman [50,51] was applied to our discussions. A detail description of the estimation of effective activation energy can be found in the supplementary materials. To demonstrate the important results of the Friedman estimation, the plots of $ln(dX/dT)$ versus $1/T_X$ for the [EPrI][TFSI]/PCL = 20/80 blend are shown in Figure 10 and the effective activation energy values of the neat PCL and [EPrI][TFSI]/PCL = 20/80 blend are demonstrated in Figure 11. In Figure 11, the plots show that the effective activation values of the [EPrI][TFSI]/PCL = 20/80 blend are smaller than that of the neat PCL, inferring that the nonisothermal crystallization of PCL would not be favored with the addition of [EPrI][TFSI] in the blends.

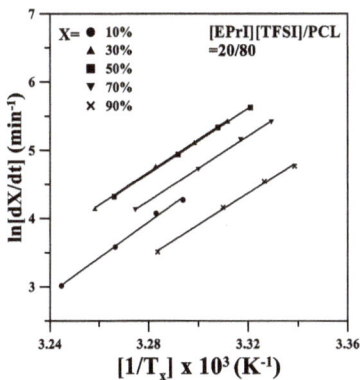

Figure 10. Friedman method plots of $ln(dX/dT)$ versus $1/T_X$ for the [EPrI][TFSI]/PCL = 20/80 blend.

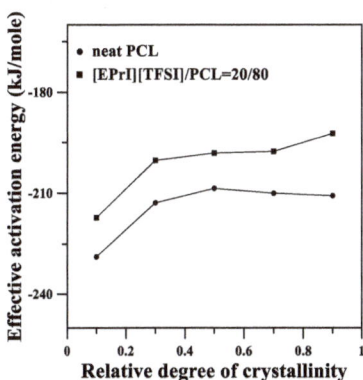

Figure 11. Effective activation energies for the neat PCL and the [EPrI][TFSI]/PCL = 20/80 blend estimated by the isoconversion Friedman method.

3.6. Wide-Angle X-ray Diffraction (WAXD) Analyzes for Crystal Structures of [EPrI][TFSI]/PCL Blends

Wide-angle X-ray diffraction (WAXD) technology was applied to discuss the crystal structures of the [EPrI][TFSI]/PCL blends. The samples for the WAXD study were isothermally crystallized at $T_c = 35\,°C$. Figure 12 demonstrates the results of WAXD for the neat PCL and the [EPrI][TFSI] = 20/80 blend. The results showed that neat PCL presented three main diffraction peaks at $2\theta = 21.3°$, $22°$, and $23.8°$, which are related to the reflection planes of (110), (111), and (200), respectively [52]. In addition, for the [EPrI][TFSI]/PCL = 20/80 blend, it showed the same reflection pattern as the neat PCL without any change in peak position or the appearance of new peaks. Relevant results revealed that the addition of [EPrI][TFSI] in PCL did not modify the crystal structures of PCL. In addition, the results obtained from the WAXD experiments also supported the abovementioned results of the isothermal crystallization kinetics (in Section 3.2), which demonstrated that the averaged Avrami n value of the neat PCL, and that of the [EPrI][TFSI]/PCL blends approached a similar value. The incorporation of [EPrI][TFSI] into PCL would not significantly influence the crystal growth mechanism of PCL, and therefore the crystal structures of PCL in the blends were not modified.

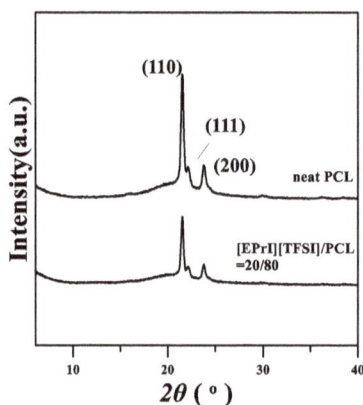

Figure 12. Wide-angle X-ray diffraction (WAXD) results of the neat PCL and the [EPrI][TFSI]/PCL = 20/80 blend. Samples subjected to the WAXD measurement were precrystallized at the T_c = 38 °C.

3.7. Possible Interactions between [EPrI][TFSI] and PCL in the Blends

Our analyses of the crystallization kinetics of the blends indicated that [EPrI][TFSI] influenced the crystallization kinetics of PCL. Furthermore, we investigated the possible interactions between [EPrI][TFSI] and PCL in the blends. FTIR was used to provide spectral evidence of the interactions between [EPrI][TFSI] and PCL. The FTIR spectra of neat PCL, neat [EPrI][TFSI], and the [EPrI][TFSI]/PCL blends are displayed in Figure 13. Figure 13a displays the part of the IR spectra indicating the carbonyl stretching region (1780–1700 cm^{-1}) of the [EPrI][TFSI]/PCL blends. The absorption peak of neat PCL occurred at 1736 cm^{-1}. Furthermore, as the [EPrI][TFSI] content increased, the peak shifted to 1733 cm^{-1}, which suggested the occurrence of an interaction between PCL and [EPrI][TFSI]. The samples were maintained in the molten amorphous state for measurements to avoid the complexity of crystallization. Figure 13b displays the part of the FTIR spectra indicating C–H stretching vibration (3225–3050 cm^{-1}) in the imidazolium cation ring. The assignment of this absorption band has been previously described in the literature [53]. For neat [EPrI][TFSI], the spectrum contains a shoulder at 3168 cm^{-1} and three peaks at 3149 cm^{-1}, 3116 cm^{-1}, and 3093 cm^{-1}. Spectral deconvolution was performed to obtain insight regarding the FTIR results for the imidazolium cation ring. Figure 14 presents the IR spectra for the deconvoluted C–H stretching vibration band of the imidazolium cation ring for the different [EPrI][TFSI]/PCL blends. The peak at 3093 cm^{-1} split into two peaks at 3087 cm^{-1} and 3097 cm^{-1}. Similar results were observed for blends comprising ILs with an imidazolium ring cation [53]. As described in the literature [53], the splitting of the peak may have been due to (i) the complexation of the IL cations with polymer chains or (ii) the noncomplexation of the IL cations. The absorption bands at low (3087 cm^{-1}) and high (3097 cm^{-1}) wavenumbers can be attributed to the complexed and noncomplexed forms of the conformation, respectively. Therefore, a complex may have formed because of the interassociation between [EPrI][TFSI] and PCL. Nevertheless, the IR results suggested the occurrence of possible intermolecular interactions between [EPrI][TFSI] and PCL. The retarded crystallization kinetics of PCL in the [EPrI][TFSI]/PCL blends might be attributed to the interactions between [EPrI][TFSI] and PCL. The retardation of the crystallization kinetics caused by the intermolecular interactions has been also reported in the works discussing the blends containing crystalline polymer [54,55].

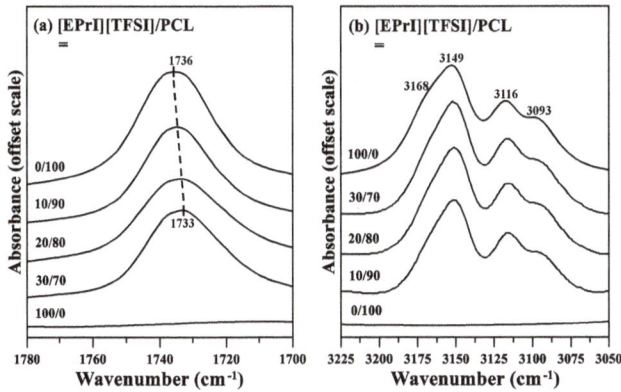

Figure 13. Fourier transform infrared (FTIR) spectra of the [EPrI][TFSI]/PCL blends indicating the (**a**) carbonyl stretching region (1780–1700 cm^{-1}) and (**b**) C–H stretching vibration band (3225–3050 cm^{-1}) of the imidazolium cation ring.

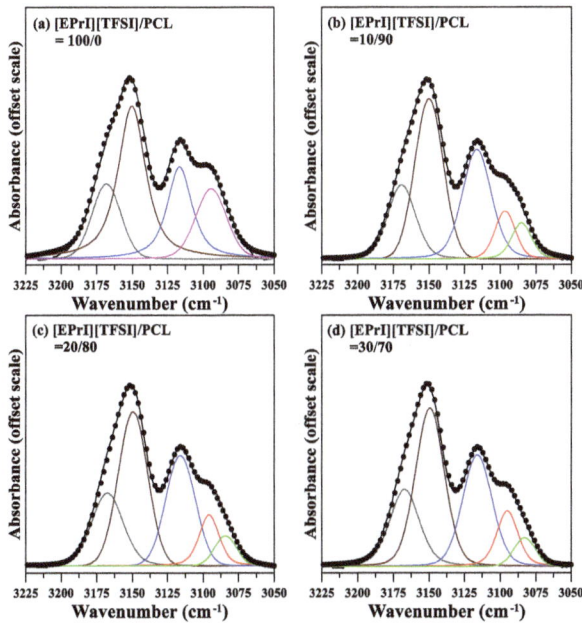

Figure 14. FTIR deconvolution results for the C–H stretching vibrational region of the imidazolium cation ring of the [EPrI][TFSI]/PCL blends. The figures show the results for the blending compositions of (**a**) 0/100; (**b**) 10/90; (**c**) 20/80; and (**d**) 30/70.

4. Conclusions

PCL is an attractive biodegradable polymer, and numerous studies have focused on its physical properties and blending system. In this study, blends comprising PCL and a novel IL, [EPrI][TFSI], were analyzed. We investigated the influence of [EPrI][TFSI] on the crystallization kinetics of PCL in the blends. The isothermal and nonisothermal crystallization kinetics of the [EPrI][TFSI]/PCL blends were evaluated. The isothermal crystallization kinetics were analyzed using the Avrami equation, and the results indicated that the presence of [EPrI][TFSI] reduced the isothermal crystallization rate of PCL. The k and $1/t_{0.5}$ values of neat PCL were smaller than those of the [EPrI][TFSI]/PCL blends. The nonisothermal crystallization of the [EPrI][TFSI]/PCL blends was suitably described using the Mo model. The $F(T)$ values for the blends, estimated using the Mo model, were larger than those for neat the PCL. $F(T)$ increased with an increase in the [EPrI][TFSI] content of the blend. This result indicated that the presence of [EPrI][TFSI] decreased the nonisothermal crystallization rate of PCL in the blends. The studies of POM showed that under a constant cooling rate, the spherulite appearance temperature of the blending sample was higher than that of the neat PCL. The analysis of the effective activation energy supposed that the nonisothermal crystallization of PCL would not be favored when the [EPrI][TFSI] was incorporated into the blends. The WAXD discussions indicated that the addition of [EPrI][TFSI] would not change the crystal structures of PCL. In this study, we determined that the addition of [EPrI][TFSI] to PCL weakened both the isothermal and nonisothermal crystallization kinetics of PCL. The FTIR results suggested the formation of intermolecular interactions between [EPrI][TFSI] and PCL in the blends. The retardation of the crystallization kinetics of PCL in the blends could be attributed to the interactions between PCL and [EPrI][TFSI]. The influence of [EPrI][TFSI] on the crystallization kinetics of biodegradable PCL is thoroughly reported in this paper.

Supplementary Materials: The following are available online at http://www.mdpi.com/2073-4360/10/5/543/s1, Theoretical Backgrounds and Methods, Figure S1: Optical microscopy (OM) images of the [EPrI][TFSI]/PCL blends; Figure S2: Scanning electron microscopy (SEM) mapping results for the [EPrI][TFSI]/PCL = 20/80 blend; Figure S3: Differential scanning calorimetry (DSC) heating scans for the [EPrI][TFSI]/PCL blends after the melting/quenching treatment. Blends showing various compositions (wt %) were detected; Figure S4: DSC heating traces at different T_c of the [EPrI][TFSI]/PCL blends with compositions (a) 0/100, (b) 10/90, (c) 20/80, and (d) 30/70; Figure S5: Hoffman–Weeks plots of the [EPrI][TFSI]/PCL blends; Table S1: Equilibrium melting points of the [EPrI][TFSI]/PCL blends with different compositions; References of Supplementary Materials.

Author Contributions: C.-T.Y. was responsible for performing the experiments and data analysis. L.-T.L. was responsible for developing the research outline, designing the experiments, and writing most of the paper. T.-Y.W. helped the synthesis of IL.

Acknowledgments: This work has been financially supported by basic research grants of MOST 105-2221-E-035-091- and MOST 106-2221-E-035-084—from Taiwan's Ministry of Science and Technology (MOST), to which the authors express their gratitude.

Conflicts of Interest: The authors declare no conflicts of interest.

References

1. Xiao, H.; Lu, W.; Yeh, J.T. Effect of plasticizer on the crystallization behavior of poly(lactic acid). *J. Appl. Polym. Sci.* **2009**, *113*, 112–121. [CrossRef]
2. Qiu, Z.; Ikehara, T.; Nishi, T. Crystallization behaviour of biodegradable poly(ethylene succinate) from the amorphous state. *Polymer* **2003**, *44*, 5429–5437. [CrossRef]
3. Chandra, R.; Rustgi, R. Biodegradable polymers. *Prog. Polym. Sci.* **1998**, *23*, 1273–1335. [CrossRef]
4. Gross, R.A.; Kalra, B. Biodegradable polymers for the environment. *Science* **2002**, *297*, 803–807. [CrossRef] [PubMed]
5. Dhanvijay, P.U.; Shertukde, V.V.; Kalkar, A.K. Isothermal and nonisothermal crystallization kinetics of poly(ε-caprolactone). *J. Appl. Polym. Sci.* **2012**, *124*, 1333–1343. [CrossRef]
6. Xing, Z.; Zha, L.; Yang, G. Thermomechanical behavior and nonisothermal crystallization kinetics of poly(ε-caprolactone) and poly(ε-caprolactone)/poly(N-vinylpyrrolidone) blends. *e-Polymers* **2010**, *121*, 1–13. [CrossRef]
7. Liu, Q.; Deng, B.; Zhu, M.; Shyr, T.W.; Shan, G. Nonisothermal crystallization kinetics of poly(ε-caprolactone)/zinc oxide nanocomposites with high zinc oxide content. *J. Macromol. Sci. Part B Phys.* **2011**, *50*, 2366–2375. [CrossRef]
8. Hua, L.; Kai, W.H.; Inoue, Y. Crystallization behavior of poly(ε-caprolactone)/graphite oxide composites. *J. Appl. Polym. Sci.* **2007**, *106*, 4225–4232. [CrossRef]
9. Piao, L.; Dai, Z.; Deng, M.; Chen, X.; Jing, X. Synthesis and characterization of PCL/PEG/PCL triblock copolymers by using calcium catalyst. *Polymer* **2003**, *44*, 2025–2031. [CrossRef]

10. Shin, K.; Dong, T.; He, Y.; Taguchi, Y.; Oishi, A.; Nishida, H.; Inoue, Y. Inclusion complex formation between ε-Cyclodextrin and biodegradable aliphatic polyesters. *Macromol. Biosci.* **2004**, *4*, 1075–1083. [CrossRef] [PubMed]

11. Zhang, Y.; Prud'homme, R.E. Crystallization of poly(ε-caprolactone)/poly(vinyl chloride) miscible blends under strain: The role of molecular weight. *Macromol. Rapid Commun.* **2006**, *27*, 1565–1571. [CrossRef]

12. Chen, X.; Gross, R.A. Versatile copolymers from [L]-lactide and [D]-xylofuranose. *Macromolecules* **1999**, *32*, 308–314. [CrossRef]

13. Seoane, I.T.; Manfredi, L.B.; Cyras, V.P. Effect of two different plasticizers on the properties of poly(3-hydroxybutyrate) binary and ternary blends. *J. Appl. Polym. Sci.* **2018**, *135*, 46016. [CrossRef]

14. Shibita, A.; Mizumura, Y.; Shibata, M. Stereocomplex crystallization behavior and physical properties of polyesterurethane networks incorporating diglycerol-based enantiomeric 4-armed lactide oligomers and a 1,3-propanediol-based 2-armed rac-lactide oligomer. *Polym. Bull.* **2017**, *74*, 3139–3160. [CrossRef]

15. Tsanaktsis, V.; Bikiaris, D.N.; Guigo, N.; Exarhopoulos, S.; Papageorgiou, D.G.; Sbirrazzuoli, N.; Papageorgiou, G.Z. Synthesis, properties and thermal behavior of poly(decylene-2,5-furanoate): A biobased polyester from 2,5-furan dicarboxylic acid. *RSC Adv.* **2015**, *5*, 74592–74604. [CrossRef]

16. Chang, L.; Woo, E.M. Crystallization of poly(3-hydroxybutyrate) with stereocomplexed polylactide as biodegradable nucleation agent. *Polym. Eng. Sci.* **2012**, *52*, 1413–1419. [CrossRef]

17. Cui, Z.; Qiu, Z. Thermal properties and crystallization kinetics of poly(butylene suberate). *Polymer* **2015**, *67*, 12–19. [CrossRef]

18. Palacios, J.K.; Zhao, J.; Hadjichristidis, N.; Müller, A.J. How the Complex Interplay between Different Blocks Determines the Isothermal Crystallization Kinetics of Triple-Crystalline PEO-b-PCL-b-PLLA Triblock Terpolymers. *Macromolecules* **2017**, *50*, 9683–9695. [CrossRef]

19. Lai, W.C.; Liau, W.B.; Lin, T.T. The effect of end groups of PEG on the crystallization behaviors of binary crystalline polymer blends PEG/PLLA. *Polymer* **2004**, *45*, 3073–3080. [CrossRef]

20. Maiz, J.; Schäfer, H.; Rengarajan, G.T.; Hartmann-Azanza, B.; Eickmeier, H.; Haase, M.; Mijangos, C.; Steinhart, M. How gold nanoparticles influence crystallization of polyethylene in rigid cylindrical nanopores. *Macromolecules* **2013**, *46*, 403–412. [CrossRef]

21. Yuan, Q.; Awate, S.; Misra, R. Nonisothermal crystallization behavior of polypropylene-clay nanocomposites. *Eur. Polym. J.* **2006**, *42*, 1994–2003. [CrossRef]

22. Joshi, M.; Butola, B. Studies on nonisothermal crystallization of HDPE/POSS nanocomposites. *Polymer* **2004**, *45*, 4953–4968. [CrossRef]

23. Choi, J.; Chun, S.W.; Kwak, S.Y. Non-Isothermal crystallization of hyperbranched poly(ε-caprolactone)s and their linear counterpart. *Macromol. Chem. Phys.* **2006**, *207*, 1166–1173. [CrossRef]

24. Long, Y.; Shanks, R.A.; Stachurski, Z.H. Kinetics of polymer crystallisation. *Prog. Polym. Sci.* **1995**, *20*, 651–701. [CrossRef]

25. Tadmor, Z.; Gogos, C.G. *Principles of Polymer Processing*, 2nd ed.; John Wiley & Sons: Hoboken, NJ, USA, 2013.

26. Seddon, K.R.; Stark, A.; Torres, M.J. Influence of chloride, water, and organic solvents on the physical properties of ionic liquids. *Pure Appl. Chem.* **2000**, *72*, 2275–2287. [CrossRef]

27. Holbrey, J.; Seddon, K. Ionic liquids. *Clean Technol. Environ. Policy* **1999**, *1*, 223–236. [CrossRef]

28. Huddleston, J.G.; Visser, A.E.; Reichert, W.M.; Willauer, H.D.; Broker, G.A.; Rogers, R.D. Characterization and comparison of hydrophilic and hydrophobic room temperature ionic liquids incorporating the imidazolium cation. *Green Chem.* **2001**, *3*, 156–164. [CrossRef]

29. Ye, Y.S.; Rick, J.; Hwang, B.J. Ionic liquid polymer electrolytes. *J. Mater. Chem. A* **2013**, *1*, 2719–2743. [CrossRef]

30. Earle, M.J.; Seddon, K.R. Ionic liquids. Green solvents for the future. *Pure Appl. Chem.* **2000**, *72*, 1391–1398. [CrossRef]

31. Earle, M.J.; Esperança, J.M.; Gilea, M.A.; Lopes, J.N.C.; Rebelo, L.P.; Magee, J.W.; Seddon, K.R.; Widegren, J.A. The distillation and volatility of ionic liquids. *Nature* **2006**, *439*, 831–834. [CrossRef] [PubMed]

32. Fredlake, C.P.; Crosthwaite, J.M.; Hert, D.G.; Aki, S.N.; Brennecke, J.F. Thermophysical properties of imidazolium-based ionic liquids. *J. Chem. Eng. Data* **2004**, *49*, 954–964. [CrossRef]

33. Chaurasia, S.K.; Singh, R.K.; Chandra, S. Ion-polymer and ion-ion interaction in PEO-based polymer electrolytes having complexing salt LiClO$_4$ and/or ionic liquid, [BMIM][PF$_6$]. *J. Raman Spectrosc.* **2011**, *42*, 2168–2172. [CrossRef]

34. Schäfer, T.; Di Paolo, R.E.; Franco, R.; Crespo, J.G. Elucidating interactions of ionic liquids with polymer films using confocal Raman spectroscopy. *Chem. Commun.* **2005**, *50*, 2594–2596.

35. Wu, T.Y.; Chen, B.K.; Kuo, C.W.; Hao, L.; Peng, Y.C.; Sun, I.W. Standard entropy, surface excess entropy, surface enthalpy, molar enthalpy of vaporization, and critical temperature of bis(trifluoromethanesulfonyl) imide-based ionic liquids. *J. Taiwan Inst. Chem. Eng.* **2012**, *43*, 860–867. [CrossRef]

36. Pizzoli, M.; Scandola, M.; Ceccorulli, G. Crystallization and melting of isotactic poly(3-hydroxy butyrate) in the presence of a low molecular weight diluent. *Macromolecules* **2002**, *35*, 3937–3941. [CrossRef]

37. Hoffman, J.D.; Weeks, J.J. Melting process and the equilibrium melting temperature of polychlorotrifluoroethylene. *Res. Natl. Bur. Stand. A* **1962**, *66*, 13–28. [CrossRef]

38. Avrami, M. Kinetics of phase change II Transformation-time relations for random distribution of nuclei. *J. Chem. Phys.* **1940**, *8*, 212–224. [CrossRef]

39. Avrami, M. Kinetics of phase change III Granulation, phase change, and microstructure. *J. Chem. Phys.* **1941**, *9*, 177–184. [CrossRef]

40. Balsamo, V.; Calzadilla, N.; Mora, G.; Müller, A.J. Thermal characterization of polycarbonate/polycaprolactone blends. *J. Polym. Sci. Part B Polym. Phys.* **2001**, *39*, 771–785. [CrossRef]

41. L'abee, R.; Van Duin, M.; Goossens, H. Crystallization kinetics and crystalline morphology of poly(ε-caprolactone) in blends with grafted rubber particles. *J. Polym. Sci. Part B Polym. Phys.* **2010**, *48*, 1438–1448. [CrossRef]

42. Remiro, P.M.; Cortazar, M.M.; Calahorra, M.E.; Calafel, M.M. Miscibility and crystallization of an amine-cured epoxy resin modified with crystalline poly(ε-caprolactone). *Macromol. Chem. Phys.* **2001**, *202*, 1077–1088. [CrossRef]

43. Wunderlich, B. *Macromolecular Physics*; Academic Press: New York, NY, USA, 1976; Volume 2, p. 147.

44. Jeziorny, A. Parameters characterizing the kinetics of the non-isothermal crystallization of poly(ethylene terephthalate) determined by d.s.c. *Polymer* **1978**, *19*, 1142–1144. [CrossRef]

45. Ozawa, T. Kinetics of non-isothermal crystallization. *Polymer* **1971**, *12*, 150–158. [CrossRef]

46. Liu, T.X.; Mo, Z.S.; Wang, S.G.; Zhang, H.F. Nonisothermal melt and cold crystallization kinetics of poly(aryl ether ether ketone ketone). *Polym. Eng. Sci.* **1997**, *37*, 568–575. [CrossRef]

47. Su, Z.; Guo, W.; Liu, Y.; Li, Q.; Wu, C. Non-isothermal crystallization kinetics of poly(lactic acid)/modified carbon black composite. *Polym. Bull.* **2009**, *62*, 629–642. [CrossRef]

48. Mohsen-Nia, M.; Memarzadeh, M.R. Characterization and non-isothermal crystallization behavior of biodegradable poly(ethylene sebacate)/SiO$_2$ nanocomposites. *Polym. Bull.* **2013**, *70*, 2471–2491. [CrossRef]

49. Auliawan, A.; Woo, E.M. Crystallization kinetics and degradation of nanocomposites based on ternary blend of poly(L-lactic acid), poly(methyl methacrylate), and poly(ethylene oxide) with two different organoclays. *J. Appl. Polym. Sci.* **2012**, *125*, E444–E458. [CrossRef]

50. Friedman, H. Kinetics of thermal degradation of char-forming plastics from thermogravimetry. Application to a phenolic plastic. *J. Polym. Sci. Part C* **1964**, *6*, 183–195. [CrossRef]

51. Vassiliou, A.A.; Papageorgiou, G.Z.; Achilias, D.S.; Bikiaris, D.N. Non-isothermal crystallisation kinetics of in situ prepared poly(ε-caprolactone)/surface-treated SiO$_2$ nanocomposites. *Macromol. Chem. Phys.* **2007**, *208*, 364–376. [CrossRef]

52. Papadimitriou, S.A.; Papageorgiou, G.Z.; Bikiaris, D.N. Crystallization and enzymatic degradation of novel poly(ε-caprolactone-co-propylene succinate) copolymers. *Eur. Polym. J.* **2008**, *44*, 2356–2366. [CrossRef]

53. Singh, V.K.; Singh, R.K. Development of ion conducting polymer gel electrolyte membranes based on polymer PVdF-HFP, BMIMTFSI ionic liquid and the Li-salt with improved electrical, thermal and structural properties. *J. Mater. Chem. C* **2015**, *3*, 7305–7318.

54. Lim, J.S.; Noda, I.; Im, S.S. Effects of metal ion-carbonyl interaction on miscibility and crystallization kinetic of poly(3-hydroxybutyrate-co-3-hydroxyhexanoate)/lightly ionized PBS. *Eur. Polym. J.* **2008**, *44*, 1428–1440. [CrossRef]

55. Papageorgiou, G.Z.; Bikiaris, D.N.; Panayiotou, C.G. Novel miscible poly(ethylene sebacate)/poly(4-vinyl phenol) blends: Miscibility, melting behavior and crystallization study. *Polymer* **2011**, *52*, 4553–4561. [CrossRef]

© 2018 by the authors. Licensee MDPI, Basel, Switzerland. This article is an open access article distributed under the terms and conditions of the Creative Commons Attribution (CC BY) license (http://creativecommons.org/licenses/by/4.0/).

polymers

MDPI

Article

Renewable, Eugenol—Modified Polystyrene Layer for Liquid Crystal Orientation

Changha Ju [†], Taehyung Kim [†] and Hyo Kang [*]

Department of Chemical Engineering, Dong-A University, 37 Nakdong-Daero 550beon-gil, Saha-gu, Busan 604-714, Korea; 1771088@donga.ac.kr (C.J.); xogud1290@donga.ac.kr (T.K.)
* Correspondence: hkang@dau.ac.kr; Tel.: +82-51-200-7720; Fax: +82-51-200-7728
† The authors contributed equally to this work.

Received: 14 December 2017; Accepted: 13 February 2018; Published: 17 February 2018

Abstract: We synthesized a series of plant-based and renewable, eugenol-modified polystyrene (PEUG#) (# = 20, 40, 60, 80, and 100, in which # is the molar content of the eugenol moiety in the side group). Eugenol is extracted from clove oil. We used polymer modification reactions to determine the liquid crystal (LC) orientation properties of the polymer films. In general, the LC cells fabricated using the polymer films with a higher molar content of eugenol side groups exhibited vertical LC orientation behavior. The vertical orientation behavior was well correlated with the surface energy value of the polymer films. The vertical LC orientation could be formed due to the low polar surface energy value on the polymer film generated by the nonpolar carbon group. Electro-optical performances (e.g., voltage holding ratio (VHR), residual DC voltage (R-DC), and thermal orientation stabilities) were good enough to be observed for LC cells using PEUG100 polymer as an eco-friendly LC orientation material.

Keywords: liquid crystal; orientation; polystyrene; eugenol

---

1. Introduction

Liquid crystal (LC) molecules have been intensively studied due to their attractive characteristics such as liquid-like fluidity and solid-like ordering [1]. LC molecules have been recognized to have anisotropic physico-chemical properties such as optical anisotropy and dielectric anisotropy induced by external stimuli due to their unique chemical structures [2]. Therefore, LC molecules have been applied for diverse fields, such as information technology [3–5], nanotechnology [6], biotechnology [7], and energy & environment technology [8,9] using interesting physico-chemical properties. For example, LC molecules have been widely utilized in the display industry, such as the transmissive one using nematic LC and the reflective one using cholesteric LC, respectively [10]. It is important to orientate the LC molecules that have anisotropy on the substrate in the same direction [10].

Orientation methods for liquid crystal molecules have been the focus of intensive development because of scientific and technical interest in flexible displays, as well as rigid panel displays [11–16]. The conventional method of mechanical rubbing of polyimide (PI) surfaces commonly used in the display industry produces stable LC orientation layers on these surfaces [17–25]. Polyimide derivatives containing hydrophobic long alkyloxy and/or alkyl side chains such as (*n*-decyloxy)biphenyloxy and *n*-octadecyl groups have been improved for use as vertical LC orientation layers [26–29]. However, PI derivatives require a high baking temperature of over 200 °C for their synthesis from the PI precursor to apply the LC orientation layers on plastic substrates for versatile displays. Therefore, new polystyrene (PS) derivatives containing long alkyl or fluoroalkyl chains, which are one of the comb-like polymers and an alternative to conventional polyimide derivatives, have been improved via a simple polymer reaction to produce vertical LC orientation layers for next-generation displays, including flexible displays, due to their advantages such as low temperature processability. For instance,

LC cells fabricated using polymer films of *n*-alkylthiomethyl- and *n*-alkylsulfonylmethyl-modified polystyrenes that contain long alkyl groups (number of alkylcarbon > 8) exhibit vertical LC orientation characteristics [30]. LC cells made using 4-alkylphenoxymethyl-modified polystyrenes also show vertical LC orientation, even at a very high rubbing strength and irrespective of the length of the alkyl side groups [31].

Eugenol (2-methoxy-4-(prop-2-en-1-yl)phenol) is the major component (80–95%) of cloves [32]. It has been used extensively in perfumes, flavorings, and essential oils for a long time [33,34]. Eugenol has a methoxyphenol structure with a short hydrocarbon chain. In this structure, the hydroxy group of the aromatic ring in the eugenol is known to have antioxidant and antimicrobial properties [35–37]. For example, eugenol moieties inhibit the growth of pathogenic bacteria, as previously reported by other research groups [38–42]. Therefore, eugenol has been used in many fields and for many applications such as dental materials [43], inflammation medicines [33,44], and antioxidants in the plastics and rubber industries [45]. Eugenol can also be used for modifying the surfaces of substrates such as metal and glass using the hydroxy group via primary and secondary bonding for various film applications. The wettability on the surface of the modified film can be controlled by changing the eugenol content [46,47].

In this paper, bio-renewable eugenol modified PSs (PEUG#) were synthesized to inform vertical orientation of the LCs and investigate the effect of the molar ratio of the modified side groups on the LC orientation behavior. The optical and electrical characteristics of the LC cells made by using PEUG# films were also determined.

2. Materials and Methods

2.1. Materials

4-Chloromethylstyrene (CMS), eugenol, and potassium carbonate were commercially supplied by Aldrich Chemical Co. (Yongin, Korea). MLC-6608 (n_o, n_e, and $\Delta\varepsilon$ indicate ordinary refractive index, extraordinary refractive index, and dielectric anisotropy, respectively, $n_o = 1.4756$, $n_e = 1.5586$, and $\Delta\varepsilon = -4.2$); nematic LC material, obtained in Merck Co. (Pyeongtaek, Korea). Tetrahydrofuran (THF), was dried using reflux method with sodium and benzophenone. Methanol and *N*,*N*′-dimethylacetamide (DMAc) were dried using molecular sieves (4 Å). 4-Chloromethylstyrene was purified using column chromatography filled with silica gel in eluting reagent, hexane, to eliminate nitroparaffin and *tert*-butylcatechol inhibitors and impurities. Poly(4-chloromethylstyrene) (PCMS) (M_n and M_w/M_n indicate number average molecular weight and polydispersity index, respectively, $M_n = 30,000$ and $M_w/M_n = 2.10$) was synthesized by free radical polymerization of 4-chloromethylstyrene using initiator, 2,2′-azoisobutyronitrile (AIBN), under a nitrogen atmosphere. AIBN was puchsed from Junsei Chemical Co., Ltd. (Tokyo, Japan) and purified using methanol by crystallization. Other reagents and solvents were used, as received.

2.2. Synthesis of Eugenol Modified Polystyrenes

The following procedure was used to synthesize all the PEUG#, in which # represents the molar content (%) of the eugenol moiety in the side group as previously reported in a similar procedure [48]. The synthesis of the PEUG100 is given as an example. Eugenol (1.61 g, 9.83 mmol, 150 mol % compared to poly(4-chloromethylstyrene)) and potassium carbonate (1.631 g, 11.81 mmol) mixture in *N*,*N*′-dimethylacetamide (30 mL) was heated up to 75 °C. Poly(4-chloromethylstyrene) (1 g, 6.56 mmol) dissolved in *N*,*N*′-dimethylacetamide (20 mL) was added to eugenol and potassium carbonate mixture and then stirred using magnetic bar in a nitrogen atmosphere at 70 °C for 24 h. The solution was cooled down to room temperature and poured slowly into methanol to yield white precipitate. The precipitate was further purified and washed with methanol to eliminate remaining *N*,*N*′-dimethylacetamide, potassium carbonate, and other used materials. The PEUG100 was dried in vacuum overnight to obtain a yield of over 80%. The substitution ratio of chloromethyl to eugenyl methyl ether was confirmed to be close to 100% within experimental error.

PEUG100 ^{1}H NMR (CDCl$_3$): δ = 0.8–1.9 (3H), 3.1–3.2 (2H), 3.7–3.8 (3H), 4.7–5.0 (2H), 5.0–5.9 (2H), 5.8–6.0 (1H), 6.3–7.2 (7H).

The other polystyrene derivatives containing eugenol side groups were synthesized using the same procedure as for the preparation of PEUG100 except for differing amounts of eugenol in the reaction. For example, PEUG80, PEUG60, PEUG40, and PEUG20 were prepared with eugenol amounts of 0.86 g (5.21 mmol), 0.64 g (3.92 mmol), 0.43 g (2.59 mmol), and 0.21 g (1.30 mmol), respectively, using slightly larger amounts of potassium carbonate (1.631 g, 11.81 mmol, 180 mol % compared with poly(4-chloromethylstyrene)).

2.3. Film Preparation and LC Cell Assembly

Solutions of PEUG# in tetrahydrofuran (1 wt %) were prepared and filtered using a polytetrafluoroethylene (PTFE) membrane with a pore size of 0.45 μm. Thin films of the polymers were spin-coated on glass substrates at 2000 rpm for 60 s. The LC cells were produced using the polymer films on glass slides and the films facing each other using spacers with thicknesses of 4.25 μm. Subsequently, the cells were filled with the nematic LC material MLC-6608 and were sealed with epoxy glue.

2.4. Instrumentation

The ^{1}H-nuclear magnetic resonance (NMR) spectroscopy measurements were performed on Bruker AVANCE spectrometer (Bruker Co., Billerica, MA, USA) with 300 MHz magnet. Gel permeation chromatography (GPC) was performed to determine M_n and M_w/M_n of PEUG# with reference to standardized polystyrenes using UV detector and THF as eluting reagent. The transmittance of the polymer films on the glass substrates was acquired using ultraviolet-visible (UV-Vis) spectroscopy (Perkin Elmer Lambda 20 spectrometer, PerkinElmer, Inc., Waltham, MA, USA). The contact angles for distilled water and diiodomethane on PEUG# films were detected with contact angle analyzer (Kruss DSA10, KRÜSS scientific instruments Inc., Hamburg, Germany)-installed drop shape analysis software. The surface energy value was measured by the Owens-Wendt's equation:

$$\gamma_{sl} = \gamma_s + \gamma_l - 2(\gamma_s^d \gamma_l^d)^{1/2} - 2(\gamma_s^p \gamma_l^p)^{1/2} \tag{1}$$

in which γ_l is the surface energy of the liquid, γ_{sl} is the interfacial energy of the solid/liquid interface, γ_s is the surface energy of the solid, γ_l^d and γ_l^p are known for the test liquids, and γ_s^d and γ_s^p can be calculated from the measured static contact angles [49]. Polarized optical microscopy (POM) images of LC cells were photographed using optical microscopy (Nikon, Eclipse E600 POL, NIKON Co., Tokyo, Japan) installed polarizer and digital camera (Nikon, Coolpix 995, NIKON Co., Tokyo, Japan). Voltage holding ratio (VHR) was determined with VHR measurement system (VHRM 105, Autronic-Melchers, Autronic-Melchers CDT. Ltd., Karlsruhe, Germany) under following conditions (data voltage, pulse width, and frame frequency were 1.0 V, 64 μs, and 60 Hz, respectively). The measuring temperatures were 25 and 60 °C. Residual DC voltage (R-DC) was assessed by capacitance-voltage (C-V) hysteresis method using Nissan Chemical Industries, Ltd. (Tokyo, Japan).

3. Results and Discussion

3.1. Synthesis and Characterization of Eugenol Modified Polystyrene

Figure 1 shows the synthetic routes to the PEUG100 and copolymers (PEUG80, PEUG60, PEUG40, and PEUG20). The copolymers that had different degrees (%) of substitution were obtained by varying the amounts of eugenol in the reaction as previously reported similar procedure [48]. Conversions of nearly 100% from chloromethyl to eugenyl methyl ether were obtained when 150 mol % of eugenol was used at 75 °C for 24 h. The chemical constituents of monomeric units in PEUG# were checked by ^{1}H NMR spectra. The ^{1}H NMR spectrum of PEUG100 shows the existence of protons from styrene backbone (δ = 6.3–7.2 ppm). The proton peaks of eugenol side chains (δ = 3.1–3.2 (2H), 3.7–3.8 (3H), 5.0–5.9 (2H), 5.8–6.0 (1H)) indicate the inclusion of eugenol moieties in the polymer. The content of eugenol was calculated by comparing the integration value of the proton peaks of eugenol (δ = 3.1–3.2 and 3.7–3.8) and the chloromethyl side chains (δ = 4.7–5.0). Similar integrations and calculations were performed for PEUG80, PEUG60, PEUG40, and PEUG20 and were typically within ±5% of the expected values.

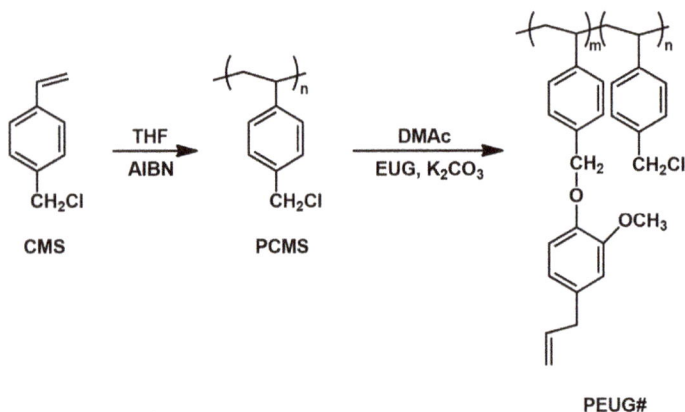

Figure 1. Synthetic route to eugenol modified polystyrene (PEUG#), where # indicates the mole percent of eugenol containing monomeric units in the polymer.

The M_n values of the polymer series synthesized from the PCMS (M_n = 30,000) were always larger than 31,000, indicating that the polymer modification from PCMS to the polymers resulted in an increase in the M_n values of the polymers, which was expected result (Table 1). PEUG# series are soluble in many intermediate polar solvents with low boiling points (e.g., chloroform and THF), and in aprotic polar solvents (e.g., *N*,*N*'-dimethylformamide (DMF), *N*,*N*'-dimethylacetamide (DMAc), *N*-methyl-2-pyrrolidone (NMP)). The solubility of all samples for various solvents indicated their suitability as thin film materials for flexible devices.

Table 1. Reaction conditions and characterizations for the synthesis of the PCMS and PEUG#.

Polymer Designation	Eugenol [mol %]	Degree of Substitution [%]	M_n [1]	M_w/M_n [1]
PCMS	-	-	3000	2.10
PEUG20	20	20	31,000	2.22
PEUG40	40	40	34,000	2.54
PEUG60	60	60	37,000	2.69
PEUG80	80	80	37,000	2.55
PEUG100	150	100	38,000	2.48

[1] Obtained from GPC using tetrahydrofuran as solvent with respect to monodisperse polystyrene as standard.

The thermal behaviors of PEUG# were investigated by differential scanning calorimetry (DSC). All PEUG# series were amorphous. This can be explained that only one glass transition was monitored from the DSC thermograms. As the molar ratio of PEUG# side group increased, PEUG20 to PEUG100, T_g decreased from 74 °C to 47 °C (Figure 2). In general, the T_g value of polymers depends on the polarity, flexibility, and bulkiness of the side group. It has been reported that the T_g values of polymers increased with increasing polarity of the side group [50], while the values can be increased or decreased according to an increase of side group bulkiness. For example, the T_g of the poly(vinylnaphthalene) with relatively bulky substituents, such as naphthalene, is higher than that of polystyrene [50]. This means that the incorporation of a bulky side group decreases the flexibility of the polymer backbone. In this case, when the bulky side chain such as eugenol was incorporated, the free volume of the polymer was increased, which increased the distance between the polymer chain segments [51]. Therefore, the intermolecular and intersegmental interactions are weaker, which decreases the T_g value [52–54].

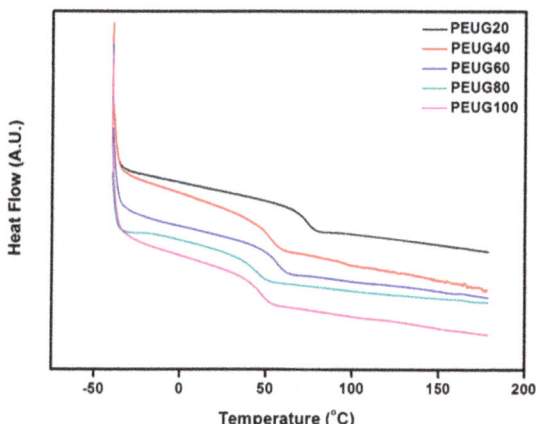

Figure 2. DSC (differential scanning calorimetry) thermogram of eugenol modified polystyrene (PEUG#).

3.2. Transmittance of Eugenol Modified Polystyrene Film

A quantitative analysis of the transparency of the PEUG# films was performed using UV-Vis spectroscopy to investigate the possibility of surface-coating applications (Figure 3). The transmittance value of the PEUG# film coated onto the glass substrate is in range of 96.5–99.0% at 550 nm, which is higher than the transmittance value (80.5%) of the widely used polyimide film, which has an intrinsic yellowish coloration issue related to the diimide fragment conjugation; this reduces the usability of the polyimide film as an LC orientation layer. The results show that the optical transparency of the PEUG# film in the visible light region is sufficiently good to enable the use of the film as an optical material for display devices.

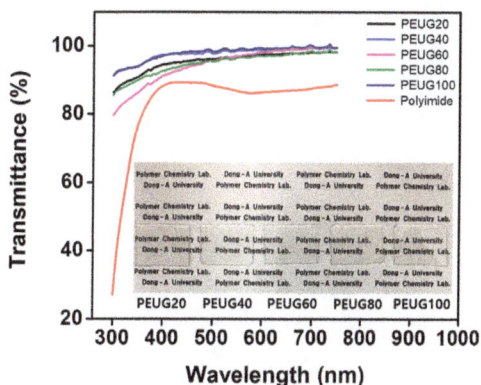

Figure 3. Optical transmittance spectra of eugenol modified polystyrene (PEUG#) and polyimide orientation layers onto quartz substrates.

3.3. LC Orientation Behavior of the LC Cells Fabricated with Eugenol Modified Polystyrene Films

Figure 4 shows the photograph images of the LC cells fabricated from the copolymers (PEUG#). The LC cell fabricated using the PEUG# film with the eugenol side group content of 20 mol % (PEUG20) shows a random planar LC orientation, while random planar and/or tilted LC orientations are observed for the LC cells fabricated using the PEUG40 and PEUG60 polymer films. Good uniformity of the vertical LC orientation behavior was observed for the LC cells fabricated with polymer films with a eugenol side group content of more than 80 mol % (PEUG80 and PEUG100). PEUG# (# = 80, and 100) films showed stabilized vertical LC orientation in whole area and lasted for more than several months. Therefore, as the molar ratio of side groups in PEUG# increases, the vertical LC orientation also increases.

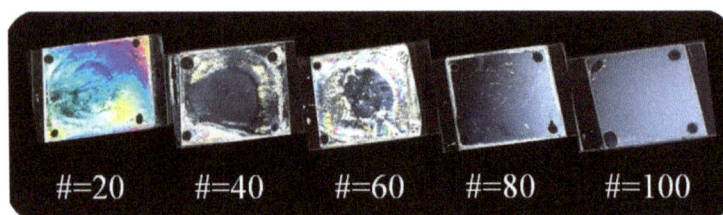

Figure 4. Photograph images of the LC cells made from PEUG# films according to the molar content of eugenol moiety.

As shown Figure 5, the LC orientation performances on PEUG# film were investigated using POM images. A random planar LC orientation on PCMS film was observed. At a molar content of 20% of the eugenol containing monomeric unit in the PEUG#, the LC cells fabricated with the PEUG# film exhibited planar LC orientation in the conoscopic POM images. At a molar content in the range of 40–60%, the LC cells fabricated with the PEUG# film exhibited a random tilted LC orientation in the conoscopic POM images. On the other hand, all the PEUG80 and PEUG100 films exhibited stable vertical LC orientation layers.

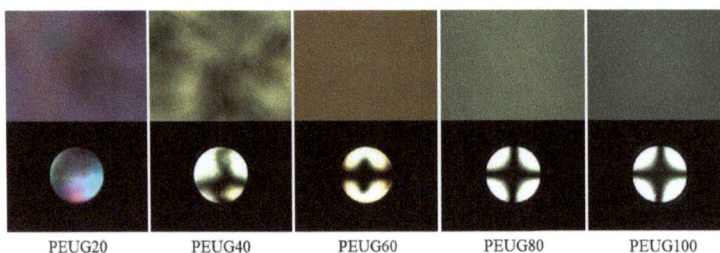

| PEUG20 | PEUG40 | PEUG60 | PEUG80 | PEUG100 |

Figure 5. Polarized optical microscopy (POM) images of the LC cells made from PEUG#.

3.4. Surface Properties of Eugenol Modified Polystyrene Films

The results obtained for the LC orientation behavior indicated a general trend of the occurrence of vertical LC orientation with increasing molar content of the eugenol side groups. It is known that high pretilt angles of LC molecules resulting in vertical orientation performance are associated with low surface energy of the orientation layer and/or sterically repulsive force between LC molecules and orientation layers [55,56]. For example, polyimide derivatives that have nonpolar and bulky groups (e.g., pentylcyclohexylbenzene [55] and 4-(n-octyloxy)phenyloxy [56]) exhibited vertical orientation behavior. Therefore, we tried to explain the LC orientation performances of the PEUG# films utilizing surface energy value measurements, one of surface characterization techniques. Surface energy values were calculated based on contact angles of (distilled) water and diiodomethane (Figure 6 and Table 2). The total surface energy values were calculated by Owens-Wendt's equation, which is summation of polar and dispersion terms. The total surface energy values of PEUG# films according to the molar content of the eugenol moiety in the side groups increased to 46.66, 47.14, 47.18, 48.23, and 48.28 mJ m^{-2}. The dispersion surface energy values according to the molar content of the eugenol moiety also increased to 44.70, 45.78, 46.10, 47.42, and 47.53 mJ m^{-2}. On the other hand, the polar surface energy values of PEUG# films according to the molar content of the eugenol moiety in the side groups decreased to 1.97, 1.36, 1.08, 0.81, and 0.75 mJ m^{-2}. We have found that the vertical LC orientation was well correlated with the polar surface energy term of the polymer film. It has been described that the polar surface energy of polymer film can affect the LC orientation behavior, as previously reported by other research [57–59]. Since eugenol has a nonpolar and bulky group, such as not only phenyl group but allyl group attached to phenyl group in the *para* position, the increase in the degree of substitution of EUG causes a decrease in the polar surface energy, in which vertical orientation could be formed. Therefore, it is reasonable to conclude that the vertical LC orientation behavior of PEUG100 and PEUG80 is due to the increased steric repulsions between the LC molecules and the polymer surfaces that result from the incorporation of the nonpolar and bulky eugenol moieties into the side group of the PS and the low polar surface energy originating from the unique chemical structure of nonpolar and long carbon groups.

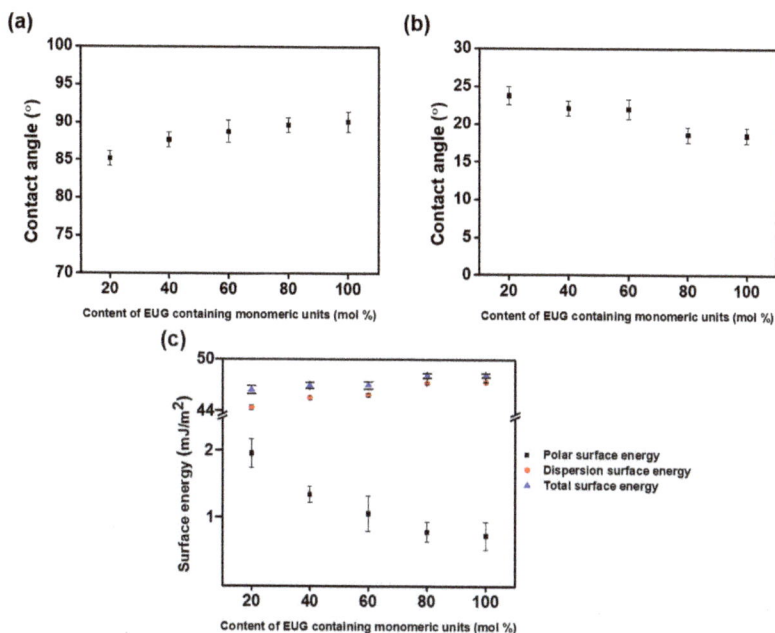

Figure 6. (a) Water, (b) diiodomethane contact angle, and (c) surface energy values of PEUG# films according to the molar content of the eugenol moiety in the side groups.

Table 2. Contact angle and surface energy values and LC orientation properties on the polymer films.

Polymer Designation	Contact Angle [o] [1]		Surface Energy [mJ/m^2] [2]			LC Aligning Ability [3]
	Water	Diiodomethane	Polar	Dispersion	Total	
PCMS	71.1	35.2	8.67	37.00	45.67	X
PEUG20	85.2	23.8	1.97	44.70	46.66	X
PEUG40	87.7	22.1	1.36	45.78	47.14	X
PEUG60	88.8	22.0	1.08	46.10	47.18	X
PEUG80	89.7	18.6	0.81	47.42	48.23	O
PEUG100	90.1	18.5	0.75	47.53	48.28	O

[1] Measured from static contact angles; [2] calculated from Owens-Wendt's equation; [3] circle (O) and cross (X) indicate polymer film have vertical and random planar, tilted LC aligning ability, respectively.

3.5. Reliability and Electro-Optical Performance of the LC Cells Fabricated with Eugenol Modified Polystyrene Films

The reliability of the LC cells fabricated from the polymer films was investigated using a stability test of the LC orientation under harsh conditions such as high temperatures. The thermal stability of the LC cell fabricated from the PEUG100 film was estimated from the POM image after heating for 10 min at 100, 150, and 200 °C, respectively. As shown in Figure 7, there are no distinguishable differences in the pretilt angle of the PEUG100 film with vertical LC orientation as shown in the Maltese cross pattern in the conoscopic POM images, indicating that the vertical LC orientation of the PEUG100 LC cell was maintained, even at a high temperature.

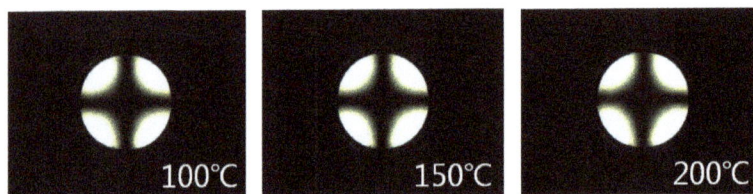

Figure 7. Conoscopic POM images of PEUG100 LC cells, after heat treatment at 100, 150, and 200 °C for 10 min, respectively.

The electro-optical (E-O) performance of the LC cell fabricated with the PEUG100 film was measured with regard to potential and practical display applications. Voltage holding ratio (VHR) of LC cell was demonstrated more than 99% at 25 °C, which was maintained at 60 °C. The VHR is enough to be high for substantive applications as the LC orientation layer in thin film transistor (TFT)-addressed display [11]. The R-DC measured by C-V hysteresis method of the LC cell was discovered to be very low, about 10 mV, which is even smaller than the R-DC of polyimides [11]. The excellent thermal stability, VHR, and R-DC of the LC cell fabricated using the PEUG100 film was ascribed to the intrinsic properties of polymers with high carbon contents, such as good thermal stability and a low dielectric constant.

Recently, considerable efforts have been devoted to the development of plastic substrates for flexible displays [60]. All the LC cells fabricated from the PEUG# films on the polyethylene terephthalate (PET) substrates exhibited similar LC orientation behavior compared with LC cells fabricated from the same polymer films on glass substrates (Figures 4 and 5). We found that the LC cells fabricated using the PEUG100 film on plastic PET substrates showed vertical LC orientation behavior. Furthermore, this type of LC cell exhibited a very good vertical LC orientation that was maintained after bending it several hundred times. Therefore, as a plant-based and renewable resource, the PEUG100 film can be considered a potential material for LC orientation layers for eco-friendly flexible displays.

Polymers **2018**, *10*, 201

4. Conclusions

A series of PS derivatives containing plant-based and renewable eugenol side groups (PEUG#) were synthesized to determine the LC orientation properties of the polymer films. The LC cells fabricated from the polymer films with more than 80 mol % of eugenol (PEUG80 and PEUG100) exhibited vertical LC orientation. However, LC cells made using PEUG# films with less than 60 mol % of eugenol exhibited random planar and/or tilted LC orientation behavior. The PEUG100 polymer films exhibited good optical transparency in the visible light region (400–700 nm). At 550 nm, the transmittance value was higher for the PEUG100 film (99.0%) than for the polyimide film (80.5%), the most commonly used LC orientation layer. The vertical LC orientation was ascribed to the steric repulsions between the LC molecules and the polymer surfaces due to the incorporation of a nonpolar and bulky eugenol moiety into the side chain. In addition, the vertical LC orientation was well correlated with the polar surface energy values of the polymer induced by the long alkyl groups. These results provide a basis for the design of eco-friendly and renewable LC orientation layers based on eugenol containing polymer films.

Acknowledgments: Financial supports by the Dong-A University Research Fund are gratefully acknowledged.

Author Contributions: Hyo Kang conceived the project. Changha Ju and Taehyung Kim contributed equally that they designed and accomplished all experiments. Changha Ju and Taehyung Kim synthesized and characterized the polymers. Changha Ju and Taehyung Kim performed the analysis of liquid crystal orientation. Changha Ju and Taehyung Kim and Hyo Kang wrote the paper. All authors participated in discussions of the research and provided the feedback for the paper.

Conflicts of Interest: The authors declare no conflict of interest.

References

1. Chandrasekhar, S. *Liquid Crystals*, 2nd ed.; Cambridge University Press: Cambridge, UK, 1992. ISBN 0-521-41747-3.
2. Meyer, R.B. Effects of electric and magnetic fields on the structure of cholesteric liquid crystals. *Appl. Phys. Lett.* **1968**, *12*, 281–282. [CrossRef]
3. Brostow, W.; Hagg Lobland, H.E. *Materials: Introduction and Applications*, 1st ed.; John Wiley & Sons: West Sussex, UK, 2017; pp. 242–246. ISBN 978-1-119-28100-9.
4. Baetens, R.; Jelle, B.P.; Gustavsen, A. Properties, requirements and possibilities of smart windows for dynamic daylight and solar energy control in buildings: A state-of-the-art review. *Sol. Energy Mater. Sol. Cells* **2010**, *94*, 87–105. [CrossRef]
5. Mori, H.; Itoh, Y.; Nishiura, Y.; Nakamura, T.; Shinagawa, T. Performance of a novel optical compensation film based on negative birefringence of discotic compound for wide-viewing-angle twisted-nematic liquid-crystal displays. *Jpn. J. Appl. Phys.* **1997**, *36*, 143–147. [CrossRef]
6. Bisoyi, H.K.; Kumar, S. Liquid-crystal nanoscience: An emerging avenue of soft self-assembly. *Chem. Soc. Rev.* **2011**, *40*, 306–319. [CrossRef] [PubMed]
7. Woltman, S.J.; Jay, G.D.; Crawford, G.P. Liquid-crystal materials find a new order in biomedical applications. *Nat. Mater.* **2007**, *6*, 929–938. [CrossRef] [PubMed]
8. Kato, T.; Yoshio, M.; Ichikawa, T.; Soberats, B.; Ohno, H.; Funahashi, M. Transport of ions and electrons in nanostructured liquid crystals. *Nat. Rev. Mater.* **2017**, *2*, 17001. [CrossRef]
9. Kumar, M.; Kumar, S. Liquid crystals in photovoltaics: A new generation of organic photovoltaics. *Polym. J.* **2017**, *49*, 85–111. [CrossRef]
10. Stöhr, J.; Samant, M.G.; Cossy-Favre, A.; Diaz, J.; Momoi, Y.; Odahara, S.; Nagata, T. Microscopic origin of liquid crystal alignment on rubbed polymer surfaces. *Macromolecules* **1998**, *31*, 1942–1946. [CrossRef]
11. Takatoh, K.; Sakamoto, M.; Hasegawa, R.; Koden, M.; Itoh, N.; Hasegawa, M. *Alignment Technology and Applications of Liquid Crystal Devices*; CRC Press: Boca Raton, FL, USA, 2005; pp. 7–8. ISBN 978-07-484-0902-0.
12. Ichimura, K. Photoalignment of liquid-crystal systems. *Chem. Rev.* **2000**, *100*, 1847–1874. [CrossRef] [PubMed]
13. Stöhr, J.; Samant, M.G. Liquid crystal alignment by rubbed polymer surfaces: A microscopic bond orientation model. *J. Electron. Spectrosc. Relat. Phenom.* **1999**, *98*, 189–207. [CrossRef]

14. Schadt, M. Liquid crystal materials and liquid crystal displays. *Annu. Rev. Mater. Sci.* **1997**, *27*, 305–379. [CrossRef]
15. Latansohn, A.; Rochon, P. Photoinduced motions in azo-containing polymers. *Chem. Rev.* **2002**, *102*, 4139–4176. [CrossRef]
16. Ree, M. High performance polyimides for applications in microelectronics and flat panel displays. *Macromol. Res.* **2006**, *14*, 1–33. [CrossRef]
17. Ghosh, M.K.; Mittal, K.L. *Polyimides: Fundamentals and Applications*; Dekker, M., Ed.; CRC Press: New York, NY, USA, 1996; pp. 806–807. ISBN 978-0824794668.
18. Feller, M.B.; Chen, W.; Shen, T.R. Investigation of surface-induced alignment of liquid-crystal molecules by optical second-harmonic generatio. *Phys. Rev. A* **1991**, *43*, 6778–6792. [CrossRef] [PubMed]
19. Van Aerle, J.N.A.; Tol, A.J.W. Molecular orientation in rubbed polyimide alignment layers used for liquid-crystal displays. *Macromolecules* **1994**, *27*, 6520–6526. [CrossRef]
20. Lee, K.W.; Peak, S.-H.; Lien, A.; Durning, C.; Fukuro, H. Microscopic molecular reorientation of alignment layer polymer surfaces induced by rubbing and its effects on LC pretilt angles. *Macromolecules* **1996**, *29*, 8894–8899. [CrossRef]
21. Weiss, K.; Wöll, C.; Böhm, E.; Fiebranz, B.; Forstmann, G.; Peng, B.; Scheumann, V.; Johannsmann, D. Molecular orientation at rubbed polyimide surfaces determined with X-ray absorption spectroscopy: Relevance for liquid crystal alignment. *Macromolecules* **1998**, *31*, 1930–1936. [CrossRef]
22. Meister, R.; Jerome, B. The conformation of a rubbed polyimide. *Macromolecules* **1999**, *32*, 480–486. [CrossRef]
23. Ge, J.J.; Li, C.Y.; Xue, G.; Mann, I.K.; Zhang, D.; Wang, S.-Y.; Harris, F.W.; Cheng, S.Z.D.; Hong, S.-C.; Zhuang, X. Rubbing-induced molecular reorientation on an alignment surface of an aromatic polyimide containing cyanobiphenyl side chains. *J. Am. Chem. Soc.* **2001**, *123*, 5768–5776. [CrossRef] [PubMed]
24. Kim, D.; Oh-e, M.; Shen, Y.R. Rubbed polyimide surface studied by sum-frequency vibrational spectroscopy. *Macromolecules* **2001**, *34*, 9125–9129. [CrossRef]
25. Vaughn, K.E.; Sousa, M.; Kang, D.; Rosenblatt, C. Continuous control of liquid crystal pretilt angle from homeotropic to planar. *Appl. Phys. Lett.* **2007**, *90*, 194102. [CrossRef]
26. Lee, S.W.; Kim, S.I.; Park, Y.H.; Ree, M.; Rim, Y.N.; Yoon, H.J.; Kim, H.C.; Kim, Y.-B. Liquid-crystal alignment on the rubbed film surface of semi-flexible copolyimides containing *n*-alkyl side groups. *Mol. Cryst. Liq. Cryst.* **2000**, *349*, 279–282. [CrossRef]
27. Lee, Y.J.; Kim, Y.W.; Ha, J.D.; Oh, J.M.; Yi, M.H. Synthesis and characterization of novel polyimides with 1-octadecyl side chains for liquid crystal alignment layers. *Polym. Adv. Technol.* **2007**, *18*, 226–234. [CrossRef]
28. Lee, S.W.; Chae, B.; Lee, B.; Choi, W.; Kim, S.B.; Kim, S.I. Rubbing-induced surface morphology and polymer segmental reorientations of a model brush polyimide and interactions with liquid crystals at the surface. *Chem. Mater.* **2003**, *15*, 3105–3112. [CrossRef]
29. Lee, S.B.; Shin, G.J.; Chi, J.H.; Zin, W.-C.; Jung, J.C.; Hahm, S.G.; Ree, M.; Chang, T. Synthesis, characterization and liquid-crystal-aligning properties of novel aromatic polypyromellitimides bearing (*n*-alkyloxy) biphenyloxy side chains. *Polymer* **2006**, *47*, 6606–6621. [CrossRef]
30. Kang, H.; Park, J.S.; Kang, D.; Lee, J.-C. Liquid crystal alignment property of *n*-alkylthiomethyl- or *n*-alkylsulfonylmethyl-substituted polystyrenes. *Polym. Adv. Technol.* **2009**, *20*, 878–886. [CrossRef]
31. Kang, H.; Kim, T.-H.; Kang, D.; Lee, J.-C. 4-Alkylphenoxymethyl-substituted polystyrenes for liquid crystal alignment layers. *Macromol. Chem. Phys.* **2009**, *210*, 926–935. [CrossRef]
32. Szabadics, J.; Erdelyi, L. Pre-and postsynaptic effects of eugenol and related compounds on *Helix pomatia* L. neurons. *Acta Biol. Hung.* **2000**, *51*, 265–273. [PubMed]
33. Prakash, P.; Gupta, N. Therapeutic uses of *Ocimum sanctum Linn* (Tulsi) with a note on eugenol and its pharmacological actions: A short review. *Indian J. Physiol. Pharmacol.* **2005**, *49*, 125–131. [PubMed]
34. Tokuoka, Y.; Uchiyama, H.; Abe, M.; Ogino, K. Solubilization of synthetic perfumes by nonionic surfactants. *J. Colloid Interface Sci.* **1992**, *152*, 402–409. [CrossRef]
35. Gülçin, İ. Antioxidant activity of eugenol: A structure-activity relationship study. *J. Med. Food* **2011**, *14*, 975–985. [CrossRef] [PubMed]
36. El-Baroty, G.S.; El-Baky, H.A.; Farag, R.; Saleh, M.A. Characterization of antioxidant and antimicrobial compounds of cinnamon and ginger essential oils. *Afr. J. Biochem. Res.* **2010**, *4*, 167–174.
37. Ultee, A.; Bennik, M.H.J.; Moezelaar, R. The phenolic hydroxyl group of carvacrol is essential for action against the food-borne pathogen *Bacillus cereus*. *Appl. Environ. Microbiol.* **2002**, *68*, 1561–1568. [CrossRef] [PubMed]

38. Bagamboula, C.; Uyttendaele, M.; Debevere, J. Inhibitory effect of thyme and basil essential oils, carvacrol, thymol, estragol, linalool and *p*-cymene towards *Shigella sonnei* and *S. flexneri*. *Food Microbiol.* **2004**, *21*, 33–42. [CrossRef]

39. Balszyk, M.; Holley, A. Interaction of monolaurin, eugenol and sodium citrate on growth of common meat spoilage and pathogenic organisms. *Int. J. Food Microbiol.* **1998**, *39*, 175–183. [CrossRef]

40. Karapinar, M.; Aktuğ, Ş.E. Inhibition of foodborne pathogens by thymol, eugenol, menthol and anethole. *Int. J. Food Microbiol.* **1987**, *4*, 161–166. [CrossRef]

41. Kim, J.; Marshall, M.R.; Wei, C.I. Antibacterial activity of some essential oil components against five foodborne pathogens. *J. Agric. Food Chem.* **1995**, *43*, 2839–2845. [CrossRef]

42. Kim, O.H.; Park, S.W.; Park, H.D. Inactivation of *Escherichia coli* O157:H7 by cinnamic aldehyde purified from *Cinnamomum cassia* shoot. *Food Microbiol.* **2004**, *21*, 105–110. [CrossRef]

43. Schmalz, G.; Arenholt-Bindslev, D. *Biocompatibility of Dental Materials*, 1st ed.; Springer: Heidelberg, Germany, 2009. ISBN 978-3-540-77782-3.

44. Grespan, R.; Paludo, M.; de Paula Lemos, H.; Barbosa, C.P.; Bersani-Amado, C.A.; de Oliveira Dalalio, M.M.; Cuman, R.K.N. Anti-arthritic effect of eugenol on collagen-induced arthritis experimental model. *Biol. Pharm. Bull.* **2012**, *35*, 1818–1820. [CrossRef] [PubMed]

45. Garkal, D.; Taralkar, S.P.; Kulkarni, P.; Jagtap, S.; Nagawade, A. Kinetic model for extraction of eugenol from leaves of *Ocimum Sanctum Linn* (Tulsi). *Int. J. Pharm. Appl.* **2012**, *3*, 267–270.

46. Ryu, D.Y.; Shin, K.; Drockenmuller, E.; Hawker, C.J.; Russell, T.P. A generalized approach to the modification of solid surfaces. *Science* **2005**, *308*, 236–239. [CrossRef] [PubMed]

47. Bain, C.D.; Whitesides, G.M. Depth sensitivity of wetting: Monolayers of ω-mercapto ethers on gold. *J. Am. Chem. Soc.* **1988**, *110*, 5897–5898. [CrossRef]

48. Ju, C.; Kim, T.; Kang, H. Liquid crystal alignment behaviors on capsaicin substituted polystyrene films. *RSC Adv.* **2017**, *7*, 41376–41383. [CrossRef]

49. Owens, D.K.; Wendt, R.C. Estimation of the surface free energy of polymers. *J. Appl. Polym. Sci.* **1969**, *13*, 1741–1747. [CrossRef]

50. Gedde, U. *Polymer Physics*; Chapman and Hall: London, UK, 1995; pp. 78–79. ISBN 978-94-011-0543-9.

51. Brostow, W. Realiability and prediction of long term performance of polymer-based materials. *Pure Appl. Chem.* **2009**, *81*, 417–432. [CrossRef]

52. Hayes, R.A. The relationship between glass temperature, molar cohesion, and polymer structure. *J. Appl. Polym. Sci.* **1961**, *15*, 318–321. [CrossRef]

53. Wesslin, B.; Lenz, R.W.; MacKnight, W.J.; Karaz, F.E. Glass transition temperatures of poly(ethyl α-chloroacrylates). *Macromolecules* **1971**, *4*, 24–26. [CrossRef]

54. Lee, J.-C.; Litt, M.H.; Rogers, C.E. Oxyalkylene polymers with alkylsulfonylmethyl side chains: Gas barrier properties. *J. Polym. Sci. Part B Polym. Phys.* **1998**, *36*, 75–83. [CrossRef]

55. Lee, J.-B.; Lee, H.-K.; Park, J.-C.; Kim, Y.-B. The structural effect on the pretilt angle of alignment materials with alkylcyclohexylbenzene as a side chain in polyimides. *Mol. Cryst. Liq. Cryst.* **2005**, *439*, 161–172. [CrossRef]

56. Lee, S.W.; Lee, S.J.; Hahm, S.G.; Lee, T.J.; Lee, B.; Chae, B.; Kim, S.B.; Jung, J.C.; Zin, W.C.; Sohn, B.H. Role of the *n*-alkyl end of bristles governing liquid crystal alignment at rubbed films of brush polymer rods. *Macromolecules* **2005**, *38*, 4331–4338. [CrossRef]

57. Paek, S.H.; Durning, C.J.; Lee, K.W.; Lien, A. A mechanistic picture of the effects of rubbing on polyimide surfaces and liquid crystal pretilt angles. *J. Appl. Phys.* **1998**, *83*, 1270–1280. [CrossRef]

58. Ban, B.S.; Kim, Y.B. Surface energy and pretilt angle on rubbed polyimide surfaces. *J. Appl. Polym. Sci.* **1999**, *74*, 267–271. [CrossRef]

59. Wu, H.Y.; Wang, C.Y.; Lin, C.J.; Pan, R.P.; Lin, S.S.; Lee, C.D.; Kou, C.S. Mechanism in determining pretilt angle of liquid crystals aligned on fluorinated copolymer films. *J. Phys. D Appl. Phys.* **2009**, *42*, 155303. [CrossRef]

60. MacDonald, B.A.; Rollins, K.; Mackerron, D.; Rakos, K.; Eveson, R.; Hashimoto, K.; Rustin, B. Engineered films for display technologies. In *Flexible Flat Panel Displays*; Crawford, G.P., Ed.; John Wiley & Sons: West Sussex, UK, 2005; pp. 11–33. ISBN 978-0-470-87048-8.

© 2018 by the authors. Licensee MDPI, Basel, Switzerland. This article is an open access article distributed under the terms and conditions of the Creative Commons Attribution (CC BY) license (http://creativecommons.org/licenses/by/4.0/).

![polymers]

MDPI

Article

Cardanol Groups Grafted on Poly(vinyl chloride)— Synthesis, Performance and Plasticization Mechanism

Puyou Jia [1], Meng Zhang [1,2,*], Lihong Hu [1,2], Rui Wang [3], Chao Sun [3] and Yonghong Zhou [1,*]

1   National Engineering Lab for Biomass Chemical Utilization, Key Lab on Forest Chemical Engineering, State Forestry Administration, and Key Lab of Biomass Energy and Materials, Institute of Chemical Industry of Forest Products, Chinese Academy of Forestry (CAF), 16 Suojin North Road, Nanjing 210042, China; jiapuyou@icifp.cn (P.J.); hlh@icifp.cn (L.H.)

2   Institute of New Technology of Forestry, Chinese Academy of Forest (CAF), Beijing 100091, China

3   College of Materials Science and Engineering, Nanjing Forestry University, 159 Longpan Road, Nanjing 210037, China; by861209@126.com (R.W.); sc930510@163.com (C.S.)

*   Correspondence: zhangmeng@icifp.cn (M.Z.); zyh@icifp.cn (Y.Z.); Tel.: +86-025-85482520 (Y.Z.)

Received: 8 September 2017; Accepted: 11 November 2017; Published: 15 November 2017

Abstract: Internally plasticized poly(vinyl chloride) (PVC) materials are investigated via grafting of propargyl ether cardanol (PEC). The chemical structure of the materials was studied by FT-IR and ^{1}H NMR. The performance of the obtained internally plasticized PVC materials was also investigated with TGA, DSC and leaching tests. The results showed that grafting of propargyl ether cardanol (PEC) on PVC increased the free volume and distance of PVC chains, which efficiently decreased the glass transition temperature (T_g). No migration was found in the leaching tests for internally plasticized PVC films compared with plasticized PVC materials with commercial plasticizer dioctyl phthalate (DOP). The internal plasticization mechanism was also disscussed according to lubrication theory and free volume theory. This work provides a meaningful strategy for designing no-migration PVC materials by introducing cardanol groups as branched chains.

Keywords: cardanol; plasticizer; poly(vinyl chloride); migration; plasticization mechanism

1. Introduction

Plasticizer is one of the most important plastic additives, and is used to improve processability, plasticity and flexility of plastics. The most widely used plasticizers are phthalate esters, which account for 80% of the total consumption of plasticizer [1]. However, the potential toxicity of these phthalate esters to the human body has been reported, which has led to their restriction in consumer products [2–4]. In order to reduce the toxicity of plasticzier, many environmentally friendly plasticizers have been investigated, such as epoxidized jatropha oil [5], cardanol derivatives [6], polymer-plasticizer [7], polyol ester [8] and phosphate plasticizer [9]. However, these alternative plasticizers will migrate from plastic products with the prolongation of aging time, thus shortening the life of the products. To avoid the migration of plasticizer, Navarro et al. and Lee et al. [10–12] have studied an internal plasticization strategy whereby phthalate-based thiol additives and hyperbranched polyglycerol, respectively, were grafted onto the polymer matrix. The glass transition temperature (T_g) of PVC materials grafting hyperbranched polyglycerol groups was kept below 0 °C. Propargyl ether triethyl citrate and castor oil-based derivative were covalently bonded to the polymer matrix as an internal plasticizer, as reported by Jia et al. [13,14]. These studies indicated that this strategy was efficient for producing flexible polymer materials, and avoiding the migration of plasticizer.

Cardanol is one of the most favorable biomass resources for plasticizer production due to its relatively low cost and similar chemical structure to conventional phthalate plasticizers such as dioctyl phthalate (DOP). Cardanol derives from an agricultural by-product that is abundantly available in many parts of the chemical industry [15,16]. There are some reactive groups on the chemical structure of cardanol, such as the unsaturated carbon chains on the branched chains, benezene ring and hydroxy groups. These groups can occur hydrogenation, polymerization, sulfuration, esterification and epoxidation reactions. [17]. Therefore, the strategy of grafting cardanol groups grafted on PVC matrices is expected to produce plasticized PVC materials without migration. Recently, Po et al. studied an approach that covalently linked cardanol to the PVC matrix via Click reaction, but the internal plasticization mechanism was not discussed in detail in terms of traditional plasticization theory [18]. A kind of internally plasticized PVC material was prepared via replacing chlorine using the mannich base of cardanol butyl ether in 2017 [19], but the plasticizing efficiency of the method was lower than for the Click reaction [10,12].

Inspired by the above-mentioned pioneer works, we herein report a strategy for developing internally plasticized PVC materials by grafting propargyl ether cardanol onto the PVC matrix. Specifically, different mass of cardanol groups were grafted onto PVC matrices to obtain no-migration, flexible PVC materials. Differential scanning calorimetry (DSC) measurements were used to detect the plasticizing efficiency of the strategy, and to compare it with pure PVC. The migration stability of the internally plasticized PVC materials in n-hexane was investigated. For comparison, the migration stability of PVC films plasticized with the traditional plasticizer DOP was also tested. In addition, the internal plasticization mechanism was discussed in terms of traditional plasticization theories, such as free volume theory and lubrication theory.

2. Materials and Methods

2.1. Materials

Cardanol (99%, acid value 5.5–6.6. Iodine value 210–250) was provided by Jining Hengtai Chemical Co., Ltd. (Jining, China) Propargyl bromide solution, tetrahydrofuran (THF), potassium carbonate, sodium azide, methanol, acetone, cuprous bromide, 5,5-dimethyl-2,2-dipyridyl, *N*,*N*-dimethylformamide (DMF) and dioctyl phthalate (DOP) were kindly provided by Nanjing Chemical Reagent Co., Ltd. (Nanjing, China) Polyvinyl chloride (PVC) was supplied by Hanwha (KM-31, Seoul, Korea).

2.2. Synthesis of Propargyl Ether Cardanol

To a 100 mL flask equipped with a condenser tube was added 30.4 g (100 mmol) of cardanol, 13.08 g (110 mmol) of propargyl bromide solution and 15.2 g (110 mmol) of potassium carbonate in 50 mL of acetone. The mixture was stirred at 65 °C for 12 h. The solution was purified by evaporating under vacuum after washing the mixture with deionized water (yield: 97%).

2.3. Synthesis of Azide-Functionalized PVC (PVC-N$_3$)

To a 100 mL flask were added 2.0 g of PVC, 2.0 g of NaN$_3$ and 100 mL of DMF. The mixture was allowed to stir at 30 °C for 24 h and precipitated into water/methanol mixture (1/1 by volume), and dried in a vacuum to obtain the PVC–N$_3$ [14]. Figure 1 showed the synthetic route of PVC–N$_3$. Elemental analysis: 35.13% C, 8.21% H, 18.47% N, and 38.19% Cl.

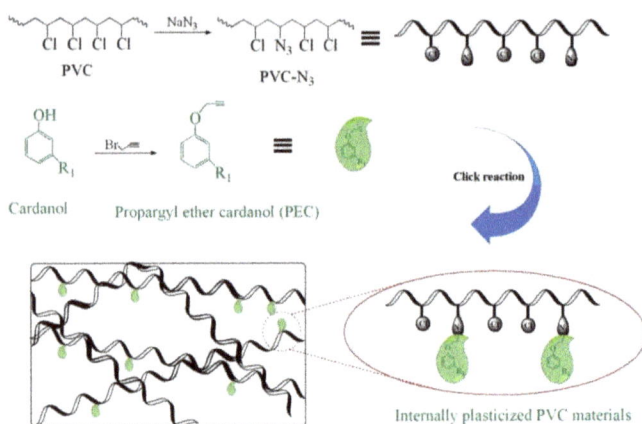

Figure 1. Synthesis of internally plasticized PVC materials.

2.4. Synthesis of Internally Plasticized PVC Materials

Internally plasticized PVC materials were prepared by dissolving a certain amount of PVC–N$_3$, PEC, cuprous bromide and 2,2'-dipyridine in 40 mL of DMF in a three-neck flask equipped with a mechanical stirrer, nitrogen pipe and thermometer. The reaction was kept at 30 °C and stirred for 24 h. Then, the mixture was precipitated into water/methanol mixture (1/1 by volume) after filtering to remove the copper salts, and dried in a vacuum to obtain PVC-0.25PEC. Table S1 (see Supplementary Materials) shows the composition of reactants. Figure 1 shows the synthesis route.

2.5. Preparation of PVC Films and PVC–DOP Films

A total of 1 g of internally plasticized PVC materials were dissolved in 20 mL of THF. The mixture was stirred at 40 °C for 20 min until the solution appeared transparent, and was then poured into a glass petri dish (5 cm diameter), and dried in a constant temperature drying box at 60 °C for 24 h to completely remove residual THF. PVC–DOP films were prepared by dissolving a total of 2 g of PVC and 1 g of DOP in 20 mL of THF using the same method.

2.6. Characterization

2.6.1. Elemental Analysis

Elemental analysis was conducted on an elemental PE-2400 analyzer (PERKINELMER Instrument Crop., Waltham, MA, USA).

2.6.2. Fourier-Transform Infrared (FT-IR)

Fourier transform infrared (FT-IR) spectra of propargyl ether cardanol and internally plasticized PVC materials were investigated on a Nicolet iS10 FTIR measurement (Nicolet Instrument Crop., Madison, WI, USA). The spectra were acquired in the range of 4000 to 500 cm^{-1} at a resolution of 4 cm^{-1}.

2.6.3. ^{1}H Nuclear Magnetic Resonance (NMR)

^{1}H NMR measurements were conducted on an AV-300 NMR spectrometer (Bruker Instrument Crop., Karlsruhe, Germany) at a frequency of 400 MHz. CDCl$_3$ was used as solvent and tetrametnylsilane (TMS) as an internal standard.

2.6.4. Gel Permeation Chromatography (GPC)

The molecular weights of PVC and internally plasticized PVC materials were measured using an efficient gel chromatograph made by Waters, Milford, MA, USA at 30 °C (flow rate: 1 mL/min, column: mixed PL gel 300 mm × 718 mm, 25 μm) using THF as solvent. The number-average molecular weight, weight-average molecular weight and polydispersity indices (M_W/M_n) were calculated by calibrating with polystyrene standards.

2.6.5. Thermogravimetric Analysis (TGA)

TGA was carried out using a TG209F1 TGA thermal analysis instrument (Netzsch Instrument Corp., Bavaria, Germany) in N_2 atmosphere (50 mL/min) at a heating rate of 10 °C/min. The temperature was scanned from 40 to 600 °C.

2.6.6. Differential Scanning Calorimetry (DSC)

Glass transition temperature (T_g) was characterized using a NETZSCH DSC 200 PC analyzer (Bavaria, Germany); the temperature ranged from −60 °C to 100 °C in N_2 atmosphere (50 mL/min) at a heating rate of 20 °C/min. T_g values reported were taken from the second heating run in order to eliminate thermal history and to correspond to the midpoint of the DSC curves measured from the extension of the pre- and post-transition baseline.

2.6.7. Leaching Tests

PVC films and internally plasticized PVC films were cut so that different groups had the same surface area. The thickness of all the films was around 2.0 mm. The films were weighted and immersed in n-hexane at 50 °C for 2 h. Then, these films were dried and reweighed. The extraction loss was calculated according to Equation (1).

$$\text{Degree of migration} = (W_1 - W_2)/W_1 \times 100 \tag{1}$$

where W_1 = initial weight of test films, and W_2 = final weight of test PVC films [20].

3. Results

3.1. Chemical Structure of PEC

Figure S1 presents the FT-IR spectra of the cardanol and PEC. The strong and broad absorption peak at 3332 cm^{-1} in the FT-IR spectra of the cardanol was attributed to stretch vibration of –OH groups. The absorption peak appeared at 3007 cm^{-1}, which was assigned to the olefinic C–H stretch of cardanol. The peaks at 2924 and 2853 cm^{-1} were associated with the =C–H and –C–H bonds, respectively. The peak at 1586 cm^{-1} was attributed to the C–C bonds. The broad absorption of the C=C stretching vibration mode from the benzene ring could be observed in the range of 1454–1152 cm^{-1} [20,21]. In comparison to the FT-IR spectra of the cardanol, the characteristic absorption peak of the –OH groups at 3007 cm^{-1} disappeared in the FT-IR spectra of PEC. A new peak appeared at 3306 cm^{-1}, which was attributed to alkyne C–H stretch, and the C≡C stretch characteristic absorption peak appeared at 2122 cm^{-1}.

^{1}H NMR of cardanol and PEC was also investigated. In ^{1}H NMR of cardanol (Figure S2, see Supplementary Materials), the signal appeared at δ 0.94 ppm, which was assigned to the protons of the methyl groups on the branched chains. Strong signals appeared at δ 1.36, δ 1.41, δ 2.09, δ 2.5 and δ 2.8 ppm, and were attributed to the protons of methylene groups. Peaks appeared at δ 5.41 ppm, which were associated with protons of hydroxyl groups. The signal of the olefin groups appeared at δ 5.44 ppm, and the signals of the protons for the benzene rings could be observed at δ 6.71, δ 6.82 and δ 7.18 ppm [19,21,22]. In comparison to the ^{1}H NMR spectrum of cardanol, a new signal appeared at δ 4.75 ppm, which was attributed to the protons of methylene groups connected to alkynyl groups.

Another new peak appeared at δ 2.61 ppm, which was associated with the protons of alkynyl groups. The FT-IR and ^{1}H NMR data indicated that PEC had been obtained.

3.2. Chemical Structure of Internally Plasticized PVC Materials

Figure 2 shows the FT-IR spectra of PVC and PVC-0.25PEC, PVC-0.50PEC and PVC-0.75PEC. In comparison to the FT-IR spectra of PVC, a new strong peak appeared at 2110 cm^{-1}, which was attributed to the characteristic absorption peak of the azide group (–N$_3$ stretch) [23–25], indicating that PVC–N$_3$ had been prepared. With more PEC grafted onto the PVC matrix, the peaks at 2910, 1586 and 1152–1454 cm^{-1}, which were attributed to C–H stretch, C–C bonds, and the C=C stretching vibration mode derived from the benzenoid ring of PEC, respectively, appeared stronger than PVC–N$_3$, indicating that the PEC had been grafted onto the PVC matrix.

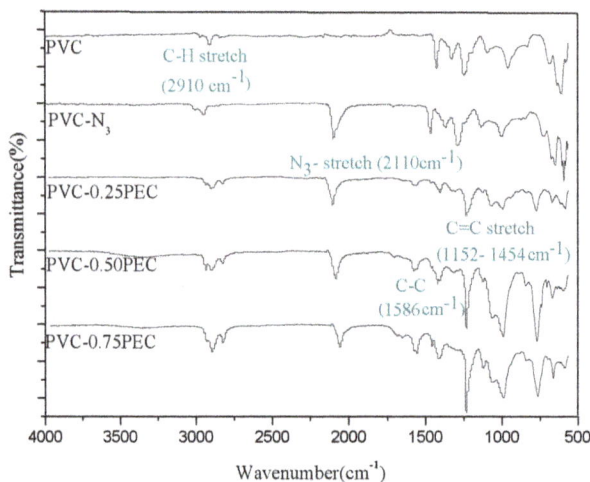

Figure 2. FT-IR spectra of PVC and internally plasticized PVC materials.

^{1}H NMR of PVC and internally plasticized PVC materials are presented in Figure 3. As can be seen from the ^{1}H NMR spectra of PVC, peak at δ 4.5 ppm was attributed to protons of –CHCl–(CH$_2$)– [14,18,26]. Peak b at δ 2.2 ppm was assigned to protons of –CHCl–(CH$_2$)–. The ^{1}H NMR of PVC–N$_3$ was similar to PVC, indicating that the chemical environment of the protons was similar. Peak a′ at δ 4.5 ppm was associated with the protons of –CHCl–(CH$_2$)–, and peak b′ at δ 2.2 ppm was attributed to the protons of –CH(N$_3$)–(CH$_2$). Peaks a″ and b″ at δ 4.5 and 2.2 ppm were associated with the protons of –CHCl–(CH$_2$)– and –CH(N$_3$)–(CH$_2$)–, respectively, which were weaker than for PVC and PVC–N$_3$, but some new peaks appeared at δ 7.18, δ 5.5, δ 2.85 and δ 0.89 ppm, which were associated with the protons of benzene rings, olefin groups, methylene groups and methyl groups of branched chains derived from PEC [26–28], respectively, indicating that the PEC had been grafted onto the PVC matrix.

The click reaction level can be evaluated by examining the molecular weight change [29,30]. The molecular weight and polydispersity indices of PVC and internally plasticized PVC materials were investigated by GPC analysis. The results are shown in Figure 4 and Table 1. As can be seen in Figure 4, the GPC peak of internally plasticized PVC materials shows a clear shift to a higher molecular weight region compared to that of PVC, which indicates that the cardanol groups had been grafted onto the poly (vinyl chloride) via click reaction. PVC and PVC-0.25PEC show a single GPC peak with a clear shift to a higher molecular weight region, which indicates that no homopolymer contamination was generated in the reactions, while PVC-0.50PEC and PVC-0.75PEC show two GPC peaks, indicating that homopolymer was produced in the reactions. In addition, the peak area of internally plasticized PVC materials decreased gradually with an increasing number of cardanol groups grafted onto the poly (vinyl chloride), which indicates that the highly branched internally plasticized PVC materials had been filtrated by the organic membrane in the dissolution process. The data for number-average molecular weight (M_n), weight-average molecular weight (M_W), and polydispersity indices are presented in Table 2. M_n and M_W of PVC materials increased gradually with greater numbers of cardanol groups grafted onto the PVC matrix. M_n of PVC, PVC-0.25PEC, PVC-0.50PEC and PVC-0.75PEC was 15,100, 21,200, 23,800 and 25,400 g/mol, respectively, which indicates that the internally plasticized PVC materials had been obtained.

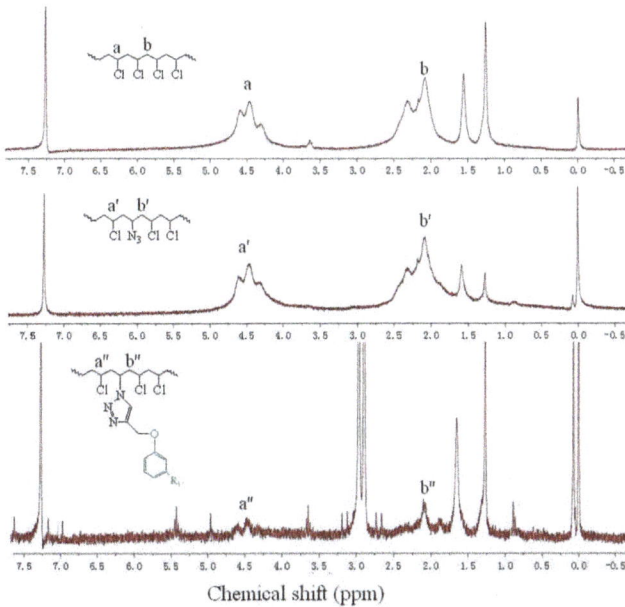

Figure 3. ^{1}H NMR of PVC and internally plasticized PVC materials.

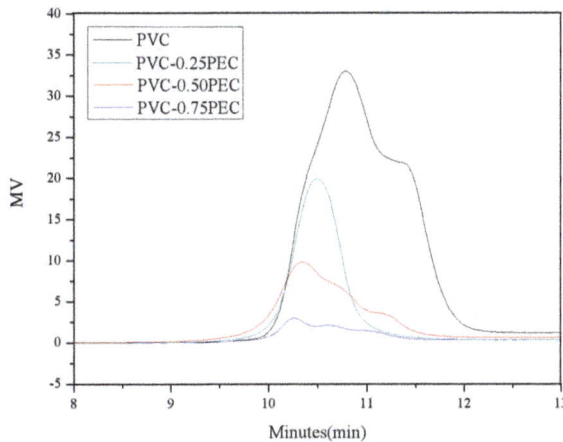

Figure 4. GPC spectra of PVC and internally plasticized PVC materials.

Table 1. Relative molecular mass and polydispersity indices of PVC materials.

PVC materials	Number average molecular weight (M_n/g·mol^{-1})	Weight-average molecular weight (M_W/g·mol^{-1})	Polydispersity indices
PVC	15,100	18,900	1.2
PVC-0.25PEC	21,200	26,700	1.2
PVC-0.50PEC	23,800	30,200	1.3
PVC-0.75PEC	25,400	31,900	1.3

Table 2. TGA and DSC data of PVC materials.

PVC materials	T_d (°C)	T_g (°C)	Char residue (%)
PVC	278.4	85	5.8
PVC-0.25PEC	211.3	67	10.0
PVC-0.50PEC	209.4	59	16.5
PVC-0.75PEC	208.6	42	17.6

3.3. Performances of Internally Plasticized PVC Materials

The thermal properties of PVC, PVC-0.25PEC, PVC-0.50PEC and PVC-0.75PEC with different cardanol group contents were investigated using DSC and TGA measurements. As shown in Figure 5, the DSC curves of PVC and internally plasticized PVC materials with increased of cardanol group content did not present any melting peaks, indicating that all of the PVC materials had amorphous characteristics. A clear broadening of the T_g is observed, which indicates that the PVC and cardanol are not miscible. Additionally, none of the PVC materials exhibited any melting peaks, indicating their amorphous characteristics. T_g data for PVC, PVC-0.25PEC, PVC-0.50PEC and PVC-0.75PEC are summarized in Table 2, and were 85, 67, 59 and 42 °C, respectively. The decline in T_g with different contents of cardanol groups was mainly caused by the internal plasticizing effect of the cardanol groups. T_g is an important evaluation criteria in Mauritz and Storey's mathematical models [31]. In this theory, plasticizing efficiency for reducing T_g can be controlled by structural features of the plasticizer, such as the branchiness of the side chains and the length of those chains. For the same molecular mass, branched plasticizers have higher plasticizing efficiency parameter values than linear structures. In this study, internally plasticized PVC macromolecules with higher branchiness side chains caused it to become a kind of internal plasticizer. The decrease in T_g illustrated that internally plasticized PVC materials induced a plasticizing effect on themselves.

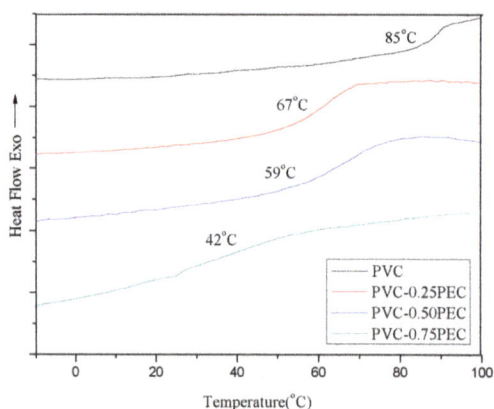

Figure 5. DSC curves of PVC and internally plasticized PVC materials.

The onset degradation temperature (T_d) and char residue are summarized in Table 2. As can be seen from Table 2, internally plasticized PVC materials had poorer thermal stability than PVC. The T_d of internally plasticized PVC materials decreased from 278.4 to around 210 °C with an increase in the number of cardanol groups grafted onto the PVC, which was caused by the unstable properties of azide groups under high temperature [18]. Char residue of PVC was 5.8% at 600 °C, the data reached 17.6% for PVC-0.75PEC. The reason for this is the fact that internally plasticized PVC materials have a higher relative carbon content than PVC, and that the thermal degradation process of internally plasticized PVC materials produces more char residue than PVC. Figure 6 presents the TGA curves of PVC and internally plasticized PVC materials with different contents of cardanol groups. PVC grafted with a variety of different functionalities showed a decrease in thermal stability below 220 °C, and increase in thermal stability above 220 °C. With an increased number of cardanol groups grafted onto the PVC chains, the thermal degradation of PVC-0.50PEC and PVC-0.75PEC at above 220 °C produced more char residue than PVC-0.25PEC. The char residue plays an important role in improving the thermal stability of internally plasticized PVC materials, because the char residue covers on the surface of internally plasticized PVC materials and prevents them from undergoing pyrolysis. One recent study has indicated that pyrolysis products of cardanol derivatives in gas phase include –OH– containing compounds, saturated hydrocarbons, CO_2, CO, and aromatic derivatives; all of these compounds were burned and produced char residue [32]. Therefore, the potential degradation products of the new plasticizer do not have any adverse effects on humans or the surrounding environment.

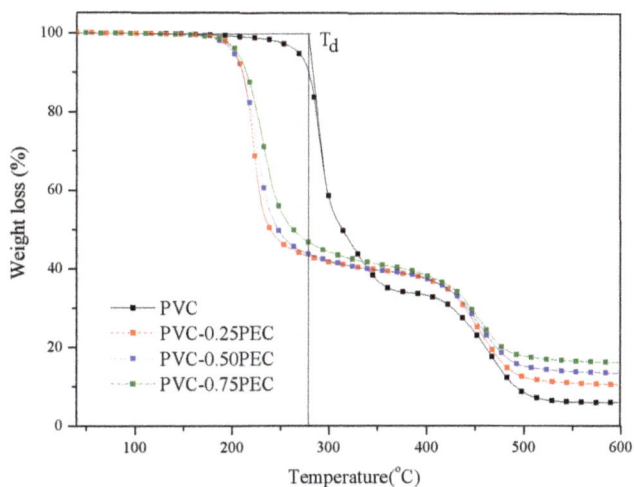

Figure 6. TGA curves of PVC and internally plasticzied PVC materials.

The common non-attached plasticizer based on cardanol acetate [6], epoxidized cardanol phenyl phosphate ester plasticizer [33], and epoxidized cardanol laurate [34] are easily leached from PVC in organic solvent, which is not conducing to make PVC products long-lasting and stable. The loss of plasticizers pollutes environments and produces a potential threat to human health. The internal plasticization strategy can effectively avoid the loss of plasticizer. The leaching tests of internally plasticized PVC films and PVC–DOP films were studied in n-hexane at 50 °C for 2 h. The results showed no migration in the leaching tests for internally plasticized PVC films, but 15.7% of the DOP leached from PVC–DOP films into n-hexane. These results indicate that the internal plasticization strategy was efficient for avoiding loss of plasticizers.

3.4. Mechanism of Internal Plasticization

Traditional plasticization mechanisms include lubricity theory, gel theory, free volume theory, kinetic theories and mathematical models. Internal plasticization mechanisms can also be illustrated based on these theories. Kilpatrick [35] and others [36,37] developed lubricity theory. This theory holds that plasticizer, which acts as a molecular lubricant, allows the polymer chains to move freely over one another when a force is applied to the plasticized polymer [34]. Figure 7 shows the structure of PVC and internally plasticized PVC materials under lubrication theory. PVC chains cannot move freely, because these PVC chains without branched chains are tangled up. These molecular characteristics of polymer chains make PVC present as stiff and difficult to process. For internally plasticized PVC materials, cardanol groups grafted onto PVC chains can serve as a kind of molecular lubricant; this structure helps them move freely, causing internally plasticized PVC chains to present as flexible and easy to process.

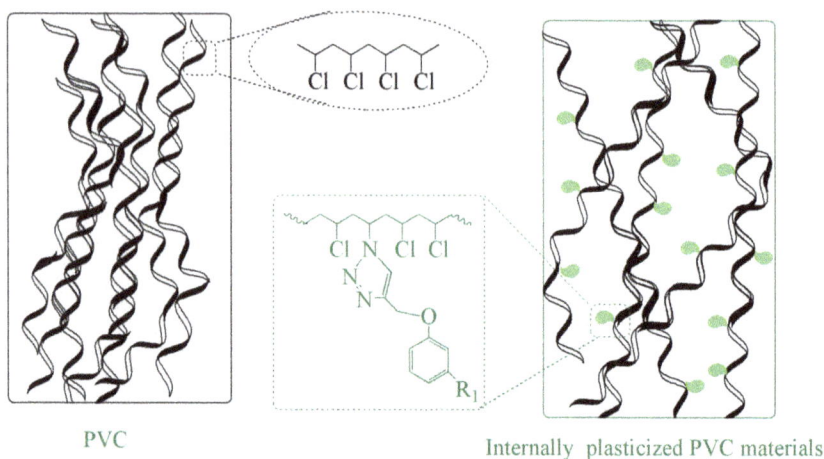

PVC   Internally plasticized PVC materials

Figure 7. Structure of PVC and internally plasticized PVC materials under lubrication theory.

Free volume theory can be used to explain internal plasticization by evaluating the three kinds of movement for chain-like macromolecules: end movement, subgroup movement and crankshaft movement [38]. As can be seen from Figure 8, there are two kinds of movement for PVC macromolecules, because there are no branched chains on the chemical structure of PVC. Meanwhile, there are three kinds of movement for internally plasticized PVC macromolecules, because cardanol groups were grafted onto PVC. The grafting of cardanol groups onto PVC increases the distance and free volume of PVC chains and makes them easier to move. The increase in distance and free volume promotes the end movement, subgroup movement and crankshaft movement of internally plasticized PVC chains, causing the materials to present as flexible and easy to process.

PVC

Internally plasticized PVC materials

⟩ :End movement

⟩ :Subgroup movement

⌣ :Crankshaft movement.

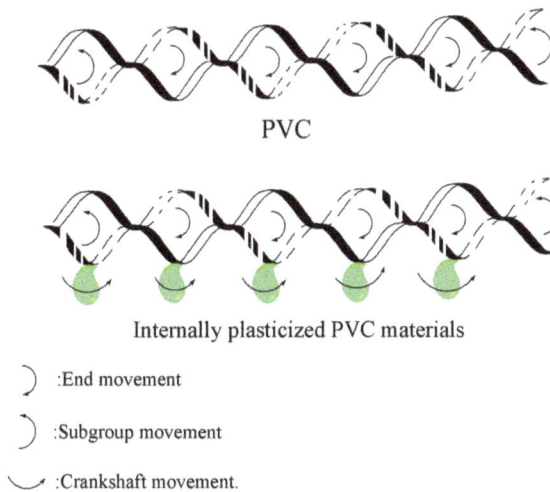

Figure 8. Three kinds of movement for chain-like macromolecules in free volume theory.

4. Conclusions

Internally plasticized PVC materials were synthesized by the grafting of PEC. The obtained PVC-0.25PEC, PVC-0.50PEC and PVC-0.75PEC exhibited efficient internal plasticization effects. Because cardanol groups grafted onto PVC chains is able to serve as a kind of molecular lubricant, the structure caused internally plasticized PVC chains to move freely, further increasing the distance and free volume of PVC chains and making them much easier to move than PVC. This will cause PVC materials to possess potentially flexible qualities, making them easy to process. The obtained internally plasticized PVC materials exhibited poorer thermal stability and more char residue than PVC. The T_g of the internally plasticized PVC materials reached 42 °C. No migration was found in leaching tests for internally plasticized PVC films, but 15.7% of DOP leached from PVC–DOP films into n-hexane. Therefore, this work provides a meaningful strategy for designing advanced PVC materials by introducing cardanol groups as branched chains.

Supplementary Materials: The following are available online at http://www.mdpi.com/2073-4360/9/11/621/s1, Figure S1: FT-IR spectra of cardanol and PEC, Figure S2: ^{1}H NMR of cardanol and PEC, Table S1: The composition of reactants.

Acknowledgments: This work was supported by the National Natural Science Foundation of China (Grant Nos. 31670578, 31670577 and 31700499), National Key Research and Development Program of China (2017YFD0601000) and the Fundamental Research Funds from Jiangsu Province Biomass and Materials Laboratory (JSBEM-S-2017010).

Author Contributions: Puyou Jia, Meng Zhang and Yonghong Zhou conceived and designed the experiments; Puyou Jia, Rui Wang, Lihong Hu and Chao Sun performed the experiments and analyzed the data; Puyou Jia wrote the paper.

Conflicts of Interest: The authors declare no conflict of interest.

References

1. Bocque, M.; Voirin, C.; Lapinte, V.; Caillol, S.; Robin, J.J. Petro-based and bio-based plasticizers: Chemical structures to plasticizing properties. *J. Polym. Sci. Part B Polym. Chem.* **2016**, *54*, 11–13. [CrossRef]
2. Boisvert, A.; Jones, S.; Issop, L.; Erythropel, H.C.; Papadopoulos, V.; Culty, M. In vitro functional screening as a means to identify new plasticizers devoid of reproductive toxicity. *Environ. Res.* **2016**, *150*, 496–512. [CrossRef] [PubMed]

3. Bui, T.T.; Giovanoulis, G.; Cousins, A.P.; Magnér, J.; Cousins, I.T.; Wit, C.C. Human exposure, hazard and risk of alternative plasticizers to phthalate esters. *Sci. Total* **2016**, *541*, 451–467. [CrossRef] [PubMed]
4. Pérez-Albaladejo, E.; Fernandes, D.; Lacorte, S.; Porte, C. Comparative toxicity, oxidative stress and endocrine disruption potential of plasticizers in JEG-3 human placental cells. *Toxicol. In Vitro* **2017**, *38*, 41–48. [CrossRef] [PubMed]
5. Chieng, B.W.; Ibrahim, N.A.; Then, Y.Y.; Loo, Y.Y. Epoxidized jatropha oil as a sustainable plasticizer to poly(lactic acid). *Polymers* **2017**, *9*, 204. [CrossRef]
6. Greco, A.; Maffezzoli, A. Cardanol derivatives as innovative bio-plasticizers for poly-(lactic acid). *Polym. Degrad. Stabil.* **2016**, *132*, 213–219. [CrossRef]
7. Jarray, A.; Gerbaud, V.; Hemati, M. Polymer-plasticizer compatibility during coating formulation: A multi-scale investigation. *Prog. Org. Coat.* **2016**, *101*, 195–206. [CrossRef]
8. Jia, P.Y.; Zhang, M.; Hu, L.; Feng, G.; Bo, C.; Zhou, Y. Design and synthesis of a castor oil based plasticizer containing THEIC and diethyl phosphate groups for the preparation of flame-retardant PVC materials. *RSC Adv.* **2017**, *7*, 897–903. [CrossRef]
9. Wang, X.; Zhou, S.; Guo, W.; Wang, P.; Xing, W.; Song, L.; Hu, Y. Renewable cardanol-based phosphate as a flame retardant toughening agent for epoxy resins. *ACS Sustain. Chem. Eng.* **2017**, *5*, 3409–3416. [CrossRef]
10. Navarro, R.; Perrino, M.P.; Tardajos, M.G.; Reinecke, H. Phthalate Plasticizers Covalently Bound to PVC: Plasticization with suppressed migration. *Macromolecules* **2010**, *43*, 2377–2381. [CrossRef]
11. Lee, K.W.; Chung, J.W.; Kwak, S. Structurally enhanced self-plasticization of poly(vinyl chloride) via click grafting of hyperbranched polyglycerol. *Macromol. Rapid Commun.* **2016**, *37*, 2045. [CrossRef] [PubMed]
12. Navarro, R.; Perrino, M.P.; García, C.; Elvira, C.; Gallardo, A.; Reinecke, H. Highly flexible PVC materials without plasticizer migration as obtained by efficient one-pot procedure using trichlorotriazine chemistry. *Macromolecules* **2016**, *49*, 2224–2227. [CrossRef]
13. Jia, P.Y.; Hu, L.; Feng, G.; Bo, C.; Zhang, M.; Zhou, Y. PVC materials without migration obtained by chemical modification of azide-functionalized PVC and triethyl citrate plasticizer. *Mater. Chem. Phys.* **2017**, *190*, 25–30. [CrossRef]
14. Jia, P.Y.; Hu, L.; Zhang, M.; Feng, G.; Zhou, Y. Phosphorus containing castor oil based derivatives: Potential non-migratory flame retardant plasticizer. *Eur. Polym. J.* **2017**, *87*, 209–220. [CrossRef]
15. Fouquet, T.; Fetzer, L.; Mertz, G.; Puchot, L.; Verge, P. Photoageing of cardanol: Characterization, circumvention by side chain methoxylation and application for photocrosslinkable polymers. *RSC Adv.* **2015**, *5*, 54899–54912. [CrossRef]
16. Gour, R.S.; Kodgire, V.V.; Badiger, M.V.J. Toughening of epoxy novolac resin using cardanol based flexibilizers. *Appl. Polym. Sci.* **2016**, *133*, 43318. [CrossRef]
17. Perdriau, S.; Harder, S.; Heeres, H.J.; de Vries, J.G. Selective conversion of polyenes to monoenes by RuCl$_3$-catalyzed transfer hydrogenation: The case of cashew nutshell liquid. *ChemSusChem* **2012**, *5*, 2427–2434. [CrossRef] [PubMed]
18. Yang, P.; Yang, J.; Sun, H.; Fan, H.; Chen, Y.; Wang, F.; Shi, B. Novel environmentally sustainable cardanol-based plasticizer covalently bound to PVC via click chemistry: Synthesis and properties. *RSC Adv.* **2015**, *5*, 16980–16985. [CrossRef]
19. Jia, P.Y.; Hu, L.H.; Shang, Q.Q.; Wang, R.; Zhang, M.; Zhou, Y. Self-Plasticization of PVC materials via chemical modification of mannich base of cardanol butyl ether. *ACS Sustain. Chem. Eng.* **2017**, *5*, 6665–6673. [CrossRef]
20. Zhang, M.; Zhang, J.; Chen, S.; Zhou, Y. Synthesis and fire properties of rigid polyurethane foams made from a polyol derived from melamine and cardanol. *Polym. Degrad. Stabil.* **2014**, *110*, 27–34. [CrossRef]
21. Chen, J.; Liu, Z.; Jiang, J.; Nie, X.; Zhou, Y.; Murray, R.E. A novel biobased plasticizer of epoxidized cardanol glycidyl ether: Synthesis and application in soft poly(vinyl chloride) films. *RSC Adv.* **2015**, *5*, 56171–56180. [CrossRef]
22. Yao, L.L.; Chen, C.; Xu, W.; Ye, Z.; Shen, Z.; Chen, M. Preparation of cardanol based epoxy plasticizer by click chemistry and its action on poly(vinyl chloride). *J. Appl. Polym. Sci.* **2017**. [CrossRef]
23. Lafarge, J.; Kebir, N.; Schapman, D.; Gadenne, V.; Burel, F. Design of bacteria repellent PVC surfaces using the click chemistry. *Cellulose* **2013**, *20*, 2779–2790. [CrossRef]
24. Kiskan, B.; Demiray, G.; Yagci, Y. Thermally curable polyvinylchloride via click chemistry. *J. Polym. Sci. Part A Polym. Chem.* **2008**, *46*, 3512–3518. [CrossRef]

25. Akat, H.; Ozkan, M. Synthesis and characterization of poly (vinylchloride) type macrophotoinitiator comprising side-chain thioxanthone via click chemistry. *eXPRESS Polym. Lett.* **2011**, *5*, 318–326. [CrossRef]
26. Liu, Z.; Chen, J.; Knothe, G.; Nie, X.; Jiang, J. Synthesis of epoxidized cardanol and its antioxidative properties for vegetable oils and biodiesel. *ACS Sustain. Chem. Eng.* **2016**, *4*, 901–906. [CrossRef]
27. Chen, J.; Nie, X.; Liu, Z.; Mi, Z.; Zhou, Y. Synthesis and application of polyepoxide cardanol glycidyl ether as biobased polyepoxide reactive diluent for epoxy resin. *ACS Sustain. Chem. Eng.* **2015**, *3*, 1164–1171. [CrossRef]
28. Huang, K.; Zhang, Y.; Li, M.; Lian, J.; Yang, X.; Xia, J. Preparation of a light color cardanol-based curing agent and epoxy resin composite: Cure-induced phase separation and its effect on properties. *Prog. Org. Coat.* **2012**, *74*, 240–247. [CrossRef]
29. Bowers, B.F.; Huang, B.; Shu, X.; Miller, B.C. Investigation of reclaimed asphalt pavement blending efficiency through GPC and FTIR. *Constr. Build. Mater.* **2014**, *50*, 517–523. [CrossRef]
30. Geng, J.; Li, H.; Sheng, Y. Changing regularity of SBS in the aging process of polymer modified asphalt binder based on GPC analysis. *Int. J. Pavement Res. Technol.* **2014**, *7*, 77–82.
31. Mauritz, K.A.; Storey, R.F.; Wilson, B.S. Efficiency of plasticization of PVC by higher-order di-alkyl phthalates and survey of mathematical models for prediction of polymer/diluent blend Tg's. *J. Vinyl Technol.* **1990**, *12*, 165–1173. [CrossRef]
32. Caiying, B.; Lihong, H.; Puyou, J.; Bingchuan, L.; Jing, Z.; Yonghong, Z. Structure and thermal properties of phosphoruscontaining polyol synthesized from cardanol. *RSC Adv.* **2015**, *5*, 106651–106660.
33. Jie, C.; Xiaoying, L.; Yigang, W.; Ke, L.; Jinrui, W.; Jianchun, J.; Xiaoan, N. Synthesis and application of a novel environmental plasticizer based on cardanol for poly(vinyl chloride). *J. Taiwan Inst. Chem. Eng.* **2016**, *65*, 488–497.
34. Xiaoying, L.; Xiaoan, N.; Jie, C.; Yigang, W.; Ke, L. Synthesis and Application of a Novel Epoxidized Plasticizer Based on Cardanol for Poly(vinyl chloride). *J. Renew. Mater.* **2017**, *5*, 154–164.
35. Daniels, P.H. A brief overview of theories of PVC plasticization and methods used to evaluate PVC-plasticizer interaction. *J. Vinyl Addit. Technol.* **2009**, *15*, 219–223. [CrossRef]
36. Cadogan, D.F.; Howick, C.J. *Plasticizers in Kirk-Othmer Encyclopedia of Chemical Technology*; John Wiley and Sons: New York, NY, USA, 1996.
37. Clark, F.W. Plasticizer. *Chem. Ind.* **1941**, *60*, 225–230.
38. Wancong, S.; Zhibo, S.; Pingping, J. *Plasticizers and Their Applications*, 2nd ed.; Chemical Industry Press: Beijing, China, 2004; pp. 536–538.

© 2017 by the authors. Licensee MDPI, Basel, Switzerland. This article is an open access article distributed under the terms and conditions of the Creative Commons Attribution (CC BY) license (http://creativecommons.org/licenses/by/4.0/).

MDPI

Article

Starch-Chitosan Polyplexes: A Versatile Carrier System for Anti-Infectives and Gene Delivery

Hanzey Yasar [1,2,†], Duy-Khiet Ho [1,2,†], Chiara De Rossi [1], Jennifer Herrmann [1], Sarah Gordon [1], Brigitta Loretz [1,*] and Claus-Michael Lehr [1,2]

[1]   Helmholtz Institute for Pharmaceutical Research Saarland (HIPS), Helmholtz Center for Infection Research (HZI), Saarland University, D-66123 Saarbrücken, Germany; Hanzey.Yasar@helmholtz-hzi.de (H.Y.); DuyKhiet.Ho@helmholtz-hzi.de (D.-K.H.); Chiara.DeRossi@helmholtz-hzi.de (C.D.R.); jennifer.herrmann@helmholtz-hzi.de (J.H.); S.C.Gordon@ljmu.ac.uk (S.G.); Claus-Michael.Lehr@helmholtz-hzi.de (C.-M.L.)
[2]   Department of Pharmacy, Saarland University, D-66123 Saarbrücken, Germany
*   Correspondence: Brigitta.Loretz@helmholtz-hzi.de; Tel.: +49-681-98806-1030
†   These authors contributed equally to this work.

Received: 8 December 2017; Accepted: 27 February 2018; Published: 1 March 2018

Abstract: Despite the enormous potential of nanomedicine, the search for materials from renewable resources that balance bio-medical requirements and engineering aspects is still challenging. This study proposes an easy method to make nanoparticles composed of oxidized starch and chitosan, both isolated from natural biopolymers. The careful adjustment of C/N ratio, polymer concentration and molecular weight allowed for tuning of particle characteristics. The system's carrier capability was assessed both for anti-infectives and for nucleic acid. Higher starch content polyplexes were found to be suitable for high encapsulation efficiency of cationic anti-infectives and preserving their bactericidal function. A cationic carrier was obtained by coating the anionic polyplex with chitosan. Coating allowed for a minimal amount of cationic polymer to be employed and facilitated plasmid DNA loading both within the particle core and on the surface. Transfection studies showed encouraging result, approximately 5% of A549 cells with reporter gene expression. In summary, starch-chitosan complexes are suitable carriers with promising perspectives for pharmaceutical use.

Keywords: polymeric nanoparticles; renewable polysaccharides; anionic starch; cationic anti-infectives; transfection

---

1. Introduction

Nanoparticulate carrier systems represent a well established platform for vaccination and treatment of severe diseases, such as infection and cancer, by protecting active agents, preventing burst release kinetics, providing the potential to enhance crossing of biological barriers and improving local drug delivery [1–4]. However, the selection of materials or excipients for nanomedical applications remains challenging due to strict requirements of the field. Such materials should be biocompatible and biodegradable, safe and at the same time provide good drug loading capacity as well as a potential to carry diverse bioactive agents [3]. Moreover, for large scale production, the used materials should be environmentally friendly, and able to be manufactured by facile processes. In recent years, a variety of polymeric materials derived from natural biopolymers have been synthesized and investigated to formulate vehicles to deliver bioactive molecules. These molecules have been embedded inside the polymeric matrix or adsorbed onto the colloidal surface [5] by either physical interaction (e.g., electrostatic complexation) or chemical modification. Nevertheless, the number of biodegradable and biocompatible polymers which are further compatible with water (as a solvent suitable for pharmaceutical use) and can form nanoparticles with a high and versatile active agent encapsulation capacity are still limited. Hence, the production of excipients for nanomedicine with a balance between pharmaceutical requirements and engineering aspects as well as a tunable potential for drug delivery has gained considerable attention. In particular, natural and modified polysaccharides such as chitosan, alginate, starch and dextrin, and their synthetic derivatives, have been considered as efficient candidates for drug carrier systems [1,6,7]. However, achieving a consistent and robust production of polysaccharide nanoparticles is challenging due to the heterogeneous physicochemical properties of natural and synthetic polymers. In addition, depending on the actives to be delivered and the route of administration, different protocols are needed [8] to prepare polysaccharide-based polymeric nanoparticles [9,10]. Thus, the chosen polymers need to be appropriately tailored, chemically modified and optimized to qualify for targeted applications [1].

Among natural polysaccharides, starch and chitosan have many promising properties. Starch is a biocompatible and biodegradable polysaccharide, which is degraded by α-amylase, and available at relatively low cost. It has been widely used in tablets and capsules, e.g., as a binder or diluent [11]. Slightly modified derivatives of starch with fractional molecular weights have previously been studied as a platform to formulate homogenous carrier systems for gene delivery [12]. Other researchers have also studied starch-based particulate systems for drug delivery [13–15]. Chitosan is similarly biodegradable and biocompatible, and has been investigated and widely used in pharmaceutical research for drug [16,17], protein [18] and nucleic acid delivery, and for vaccination purposes [19–21]. It has also been used as a biomedical material for artificial skin and wound healing bandages [22] as a biodegradable polysaccharide [23]. Moreover, chitosan has good biocompatibility as tested in humans [24]. Yamada et al. [12] has reported the preparation of anionic starch derivatives by mild chemical modification, and the separation of different molecular weights by a fractional cut-off protocol, which was later aimed for transfection study. The research showed promising perspectives of starch derivatives as drug carrier system. However, the charge mediated complexation of fractional starch derivatives was not fully explored in that study; the carrier capacity of such system thus remains to be investigated.

In light of these advantages, the aim of this work was to produce versatile and flexible nanocarriers using both starch and chitosan, with a facile and organic solvent-free preparation method combining the advantages of these two polymers into a carrier system. The investigated systems were composed of starch derivatives of molecular weight (M_w) >100 kDa or with M_w range of 30–100 kDa, and oligochitosan M_w 5 kDa or Protasan M_w 90 kDa as chitosan derivatives. A wide range of molecular weights was used to achieve complex stability. We also explored the design space of the system to obtain particles with high colloidal stability as well as tunable surface charge and size. Thus, the varied production parameters of starch-chitosan polyplexes (Scheme 1A) were: (i) molar ratio of carboxylate and amine functional groups (C/N ratio) of starch and chitosan, respectively; (ii) polymer concentration; and (iii) counter polymer type. The loading capacity and versatility of these simple carriers was then investigated using tobramycin and colistin as clinically relevant models of small molecule and peptide anti-infectives respectively [25,26], as well as nucleic acids (plasmid DNA). Furthermore, to improve encapsulation capacity, we coated the starch-chitosan polyplexes with an additional chitosan (Protasan) layer (Scheme 1B), and explored the loading capacity of the resulting nanoparticles. Coating the polyplexes enabled drug loading on the surface of particles, which led to a better encapsulation particularly in the case of the utilized nucleic acids. This approach also creates the further potential for formulating a multifunctional delivery system. The novel approach of starch-chitosan-based complex-coacervation suggested in this study is a straightforward and promising technique to prepare versatile carrier systems with potential in nanomedicine applications. Therefore, we undertook preliminary studies of design, synthesis, and formulation of such carrier systems, and explored their flexibility and capacity for encapsulating selected model macromolecular drugs.

A) Preparation of anionic starch-chitosan core polyplexes (anCP)

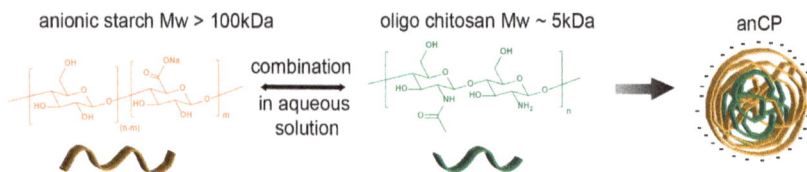

anionic starch Mw > 100kDa

oligo chitosan Mw ~ 5kDa

anCP

combination

in aqueous

solution

B) Preparation of Protasan coated CP (cCP)

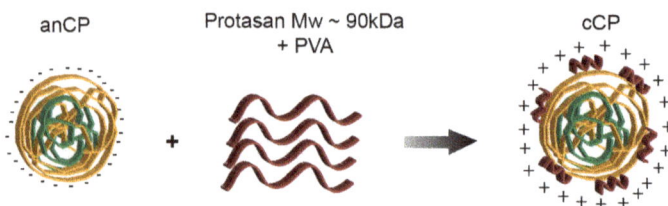

anCP

Protasan Mw ~ 90kDa

+ PVA

cCP

+

Scheme 1. Illustration of drug-free (plain) starch-chitosan polyplex-preparation.

2. Experimental Section

2.1. Materials

As raw material, partially hydrolyzed potato starch (M_w of 1300 kDa), which was a kind gift from AVEBE (Veendam, The Netherlands), was used. Selective oxidation of the primary alcohol on starch was performed to increase water solubility and obtain an anionic charge. The oxidation procedure and molecular weight fractionation of three M_w samples (5, 30–100, and >100 kDa) was conducted in accordance with the protocol of Yamada et al. [12]. The obtained starch derivatives had an oxidation degree of 45%. The M_w fraction >100 kDa is used unless stated otherwise and is termed "anionic starch" in all further descriptions.

Chitosan oligosaccharide lactate (oligochitosan; M_w 5 kDa), polyvinyl alcohol (PVA; Mowiol® 4-88), sodium hydroxide, trifluoroacetic acid (TFA), acetonitrile and acetic acid were purchased from Sigma-Aldrich (Darmstadt, Germany). Tobramycin sulfate salt and colistin sulfate salt were used as received also from Sigma-Aldrich. Ultrapure chitosan chloride salt (Protasan UP CL113; M_w ~90 kDa, deacetylation degree 75–90%) was obtained from FMC Biopolymer AS NovaMatrix (Sandvika, Norway). Purified water was produced by a Milli-Q water purification system from Merck Millipore (Darmstadt, Germany). O-Phthalaldehyde (OPA), 2-mercaptoethanol, phosphotungstic acid (PTA) and boric acid were used as purchased from Sigma-Aldrich.

Agarose SERVA for DNA Electrophoresis of research grade was bought from Serva (Heidelberg, Germany). Ethidium bromide solution (10 mg/mL), heparin sodium salt from porcine intestinal mucosa, 3-(4,5-dimethylthiazol-2-yl)-2,5-diphenyltetrazolium bromide) (MTT reagent), Triton™ X-100, dimethyl sulfoxide (DMSO) and Dulbecco's phosphate buffered saline solution (PBS) were obtained from Sigma-Aldrich. Gibco Hanks' balanced salt solution (HBSS) buffer was purchased from Thermo Fisher Scientific (Darmstadt, Germany). A549 cells (human lung carcinoma cell line, No. ACC 107) were obtained from DSMZ GmbH (Braunschweig, Germany). Cell culture medium (RPMI 1640) was purchased from PAA laboratories GmbH (Pasching, Austria) and supplemented with 10% fetal calf serum (FCS, Sigma-Aldrich). Plasmid DNA (pDNA) encoding for the fluorescent protein AmCyan was bought from Clontech Laboratories, Inc. (pAmCyan 1-N1, Mountain View, CA, USA). The plasmid was propagated in *Escherichia coli* DH5α and isolated with Qiagen EndoFree Plasmid Mega Kit (Qiagen, Hilden, Germany) to obtain pDNA of cell culture quality. jetPRIME® transfection reagent was purchased from Polyplus-transfection (Illkirch, France). Rhodamine *Ricinus communis* agglutinin I (RGA I) was obtained from Vector Laboratories. 4′,6-diamidino-2-phenylindole (DAPI) was purchased from Life Technologies (Darmstadt, Germany).

2.2. Preparation, Optimization and Characterization of Starch–Chitosan Core Polyplexes

2.2.1. Preparation and Optimization of Starch-Chitosan Core Polyplexes (CP)

Starch-chitosan core polyplexes (CP) were prepared by self-assembly of anionic starch derivatives and chitosan derivatives in aqueous medium. CP characteristics, including their: (i) surface properties; (ii) size; and (iii) physicochemical stability were varied by: (i) the molecular weight of utilized anionic starch and chitosan derivatives; (ii) polymer concentration; and (iii) molar ratio of carboxylate (COONa) to amine (NH$_2$) groups (C/N ratio) in oxidized starch and chitosan, respectively. The polyplex formulation procedure is described in Scheme 1A. Briefly, a solution of anionic starch was prepared in Milli-Q water at a defined concentration, while the utilized chitosan derivative was solubilized in 0.02 M acetic acid, followed by pH adjustment to 5.5. The assembly into CP of oxidized starch and its counter excipient occurred by the addition of an appropriate amount of starch polymer solution into the pre-warmed solution of chitosan derivative, followed by 2 min of vortexing and 1 h incubation at room temperature. To prepare anionic core polyplexes (anCP), anionic starch (M_w of >100 kDa) and oligochitosan (M_w of 5 kDa) were employed at various C/N ratios, ranging from 50:1 to 10:1 and further to 1:1, designed to optimize the formulation and stability of the polyplexes. Cationic core polyplexes (cationic CP) were prepared by co-assembly of negative starch (M_w of 30–100 kDa) and Protasan (M_w of 90 kDa) having a higher amount of positively charged amine groups. The optimal C/N ratio was identified by investigating the ratios of 1:30, 1:10 and 1:1. All samples with a solvent pH-value of 5.5 were characterized by dynamic light scattering (DLS), using a Zetasizer Nano from Malvern Instruments (UK) to obtain hydrodynamic size, polydispersity index (PDI), and using laser Doppler velocimetry to obtain ζ-potential. All samples were prepared at least in three different batches.

2.2.2. Preparation and Optimization of Protasan Coated CP (cCP)

Another approach taken to further improve the loading capacity of starch-chitosan carriers was to prepare coated polyplexes with a further layer of Protasan on anCP. The optimized coating method is described briefly as following: anCP were prepared as described and then coated with an additional layer of positively charged Protasan, by an association of amine functional groups of the chitosan and the anionic surface of the anCP (Scheme 1B). The coating solution was prepared by dissolving 3 mg of Protasan in 1 mL PVA 2% (*w/v*) solution, which was then diluted with Milli-Q water to a 1.5 mg/5 mL concentration for coating. A 500 μL volume (6.6 mg/mL) of anCP was added dropwise to the prepared Protasan solution, which was continuously stirred for 30 min at 150 rpm. This was followed by incubation at room temperature for 3 h prior to characterization. The resulting Protasan-coated anCP (cCP, *c* = 0.87 mg/mL) were kept for further studies. Samples were prepared in at least three different batches. All particle samples were characterized for their hydrodynamic size, PDI and ζ-potential. This method was also applied to investigate the physicochemical stability of anCP and cationic CP under storage conditions of 4 °C for 27 days.

2.2.3. pH-Stability of Drug-Free CP and cCP

The colloidal stability of anCP and cCP at different pH values was investigated by incubating particle suspensions at pH values of 3.5, 4.0, 4.5, 5.5, 6.0, 7.5 and 8.0, all within the physiologically-relevant range. Samples were analyzed to obtain hydrodynamic size, PDI, and ζ-potential, after predetermined incubation times (30 min, 1 h, 3 h and 24 h). The pH-value was adjusted following polyplex preparation at pH 5.5 (as described above) by using either 0.02 M acetic acid solution or 1 M NaOH solution.

All experiments were conducted in triplicates with $n = 3$, and results expressed as mean ± standard deviation (SD).

2.2.4. Morphology

The morphology of all produced polyplexes was visualized by transmission electron microscopy (TEM, JEM 2011, JEOL, St Andrews, UK). Before the TEM visualization, 8.7 µg/10 µL of polyplexes were added on a copper grid (carbon films on 400 mesh copper grids, Plano GmbH, Wetzlar, Germany) and incubated for 10 min to allow an adhesion of polyplexes to the surface. The excess was removed, and polyplexes were further stained with 0.5% (*w/v*) PTA to improve the contrast of TEM images.

2.2.5. Cytotoxicity Study: MTT Assay

A549 cells were seeded in a 96 well plate at a density of 1×10^5 cells per well, in 200 µL of RPMI cell culture medium supplemented with 10% FCS. Cells were grown for 4 days prior to the conduction of the assay to allow for approximately 95% cell confluency. On Day 4, CP and cCP samples were diluted with a suitable amount of RPMI medium (without FCS) to achieve test concentrations of 5, 10, 40, 70, 100, 200 and 500 µg/mL. Cells were then washed twice with 200 µL HBSS buffer (pH 7.4), and polyplex samples were added to cells in triplicate. Cells incubated with only RPMI medium were used as a negative control (determined to result in 100% cell viability) and cells treated with 1% Triton™ X-100 in RPMI medium were used as positive control (designated as 0% cell viability). All samples were incubated with cells for 4 h, on a horizontal shaker with careful shaking at 150 rpm at 37 °C and 5% CO_2. Subsequently, the supernatant was removed, and cells were washed once with HBSS. Then, 200 µL of the MTT-reagent (5 mg/mL) in HBSS was applied to each well and further incubated for 4 h with gentle shaking. The supernatant was then removed and DMSO was immediately added to achieve cell lysis. Cells were incubated in DMSO for 15 min under careful shaking and protected from light. The absorbance of each well at 550 nm was then measured with a plate reader (Infinite® 200 Pro, TECAN, Männedorf, Switzerland). The percentage of viable cells was calculated in comparison to negative and positive controls as described by Nafee et al. [27].

2.3. Cationic Anti-Infective Loaded anCP

2.3.1. Preparation and Optimization of Cationic Anti-Infective Loaded anCP

Isothermal Titration Calorimetry

Two relevant anti-infectives were used to test the loading capacity of anCP. Tobramycin was used as an example of a cationic small molecule antibiotic having a molecular weight of 467.5 Da, and colistin (polymyxin E) was used as an example of a peptide antibiotic with a molecular weight of 1267.5 Da (Scheme 2A).

Interaction between anionic starch and the cationic anti-infectives tobramycin and colistin was investigated by isothermal titration calorimetry (ITC) using a NanoITC 2G (TA Instruments, New Castle, DE, USA). The purpose of such measurement was to optimize excipient to cargo ratio in drug loaded carrier production. Briefly, all drug and anionic starch solutions were prepared in milli-Q water. A 25 mM solution of tobramycin or colistin was prepared in a 250 µL syringe and used to saturate 1.5 mL of anionic starch at a concentration of 0.1 mM filled in the sample cell. Following an initial delay of 300 s, 250 µL of drug solution was repeatedly injected into the sample cell with a spacing of 500 s between injections, and at a reference power of 10 µCal/s. The final thermogram and thermodynamic parameters were produced by subtracting the heat of dilution of either tobramycin or colistin (25 mM in 1.5 mL milli-Q water), followed by fitting using the One Set of Sites model in the data analysis software NanoAnalyze. The free energy of binding (ΔG) was calculated using the equation $\Delta G = \Delta H - T\Delta S$, where ΔH is the enthalpy change, T is temperature (Kelvin), and ΔS is the change in entropy. All measurements were performed at 25 °C.

Preparation and Optimization of Cationic Anti-Infective Loaded anCP

Both tobramycin and colistin were loaded using the same procedure, during formation of anCP, employing various C/N ratios, as follows: (i) 1 mg tobramycin or 3 mg colistin was incubated with an appropriate amount of anionic starch solution for 2 h; and (ii) pre-warmed chitosan solution at pH 5.5 was added, and coacervation was achieved by vortex mixing (2 min).

The anti-infective loaded anCP suspension was then centrifuged at 13,000× *g* and 4 °C for 20 min at least twice and allowed to equilibrate at 4 °C overnight before conducting further experiments. In all experiments the supernatant produced by centrifugation was collected for drug loading quantification.

Loading Quantification

The degree of anti-infective loading in anCP was determined using an indirect quantification method (drug amount inside anCP = initial drug amount − drug amount in the supernatant). Colistin was quantified by high-performance liquid chromatography (HPLC), while tobramycin was quantified based on a protocol for detection of aminoglycosides [28], as detailed below.

HPLC Analysis

The HPLC analysis was performed on a Dionex UltiMate 3000 system (Thermo-Fischer Scientific, Dreieich, Germany) equipped with LPG-3400 SD pump, WPS-3000 autosampler, DAD3000 detector, and TCC-3000 column oven. Chromeleon software (Chromeleon 6.80 SP2 build 9.68, Thermo Scientific Dionex, Dreieich, Germany) was used for data analysis. A column set of LiChrospher® 100 RP-18 (5 µm) LiChroCART® 125-4, consisting of a 125 mm × 4 mm LiChrospher 100/RP-18 column (Merck-Hitachi, Darmstadt, Germany) with a LiChrospher 100/RP-18 guard column (5 µm) (Merck-Hitachi, Darmstadt, Germany) at 30 °C was used as stationary phase for all substances. A gradient method was used starting with 20% A, increasing to 50% A within 2 min, and holding for 1.5 min (A = acetonitrile, B = 0.1% TFA solution in water). Before injection, the samples were filtered through a cellulose acetate 0.2 µm membrane. The flow rate was 1.0 mL/min, and the injection volume was 50 µL. A calibration curve was constructed using eight different concentrations of colistin in water, ranging from 0.2 mg/mL to 0.005 mg/mL ($r^2 = 0.9955$). All 8 standards were measured 5 times, and a percent relative standard deviation (% RSD) of less than 3.9% was calculated. The run time was 6 min, and a retention time of 3.6 min and 3.9 min was observed for colistin A and colistin B, respectively. As colistin is a mixture of two main fractions, colistin A and colistin B, both were quantified to determine colistin loading. The detection wavelength was 210 nm for colistin A and 214 nm for colistin B.

Aminoglycoside Detection Protocol

The product fluorescence of tobramycin reacted with a fluorescent reagent was measured at 344/450 nm (Ex/Em) using a Tecan microplate reader following a published method [28]. To prepare the reagent solution, a 0.2 g amount of OPA reagent was dissolved in a mixture of 1 mL methanol, 19 mL boric acid 0.4 M at pH 10.4, and 0.4 mL of 14.3 M 2-mercaptoethanol. A 2 mL of the resulting mixture was then diluted with 16 mL methanol before use. A calibration curve was constructed using five different concentrations of tobramycin in water (0.04–0.005 mg/mL, $r^2 = 0.9976$).

In both cases, the encapsulation efficiency (EE) and the drug loading rate (LR) were calculated according to the following equations:

$$EE = \frac{\text{Weight of encapsulated drug in nanoparticles}}{\text{Initial amount of drug in the system}} \times 100$$
$$LR = \frac{\text{Weight of drug in nanoparticles}}{\text{Weight of nanoparticles}} \times 100$$

(1)

where "weight of nanoparticles" was calculated as weight of polymeric material + weight of encapsulated drug in nanoparticles.

Each sample was assayed at least in triplicate, and results are reported as the mean $\pm$ SD.

Drug Release Study

Tobramycin or colistin release profiles from tobramycin loaded anCP or colistin loaded anCP was performed in PBS (pH 7.4) at 37 °C. Briefly, either tobramycin loaded anCP or colistin loaded anCP was diluted in PBS to have final tobramycin or colistin concentration at 10% (*w/w*) and loaded into dialysis membrane (MWCO 1 kDa, Spectrum Labs, Rancho Domiguez, CA, USA) in the case of tobramycin, or dialysis membrane (MWCO 3.5–5 kDa, Spectrum Labs, USA) in the case of colistin. After that, the whole system was put into 20 mL PBS and placed on a shaker at 400 rpm at 37 °C. The concentration of released drug was analyzed by collecting samples from the supernatant during the period from 1 h to 24 h. The amount of colistin and tobramycin were determined by HPLC and aminoglycoside detection protocol, respectively. The volume was kept constant by refilling with an identical volume of PBS. The cumulative released drug (%) was calculated (mean $\pm$ SD of $n = 3$). Three independent experiments were conducted in triplicates, and results expressed as the mean $\pm$ standard deviation (SD).

2.3.2. Minimum Inhibitory Concentration (MIC) Assay

The antimicrobial properties of anCP, anti-infective loaded anCP, and free drugs were performed by standard microbroth dilution assays with *Escherichia coli* (DH5α) and *Pseudomonas aeruginosa* (PA14) in 96 well plates. A suspension of *E. coli* or *P. aeruginosa* prepared from mid log cultures in Mueller-Hinton broth or Lysogeny Broth medium (at 25 °C) was first diluted to OD_{600} (absorption at 600 nm) 0.01, which corresponds to approximately 5×10^6 CFU/mL (CFU, colony-forming units). Polyplex samples (anCP, drug-loaded anCP), free drug solution and PBS as control were then added to bacteria-containing wells by serial dilution over a range of 0.03–64 µg/mL. After incubation for 16 h at 37 °C, inhibitory concentration (IC) IC_{90} values were determined by sigmoidal curve fitting of absorption values (600 nm) that were measured on a Tecan microplate reader. The experiments were conducted in duplicate.

2.4. Preparation of pDNA Loaded cCP

Plasmid DNA pAmCyan was incorporated into the polyplexes to evaluate the potential of the carrier system with respect to nucleic acid actives. A ratio of amine groups (chitosan) to phosphate groups (pDNA) of 20/1 was chosen and is referred to as N/P ratio. The preparation was performed in three steps: first, an appropriate amount of pAmCyan was added to a solution of anionic starch and mixed thoroughly. A 1 mL volume of this pAmCyan-starch solution was added to 1 mL of oligochitosan solution (650 µg/mL) and mixed immediately by vortex for 15 s. A further incubation for 1 h at room temperature was then carried out, leading to the formation of pAmCyan-loaded anCP. In the second step, the pAmCyan loaded anCP were coated by Protasan as described in Section 2.2, to form pAmCyan-loaded cCP. In the third step, a further layer of pAmCyan was applied to pAmCyan-loaded cCP (1:30 *w/w*) resulting in pAmCyan double loaded cCP (Scheme 2B). The pDNA encapsulation efficiency of each step was analyzed by pelleting the samples down and measuring the absorbance of unbound pDNA (at 260/280 nm with NanoDrop Spectrophotmeter) remaining in the supernatant after centrifugation for 30 min at 24,400× *g*. Thus, the amount of bound pDNA was examined indirectly. The products of each step were characterized to obtain hydrodynamic size, PDI, and ζ-potential, and their morphology was observed by TEM.

2.4.1. Determination the Complexation of pAmCyan in Starch-Chitosan Polyplexes

Complexation and stability of pAmCyan in starch-chitosan polyplexes was evaluated by a gel retardation assay using agarose gel electrophoresis. Further, to facilitate DNA fragmentation, the endonuclease BamHI was used, which linearizes the plasmid, and heparin addition to cause the release of pDNA from the complex. Polyplexes containing 500 ng of pDNA per sample from each step of the formulation process were first digested with 0.5 µL BamHI for 2 h at 37 °C with shaking. Afterward, 3 µL (30 mg/mL) heparin was added to solutions of digested polyplexes, incubated for 15 min at room temperature and then mixed with 2 µL of orange DNA loading dye (6×; Thermo Fisher Scientific, Waltham, MA, USA). These mixtures were then loaded into 0.75% (*w/v*) agarose gel containing 5 µL of ethidium bromide and run for 60 min at 50 V in 0.5× TBE-buffer. The visualization of the bands was performed with a UV illuminator, Fusion FX7 imaging system from Peqlab (Erlangen, Germany).

2.4.2. In Vitro Transfection Studies in A549 Cells

To test the efficiency of the pAmCyan loaded polyplexes, in vitro transfection studies were performed in A549 cells. Briefly, A549 cells were seeded in 24-well plates, at a density of 25×10^4 cells per well in 500 µL of RPMI cell culture medium with 10% FCS. Cells were grown for 2 days to reach a cell confluency of 60–70%. Polyplexes of the pAmCyan double loaded carrier system (see Section 2.4) containing 1 µg of pAmCyan (polyplex concentration ~60 µg/mL) were prepared with a ratio of 1:30, 1:50 and 1:100 between pDNA:polyplexes in 500 µL of HBSS buffer. Then, cells were washed twice with HBSS buffer and incubated with the polyplexes for 6 h. After 6 h of incubation, polyplexes were removed and replaced with RPMI containing 10% FCS. Cells were further grown for 2, 3 and 4 days to identify the time point of maximum reporter gene expression. For comparison, the commercially available transfection reagent jetPRIME® was used as positive control. Cells treated with pAmCyan-free cCP and cell culture medium alone were used as negative controls. For confocal laser scanning microscope (CLSM; Leica TCS SP 8, Leica, Wetzlar, Germany) visualization, cell membranes were stained using RGA I (15 µg/mL), and cell nuclei were stained with DAPI (0.1 µg/mL). Samples were then fixed with 3% paraformaldehyde and stored at 4 °C until analysis. All images were acquired using a 25× water immersion objective at 1024 × 1024 resolution and further processed with LAS X software (LAS X 1.8.013370, Leica Microsystems, Leica, Germany). The percentage efficiency of transfected cells was quantified using flow cytometry (BD LSRFortessa™ Cell Analyzer, Biosciences, Heidelberg, Germany). Fifty thousand cells per sample were counted by the cytometer and data were analyzed using FlowJo software (FlowJo 7.6.5, FlowJo LLC, Ashland, OR, USA). Three independent experiments were performed in triplicates, and results expressed as the mean ± standard deviation (SD).

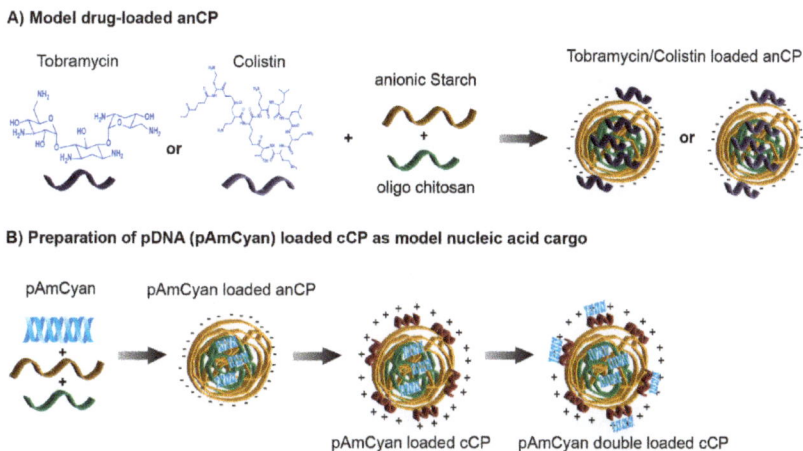

Scheme 2. Illustration of starch-chitosan polyplex-preparation for drug-loaded polyplexes.

3. Results and Discussion

3.1. Preparation and Characterization of Drug-Free Starch-Chitosan Polyplexes

This study represents an extension in comparison to the particle preparation approach of Barthold et al. [29], in which the large poly dispersity index modified starch was employed for colloidal formation. Furthermore, the chemical modification in that reported study, which was used to produce cationic starch derivative, resulted in unfavorably additional synthesis step. Although the particle preparation was well established, the lack of cationic strength due to an obviously low converting yield of cationic starch synthesis limited the carrier capacity for anionic net charge actives of such system. Thus, we used fractionally modified starch derivatives to have better control of colloidal stability, and different molecular weight chitosan derivatives as strong counter excipient for the polyplexes produce. Both excipients are polysaccharides and therefore have favorable characteristics with respect to biological safety, biocompatibility and biodegradability. The simple production of polyplexes using these excipients has the perspective to be readily up-scaled. In the first series of preparations, we studied the plain polymeric complexes by combining both excipients in aqueous solution, with the electrostatic interaction between opposite charges of the individual polymers resulting in polyplex self-assembly. During the optimization of this process, various combinations of types of polymers, C/N ratio, and initial polymer solution concentration were investigated to find a stable and narrow size distribution of the produced colloidal structures (details of the optimization can be found in the Supplementary Materials, Tables S1 and S2). The best of several stable polyplex formulations was produced using a C/N ratio of 10:1, utilizing anionic starch and oligochitosan. Starch-chitosan polyplexes were obtained with an anionic surface charge evidenced by a ζ-potential of around -30 mV. The size of polyplexes could be varied from 150 nm to 350 nm by changing of polymer concentration, with a narrow PDI (<0.3) in all cases. The impact of polymer concentration on polyplex size was expected and already described for comparable systems [30,31]. Spherical polyplex morphology was visualized using TEM (Figure 1A).

Drug-free anCP   Drug-free cCP

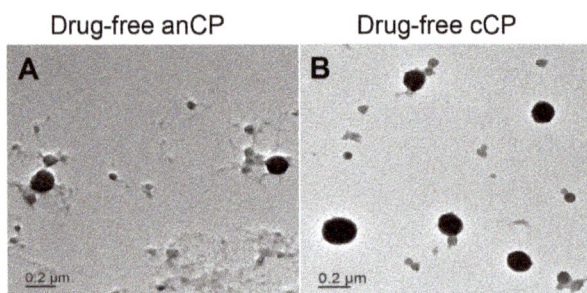

Figure 1. Transmission electron microscope (TEM) images of drug-free starch-chitosan polyplexes stained by 0.5% phosphotungstic acid solution: (**A**) drug-free anCP; and (**B**) drug-free cCP.

Reversing the C/N ratio to 1:10, and using starch (M_w 30–100 kDa) and Protasan (M_w ~90 kDa) resulted in a switch of the surface charge from anionic to cationic (further termed as cationic CP), with a ζ-potential of around $+40$ mV. The size of particles varied from 214.3 nm to approximately 400 nm depending on the polymer concentration and C/N ratio (Supplementary Materials, Tables S1 and S2). As both anCP and cationic CP systems formed as a result of attractive forces of polymer functional groups, further aggregation of systems over time may potentially occur; the physical stability of the polyplexes was therefore studied over a time course with storage at 4 °C. The colloidal characteristics of both, anCP and cationic CP, remained stable for 27 days with a PDI of ~0.18 and a ζ-potential of -30 mV and $+35$ mV for anCP and cationic CP, respectively (Supplementary Materials,

Figure S1). Consequently, the utilized preparation process represents a straightforward approach for the formulation of versatile nanoparticles.

The possibility to control the surface charge of a nanocarrier is advantageous for both improving the interaction with the drug to be encapsulated as well as in a later stage the interaction with the target cell [32]. Therefore, the ability to tune surface charge by changing the C/N ratio and molecular weight of starch and chitosan derivatives is a distinct advantage of this novel type of carrier. The ability to load drug molecules of differing structure size and charge, such as e.g., low-M_w anti-infectives as well as high-M_w plasmid DNA, into these carriers was then investigated. Furthermore, a simple coating process was employed to minimize the use of cationic polymer, while still allowing for positive surface charge tuning of particles. The anCP were coated with an additional layer of Protasan (M_w of 90 kDa) resulting in cationic coated polyplexes, cCP. The organic solvent-free procedure was performed in aqueous solution in the presence of PVA as a stabilizer, and led to stable cationic particles with a ζ-potential of +27.1 mV, and a spherical morphology (Figure 1B). The hydrodynamic size and PDI decreased in comparison to anCP (Table 1, Supplementary Materials Table S3) due to the improved electrostatic interaction between the excipients. Furthermore, the anionic, cationic and coated polyplexes overall indicated ζ-potential values of around ±30 mV at which the value ensures improved colloidal stability [33–35], giving the polyplexes the possibility to survive and overcome various biological barriers and reach a specific site of interest.

Table 1. Summary of characteristics of representative drug-free (plain) polyplexes. All measurements were conducted in triplicates. $n = 3$, mean ± SD.

Polyplexes	Size (nm)	PDI	ζ-potential (mV)
Drug-free anCP	287.9 ± 5.0	0.22 ± 0.01	−29.7 ± 0.4
Drug-free cCP	205.4 ± 3.9	0.14 ± 0.02	27.1 ± 1.0

3.1.1. Colloidal Stability of Drug-Free anCP and cCP

To explore the potential to administer anCP and cCP by various routes, the physicochemical stability of these systems was investigated at pH values ranging from 3.5 to 8.0, as relevant for various drug administration pathways. The stability of the polyplexes in different conditions of pH was investigated following 30 min, 1 h, 3 h and 24 h of incubation. As the assembly of the polysaccharide nanoparticulate systems was based on electrostatic interaction, stability of such systems mainly depends on its surface properties which are, in turn, influenced by surrounding environmental factors, e.g., ionic strength and pH values [35,36]. anCP showed stable characteristics regarding size, PDI and ζ-potential, even at the lowest investigated pH value of 3.5 after 3 h incubation (Supplementary Materials, Figure S2). In agreement with the results of Yamada et al. [12], the relatively high M_w (>100 kDa) of the anionic starch clearly aids in stabilization of the particles.

However, the possible dissociation of carboxylate groups on particle surfaces may have eventually led to colloidal aggregation [36] at pH 3.5 and hence destabilized the polyplexes, as indicated by stability data after 24 h of incubation. By contrast the anCP remained stable at all other, higher, pH values after a 24 h incubation (Figure 2), which could be explained by an enhanced repulsive force among anionic particles due to increasing deprotonation of surface carboxylate groups at high pH values. The stability test performed on cCP revealed a stable particle size and PDI at all pH values after 24 h, however a reduction in cCP ζ-potential was seen from pH 3.5 to 8.0 (Figure 2). This behavior is explained by protonation of chitosan molecules, which, being a weak polyelectrolyte with a pK_a of approximately 6.5, has a changing protonation degree depending on the pH of the surrounding solution [37]. An increase in pH value up to 8.0 resulted in a diminishing protonation degree of the chitosan polymer [37,38], thereby resulting in a decrease ζ-potential. Nevertheless, a continued stability of cCP at all tested pH values, especially at pH 7.5 and 8, which are higher than the chitosan pK_a value, could be conferred by the presence of PVA, as a stabilizer that interrupts colloid interaction and aggregation. The stability of both anCP and cCP over a broad range of pH values clearly indicated flexibility in the potential application of such a tunable carrier system, for drug delivery via various routes of administration. The system could be considered for use in pulmonary delivery, where the local pH is nearly neutral; for gastrointestinal and vaginal delivery, where a low pH environment is encountered [39–41]; and in other applications.

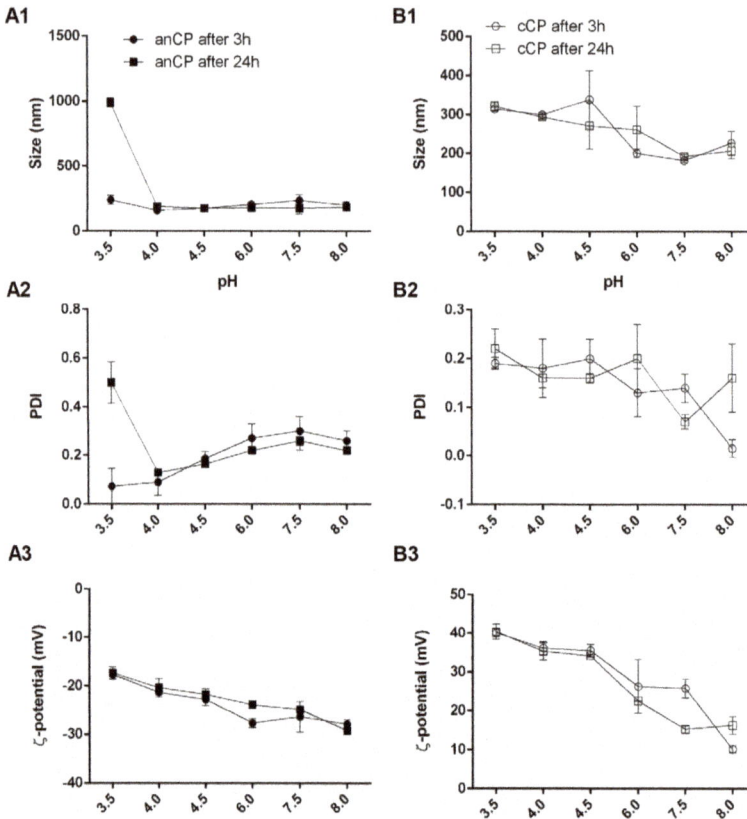

Figure 2. Characteristics of anionic core polyplexes (anCP, **A1**, **A2** and **A3**) and Protasan-coated core polyplexes (cCP, **B1**, **B2** and **B3**) after 3 h and 24 h incubation in pH conditions ranging from 3.5 to 8.0. The pH-values changed accordingly starting from an initial pH-value of the samples of 5.5. $n = 3$, mean ± SD.

3.1.2. Cytotoxicity Assessment

A549 cells were used in our study as a model cell line to test the potential of our carrier system. Figure 3 shows the viability of A549 cells exposed to anCP and cCP with concentrations up to 500 µg/mL, with the light grey area marking the concentration used for later MIC assays and the dark grey showing the concentration employed in subsequent transfection studies. The anCP demonstrated almost no cytotoxicity over the tested concentration range, with an observed cell viability of nearly 100% at all concentrations. However, in contrast, cell viability decreased markedly following treatment with increasing concentrations of cCP. This may be due to their cationic surface charge [42], which, on the other hand, could potentially facilitate a higher cellular uptake of cCP [43], as is particularly relevant for pDNA delivery applications. Ultimately, the cationic surface charge of such pDNA polyplex must be carefully tuned towards an acceptable compromise between transfection efficacy and biocompatibility.

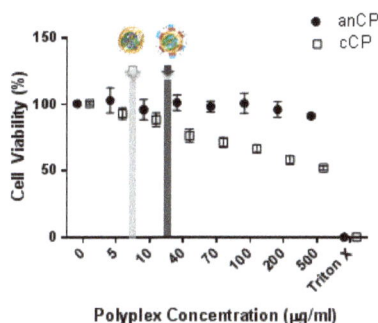

Figure 3. Cell viability assayed by MTT after 4 h incubation (mean ± SD, *n* = 9 from three independent experiments).

3.2. Loading of anCP with Low-Mw Anti-Infectives

3.2.1. Optimization of the Preparation Process and Drug-Loading

To explore the potential of anCP as a carrier for diverse drug cargos, the aminoglycoside tobramycin (M_w = 467.5 Da), and the oligopeptide colistin (M_w = 1267.5 Da) were chosen as low molecular weight cargos. Tobramycin and colistin are active against Gram negative bacteria, and are two of the four drugs specifically approved in Europe for application as inhaled therapies for chronic bronchopulmonary *P. aeruginosa* infection in cystic fibrosis patients [25,26,44]. Fast elimination and poor permeability however often limit the delivery of hydrophilic anti-infectives, such as tobramycin and colistin, requiring frequent and high dosing with the risk of adverse drug effects and the development of bacterial resistance. Approaches to encapsulate these essential anti-infectives within drug carrier systems to avoid such delivery problems and preserve their activity have therefore been described [30,45]. Tobramycin and colistin both have net positive charges at a neutral pH value due to the presence of amine functional groups in their structures. Thus, it was hypothesized that their properties would be conducive to incorporation into anionic starch-based particles. Consequently, the interaction of further applied chitosan molecules and anionic starch would be affected, which would eventually lead to unstable colloids and aggregation of the resulting system. Therefore, before coacervation, the potential binding of anti-infective molecules to oxidized starch polymer (M_w of >100 kDa), was investigated by isothermal titration calorimetry (ITC), which revealed the thermodynamics of the binding and helped to estimate the optimal drug amount for loading in anCP. Tobramycin and colistin respectively were injected as aqueous solutions to saturate an anionic starch solution, as shown in Figure 4. Values in the inset tables were calculated by the software NanoAnalyze, yielding the same Gibbs free energy (Δ*G*) for the interaction of around −17.12 kJ/mol for both tobramycin and colistin respectively with the anionic starch polymer. Moreover, based on the thermograms from the ITC analysis, the amount of tobramycin or colistin needed to completely saturate the anionic starch polymer is known. To completely saturate the fixed amount of anionic starch (e.g., 5 mg), there is a need of 1 mg tobramycin, while the needed amount of colistin is 3 mg. The interaction between drug molecule-anionic starch, as well as the number of amine groups on each drug molecule are similar; their molecular weight, however, are nearly three times different. Thus, the amount of the used colislin was three times higher than that of tobramycin. With the aforementioned optimization, the amounts of drugs were selected and for further investigation of drug-loaded anCP.

Having illustrated a clear interaction of tobramycin and colistin with anionic starch, preparation of drug-loaded anCP using chitosan as a counter polymer was investigated. Anionic starch and the selected anti-infective were first incubated, followed by the addition of an appropriate amount of pre-warmed chitosan solution, leading to the formation of polyplexes by self-assembly of these polyelectrolytes. A comparable particle preparing procedure was described by Deacon et al. for tobramycin and alginate [30]. The C/N ratio and the initial concentrations of the three components (anionic starch, chitosan, and tobramycin or colistin) were varied, and the characteristics of the resulting polyplexes were investigated in order to achieve an optimal formulation. The results of this optimization work are highlighted in Table 2, with additional data shown in the Supplementary Materials (Tables S4 and S5).

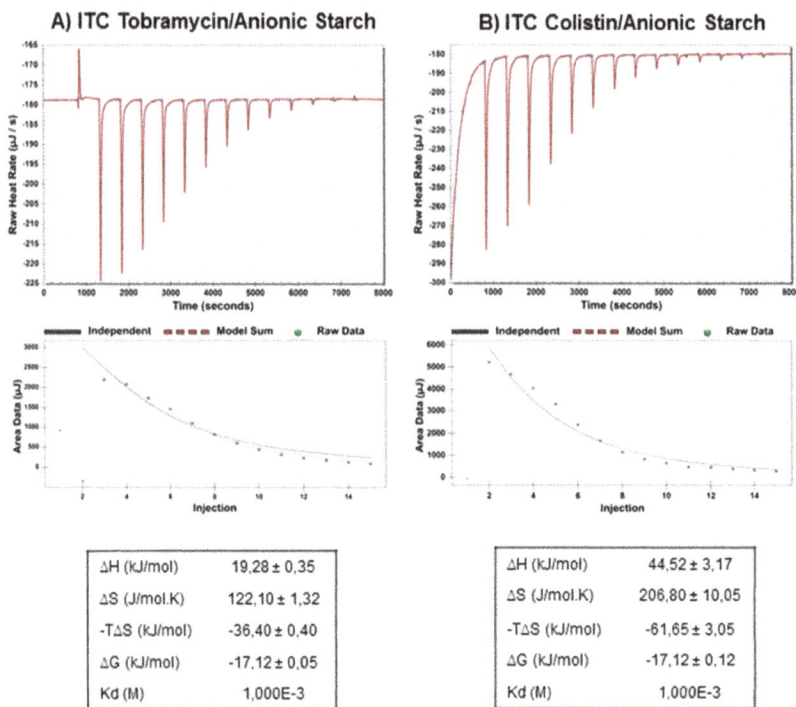

A) ITC Tobramycin/Anionic Starch		B) ITC Colistin/Anionic Starch	
ΔH (kJ/mol)	19,28 ± 0,35	ΔH (kJ/mol)	44,52 ± 3,17
ΔS (J/mol.K)	122,10 ± 1,32	ΔS (J/mol.K)	206,80 ± 10,05
-TΔS (kJ/mol)	-36,40 ± 0,40	-TΔS (kJ/mol)	-61,65 ± 3,05
ΔG (kJ/mol)	-17,12 ± 0,05	ΔG (kJ/mol)	-17,12 ± 0,12
Kd (M)	1,000E-3	Kd (M)	1,000E-3

Figure 4. Isothermal titration calorimetry (ITC): (**A**) titration of tobramycin (25 mM) into anionic starch derivative (M_W of >100 kDa) (0.1 mM); and (**B**) titration of colistin (25 mM) into anionic starch derivative (M_W of >100 kDa) (0.1 mM).

The loading capacity of tobramycin and colistin in anCP was then evaluated. Encapsulation of these molecules into anCP was based on the association of anionic ions of oxidized starch and cationic ions of drug molecules. Hence, by using a fixed amount of drug molecules, and varying the C/N ratio (by varying the amount of added chitosan derivative) as well as the initial concentration of polymer solution, stable drug-loaded colloids could be produced. As shown in Table S4, tobramycin-loaded anCP ranging in size from 165.8 ± 0.8 nm to 375.9 ± 1.8 nm with a homogenous distribution (PDI < 0.3) were formed. The ζ-potential of tobramycin-loaded anCP generally increased from nearly −30 mV to average −17 mV, which suggested the presence of cationic drug molecules not only within the polyplex matrix, but also on the surface of polyplexes. An increasing C/N ratio also resulted in a tendency for decreasing particle size from 375.9 ± 1.8 nm to 175.2 ± 2.8 nm. This decrease in size could be due to a condensing effect when using a higher amount of starch, which introduced an excess of available anionic ions for interaction with chitosan, even after incubation with tobramycin. As a result, the colloidal characteristics of tobramycin-loaded nanoparticles were not significantly different to those of unloaded systems. To obtain colistin-loaded anCP, an amount of colistin three times higher in comparison to tobramycin was employed for pre-incubation with anionic starch, due to the molecular weight difference between the two drugs. Colistin-loaded anCP were prepared again using varying polymer concentrations. As the final concentration decreased, while C/N ratio was maintained at 40/1, the particle size decreased from 324.4 ± 3.6 nm to 266.3 ± 6.5 nm (Table S5). The colistin loaded systems were also stable and homogenous with PDI values lower than 0.3. The results, therefore, clearly show that a reduction in polyplex size resulted from a decrease in employed polymer concentration; this is also in accordance with observations in previous studies [31]. Overall, there was an increase in the ζ-potential of drug loaded carriers as compared to unloaded, which is evidence for the presence of positive net charge anti-infective molecules on carrier surfaces. The size of colistin-loaded anCP was generally larger than the corresponding tobramycin-loaded anCP, which could be explained by the possible formation of colistin micelles during incubation with starch solution. This is made possible by the amphiphilic molecular structure of colistin, which possesses a lipophilic fatty acyl tail and a hydrophilic head group [46]. Consequently, the addition of chitosan supported the colloidal stability of the polyplex system. The morphology of anti-infective loaded anCP was spherical as investigated by TEM (Figure 5A,B). The encapsulation efficiency (EE) and loading rate (LR) of colistin- and tobramycin-loaded anCP are highlighted in Table 2 and Tables S6 and S7. The EE and LR values were indirectly calculated by collecting supernatants after two washing steps. As determined using HPLC, colistin encapsulated within anCP showed maximum values of 96.57 ± 0.19% and 22.70 ± 0.33% for EE and LR, respectively. Incorporation of tobramycin, determined by product fluorescence at 344/450 nm, also showed an EE higher than 98% in all cases, but comparatively lower LR values (2.9 ± 0.0% maximum). The high EE of both model drugs (>90% in all cases) was a result of pre-determination of the interaction between drug molecules and anionic starch, which allowed estimating the amount of used drug in encapsulation and thereby maximization of the encapsulation efficiency. The LR of tobramycin-loaded anCP showed a rational loading capacity for polymeric nanoparticles with a size of approximately 200 nm, while the LR of colistin-loaded anCP was surprisingly high. This might be due to the aforementioned micelle formation of colistin molecules, stabilized by the starch polymer solution. Hence, colistin could be localized in the core of nanoparticles, covered by starch polymer molecules, and could also be loaded on the surface of the system due to charge interaction. The results clearly demonstrate the capacity of the anCP carrier system to be loaded with either type of low-M_w anti-infectives.

Table 2. Summary of characteristics of drug-loaded anCP, %EE = encapsulation efficiency and %LR = loading rate. All measurements were conducted in triplicates. n = 3, mean ± SD.

Polyplexes	Size (nm)	PDI	ζ-potential (mV)	%EE	%LR
Tobramycin loaded anCP	175.2 ± 2.8	0.18 ± 0.00	−16.8 ± 1.0	98.7 ± 0.1	2.9 ± 0.0
Colistin loaded anCP	266.3 ± 6.5	0.27 ± 0.01	−14.6 ± 0.5	96.6 ± 0.2	17.2 ± 0.1
pAmCyan loaded anCP	271.8 ± 2.4	0.25 ± 0.01	−29.8 ± 0.6	76.6 ± 0.6	0.3 ± 0.002
pAmCyan loaded cCP	214.0 ± 3.5	0.17 ± 0.01	28.0 ± 0.6	67.7 ± 14.1	0.2 ± 0.036
pAmCyan double loaded cCP	204.6 ± 3.5	0.16 ± 0.02	25.5 ± 0.6	93.9 ± 4.5	3.3 ± 0.150

Furthermore, the cumulative release profile of both tobramycin and colistin from drug-loaded anCP were studied in PBS at 37 °C, the results are shown in Figure S3. Clearly, the controlled release of anti-infective in PBS could be observed in both cases, with over 40% and 20% of drug released over the period 16–24 h for tobramycin and colistin, respectively. The initial burst after 4–6 h incubation was recorded as on average nearly 30% for tobramycin and 20% for colistin. The percentage of initial anti-infective released from the anCP would represent the amount of drug molecule loaded on the particles surface. Interestingly, the release of colistin at all time points are relatively lower than that of tobramycin, which would again be explained by the aforementioned micelle formation of colistin molecules that are stabilized and maybe then embedded inside the polymeric polyplex. The release results would help predict the drug carriers behavior in further in vitro experiments. To evaluate the release of the anti-infectives from drug-loaded anCP, and have better insight into the controlled release in vitro or in vivo, in which other components exist, e.g., bacteria, would require more complex biologically simulated tests that were beyond the scope of the present study.

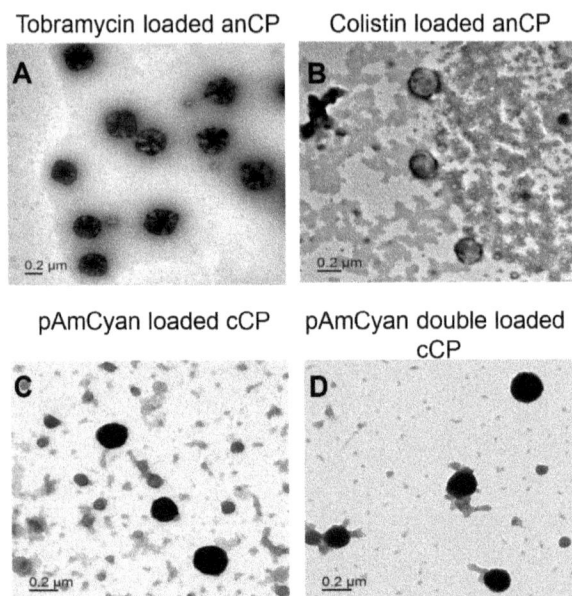

Figure 5. Transmission electron microscope (TEM) images of drug-loaded starch-chitosan polyplexes stained by 0.5% phosphotungstic acid solution: (**A**) tobramycin loaded anCP; (**B**) colistin loaded anCP; (**C**) pAmCyan loaded cCP; and (**D**) pAmCyan double loaded cCP.

3.2.2. Efficacy of Anti-Infective Loaded anCP

While anCP have the capacity to load different types of anti-infectives including small molecule and peptide drugs, it is important that the particle excipients do not interfere with action of active agents which might confound the further evaluation of drug delivery systems. Hence, the anti-microbial activity of blank anCP and anti-infective loaded anCP were studied against *E. coli* and *P. aeruginosa* in comparison to the use of free drugs. As shown in Table 3, the antibacterial activity of drug-loaded anCP was relatively similar to that of the corresponding free drug. MIC values obtained show that blank anCP were not active against *E. coli* and *P. aeruginosa* at the highest tested concentrations, which means the formation of polyplexes with anCP did not compromise the intrinsic anti-microbial efficiency of either antibiotic. To demonstrate a superior safety and efficiency profile of such nanocarriers in comparison to the free drug would require some more complex biological test systems that were beyond the scope of the present study.

Table 3. MIC assay results against *E. coli* and *P. aeruginosa*.

Samples	IC90 against *E. coli* (µg/mL)	IC90 against *P. aeruginosa* (µg/mL)
Tobramycin	0.2–0.3	1.56
Tobramycin loaded anCP	0.2–0.3 *	1.56 *
Colistin	0.4–0.5	3.125
Colistin loaded anCP	0.5 *	3.125–6.25 *
anCP	>64	>64
PBS buffer	no inhibition	no inhibition

* Drug content in anCP.

3.3. Loading of anCP with High-Mw pDNA

A model plasmid DNA encoding a fluorescent dye (pAmCyan) was further incorporated into the carrier system in a three-step procedure (core formation, Protasan coating and pDNA complexation), to demonstrate the ability of the polyplexes to deliver a broad spectrum of cargos. The three-step procedure also lead to an increase of nucleic acid encapsulation within the polyplexes, protecting nucleic acids from enzymatic degradation. Produced pAmCyan loaded polyplexes were again found to have a spherical structure (Figure 5C,D). The physiochemical characteristics of all intermediate and final polyplexes in this stepwise production can be found in Table 2. Each subsequent step in the preparation procedure results in a denser complexation, with the pAmCyan double loaded cCP showing the smallest size and most narrow size distribution (lowest PDI value). Furthermore, the ζ-potential was observed to switch from negative to positive after coating with Protasan, with a further slight decrease after complexation with negatively charged pAmCyan. The additional complexation with pAmCyan resulted in a 15% higher encapsulation efficiency in comparison to the intermediate step 2 (Table 2). Additionally, agarose gel electrophoresis (Figure 6, left) elucidates that no pDNA could run through the gel, which indicates that pDNA is strongly complexed within the polyplexes. Only further treatment with BamHI and heparin causes pDNA release as seen through the bands (Figure 6, right). Furthermore, pDNA loaded anCP and pDNA associated on the surface of polyplexes (pDNA double loaded cCP) allow an easier intercalation of EtBr and faster release with heparin, whereas pDNA loaded cCP is densely packed impeding pDNA release as no free pDNA bands can be observed in the gel.

1. Marker	5. pDNA loaded anCP	8. pDNA + BamHI
2. Naked pDNA	6. pDNA loaded cCP	9. pDNA loaded anCP + BamHI + Heparin
3. Plain anCP	7. pDNA double loaded cCP	10. pDNA loaded cCP + BamHI + Heparin
4. Plain cCP		11. pDNA double loaded cCP + BamHI + Heparin

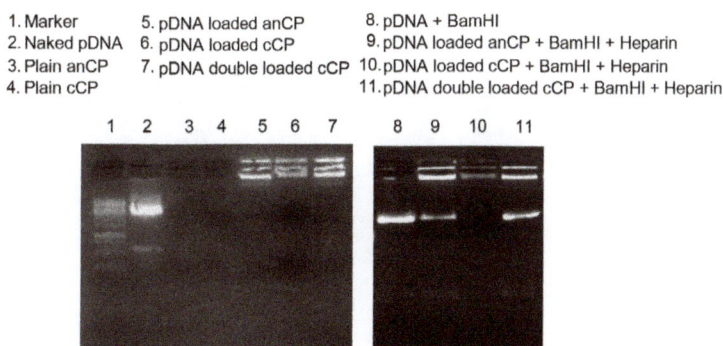

Figure 6. Gel retardation assay using agarose gel electrophoresis of plain and pDNA (pAmCyan) incorporated polyplexes for all three preparation steps in comparison with naked pDNA (undigested pDNA) and digested pDNA (pDNA + BamHI).

Potential of Polyplexes for pDNA Delivery

Using nanoparticles as a non-viral delivery system for gene therapy represents a significant challenge, as nanocarriers need to cross several biological barriers while preserving the functionality of carried pDNA. pDNA condensed inside the nanocarriers must survive the acidic conditions inside the lysosomes and escape the lysosomal compartment in order to cross the nuclear membrane [43]. Current knowledge of polymeric transfection systems suggests that a good pH-buffering capacity (a process known as the "proton sponge effect") [47] is an important factor in the achievement of endosomal escape. Here, the potential of starch–chitosan polyplexes for nucleic acid delivery was explored by in vitro transfection studies using A549 cells. Three different ratios between pDNA:polyplexes have been studied to investigate the best transfection rate. While 1:50 and 1:100 show no significant transfection (data not shown), 1:30 mediated successful transfection, with the highest reporter gene expression observed after 48 h with 5% of transfected cells. In comparison, jetPRIME® as positive control had a higher transfection efficiency (45%) after 48 h, which rapidly decreased to 30% after 72 h and to 25% after 96 h (Figure 7). The comparatively lower transfection efficiency of the polyplexes may be due to a high stability of condensed pDNA, leading to an incomplete release of pDNA inside the cytoplasmic compartment [48,49]. Further improvement of the transfection efficiency would presumably be achievable by addition of endosomal escape moieties [50,51], or with chitosan derivatives (e.g., trimethylation or amino acid conjugation) [52,53]. However, such efficacy improvements often impact the biocompatibility. Thus, optimization between safety and efficacy should be performed for a selected nucleotide type, target application, and delivery route, since carrier stability, cellular uptake, and functional efficacy are highly dependent on all these factors.

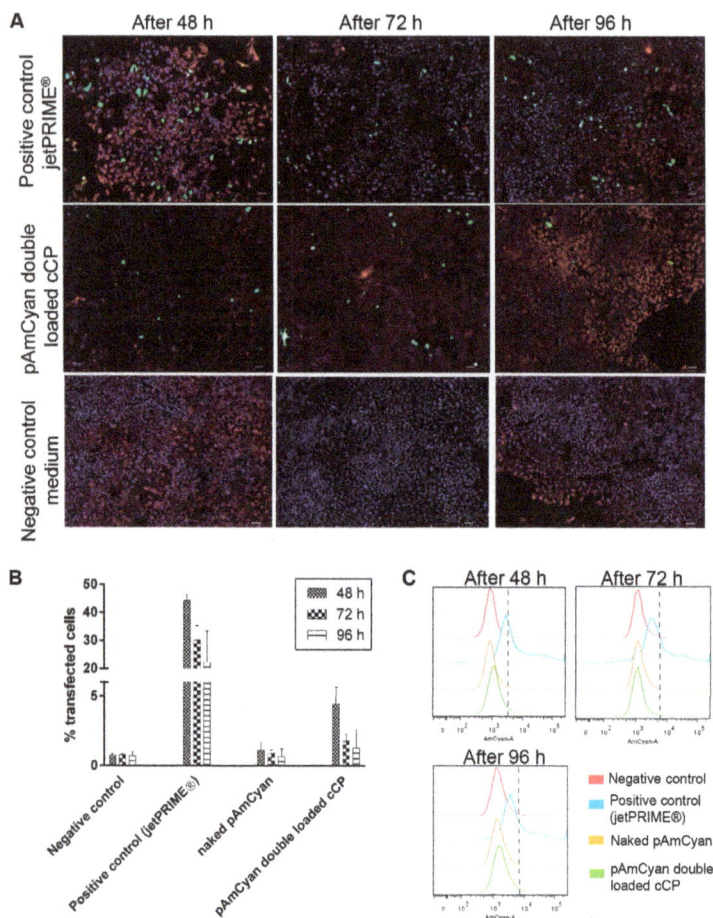

Figure 7. (**A**) Representative confocal images of A549 cells transfected with pAmCyan double loaded pAmCyan by using jetPRIME® as positive control and only cell culture medium as negative control. Transfection was analyzed with CLSM after 48 h, 72 h, and 96 h. Green fluorescence reveals cells successfully transfected with the polyplexes while their morphology remains consistent with non-transfected cells (red: cell membrane; blue: cell nucleus; scale bar 50 μm). (**B**) The transfection efficiency was further quantified using flow cytometry, which indicated the highest amount of transfection after 48 h for pAmCyan double loaded cCP. (**C**) Representative graphs obtained with flow cytometer.

4. Conclusions

In this work, we produced a flexible, straightforward and organic solvent-free procedure for the manufacture of nanocarrier systems based on the natural, biodegradable and biocompatible polysaccharides starch and chitosan. Starch and chitosan derivatives of different M_w ranges were combined by adjusting the molar ratio of carboxyl and amine functional groups, polymer concentration and counter polymer type to obtain a delivery system with tunable properties including surface charge and size. Core polyplexes (CP) were built by complex coacervation of anionic starch (M_w ~100 kDa) with positively charged chitosan derivatives (M_w ~5 kDa) in aqueous solution. The polyplexes with the best colloidal properties were obtained at a molar ratio of carboxyl and amine groups of 10:1. The negatively charged core polyplexes remained stable on storage for over 27 days. We further focused on optimizing anionic CPs by coating them with an additional layer of chitosan (Protasan, M_w ~90 kDa). Cell viability testing of anCPs and cCPs indicated a low level of cytotoxicity acceptable for use in biological systems, and colloidal stability at different tested pH values. The developed anCP system further showed good carrier properties, allowing for high encapsulation efficiency (>90%) of cationic peptide (colistin) and small molecule (tobramycin) anti-infectives without compromising antimicrobial activity. Moreover, the cationic polyplexes, cCP, allowed for double encapsulation of plasmid DNA (pAmCyan) for intracellular delivery as confirmed by gel retardation assay, and facilitating in-vitro transfection in A549 cells.

Starch-chitosan polyplexes show high flexibility for designing multifunctional carriers, in which for example the core polyplexes can encapsulate anti-infectives, while the outer coating layer could be used to incorporate other components like enzymes or nucleases (e.g., deoxyribonuclease I) to enhance drug penetration through biofilms or mucus [30]. For gene therapy purposes the inner polyplex can be used to carry and protect plasmid DNA, while the surface could be decorated with a second polynucleotide.

Supplementary Materials: The following are available online at http://www.mdpi.com/2073-4360/10/3/252/s1, Figure S1: Physicochemical stability of starch-chitosan CP, in which anCP was produced with C/N ratio 10/1, and catCP was produced with C/N ratio 1/10, upon storage (4 °C). The particles were diluted into Milli-Q water at each time point for the measurement of size, PDI and ζ-potential. $N = 3$, $n = 3$, mean ± SD; Figure S2: Physicochemical stability of starch-chitosan anCP and cCP at different pH values ranging from 3.5 to 8.0, after 30 min and 1 h incubation. The initial pH-value of the samples was 5.5. $N = 3$, $n = 3$, mean ± SD; Figure S3: Cumulative release of tobramycin from tobramycin loaded anCP, and colistin from colistin loaded anCP performed in PBS at 37 °C. $N = 3$, $n = 3$, mean ± SD; Table S1: Summary of starch-chitosan CP characteristics obtained by varying polymer types, polymer concentration, and C/N molar ratio. $N > 3$, $n = 3$, mean ± SD; Table S2: Summary of starch-chitosan CP characterization with optimal C/N ratio varied by change of polymer concentration. $N > 3$, $n = 3$, mean ± SD; Table S3: Summary of anionic CP (anCP) and Protasan coated anCP (cCP) characteristics, in which anCP was produced with parameters, namely C/N ratio 10/1, and polymer concentration at 6.5 mg/mL. $N > 3$, $n = 3$, mean ± SD; Table S4: Summary of tobramycin-loaded anCP characteristics achieved by variation of C/N ratio and polymer concentration. $N > 3$, $n = 3$, mean ± SD; Table S5: Summary of colistin-loaded anCP characteristics resulting from variation of polymer concentration. $N > 3$, $n = 3$, mean ± SD; Table S6: Summary of drug loading quantification of tobramycin-loaded anCP. $N > 3$, $n = 3$, mean ± SD; Table S7: Summary of drug loading quantification of colistin-loaded anCP. $N > 3$, $n = 3$, mean ± SD.

Acknowledgments: The authors thank Xabier Murgia Esteve, Florian Gräf, and Arianna Castoldi for fruitful discussions; Petra König and Jana Westhues for support and handling of cell cultures; and Viktoria Schmitt for bacteria culture. This project has received funding from the European Union Framework Programme for Research and Innovation Horizon 2020 (2014–2020) under the Marie Sklodowska-Curie Grant Agreement No. 642028.

Author Contributions: Hanzey Yasar and Duy-Khiet Ho contributed equally in this study in which they initiated the research idea, conceived and designed all experiments. Hanzey Yasar and Duy-Khiet Ho synthesized and prepared molecular weight fractionalized anionic starch, and optimized particles preparing process of starch and chitosan (including negative, positive and coated surface particles), as well as investigated stability of all particles at storage condition and different pH environments. Hanzey Yasar further studied cytotoxicity (by MTT assay) and plasmid (pAmCyan) loading capacity (by a gel retardation assay) of the particles, and performed transfection study on A549 cell line by CLSM and quantification method using flow cytometry. Duy-Khiet Ho further studied and optimized cationic anti-infectives (tobramycin and colistin) loading capacity (by isothermal titration calorimetry) of the particles, and performed the minimum inhibitory concentration (MIC) assay on *E. coli* and *P. aeruginosa*. Hanzey Yasar and Duy-Khiet Ho analyzed all the data and wrote the manuscript with an equal manner. Chiara De Rossi visualized the developed particles by using TEM, performed HPLC, and contributed

Polymers **2018**, *10*, 252

her expertise in imaging and analyzing the transfection study by using CLSM. Jennifer Herrmann contributed her expertise in bacteria study and analyzed the data. Sarah Gordon, Brigitta Loretz and Claus-Michael Lehr supervised Hanzey Yasar and Duy-Khiet Ho, initiated the project and have been responsible for the overall scientific approach. All authors contributed with their scientific input to the written manuscript.

Conflicts of Interest: The authors declare no competing financial interest.

Abbreviations

CP	core polyplexes
anCP	anionic core polyplexes
cationic CP (or catCP)	cationic core polyplexes
cCP	coated polyplexes

References

1. Kang, B.; Opatz, T.; Landfester, K.; Wurm, F.R. Carbohydrate nanocarriers in biomedical applications: Functionalization and construction. *Chem. Soc. Rev.* **2015**, *44*, 8301–8325. [CrossRef] [PubMed]
2. Bachmann, M.F.; Jennings, G.T. Vaccine delivery: A matter of size, geometry, kinetics and molecular patterns. *Nat. Rev. Immunol.* **2010**, *10*, 787–796. [CrossRef] [PubMed]
3. Abed, N.; Couvreur, P. Nanocarriers for antibiotics: A promising solution to treat intracellular bacterial infections. *Int. J. Antimicrob. Agents* **2014**, *43*, 485–496. [CrossRef] [PubMed]
4. D'Angelo, I.; Conte, C.; Miro, A.; Quaglia, F.; Ungaro, F. Pulmonary drug delivery: A role for polymeric nanoparticles? *Curr. Top. Med. Chem.* **2015**, *15*, 386–400. [CrossRef] [PubMed]
5. Mahapatro, A.; Singh, D.K. Biodegradable nanoparticles are excellent vehicle for site directed in-vivo delivery of drugs and vaccines. *J. Nanobiotechnol.* **2011**, *9*, 55. [CrossRef] [PubMed]
6. Baldwin, A.D.; Kiick, K.L. Polysaccharide-modified synthetic polymeric biomaterials. *Biopolymers* **2010**, *94*, 128–140. [CrossRef] [PubMed]
7. Wikström, J.; Elomaa, M.; Syväjärvi, H.; Kuokkanen, J.; Yliperttula, M.; Honkakoski, P.; Urtti, A. Alginate-based microencapsulation of retinal pigment epithelial cell line for cell therapy. *Biomaterials* **2008**, *29*, 869–876. [CrossRef] [PubMed]
8. Hans, M.L.; Lowman, A.M. Biodegradable nanoparticles for drug delivery and targeting. *Curr. Opin. Solid State Mater. Sci.* **2002**, *6*, 319–327. [CrossRef]
9. Azzam, T.; Eliyahu, H.; Shapira, L.; Linial, M.; Barenholz, Y.; Domb, A.J. Polysaccharide? Oligoamine Based Conjugates for Gene Delivery. *J. Med. Chem.* **2002**, *45*, 1817–1824. [CrossRef] [PubMed]
10. Sim, H.J.; Thambi, T.; Lee, D.S. Heparin-based temperature-sensitive injectable hydrogels for protein delivery. *J. Mater. Chem. B* **2015**, *3*, 8892–8901. [CrossRef]
11. Builders, P.F.; Arhewoh, M.I. Pharmaceutical applications of native starch in conventional drug delivery. *Starch Stärke* **2016**, *68*, 864–873. [CrossRef]
12. Yamada, H.; Loretz, B.; Lehr, C.-M. Design of starch-graft-PEI polymers: An effective and biodegradable gene delivery platform. *Biomacromolecules* **2014**, *15*, 1753–1761. [CrossRef] [PubMed]
13. Mahmoudi Najafi, S.H.; Baghaie, M.; Ashori, A. Preparation and characterization of acetylated starch nanoparticles as drug carrier: Ciprofloxacin as a model. *Int. J. Biol. Macromol.* **2016**, *87*, 48–54. [CrossRef] [PubMed]
14. Balmayor, E.R.; Baran, E.T.; Azevedo, H.S.; Reis, R.L. Injectable biodegradable starch/chitosan delivery system for the sustained release of gentamicin to treat bone infections. *Carbohydr. Polym.* **2012**, *87*, 32–39. [CrossRef]
15. Santander-Ortega, M.J.; Stauner, T.; Loretz, B.; Ortega-Vinuesa, J.L.; Bastos-González, D.; Wenz, G.; Schaefer, U.F.; Lehr, C.M. Nanoparticles made from novel starch derivatives for transdermal drug delivery. *J. Control. Release* **2010**, *141*, 85–92. [CrossRef] [PubMed]
16. Bernkop-Schnurch, A.; Dunnhaupt, S. Chitosan-based drug delivery systems. *Eur. J. Pharm. Biopharm.* **2012**, *81*, 463–469. [CrossRef] [PubMed]
17. Grenha, A.; Gomes, M.E.; Rodrigues, M.; Santo, V.E.; Mano, J.F.; Neves, N.M.; Reis, R.L. Development of new chitosan/carrageenan nanoparticles for drug delivery applications. *J. Biomed. Mater. Res. Part A* **2010**, *92*, 1265–1272. [CrossRef] [PubMed]

18. Gan, Q.; Wang, T. Chitosan nanoparticle as protein delivery carrier—Systematic examination of fabrication conditions for efficient loading and release. *Colloids Surf. B Biointerfaces* **2007**, *59*, 24–34. [CrossRef] [PubMed]

19. Wen, Z.-S.; Xu, Y.-L.; Zou, X.-T.; Xu, Z.-R. Chitosan nanoparticles act as an adjuvant to promote both Th1 and Th2 immune responses induced by ovalbumin in mice. *Mar. Drugs* **2011**, *9*, 1038–1055. [CrossRef] [PubMed]

20. De Campos, A.M.; Sánchez, A.; Alonso, M.J. Chitosan nanoparticles: A new vehicle for the improvement of the delivery of drugs to the ocular surface. Application to cyclosporin A. *Int. J. Pharm.* **2001**, *224*, 159–168. [CrossRef]

21. Van der Lubben, I.M.; Verhoef, J.C.; van Aelst, A.C.; Borchard, G.; Junginger, H.E. Chitosan microparticles for oral vaccination: Preparation, characterization and preliminary in vivo uptake studies in murine Peyer's patches. *Biomatrials* **2001**, *22*, 687–694. [CrossRef]

22. Dodane, V.; Vilivalam, V.D. Pharmaceutical applications of chitosan. *Pharm. Sci. Technol. Today* **1998**, *1*, 246–253. [CrossRef]

23. Onishi, H.; Machida, Y. Biodegradation and distribution of water-soluble chitosan in mice. *Biomaterials* **1999**, *20*, 175–182. [CrossRef]

24. Aspden, T.J.; Mason, J.D.; Jones, N.S.; Lowe, J.; Skaugrud, O.; Illum, L. Chitosan as a nasal delivery system: The effect of chitosan solutions on in vitro and in vivo mucociliary transport rates in human turbinates and volunteers. *J. Pharm. Sci.* **1997**, *86*, 509–513. [CrossRef] [PubMed]

25. Elborn, J.S.; Vataire, A.-L.; Fukushima, A.; Aballea, S.; Khemiri, A.; Moore, C.; Medic, G.; Hemels, M.E.H. Comparison of Inhaled Antibiotics for the Treatment of Chronic *Pseudomonas aeruginosa* Lung Infection in Patients with Cystic Fibrosis: Systematic Literature Review and Network Meta-analysis. *Clin. Ther.* **2016**, *38*, 2204–2226. [CrossRef] [PubMed]

26. Elborn, S.; Vataire, A.-L.; Fukushima, A.; Aballéa, S.; Khemiri, A.; Moore, C.; Medic, G.; Hemels, M. Efficacy and safety of inhaled antibiotics for chronic pseudomonas infection in cystic fibrosis: Network meta-analysis. *Eur. Respir. J.* **2016**, *48*, PA4863. [CrossRef]

27. Nafee, N.; Schneider, M.; Schaefer, U.F.; Lehr, C.-M. Relevance of the colloidal stability of chitosan/PLGA nanoparticles on their cytotoxicity profile. *Int. J. Pharm.* **2009**, *381*, 130–139. [CrossRef] [PubMed]

28. Benson, J.R.; Hare, P.E. O-phthalaldehyde: Fluorogenic detection of primary amines in the picomole range. Comparison with fluorescamine and ninhydrin. *Proc. Natl. Acad. Sci. USA* **1975**, *72*, 619–622. [CrossRef] [PubMed]

29. Barthold, S.; Kletting, S.; Taffner, J.; de Souza Carvalho-Wodarz, C.; Lepeltier, E.; Loretz, B.; Lehr, C.-M. Preparation of nanosized coacervates of positive and negative starch derivatives intended for pulmonary delivery of proteins. *J. Mater. Chem. B* **2016**, *4*, 2377–2386. [CrossRef]

30. Deacon, J.; Abdelghany, S.M.; Quinn, D.J.; Schmid, D.; Megaw, J.; Donnelly, R.F.; Jones, D.S.; Kissenpfennig, A.; Elborn, J.S.; Gilmore, B.F.; et al. Antimicrobial efficacy of tobramycin polymeric nanoparticles for *Pseudomonas aeruginosa* infections in cystic fibrosis: Formulation, characterisation and functionalisation with dornase alfa (DNase). *J. Control. Release* **2015**, *198*, 55–61. [CrossRef] [PubMed]

31. Dul, M.; Paluch, K.J.; Kelly, H.; Healy, A.M.; Sasse, A.; Tajber, L. Self-assembled carrageenan/protamine polyelectrolyte nanoplexes-Investigation of critical parameters governing their formation and characteristics. *Carbohydr. Polym.* **2015**, *123*, 339–349. [CrossRef] [PubMed]

32. Radovic-Moreno, A.F.; Lu, T.K.; Puscasu, V.A.; Yoon, C.J.; Langer, R.; Farokhzad, O.C. Surface charge-switching polymeric nanoparticles for bacterial cell wall-targeted delivery of antibiotics. *ACS Nano* **2012**, *6*, 4279–4287. [CrossRef] [PubMed]

33. Mandzy, N.; Grulke, E.; Druffel, T. Breakage of TiO_2 agglomerates in electrostatically stabilized aqueous dispersions. *Powder Technol.* **2005**, *160*, 121–126. [CrossRef]

34. Jonassen, H.; Kjoniksen, A.-L.; Hiorth, M. Stability of chitosan nanoparticles cross-linked with tripolyphosphate. *Biomacromolecules* **2012**, *13*, 3747–3756. [CrossRef] [PubMed]

35. Honary, S.; Zahir, F. Effect of Zeta Potential on the Properties of Nano-Drug Delivery Systems—A Review (Part 2). *Trop. J. Pharm. Res.* **2013**, *12*, 265–273. [CrossRef]

36. Yoo, M.K.; Sung, Y.K.; Chong, S.C.; Young, M.L. Effect of polymer complex formation on the cloud-point of poly(N-isopropyl acrylamide) (PNIPAAm) in the poly(NIPAAm-co-acrylic acid): Polyelectrolyte complex between poly(acrylic acid) and poly(allylamine). *Polymer* **1997**, *38*, 2759–2765. [CrossRef]

37. Fan, W.; Yan, W.; Xu, Z.; Ni, H. Formation mechanism of monodisperse, low molecular weight chitosan nanoparticles by ionic gelation technique. *Colloids Surf. B Biointerfaces* **2012**, *90*, 21–27. [CrossRef] [PubMed]

38. Shu, X.; Zhu, K. The influence of multivalent phosphate structure on the properties of ionically cross-linked chitosan films for controlled drug release. *Eur. J. Pharm. Biopharm.* **2002**, *54*, 235–243. [CrossRef]

39. Ensign, L.M.; Cone, R.; Hanes, J. Nanoparticle-based drug delivery to the vagina: A review. *J. Control. Release* **2014**, *190*, 500–514. [CrossRef] [PubMed]

40. Evans, D.F.; Pye, G.; Bramley, R.; Clark, A.G.; Dyson, T.J.; Hardcastle, J.D. Measurement of gastrointestinal pH profiles in normal ambulant human subjects. *Gut* **1988**, *29*, 1035–1041. [CrossRef] [PubMed]

41. Melis, G.B.; Ibba, M.T.; Steri, B.; Kotsonis, P.; Matta, V.; Paoletti, A.M. Ruolo del pH come modulatore dell'equilibrio fisiopatologico vaginale. *Min. Ginecol.* **2000**, *52*, 111–121.

42. Fischer, D.; Li, Y.; Ahlemeyer, B.; Krieglstein, J.; Kissel, T. In vitro cytotoxicity testing of polycations: Influence of polymer structure on cell viability and hemolysis. *Biomatrials* **2003**, *24*, 1121–1131. [CrossRef]

43. Frohlich, E. The role of surface charge in cellular uptake and cytotoxicity of medical nanoparticles. *Int. J. Nanomed.* **2012**, *7*, 5577–5591. [CrossRef] [PubMed]

44. Maiz, L.; Giron, R.M.; Olveira, C.; Quintana, E.; Lamas, A.; Pastor, D.; Canton, R.; Mensa, J. Inhaled antibiotics for the treatment of chronic bronchopulmonary *Pseudomonas aeruginosa* infection in cystic fibrosis: Systematic review of randomised controlled trials. *Expert Opin. Pharmacother.* **2013**, *14*, 1135–1149. [CrossRef] [PubMed]

45. Bargoni, A.; Cavalli, R.; Zara, G.P.; Fundaro, A.; Caputo, O.; Gasco, M.R. Transmucosal transport of tobramycin incorporated in solid lipid nanoparticles (SLN) after duodenal administration to rats. Part II—Tissue distribution. *Pharmacol. Res.* **2001**, *43*, 497–502. [CrossRef] [PubMed]

46. Wallace, S.J.; Li, J.; Nation, R.L.; Prankerd, R.J.; Velkov, T.; Boyd, B.J. Self-assembly behavior of colistin and its prodrug colistin methanesulfonate: Implications for solution stability and solubilization. *J. Phys. Chem. B* **2010**, *114*, 4836–4840. [CrossRef] [PubMed]

47. Akinc, A.; Thomas, M.; Klibanov, A.M.; Langer, R. Exploring polyethylenimine-mediated DNA transfection and the proton sponge hypothesis. *J. Gene Med.* **2005**, *7*, 657–663. [CrossRef] [PubMed]

48. Truong, N.P.; Jia, Z.; Burgess, M.; Payne, L.; McMillan, N.A.J.; Monteiro, M.J. Self-catalyzed degradable cationic polymer for release of DNA. *Biomacromolecules* **2011**, *12*, 3540–3548. [CrossRef] [PubMed]

49. Hartono, S.B.; Phuoc, N.T.; Yu, M.; Jia, Z.; Monteiro, M.J.; Qiao, S.; Yu, C. Functionalized large pore mesoporous silica nanoparticles for gene delivery featuring controlled release and co-delivery. *J. Mater. Chem. B* **2014**, *2*, 718–726. [CrossRef]

50. Truong, N.P.; Gu, W.; Prasadam, I.; Jia, Z.; Crawford, R.; Xiao, Y.; Monteiro, M.J. An influenza virus-inspired polymer system for the timed release of siRNA. *Nat. Commun.* **2013**, *4*, 1902. [CrossRef] [PubMed]

51. Sanz, V.; Coley, H.M.; Silva, S.R.P.; McFadden, J. Protamine and Chloroquine Enhance Gene Delivery and Expression Mediated by RNA-Wrapped Single Walled Carbon Nanotubes. *J. Nanosci. Nanotechnol.* **2012**, *12*, 1739–1747. [CrossRef] [PubMed]

52. Kean, T.; Roth, S.; Thanou, M. Trimethylated chitosans as non-viral gene delivery vectors: Cytotoxicity and transfection efficiency. *J. Control. Release* **2005**, *103*, 643–653. [CrossRef] [PubMed]

53. Zheng, H.; Tang, C.; Yin, C. Exploring advantages/disadvantages and improvements in overcoming gene delivery barriers of amino acid modified trimethylated chitosan. *Pharm. Res.* **2015**, *32*, 2038–2050. [CrossRef] [PubMed]

© 2018 by the authors. Licensee MDPI, Basel, Switzerland. This article is an open access article distributed under the terms and conditions of the Creative Commons Attribution (CC BY) license (http://creativecommons.org/licenses/by/4.0/).

polymers

MDPI

Article

Thiolated Chitosan Masked Polymeric Microspheres with Incorporated Mesocellular Silica Foam (MCF) for Intranasal Delivery of Paliperidone

Stavroula Nanaki [1], Maria Tseklima [1], Evi Christodoulou [1], Konstantinos Triantafyllidis [2], Margaritis Kostoglou [2] and Dimitrios N. Bikiaris [1,*]

[1] Laboratory of Polymer Chemistry and Technology, Aristotle University of Thessaloniki, GR-54124 Thessaloniki, Greece; sgnanaki@chem.auth.gr (S.N.); mtseklima@pharmathen.com (M.T.); evicius@gmail.com (E.C.)

[2] Laboratory of General and Inorganic Chemical Technology, Aristotle University of Thessaloniki, GR-54124 Thessaloniki, Greece; ktrianta@chem.auth.gr (K.T.); kostoglu@chem.auth.gr (M.K.)

* Correspondence: dbic@chem.auth.gr; Tel.: +30-2310-997812

Received: 30 October 2017; Accepted: 13 November 2017; Published: 15 November 2017

Abstract: In this study, mesocellular silica foam (MCF) was used to encapsulate paliperidone, an antipsychotic drug used in patients suffering from bipolar disorder. MCF with the drug adsorbed was further encapsulated into poly(lactic acid) (PLA) and poly(lactide-*co*-glycolide) (PLGA) 75/25 *w*/*w* microspheres and these have been coated with thiolated chitosan. As found by TEM analysis, thiolated chitosan formed a thin layer on the polymeric microspheres' surface and was used in order to enhance their mucoadhesiveness. These microspheres aimed at the intranasal delivery of paliperidone. The DSC and XRD studies showed that paliperidone was encapsulated in amorphous form inside the MCF silica and for this reason its dissolution profile was enhanced compared to the neat drug. In coated microspheres, thiolated chitosan reduced the initial burst effect of the paliperidone dissolution profile and in all cases sustained release formulations have been prepared. The release mechanism was also theoretically studied and three kinetic models were proposed and successfully fitted for a dissolution profile of prepared formulations to be found.

Keywords: PLA; PLGA; mesoporous cellular foam; paliperidone; thiolated chitosan; microspheres; drug encapsulation

---

1. Introduction

Mesoporous silica nanoparticles (MSN) are among a variety of drug delivery systems that have been tested extensively in recent years for the delivery of water-insoluble drugs [1]. Due to their large surface areas and porous interiors they can be used as reservoirs to store hydrophobic drugs. The pore size and environment can be tailored to selectively store different molecules of interest [2,3], while the size and shape of the particles can be tuned to maximize cellular uptake. Such silica-based materials have been successfully used as drug-delivery vectors [4,5], gene transfection reagents [6], cell markers [7], and carriers of molecules [8]. Mesocellular foam (MCF) silica particles have an almost spherical morphology and a continuous three-dimensional pore system, with pore size in the range of ca. 10–30 nm, having already been used for delivery of drugs [9,10] through oral [11–13] and intravenous [14,15] routes.

Paliperidone is a second-generation antipsychotic drug and is effective in treating both positive and negative symptoms of schizophrenia with an increased safety effect towards extrapyramidal symptoms. For that, paliperidone is administrated in two forms: in oral form, present on the market as a 24-h extended-release tablet [16], and in injectable form [17], marketed by Janssen (Invega

Sustenna, Titusville, NJ, USA). Such long active injectables (LAI) have gained attention recently for their promise to treat diseases like schizophrenia. However, paliperidone is a poorly water soluble drug, which limits its effectiveness. For this reason, in our previous study [18] it was first adsorbed into MCF silica and then encapsulated in polymeric microparticles. Poly(L-lactic acid) (PLA) [19,20] and poly(D,L-lactide-*co*-glycolide) (PLGA) [21] are two Food and Drug Administration (FDA)-approved polymers that are used particularly in nasal application of active compounds. These polymers are transformed into lactic and/or glycolic acid in the body due to hydrolysis. However, when Darville et al. [17] investigated the local disposition of paliperidone in rats, they found that drug release caused a reaction leading to chronic inflammation. Also, large amounts of crystalline paliperidone-LAI particles were found within the infiltrating macrophages, supporting the hypothesis that the sudden drop in dissolution rate of the drug and its potential absorption lead to accumulation of the drug in macrophages. The same group investigated the co-administration of liposomal clodronate and sunitinib in order to inhibit the depot infiltration and nano-/microparticle phagocytosis by macrophages, and the neovascularization of the depot, respectively [22].

In recent years, the nasal route has gained importance as a non-invasive drug application route that offers many advantages for the introduction of drugs into systemic circulation, such as rapid absorption of drugs and therefore quick onset of their effect, and avoidance of the hepatic first-pass effect. However, disadvantages of the nasal route include enzymatic barriers, particularly in the case of macromolecular drugs, and the low permeability of the nasal epithelia. Furthermore, it is necessary that any formulation intended for nasal delivery must have strong mucoadhessive properties. Chitosan is a positively charged polymer [23] frequently used in nasal application of macromolecules [24,25], due to its strong mucoadhesive properties and its ability to transiently open the tight junctions in the nasal mucosa. Mucoadhesion is achieved by the ionic interaction of positively charged amine groups of D-glucosamine units of chitosan with negatively charged sialic acid groups of musin or other negatively charged groups of the mucosal membrane [26]. It has been reported that chitosan does not lead to any histological changes in the nasal mucosa [27–29]. So in order to increase the mucoadhesive properties of PLA or PLGA microspheres, their surface could be coated with chitosan. Moreover, from the literature it was found that thiolation of chitosan shows significant improvement in permeation and better mucoadhesive properties than neat chitosan due to the fact that the covalent bonds formed between the thiol group and the mucus glycoprotein are stronger than the noncovalent bonds between chitosan and mucus glycoprotein [30–35].

The aim of the present study was to enhance the solubility of paliperidone by its adsorption in MCF particles and to prepare proper microspheres for intranasal delivery of the drug. Microparticles prepared expected to act as extended release formulations of the drug, lasting for about one month. In order to be administered by the intranasal route, thiolated chitosan was used as a coating membrane. Thiolated chitosan, apart from mucoadhesion enhancement, has the additional advantage of increasing drug penetration into the nasal mucous membrane. As far as we know, no other study has been published concerning the intranasal delivery of paliperidone via thiolated masked polymeric microparticles.

2. Materials and Methods

2.1. Materials and Reagents

Pluronic P-123 (Poly(ethylene glycol)-block-poly(propylene glycol)-block-poly(ethylene glycol)) triblock copolymer with average M_n~5800 was acquired from Sigma-Aldrich (St. Louis, MO, USA) and was used as the MCF mesostructure-directing agent, together with 1,2,3-trimethylbenzene (TMB, Fluka, Munich, Germany), which was utilized as a co-surfactant and swelling agent, as well as ammonium fluoride (NH_4F, Merck, Kenilworth, NJ, USA) serving as a mineralizing agent [36]. Tetraethyl orthosilicate (TEOS) was acquired from Merck and used as the silica source of MCF. Chitosan low molecular weight (50,000–190,000 Da, degree of deacetylation $\geq$ 75%) was purchased from

Sigma-Aldrich. Thioglycolic acid (purum $\geq$ 98%), *N*-Ethyl-*N'*-(3-dimethylaminopropyl) carbodiimide hydrochloride (EDAC·HCl) (purum $\geq$ 99%) and *N*-Hydroxysuccinimide (NHS) (purum 98%) were purchased from Sigma-Aldrich (St. Louis, MO, USA). Poly(lactic acid) (PURASORB PDL 02 with an inherent viscosity midpoint of 0.2 dL/g), and Poly(lactide-*co*-glycolide) 75/25 *w/w* copolymer (PURASORB PDLG 7502 with an inherent viscosity midpoint of 0.2 dL/g) were kindly donated by Corbion (Montmelo, Spain). Paliperidone was kindly donated by Pharmathen S.A (Athens, Greece). All other reagents were of analytical grade.

2.2. Synthesis of MCF

MCF was previously synthesized in our lab using the self-assembly method described by Winkel et al. [18,36]. In brief, Pluronic P-123 was used as the structure directing agent, 1,2,3-trimethylbenzene (TMB) was used as the swelling agent, ammonium fluoride (NH_4F) as the mineralizing agent, and tetraethyl orthosilicate (TEOS) as the silica source, in acidic pH conditions. P-123 was dissolved in aqueous HCl 1.6 M followed by the addition of NH_4F and TMB, and the mixture was stirred for 1 h at 40 °C. TEOS was then added to the solution and stirring continued for 20 h at 40 °C. The resulting mixture was transferred into an autoclave and heated at 100 °C for 24 h. Filtration was used to recover the solid products, followed by a washing step with water, and calcination was conducted in air at 500 °C, for 8 h, with a heating rate of 1 °C/min, in order to combust the organic templates.

2.3. Synthesis of Thiolated Chitosan

Thiolated chitosan (TMC) was synthesized by a two-stage procedure reported by Zhu et al. [37]. In brief, 1 mL thioglycolic acid, 3.5 mg EDAC·HCl, and 2.0 mg NHS were inserted into a flask containing 2 mL DMF and the mixture was left under magnetic stirring overnight, resulting in the synthesis of NHS–ester as an intermediate product. In the second stage, 500 mg of low molecular weight chitosan was added to 4 mL hydrochloric solution, 1 M in concentration, and diluted with water to a final concentration of 2.5%. After that, NHS–ester was inserted dropwise into the chitosan solution, and the pH value was adjusted to 5. The resulting mixture was left under magnetic stirring overnight. Synthesized TMC was lyophilized and washed in a Soxhlet extractor until total elimination of unreacted monomers was achieved. After lyophilization, TMS appeared as a white, odorless solid with a fibrous structure. Its successful synthesis was evaluated by FTIR.

2.4. Paliperidone Loading Procedure on MCF

Paliperidone was loaded on MCF by adsorption [18]. In brief, the proper quantity of the drug was dissolved in a mixture of organic solvents isopropanol/dichloromethane 75/25 *v/v*, resulting in a solution of 0.1% *w/v*. Mesoporous silica MCF (100 mg) was added to the organic solution and the resulting dispersion was left under magnetic stirring for 24 h, in a nitrogen atmosphere. MCF with paliperidone adsorbed was isolated by centrifugation at 4000 rpm for 10 min and at room temperature until total evaporation of the solvent was achieved. The dried material was washed with acetone to remove the quantity of paliperidone that was deposited onto MCF's surface. The amount of the adsorbed drug was determined by TGA and BET.

2.5. Preparation of PLA and PLGA Microspheres Loaded with MCF/Paliperidone

Polymeric microspheres (PLA and PLGA) containing MCF with adsorbed paliperidone were prepared by the solid-oil-water (s/o/w) modified double emulsification method [38]. According to this procedure, 100 mg of polymer, PLA or PLGA 75/25 *w/w* were dissolved in 5 mL of dichloromethane. Ten milligrams of MCF containing adsorbed paliperidone were added to the polymeric solution and dispersed using a probe sonicator for 1 min. The dispersion was inserted dropwise in 100 mL of PVA solution, 1% *w/v* in concentration, and homogenized using a homogenizer for 2 min. After that, 100 mL water were added and the mixture was left under magnetic stirring until the total evaporation of dichloromethane was achieved. After centrifugation at 8000 rpm for 10 min, the microspheres were collected and washed three times with distilled water in order to remove traces of the residual solvent and PVA. The resulting microspheres were finally freeze-dried and stored at 4 °C for further evaluation.

2.6. Preparation of Thiolated Coated PLA and PLGA Microspheres

Microspheres prepared via the above stages were further modified with thiolated chitosan according to the procedure described by Jiang et al. [39]. In brief, pre-weighted thiolated chitosan was dissolved in deionized water at a concentration of 0.025 mg/mL. The prepared PLA and PLGA 75/25 *w*/*w* microspheres loaded with MCF/paliperidone were suspended in a thiolated chitosan solution at a concentration of 9.5 mg/mL by sonication for 1 min at 30 W power output. They were left under magnetic stirring for 15 min and then collected by centrifugation at 80,000× *g* for 15 min. The modified microspheres were washed with water once and then collected by freeze drying. They were stored in a vacuum until further use.

MCF with loaded paliperidone was also masked with thiolated chitosan for comparison reasons following the procedure described above. All the formulations were fully characterized.

2.7. Characterization of Prepared Formulations

2.7.1. Morphology, Thermal Properties and Crystallinity of Prepared Formulations

Fourier Transform-Infrared spectroscopy (FTIR) spectra were obtained on a Perkin-Elmer FTIR spectrometer (Spectrum 1, Waltham, MA, USA) using pellets of MCF and MCF-Paliperidone diluted in KBr. Infrared (IR) absorbance spectra were obtained between 450 and 4000 cm^{-1} at a resolution of 4 cm^{-1} using 20 co-added scans. All spectra presented are baseline-corrected and normalized.

X-ray diffraction (XRD) analysis was performed on MCF silica and MCF-paliperidone over the 5–45° 2θ range, using a MiniFlex II diffractometer (Rigaku Co., Tokyo, Japan) with Bragg–Brentano geometry (θ, 2θ) and Ni-filtered Cu Kα radiation (λ = 0.154 nm).

Thermogravimetric analysis (TGA) was carried out with a SETARAM SETSYS TG-DTA 16/18 (Caluire, France). Samples (6.0 ± 0.2 mg) were placed in alumina crucibles. An empty alumina crucible was used as a reference. Paliperidone, MCF, and MCF-Paliperidone were heated from ambient temperature to 600 °C in a 50 mL/min flow of N$_2$ at heating rate of 20 °C/min.

A Perkin-Elmer Pyris 1 differential scanning calorimeter (DSC) (Waltham, MA, USA), calibrated with Indium and Zinc standards was employed. A sample of about 10 mg was used for each test, placed in scaled aluminum pan and heated to 300 °C at a heating rate of 20 °C/min. The sample was held at that temperature for 5 min in order to erase any thermal history. After that it was quenched to 30 °C with liquid nitrogen and scanned immediately to 300 °C at a heating rate of 20 °C/min.

The morphology of the prepared microspheres was examined using a scanning electron microscope (SEM), type Jeol (JMS-840, Peabody, MA, USA). All the studied surfaces were coated with carbon black to avoid charging under the electron beam. SEM photos were also received after a dissolution study.

Transmittance electron microscopy (TEM) experiments were carried out on a JEOL 2011 TEM (Peabody, MA, USA) with a LaB6 filament and an accelerating voltage of 200 kV. The specimens were prepared by evaporating drops of SBA-15 silica–ethanol suspension after sonication onto a carbon-coated lacy film supported on a 3 mm diameter, 300 mesh copper grid.

2.7.2. High-Pressure Liquid Chromatography (HPLC) Quantitative Analysis and Drug Loading

Quantitative analysis and drug loading was performed using a Shimadzu HPLC (Kyoto, Japan) prominence system consisting of a degasser (DGU-20A5), a liquid chromatograph (LC-20 AD), an auto sampler (SIL-20AC), a UV/Vis detector (SPD-20A) and a column oven (CTO-20AC). For the analysis a validated method was used [40]. In detail, a C18 reversed-phase column (250 mm × 4.6 mm i.d., 5-μm particle) was used; the mobile phase was water (pH = 3.5):methanol 70:30 *v*/*v* and the flow rate was

1 mL·min^{-1}. UV detection was performed at 275 nm. In brief, the proper amount of microspheres was dissolved in dichloromethane and left under magnetic stirring overnight. The samples were filtered through a 0.45-μm membrane and quantified by HPLC.

Microparticle yield (%), drug loading (%), and entrapment efficiency (%) were calculated by the following equations:

$$\text{Microparticles yield (\%)} = \frac{\text{weight of microparticles}}{\text{weight of polymer and drug fed initially}} \cdot 100 \tag{1}$$

$$\text{Drug loading content (\%)} = \frac{\text{weight of drug in microparticles}}{\text{weight of microparticles}} \cdot 100 \tag{2}$$

$$\text{Entrapment efficiency (\%)} = \frac{\text{weight of drug in microparticles}}{\text{weight of drug fed initially}} \cdot 100 \tag{3}$$

2.7.3. In Vitro Release Profile

In vitro release rates of paliperidone from the prepared formulations were measured in USP dissolution apparatus I (basket apparatus). The dissolution apparatus used was a DISKTEK 2100 C (Markham, ON, Canada) with an auto sampler DISTEK EVOLUTION 4300 and a DISKTEK syringe pump. Dissolution tests were performed in 900 mL simulated body fluid used as dissolution medium having pH 7.2 and temperature 37 ± 1 °C. The rotation speed was set at 50 rpm.

3. Results and Discussion

3.1. Characterization of MCF/Paliperidone-Loaded Nanoparticles

MCF silica was previously synthesized in our lab and porosimetry studies showed that the N$_2$ adsorption isotherms of the parent MCF were of type IV according to the IUPAC classification [18], being characteristic of this type of mesoporous materials. The BET method showed that MCF has a relative high specific surface area, 445 m^2/g, compared to classical synthesized silica particles, and its average mesopore size (diameter) was about 15 nm, a pore size ideal for paliperidone adsorption. TEM images (shown in our previous work) verified its cellular pore morphology.

MCF was also characterized after the adsorption of paliperidone [18]. In brief, the BET method showed that MCF, after paliperidone's adsorption, showed a reduced specific surface area of 232 m^2/g, with its total pore volume being reduced from 1.216 to 0.873 cc/g, showing that paliperidone had been successfully inserted into its pores. Furthermore, thermogravimetric analysis (TGA) was used to determine the drug content loaded on the MCF particles, which was found to be 23.78 wt % [18].

3.2. Characterization of Thiolated Chitosan

Synthesis of thiolated chitosan was evaluated using FTIR. As can be seen in Figure 1, thiolated chitosan showed some characteristic peaks from the newly formed amide bond at 1590 cm^{-1} and peaks of thiol groups; one at 1233 cm^{-1} owning to the S–C bond and one small peak (which presents as a shoulder) in 2680 cm^{-1} owning to the H–S bond [37]. The existence of these peaks showed the successful synthesis of thiolated chitosan.

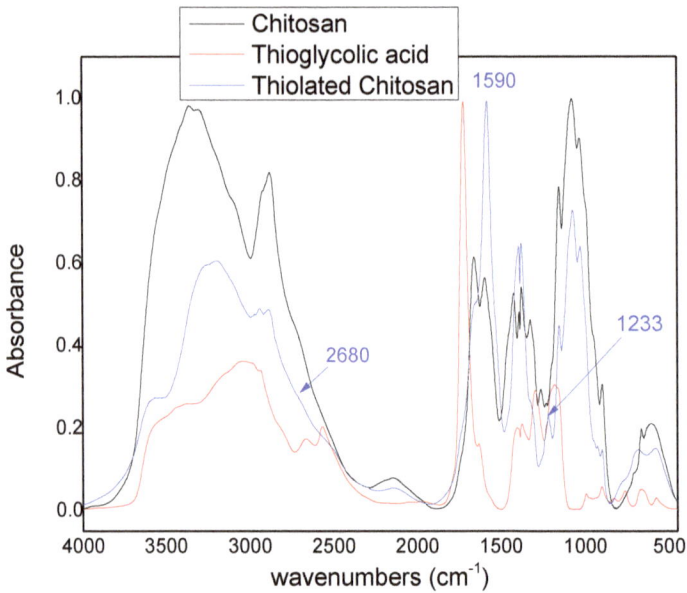

Figure 1. FTIR spectra of chitosan, thioglycolic acid, and thiolated chitosan.

These changes in the chemical form of chitosan affected its physical stage. XRD patterns (Figure 2) showed that chitosan has a wide angle with high intensity at 19.78°, attributed to its crystalline nature. This peak disappeared in thiolated chitosan, probably due to the decrease in the amount of the free amino groups after the introduction of thiol groups, which significantly reduce intermolecular hydrogen bonds. This reduces the ability of chitosan to be crystallized. The absence of a specific peak also illustrates that thiolated chitosan was successfully synthesized and is completely amorphous, as was already reported in the literature [41].

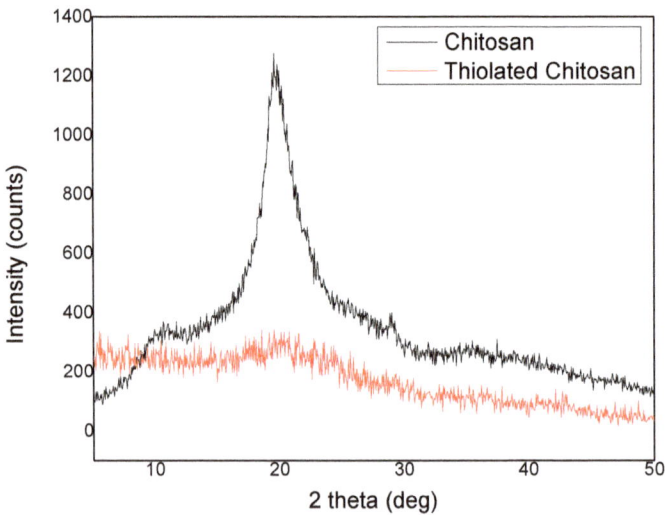

Figure 2. XRD patterns of chitosan and thiolated chitosan.

A TGA thermogram of chitosan and thiolated chitosan (Figure 3) showed a first stage weight loss up to 105 °C, attributed to the water adsorbed or bound to the polymers, showing weight losses of 10% and 8%, respectively. No mass loss was observed up to 255 °C for chitosan, while after that, a second stage weight loss of 45% was recorded up to 350 °C. This stage is probably due to the dehydration of the saccharide rings, depolymerization, and decomposition of the acetylated and deacetylated units of the chitosan [42,43]. A third stage was recorded up to 600 °C, at which chitosan was completely decomposed. Thiolated chitosan showed a second stage that ranged between 105 and 255 °C, with a weight loss equal to 15% probably due to the degradation of thiol groups in thiolated chitosan. From that temperature to 350 °C, a third stage of thermal decomposition, similar to the second stage decomposition of chitosan, was recorded, corresponding to 40% mass loss and attributed to the dehydration of the saccharide rings, depolymerization, and decomposition of the acetylated and deacetylated units of the chitosan. A final stage up to 600 °C was recorded and led to complete decomposition of thiolated chitosan.

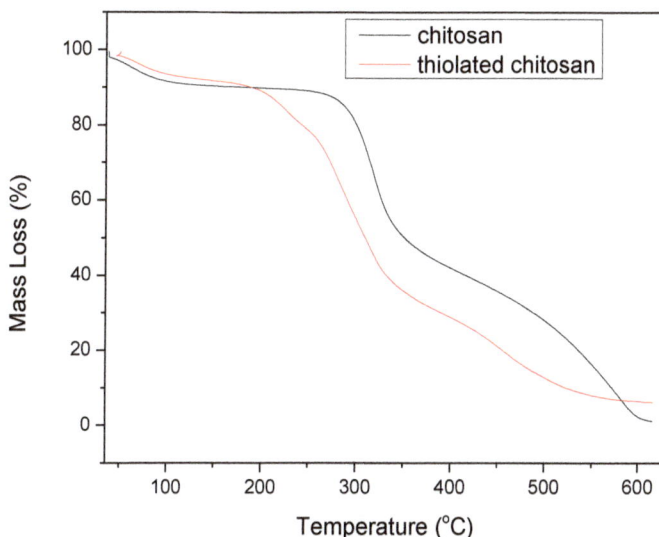

Figure 3. TGA thermograms of chitosan and thiolated chitosan.

3.3. Characterization of MCF-Paliperidone Nanoparticles Coated with Thiolated Chitosan

The synthesized thiolated chitosan was first used as a coating polymer for MCF/paliperidone-loaded particles. SEM and TEM micrographs were taken in order to examine their morphology. As can be seen in Figure 4a, the primary nanoparticles are aggregated in microparticles with irregular shapes and from EDX analysis (Figure 4b) it is detected that they have both Si and S elements, confirming the presence of MCF and thiolated chitosan, respectively. From the TEM images (Figure 4c) it is evident that the morphology of the primary MCF silica nanoparticles was similar to that of the parent MCF and thiolated chitosan covered these nanoparticles [18].

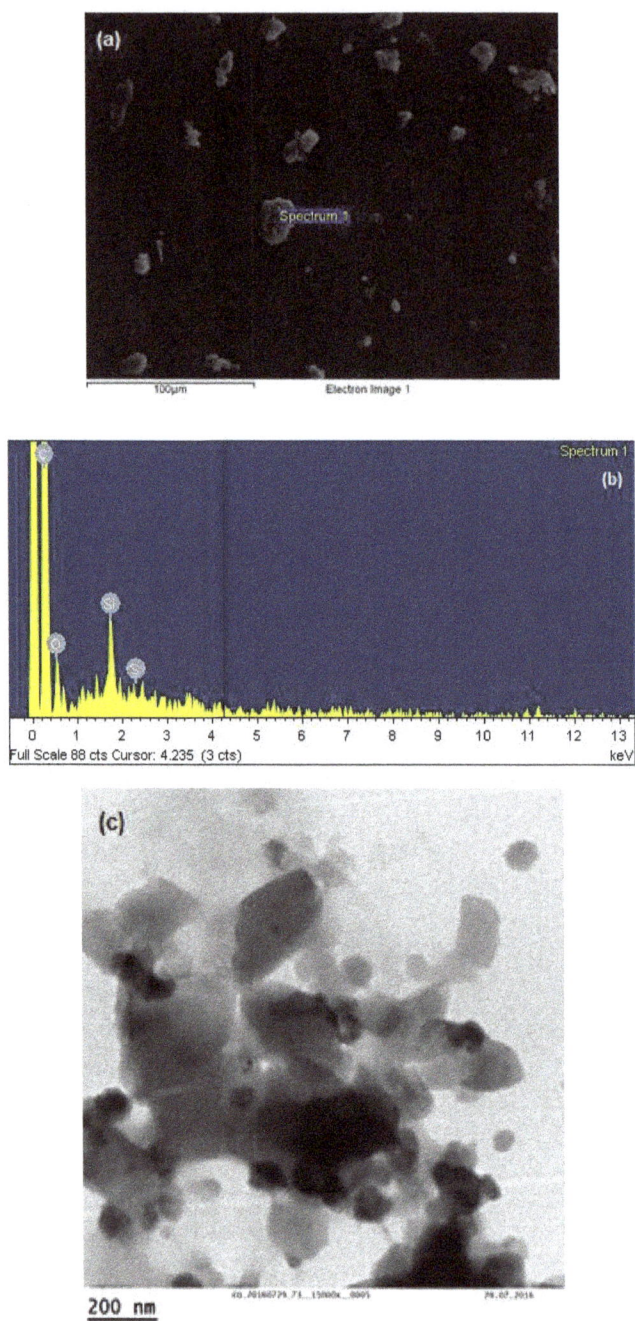

Figure 4. MCF/paliperidone-loaded nanoparticles coated with thiolated chitosan: (**a**) SEM micrograph, (**b**) EDX analysis, and (**c**) TEM micrograph.

As was found from XRD studies, the incorporation of paliperidone into MCF pores leads to drug amorphization [18]. This also happened when the loaded particles were coated with thiolated chitosan and in XRD patterns no characteristic crystalline peaks were observed (data not shown) [18]. This amorphization was also verified by DSC. As can be seen from Figure 5, neat paliperidone has a melting point of 180.16 °C. After its encapsulation into MCF particles and coating with thiolated chitosan, this peak has disappeared, showing that paliperidone is in a totally amorphous state. A small peak was recorded at 103.6 °C, maybe due to the water removal, which was also recorded in neat MCF particles but at a much lower temperature (72.95 °C). The small difference between the two materials could be due to the existence of thiolated chitosan in the coated MCF, which is also a hydroscopic material and absorbs more water than neat MCF.

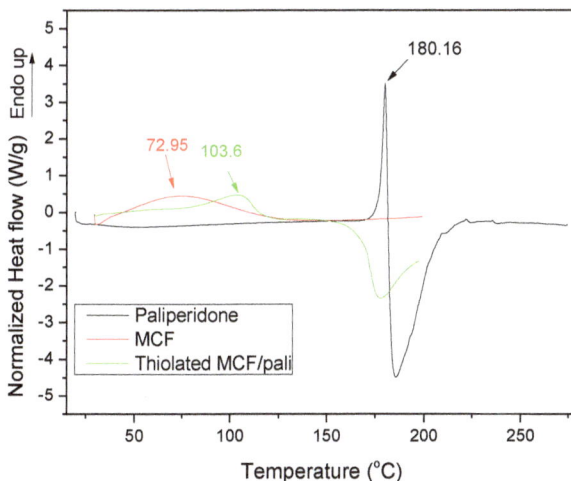

Figure 5. DSC thermograms of paliperidone, MCF, and thiolated MCF with adsorbed paliperidone.

3.4. Characterization of Polymeric PLA and PLGA Microspheres before and after Coating with Thiolated Chitosan

The incorporation of paliperidone into MCF nanoparticles was done in order to prepare a completely amorphous drug and thus increase its dissolution profile [18]. It is well known that amorphous drugs have in some cases 1000 times higher solubility compared with their crystalline forms [44,45]. However, these systems lead to immediate release formulations, which are inappropriate for paliperidone drug release. For this reason, MCF/paliperidone-loaded nanoparticles have been microencapsulated into PLA and PLGA microspheres and given controlled release formulations [18]. In the present work we have extended our previous study by preparing the same microspheres but coated with thiolated chitosan, in order to increase the mucoadhesive properties of their surfaces. Also, the previous microspheres were appropriate for injectable formulations and have recently been prepared for intranasal delivery. The successful coating, as well as the morphology of neat and modified microparticles with thiolated chitosan, was verified with SEM and TEM techniques. As can be seen in Figure 6, in all cases microspheres with spherical shapes formed. Thiolated/PLA/Pal microspheres have sizes varying between 1 and 3 μm, while after MCF incorporation the proper microparticles, i.e., Thiolated_PLA_MCF_Pal, showed bigger sizes that varied between 6 and 10 μm. Analogous observations were made for microspheres prepared with PLGA copolymer, since Thiolated_PLGA75/25_Pal and Thiolated_PLGA75/25_MCF_Pal have sizes between 2–4 μm and 3–6 μm, respectively. This observation is probably due to MCF incorporation, which leads to microspheres with bigger sizes. Another observation is that microspheres prepared with PLGA have smaller sizes than those prepared with PLA, probably due to glycolic acid entanglements resulting in the formation of tighter segments.

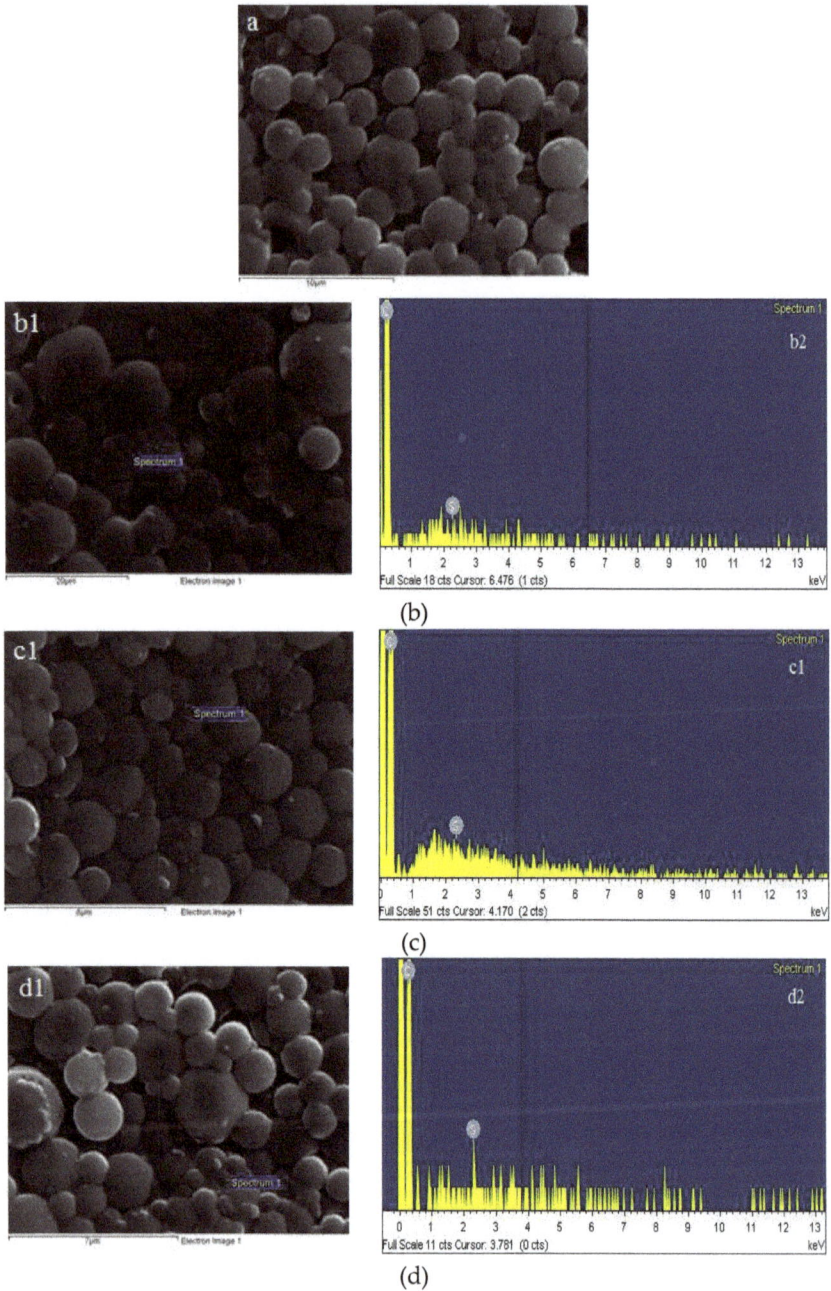

Figure 6. SEM images of prepared microspheres: (**a**) Thiolated_PLA_Pal, (**b**) Thiolated_PLA_MCF_Pal, (**c**) Thiolated_PLGA75/25_Pal, and (**d**) Thiolated_PLGA75/25_MCF_Pal.

TEM images verified the successful incorporation of MCF/paliperidone nanoparticles inside of formed microspheres. As can be seen from Figure 7a, MCF with adsorbed paliperidone were detected inside the microspheres, while some of them were also located on the microsphere's surface (Figure 7b). Also, it is clear that thiolated chitosan forms a thin coating layer on the microsphere surface, which ranged between 20 and 50 nm (Figure 7b).

The physical state of encapsulated MCF/paliperidone nanoparticles in PLA and PLGA microspheres was studied with XRD. From the recorded patterns it can be seen that all microparticles are amorphous, since only a very broad peak was recorded (Figure 8). This was expected since both PLA and PLGA form amorphous microspheres, while paliperidone was also incorporated in an amorphous form inside the MCF pores [18].

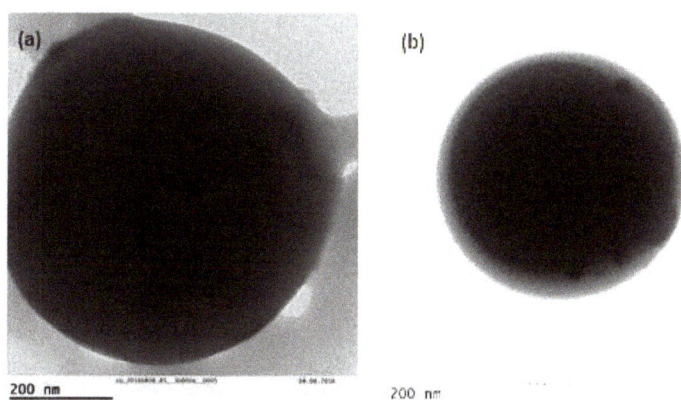

Figure 7. TEM images of microspheres: (a) Thiolated_PLA_MCF_Pal and (b) Thiolated_PLGA75/25_MCF_Pal.

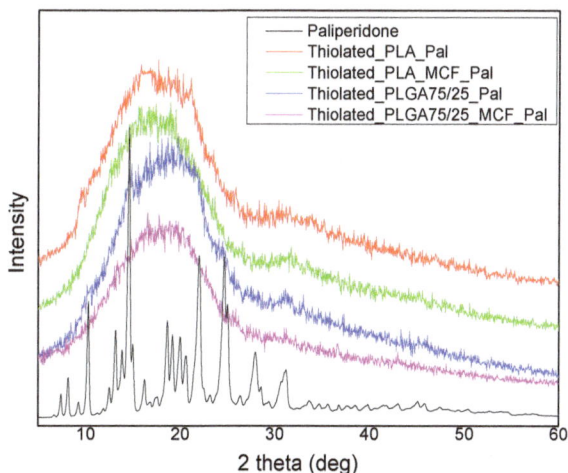

Figure 8. XRD patterns of thiolated microspheres loaded with paliperidone.

DSC studies were also conducted in order to examine the crystallinity and structure behavior of prepared PLA and PLGA microspheres loaded with paliperidone and MCF/paliperidone and coated with thiolated chitosan. Figure 9a shows a DSC thermogram of neat PLA and thiolated microspheres of PLA_Pal and PLA_MCF_Pal. As can be seen, PLA exhibits a glass transition temperature of 61.16 °C. Microspheres masked with thiolated chitosan showed some changes concerning the recorded T_g value, indicating that the addition of both paliperidone and MCF loaded with paliperidone can affect the mobility of PLA chains. For this reason, the T_g value in Thiolated_PLA_Pal changed to 62.91 °C, while this value shifted to 64.32 °C in Thiolated_PLA_MCF_Pal. However, these chances must be attributed to the incorporation of inorganic nanoparticles into a polymer matrix rather than the addition of a thiolated layer on the microparticle surface. So it is clear that the addition of inorganic nanoparticles causes a higher reduction of chain mobility, which is in agreement with the literature [18]. At higher temperatures PLA can be crystallized; a cold crystallization (T_{cc}) peak was recorded at 128.13 °C and formed crystals are melted (T_m) at 152.36 °C. These temperatures were also affected by MCF/drug incorporation. T_{cc} was shifted to 115.17 °C and in Thiolated_PLA_Pal and to 113.39 °C in Thiolated_PLA_MCF_Pal microspheres. This can be attributed mainly to the crystallization effect of both the drug and MCF on PLA crystallization, which can act as nucleating agents [46] and not to coating layer of thiolated chitosan. A small effect was also found on the melting temperature of PLA: two peaks are present in thiolated microspheres loaded with the drug and MCF/paliperidone, probably due to the formation of two types of crystals. In addition, no further peak owing to MCF and/or paliperidone was present in the DSC thermogram, indicating that MCF with paliperidone adsorbed is encapsulated in an amorphous state in microspheres. This observation is in accordance with the results found by XRD analysis.

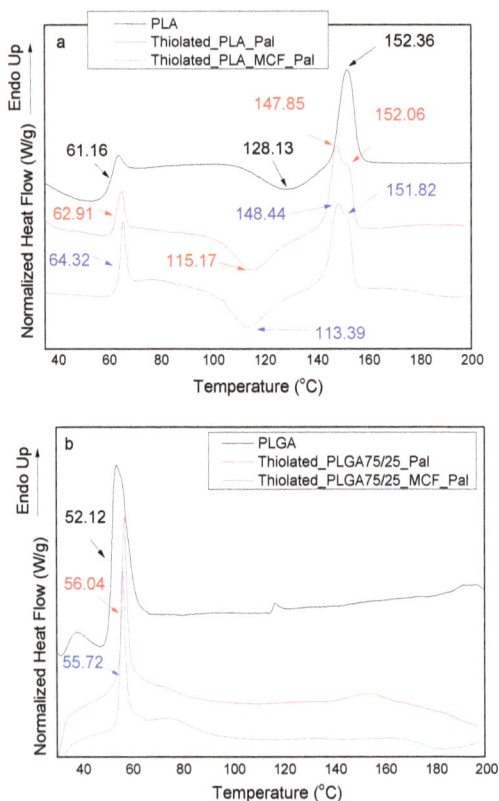

Figure 9. DSC thermograms of thiolated chitosan coated microspheres consisted from (**a**) PLA and (**b**) PLGA with encapsulated paliperidone and MCF_paliperidone.

Some changes can also be seen in the DSC thermograms of thiolated microspheres prepared with PLGA 75/25 w/w copolymer (Figure 9b). As can be seen, PLGA is amorphous, showing a T_g value at 52.12 °C, while coated microspheres with encapsulated paliperidone and MCF/paliperidone showed a slight increase in the T_g value up to about 56 °C, indicating a slight decrease in the mobility of PLGA chains. A similar shift was also recorded in PLA microspheres, as discussed previously. However, as can be seen, PLGA cannot be crystallized and remains in an amorphous form without recording any T_{cc} and T_m temperatures, as in the case of PLA.

3.5. Dissolution Study of Prepared Microspheres

Table 1 shows the drug loading, entrapment efficiency, and microparticle yield of all prepared formulations. As can be seen, all prepared microspheres showed particularly high yield values higher than 86%. Microparticles containing MCF appeared to have higher values than ones prepared with paliperidone alone, maybe due to the incorporation of MCF inside the microparticles. Drug loading values showed a similar behavior, and those with MCF have higher values compared to microparticles with an encapsulated neat drug. Finally, the entrapment efficiency was higher in copolymers, compared to PLA microparticles.

Table 1. Microparticles yield (%), drug loading (%), and entrapment efficiency (%) of prepared thiolated coated microparticles.

Sample	Microparticle Yield (%)	Drug Loading (%)	Entrapment Efficiency (%)
Thiolated_PLA_Pal	85.95 ± 4.94	8.65 ± 1.90	36.25 ± 3.03
Thiolated_PLA_MCF_Pal	93.64 ± 6.17	18.09 ± 4.45	33.62 ± 2.43
Thiolated_PLGA75/25_Pal	90.43 ± 1.99	9.47 ± 2.79	47.77 ± 5.98
Thiolated_PLGA75/25_MCF_Pal	94.70 ± 3.12	21.92 ± 1.35	41.27 ± 5.06

Figure 10 shows the dissolution profiles of MCF with adsorbed paliperidone and thiolated coated nanoparticles. As can be seen, the dissolution profile of neat paliperidone reached its maximum at about 10% in the first hour, without any further dissolution up to 20 days. This is clear proof that paliperidone is a poorly water soluble drug. When it is incorporated in MCF nanopores, the dissolution rate is substantially enhanced and it seems that after 10 days almost 90% of the encapsulated drug was released. This is due to the drug amorphization [44,45]. Thiolated MCF/paliperidone also showed an enhanced dissolution drug profile compared with neat paliperidone, and after 24 days about 80% of the drug was released. From these release studies it is clear that thiolation slightly reduces paliperidone release compared to MCF/paliperidone nanoparticles. This delay is also recorded from the beginning since in the first 6 h no significant drug released was observed; i.e., only 2.8%, in thiolated MCF/paliperidone nanoparticles. This is probably due to the entrance of thiolated chitosan into MCF pores during coating, which can close the gates of pores and prevent drug release from MCF. Also, the thiolated coating layer that formed on the microparticles' surface (Figure 4c) can further delay the drug release.

Figure 10b shows the dissolution of paliperidone from thiolated microspheres using PLA and PLGA75/25 w/w polymers as matrices. It seems that the masking of microparticles with thiolated chitosan reduces paliperidone's release, compared to uncoated ones [18], probably due to the addition of a coated layer that acts as a barrier. Dissolution profiles showed that in the first 6 h no significant drug release occurred, probably because at that period swelling of thiolated chitosan may take place, delaying the drug release. This also causes the disappearance of the burst effect that appears in loaded microspheres without the thiolated chitosan coating layer [18]. After that, there is a continuous and sustained release paliperidone from all microspheres up to 14–15 days. A decrease in the dissolution rate was observed after that, depending on the microsphere used. As can be seen, the release is lower in PLA microspheres than in PLGA. As was found in our previous study, the drug release rate from polymeric microspheres depends on the melting point and glass transition of the polymers used [18].

So, this lower rate of paliperidone release from PLA microspheres should be attributed to its high T_g (61–64 °C) compared with the lower T_g of PLGA (52–55 °C). Furthermore, in both type of microspheres those with MCF have higher release rates, compared with microspheres where only paliperidone was dispersed. This could be attributed to the higher available surface on which the drug is dispersed, since MCF have a very high specific surface area (445 m^2/g).

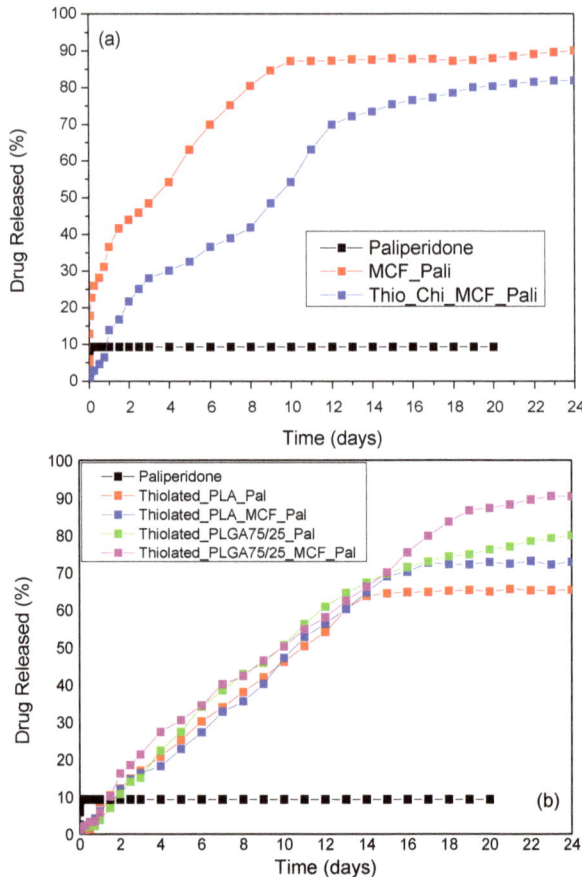

Figure 10. Dissolution profile of (**a**) Thiolated_ MCF_paliperidone and (**b**) Thiolated microspheres.

3.6. Kinetic Analysis of Drug Release

An extensive discussion of the release mechanisms and description of corresponding kinetics for the systems examined here in the absence of the thiolated chitosan layer has been given in a previous work [18]. So the analysis in the following will be mainly focused on the influence of the coating layer. Starting with the case of bulk paliperidone release, it was found that it is a simple dissolution process dominated by paliperidone solubility in water. The fast release curve shown in Figure 10a was very accurately described by considering an appropriate combination of dissolution rate constant and mass transfer coefficient. Before we proceed to the discussion of release from other materials, let us discuss what is expected to be the effect of the thiolated chitosan layer.

There are some studies on swelling and diffusion in chitosan derivatives [47,48]. At first the water diffuses into the chitosan layer, inducing its swelling. The swelled chitosan is a gel containing pores with water allowing pore diffusion of solutes. So the transient diffusion equation for paliperidone in the swelled layer combined with the diffusion of water in the dry layer determines the additional transport of paliperidone. It is noticed that the coating layer is initially free of paliperidone, so its dynamic characteristics can be deduced from the delay in the release process. In the present experiments this delay is always less than 0.7 days, which is relatively small compared to the characteristic overall release time (more than 10 days). This difference between the time scales suggests that the dynamic phenomena in the coating layer can be ignored and the effect of coating can be simply accounted as an additional resistance to paliperidone diffusion through the particles.

The next step is to analyze the release from coated MCF particles. Here, the process is desorption accompanied by diffusion, first through the MCF particle and then through the chitosan layer. In the previous work it was shown that the release curves from MCF particles cannot be described assuming structural homogeneity (i.e., a single apparent diffusion coefficient for the whole particle). Instead, three different diffusion coefficients corresponding to different fractions of the adsorbed paliperidone were considered. The first fraction is assumed to extend to the whole volume of the particles so the complete series solution of the transient diffusion equation is employed (necessary for the capture of the initial diffusion burst). The other two fractions are assumed to be initially located in the interior of the particles so the approximating linear driving force equation is used to account for their diffusion [49]. As is obvious from Figure 10a, the addition of a chitosan layer considerably delays the paliperidone release. The initial burst is completely eliminated since there is no initial paliperidone in the outer (chitosan) layer. This suggests that the fit can be done using solely the linear driving force formula (LDF) (instead of the complete diffusion equation). An acceptable fitted curve results from assuming two different diffusion constants (instead of three in the absence of a chitosan layer). This is quite reasonable since the additional diffusion resistance of the coating layer shadows the internal inhomogeneities. The equation used for fitting is

$$\frac{C}{C_\infty} = \varphi_1(1 - \exp(-K_1 t)) + \varphi_2(1 - \exp(-K_2 U(t - \tau))) \tag{4}$$

where $U(x)$ is a function defined as $U = 0$ for $x \le 0$ and $U = x$ for $x > 0$. The coefficient values are $\varphi_1 = 0.48$, $K_1 = 0.37$ days^{-1}, $\varphi_2 = 0.52$, $K_2 = 0.24$ days^{-1}, $\tau = 7$ days. The time shifts in the absence of chitosan layer were 0.75 and 3.7 days [18]. The first one is eliminated and the second is shifted to seven days due to the mass transfer resistance imposed by the chitosan layer. The extension of the LDF to account for a coating layer leads to the relation $K = \left(\frac{R^2}{15 D_p} + \frac{R\delta}{3 D_c}\right)^{-1}$, where R is the MCF particles' radius (an average value is considered), δ is the chitosan layer thickness, and D_p, D_c are the apparent diffusion coefficients in the particle and in the chitosan layer, respectively. The particle diffusivity should not depend on the existence of the coating layer so the values of D_p from fractions 2 and 3 found in the previous work can be employed in order to find the chitosan layer diffusivity D_c. Assuming a layer thickness of 40 nm, the result is $D_c = 1.84 \times 10^{-20}$ m^2/s. This value is on the same order as the diffusivities in the MCF particles [18], which means that the primary effect of the coating layer in this case is the creation of an initial solute-free zone outside the particles and the secondary effect is the additional mass transfer resistance. The comparison between the experimental and fitted release curves is shown in Figure 11.

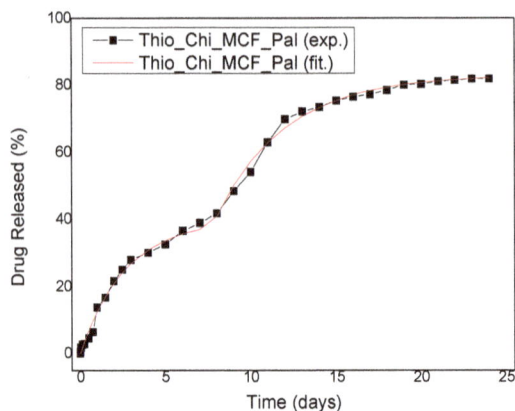

Figure 11. Paliperidone release profile for thiolated MCF particles. The experimental data are shown as symbols and the fitting data as a continuous line.

The next step is the analysis of the release curves from the coated polymer microspheres. In our previous work we found that the release from the uncoated microspheres can be described as a Fickian diffusion process [50]. The complete solution of the transient diffusion equation was successfully used to fit the data (including the initial release burst). In the presence of chitosan coating for PLGA, the initial burst disappears and the whole release process is delayed (similar to the MCF chitosan-coated particles). The process can be described in this case using a simple LDF formula:

$$\frac{C}{C_\infty} = 1 - \exp(-KU(t - \tau)), \tag{5}$$

where τ accounts for the initial dead time due to the existence of the (initially empty of solute) chitosan layer. The fit to the data gives $\tau = 0.7$ days and $K = 0.09$ days^{-1}. This value is compatible with the relation $K = \left(\frac{R^2}{15D_p} + \frac{R\delta}{3D_c}\right)^{-1}$, where R is the microsphere radius (the typical value of 1.3 μm is used here), δ is the chitosan layer thickness of 40 nm, D_p is the diffusion coefficient of paliperidone in PLGA microspheres found in the previous work ($D_p = 2.64 \times 10^{-18}$ m^2/s), and D_c is the diffusion coefficient of paliperidone in the coating layer found by analyzing the coated MCF particles' release data. Whereas the coated PLGA data are compatible with the uncoated PLGA and the coated MCF release data, the situation is completely different for the coated PLA microspheres data. The release curve can be described as a simple line, which means that the release rate is constant up to the disappearance of the paliperidone from the particles. However, it has been shown [18] that the dominant mechanism for release from uncoated PLA particles is diffusion. In addition, in the present work it is shown that the coating layer acts as an additional diffusion resistance. The only explanation for having a constant rate process for the coated particles (instead of a diffusion one as in the PLGA case) is an interfacial transport barrier in the boundary between PLA and coating. This barrier exhibits a zero-order kinetic, i.e., the rate of transport from PLA to coating does not depend on the solute concentration in PLA. Such a barrier does not exist in the PLGA–coating interface. From a practical point of view the achievement of a constant release rate in the PLA-coating system is a major issue since release rate uniformity is almost impossible to achieve for a diffusion-dominated system [51]. The comparison between fitting and experimental coated polymer microspheres release curves is shown in Figure 12.

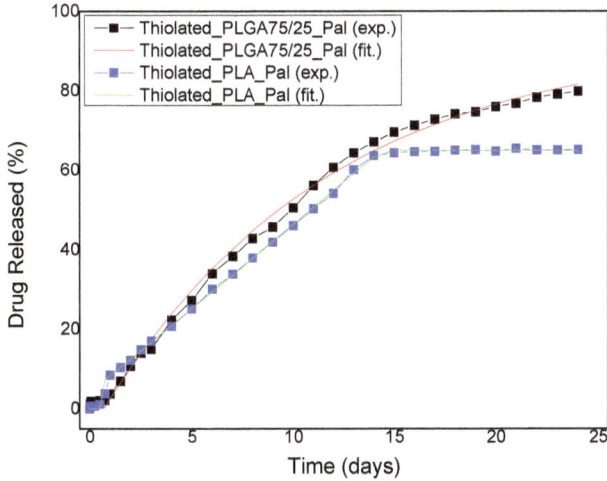

Figure 12. Paliperidone release profile for thiolated PLGA and PLA microspheres. The experimental data are shown as symbols and the fitting data as a continuous line.

Let us finally discuss the release curves for the combined MCF/polymer/coating system. In the absence of a coating, the composite particle release showed some resemblance to MCF particle release, which was physically and mathematically described by the presence of MCF particles on the surface of composite particles [18]. Here the coating layer excludes this possibility and this is evident in the release curves that have no similarity with the MCF particle release profile. The effect of the coating layer is dominant, such that the addition of MCF particles into the polymer particles has no actual influence on the shape of the controlled release curves. The fitting equations are the same as those used for the coated polymer particles and the fitting results are shown in Figure 13. As a confirmation of the dominance of the PLA–coating layer interface transport on the release rate, it is noted that the release rate is the same in the absence and presence of MCF particles in the polymer particle.

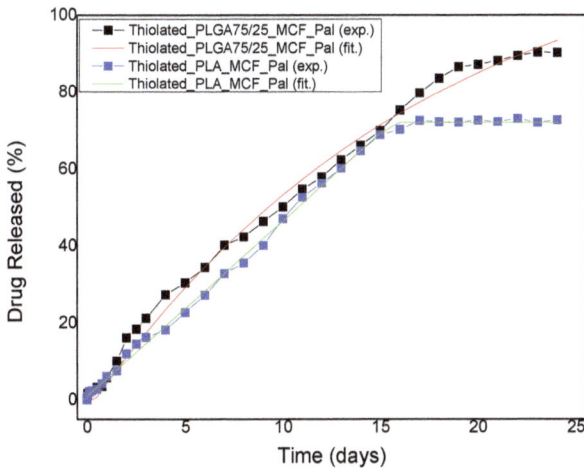

Figure 13. Paliperidone release profile for thiolated PLGA/MCF and PLA/MCF microspheres. The experimental data are shown as symbols and the fitting data as a continuous line.

4. Conclusions

In this study thiolated chitosan was used to mask microspheres of PLA and PLGA 75/25 w/w containing MCF silica loaded with paliperidone drug. The main scope was to evaluate any possible changes that happened in microspheres after their masking with thiolated chitosan, a mucoadhesive polymer that is used for intranasal delivery formulations. In all cases polymeric microspheres with spherical shapes were formed, with sizes varying between 1 and 3 μm. In XRD studies it was found that paliperidone was encapsulated in an amorphous form inside of MCF pores. SEM and TEM images showed that thiolated chitosan formed a film that coated the microspheres. As was found from drug release studies and model fitting of release rates, this film coating resulted in the delay of the dissolution release profile of paliperidone. Also, it was found that, due to this coating layer, the addition of MCF particles into the polymer microspheres had no influence on the drug release rate. However, the encapsulation of paliperidone into MCF nanoparticles is essential for drug dissolution enhancement.

Acknowledgments: Support for this study was received in the framework of the Hellenic Republic–Siemens Settlement Agreement from the State Scholarships Foundation through Operational Program "IKY FELLOWSHIPS OF EXCELLENCE FOR POSTGRADUATE STUDIES IN GREECE—SIEMENS PROGRAM," which is gratefully acknowledged.

Author Contributions: Stavroula Nanaki, Maria Tseklima, and Evi Christodoulou performed the experiments; Dimitrios N. Bikiaris conceived and designed the experiments; Konstantinos Triantafyllidis prepared MCF materials; Margaritis Kostoglou conducted the kinetic studies. All authors participated to paper writing.

Conflicts of Interest: The authors declare no conflict of interest. The founding sponsors had no role in the design of the study; in the collection, analyses, or interpretation of data; in the writing of the manuscript, and in the decision to publish the results.

References

1. Kresge, C.T.; Leonowilz, M.E.; Roth, W.J.; Vartuli, J.C.; Beck, J.S. Ordered mesoporous molecular sieves synthesized by a liquid-crystal template mechanism. *Nature* **1992**, *359*, 710–712. [CrossRef]
2. Munoz, B.; Ramila, A.; Perez-Pariente, J.; Diaz, I.; Vallet-Regi, M. MCM-41 Organic modification as drug delivery rate regulator. *Chem. Mater.* **2003**, *15*, 500–503. [CrossRef]
3. Han, Y.J.; Stucky, G.D.; Butler, A. Mesoporous silicate sequestration and release of proteins. *J. Am. Chem. Soc.* **1999**, *121*, 9897–9898. [CrossRef]
4. Arruebo, M.; Galan, M.; Navascues, N.; Tellez, C.; Marquina, C.; Ibarra, M.R.; Santamaria, J. Development of magnetic nanostructured silica-based materials as potential vectors for drug-delivery applications. *Chem. Mater.* **2006**, *18*, 1911–1919. [CrossRef]
5. Arruebo, M.; Fernandez-Pacheco, R.; Irusta, S.; Arbiol, J.; Ibarra, M.R.; Santamaria, J. Sustained release of doxorubicin from zeolite-magnetite nanocomposites prepared by mechanical activation. *Nanotechnology* **2006**, *17*, 4057–4064. [CrossRef] [PubMed]
6. Radu, D.R.; Lai, C.Y.; Jeftinija, K.; Rowe, E.W.; Jeftinija, S.; Lin, V.S.Y. A polyamidoamine dendrimer-capped mesoporous silica nanosphere-based gene transfection reagent. *J. Am. Chem. Soc.* **2004**, *126*, 13216–13217. [CrossRef] [PubMed]
7. Lin, Y.S.; Tsai, C.P.; Huang, H.Y.; Kuo, C.T.; Hung, Y.; Huang, D.M.; Chen, Y.C.; Mou, C.Y. Well-ordered mesoporous silica nanoparticles as cell markers. *Chem. Mater.* **2005**, *17*, 4570–4573. [CrossRef]
8. Lai, C.Y.; Trewyn, B.G.; Jeftinija, D.M.; Jeftinija, K.; Xu, S.; Jeftinija, S.; Lin, V.S.Y. A mesoporous silica nanosphere-based carrier system with chemically removable CdS nanoparticle caps for stimuli-responsive controlled release of neurotransmitters and drug molecules. *J. Am. Chem. Soc.* **2003**, *125*, 4451–4459. [CrossRef] [PubMed]
9. Zhang, Y.; Zhang, J.; Jiang, T.; Wang, S. Inclusion of the poorly water-soluble drug simvastatin in mesocellular foam nanoparticles: Drug loading and release properties. *Int. J. Pharm.* **2011**, *410*, 118–124. [CrossRef] [PubMed]

10. Zhang, Y.; Jiang, T.; Zhang, Q.; Wang, S. Inclusion of telmisartan in mesocellular foam nanoparticles: Drug loading and release property. *Eur. J. Pharm. Biopharm.* **2010**, *76*, 17–23. [CrossRef] [PubMed]

11. Lee, C.H.; Lo, L.W.; Mou, C.Y.; Yang, C.S. Synthesis and characterization of positive-charge functionalized mesoporous silica nanoparticles for oral drug delivery of an anti-inflammatory drug. *Adv. Funct. Mater.* **2008**, *18*, 3283–3292. [CrossRef]

12. Slowing, I.I.; Vivero-Escoto, J.L.; Wu, C.W.; Lin, V.S.Y. Mesoporous silica nanoparticles as controlled release drug delivery and gene transfection carriers. *Adv. Drug Deliv. Rev.* **2008**, *60*, 1278–1288. [CrossRef] [PubMed]

13. Zhang, Y.; Zhi, Z.; Jiang, T.; Zhang, J.; Wang, Z.; Wang, S. Spherical mesoporous silica nanoparticles for loading and release of the poorly water-soluble drug telmisartan. *J. Control. Release* **2010**, *145*, 257–263. [CrossRef] [PubMed]

14. Lu, J.; Liong, M.; Zink, J.I.; Tamanoi, F. Mesoporous silica nanoparticles as a delivery system for hydrophobic anticancer drugs. *Small* **2007**, *3*, 1341–1346. [CrossRef] [PubMed]

15. Rosenholm, J.M.; Sahlgren, C.; Linden, M. Towards multifunctional, targeted drug delivery systems using mesoporous silica nanoparticles—Opportunities & challenges. *Nanoscale* **2010**, *2*, 1870–1883. [CrossRef] [PubMed]

16. Meltzer, H.; Kramer, M.; Gassmann-Mayer, C.; Lim, P.; Bobo, W.; Eerdekens, M. Efficacy and tolerability of oral paliperidone extended-release tablets in the treatment of acute schizophrenia: Pooled data from three 6-week placebo-controlled studies. *J. Clin. Psychiatry* **2006**, *69*, 817–829. [CrossRef]

17. Darville, N.; van Heerden, M.; Vynckier, A.; De Meulder, M.; Sterkens, P.; Annaert, P.; Van den Mooter, G. Intramuscular administration of paliperidone palmitate extended-release injectable microsuspension induces a subclinical inflammatory reaction modulating the pharmacokinetics in rats. *J. Pharm. Sci.* **2014**, *103*, 2072–2087. [CrossRef] [PubMed]

18. Nanaki, S.; Tseklima, M.; Terzopoulou, Z.; Nerantzaki, M.; Giliopoulos, D.J.; Triantafyllidis, K.; Kostoglou, M.; Bikiaris, D.N. Use of mesoporous cellular foam (MCF) in preparation of polymeric microspheres for long acting injectable release formulations of paliperidone antipsychotic drug. *Eur. J. Pharm. Biopharm.* **2017**, *117*, 77–90. [CrossRef] [PubMed]

19. Gao, X.; Tao, W.; Lu, W.; Zhang, Q.; Zhang, Y.; Jiang, X.; Fu, S. Lectin-conjugated PEG–PLA nanoparticles: Preparation and brain delivery after intranasal administration. *Biomaterials* **2006**, *27*, 3482–3490. [CrossRef] [PubMed]

20. Vila, A.; Sanchez, A.; Evora, C.; Soriano, I.; McCallion, O.; Alonso, M.J. PLA-PEG particles as nasal protein carriers: The influence of the particle size. *Int. J. Pharm.* **2005**, *292*, 43–52. [CrossRef] [PubMed]

21. Csaba, N.; Sánchez, A.; Alonso, M.J. PLGA: Poloxamer and PLGA: Poloxamine blend nanostructures as carriers for nasal gene delivery. *J. Control. Release* **2006**, *113*, 164–172. [CrossRef] [PubMed]

22. Darville, N.; Heerden, M.; Mariën, D.; DeMeulder, M.; Rossenu, S.; Vermeulen, A.; Vynckier, A.; De Jonghe, S.; Sterkens, P.; Annaert, P.; et al. The effect of macrophage and angiogenesis inhibition on the drug release and absorption from an intramuscular sustained-release paliperidone palmitate suspension. *J. Control. Release* **2016**, *230*, 95–108. [CrossRef] [PubMed]

23. Koukaras, E.N.; Papadimitriou, S.A.; Bikiaris, D.N.; Froudakis, G.E. Properties and energetics for design and characterization of chitosan nanoparticles used for drug encapsulation. *RSC Adv.* **2014**, *4*, 12653–12661. [CrossRef]

24. Van der Lubben, I.M.; Verhoef, J.C.; Borchard, G.; Junginger, H.E. Chitosan and its derivatives in mucosal drug and vaccine delivery. *Eur. J. Pharm. Sci.* **2001**, *14*, 201–207. [CrossRef]

25. Wong, T.W. Chitosan and its use in design of insulin delivery system. *Recent Pat. Drug Deliv. Formul.* **2009**, *3*, 8–25. [CrossRef] [PubMed]

26. Henriksen, I.; Green, K.L.; Smart, J.D.; Smistad, G.; Karlsen, J. Bioadhesion of hydrated chitosan: An in vitro and in vivo study. *Int. J. Pharm.* **1996**, *145*, 231–240. [CrossRef]

27. Schipper, N.G.; Olsson, S.; Hoogstraate, J.A.; deBoer, A.G.; Varum, K.M.; Artursson, P. Chitosans as absorption enhancers for poorly absorbable drugs 2: Mechanism of absorption enhancement. *Pharm. Res.* **1997**, *14*, 923–929. [CrossRef] [PubMed]

28. Lehr, C.M.; Bouwstra, J.A.; Schacht, E.H.; Junginger, H.E. In vitro evaluation of mucoadhesive properties of chitosan and some other natural polymers. *Int. J. Pharm.* **1992**, *78*, 43–48. [CrossRef]

29. Artursson, P.; Lindmark, T.; Davis, S.S.; Illum, L. Effect of chitosan on the permeability of monolayers of intestinal epithelial cells (Caco-2). *Pharm. Res.* **1994**, *11*, 1358–1361. [CrossRef] [PubMed]

30. Zahir-Jouzdani, F.; Mahbod, M.; Soleimani, M.; Vakhshiteh, F.; Arefian, E.; Shahosseini, S.; Dinarvand, R.; Atyabi, F. Chitosan and thiolated chitosan: Novel therapeutic approach for preventing corneal haze after chemical injuries. *Carbohydr. Polym.* **2018**, *17*, 42–49. [CrossRef] [PubMed]

31. Anitha, A.; Deepa, N.; Chennazhi, K.P.; Nair, S.V.; Tamura, H.; Jayakumar, R. Development of mucoadhesive thiolated chitosan nanoparticles for biomedical applications. *Carbohydr. Polym.* **2011**, *83*, 66–73. [CrossRef]

32. Sreenivas, S.A.; Pai, K.V. Thiolated Chitosans: Novel polymers for mucoadhesive drug delivery—A Review. *Trop. J. Pharm. Res.* **2008**, *7*, 1077–1088. [CrossRef]

33. Mahmood, A.; Lanthaler, M.; Laffleur, F.; Huck, C.W.; Bernkop-Schnürch, A. Thiolated chitosan micelles: Highly mucoadhesive drug carriers. *Carbohydr. Polym.* **2017**, *167*, 250–258. [CrossRef] [PubMed]

34. Duggan, S.; Cummins, W.; O'Donovan, O.; Hughes, H.; Owens, E. Thiolated polymers as mucoadhesive drug delivery systems. *Eur. J. Pharm. Sci.* **2017**, *100*, 64–78. [CrossRef] [PubMed]

35. Menzel, C.; Jelkmann, M.; Laffleur, F.; Bernkop-Schnürch, A. Nasal drug delivery: Design of a novel mucoadhesive and in situ gelling polymer. *Int. J. Pharm.* **2017**, *517*, 196–202. [CrossRef] [PubMed]

36. Schmidt-Winkel, P.; Lukens, W.W.; Zhao, J.D.; Yang, P.; Chmelka, B.F.; Stucky, G.D. Mesocellular siliceous foams with uniformly, sized cells and windows. *J. Am. Chem. Soc.* **1999**, *121*, 254–255. [CrossRef]

37. Zhu, X.; Su, M.; Tang, S.; Wang, L.; Liang, X.; Meng, F.; Hong, Y.; Xu, Z. Synthesis of thiolated chitosan and preparation nanoparticles with sodium alginate for ocular drug delivery. *Mol. Vis.* **2012**, *18*, 1973–1982. [PubMed]

38. Filippousi, M.; Siafaka, P.I.; Amanatiadou, E.P.; Nanaki, S.G.; Nerantzaki, M.; Bikiaris, D.N.; Vizirianakis, I.S.; Van Tendeloo, G. Modified chitosan coated mesoporous strontium hydroxyapatite nanorods as drug carriers. *J. Mater. Chem. B* **2015**, *3*, 5991–6000. [CrossRef]

39. Jiang, L.; Li, X.; Liu, L.; Zhang, Q. Thiolated chitosan-modified PLA-PCL-TPGS nanoparticles for oral chemotherapy of lung cancer. *Nanoscale Res. Lett.* **2013**, *8*, 66. [CrossRef] [PubMed]

40. Azeem, A.; Iqbal, Z.; Ahmad, F.; Khar, K.; Talegaonkar, S. Development and validation of a stability-indicating method for determination of ropinirole in the bulk drug and in pharmaceutical dosage forms. *Acta Chromatogr.* **2008**, *20*, 95. [CrossRef]

41. Li, J.; Liu, D.; Tan, G.; Zhao, Z.; Yang, X.; Pan, W. A comparative study on the efficiency of chitosan-*N*-acetylcysteine, chitosan oligosaccharides or carboxymethyl chitosan surface modified nanostructured lipid carrier for ophthalmic delivery of curcumin. *Carbohydr. Polym.* **2016**, *146*, 435–444. [CrossRef] [PubMed]

42. Yin, L.; Ding, J.; He, C.; Cui, L.; Tang, C.; Yin, C. Drug permeability and mucoadhesion properties of thiolated trimethyl chitosan nanoparticles in oral insulin delivery. *Biomaterials* **2009**, *30*, 5691–5700. [CrossRef] [PubMed]

43. Chrissafis, K.; Paraskevopoulos, K.M.; Papageorgiou, G.Z.; Bikiaris, D.N. Thermal and dynamic mechanical behaviour of bionanocomposites: Fumed silica nanoparticles dispersed in poly(vinyl pyrrolidone), chitosan and poly(vinyl alcohol). *J. Appl. Polym. Sci.* **2008**, *110*, 1739–1749. [CrossRef]

44. Bikiaris, D.N. Solid dispersions, Part I: Recent evolutions & future opportunities in manufacturing methods for dissolution rate enhancement of poorly water soluble drugs. *Expert Opin. Drug Deliv.* **2011**, *8*, 1501–1519. [CrossRef] [PubMed]

45. Bikiaris, D.N. Solid dispersions, Part II: New strategies in manufacturing methods for dissolution rate enhancement of poorly water soluble drugs. *Expert Opin. Drug Deliv.* **2011**, *8*, 1663–1689. [CrossRef] [PubMed]

46. Papageorgiou, G.Z.; Achilias, D.S.; Nanaki, S.; Beslikas, T.; Bikiaris, D. PLA nanocomposites: Effect of filler type on non-isothermal crystallization. *Thermochim. Acta* **2010**, *511*, 129–139. [CrossRef]

47. Kyzas, G.Z.; Kostoglou, M. Swelling-adsorption interactions during mercury and nickel ions removal by chitosan derivatives. *Sep. Purif. Technol.* **2015**, *149*, 92–102. [CrossRef]

48. Kyzas, G.Z.; Lazaridis, N.K.; Kostoglou, M. Modelling the effect of pre-swelling on adsorption dynamics of dyes by chitosan derivatives. *Chem. Eng. Sci.* **2012**, *81*, 220–230. [CrossRef]

49. Tien, C. *Adsorption Calculations and Modeling*; Butterworth-Heinemann: Boston, MA, USA, 1994.

50. Crank, J. *The Mathematics of Diffusion*; Oxford University Press: Oxford, UK, 1975.
51. Georgiadis, M.C.; Kostoglou, M. On the optimization of drug release from multi-laminate polymer matrix devices. *J. Control. Release* **2002**, *77*, 273–285. [CrossRef]

© 2017 by the authors. Licensee MDPI, Basel, Switzerland. This article is an open access article distributed under the terms and conditions of the Creative Commons Attribution (CC BY) license (http://creativecommons.org/licenses/by/4.0/).

![polymers logo] *polymers*

MDPI

Article

Tailoring Drug Release Properties by Gradual Changes in the Particle Engineering of Polysaccharide Chitosan Based Powders

Ednaldo G. do Nascimento [1], Lilia B. de Caland [1], Arthur S.A. de Medeiros [1], Matheus F. Fernandes-Pedrosa [1], José L. Soares-Sobrinho [2], Kátia S.C.R. dos Santos [3] and Arnóbio Antonio da Silva-Júnior [1,*]

[1] Laboratory of Pharmaceutical Technology and Biotechnology, Department of Pharmacy, Federal University of Rio Grande do Norte, UFRN, Gal. Gustavo Cordeiro de Farias, Petropolis, Natal 59072-570, RN, Brazil; ednaldogn40@gmail.com (E.G.d.N.); liliabasiliocaland@gmail.com (L.B.d.C.); arthursergiomedeiros@gmail.com (A.S.A.d.M.); mpedrosa@ufrnet.br (M.F.F.-P.)
[2] Department of Pharmacy, Center of Health Sciences, Federal University of Pernambuco, Professor Moraes Rego 1235, Recife 50670-901, PE, Brazil; joselamartine@hotmail.com
[3] School of Pharmaceutical Sciences, Federal University of Amazonas, UFAM, General Rodrigo Octávio Jordão Ramos, 6200, South Sector, Manaus 69077-000, AM, Brazil; katiasolange@hotmail.com
* Correspondence: arnobiosilva@gmail.com; Tel.: +55-084-3342-9820

Academic Editor: George Papageorgiou
Received: 14 June 2017; Accepted: 24 June 2017; Published: 29 June 2017

Abstract: Chitosan is a natural copolymer generally available in pharmaceutical and food powders associated with drugs, vitamins, and nutraceuticals. This study focused on monitoring the effect of the morphology and structural features of the chitosan particles for controlling the release profile of the active pharmaceutical ingredient (API) propranolol hydrochloride. Chitosan with distinct molecular mass (low and medium) were used in the formulations as crystalline and irregular particles from commercial raw material, or as spherical, uniform, and amorphous spray-dried particles. The API–copolymer interactions were assessed when adding the drug before (drug-loaded particles) or after the spray drying (only mixed with blank particles). The formulations were further compared with physical mixtures of the API with chitin and microcrystalline cellulose. The scanning electron microscopy (SEM) images, surface area, particle size measurements, X-ray diffraction (XRD) analysis and drug loading have supported the drug release behavior. The statistical analysis of experimental data demonstrated that it was possible to control the drug release behavior (immediate or slow drug release) from chitosan powders using different types of particles.

Keywords: polysaccharides; chitosan; microparticles; spray drying; structural properties; polymer characterization; drug release

---

1. Introduction

Polysaccharides have been extensively used as ingredients in the pharmaceutical and food industries. The chitosan is a cationic polysaccharide copolymer composed of glucosamine (β (1–4)-linked 2-amino-2-D-glucose) and N-acetylglucosamine (2-acetamido-2-deoxy-D-glucose) (Figure 1). This raw material is generally produced by the partial deacetylation of chitin, the second most abundant polymer in nature, present in exoskeletons of crustaceans [1,2]. In solid dosage forms, the chitosan has been used as excipient for tablets [3,4], diluent or granulating agent [5,6]. In liquid dispersions, this copolymer is also used as an emulsifying or thickness agent [7,8]. Furthermore, it has been applied in drug delivery systems for modulating the release or enhancing the solubility of pharmaceutical ingredients [9–11].

Polymers **2017**, *9*, 253

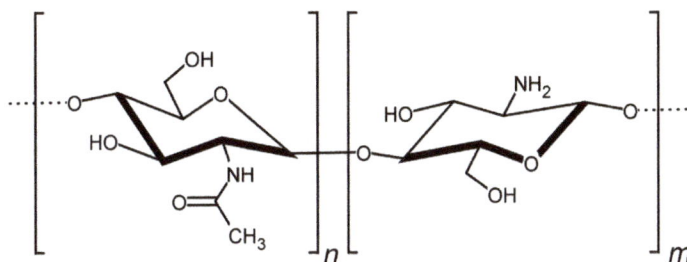

Figure 1. Schematic presentation of chemical structure of the copolymer chitosan composed of N-acetylglucosamine (2-acetamido-2-deoxy-D-glucose) [n] and glucosamine (β (1–4)-linked 2-amino-2-D-glucose) [m].

Chitosan based microparticles have also been used for drug targeting in specific tissues such as gastric or colon mucosa [12,13]. This copolymer is responsible for enhancement of drug permeability and bioavailability, due to the mucoadhesive properties [14]. Added to their versatile properties, chitosan is a biomaterial biodegradable with safety established for the oral rout in humans [15–17]. The chitosan particles can be produced using several methods, such as freeze-drying [18], spray-drying [19], ionotropic gelation, coacervation or co-precipitation with solvent evaporation [18,20,21]. The desired properties of particles such as size, shape, density, porosity and the drug-loading ability are certainly considered in the experimental design [22]. The spray drying is a single step method, widely used in foods, able to produce small, uniform and spherical polymeric particles with the drug generally homogeneously dispersed into the polymeric matrix [19,23].

Considering the active pharmaceutical ingredient (API) release rate for oral rout, the pharmaceutical dosage forms are classified as the immediate release, in which generally 80% w/w of the ingredient is released until 45 min in acid medium. The modified drug delivery devices can present the drug release as being pH-tunable or slow drug release in a superior interval time. Taking into account the hydrogel behavior of chitosan particles, the drug release rate can be modulated by the pH of medium or by inherent copolymer properties such as degrees of deacetylation (DD) and molecular mass [24,25]. Other aspects involve the structural properties of particles, drug crystallinity, and drug–copolymer interactions.

In this study, chitosan with different viscosimetric molecular mass (low and medium) were used to evaluate how the morphology of particles (commercial raw material or spray-dried particles) and drug–copolymer interaction in physical mixtures or in drug-loaded spray-dried microparticles affected the drug release rate. For this purpose, the anti-hypertensive propranolol hydrochloride was used as a pharmaceutical ingredient due to high water solubility. The physicochemical and structural features of different powder samples were carefully monitored and the drug dissolution data subjected to comparisons using a suitable mathematical modeling approach.

2. Materials and Methods

2.1. Materials

Low and medium molecular mass chitosan (LMC and MMC, respectively) were purchased from Sigma-Aldrich (São Paulo, SP, Brazil). Propranolol hydrochloride (PPHy) and microcrystalline cellulose (MCC) were from SM pharmaceutical enterprises (São Paulo, SP, Brazil). Chitin (CH) was from Polymar (Fortaleza, CE, Brazil). All other reagents were of analytical grade. The purified water (1.2 μS) was prepared from an OS50 LX reverse osmosis purification apparatus Gehaka (São Paulo, SP, Brazil).

2.2. Physicochemical Characterization of Chitosan

The viscosimetric molecular mass (Mv) of chitosan was determined using the flow time of six chitosan solutions at different concentrations (0.17; 0.18; 0.19; 0.21 and 0.23% w/v) diluted in acetic acid 0.5 M/sodium acetate 0.2 M buffer. The dispersions passed through capillary viscosimeter CFRC-100 model (Cannon-Fenske Routine, USA), with size 100 at 25 ± 0.1 °C ($n = 6$). The relative (ηr), specific (ηsp), reduced (ηred) and inherent (ηinh) viscosities were determined, as shown below in Equations (1)–(4).

$$\eta r = t/t0 \tag{1}$$

$$\eta sp = \eta r - 1 \tag{2}$$

$$\eta red = \eta sp/C \tag{3}$$

$$\eta inh = \ln(\eta r)/C \tag{4}$$

where t and t0 represent flow time of chitosan solutions and buffer solutions, respectively, C represents the natural logarithmic chitosan concentration.

The ηred and ηinh values were plotted versus chitosan concentration and the intrinsic viscosity (η) were determined by the curve extrapolation for the absence of chitosan. Therefore, the correlation of molecular mass with the η value were determined by the empirical equation of Mark-Houwink-Sakurada Equation (5), where [η] is the intrinsic viscosity, k and a are constants for a specific polymer solution with the used solvent at a specific temperature, that is $k = 3.5 \times 10^{-4}$ and $a = 0.76$ [26].

$$[\eta] = kMva \tag{5}$$

The degree of deacetylation was determined using the conductometric titration method [27]. The protonated amine groups were determined in chitosan solution of 0.5% w/v dissolved with aqueous hydrochloric acid of 0.06 M, adding in the following NaOH solution (0.15 M) to allow the deprotonation of chitosan. The degree of deacetylation was measured according the following Equation (6):

$$DD = 100 \, MA \, (\Delta V \cdot C_{NaOH}/\Delta V \cdot C_{NaOH} \cdot \Delta M + WCHIT \cdot ms) \tag{6}$$

where DD is the degree of deacetylation, MA is the molecular mass of acetylated copolymer, ΔV is the variation of NaOH volume, ΔM is the difference between the molecular mass between acetylated and deacetylated copolymers, WCHIT is the solid mass fraction of chitosan and ms is the mass of the sample.

2.3. Preparation of Powder Samples

2.3.1. Spray-Dried Microparticles

The aqueous solution (acetic acid 1.0% w/v) containing chitosan at 0.5% (w/v) was prepared under magnetic stirring for 24 h and dried in an ADL311S spray dryer (Yamato Scientific Co., Tokyo, Japan) to obtain the blank microparticles. For the drug-loaded chitosan microparticles, the propranolol was dissolved with the copolymer at mass ratio drug-chitosan of 1:2. The drying conditions were previously established in previous studies [28]. The inlet temperature of 140 °C, outlet temperature of about 90 °C, air pressure of 0.1 MPa, air flow of 0.32 m^3/min and a feed flow rate of 5 mL/min through a nozzle of 0.4 mm were used during the experiments.

2.3.2. Powder Formulations

The hard gelatin capsules n° 2 were filled with powder mixtures containing propranolol hydrochloride. The drug was physically mixed with the different excipients using mortar, or loaded in the spray-dried chitosan microparticles. The final composition of different samples are shown in Table 1.

Table 1. Composition of powders used in the hard gelatin capsules.

Samples	Powder (Composition)	Mass of Diluent (mg)
LMC	Low molecular mass chitosan	76
MMC	Medium molecular mass chitosan	69
CH	Chitin	76
MCC	Microcystalline cellulose	76
Blank LMC-MPs	Microparticles of LMC	76
Blank MMC-MPs	Microparticles of MMC	76
PPHy LMC-MPs	Drug-loaded LMC-MPs	116 *
PPHy MMC- MPs	Drug-loaded MMC-MPs	116 *

* 116 mg (76 mg of Chitosan + 40 mg of PPHy). All capsules also contain the addition of 21 mg of talc (diluent) and 40 mg of PPHy.

2.4. Morphology, Particle Size and Surface Area Analysis

The shape and surface aspect of particles were accessed using the scanning electron microscopy (SEM) images taken in TM 3000 Microscope Hitachi (Tokyo, Japan). The particles were dried and mounted on metal stubs using double-sided adhesive carbon tape and analyzed at the voltage of 20.0 kV. The mean diameter and the size distribution of the microparticles was determined using dynamic light scattering (DLS) in a Nanotrac NPA252 (Montgomeryville, PA, USA) with Flex software 10.4.3. The amount of 4.0 mg of powder was dispersed in 15.0 mL of aqueous solution of polysorbate 80 at 0.5% w/v. The cumulative diameter of 10, 50 and 90% in the particle size distribution were determined in triplicate. The index span was calculated by the equation: SPAN = D90 − D10/D50 [28].

The surface area, pore volume and pore size of the chitosan microparticles were determined following the method of Brunauer-Emmett-Teller (BET), using the liquid N2 adsorption and desorption isotherms, measured at 77 K temperature with an ASAP 2420 surface area analyzer (Micromeritics Instrument, Norcross, USA). All samples were degassed and stored at vacuum at room temperature overnight prior to measurements. The experiments were repeated at least three times using fresh powder.

2.5. X-ray Diffraction Analysis

The X-ray diffraction (XRD) analysis was performed for pure compounds, physical mixtures and spray-dried microparticles in a D2 Phaser diffractometer (Bruker corporation, Billerica, USA) using CuK radiation (λ = 1.54 Å) with a Ni filter. The measurements were performed in a 2-Theta angle variation of 5–45°, angular step of 0.004° and 30 kV voltage.

2.6. Fourier Transform Infrared Spectrophotometry (FT-IR) Studies

The FT-IR was recorded in an IR Prestige-21 equipment (Shimadzu, Kyoto, Japan). Samples of 2 mg of powder were mixed with 300 mg of potassium bromide (KBr) for pellets confection, using a hydraulic press at 10 KgF. The middle infrared region (MID) was analyzed in the range of 400–4000 cm^{-1}.

2.7. Drug Loading Analysis

The spray-dried PPHy-loaded chitosan microparticles were massed to contain 20 mg of drug and then dissolved with 20 mL of purified water. Then, the methanol was added to a final volume of 100 mL with a volumetric flask. The measurements were done in triplicate by UV/Vis spectrophotometry at a wavelength of 290 nm. The absorbance results were correlated with drug concentration through a previously determined calibration curve. The drug loading efficiency was calculated from the relationship between the analytical and the theoretical drug contents.

2.8. Drug Dissolution Assays

The dissolution profile of samples was performed in the simulated gastrointestinal medium [29], using a dissolution equipment 299 model (Ethiktechnology, Sao Paulo, Brazil). The hard gelatin capsules containing the formulations were placed in basket apparatus, using 1000 mL of hydrochloric acid solution 1% (v/v) as dissolution medium at 37 ± 0.5 °C, under 100 rpm agitation. At specific intervals, aliquots of 10 mL of dissolution medium were removed, filtered through a 0.45 µm cellulose acetate, and analyzed by UV spectrophotometry at 290 nm, previously validated in a spectrophotometer 60S Evolution (Thermo Fisher Scientific Inc, Madison, WI, USA). All experiments were performed six times ($n = 6$) and the cumulative percentage of released propranolol was plotted versus time.

The effect of the experimental variations on the drug release profile of the different samples was evaluated by comparisons using the independent-model of similarity factor (f2), according to Equation (7).

$$F2 = 50 \cdot \log \left[1 + (1/n) \cdot \sum (D1t - D2t) \, 2 \right] -0.5] \cdot 100 \tag{7}$$

where, n is the number of experimental intervals, t is the interval time, D1t is the drug ratio dissolved at a specific interval time for formulation (1), and D2t is the drug ratio dissolved at specific interval time for formulation (2). Formulations with drug dissolution profiles that are statistically similar present F2 values in the range of 50 to 100.

2.9. Statistical Analysis

The analytical data were expressed as the mean ± standard deviation. The Mann Whitney test was applied for pairwise comparisons, while multiple comparisons were assessed by the analysis of variance (ANOVA) on ranks (Kruskal Wallis test) at 0.05 of significance level.

3. Results

3.1. Physicochemical Characterization of Chitosan

The different used chitosan (LMC and MMC) have the viscosimetric molecular mass (Mv) and degree of deacetylation assessed to control the behavior of copolymers. The intrinsic viscosity [η] was determined extrapolating the intersection in the plot of the reduced (ηred) and inherent viscosities (ηinh) as a function of the chitosan concentration (Figure 2). The calculated Mv values for LMC and MMC were of 4.61×10^5 and 7.15×10^5 g·mol^{-1}, respectively, which were estimated using the Mark-Houwink-Sakurada equation (Equation (1)).

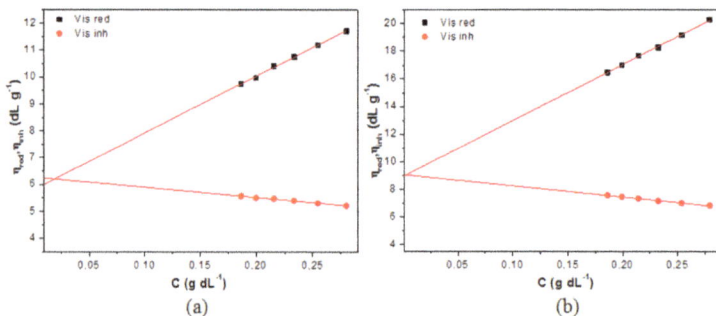

(a)　(b)

Figure 2. The reduced (ηred) and inherent viscosities (ηinh) as function of chitosan concentration (C) for the two used copolymers: (**a**) low molecular mass chitosan (LMC) and (**b**) medium molecular mass chitosan (MMC).

Figure 3 shows the conductivity curves for the MMC and LMC. The first linear descending branch of the curve is the neutralization of excess acid used to solubilize the copolymer. The second was due to the neutralization of NH_3^+ of chitosan amino groups. Finally, the ascending points represent the excess alkaline solution (OH$^-$) after the equivalence point. The extrapolation of these three lines made it possible to observe two inflection points, corresponding respectively to the volume of initial and final NaOH volume (VNaOHi and VNaOHf) required to neutralize the protonated amino group of chitosan [30]. The degrees of deacetylation of 88.16% (±0.18%) and 88.74% (±2.63%) were determined for the LMC and MMC, respectively.

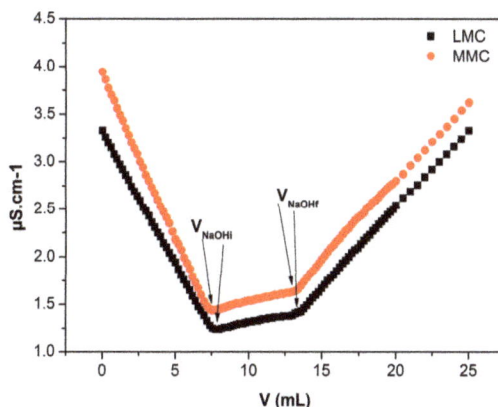

Figure 3. Conductivity titration curves as a function of added volume of NaOH solution (0.15 M) for each used chitosan.

3.2. Physicochemical Properties of Particles

The microparticles were successfully produced using the selected spray drying parameters. They were fine and dried powders, with yield levels of about 56.49 ± 0.06% for the blank microparticles and of about 57.29 ± 0.09% for drug-loaded microparticles. The encapsulation efficiency ranged between 90.1 ± 0.7 and 91.0 ± 3.0 for the PPHy-LMC-MPs and PPHy-MMC-MPs, respectively. Table 2 shows the data of particle size and BET analysis for the spray dried microparticles. The mean diameter ranged between 1.72 to 3.45 µm (SPAN index between 1 and 3).

Table 2. Particle size and BET analysis for the spray-dried microparticles.

Samples	Particle Size Analysis					BET Analysis		
	D10 (µm)	D50 (µm)	D90 (µm)	Mean (µm)	SPAN	Surface Area (m²/g)	Pore Volume (cm³/g)	Pore Size (Å)
Blank LMC-MPs	0.7	2.3	5.2	2.3	2.0	25.3	0.056	42.4
Blank MMC-MPs	1.4	3.5	5.8	3.5	1.3	25.8	0.059	39.7
PPHy-LMC-MPs	0.4	1.7	5.0	1.7	2.7	19.2	0.046	48.7
PPHy-MMC-MPs	1.4	3.4	5.8	3.4	1.3	43.5	0.098	37.5

Table 2 also shows the BET parameters calculated from the nitrogen adsorption–desorption isotherms for different samples of chitosan powders (Figure 4). No hysteresis was observed and the adsorption increased linearly at relative pressure in the range of 0.0–1.0. The drug loading seems to have slightly decreased the pore size for the two tested copolymers.

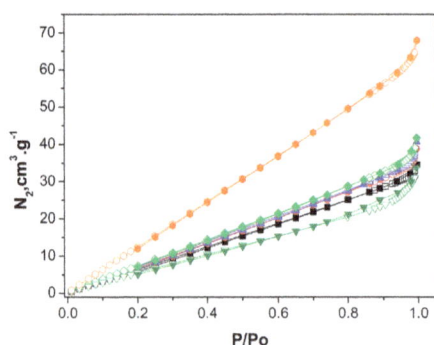

Figure 4. The N_2 adsorption-desorption isotherms for different chitosan powder samples LMC (■); MMC (●); blank LMC-MPs (▲); blank MMC-MPs (♦); PPHy-LMC-MPs (▼); PPHy-MMC-MPs (●).

Figure 5 shows the SEM images for the distinct chitosan particles. The irregular shape with wide particle-size distribution for the two commercial chitosan raw materials LMC and MMC are shown in Figure 5a,b, respectively. In contrast, the blank (Figure 5c,d) and drug-loaded spray-dried microparticles (Figure 5e,f) show regular spherical particles with narrow particle size distribution, a signature of spray dried particles [31]. This fact is important for the release rate that the hydrogel powder can supply, consequently affecting the biological activity. Another structural feature that controls the drug release from the particles is the crystallinity of the compounds. This aspect was assessed using XRD analysis (Figure 6a).

Figure 5. Scanning electron microscopy (SEM) images of different chitosan particles: (**a**) LMC, (**b**) MMC, (**c**) Blank LMC-MPs, (**d**) Blank MMC-MPs, (**e**) PPHy-LMC-MPs, and (**f**) PPHy-MMC-MPs.

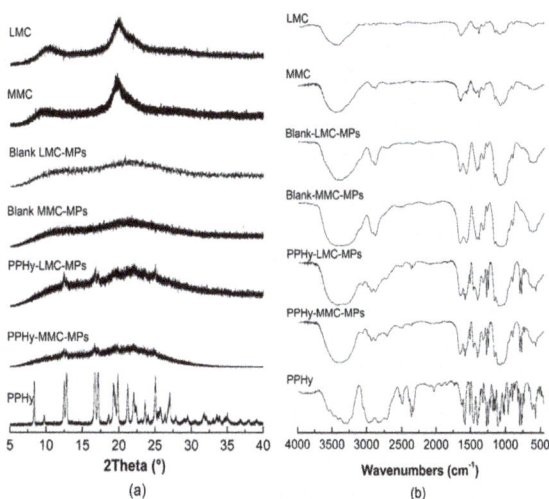

Figure 6. (**a**) X-ray diffraction (XRD) patterns and (**b**) Fourier transform infrared (FT-IR) spectra for different chitosan powders.

The FT–IR spectra (Figure 6b) of the commercial chitosan raw materials showed that the characteristic absorption bands of O–H and N–H stretch appeared overlapped in a large and broad band at 3450 cm^{-1}. The C–H stretch of methyl groups at 2890 cm^{-1}, C=O amide stretch at 1655 cm^{-1}, the N–H angular deformation bands at 1583–1594 cm^{-1}, the C–H symmetric angular deformation bands of methyl at 1380–1383 cm^{-1}, the C–N amino stretch at 1308–1380 cm^{-1} and finally the C–O secondary alcohol stretch at 1103 cm^{-1}. The propranolol spectra showed O–H stretch at 3283 cm^{-1}, the C=C aromatic stretching at 1705 cm^{-1}, the N–H angular deformation at 1590 cm^{-1}, the C–O ether stretch at 1270–1240 cm^{-1} and the C–O secondary alcohol stretch at 1103 cm^{-1}.

The FT–IR of drug-loaded particles did not show any relevant shift of recorded stretches for the pure compounds. The bands of drug and copolymer appeared overlapped, suggesting that the drying process did not induce any chemical reaction of the drug with chitosan.

3.3. Drug Dissolution Assays

Before the dissolution assay, all produced formulations of the capsule have the weight variation and drug content analytically controlled. According to the quality control of pharmaceutical and food products, a variation within the ±10% for average weight and drug content, with relative standard deviations of less than 4%, were given as acceptable. Figure 7 shows the drug dissolution profile from different formulations of powders. Figure 7a plots the formulations that exhibited the immediate drug release profile, which released more than 80% before 40 min [29]. The drug release profile from powders containing the two distinct commercial chitosan microparticles were further compared with powder mixtures containing chitin or microcrystalline cellulose due to their use as diluent for solid dosage forms. The mixtures with microcrystalline cellulose (MCC) and chitin showed similar behavior, releasing the full drug content before 10 min. The mixtures with chitosan raw materials as diluents with distinct molecular mass (LMC and MMC) exhibited a slight delay, explained due to the hydrogel behavior of this compound compared with the MCC and chitin that are water insoluble excipients. After 30 min, the capsules containing LMC and MMC dissolved the drug content of 87.0 ± 1.0% and 87.1 ± 2.7%, respectively.

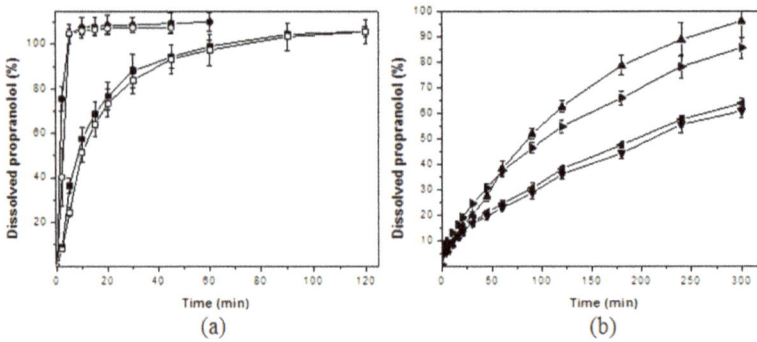

Figure 7. Dissolution profile of propranolol from different chitosan powder formulations: (a) LMC (■), MMC (□), Chitin (•), and microcrystalline cellulose (○), and (b) blank LMC-MPs (▲), blank MMC-MPs (►), PPHy-LMC-MPs (▼), and PPHy-MMC-MPs (◄).

Figure 7b shows the powder formulations with slow drug release profiles. The simple mixture of drug with distinct blank microparticles was able to modulate the drug release, enhancing about 3 times the necessary time to release 100% of drug in the dissolution medium. The drug release rate was further slowed for the drug-loaded spray-dried chitosan microparticles. These samples release about 50% of drug after 300 min. The formulations prepared with the distinct chitosan were also compared. The different molecular mass (LMC and MMC) did not induce relevant changes in the drug dissolution profiles. The comparisons of the experimental data using the similarly factored treatment (f2 > 50) showed the statistical similarity for the pairs LMC with MMC, blank LMC-MPs with MMC-MPs, and PPHy-LMC-MPs with PPHy-MMC-MPs, observed in Table 3.

Table 3. Statistical analysis of the drug release behavior.

GROUPS	SAMPLES	Molecular Weight	$Q_{30min} \pm SD$ (%)	f2 **
Immediate release	LMC	4.61×10^5	86.95	60.07
	MMC	7.15×10^5	87.15	
	CH	ND	107.60	
	MCC	3.6×10^3 *	107.04	-
Slow release	blank LMC-MPs	ND	20.27	50.35
	blank MMC-MPs	ND	24.58	
	PPHy-LMC-MPs	ND	16.95	77.98
	PPHy-MMC-MPs	ND	14.15	

* Handbook of Pharmaceutical Excipients—Fifty edition; ** similarity factor; ND = not determined.

4. Discussion

In this study, we have used two copolymers with the same deacetylation degree, but with medium and low molecular mass, that governed important physicochemical and functional properties of chitosan particles. These aspects of the chitosan were successfully characterized before to explore the physical aspects of particles (Figures 2 and 3). Moreover, we considered morphological and physical aspects of particles, e.g., the irregular commercial chitosan, or spherical spray-dried chitosan microparticles. The drug–copolymer interactions were also assessed in the simple mixtures or from the drug-loaded spray-dried microparticles, as a strategy to modulate the drug release properties for application in foods, cosmetics, chemicals and pharmaceuticals.

The data in Table 1 demonstrated the small and narrow-sized spray dried microparticles compared with previous studies [32–34]. The MMC induced the formation of larger particles compared with LMC, due to the higher packing of smaller polymer chains [35]. In addition, the particle size is highly affected by the viscosity of feed solution, consequently affecting the droplet formation during the drying process. However, these differences did not affect the high drug loading for both samples of spray-dried microparticles. Previous studies also reported spray-dried chitosan microparticles with mean diameters of approximately 3 μm, using low molecular chitosan solutions at 0.5 to 1.0% (w/v) [36,37].

The experimental data of isotherms (Figure 4) were applied to BET calculations to determine surface area of particles (Table 2). Curves typical of mesopore-rich materials were observed, with pore diameters between 20 and 50 Å [38]. The BET analysis showed that both LMC-MPs and MMC-MPs showed equivalent surface area and pore volume. Regarding the drug-loaded particles, PPHy-LMC-MPs exhibited lower surface area and pore volume than PPHy-MMC-MPs, demonstrating a more efficient packing of the copolymer with lower chains in the particles. The greater surface area of PPHy-MMC-MPs compared with PPHy-LMC-MPs did not affect the high EE% of microparticles, since similar values were approximately 90%. This parameter was previously affected by the surface characteristics of chitosan particles produced using different microencapsulation methods [39]. This fact reinforces the amazing performance of the spray drying to produce chitosan microparticles for slow release of small molecules, especially due to the water penetration and swelling of polymeric matrix.

Figure 5 shows the SEM images for the distinct chitosan particles. The irregular shape with wide particle-size distribution for the two commercial chitosan raw materials LMC and MMC are shown in Figure 5a,b. In contrast, the blank (Figure 5c,d) and drug-loaded spray-dried microparticles (Figure 5e,f) have shown regular spherical particles with narrow particle size distribution, a signature of spray dried particles [31]. This fact is important for the release rate that powders with hydrogel character can supply, consequently affecting the biological activity. Mys et al. obtained the technique of spray-dried microparticles in the spherical form of syndiotactic polystyrene [40,41].

Another structural feature that controls the drug release from the particles is the crystallinity of the compounds into the particles. This aspect was assessed using XRD analysis (Figure 6a).

The two characteristic reflections from crystalline samples of commercial chitosan raw materials were observed at 10° and 20° for the LMC, and at 10° and 19° for the MMC, respectively, in accordance with previous studies [42,43]. The pure drug also presented the characteristic crystalline diffraction pattern [44]. The spray drying produced amorphous chitosan microparticles. Considering the drug-loaded microparticles, the PPHy was homogeneously dispersed in the polymeric matrix in the amorphous state. The random drug distribution in the particles induces an energetic favorable drug dissolution compared with crystalline particles of drug into polymeric devices. In this approach, the drug dissolution in the chitosan loaded-microparticles offered freely drug–chitosan interactions, which were able to control the drug dissolution. This kind of interaction does not occur when the drug is physically mixed with the chitosan particles. Thus, we can observe interesting differences in drug dissolution behavior. The interactions of drug with the chitosan were assessed using FT-IR analysis (Figure 6b).

The amorphization of chitosan in the blank microparticles caused wavenumber shifts and changes in the transmittance intensity for the C=O amide stretch at approximately 1650 cm^{-1} and for the C–O secondary alcohol stretch at 1103 cm^{-1}. This fact was also previously demonstrated [45]. The FT-IR of drug-loaded particles did not show any relevant shift of recorded stretches for the pure compounds. The bands of drug and copolymer appeared overlapped, suggesting that the drying process did not induce any chemical reaction of drug with chitosan.

The adjuvant functionality of chitin, microcrystalline cellulose, chitosan and chitosan microparticles was evaluated, using these components as diluents in propranolol capsules. The drug-loaded microparticles was also encapsulated to observe the drug delivery profile. Before dissolution assay, all produced capsule batches were confronted in weight variation and content uniformity tests (data not shown). All capsule formulations presented the average weight within the ±10% range. Furthermore, the relative standard deviations were less than 4%. Regarding uniformity of content, the results showed that propranolol was assayed within the range 90–110%.

The drug release behavior also showed that the morphology and structural features of the particles also modulate the hydrogel behavior of chitosan, which is pH dependent. The chitosan is soluble at an acidic pH, but the different shape, size, and crystallinity of the spray−dried particles (blank LMC and blank MMC) enhanced the hydrogel behavior of chitosan, decreasing the drug dissolution rate from the powders. Other researchers have observed similar behavior using another polymer (hydroxypropyl methyl cellulose phthalate) in the release of API, verifying that it depends directly on the physical property of the particle and the pH of the medium [46]. The amorphous character and great surface area of these particles compared with the commercial raw chitosan changes the solubility and the swelling rate of the copolymer, changing its interaction with the drug in the aqueous medium [47,48]. In the case of drug-loaded spray-dried chitosan microparticles (PPHy-LMC-MPs and PPHy-MMC-MPs), the drug entrapment in the polymeric matrix further enhanced the drug–copolymer interaction, improving the slow release effect of the powder formulations. The similar dissolution between the pairwise formulations also evidence that the neglectable particle size variations, within 1 μm range between the samples, did not influence the dissolution rate. The particle size directly affects the drug dissolution, in which higher surface area or lower particle size leads to a faster dissolution of the entire particle, as shown in the Noyes-Whitney equation [49]. In the case of polymeric particles with hydrogel character, the slow drug dissolution followed by diffusion from the particle is expected. Therefore, the largest differences in the particle size of hydrogel particles can induce significant differences in drug release kinetic behavior [50].

In this approach, different deacetylation degrees were not tested, because the chitosan commercially ranged from 80 to 85%. Thus, drug release performance from the chitosan powders was dependent on its hydrogel behavior. In addition, the experimental data proved that it was possible to control this property, and consequently the drug–copolymer interaction manipulating the morphology and structural features of the chitosan particles.

5. Conclusions

In this study, we have demonstrated the versatility of polysaccharides such as chitosan for food, chemistry and pharmaceutical industry as an ingredient capable of modulating the release behavior of additives, vitamins and active pharmaceutical ingredients. The experimental data discussed in this approach demonstrated that it was possible to modulate the active pharmaceutical ingredient release in chitosan powders to induce an immediate or slow release behavior. This behavior was dependent on the hydrogel behavior of the copolymer, which we have controlled using different types of unprocessed chitosan or spray-dried particles. The mixture of API in the particles also affected the release rate.

Acknowledgments: The authors wish to thank the Brazilian National Council for Scientific and Technological Development (CNPq) for financial support (grant numbers: 483073/2010-5; 481767/2012-6) and Coordination for the Improvement of Higher Level (CAPES) (grant number: AUXPE PNPD 23038.007487/2011-91 and scholarship of de Caland, LB). The authors also acknowledge the help extended by the Multifunctional Materials Laboratory and Numerical Experimentation (LAMMEN—UFRN), Electron Microscopy Laboratory Scanning (DEMAT—UFRN) and Surfactants Technology Laboratory (LTT—UFRN) for sharing their resources.

Author Contributions: Ednaldo G. do Nascimento planned, executed, and analyzed the experiments in cooperation with Kátia S. C. R. dos Santos and Arnóbio Antonio da Silva-Júnior and wrote the paper. Lilia B. de Caland, Arthur S. A. de Medeiros, Matheus F. Fernandes-Pedrosa and José L. Soares-Sobrinho analyzed the data and supported the research.

Conflicts of Interest: The authors declare no conflict of interest.

References

1. Aider, M. Chitosan application for active bio-based films production and potential in the food industry: Review. *LWT Food Sci. Technol.* **2010**, *43*, 837–842. [CrossRef]
2. Ngo, D.-H.; Vo, T.-S.; Ngo, D.-N.; Kang, K.-H.; Je, J.-Y.; Pham, H.N.-D.; Byun, H.-G.; Kim, S.-K. Biological effects of chitosan and its derivatives. *Food Hydrocoll.* **2015**, *51*, 200–216. [CrossRef]

3. Mir, V.G.; Heinämäki, J.; Antikainen, O.; Revoredo, O.B.; Colarte, A.I.; Nieto, O.M.; Yliruusi, J. Direct compression properties of chitin and chitosan. *Eur. J. Pharm. Biopharm.* **2008**, *69*, 964–968. [CrossRef] [PubMed]

4. Lopez, O.; Garcia, M.A.; Villar, M.A.; Gentili, A.; Rodriguez, M.S.; Albertengo, L. Thermo-compression of biodegradable thermoplastic corn starch films containing chitin and chitosan. *LWT Food Sci. Technol.* **2014**, *57*, 106–115. [CrossRef]

5. Yang, Y.; Hadinoto, K. A highly sustainable and versatile granulation method of nanodrugs via their electrostatic adsorption onto chitosan microparticles as the granulation substrates. *Int. J. Pharm.* **2013**, *452*, 402–411. [CrossRef] [PubMed]

6. Phaechamud, T.; Darunkaisorn, W. Drug release behavior of polymeric matrix filled in capsule. *Saudi Pharm. J.* **2016**, *24*, 627–634. [CrossRef] [PubMed]

7. Zinoviadou, K.G.; Scholten, E.; Moschakis, T.; Biliaderis, C.G. Engineering interfacial properties by anionic surfactant-chitosan complexes to improve stability of oil-in-water emulsions. *Food Funct.* **2012**, *3*, 312–319. [CrossRef] [PubMed]

8. Zhang, S.; Zhou, Y.; Yang, C. Pickering emulsions stabilized by the complex of polystyrene particles and chitosan. *Colloids Surf. A Physicochem. Eng. Asp.* **2015**, *482*, 338–344. [CrossRef]

9. Popat, A.; Karmakar, S.; Jambhrunkar, S.; Xu, C.; Yu, C. Curcumin-cyclodextrin encapsulated chitosan nanoconjugates with enhanced solubility and cell cytotoxicity. *Colloids Surf. B. Biointerfaces* **2014**, *117*, 520–527. [CrossRef] [PubMed]

10. Elwerfalli, A.M.; Al-Kinani, A.; Alany, R.G.; ElShaer, A. Nano-engineering chitosan particles to sustain the release of promethazine from orodispersables. *Carbohydr. Polym.* **2015**, *131*, 447–461. [CrossRef] [PubMed]

11. Ganesh, M.; Jeon, U.J.; Ubaidulla, U.; Hemalatha, P.; Saravanakumar, A.; Peng, M.M.; Jang, H.T. Chitosan cocrystals embedded alginate beads for enhancing the solubility and bioavailability of aceclofenac. *Int. J. Biol. Macromol.* **2015**, *74*, 310–317. [CrossRef] [PubMed]

12. Fernandes, M.; Gonçalves, I.C.; Nardecchia, S.; Amaral, I.F.; Barbosa, M.A.; Martins, M.C.L. Modulation of stability and mucoadhesive properties of chitosan microspheres for therapeutic gastric application. *Int. J. Pharm.* **2013**, *454*, 116–124. [CrossRef] [PubMed]

13. Newton, A.M.J.; Indana, V.L.; Kumar, J. Chronotherapeutic drug delivery of Tamarind gum, chitosan and okra gum controlled release colon targeted directly compressed propranolol HCl matrix tablets and in vitro evaluation. *Int. J. Biol. Macromol.* **2015**, *79*, 290–299. [CrossRef] [PubMed]

14. Shrestha, N.; Shahbazi, M.-A.; Araújo, F.; Zhang, H.; Mäkilä, E.M.; Kauppila, J.; Sarmento, B.; Salonen, J.J.; Hirvonen, J.T.; Santos, H.A. Chitosan-modified porous silicon microparticles for enhanced permeability of insulin across intestinal cell monolayers. *Biomaterials* **2014**, *35*, 7172–7179. [CrossRef] [PubMed]

15. Tapola, N.S.; Lyyra, M.L.; Kolehmainen, R.M.; Sarkkinen, E.S.; Schauss, A.G. Safety aspects and cholesterol-lowering efficacy of chitosan tablets. *J. Am. Coll. Nutr.* **2008**, *27*, 22–30. [CrossRef] [PubMed]

16. Baldrick, P. The safety of chitosan as a pharmaceutical excipient. *Regul. Toxicol. Pharmacol.* **2010**, *56*, 290–299. [CrossRef] [PubMed]

17. Gomathi, T.; Govindarajan, C.; Rose, H.R.M.H.; Sudha, P.N.; Imran, P.K.M.; Venkatesan, J.; Kim, S.-K. Studies on drug-polymer interaction, in vitro release and cytotoxicity from chitosan particles excipient. *Int. J. Pharm.* **2014**, *468*, 214–222. [CrossRef] [PubMed]

18. Diop, M.; Auberval, N.; Viciglio, A.; Langlois, A.; Bietiger, W.; Mura, C.; Peronet, C.; Bekel, A.; Julien David, D.; Zhao, M.; et al. Design, characterisation, and bioefficiency of insulin-chitosan nanoparticles after stabilisation by freeze-drying or cross-linking. *Int. J. Pharm.* **2015**, *491*, 402–408. [CrossRef] [PubMed]

19. Prata, A.S.; Grosso, C.R.F. Production of microparticles with gelatin and chitosan. *Carbohydr. Polym.* **2015**, *116*, 292–299. [CrossRef] [PubMed]

20. Karnchanajindanun, J.; Srisa-ard, M.; Baimark, Y. Genipin-cross-linked chitosan microspheres prepared by a water-in-oil emulsion solvent diffusion method for protein delivery. *Carbohydr. Polym.* **2011**, *85*, 674–680. [CrossRef]

21. Jia, R.; Jiang, H.; Jin, M.; Wang, X.; Huang, J. Silver/chitosan-based Janus particles: Synthesis, characterization, and assessment of antimicrobial activity in vivo and vitro. *Food Res. Int.* **2015**, *78*, 433–441. [CrossRef] [PubMed]

22. Passos, J.J.; De Sousa, F.B.; Mundim, I.M.; Bonfim, R.R.; Melo, R.; Viana, A.F.; Stolz, E.D.; Borsoi, M.; Rates, S.M.K.; Sinisterra, R.D. Double continuous injection preparation method of cyclodextrin inclusion compounds by spray drying. *Chem. Eng. J.* **2013**, *228*, 345–351. [CrossRef]

23. Singh, K.; Tiwary, A.K.; Rana, V. Spray dried chitosan-EDTA superior microparticles as solid substrate for the oral delivery of amphotericin B. *Int. J. Biol. Macromol.* **2013**, *58*, 310–319. [CrossRef] [PubMed]

24. Ray, S.D. Potential aspects of chitosan as pharmaceutical excipient. *Acta Pol. Pharm. Drug Res.* **2011**, *68*, 619–622.

25. Szymańska, E.; Winnicka, K. Stability of chitosan—A challenge for pharmaceutical and biomedical applications. *Mar. Drugs* **2015**, *13*, 1819–1846. [CrossRef] [PubMed]

26. Kasaai, M.R. Calculation of Mark-Houwink-Sakurada (MHS) equation viscometric constants for chitosan in any solvent-temperature system using experimental reported viscometric constants data. *Carbohydr. Polym.* **2007**, *68*, 477–488. [CrossRef]

27. Raymond, L.; Morin, F.G.; Marchessault, R.H. Degree of deacetylation of chitosan using conductometric titration and solid-state NMR. *Carbohydr. Res.* **1993**, *246*, 331–336. [CrossRef]

28. Mesquita, P.C.; Oliveira, A.R.; Pedrosa, M.F.F.; De Oliveira, A.G.; Da Silva-Júnior, A.A. Physicochemical aspects involved in methotrexate release kinetics from biodegradable spray-dried chitosan microparticles. *J. Phys. Chem. Solids* **2015**, *81*, 27–33. [CrossRef]

29. *Farmacopeia Brasileira*, 5th ed.; ANVISA: Brasília, Brazil, 2010. (In Portuguese)

30. Dos Santos, Z.M.; Caroni, A.L.P.F.; Pereira, M.R.; da Silva, D.R.; Fonseca, J.L.C. Determination of deacetylation degree of chitosan: A comparison between conductometric titration and CHN elemental analysis. *Carbohydr. Res.* **2009**, *344*, 2591–2595. [CrossRef] [PubMed]

31. Cal, K.; Sollohub, K. Spray drying technique. I: Hardware and process parameters. *J. Pharm. Sci.* **2010**, *99*, 575–586. [CrossRef] [PubMed]

32. Alvim, I.D.; Stein, M.A.; Koury, I.P.; Dantas, F.B.H.; Cruz, C.L.d.C.V. Comparison between the spray drying and spray chilling microparticles contain ascorbic acid in a baked product application. *LWT Food Sci. Technol.* **2016**, *65*, 689–694. [CrossRef]

33. Consoli, L.; Grimaldi, R.; Sartori, T.; Menegalli, F.C.; Hubinger, M.D. Gallic acid microparticles produced by spray chilling technique: Production and characterization. *LWT Food Sci. Technol.* **2016**, *65*, 79–87. [CrossRef]

34. Palazzo, F.; Giovagnoli, S.; Schoubben, A.; Blasi, P.; Rossi, C.; Ricci, M. Development of a spray-drying method for the formulation of respirable microparticles containing ofloxacin-palladium complex. *Int. J. Pharm.* **2013**, *440*, 273–282. [CrossRef] [PubMed]

35. Al-Qadi, S.; Grenha, A.; Carrión-Recio, D.; Seijo, B.; Remuñán-López, C. Microencapsulated chitosan nanoparticles for pulmonary protein delivery: In vivo evaluation of insulin-loaded formulations. *J. Control. Release* **2012**, *157*, 383–390. [CrossRef] [PubMed]

36. Kašpar, O.; Jakubec, M.; Štěpánek, F. Characterization of spray dried chitosan—TPP microparticles formed by two- and three-fluid nozzles. *Powder Technol.* **2013**, *240*, 31–40. [CrossRef]

37. Kašpar, O.; Tokárová, V.; Nyanhongo, G.S.; Gübitz, G.; Štěpánek, F. Effect of cross-linking method on the activity of spray-dried chitosan microparticles with immobilized laccase. *Food Bioprod. Process.* **2013**, *91*, 525–533. [CrossRef]

38. Santos, J.V. Otimiza çã o dos par â metros de secagem por aspers ã o de micropart í culas de quitosana como carreadores de insulina. *Lat. Am. J. Pharm.* **2003**, *22*, 327–334.

39. Yang, H.-C.; Hon, M.-H. The effect of the degree of deacetylation of chitosan nanoparticles and its characterization and encapsulation efficiency on drug delivery. *Polym. Plast. Technol. Eng.* **2010**, *49*, 1292–1296. [CrossRef]

40. Mys, N.; Verberckmoes, A.; Cardon, L. Processing of syndiotactic polystyrene to microspheres for part manufacturing through selective laser sintering. *Polymers (Basel)* **2016**, *8*, 383. [CrossRef]

41. Mys, N.; Van De Sande, R.; Verberckmoes, A.; Cardon, L. Processing of polysulfone to free flowing powder by mechanical milling and spray drying techniques for use in selective laser sintering. *Polymers (Basel)* **2016**, *8*, 150. [CrossRef]

42. Li, B.; Shan, C.-L.; Zhou, Q.; Fang, Y.; Wang, Y.-L.; Xu, F.; Han, L.-R.; Ibrahim, M.; Guo, L.-B.; Xie, G.-L.; et al. Synthesis, characterization, and antibacterial activity of cross-linked chitosan-glutaraldehyde. *Mar. Drugs* **2013**, *11*, 1534–1552. [CrossRef] [PubMed]

43. Zhang, Z.-H.; Han, Z.; Zeng, X.-A.; Xiong, X.-Y.; Liu, Y.-J. Enhancing mechanical properties of chitosan films via modification with vanillin. *Int. J. Biol. Macromol.* **2015**, *81*, 638–643. [CrossRef] [PubMed]

44. Bartolomei, M.; Bertocchi, P.; Cotta Ramusino, M.; Santucci, N.; Valvo, L. Physico-chemical characterisation of the modifications I and II of (R,S) propranolol hydrochloride: Solubility and dissolution studies. *J. Pharm. Biomed. Anal.* **1999**, *21*, 299–309. [CrossRef]

45. Florey, K. *Analytical Profiles of Drug Substances*; Florey, K., Ed.; Academic Press: New York, NY, USA; London, UK, 1972.

46. Lei, H.; Gao, X.; Wu, W.D.; Wu, Z.; Chen, X.D. Aerosol-assisted fast formulating uniform pharmaceutical polymer microparticles with variable properties toward pH-sensitive controlled drug release. *Polymers (Basel)* **2016**, *8*, 195. [CrossRef]

47. Gómez-Burgaz, M.; García-Ochoa, B.; Torrado-Santiago, S. Chitosan-carboxymethylcellulose interpolymer complexes for gastric-specific delivery of clarithromycin. *Int. J. Pharm.* **2008**, *359*, 135–143. [CrossRef] [PubMed]

48. Ragelle, H.; Vandermeulen, G.; Préat, V. Chitosan-based siRNA delivery systems. *J. Control. Release* **2013**, *172*, 207–218. [CrossRef] [PubMed]

49. Dokoumetzidis, A.; Macheras, P. A century of dissolution research: From noyes and whitney to the biopharmaceutics classification system. *Int. J. Pharm.* **2006**, *321*, 1–11. [CrossRef] [PubMed]

50. Tran, T.T.D.; Tran, K.A.; Tran, P.H.L. Modulation of particle size and molecular interactions by sonoprecipitation method for enhancing dissolution rate of poorly water-soluble drug. *Ultrason. Sonochem.* **2015**, *24*, 256–263. [CrossRef] [PubMed]

© 2017 by the authors. Licensee MDPI, Basel, Switzerland. This article is an open access article distributed under the terms and conditions of the Creative Commons Attribution (CC BY) license (http://creativecommons.org/licenses/by/4.0/).

polymers

MDPI

Article

Dual Drug Delivery of Sorafenib and Doxorubicin from PLGA and PEG-PLGA Polymeric Nanoparticles

György Babos [1,2], Emese Biró [1,2], Mónika Meiczinger [2] and Tivadar Feczkó [1,2,*]

[1]   Institute of Materials and Environmental Chemistry, Research Centre for Natural Sciences, Hungarian Academy of Sciences, Magyar tudósok körútja 2., H-1117 Budapest, Hungary; babos@mukki.richem.hu (G.B.); biro@mukki.richem.hu (E.B.)

[2]   Research Institute of Biomolecular and Chemical Engineering, University of Pannonia, Egyetem u. 10, H-8200 Veszprém, Hungary; meiczinger@mukki.richem.hu

*   Correspondence: tivadar.feczko@gmail.com; Tel.: +36-88-624000 (ext. 3508)

Received: 9 July 2018; Accepted: 6 August 2018; Published: 9 August 2018

Abstract: Combinatorial drug delivery is a way of advanced cancer treatment that at present represents a challenge for researchers. Here, we report the efficient entrapment of two clinically used single-agent drugs, doxorubicin and sorafenib, against hepatocellular carcinoma. Biocompatible and biodegradable polymeric nanoparticles provide a promising approach for controlled drug release. In this study, doxorubicin and sorafenib with completely different chemical characteristics were simultaneously entrapped by the same polymeric carrier, namely poly(D,L-lactide-*co*-glycolide) (PLGA) and polyethylene glycol-poly(D,L-lactide-*co*-glycolide) (PEG-PLGA), respectively, using the double emulsion solvent evaporation method. The typical mean diameters of the nanopharmaceuticals were 142 and 177 nm, respectively. The PLGA and PEG-PLGA polymers encapsulated doxorubicin with efficiencies of 52% and 69%, respectively, while these values for sorafenib were 55% and 88%, respectively. Sustained drug delivery under biorelevant conditions was found for doxorubicin, while sorafenib was released quickly from the PLGA-doxorubicin-sorafenib and PEG-PLGA-doxorubicin-sorafenib nanotherapeutics.

Keywords: sorafenib; doxorubicin; polymeric nanoparticles; drug delivery

1. Introduction

Hepatocellular carcinoma (HCC) is one of the most destructive cancers. At present, sorafenib is the only drug available that prolongs the life of patients with HCC. However, non-specific uptake leads to high toxicity and serious side effects. Sorafenib is a multikinase inhibitor that targets various receptor tyrosine kinases and RAF kinases; hence, it hampers tumor growth and exerts cytostatic effects and thus demonstrates a significant overall survival rate of patients, e.g., with HCC. However, its water immiscibility results in low bioavailability [1]; thus, a high dosage is required. Doxorubicin is a common chemotherapeutic agent in numerous cancer therapies [2]. It is an anthracycline antibiotic. Doxorubicin hydrochloride salt is a water-soluble, hygroscopic, crystalline form of the drug, which possesses better bioavailability. Doxorubicin activation on the nucleic acids of dividing cells can occur by intercalation between the base pairs of the DNA strands, thus inhibiting the synthesis of DNA and RNA by impeding the replication and transcription in the cells and producing iron-mediated free radicals that destroy cell membranes, proteins, and DNA. The most disadvantageous side effects of doxorubicin are myelosuppression and cardiotoxicity.

The drawbacks of the use of anticancer agents could be decreased by the application of a nanocarrier that supports the targeted drug delivery and controls the release of effective agents. Polymeric nanoparticulate drug delivery systems have been shown to be a valid approach to sustain drug liberation and to enable a targeting function. There are some existing papers on sorafenib or doxorubicin microencapsulation using PLGA copolymers. Nevertheless, the sorafenib loading in PLGA nanoparticles is generally rather low. E.g., a 1.4% sorafenib loading in PLGA nanoparticles with an oil-in-water single-emulsion solvent evaporation method was achieved in [3]. Multiblock polymer nanoparticles consisting of (poly(lactic acid)-poly(ethylene glycol)-poly(L-lysine)-diethylenetriamine pentaacetic acid and the pH-sensitive material poly(L-histidine)-poly(ethylene glycol)-biotin could encapsulate 2.4% sorafenib [4]; however, by a nanoprecipitation-dialysis method using a block copolymer of dextran and poly(D,L-lactide-*co*-glycolide) the realized drug content was substantially higher, with a maximum of 5.3% [5]. Doxorubicin-loaded PEG-PLGA-Au nanoparticles with a cytostatic drug content of 3.9% were prepared to enable combined treatment based on chemotherapy and heat-therapy by near-infrared radiation in [6].

By simultaneous delivery of anticancer drugs to tumor cells, a synergistic effect can be realized by an appropriate composition [7]. In some studies, co-delivery of sorafenib and doxorubicin has already been successfully done. E.g., a nanocomposite composed of doxorubicin containing a polyvinyl alcohol core and a human serum albumin-sorafenib shell was manufactured by a sequential freeze-thaw method followed by ethanol coacervation in [8]. The drug loading and the encapsulation efficiency of doxorubicin were 3.0% and 82.0%, respectively, in the nanocore; these values for sorafenib were 2.4% and 91%, respectively, in the albumin nanoshell. Lipid-polymer hybrid nanoparticles decorated with the tumor-homing peptide iRGD were prepared in [9]. The hybrid nanocomposites possessed synergistic cytotoxicity, a pro-apoptotic ability, and improved uptake by HepG2 human hepatocellular carcinoma cells. The blood circulation time and bioavailability and antitumor effects were also significantly increased in HCC xenograft mouse models. Although the drug loading for sorafenib was rather low (3.6%), high doxorubicin content (13.6%) was realized in this work. Very recently, Xiong et al. [10] entrapped sorafenib adamantine-terminated doxorubicin using poly(ethylene glycol)-β-cyclodextrin. Their reduction-responsive supramolecular nanosystem was manufactured through host-guest interaction between cyclodextrin and adamantine moieties, which then self-assembled into regular spherical nanoparticles that showed an inhibitory effect against HepG2 hepatocellular carcinoma cells.

In our study, PLGA and PEG-PLGA carriers, respectively, were used to entrap doxorubicin and sorafenib together in nanotherapeutics in order to enable the anticancer drugs to exert a synergistic influence. The double emulsion solvent evaporation method was applied for the simultaneous entrapment of the drugs. After the optimization of size and encapsulation efficiency, the drug release profile was investigated in human blood plasma. In vitro cellular studies in HT-29 cancer cells were performed to study the cellular uptake and cytotoxicity of the drug-loaded nanocomposites.

2. Materials and Methods

2.1. Materials

Poly(D,L-lactide-*co*-glycolide) (PLGA) polymer, Resomer RG 752H (lactide:glycolide: 75:25, inherent viscosity 0.14–0.22 dL/g, M_w = 4000–15,000), and Resomer RG 502H (lactide:glycolide: 50:50, inherent viscosity 0.16–0.24 dL/g, M_w = 7000–17,000 g/mol) were produced by Boehringer Ingelheim (Ingelheim am Rhein, Germany). PEGylated-PLGA (PEG-PLGA) polymer, Resomer, RGP d 5055 (PEG-PLGA) (PEG content: 3–7% (m/m), inherent viscosity: 0.93 dL/g, M_w = 33,500 g/mol) was obtained from Evonik (Essen, Germany). Polyvinyl alcohol (PVA, M_w = 30,000–70,000 g/mol, 87–90% hydrolysed), dichloromethane (DCM), acetone, glacial acetic acid, dimethyl sulfoxide (DMSO), sodium azide, 1-ethyl-3(3-dimethylaminopropyl) carbodiimide (EDC), N-hydroxysuccinimide (NHS), sodium dodecyl sulphate (SDS), 3-(4,5-dimethylthiazol-2-yl)-2,5-diphenyltetrazolium bromide (MTT), piroxicam, and RPMI-1640 medium were obtained from Sigma Aldrich (St. Louis, MO, USA). Sorafenib (free base) and doxorubicin HCl were purchased from Active Biochem (Hong Kong, China). Cyanine 5 amine was produced by Lumiprobe GmBH (Hannover, Germany).

2.2. Preparation of Nanocomposites

For the preparation of our dual-agent nanocomposites, the water-in-oil-in-water double emulsion solvent evaporation process was found to be appropriate. Briefly, the inner water phase was composed of 0.2 mL 0.5% (*w/v*) doxorubicin HCl solution in MilliQ water, which was added to the organic phase that consisted of 20 mg encapsulating polymer (Resomer RG 752H, Resomer RG 502H, or Resomer RGP d5055) dissolved in 1.0 mL DCM combined with 1.0 mg sorafenib dissolved in 0.1 mL acetone. The first emulsification was performed by sonication using a sonicator (Sonics Vibra Cell VCX 130, 130 W, Newtown, CT, USA) at an amplitude of 30% for 30 s. Then, the prepared water-in-oil emulsion was pipetted into the outer water phase that consisted of 1% (*w/v*) PVA in 5 mL phosphate buffer (pH 8). The water-in-oil-in-water emulsion was formed by another sonication at an amplitude of 50% for 60 s. The organic solvents were evaporated by magnetic stirring for 3 h under atmospheric pressure at room temperature. Nanoparticles were centrifuged by a Hermle Z216 MK microcentrifuge (Schwerte, Germany) at 15,000 rpm for 20 min, washed thrice, and redispersed in MilliQ water or phosphate-buffered saline (PBS, pH 7.4).

2.3. Investigation of Nanocapsules

2.3.1. Morphology and Size Analysis

The morphology of nanocapsules was monitored after centrifuging and redispersing them in distilled water, dropping them onto a grid, and drying them under room temperature. Then, they were examined with an FEI Apreo scanning electron microscope (SEM, Thermofisher, Waltham, MA, USA) at 20 kV.

The size distribution of the obtained nanoparticles was determined by a Zetasizer Nano ZS (Malvern Instruments, Malvern, UK) operated with dynamic light scattering. The particles were characterized by their intensity mean diameter and polydispersity index (PDI).

2.3.2. Nanoparticle Yield and Encapsulation Efficiency

The yield of the nanocomposites was determined by gravimetry after washing and drying of a known volume of nanoparticle suspension. The drug loading and encapsulation efficiency were investigated directly by dissolving 10 mg nanoparticles in 1 mL DMSO, and the solution was diluted to be detectable in the linear calibration range (1–20 mg/L). The absorbance of the solutions was measured spectrophotometrically (PG Instruments T80, Leicestershire, UK) at the absorbance maxima of doxorubicin (480 nm) and sorafenib (270 nm) in DMSO. The encapsulation efficiency of the active agents was calculated as follows:

Encaps. efficiency (%) = (mass of drug in nanocomposite/mass of total loaded drug) × 100

2.3.3. In Vitro Drug Release Experiment

The in vitro drug release of the nanocomposites was investigated in human blood plasma and in ammonium acetate buffer (pH 5.5) because of the acidic tumor microenvironment. For in vitro release experiments after the washing steps, a 2 mL suspension including 4.4 mg PLGA or 6.4 mg PEG-PLGA nanocomposites was resuspended in 15 mL human blood plasma containing 0.03% sodium azide bactericide. Five milliliters (5 mL) of nanoparticle suspension in the release medium were pipetted to 5 mL non-transparent Eppendorf tubes, incubated at 37 °C in a G24 Environmental Incubator Shaker (New Brunswick Scientific Co. Inc., Edison, NJ, USA), and shaken by a BIO RS-24 Mini-rotator (Biosan, Rīga, Latvia) for 7 days at 700 rpm. Three parallel samples per nanocomposite were investigated. After 1 h and every 24 h, 0.5 mL from each sample were centrifuged (Hermle Z216 MK microcentrifuge, Gosheim, Germany) for 20 min at 15,000 rpm, washed three times, and the pellet was dissolved in 0.5 mL DMSO.

The sorafenib and doxorubicin concentration was measured by a Young Lin YL 9100 HPLC instrument (YL Instruments Co., Ltd., Gyeonggi-do, Korea) at 30 °C. The active agents were separated by a Zorbax SB-Aq column (150 mm × 4.6 mm, 5 μm; Agilent, Santa Clara, CA, USA). The mobile phase composition is given in Table 1. The flow rate was adjusted to 1 mL/min. The detection wavelength of sorafenib and doxorubicin was 280 and 480 nm, respectively. Piroxicam was used as an internal standard during the measurements.

Table 1. Mobile phase composition in HPLC analysis of doxorubicin (DOX) and sorafenib (SOR) co-loaded nanocomposites.

Time (min)	Methanol (%)	0.1% Tetrafluoroacetic Acid in H_2O (%)
0.0	30.0	70.0
5.0	30.0	70.0
8.00	40.0	60.0
11.00	50.0	50.0
14.00	60.0	40.0
17.00	70.0	30.0
20.00	80.0	20.0
23.00	90.0	10.0
27.00	100.0	0.0
30.00	100.0	0.0
35.00	30.0	70.0

The concentration of sorafenib and doxorubicin was calculated using calibration curves and the encapsulation efficiencies were calculated as follows:

$$\text{Encapsulation efficiency (\%)} = (\text{mass of drug in NP/mass of total loaded drug}) \times 100$$

2.4. Attachment of Fluorescent Dye

A 1 mL nanoparticle suspension (2.2 mg/mL PLGA and 3.2 mg/mL PEG-PLGA) was centrifuged and washed with MilliQ water, redispersed in 0.5 mL PBS (pH 7.4), mixed with 0.1 mL PBS (pH 7.4) involving a 50× molar excess of EDC and NHS related to the (PEG-)PLGA concentration, then incubated for 60 min at 25 °C, centrifuged and washed with MilliQ water, and redispersed in 1.0 mL PBS (pH 7.4). The obtained carbodiimide-activated nanoparticle dispersion was pipetted to a 0.02 mL PBS (pH 7.4) solution containing 0.5 mg/mL Cyanine 5 amine fluorescent dye and incubated for 1 h at 25 °C. Then, the nanocomposite dispersion was centrifuged, washed three times, and redispersed in 1 mL PBS.

2.5. Cell Cultures

The human cancer cell line HT-29 was grown in RPMI-1640 medium supplemented with 10% fetal calf serum (FCS) and 100 U/mL penicillin. The cells were cultured at 37 °C in a humidified atmosphere containing 5% CO_2. They were trypsinized, resuspended, and precultured before use.

2.6. In Vitro Cellular Uptake and Cytotoxicity Studies

The HT-29 cellular uptake of the nanoparticles was evaluated using flow cytometry. The cells were cultured in 24-well plates at a cell density of 2×10^5 cells per well at 37 °C for 24 h. After cultivation, 100 mg of fluorescently labelled nanoparticles/well were added to the cells and incubated for 24 h. Cells grown without nanoparticles were used as a negative control. The cells were washed by PBS, trypsinized, and redispersed in PBS containing 2% BSA. Flow cytometry was performed on a BD FACSAria III Cell sorter (BD Biosciences, San Jose, CA, USA) at an Ex/Em wavelength of 633/660 nm. Every sample was analyzed in triplicate.

The in vitro cytotoxicity caused in HT-29 cells was assayed using MTT reagent. Cells were seeded (10,000 cells/well) in 96-well plates. After 24 h of pre-incubation, the growth media were replaced with 200 μL of fresh RPMI-1640 medium containing 10% FCS and PLGA- and PEG-PLGA blank nanoparticles or the dual-drug-entrapping nanocomposites. Three different doxorubicin concentration levels of the added nanopharmaceuticals were applied: 0.5, 1.0, and 2.5 μg per well. The nanocomposites also contained sorafenib; however, it was a higher amount of sorafenib (with 6% and 28% using PLGA and PEG-PLGA, respectively) since its entrapment was more efficacious. The positive control samples were also supplied with the same amount of free doxorubicin and sorafenib in DMSO solution. DMSO cytotoxicity (without drugs) was also investigated. After 48 h of incubation, 20 mL/well MTT solution (5 mg MTT/mL) and 0.2 mL/well supplemented culture media were added followed by further incubation for 2 h. The supernatant was removed, and MTT lysis solution (DMSO, 1% acetic acid, 10% SDS) was added into each well to dissolve the cells with MTT formazan crystals. The absorbance was determined at 492 nm by a Robonik Readwell Touch (Navi Mumbai, India) plate reader. The percentage of viable cells was calculated by comparing the absorbance of treated cells against that of the untreated cells (negative control). The data were presented as the mean and standard deviation with eight replicates.

3. Results and Discussion

3.1. Preliminary Method Development

Our aim was to synthesize nanocomposites that are capable of encapsulating doxorubicin and sorafenib with high loading, high yield, and high encapsulation efficiency as well as a small size. Though PLGA copolymers are very frequently applied nanocarriers, to our knowledge co-loaded sorafenib and doxorubicin PLGA and PEG-PLGA nanotherapeutics have not been prepared so far. This fact is not surprising, because the solubility of the two drug molecules differs substantially; hence, it is not an easy task to involve them in a matrix comprised of one polymer. As was shown in the literature survey, hybrid nanosystems have mostly been used for their co-entrapment.

It must be emphasized that the conditions described in the experimental section were selected after extensive process-optimizing experiments, which also included some trials of nanoprecipitation and single emulsion methods. Nevertheless, these procedures are not described here in detail, since it was proved early in these examinations that they were not suitable for the efficient co-encapsulation of our active agents. Very briefly, for nanoprecipitation, a joint solvent of the active agents and the encapsulating polymer would be necessary; thus, doxorubicin HCl was converted to a free base using trimethylamine [11] before the process. However, the desalted doxorubicin could not be completely dissolved by the applicable solvents (ethanol, acetone, THF); thus, the nanoprecipitation process did not result in desirable encapsulation. In the oil-in-water emulsion solvent evaporation probe, the desalting of doxorubicin HCl was done during the emulsification; however, the doxorubicin encapsulation efficiency was too low in this case.

Since the solubility of doxorubicin decreases with increasing pH, the double emulsion solvent evaporation technique can be a suitable tool to microencapsulate doxorubicin effectively using an outer water phase with a pH higher than 7 [12]. In this approach, doxorubicin is included in the inner water phase. Because sorafenib is an organic soluble drug, it could be entrapped by the polymers after dissolving it in the organic phase. The nanocomposites formed by the double emulsion solvent evaporation method are characterised in the following sections.

3.2. Size, Yield, Drug Encapsulation Efficiency, and Drug Encapsulation Content

Particles smaller than 10 nm are quickly cleared by the renal filtration [13], while the ones bigger than 300 nm can be easily recognized by the reticuloendothelial system (RES) and removed from the blood circulation [14]. Thus, nanoparticles ranging from 10 to 200 nm could extravasate from the disorganized tumor vasculature to the tumor microenvironment due to tumor angiogenesis. Therefore, the manufacture of nanoparticles less than 200 nm in size and with a negative surface charge are desirable to prevent protein adsorption and promote accumulation in tumors. From low molecular weight PLGAs, such as Resomer RG502H and Resomer RG752H, the typical available particle sizes range from 60 to 200 nm [15].

Although the Resomer RG 502H PLGA provided substantially higher encapsulation efficiencies than the Resomer RG 752H PLGA for both of the drugs, the size (164.6 nm) and PDI (0.203) of their nanocomposites were significantly larger (Table 2). The relatively high PDI indicates the presence of bigger particles formed by the separate precipitation of the drug during the solvent's evaporation from the polymer nanoparticles, which is also supported by the size distributions (Figure 1). Such high PDI values were also found in our preliminary nanoprecipitation experiments or, e.g., by Lin et al. [16], who prepared PLGA-sorafenib nanocomposites with nanoprecipitation (PDI 0.21–0.35).

Table 2. Yield, size, polydispersity index (PDI), and encapsulation efficiency (EE) of doxorubicin (DOX) and sorafenib (SOR) co-loaded nanocomposites.

Polymer	PLGA RG 502H	PLGA RG 752H	PEG-PLGA
Yield (%)	70.0	49.7	75.3
Intensity mean diameter (nm)	164.6	142.2	177.2
PDI	0.203	0.123	0.076
EE (DOX) (%)	74	52	69
EE (SOR) (%)	67	55	88
DOX loading (%)	4.81	4.76	4.17
SOR loading (%)	4.35	5.03	5.31

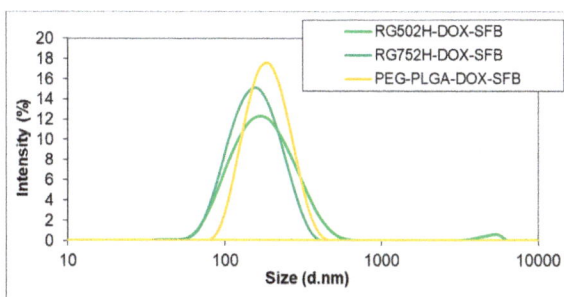

Figure 1. Size distribution by intensity of doxorubicin (DOX)- and sorafenib (SOR)-containing PLGA (Resomer RG502H and Resomer RG752H) and PEG-PLGA nanoparticles.

SEM images of the nanocomposites (Figure 2) suggested significantly smaller nanoparticles than found by Dynamic Light Scattering (DLS) measurements (Figure 1). This can be interpreted as the fact that the DLS method displays the hydrodynamic diameter of the nanocomposites while SEM shows them in a dry state. We cannot exclude that the aggregation of some smaller particles occurred, which can also indicate a higher size during the DLS study. This latter hypothesis might be supported by the zeta potential measurements, which provided relatively low negative values, which means that their aggregation might have taken place. Neither of them showed variation as a function of encapsulating polymer, but varied in the narrow range between -17.6 mV (PLGA) and -18.8 mV (PEG-PLGA). Since the PVA surfactant cannot be completely removed from the surface of the nanoparticles due to its strong adsorption, they are sterically stabilized, which cannot be characterized by zeta potential measurements. This may be the main reason why a difference between the zeta potential values of the two types of nanocomposites was not found.

Doxorubicin has absorbance in the visible range; thus, its concentration could be measured by UV–Vis spectrophotometry. Sorafenib absorbs exclusively in the UV region, while the absorbance of doxorubicin in the same UV region is also considerable although substantially lower than that of sorafenib. Because of their overlapping in the UV region, the HPLC method was applied to determine the concentration of both active agents after the dissolution of the nanocomposites. Sorafenib concentration was measured at a wavelength of 280 nm (Channel 1, Figure 3), while doxorubicin was analysed at 480 nm (Channel 2, Figure 3).

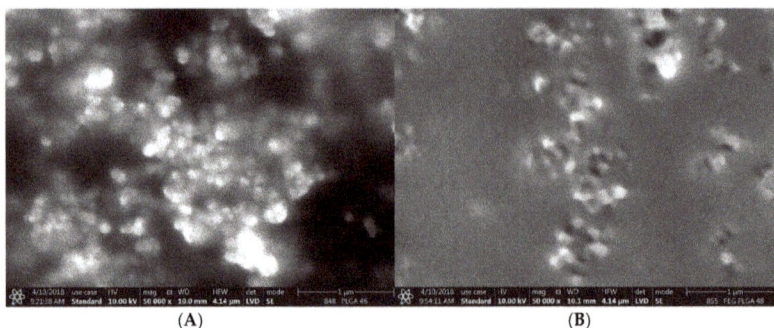

Figure 2. Scanning electron microscopic images of doxorubicin- and sorafenib-containing PLGA (**A**) and PEG-PLGA (**B**) nanocomposites.

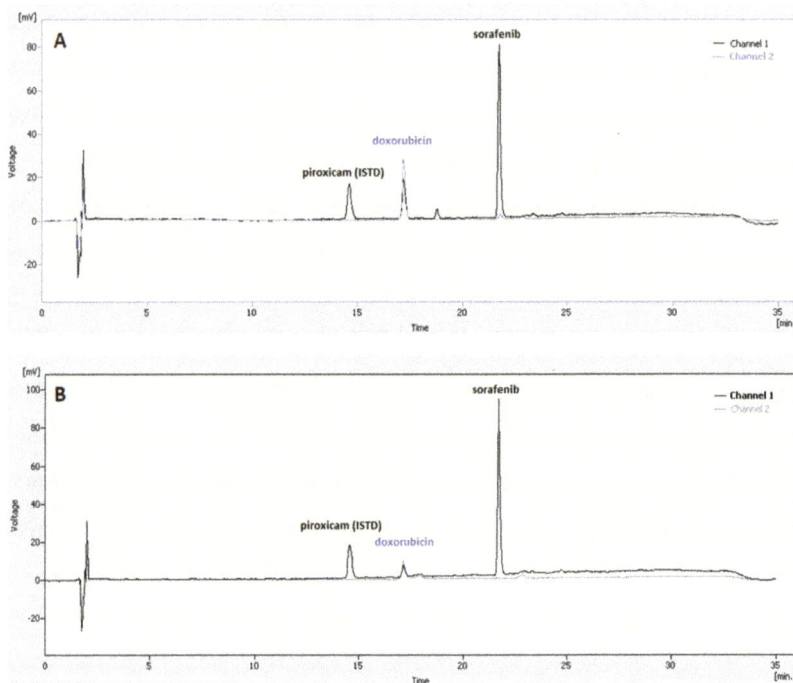

Figure 3. HPLC plot of dissolved sorafenib- and doxorubicin-loaded PLGA (**A**) and PEG-PLGA (**B**) nanocomposites.

The highest yield (75.3%) and sorafenib encapsulation efficiency (88%) were achieved by the PEG-PLGA polymer (Table 2). Its doxorubicin encapsulation efficiency is also satisfactory (69%). A clear correlation can be observed between the particle yield and encapsulation efficiency at each of the nanocomposites, which means the higher the yield, the higher the drug entrapment that can be achieved, which resulted in similar drug contents in the nanomedicines: 4.81%, 4.76%, and 4.17% for doxorubicin and 4.35%, 5.03%, and 5.31% for sorafenib in the case of the Resomer RG 502H PLGA, the Resomer RG 752H PLGA, and PEG-PLGA copolymers, respectively.

3.3. In Vitro Sorafenib Release

In preliminary studies, we found that the release profile of the Resomer RG 752H PLGA was much more beneficial than that of the Resomer RG 502H PLGA because the Resomer RG 502H PLGA showed a high initial burst and then the release rate became very low. Thus, the nanocomposites of the Resomer RG 752H and the PEG-PLGA copolymers were investigated in biorelevant drug release studies. The released ratios of sorafenib and doxorubicin were determined by HPLC indirectly after the dissolution of washed nanocomposites was sampled at a predetermined time. The released amount was calculated from the remaining drug content in the nanoparticles.

For the water-soluble drug doxorubicin HCl, the more hydrophilic carrier PEG-PLGA provided for a quicker release than PLGA with an initial burst of $54 \pm 10\%$, while it was $23 \pm 4\%$ for PLGA-based nanoparticles (Figure 4A). After the burst release, both types of nanocomposites showed sustained release until the end of the study (1 week) and provided almost the complete liberation of doxorubicin ($96 \pm 6\%$ for PLGA and $97 \pm 19\%$ for PEG-PLGA).

Sorafenib was released much more quickly from both of the carriers (Figure 4B). The initial burst of PLGA and PEG-PLGA was $88 \pm 12\%$ and $48 \pm 5\%$, respectively.

The ratio of glycolide to lactide at different compositions enables control of the degree of crystallinity of the PLGA polymers. When the crystalline poly(glycolic acid) is co-polymerized with poly(lactic acid), the crystallinity degree decreases; consequently, the hydration rate and hydrolysis are enhanced. Thus, the degradation time of the copolymer is related to the ratio of monomers used in the synthesis. In general, the higher the content of glycolide, the quicker the rate of degradation [17]. In the Resomer RG 752H polymer, which was used as the PLGA matrix for the release studies, the lactide:glycolide ratio was 75:25; hence, a slower release was expected, especially for sorafenib. However, the extremely quick sorafenib release can be interpreted as the substantial influence of doxorubicin on sorafenib microencapsulation. It is hypothesized that, due to the doxorubicin incorporation, most of the sorafenib must be precipitated on the surface of the nanocomposites, which can be easily dissolved in the blood plasma due to the strong interaction of sorafenib and serum proteins [18].

The release of the active agents was investigated also under an acidic condition since the tumor microenvironment is generally acidic and nanoparticles accumulate generally in lysosomes that can be characterized by an internal acidic pH [19]. The release characteristics in acidic buffer were found to be opposite compared to those in human blood plasma; that is, doxorubicin was liberated within 1 day in both encapsulating polymers (Figure 5). Sorafenib release was completed in 6 days. The PEG-PLGA carrier displayed considerably faster release compared to the PLGA carrier, especially for sorafenib.

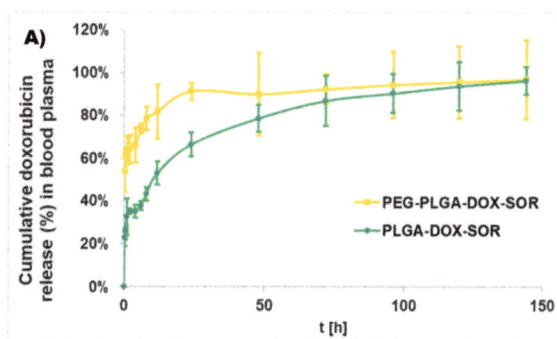

Figure 4. *Cont.*

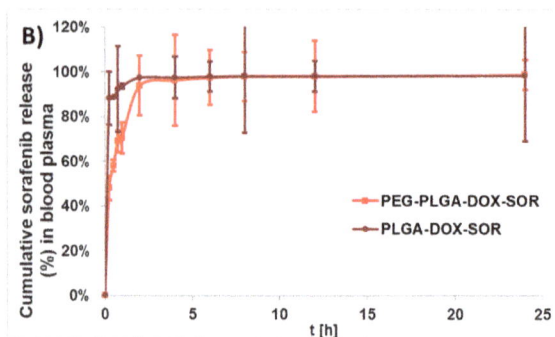

Figure 4. Doxorubicin (**A**) and sorafenib (**B**) release from PLGA-doxorubicin-sorafenib (PLGA-DOX-SFB) and PEG-PLGA-doxorubicin-sorafenib (PEG-PLGA-DOX-SFB) nanocomposites in human blood plasma. Data are presented as mean ± SD from three replicates for each concentration.

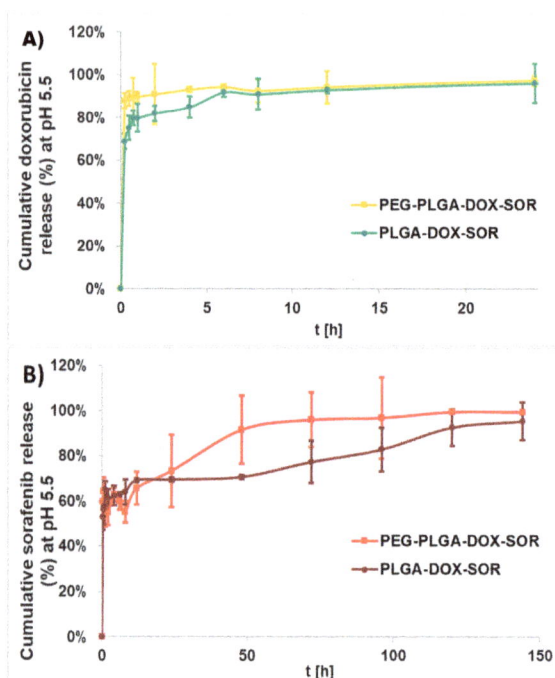

Figure 5. Doxorubicin (**A**) and sorafenib (**B**) release from PLGA-doxorubicin-sorafenib (PLGA-DOX-SFB) and PEG-PLGA-doxorubicin-sorafenib (PEG-PLGA-DOX-SFB) nanocomposites in ammonium acetate buffer (pH 5.5). Data are presented as mean ± SD from three replicates for each concentration.

3.4. Cellular Uptake

Flow cytometry was used to study the in vitro cellular uptake by the HT-29 human cancer cell line. After 1 day of incubation with the Cyanine-5-conjugated nanocomposites, all of the living cells seemed to engulf nanoparticles according to the fluorescence activated cell sorting (FACS) experiments in each of the examined wells. There was a difference only among the amount of cells that was engulfed and expressed in different fluorescent intensity values (Table 3). A significantly higher amount of the nanocomposites prepared using PEG-PLGA was taken up by the cells than that using PLGA. It is also noted that the drug-containing nanoparticles were engulfed to a substantially higher extent than the blank nanoparticles, which might be the result of the changed surface characteristics.

Table 3. Fluorescent intensity values of blank as well as doxorubicin (DOX) and sorafenib (SOR) co-loaded PLGA and PEG-PLGA nanoparticles in HT-29 cellular uptake studies.

Sample	Negative Control	PLGA Blank	PLGA-DOX-SOR	PEG-PLGA Blank	PEG-PLGA-DOX-SOR
mean fluorescent intensity	22	17,105	22,243	21,846	45,765
SD (%)	4.5	1.7	30.0	3.9	24.8

3.5. Cytotoxicity

To determine the cytotoxic effect of active agents in solution and nanocomposites on HT-29 cancer cells, an MTT assay was performed. As shown in Figure 6, the cells remained viable in the negative control, DMSO-, and blank-nanoparticles-treated wells. The increasing concentration of drugs in solution decreased the cell viability to 15%. Drug-containing PLGA nanoparticles caused similar cytotoxicity at all the three concentration levels (viability: 49–45%). Drug-loaded PEG-PLGA nanocomposites reduced the cell viability more substantially (38–23%). The higher cytotoxicity of the PEG-PLGA nanoparticles compared to the PLGA nanocomposites is in accordance with their quicker drug release.

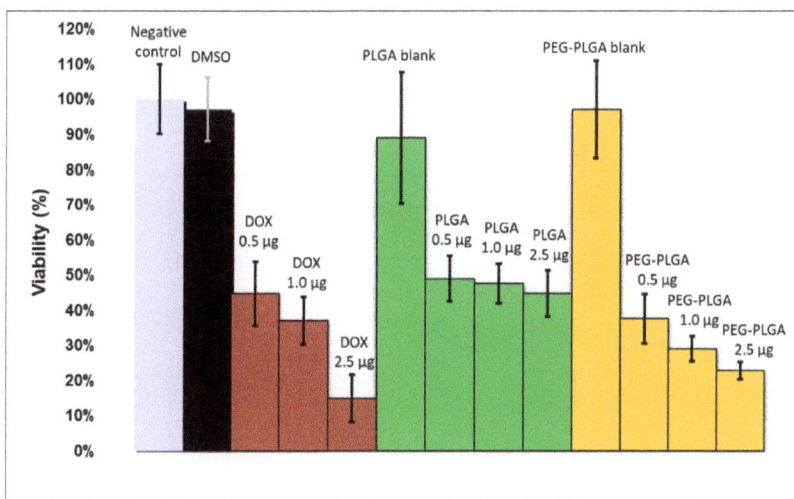

Figure 6. Viability of HT-29 cells by MTT assay due to different concentrations of doxorubicin (DOX) and sorafenib in solution (red columns), PLGA nanoparticles (green columns), and PEG-PLGA nanoparticles (yellow columns).

4. Conclusions

Doxorubicin and sorafenib co-loaded therapeutic nanocomposites were developed using PLGA- and PEG-PLGA-encapsulating polymers, respectively, by the double emulsion solvent evaporation method. The nanoparticles possessed promising physical and chemical properties; that is, a small size, high yield, high drug encapsulation efficiency, and high drug loading. The doxorubicin was released continuously within 6 days, while the sorafenib was released quickly during 24 h under biorelevant conditions. In an acidic tumor-simulated condition, the two agents presented opposing characteristics with accelerated doxorubicin and sustained sorafenib release. The hydrolysis of PEG-PLGA was quicker in both of the media. PEG-PLGA nanocomposites displayed higher cellular uptake, which harmonizes with the higher cytotoxicity experienced.

Author Contributions: G.B. took part in the design of the experiments, and he performed them with E.B. M.M. did the HPLC analysis, while T.F. conceived the experiments and wrote the paper.

funding: This research was funded by the Ministry of National Economy (Hungary) via the Economic Development and Innovation Operation Programme (project No. BIONANO_GINOP-2.3.2-15-2016-00017).

Acknowledgments: Tivadar Feczkó is grateful for the support of the Alexander von Humboldt Foundation (Ref. No.: 3.3-UNG/1161203 STP and 3.3-1161203-HUN-HFST-E). The funding of the BIONANO_GINOP-2.3.2-15-2016-00017 project is acknowledged.

Conflicts of Interest: The authors declare no conflicts of interest.

References

1. Heidarinasab, A.; Panahi, H.A.; Faramarzi, M.; Farjadian, F. Synthesis of thermosensitive magnetic nanocarrier for controlled sorafenib delivery. *Mater. Sci. Eng. C* **2016**, *67*, 42–50. [CrossRef] [PubMed]
2. Cagel, M.; Grotz, E.; Bernabeu, E.; Moretton, M.A.; Chiappetta, D.A. Doxorubicin: Nanotechnological overviews from bench to bedside. *Drug Discov. Today* **2017**, *22*, 270–281. [CrossRef] [PubMed]
3. Liu, J.; Boonkaew, B.; Arora, J.; Mandava, S.H.; Maddox, M.M.; Chava, S.; Callanghan, C.; He, J.; Dash, S.; John, V.T.; et al. Comparison of sorafenib-loaded poly(lactic/glycolic) acid and DPPC liposome nanoparticles in the in vitro treatment of renal cell carcinoma. *J. Pharm. Sci.* **2015**, *104*, 1187–1196. [CrossRef] [PubMed]
4. Liu, Y.; Feng, L.; Liu, T.; Zhang, L.; Yao, Y.; Yu, D.; Wang, L.; Zhang, N. Multifunctional pH-sensitive polymeric nanoparticles for theranostics evaluated experimentally in cancer. *Nanoscale* **2014**, *6*, 3231–3242. [CrossRef] [PubMed]
5. Kim, D.H.; Kim, M.D.; Choi, C.W.; Chung, C.W.; Ha, S.H.; Kim, C.H. Antitumor activity of sorafenib-incorporated nanoparticles of dextran/poly(dl-lactide-co-glycolide) block copolymer. *Nanoscale Res. Lett.* **2012**, *7*, 91. [CrossRef] [PubMed]
6. Lee, S.M.; Park, H.; Yoo, K.H. Synergistic cancer therapeutic effects of locally delivered drug and heat using multifunctional nanoparticles. *Adv. Mater.* **2010**, *22*, 4049–4053. [CrossRef] [PubMed]
7. Wen, Y.H.; Lee, T.Y.; Fu, P.C.; Lo, C.L.; Chiang, Y.T. Multifunctional polymer nanoparticles for dual drug release and cancer cell targeting. *Polymers* **2017**, *9*, 213. [CrossRef]
8. Malarvizhi, G.L.; Retnakumari, A.P.; Nair, S.; Koyakutty, M. Transferrin targeted core-shell nanomedicine for combinatorial delivery of doxorubicin and sorafenib against hepatocellular carcinoma. *Nanomed. Nanotechnol.* **2014**, *10*, 1649–1659. [CrossRef] [PubMed]
9. Zhang, J.; Hu, J.; Chan, H.F.; Skibba, M.; Liang, G.; Chen, M. iRGD decorated lipid-polymer hybrid nanoparticles for targeted co-delivery of doxorubicin and sorafenib to enhance anti-hepatocellular carcinoma efficacy. *Nanomed. Nanotechnol.* **2016**, *12*, 1303–1311. [CrossRef] [PubMed]
10. Xiong, Q.; Cui, M.; Yu, G.; Wang, J.; Song, T. Facile fabrication of reduction-responsive supramolecular nanoassemblies for co-delivery of doxorubicin and sorafenib toward hepatoma cells. *Front. Pharm.* **2018**, *9*, 61. [CrossRef] [PubMed]
11. Yang, Q.; Tan, L.; He, C.; Liu, B.; Xu, Y.; Zhu, Z.; Shao, Z.; Gong, B.; Shen, Y.-M. Redox-responsive micelles self-assembled from dynamic covalent block copolymer rsfor intracellular drug delivery. *Acta Biomater.* **2015**, *17*, 193–200. [CrossRef]
12. Khemani, M.; Sharon, M.; Sharon, M. pH Dependent Encapsulation of Doxorubicin in PLGA. *Ann. Biol. Res.* **2012**, *3*, 4414–4419.
13. Fox, M.E.; Szoka, F.C.; Frechet, J.M.J. Soluble Polymer Carriers for the Treatment of Cancer: The Importance of Molecular Architecture. *Acc. Chem. Res.* **2009**, *42*, 1141–1151. [CrossRef] [PubMed]
14. Kobayashi, H.; Watanabe, R.; Choyke, P.L. Improving conventional enhanced permeability and retention (EPR) effectts; what is the appropriate target? *Theranostics* **2013**, *4*, 81–89. [CrossRef] [PubMed]
15. Halayqa, M.; Domanska, U. PLGA Biodegradable nanoparticles containing perphenazine or chlorpromazine hydrochloride: Effect of formulation and release. *Int. J. Mol. Sci.* **2014**, *15*, 23909–23923. [CrossRef] [PubMed]

16. Lin, T.T.; Gao, D.Y.; Liu, Y.C.; Sung, Y.C.; Wan, D.; Liu, J.Y.; Chiang, T.; Wang, L.; Chen, Y. Development and characterization of sorafenib-loaded PLGA nanoparticles for the systemic treatment of liver fibrosis. *J. Control. Release* **2016**, *221*, 62–70. [CrossRef] [PubMed]
17. Park, T.G.; Lu, W.; Crotts, G. Importance of in vitro experimental conditions on protein release kinetics, stability and polymer degradation in protein encapsulated poly(D,L-lactic acid-co-glycolic acid) microspheres. *J. Control. Release* **1995**, *33*, 211–222. [CrossRef]
18. Shi, J.H.; Chen, J.; Wang, J.; Zhu, Y.Y.; Wang, Q. Binding interaction of sorafenib with bovine serum albumin: Spectroscopic methodologies and molecular docking. *Spectrochim. Acta Part A* **2015**, *149*, 630–637. [CrossRef] [PubMed]
19. Yin, T.; Liu, J.; Zhao, Z.; Zhao, Y.; Dong, L.; Yang, M.; Zhou, J.; Huo, M. Redox sensitive hyaluronic acid-decorated graphene oxide for photothermally controlled tumor-cytoplasm-selective rapid drug delivery. *Adv. Funct. Mater.* **2017**, *14*, 1604620. [CrossRef]

© 2018 by the authors. Licensee MDPI, Basel, Switzerland. This article is an open access article distributed under the terms and conditions of the Creative Commons Attribution (CC BY) license (http://creativecommons.org/licenses/by/4.0/).

polymers

[MDPI]

Article

Inhalable Fucoidan Microparticles Combining Two Antitubercular Drugs with Potential Application in Pulmonary Tuberculosis Therapy

Ludmylla Cunha [1,2], Susana Rodrigues [1,2], Ana M. Rosa da Costa [3], M Leonor Faleiro [1], Francesca Buttini [4] and Ana Grenha [1,2,*]

[1]   Centre for Biomedical Research, University of Algarve, 8005-139 Faro, Portugal;
       ludmyllacc@gmail.com (L.C.); susananasus@gmail.com (S.R.); mfaleiro@ualg.pt (M.L.F.)
[2]   Centre for Marine Sciences, University of Algarve, 8005-139 Faro, Portugal
[3]   Algarve Chemistry Research Centre and Department of Chemistry and Pharmacy, University of Algarve,
       8005-139 Faro, Portugal; amcosta@ualg.pt
[4]   Food and Drug Department, University of Parma, 43124 Parma, Italy; francesca.buttini@unipr.it
*   Correspondence: amgrenha@ualg.pt; Tel.: +351-289-244-441; Fax: +351-289-800-066

Received: 12 April 2018; Accepted: 31 May 2018; Published: 8 June 2018

Abstract: The pulmonary delivery of antitubercular drugs is a promising approach to treat lung tuberculosis. This strategy not only allows targeting the infected organ instantly, it can also reduce the systemic adverse effects of the antibiotics. In light of that, this work aimed at producing fucoidan-based inhalable microparticles that are able to associate a combination of two first-line antitubercular drugs in a single formulation. Fucoidan is a polysaccharide composed of chemical units that have been reported to be specifically recognised by alveolar macrophages (the hosts of *Mycobacterium*). Inhalable fucoidan microparticles were successfully produced, effectively associating isoniazid (97%) and rifabutin (95%) simultaneously. Furthermore, the produced microparticles presented adequate aerodynamic properties for pulmonary delivery with potential to reach the respiratory zone, with a mass median aerodynamic diameter (MMAD) between 3.6–3.9 μm. The formulation evidenced no cytotoxic effects on lung epithelial cells (A549), although mild toxicity was observed on macrophage-differentiated THP-1 cells at the highest tested concentration (1 mg/mL). Fucoidan microparticles also exhibited a propensity to be captured by macrophages in a dose-dependent manner, as well as an ability to activate the target cells. Furthermore, drug-loaded microparticles effectively inhibited mycobacterial growth in vitro. Thus, the produced fucoidan microparticles are considered to hold potential as pulmonary delivery systems for the treatment of tuberculosis.

Keywords: alveolar macrophages; fucoidan; isoniazid; inhalable microparticles; rifabutin; spray-drying; tuberculosis therapy

---

1. Introduction

Tuberculosis (TB) is a leading infectious cause of death worldwide, even though a vaccine and several effective antibiotics are available for its prevention and treatment. According to the World Health Organisation (WHO), there were 10.4 million new TB cases globally and 1.3 million TB-related deaths in 2016 [1]. Global TB control is very difficult due to many factors, including late diagnosis and patient nonadherence to long-term treatments, which leads to a high incidence of extensive resistance to effective antitubercular drugs [2]. Besides drug resistance, the current therapy faces serious challenges, such as multi-drug interactions—especially with antiretroviral agents in cases of TB and HIV *co*-infection—long-term treatment, and antibiotic toxicity, which lead to adverse effects, resulting in patient non-compliance [3]. In this sense, accurate and early diagnosis are essential in TB therapy, in addition to patient adherence to the treatment and re-examination to verify the development of the active form of the disease. Moreover, in the treatment of TB, it must be considered the ability of *Mycobacterium tuberculosis* to survive intracellularly in the host alveolar macrophage for a long period. For that reason, therapy should ideally involve the intracellular delivery of antitubercular agents [4]. This could be achieved by the design of inhalable drug formulations with suitable aerodynamic properties to reach the alveoli, where macrophages infected with *M. tuberculosis* reside [5]. This strategy would potentially reduce the dosage and frequency of administration, and perhaps shorten treatment duration. As a result, systemic side effects could be avoided, improving patient adherence to the treatment [6].

In this context, fucoidan (FUC) could be used as the matrix material of the inhalable carriers to be designed. FUC is a promising biomaterial in this regard, because it presents sulphated fucose and other sugar residues [7], which can be recognised by the surface receptors of alveolar macrophages [8]. This can favour the internalisation of microparticles by alveolar macrophages, and subsequently, the delivery of drugs at the infection site. Furthermore, FUC can possibly intensify macrophage activation mediated by membrane receptors [9–11]. The working mechanism herein proposed for FUC microparticles is illustrated in Figure 1.

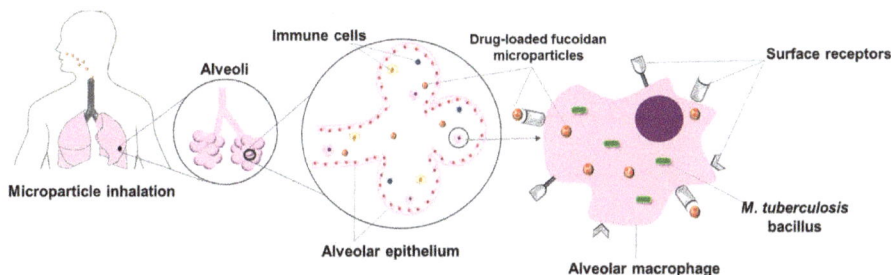

Figure 1. Illustration of microparticle uptake by alveolar macrophages, assuming targeted drug delivery mediated by fucoidan. Drug-loaded fucoidan microparticles reach the alveoli upon dry powder aerosolisation. Next, alveolar macrophages, infected with *M. tuberculosis*, engulf the microparticles. Fucoidan is expected to facilitate phagocytosis, because it possesses chemical moieties that are recognisable by the macrophage surface receptors.

The purpose of this work is to produce inhalable FUC-based microparticles combining both isoniazid (INH) and rifabutin (RFB) in a single formulation, with a potential affinity for alveolar macrophages mediated by FUC. The proposed combined therapy of INH and RFB complies with the specific WHO guidelines on TB therapy [12]. The microparticles were characterised and their respirability was evaluated to determine the ability to reach the deep lung. The uptake of FUC microparticles by macrophages was assessed by flow cytometry, and their capacity to activate macrophages was determined. The potential antimicrobial activity of the produced carriers was also evaluated, along with the determination of their cytocompatibility.

2. Materials and Methods

2.1. Materials

Fucoidan (FUC, *Laminaria japonica*) and rifabutin (RFB, M_w 847.00 g/mol) were purchased from Chemos GmbH (Regenstauf, Germany). Isoniazid (INH, M_w 137.14 g/mol), buffer solution pH 5 (citric acid ~0.096 M, sodium hydroxide ~0.20 M), dimethylformamide (DMF), Dulbecco's modified Eagle's medium (DMEM), lipopolysaccharide (LPS), N-(3-dimethylaminopropyl)-N'-ethylcarbodiimide hydrochloride (EDAC), non-essential amino acids solution and penicillin/streptomycin (10,000 units/mL,

10,000 g/mL), sodium dodecyl sulphate (SDS), Triton-X 100, trypsin-EDTA solution (2.5 g/L trypsin, 0.5 g/L EDTA), trypan blue solution (0.4%), and HCl were supplied by Sigma-Aldrich (Munich, Germany). Thiazolyl blue tetrazolium bromide (MTT), phosphate buffer saline (PBS) tablets pH 7.4 and Tween 80® were supplied by Amresco (Solon, OH, USA). Dimethyl sulfoxide (DMSO) was provided by VWR (Fontenay-sous-Bois, France) and phorbol 12-myristate 13-acetate (PMA) was provided by Cayman Chemicals (Ann Arbor, MI, USA). A lactate dehydrogenase (LDH) kit was obtained from Takara Bio (Tokyo, Japan) and L-glutamine solution (200 mM), as well as fetal bovine serum (FBS) from Gibco (Life Technologies, Waltham, MA, USA). RPMI 1640 and Ham's F12 media were supplied by Lonza Group AG (Basel, Switzerland). Middlebrook 7H9 (M7H9; 4.7 g/L) and OADC (oleic acid, albumin, dextrose and catalase) were purchased from Remel (Lenexa, KS, USA). Ultrapure water (MilliQ, Millipore, UK) was used throughout the studies and other chemicals were reagent grade.

2.2. Preparation of Microparticles

Microparticles with and without drugs were obtained from 2% (w/v) FUC solutions, which were prepared by dissolving the polymer in ultrapure water under stirring (MS-3000 Biosan, Riga, Latvia). The drugs INH and RFB were incorporated into the solution in two steps. INH was ground in a porcelain mortar, solubilised in water and then added dropwise to the prepared polymeric solution. RFB was ground in a glass mortar, dissolved in 4% (v/v) ethanol, and then incorporated drop by drop into the polymeric solution containing INH. The final solution (50 mL) was stirred for 1 h before spray-drying. Drugs were included in the formulation to obtain FUC/INH/RFB mass ratios of 10/1/0.5.

Microparticles were produced on a laboratory scale spray-dryer (Büchi B-290 Mini Spray Dryer, Büchi Labortechnik AG, Flawil, Switzerland), equipped with a high-performance cyclone. The equipment operated in open mode configuration, using compressed air. The spray flow rate was set at 473 L/h, and the operating parameters were set as indicated in Table 1.

Table 1. Operating parameters of the spray-drying process.

Microparticles	Inlet T (°C)	Aspirator (%)	Feed rate (mL/min)
Unloaded FUC	125 ± 1	80	2.0
FUC/INH/RFB	145 ± 1	85	1.0

FUC: fucoidan; INH: isoniazid; RFB: rifabutin.

The production yield of the spray-drying process was calculated comparing the weight of collected microparticles and the total amount of solids initially added to produce microparticles. Dry powders were stored in desiccators until further use.

Fluorescent microparticles of FUC were also produced to be used in the assay of microparticle uptake by macrophages. To produce fluorescently labelled FUC, the covalent attachment of the dye to the polymer was carried out by reacting it with fluorescein sodium salt in the presence of N-(3-dimethylaminopropyl)-N'-ethylcarbodiimide hydrochloride (EDAC) as activator of the fluorescein carboxyl group. In short, 1 g of FUC was dissolved at 2% (w/v) in water. Fluorescein sodium salt (24.4 mg) dissolved in 4 mL of 96% (v/v) ethanol and EDAC (9.6 mg dissolved in 16 mL of milli-Q water) were added to the former solution. The reaction mixture was kept under stirring in the dark overnight and afterwards dialysed (2000 Da M_w cut-off) against distilled water, which was also protected from the light. The dialysate was frozen and freeze-dried (FreeZone Benchtop Freeze Dry System, Labconco, Kansas City, MO, USA). Fluorescent (unloaded) FUC microparticles were prepared under the same conditions as displayed in Table 1.

2.3. Microparticle Characterisation

2.3.1. Morphology

Microparticle morphology was visualised by field emission scanning electron microscopy (FESEM Ultra Plus, Zeiss, Jena, Germany). Briefly, the dry powder was placed onto metal plates and 5-nm thick iridium film was sputter-coated (model Q150T S/E/ES, Quorum Technologies, Lewes, UK) on the sample before visualisation.

2.3.2. Feret's Diameter

The microparticle size was estimated as the Feret's diameter and measured as the mean of 300 microparticles ($n = 3$). The measurements were performed by optical microscopy (Microscope TR 500, VWR international, Leuven, Belgium).

2.3.3. Particle Size Distribution

Drug-loaded FUC microparticles were characterised in terms of median volume diameter by laser light scattering. In short, an amount of dry powder (15 mg) was dispersed in 15 mL of 2-propanol and sonicated for 5 min. Measurements were performed with a SprayTec® (Malvern, UK) and data are expressed as 50% (D_v50) of aerosol droplets. The analyses were performed three times with an obscuration threshold of 10% [13].

2.3.4. Density

A helium pycnometer (Micromeritics AccuPyc 1330, Aachen, Germany) was used to determine microparticle real density (g/cm^3; $n = 3$). Tap density (g/cm^3) was determined using a tap density tester ($n = 3$; 30 tapping/min, Densipro 250410, Deyman, Santiago de Compostela, Spain).

2.4. Drug Association Efficiency and Loading

In order to determine the drug content, the dry powder (20 mg) was solubilised in HCl 0.1 M (10 mL) under magnetic stirring for 20 min. Then, the solution was filtered (0.45 µm, RC, Sartorius, Concord, CA, USA), and a sample was analysed by high-performance liquid chromatography (HPLC—Agilent 1100 series, Concord, Germany). A LiChrospher® 100 RP-18 (4.6 µm) column of 4 mm i.d. × 250 mm length with a security guard cartridge was used, and detection was performed by a diode array detector set at a wavelength of 275 nm. For the analysis, the mobile phase consisted of phosphate buffer 20 mM pH = 7 (A) and acetonitrile (B), flowing at a rate of 1.0 mL/min. The elution was conducted with a gradient starting with A/B = 95%/5% (0–5 min), which further reached a 30%/70% A/B ratio (5–8 min), which was kept for 19 min. Under these conditions, retention times of INH and RFB were 5 min and 20 min, respectively. Calibration curves (10–400 mg/mL) were plotted using INH and RFB standard solutions (HCl 0.1M). Drug association efficiency and microparticle loading capacity were calculated ($n = 3$) by the following equations [14,15]:

$$AE\ (\%) = (\text{Real drug content/Theoretical drug content}) \times 100 \tag{1}$$

$$LC\ (\%) = (\text{Real amount of drug/Weight of MP}) \times 100 \tag{2}$$

2.5. In Vitro Drug Deposition

The in vitro aerosolisation of the dry powder was evaluated using the Andersen cascade impactor (ACI, Copley Scientific Ltd., Nottingham, UK). The used methodology respected the USP38 guidelines for dry powder inhalers (Apparatus 1, United States Pharmacopoeia, Chapter 601). ACI separates particles according to their aerodynamic diameter, and it was assembled using the appropriate adaptor kit for the 60 L/min air flow test. Cut-offs of the stages from −1 to 6 are the following: 8.60, 6.50, 4.40, 3.20, 1.90, 1.20, 0.55 and 0.26 µm. A glass fiber filter (Whatman, Milano, Italy) was placed right below stage six in order to collect particles with a diameter lower than that of the stage six cut-off. Collection plates were coated with 1% (*v/v*) Tween® 20 in ethanol to prevent particle bounce.

A powder amount of 30 mg was loaded into a size three HPMC capsule (Quali-V-I, Qualicaps, Madrid, Spain) and aerosolised using an RS01 device (IFR = 0.033 kPa$^{0.5}$/LPM, Plastiape, Lecco, Italy). The content of three capsules was discharged for each experiment, and the experiments were performed in triplicate.

The flow rate that was used during each test was adjusted at 60 L/min with a critical flow controller TPK (Copley Scientific, Nottingham, UK) in order to produce a pressure drop of 4 kPa across the inhaler. The flow rate corresponding to these pressures was measured before each experiment using a DFM 2000 Flow Meter (Copley Scientific, Nottingham, UK). The test duration time was adjusted at 4 s, so that a volume of 4 L of air was drawn through each inhaler during each test.

A mixture of water/acetonitrile (50/50, *v/v*) was used to rinse off the powder from the apparatus. Samples were then sonicated for 5 min, filtered (0.45 µm, RC, Sartorius, Concord, CA, USA), and analysed by HPLC (Agilent 1200 series, Waldbronn, Germany), following the analytical protocol previously described.

The measurement of the INH and RFB that was deposited in the impactor allowed the calculation of deposition parameters. Mass median aerodynamic diameter (MMAD) was determined by plotting the cumulative percentage of mass less than the stated aerodynamic diameter for each stage on a probability scale versus the aerodynamic diameter of the stage on a logarithmic scale. The mass of drug particles with size <5 µm (calculated from log-probability plots) was defined as a fine particle dose (FPD). Yet, the amount of drug leaving the device and reaching the impactor was considered as the emitted dose (ED). Subsequently, the fine particle fraction (FPF) was calculated as the percentage ratio between FPD and ED. Finally, the metered dose (MD) was accounted as the mass of drug recovered and quantified by HPLC, and was calculated by summing the drug recovered from the inhaler (device and capsule) and the impactor (induction port, stages −1 to 6 and F).

2.6. Drug Release Profiles

The in vitro drug release studies were carried out in PBS (pH 7.4) and in buffer solution at pH 5 (citric acid ~0.096 M, sodium hydroxide ~0.20 M, Sigma-Aldrich, Munich, Germany), both containing 1% (*v/v*) Tween® 80. Assays were performed respecting sink conditions, as the maximum amount of drug was always below 30% of its maximum solubility [16]. The drug release rate was determined by incubating the dry powder (20 mg) with a release medium (10 mL) in a test tube, which was kept at 37 °C (Dry line; VWR, Tempe, AZ, USA) under mild shaking (100 rpm; Orbital Shaker OS 10, Biosan, Riga, Latvia). Then, at pre-established time intervals, aliquots (1 mL) were withdrawn, centrifuged (16,000× *g*, 15 min; Heraeus Fresco 17 Centrifuge, Thermo Scientific, Waltham, MA, USA), and filtered (0.45 µm). In the end, the drug content in the samples was quantified by HPLC (*n* = 3) by interpolation from calibration curves obtained with standard solutions of drugs diluted in the release media.

2.7. In Vitro Biocompatibility Studies

2.7.1. Cell Culture

Human alveolar epithelium cells (A549) were purchased from the American Type Culture Collection (ATCC, Middlesex, UK). The cell line was cultured in DMEM supplemented with 10% (*v/v*) of FBS, 1% (*v/v*) L-glutamine solution 200 mM, 1% (*v/v*) non-essential amino acids, and 1% (*v/v*) penicillin/streptomycin. For the experiments, cells were used in passages 25–36.

THP-1 human monocytic cells were obtained from the Leibniz-Institute DSMZ (Braunschweig, Germany). The cell line was maintained in suspension at 0.2–0.8×10^6 cells/mL in RPMI 1640 medium supplemented with 10% (*v/v*) FBS, 1% (*v/v*) L-glutamine, and 1% (*v/v*) penicillin/streptomycin. For the assays, THP-1 cells (0.35×10^6 cells/mL) were differentiated to acquire macrophage phenotype (50 nM PMA, 48 h) before performing the cytotoxicity tests. Cells were used between passages 10–17.

In general, cell cultures were grown using 75 cm^2 flasks in a humidified 5% CO_2/95% atmospheric air incubator at 37 °C (HerAcell 150, Heraeus, Hanau, Germany).

2.7.2. Determination of Metabolic Activity

The effect of FUC microparticles on cell viability was evaluated on A549 and macrophage-differentiated THP-1 cells by MTT assay. Briefly, A549 cells were seeded in 96-well plates (Orange Scientific, Braine-l'Alleud, Belgium) at a density of 1.0×10^4 cells/well in complete medium (100 µL) and allowed to attach overnight at 37 °C in 5% CO_2 atmosphere. After that, cells were exposed to test solutions for 3 and 24 h. After the exposure time, the medium was removed, and 30 µL of MTT (0.5 mg/mL in PBS, pH 7.4) were added in each well, followed by 2 h of incubation at 37 °C. Next, purple crystals were dissolved with DMSO (50 µL), and the absorbance was determined by spectrophotometry (Infinite M200, Tecan, Grödig, Austria) at 540 nm, subtracting the background absorbance (640 nm).

Similarly, THP-1 cells were seeded in 96-well plates (0.35×10^6 cells/well) in 100 µL of complete medium and differentiated as described before. After differentiation, the cell culture medium (CCM) was renewed for another 24 h, before performing the experiments. As for A549 cells, exposure times of 3 and 24 h were applied. After exposure, MTT solution (30 µL) was added in each well (no media removal was applied), and incubation was allowed for 2 h, after which formazan crystals were solubilised with 10% SDS in a 1:1 mixture of DMF. The absorbance was determined by spectrophotometry, as described above.

Overall, test solutions were previously prepared by dissolving the microparticles (unloaded and drug-loaded) in the proper CCM without FBS at three concentrations: 0.1, 0.5, and 1.0 mg/mL. INH and RFB were also tested as free drugs at concentrations equivalent to their theoretical loadings in microparticles, i.e., 0.01, 0.05 and 0.1 mg/mL for INH and 0.005, 0.025 and 0.05 mg/mL for RFB.

CCM and 2% (w/v) SDS were used as positive and negative controls of cell viability, respectively. Cell viability of treated cells was expressed as a percentage of that observed for the positive control (CCM). The assay was performed at least three times, with six replicates at each concentration of test solutions.

2.7.3. Evaluation of Cell Membrane Integrity

The integrity of cell membrane was assessed by the quantification of LDH release upon exposure to test samples. Both A549 (1.0×10^4 cells/well) and macrophage-differentiated THP-1 cells (0.35×10^6 cells/well) were exposed to sample solutions at a concentration of 1.0 mg/mL (corresponding to the maximum concentration used in MTT assays). Solutions of free INH (0.1 mg/mL) and RFB (0.05 mg/mL) were also tested. After 24 h of exposure, aliquots (100 µL) of cell supernatant samples were centrifuged ($16,000 \times g$, 5 min) and processed using a commercial LDH kit (Takara Bio, Tokyo, Japan). The concentration of LDH was measured by spectrophotometry (Infinite M200, Tecan, Grödig, Austria) at a wavelength of 490 nm with background correction at 690 nm. Cells incubated with CCM only were considered the negative control; those treated with Triton-X100 (10%) were used as the positive control, the latter being assumed as 100% of LDH release. Thus, released LDH values were expressed as a percentage of the positive control. All of the measurements were performed in triplicate.

2.8. Macrophage Activation by Microparticles

In order to evaluate the capability of microparticles to activate macrophage-like cells, differentiated THP-1 cells (0.350×10^6 cells/well) were incubated with drug-loaded FUC microparticles. After 24 h of incubation, cell-free supernatants were collected and TNF-α and IL-8 quantified with Quantikine® HS ELISA kits (R&D Systems, Minneapolis, MN, USA). The amount of each cytokine was expressed in pg/mL based on reference standard curves. Cytokines released from cells treated with FUC solution and LPS, and untreated cells were used as controls. The absorbance of samples was determined at 450 nm in a microplate reader and corrected for background absorbance at 540 nm.

2.9. Preliminary Evaluation of Microparticle Uptake by Macrophages

The uptake of FUC microparticles by macrophage-like cells was performed by flow cytometry (FacScalibur cell analyser, BD Biosciences, Erembodegem, Belgium) and involved the exposure of cells to fluorescently-labelled microparticles. The assay was performed on human macrophage-differentiated THP-1 cells and on rat alveolar macrophages (NR8383 cells). Macrophage-differentiated THP-1 (0.35×10^6 cells/mL) and NR8383 (0.2×10^6 cells/mL) cells were seeded in 35 mm-diameter dishes, containing 5 mL of the respective complete medium. Next, cells were maintained for 24 h at 37 °C to ensure the adhesion of 50%–75% of the original population. Then, media were removed, and fluorescent microparticles (50 and 200 μg/cm^2) were aerosolised onto the macrophage layer using a Dry Powder Insufflator™ (Model DP-4, Penn-Century™, Wyndmoor, PA, USA). Cells unexposed to microparticles were considered as negative control. The phagocytic process was allowed for 2 h (incubation at 37 °C) and was stopped by the addition of a cold solution of PBS.3% FBS (5 mL, two applications). Then, cells were scraped, re-suspended in 3 mL of PBS.3% and centrifuged (1500 rpm, 2 min, room temperature, centrifuge MPW-223e, MedInstruments, Warsaw, Poland). Cells were washed with PBS.3% (5 mL) thrice and finally re-suspended in 1 mL of buffer for flow cytometry analysis (BD Biosciences FACSCalibur, Erembodegem, Belgium). For this purpose, side scatter light was used to distinguish cell viable population, whereas FSC-H and SSC-H channels were applied to measure the size and granularity of cells, respectively. A total of 10,000 events were counted within a gated region, and the data were presented as mean fluorescence (FL) intensity. The number of cells associated with fluorescence was considered the definition for uptake. The assay was replicated at least three times for each dose.

2.10. Determination of Minimum Inhibitory Concentration (MIC)

2.10.1. Culture of Mycobacteria

The in vitro efficacy of microparticles was evaluated against *Mycobacterium bovis* BCG (DSMZ 43990), provided, as a gift, by Centro de Estudos de Doenças Crónicas da Faculdade de Ciências Médicas da Universidade Nova de Lisboa (CEDOC/FCM-UNL). The stocks of mycobacteria were preserved and stored in −80 °C ultralow temperature freezers (U725 Innova New Brunswick Scientific, Edison, NJ, USA). Mycobacteria were cultivated in M7H9 broth, supplemented with 10% OADC and 0.05% of Tween® 80. Mycobacteria was handled inside a laminar flow hood (Bio48 Faster, Cornaredo, Italy), respecting the guidance and safety requirements to prevent contamination. That includes autoclave sterilisation (Uniclave88, Sintra, Portugal) of infectious materials.

2.10.2. MIC Measurements

The lowest concentration of free antibiotics required to inhibit mycobacteria growth was determined by MTT assay [17]. Firstly, stock solutions (1 mg/mL) of INH and RFB were prepared by dissolving the drugs in the M7H9 supplemented medium. RFB was previously solubilised in ethanol, and then diluted with M7H9 broth. Stock solutions were filtered (0.22-μm sterile syringe filter) and mixed at concentrations proportional to the drug mass ratio contained in the microparticles (FUC/INH/RFB = 10/1/0.5, w/w). The susceptibility of the *M. bovis* strain was then evaluated by incubating mycobacteria with a drug solution combining INH/RFB, followed by serial dilutions.

Similarly, the growth inhibition of mycobacteria promoted by FUC/INH/RFB microparticles was also assessed. A solution of dry powder was prepared at 1 mg/mL and then diluted, based on the calculations, to meet the desired drug concentrations. Generally, 96-well flat-bottom microplates (Orange Scientific, Braine-l'Alleud, Belgium) were filled with test samples, according to the scheme displayed in Figure S1 (Supplementary Materials). Three bacterial suspensions were prepared, and the assays were conducted after achieving an optical density value (OD_{600nm}) of approximately 0.2, as measured by spectrophotometry (Infinite M200, Tecan, Austria). For the experiment, solutions (180 µL) of free drugs or microparticles were added into the wells (column 4), making continuously serial two-fold dilutions with M7H9 supplemented broth (columns 5–11). Next, 20 µL of bacterial suspension were added into the respective well, completing the final volume of 200 µL per well. Wells filled only with M7H9 supplemented medium (column 2; 200 µL) were used as a negative control. Similarly, bacterial suspensions (20 µL) were introduced in wells (column 3) containing M7H9 medium (180 µL) in the absence of free drugs/microparticles, which were assumed as positive controls. The bacterial growth of each bacterial suspension—1, 2, and 3—was evaluated in the B–C, D–E, and F–G lines, respectively. Assays were performed in triplicate.

The outside lane of wells (a frame-like) were filled with sterile distilled water to avoid the evaporation of microplate content. The plates were covered with the lid, sealed with parafilm, and incubated at 37 °C (Binder, Tempe, AZ, USA) for seven days. After that time, 30 µL of MTT sterile solution was added to each well, followed by 1 h of incubation at 37 °C. Then, 50 µL of DMSO was added into the wells, resulting in a colour change from yellow to dark gold proportional to the growth of mycobacteria. The absorbance was measured by spectrophotometry (Infinite M200, Tecan, Austria) at 540 nm. The minimum inhibitory concentration was considered the one that inhibited mycobacteria growth by 95% to 100%.

2.11. Statistical Analysis

Statistical significance was determined with the student *t*-test and one-way analysis of variance (ANOVA) with the pairwise multiple comparison procedures (Holm-Sidak method). A *p*-value of less than 0.05 was considered significant. Analysis was run using Sigmaplot software (version 12.5, Systat Software Inc., London, UK).

3. Results and Discussion

3.1. Preparation and Characterisation of Fucoidan Microparticles

Spray-dried FUC microparticles loaded with a combination of two first-line antitubercular drugs (INH and RFB) were produced with a yield around 81%, indicating the effectiveness of the process. Drugs were efficiently associated to microparticles, with INH registering 97% ± 4% and RFB 95% ± 4% association efficiency. In this manner, loading capacities reached 8.5% ± 0.4% (INH) and 4.1% ± 0.2% (RFB), which were close to the theoretical values. This approach of associating antitubercular drugs in a single formulation meets the recommendations of the WHO regarding the need to establish a combined therapy for TB [12]. The selected theoretical drug loadings (8.7% for INH and 4.4% for RFB) are similar to those reported in other works [18]. INH is present in a higher amount than RFB, because the latter is a more potent drug [19], and also has a more toxic profile, as demonstrated in Section 3.4. Moreover, fucoidan proportion was kept purposely high in order to favour macrophage internalisation [8].

The fact that both drugs were associated with similar efficiencies demonstrates that the process was independent of the aqueous solubility of the drugs (125 mg/mL for INH and 0.19 mg/mL for RFB) [20,21].

The morphological analysis performed with electronic microscopy revealed that the unloaded microparticles exhibited a slightly convoluted shape but smooth surface (Figure 2a). However, the incorporation of drugs in the microparticles produced important modifications on their morphology, which became more irregular and acquired corrugated surfaces (Figure 2b). In this case, the presence of RFB may have contributed to the morphological alterations perceived on the produced microparticles, as the observed surface wrinkles have been reported for other spray-dried microparticles loaded with RFB [22]. The wrinkled surface can also be thought of as a result of the drying process, in which the removal of ethanol from the microparticles' surface occurs faster than water evaporation. The higher volatility of ethanol induces the formation of a primary shell that collapses as the water content in the core evaporates [23].

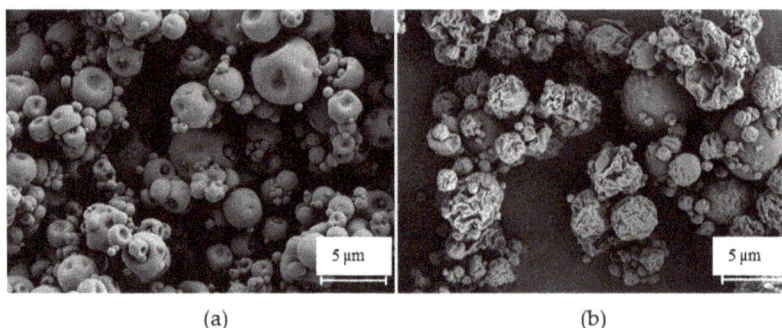

(a)   (b)

Figure 2. Scanning electron microphotographs of FUC-based microparticles: (a) unloaded microparticles and (b) FUC/INH/RFB microparticles. FUC: fucoidan, INH: isoniazid, RFB: rifabutin.

Drug-loaded microparticles were found to have a Feret's diameter of 1.4 ± 0.8 μm; no significant differences were detected after drug incorporation (1.6 ± 0.8 μm for unloaded FUC microparticles). Additionally, volume diameter measurement by laser light scattering indicated median diameters (D_v50) of 2.77 ± 0.03 μm for drug-loaded microparticles. The latter indicates that the produced microparticles present a suitable size for pulmonary deposition [24]. The particle size distribution reflects the productive efficiency of spray-drying to provide microparticles with favourable aerodynamic properties for inhalation purposes [25]. To the best of our knowledge, this is the first report describing the use of spray-drying to produce respirable FUC-based microparticles loaded with two antitubercular drugs.

Density further influences the properties of inhalable dry powders [26]. Thus, real and tap densities of FUC microparticles were determined, showing values around 1.742 ± 0.004 g/cm³ and 0.346 ± 0.019 g/cm³, respectively. These results are similar to others reported for spray-dried polysaccharide microparticles [27,28].

3.2. Aerodynamic Characterisation of Fucoidan Microparticles

The design of inhalable microparticles requires adequate flowability to promote lung deposition. In light of this, the in vitro aerosol performance was assessed in the ACI by using an RS01 dry powder inhaler. The obtained data are presented in Table 2.

Table 2. Aerodynamic parameters of fucoidan microparticles loaded with isoniazid (INH) and rifabutin (RFB) in the combined formulation. Loaded amount of powder in the capsule was 30 mg, corresponding to 2.6 mg of INH and 1.4 mg of RFB, according to the drug content of formulation ($n = 3$, mean ± SD).

Drug	Metered Dose (mg)	Emitted Dose (mg)	MMAD (μm)	FPD < 5 μm (mg)	FPF < 5 μm (%)
INH	1.91 ± 0.26	1.64 ± 0.23	3.90 ± 0.01	0.82 ± 0.02	50.2 ± 2.4
RFB	1.29 ± 0.03	1.10 ± 0.02	3.64 ± 0.32	0.53 ± 0.01	45.4 ± 1.4

FPD: fine particle dose; FPF: fine particle fraction; MMAD: mass median aerodynamic diameter.

The results demonstrated that more than 85% of the drugs were emitted from the device, indicating the suitability of FUC to be used as a matrix material in spray-dried inhalable microparticles. The adequate flowability is possibly a result of the wrinkled surfaces observed in the produced microparticles. Surface irregularities can reduce the cohesion forces between particles that lead to agglomeration, enhancing powder dispersibility, as well as improving the respirable fraction of the formulation [23,29]. This aspect is crucial when a lactose carrier is not included in the formulation, and spray-dried microparticles are aerosolised alone [16]. The FPF ($\leq$5 μm) was around 50%, which represents the respirable fraction of microparticles with the potential ability to reach the respiratory zone. The results suggest as well that there was low cohesion among microparticles, which led to high deaggregation during aerosolisation and in agreement with the in vitro respirability usually exhibited by DPI formulation [24]. Additionally, the use of the RS01 device may have contributed to the maximisation of the performance, as it is described that the spinning capsule rotation provided by this inhaler is more efficient compared with the other capsule motion in the powder deaggregation [30,31].

Figure 3 illustrates the stage-by-stage deposition profiles of INH and RFB in the ACI after aerosolisation. The drug recovery varied between 82%–85%, being in accordance with the values established by the European Pharmacopeia [32]. According to drug mass deposition on ACI stages, the determined MMAD values were 3.9 μm (INH) and 3.6 μm (RFB), suggesting a suitable size for lung deposition. In fact, particles with an aerodynamic diameter within the range of 1–5 μm display a greater tendency for reaching the respiratory zone, and if with extrafine size (<2 μm), have more peripheral deposition within the area [33]. Moreover, they are in the adequate size range (1–6 μm) that allows phagocytosis by macrophage, the target cells [34].

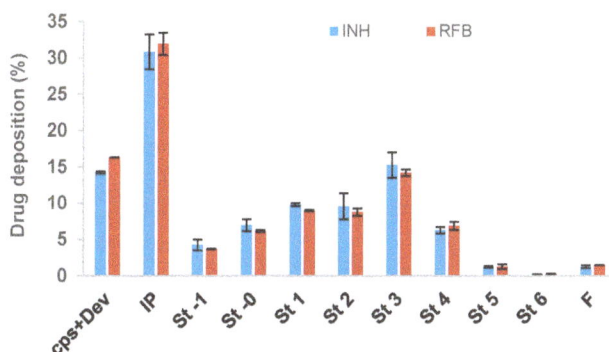

Figure 3. In vitro aerodynamic deposition of antitubercular drugs (INH and RFB) in the Andersen cascade impactor. Drugs were associated with spray-dried fucoidan microparticles. Values are mean ± SD, n = 3. Cps: capsule; Dev: inhaler device; IP: induction port; F: filter, INH: isoniazid; RFB: rifabutin St: stage.

The similarity between the profiles of both drugs demonstrates that INH and RFB were equally co-deposited on the several stages, thus supporting the decision of developing a delivery system with a combination of the two drugs. Furthermore, this indicates that the drugs have even distribution within the microparticles. In summary, the produced microparticles were shown to have adequate aerodynamic characteristics for the pulmonary delivery of antitubercular drugs with great propensity to reach the respiratory zone.

3.3. In Vitro Drug Release Profiles

The release of drugs was evaluated in PBS (pH 7.4) with the addition of 1% (*v/v*) Tween® 80, resembling the lung lining fluid in terms of pH and the presence of surfactant [35,36]. The latter also enables the dissolution of RFB, which is sparingly soluble in aqueous media. Considering that alveolar macrophages are the target cells in this study, drug release studies were also conducted in more acidic medium (pH 5), simulating the phagolysosomal environment [37]. The obtained results are shown in Figure 4.

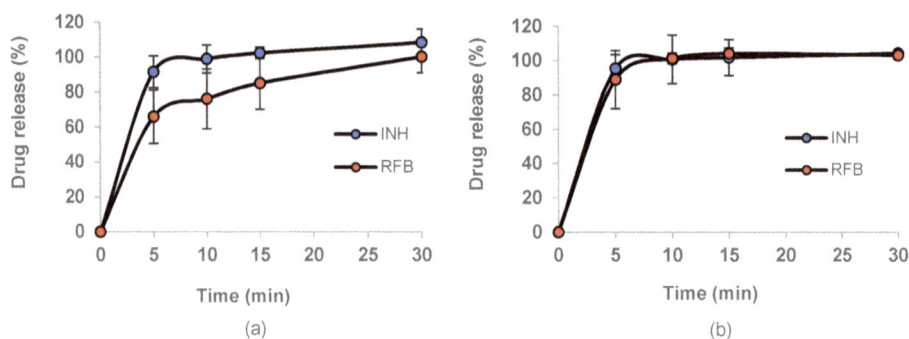

Figure 4. In vitro release profile of isoniazid (INH) and rifabutin (RFB) from FUC/INH/RFB (10/1/0.5, *w/w*) in (**a**) PBS pH 7.4-Tween 80® and (**b**) acidic medium pH 5.0-Tween 80®. FUC: fucoidan; mean ± SD, $n = 3$.

In general, the two drugs released rapidly from the microparticles, especially INH, which is completely available after 10 min, disregard the pH of the release medium (Figure 4a,b). The rapid release of this drug was expected, owing to its high solubility in water. However, the RFB release profile was also similar to that of INH, despite its lower solubility. At pH 5, RFB released exactly at the same rate as INH. In turn, at pH 7.4, the release was a little bit slower, although not to a statistically significant level. In this medium, approximately 75% of the RFB was released within 10 min, and 100% was released within 30 min. The faster release in acidic medium is possibly a consequence of a certain protonation of the drug, which increases its solubility. However, no significant difference was perceived at any time point, indicating that the release of the two drugs was not significantly influenced by pH. Other works also reported the rapid release of INH and RFB from polysaccharide microparticles [20,38].

The rapid release could be explained by two main factors: the surface irregularities, which increase the contact with the medium, and the high solubility of the polymeric matrix, i.e., FUC rapidly dissolves in the media, releasing the drugs. It should be stressed that these observations do not reflect in vivo occurrences, considering that a lower amount of liquid is present in the alveoli compared with that involved in the assays. It is well known that the alveolar epithelium is covered by a thin layer (0.01–0.1 μm) of lung lining fluid, and thus, microparticles are expected to be only partially in contact with this fluid, and not immersed in it [39,40]. Consequently, in vivo drug release will probably occur more slowly, allowing microparticle internalisation by macrophage cells before particle dissolution and complete drug release.

3.4. In Vitro Cytotoxicity

The cytotoxicity of FUC/INH/RFB microparticles was evaluated by performing two complementary tests: metabolic assay (MTT) and the LDH release assay, which assesses cell membrane integrity. In both cases, alveolar epithelial cells (A549) and macrophage-differentiated THP-1 cells were used. Drug-loaded microparticles were tested along with an unloaded formulation and free drugs, which were considered as controls. Free drugs were tested at concentrations corresponding to the respective theoretical loading in the microparticles.

3.4.1. Metabolic Activity by MTT Assay

The MTT assay was performed upon 3 and 24 h of exposure, and revealed that the viability of the A549 cells exposed to the produced microparticles, in all of the tested conditions, remained above 70%, the threshold below which it is considered the occurrence of cytotoxic effect according to ISO [41]. In fact, the exposure of A549 cells to FUC/INH/RFB microparticles induced mild effects on cell viability; meanwhile, no significant differences were observed in terms of time and microparticle concentrations (Figure 5a,b lighter colours). However, macrophage-differentiated THP-1 cells were slightly more sensitive to the contact with FUC/INH/RFB microparticles after long-time exposure. At 24 h, cell viability decreased to 65% upon exposure to the highest dose of microparticles (1.0 mg/mL) (Figure 5b, darker colours). Comparatively, A549 cells showed 76% viability in the same conditions ($p < 0.05$). Concerning macrophage-like THP-1 cells, no significant time-dependent or dose-dependent effects were observed.

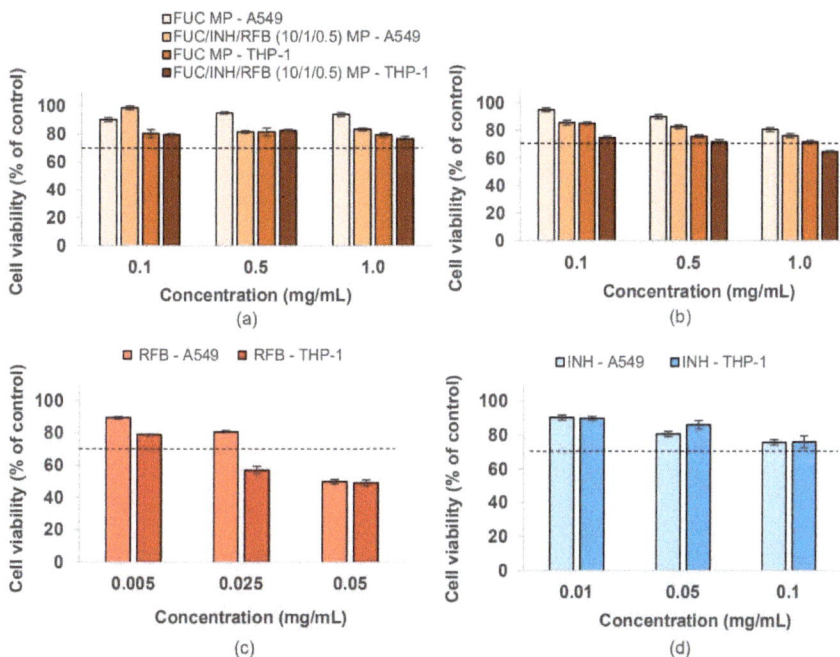

Figure 5. A549 and macrophage-differentiated THP-1 cell viability upon (**a**) 3 h and (**b**) 24 h of exposure to unloaded FUC and FUC/INH/RFB (10/1/0.5 *w/w*) microparticles; (**c**) 24 h exposure to RFB as a free drug; and (**d**) 24 h exposure to INH as a free drug. Cell viability was calculated as a percentage of positive control (untreated cells). Data represent mean ± SEM (*n* = 3, six replicates per experiment at each concentration). Dashed line indicates 70% cell viability. FUC: fucoidan; INH: isoniazid; MP: microparticles; RFB: rifabutin.

The mild toxicity observed in THP-1 cells is clearly due to the RFB content, as was already described in a recent work of our group [42]. Figure 5c,d depicts the effects of the free drugs after 24 h. While INH induced cell viabilities between 75%–90% in both cell lines, thus not exhibiting a cytotoxic behaviour, RFB generated different responses. In fact, RFB decreased cell viability to around 50% in both cell types when tested at the highest concentration (0.05 mg/mL), while the viability of differentiated THP-1 cells also decreased to 57% at the intermediate concentration (0.025 mg/mL, Figure 5c). These observations demonstrated the drug toxicity, which has been reported in vivo [43,44], and indicated that the toxic effect is due to RFB. In order to further confirm this, drug-loaded microparticles were produced with lower amounts of RFB (FUC/INH/RFB = 10/1/0.2, *w/w*) to be purposely tested regarding cytotoxicity. It was verified that this decrease in RFB content resulted in the viability of differentiated THP-1 cells over 70% in all of the tested concentrations (data not shown). Free drugs were also tested for 3 h (data not shown). INH induced the viability of 89%–99% in all of the cases. In turn, RFB decreased viability to 68% (A549 cells) and 56% (differentiated THP-1 cells) at the highest tested concentration (0.05 mg/mL). It should be stressed that at the same time point (3 h), viability remained at 84% (A549) and 77% (differentiated THP-1 cells) when cells were exposed to FUC/INH/RFB microparticles (1.0 mg/mL), as shown in Figure 5a. In this way, it is suggested that the microencapsulation had a beneficial effect on RFB cytotoxicity ($p < 0.05$). Unloaded FUC microparticles were also tested as control in both cell lines (Figure 5a,b), resulting in a cell viability above 80% in all of the cases, thus evidencing the absence of detrimental effects of the polysaccharide under the tested conditions.

Overall, it can be considered that both A549 and macrophage-differentiated cells tolerated the exposure to fucoidan-based microparticles well. A single exception was registered in macrophage-like THP-1 cells when exposed to 1.0 mg/mL microparticles for 24 h, with the toxicity being attributed to the RFB content. However, it should be highlighted that this dose is much higher than the one that was expected to occur in vivo, as after the delivery, the dry powder will be distributed through a large surface. Therefore, the effects will probably be more similar to those of the lower doses (0.1 and 0.5 mg/mL), which were concentrations where no cytotoxicity was perceived.

3.4.2. Cell Membrane Integrity

As a complement to the MTT assay, the amount of the cytoplasmic enzyme LDH released after cell exposure (24 h) to microparticles (1.0 mg/mL) and free drugs was determined (Figure 6). The results essentially confirm those of MTT, with free RFB having a more intense effect on the release of LDH. The incubation with CCM induced 21% of LDH release in A549 cells (Figure 6a) and 34% in macrophage-differentiated THP-1 cells (Figure 6b). The exposure to free RFB significantly increased these values to 42% and 57%, respectively ($p < 0.05$). Despite RFB cytotoxicity, FUC/INH/RFB microparticles generated similar LDH release compared with CCM in both cell lines, which reinforces the beneficial effect of microencapsulation regarding RFB cytotoxicity.

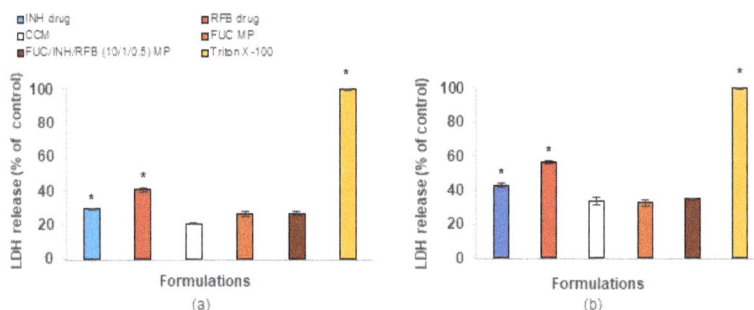

Figure 6. Release of lactate dehydrogenase (LDH) from (**a**) A549 cells and (**b**) macrophage-differentiated THP-1 cells exposed to fucoidan-based microparticles (1.0 mg/mL), free rifabutin (RFB, 0.05 mg/mL), and free isoniazid (INH, 0.1 mg/mL). Cell culture medium (CCM) and Triton X-100 were used as negative and positive controls, respectively. The released LDH calculated was based on 100% assumed for positive control. Data represent mean ± SEM ($n = 3$, six replicates per experiment at each concentration). * $p < 0.05$ compared to CCM.

Curiously, free INH also induced the release of a significantly higher amount of LDH compared with the control CCM (29% for A549 cells and 43% for macrophage-differentiated THP-1 cells, $p < 0.05$). However, microencapsulation reverted the toxicological effect ($p < 0.05$). Unloaded FUC microparticles also evidenced an absence of effect on LDH release.

The overall evaluation of in vitro cytotoxicity indicates a very acceptable toxicological profile of FUC-based microparticles. Nevertheless, it is considered beneficial to widen the toxicological studies of these microparticles to certify their use for the proposed application.

3.5. Macrophage Activation

The ability of FUC/INH/RFB microparticles to induce macrophage activation was assessed by determining the amount of TNF-α and IL-8 secreted by macrophage-like THP-1 cells upon contact with the formulation. The referred cytokines are two pro-inflammatory molecules released by human alveolar macrophages upon infection with M. tuberculosis [45,46]. The synthesis of pro-inflammatory cytokines such as TNF-α and IL-8 by macrophages contributes to the effective control of the proliferation and dissemination of pathogens [47]. The amount of secreted cytokines was compared with the levels produced upon stimulation with LPS (positive control) and untreated cells (negative control).

Figure 7 depicts the obtained results. The contact with drug-loaded FUC microparticles induced TNF-α concentrations of 1.5×10^3 pg/mL (Figure 7a), which did not differ statistically from the value registered upon LPS stimulation, although the nominal value was higher in that case (2.5×10^3 pg/mL). This observation is in agreement with a recent report showing that that FUC induces TNF-α secretion from macrophage-differentiated THP-1 cells [48]. FUC was also tested as a solution, which gives an indication of the effect of the polymer itself. It was shown to induce the production of TNF-α (1.9×10^3 pg/mL), which is not statistically different from the effect of microparticles. Importantly, when comparing with CCM, the induced TNF-α production was higher for both FUC microparticles and FUC solution ($p < 0.05$).

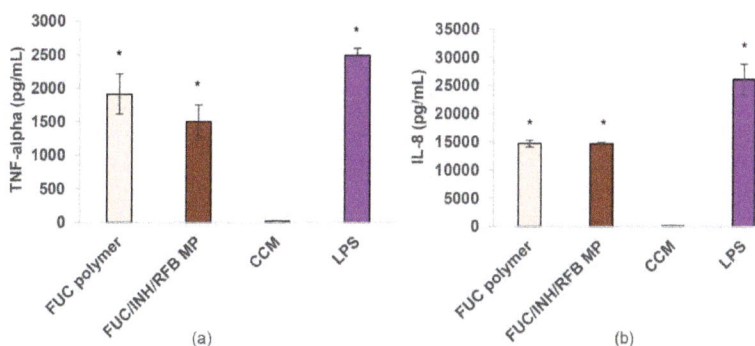

Figure 7. Assessment of (**a**) TNF-α and (**b**) IL-8 secretion induced by FUC/INH/RFB microparticles and FUC as the raw material. Lipopolysaccharide (LPS) and cell culture medium (CCM) were used as controls. FUC: fucoidan; INH: isoniazid; RFB: rifabutin. Data represent mean ± SEM (n = 3). * $p < 0.05$ compared to CCM.

Similar effects were observed regarding the production of IL-8 (Figure 7b). Although in this case, LPS generated a significantly higher amount of cytokine (26×10^3 pg/mL, $p < 0.05$), drug-loaded FUC microparticles also revealed an ability to induce its secretion, which reached 14.7×10^3 pg/mL, more than half of the amount corresponding to LPS. Moreover, IL-8 secretion stimulated by the produced microparticles and raw material were much higher than that of the control CCM ($p < 0.05$).

Overall, no significant differences were observed in the production of each interleukin upon stimulation by FUC/INH/RFB microparticles and the raw material FUC. In this way, the results suggest that FUC is the agent responsible for the observed activation of macrophage-like cells. Actually, the immune modulatory activity of FUC has been already reported, and is mediated through the regulation of immune cells, including macrophages [49,50]. Although FUC has been shown to induce the production of TNF-α [11,48] and IL-8 [51] from macrophages and other immune cells [9], the polymer seems to be endowed with anti-inflammatory activity as well. In truth, the mechanism of action of FUC as a bioactive agent is yet to be unveiled [52,53], and this is an area worth researching.

3.6. Preliminary Evaluation of Microparticle Uptake by Macrophage-Like Cells

Taking into consideration the intended application of the developed microparticles, the ability of macrophage cells to internalise the carriers was assessed by flow cytometry. Macrophage-differentiated THP-1 cells and rat alveolar macrophages NR8383 were exposed to two different doses of microparticles. As depicted in Figure 8, the percentage of THP-1 cells taking up FUC microparticles were around 23% (50 μg/cm^2) and 87% (200 μg/cm^2), demonstrating a dose-dependent uptake ($p < 0.05$). Similarly, the cellular uptake of carriers by rat macrophages varied between 68%–86% as the concentration raised from 50 to 200 μg/cm^2, also revealing a dose-dependent effect ($p < 0.05$). Significant differences ($p < 0.05$) were observed between the uptake by both cell lines, particularly at the lowest tested dose (50 μg/cm^2). In this case, 23% and 68% of FUC microparticles were taken up by THP-1 cells and NR8383 cells, respectively. Therefore, the obtained results are a preliminary demonstration of the macrophage's ability to uptake FUC microparticles at a considerable level, depending on the dose. Considering that FUC exhibits in its structure of chemical motifs (sulphate groups and sugar units) that can be recognised by macrophage surface receptors [8], a more accurate determination of preferential macrophage phagocytosis would be provided by comparing the uptake of FUC microparticles with a material devoid of recognisable moieties. Additionally, complementing cytometry data with confocal microscopy images would be beneficial in corroborating the data obtained so far.

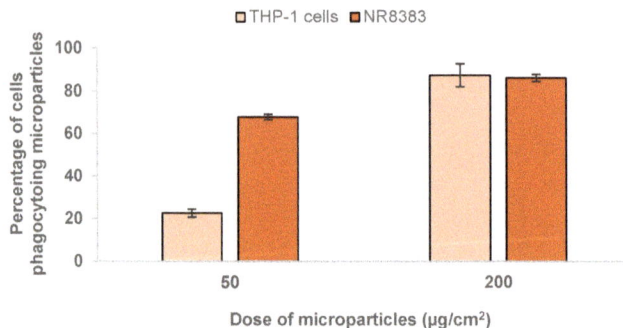

Figure 8. Uptake of fluorescently-labelled fucoidan (FUC) microparticles by macrophage-differentiated THP-1 cells and NR8383 cells exposure to 50 μg/cm^2 and 200 μg/cm^2, for a period of 2 h. Results are expressed as mean $\pm$ SEM ($n \geq 3$).

3.7. Determination of Minimum Inhibitory Concentration

In order to measure the minimum inhibitory concentration of the produced systems, *M. bovis* cells were treated with free drugs and FUC/INH/RFB microparticles. The viability of mycobacteria upon exposure to tested samples was calculated as a percentage of bacterial culture (control), which was considered as 100% of bacterial growth.

The MIC value determined for INH as free drug was 0.125 μg/mL, which is in the value range reported in the literature [54,55]. Comparatively, free RFB at a concentration of 0.004 μg/mL was sufficient to inhibit the growth of *M. bovis*, suggesting that RFB is a stronger antimycobacterial agent than INH, probably due to the higher lipophilicity that facilitates its internalisation through the cell membrane [56,57]. The literature reports variable MIC values of RFB, depending on the strain of *M. bovis* and on the method of susceptibility testing [58,59]. By combining both INH and RFB as free drugs in a single solution, the determined MIC values were 0.008 μg/mL (INH) and 0.004 μg/mL (RFB). It is worth noting that the MIC of INH reduced from 0.125 to 0.008 μg/mL when combined with RFB. Differently, either alone or in combination, RFB displayed the same MIC value (0.004 μg/mL). This indicates that the in vitro susceptibility of mycobacteria to INH is potentiated when the two antitubercular drugs were applied together.

By exposing *M. bovis* to 0.08 µg/mL of FUC/INH/RFB microparticles, the viability of mycobacteria decreased to the minimum level, therefore being the referred concentration of the MIC value of the produced systems. It is important to highlight that, at this microparticle concentration, the drug content, considering their respective association efficiency, is approximately 0.008 µg/mL (INH) and 0.004 µg/mL (RFB). These concentrations correspond to the MIC values determined for the two drugs when tested in combination (as free drugs). Therefore, the inhibition effect on the growth of *M. bovis* was very similar when comparing drug-loaded FUC microparticles and the mixed solution of INH/RFB as free drugs. This observation indicates that the microencapsulation process had no effect on the antibacterial activity of the drugs. In summary, an in vitro susceptibility of *M. bovis* has been observed towards drug-loaded FUC microparticles with considerable growth inhibition, as expected.

4. Conclusions

Inhalable dry powders based on FUC were produced to associate two first-line antitubercular drugs (INH and RFB) in a single formulation. Spray-dried microparticles were produced with high drug association efficiencies (>95%) and displayed MMADs between 3.6 and 3.9 µm, and a FPF around 45%–50%, thus evidencing the suitable aerodynamic properties for pulmonary delivery with great potential to reach the respiratory zone. FUC-based microparticles had no effect on the cell viability of alveolar epithelial cells (A549), although a slight reduction in the viability of macrophage-differentiated THP-1 cells (65% viable cells) was observed when exposed for 24 h to concentrations as high as 1.0 mg/mL. However, this dose is expected to largely overpass that to be observed in vivo. The produced microparticles were further demonstrated to be captured by macrophage-like cells (23%–87% uptake) in a dose-dependent manner. Moreover, they were able to induce macrophage activation by potentiating the secretion of cytokines. Furthermore, drug-loaded microparticles showed potential activity against a strain of mycobacteria (95% growth inhibition). According to the obtained data, the proposed delivery carrier, combining two antitubercular drugs in a single formulation, is a promising tool for the inhalable treatment of pulmonary TB. Nevertheless, unveiling the effect of a long-term administration of FUC microparticles in vivo, as well as the in vivo antimicrobial efficacy, are very relevant aspects to address in the future.

Supplementary Materials: The following are available online at http://www.mdpi.com/2073-4360/10/6/636/s1, Figure S1: Scheme of the 96-well microplate showing columns 4–11 filled with solutions of free drugs or microparticles serially diluted with M7H9 broth, containing mycobacteria in triplicate: lines B–C (suspension 1), lines D–E (suspension 2) and lines F–G (suspension 3). Contents of column 2 (only M7H9 medium) and column 3 (bacterial suspensions in broth) were considered negative and positive control, respectively.

Author Contributions: L.C., S.R. and A.G. conceived and designed the experiments; S.R. performed the experiments concerning microparticle aerosolisation properties, supervised by F.B.; M.L.F. supervised the microbiological experiments. A.M.R.d.C. collaborated with the experiments for preparing fluorescently labelled polymers and supervised chemical analytical experiments. L.C. performed the experiments, analyzed the data and wrote the paper; A.G. supervised and directed the project. All authors reviewed, edited and approved the manuscript.

Acknowledgments: Funding from the Portuguese Foundation for Science and Technology (PTDC/DTP-FTO/0094/2012, UID/Multi/04326/2013 and UID/BIM/04773/2013) is acknowledged. Ludmylla Cunha acknowledges Ph.D. grant (BEX 1168/13-4) supported by CAPES—Brazil.

Conflicts of Interest: The authors declare no conflict of interest.

References

1.	World Health Organization (WHO). *Global Tuberculosis Report 2017*; World Health Organization: Geneva, Switzerland, 2017.
2.	Garcia-Contreras, L.; Padilla-Carlin, D.J.; Sung, J.; VerBerkmoes, J.; Muttil, P.; Elbert, K.; Peloquin, C.; Edwards, D.; Hickey, A. Pharmacokinetics of ethionamide delivered in spray-dried microparticles to the lungs of guinea pigs. *J. Pharm.* **2017**, *106*, 331–337. [CrossRef] [PubMed]
3.	Zumla, A.; Raviglione, M.; Hafner, R.; Reyn, C.F. von Tuberculosis. *N. Engl. J. Med.* **2013**, *368*, 745–755. [CrossRef] [PubMed]

4. Aparna, V.; Shiva, M.; Biswas, R.; Jayakumar, R. Biological macromolecules based targeted nanodrug delivery systems for the treatment of intracellular infections. *Int. J. Biol. Macromol.* **2018**, *110*, 2–6. [CrossRef] [PubMed]

5. Tukulula, M.; Gouveia, L.; Paixao, P.; Hayeshi, R.; Naicker, B.; Dube, A. Functionalization of PLGA nanoparticles with 1,3-β-glucan enhances the intracellular pharmacokinetics of rifampicin in macrophages. *Pharm. Res.* **2018**, *35*. [CrossRef] [PubMed]

6. Eedara, B.B.; Rangnekar, B.; Sinha, S.; Doyle, C.; Cavallaro, A.; Das, S.C. Development and characterization of high payload combination dry powders of anti-tubercular drugs for treating pulmonary tuberculosis. *Eur. J. Pharm. Sci.* **2018**, *118*, 216–226. [CrossRef] [PubMed]

7. Cunha, L.; Grenha, A. Sulfated seaweed polysaccharides as multifunctional materials in drug delivery applications. *Mar. Drugs* **2016**, *14*, 42. [CrossRef] [PubMed]

8. Rodrigues, S.; Grenha, A. Activation of macrophages: Establishing a role for polysaccharides in drug delivery strategies envisaging antibacterial therapy. *Curr. Pharm. Des.* **2015**, *21*, 4869–4887. [CrossRef] [PubMed]

9. Jin, J.O.; Yu, Q. Fucoidan delays apoptosis and induces pro-inflammatory cytokine production in human neutrophils. *Int. J. Biol. Macromol.* **2015**, *73*, 65–71. [CrossRef] [PubMed]

10. Cho, M.; Lee, D.-J.; Kim, J.-K.; You, S. Molecular characterization and immunomodulatory activity of sulfated fucans from Agarum cribrosum. *Carbohydr. Polym.* **2014**, *113*, 507–514. [CrossRef] [PubMed]

11. Borazjani, N.J.; Tabarsa, M.; You, S.; Rezaei, M. Purification, molecular properties, structural characterization, and immunomodulatory activities of water soluble polysaccharides from Sargassum angustifolium. *Int. J. Biol. Macromol.* **2018**, *109*, 793–802. [CrossRef] [PubMed]

12. World Health Organization. *WHO Treatment Guidelines for Drug-Resistant Tuberculosis 2016*; WHO Library Cataloguing-in-Publication Data; World Health Organization: Geneva, Switzerland, 2016.

13. Martinelli, F.; Balducci, A.G.; Kumar, A.; Sonvico, F.; Forbes, B.; Bettini, R.; Buttini, F. Engineered sodium hyaluronate respirable dry powders for pulmonary drug delivery. *Int. J. Pharm.* **2017**, *517*, 286–295. [CrossRef] [PubMed]

14. Ota, A.; Istenič, K.; Skrt, M.; Šegatin, N.; Žnidaršič, N.; Kogej, K.; Ulrih, N.P. Encapsulation of pantothenic acid into liposomes and into alginate or alginate–pectin microparticles loaded with liposomes. *J. Food Eng.* **2018**, *229*, 21–31. [CrossRef]

15. Prabu, C.; Latha, S.; Selvamani, P.; Ahrentorp, F.; Johansson, C.; Takeda, R.; Takemura, Y.; Ota, S. Layer-by-layer assembled magnetic prednisolone microcapsules (MPC) for controlled and targeted drug release at rheumatoid arthritic joints. *J. Magn. Magn. Mater.* **2017**, *427*, 258–267. [CrossRef]

16. European Medicines Agency. *Guideline on Quality of Oral Modified Release Products*; EMA: Volume EMA/CHMP/QWP/428693/2013; European Medicines Agency: London, UK, 2014.

17. Ahmadi, Z.; Verma, G.; Jha, D. Evaluation of antimicrobial activity and cytotoxicity of pegylated aminoglycosides. *J. Bioact. Compat. Polym.* **2018**, *333*, 295–309. [CrossRef]

18. Zhang, L.; Li, Y.; Zhang, Y.; Zhu, C. Sustained release of isoniazid from polylactide microspheres prepared using solid/oil drug loading method for tuberculosis treatment. *Sci. China Life Sci.* **2016**, *59*, 724–731. [CrossRef] [PubMed]

19. Jabes, D.; Della Bruna, C.; Rossi, R.; Olliaro, P. Effectiveness of rifabutin alone or in combination with isoniazid in preventive therapy of mouse tuberculosis. *Antimicrob. Agents Chemother.* **1994**, *38*, 2346–2350. [CrossRef] [PubMed]

20. Alves, A.; Cavaco, J.; Guerreiro, F.; Lourenço, J.; Rosa da Costa, A.; Grenha, A. Inhalable antitubercular therapy mediated by locust bean gum microparticles. *Molecules* **2016**, *21*, 702. [CrossRef] [PubMed]

21. Anshakova, A.V.; Yu Konyukhov, V. Study by inverse gas chromatography of the solubility of rifabutin in water in the presence of cyclodextrin. *Russ. J. Appl. Chem.* **2017**, *90*, 209–213. [CrossRef]

22. Pai, R.V.; Jain, R.R.; Bannalikar, A.S.; Menon, M.D. Development and evaluation of chitosan microparticles based dry powder inhalation formulations of rifampicin and rifabutin. *J. Aerosol Med. Pulm. Drug Deliv.* **2016**, *29*, 179–195. [CrossRef] [PubMed]

23. Daman, Z.; Gilani, K.; Rouholamini Najafabadi, A.; Eftekhari, H.; Barghi, M. Formulation of inhalable lipid-based salbutamol sulfate microparticles by spray drying technique. *DARU J. Pharm. Sci.* **2014**, *22*, 1–9. [CrossRef] [PubMed]

24. Ni, R.; Zhao, J.; Liu, Q.; Liang, Z.; Muenster, U.; Mao, S. Nanocrystals embedded in chitosan-based respirable swellable microparticles as dry powder for sustained pulmonary drug delivery. *Eur. J. Pharm. Sci.* **2017**, *99*, 137–146. [CrossRef] [PubMed]

25. Belotti, S.; Rossi, A.; Colombo, P.; Bettini, R.; Rekkas, D.; Politis, S.; Colombo, G.; Balducci, A.G.; Buttini, F. Spray dried amikacin powder for inhalation in cystic fibrosis patients: A quality by design approach for product construction. *Int. J. Pharm.* **2014**, *471*, 507–515. [CrossRef] [PubMed]

26. Takeuchi, I.; Taniguchi, Y.; Tamura, Y.; Ochiai, K.; Makino, K. Effects of L-leucine on PLGA microparticles for pulmonary administration prepared using spray drying: Fine particle fraction and phagocytotic ratio of alveolar macrophages. *Colloids Surf. A Physicochem. Eng. Asp.* **2018**, *537*, 411–417. [CrossRef]

27. Sinsuebpol, C.; Chatchawalsaisin, J.; Kulvanich, P. Preparation and in vivo absorption evaluation of spray dried powders containing salmon calcitonin loaded chitosan nanoparticles for pulmonary delivery. *Drug Des. Dev. Ther.* **2013**, *7*, 861–873. [CrossRef] [PubMed]

28. Dalpiaz, A.; Fogagnolo, M.; Ferraro, L.; Capuzzo, A.; Pavan, B.; Rassu, G.; Salis, A.; Giunchedi, P.; Gavini, E. Nasal chitosan microparticles target a zidovudine prodrug to brain HIV sanctuaries. *Antivir. Res.* **2015**, *123*, 146–157. [CrossRef] [PubMed]

29. Yang, M.Y.; Chan, J.G.Y.; Chan, H.K. Pulmonary drug delivery by powder aerosols. *J. Control. Release* **2014**, *193*, 228–240. [CrossRef] [PubMed]

30. Martinelli, F.; Balducci, A.G.; Rossi, A.; Sonvico, F.; Colombo, P.; Buttini, F. "Pierce and inhale" design in capsule based dry powder inhalers: Effect of capsule piercing and motion on aerodynamic performance of drugs. *Int. J. Pharm.* **2015**, *487*, 197–204. [CrossRef] [PubMed]

31. Buttini, F.; Hannon, J.; Saavedra, K.; Rossi, I.; Balducci, A.G.; Smyth, H.; Clark, A. Accessorized DPI: A shortcut towards flexibility and patient adaptability in dry powder inhalation. *Pharm. Res.* **2016**, 3012–3020. [CrossRef] [PubMed]

32. Buttini, F.; Colombo, G.; Kwok, P.C.L.; Wui, W.T. Aerodynamic assessment for inhalation products: Fundamentals and current pharmacopoeial methods. In *Inhalation Drug Delivery: Techniques and Products*; Colombo, P., Traini, D., Buttini, F., Eds.; Wiley-Blackwell: West Sussex, UK, 2013; pp. 91–119.

33. Buttini, F.; Brambilla, G.; Copelli, D.; Sisti, V.; Balducci, A.G.; Bettini, R.; Pasquali, I. Effect of flow rate on in vitro aerodynamic performance of NEXThaler® in comparison with Diskus® and Turbohaler® Dry Powder Inhalers. *J. Aerosol Med. Pulm. Drug Deliv.* **2016**, *29*, 167–178. [CrossRef] [PubMed]

34. Hirota, K.; Hasegawa, T.; Hinata, H.; Ito, F.; Inagawa, H.; Kochi, C.; Soma, G.I.; Makino, K.; Terada, H. Optimum conditions for efficient phagocytosis of rifampicin-loaded PLGA microspheres by alveolar macrophages. *J. Control. Release* **2007**, *119*, 69–76. [CrossRef] [PubMed]

35. Eleftheriadis, G.K.; Akrivou, M.; Bouropoulos, N.; Tsibouklis, J.; Vizirianakis, I.S.; Fatouros, D.G. Polymer−lipid microparticles for pulmonary delivery. *Langmuir* **2018**, *34*, 3438–3448. [CrossRef] [PubMed]

36. Mulla, J.A.S.; Mabrouk, M.; Choonara, Y.E.; Kumar, P.; Chejara, D.R.; du Toit, L.C.; Pillay, V. Development of respirable rifampicin-loaded nano-lipomer composites by microemulsion-spray drying for pulmonary delivery. *J. Drug Deliv. Sci. Technol.* **2017**, *41*, 13–19. [CrossRef]

37. Vieira, A.C.; Magalhães, J.; Rocha, S.; Cardoso, M.S.; Santos, S.G.; Borges, M.; Pinheiro, M.; Reis, S. Targeted macrophages delivery of rifampicin-loaded lipid nanoparticles to improve tuberculosis treatment. *Nanomedicine* **2017**, *12*, 2721–2736. [CrossRef] [PubMed]

38. Oliveira, P.M.; Matos, B.N.; Pereira, P.A.T.; Gratieri, T.; Faccioli, L.H.; Cunha-Filho, M.; Gelfuso, G.M. Microparticles prepared with 50-190 kDa chitosan as promising non-toxic carriers for pulmonary delivery of isoniazid. *Carbohydr. Polym.* **2017**, *17*, 1–35. [CrossRef] [PubMed]

39. Haghi, M.; Ong, H.X.; Traini, D.; Young, P. Across the pulmonary epithelial barrier: Integration of physicochemical properties and human cell models to study pulmonary drug formulations. *Pharmacol. Ther.* **2014**, *144*, 235–252. [CrossRef] [PubMed]

40. Bur, M.; Huwer, H.; Muys, L.; Lehr, C.-M. Drug transport across pulmonary epithelial cell monolayers: Effects of particle size, apical liquid volume, and deposition technique. *J. Aerosol Med. Pulm. Drug Deliv.* **2010**, *23*, 119–127. [CrossRef] [PubMed]

41. ISO 10993-5: 2009. *Biological Evaluation of Medical Devices Part 5: Tests for In Vitro Cytotoxicity*; International Organization for Standardization: Geneva, Switzerland, 2009.

42. Rodrigues, S.; Alves, A.D.; Cavaco, J.S.; Pontes, J.F.; Guerreiro, F.; Rosa da Costa, A.M.; Buttini, F.; Grenha, A. Dual antibiotherapy of tuberculosis mediated by inhalable locust bean gum microparticles. *Int. J. Pharm.* **2017**, *529*, 433–441. [CrossRef] [PubMed]

43. Barluenga, J.; Aznar, F.; García, A.B.; Cabal, M.P.; Palacios, J.J.; Menéndez, M.A. New rifabutin analogs: Synthesis and biological activity against Mycobacterium tuberculosis. *Bioorganic Med. Chem. Lett.* **2006**, *16*, 5717–5722. [CrossRef] [PubMed]

44. Chien, J.-Y.; Chien, S.-T.; Huang, S.-Y.; Yu, C.-J. Safety of rifabutin replacing rifampicin in the treatment of tuberculosis: A single-centre retrospective cohort study. *J. Antimicrob. Chemother.* **2014**, *69*, 790–796. [CrossRef] [PubMed]

45. Chakraborty, P.; Kulkarni, S.; Rajan, R.; Sainis, K. Mycobacterium tuberculosis strains from ancient and modern lineages induce distinct patterns of immune responses. *J. Infect. Dev. Ctries.* **2017**, *11*, 904–911. [CrossRef]

46. Mwandumba, H.C.; Squire, S.B.; White, S.A.; Nyirenda, M.H.; Zijlstra, E.E.; Molyneux, M.E.; Russell, D.G.; Rhoades, E.R. Alveolar macrophages from HIV-infected patients with pulmonary tuberculosis retain the capacity to respond to stimulation by lipopolysaccharide. *Microbes Infect.* **2007**, *9*, 1053–1060. [CrossRef] [PubMed]

47. Duque, G.A.; Descoteaux, A. Macrophage cytokines: Involvement in immunity and infectious diseases. *Front. Immunol.* **2014**, *5*, 1–12. [CrossRef]

48. Stefaniak-Vidarsson, M.M.; Gudjónsdóttir, M.; Marteinsdottir, G. Evaluation of bioactivity of fucoidan from laminaria with in vitro human cell culture (THP-1). *Funct. Foods Health Dis.* **2017**, *7*, 688–701.

49. Zhang, W.; Oda, T.; Yu, Q.; Jin, J.-O. Fucoidan from macrocystis pyrifera has powerful immune-modulatory effects compared to three other fucoidans. *Mar. Drugs* **2015**, *13*, 1084–1104. [CrossRef] [PubMed]

50. Telles, C.B.S.; Mendes-Aguiar, C.; Fidelis, G.P.; Frasson, A.P.; Pereira, W.O.; Scortecci, K.C.; Camara, R.B.G.; Nobre, L.T.D.B.; Costa, L.S.; Tasca, T. Immunomodulatory effects and antimicrobial activity of heterofucans from Sargassum filipendula. *J. Appl. Phycol.* **2017**, *30*, 569–578. [CrossRef]

51. Jin, J.O.; Park, H.Y.; Xu, Q.; Park, J.I.; Zvyagintseva, T.; Stonik, V.A.; Kwak, J.Y. Ligand of scavenger receptor class A indirectly induces maturation of human blood dendritic cells via production of tumor necrosis factor-α. *Blood* **2009**, *113*, 5839–5848. [CrossRef] [PubMed]

52. Li, J.; Chen, K.; Li, S.; Liu, T.; Wang, F.; Xia, Y.; Lu, J.; Zhou, Y.; Guo, C. Pretreatment with fucoidan from fucus vesiculosus protected against conA-induced acute liver injury by inhibiting both intrinsic and extrinsic apoptosis. *PLoS ONE* **2016**, *11*, 1–16. [CrossRef] [PubMed]

53. Park, J.; Cha, J.D.; Choi, K.M.; Lee, K.Y.; Han, K.M.; Jang, Y.S. Fucoidan inhibits LPS-induced inflammation in vitro and during the acute response in vivo. *Int. Immunopharmacol.* **2017**, *43*, 91–98. [CrossRef] [PubMed]

54. Marianelli, C.; Armas, F.; Boniotti, M.B.; Mazzone, P.; Pacciarini, M.L.; Di Marco Lo Presti, V. Multiple drug-susceptibility screening in Mycobacterium bovis: New nucleotide polymorphisms in the embB gene among ethambutol susceptible strains. *Int. J. Infect. Dis.* **2015**, *33*, 39–44. [CrossRef] [PubMed]

55. Sturegård, E.; Ängeby, K.A.; Werngren, J.; Juréen, P.; Kronvall, G.; Giske, C.G.; Kahlmeter, G.; Schön, T. Little difference between minimum inhibitory concentrations of Mycobacterium tuberculosis wild-type organisms determined with BACTEC MGIT 960 and Middlebrook 7H10. *Clin. Microbiol. Infect.* **2015**, *21*, 148.e5–148.e7. [CrossRef] [PubMed]

56. Pinheiro, M.; Silva, A.S.; Reis, S. Molecular interactions of rifabutin with membrane under acidic conditions. *Int. J. Pharm.* **2015**, *479*, 63–69. [CrossRef] [PubMed]

57. Pinheiro, M.; Nunes, C.; Caio, J.M.; Moiteiro, C.; Lúcio, M.; Brezesinski, G.; Reis, S. The influence of Rifabutin on human and bacterial membrane models: Implications for its mechanism of action. *J. Phys. Chem. B* **2013**, *117*, 6187–6193. [CrossRef] [PubMed]

58. Ritz, N.; Tebruegge, M.; Connell, T.G.; Sievers, A.; Robins-Browne, R.; Curtis, N. Susceptibility of Mycobacterium bovis BCG vaccine strains to antituberculous antibiotics. *Antimicrob. Agents Chemother.* **2009**, *53*, 316–318. [CrossRef] [PubMed]

59. Sirgel, F.A.; Warren, R.M.; Böttger, E.C.; Klopper, M.; Victor, T.C.; van Helden, P.D. The Rationale for Using Rifabutin in the Treatment of MDR and XDR Tuberculosis Outbreaks. *PLoS ONE* **2013**, *8*, 1–5. [CrossRef] [PubMed]

© 2018 by the authors. Licensee MDPI, Basel, Switzerland. This article is an open access article distributed under the terms and conditions of the Creative Commons Attribution (CC BY) license (http://creativecommons.org/licenses/by/4.0/).

polymers

MDPI

Article

Novel Isocyanate-Modified Carrageenan Polymer Materials: Preparation, Characterization and Application Adsorbent Materials of Pharmaceuticals

Myrsini Papageorgiou [1], Stavroula G. Nanaki [2], George Z. Kyzas [3], Christina Koulouktsi [2], Dimitrios N. Bikiaris [2] and Dimitra A. Lambropoulou [1,*]

[1] Laboratory of Environmental Pollution Control, Department of Chemistry,
   Aristotle University of Thessaloniki, GR-541 24 Thessaloniki, Greece; myrsinipapag@gmail.com
[2] Laboratory of Polymer Chemistry and Technology, Department of Chemistry,
   Aristotle University of Thessaloniki, GR-541 24 Thessaloniki, Greece; sgnanaki@chem.auth.gr (S.G.N.);
   ckoulouktsi@pharmathen.com (C.K.); dbic@chem.auth.gr (D.N.B.)
[3] Hephaestus Advanced Laboratory, Eastern Macedonia and Thrace Institute of Technology,
   GR-654 04 Kavala, Greece; georgekyzas@gmail.com
* Correspondence: dlambro@chem.auth.gr; Tel.: +30-2310-997-859

Received: 27 September 2017; Accepted: 6 November 2017; Published: 10 November 2017

Abstract: The present study focused on the synthesis and application of novel isocyanate-modified carrageenan polymers as sorbent materials for pre-concentration and removal of diclofenac (DCF) and carbamazepine (CBZ) in different aqueous matrices (surface waters and wastewaters). The polymer materials were characterized using Fourier transform infrared spectroscopy (FTIR), X-ray diffraction (XRD), Thermal Gravimetric Analysis (TGA) and Scanning Electron Microscopy (SEM). The effects on the adsorption behavior were studied, and the equilibrium data were fitted by the Langmuir and Freundlich models. The maximum adsorption capacity (Q_{max}) was determined by Langmuir–Freundlich model and was ranged for iota-carrageenan (iCAR) from 7.44 to 8.51 mg/g for CBZ and 23.41 to 35.78 mg/g for DCF and for kappa-carrageenan (kCAR) from 7.07 to 13.78 mg/g for CBZ and 22.66 to 49.29 mg/g for DCF. In the next step, dispersive solid phase extraction (D-SPE) methodology followed by liquid desorption and liquid chromatography mass spectrometry (LC/MS) has been developed and validated. The factors, which affect the performance of D-SPE, were investigated. Then, the optimization of extraction time, sorbent mass and eluent's volume was carried out using a central composite design (CCD) and response surface methodology (RSM). Under the optimized conditions, good linear relationships have been achieved with the correlation coefficient (R^2) varying from 0.9901 to 0.995. The limits of detections (LODs) and limits of quantifications (LOQs) ranged 0.042–0.090 μg/L and 0.137–0.298 μg/L, respectively. The results of the recoveries were 70–108% for both analytes, while the precisions were 2.8–17.5% were obtained, which indicated that the method was suitable for the analysis of both compounds in aqueous matrices.

Keywords: carbamazepine; carrageenan polymers; diclofenac; dispersive solid phase extraction; response surface methodology

---

1. Introduction

Detection of pharmaceutically active compounds in the aquatic environment has raised concerns over their potential adverse effects on the environment [1–3]. Among them, carbamazepine (CBZ), a well-known antiepileptic compound, and diclofenac (DCF), a common non-steroidal anti-inflammatory drug, owing to ever-increasing consumption, inappropriate disposal and their incomplete removal in wastewater treatment plants (WWTPs) have been found ubiquitously in wastewater effluents [4–6] and different waters including surface water, ground water, and drinking water worldwide [7]. The release rate of CBZ into water bodies is estimated to be around 30 tons per year and, therefore, CBZ has been proposed as an anthropogenic marker for water contamination in the environment. According to several reports, environmentally-relevant concentrations of CBZ could negatively influence aquatic life (i.e., bacteria, algae, invertebrates, and fish) [8,9]. On the other hand, the anti-inflammatory DCF was identified for priority investigation because of risk perception and has recently been included on the "watch list" of priority substances under the Water Framework Directive [10]. The global consumption of DCF is estimated to be 940 tons per year, with a defined daily dose of 100 mg [9,11].

These facts have drawn extensive interest into the investigation of both compounds. Hence, it is important to develop reliable techniques for the detection and quantification of trace concentrations of these compounds in water samples in order to assess environmental exposure to CBZ and DCF. In general, the quantification of pharmaceutical compounds is usually carried out with chromatographic techniques after their extraction from aqueous matrices with a suitable sample preparation method. Among the various sample preparation methods, sorptive microextraction techniques (SμE) (e.g., Solid Phase Microextraction (SPME); Dispersive solid phase extraction, etc.) [12–16], for pre-concentration and/or cleanup of drugs offer new possibilities in sample treatment and superior advantages compared to conventional extraction methods. Accordingly, novel sorptive microextraction methods are being developed and examined for extraction of pharmaceuticals and other emerging contaminants from aqueous media with a focus on new materials with remarkable properties. The application of new sorbents has achieved a sharp increase in recent years since they can play an important role in sample and pre-concentration processes [17–22].

Given this background, in the present study, two isocyanate-modified carrageenan polymers based on kappa-carrageenan (kCAR) and iota-carrageenan (iCAR) have been synthesized and evaluated as sorbents for extraction of CBZ and DCF. Due to non-toxicity, biodegradability, and biocompatibility of carrageenan biopolymers there are considerable research efforts to provide new carrageenan-based materials [23]. In our previous article, iCAR and kCAR carrageenan microparticles were synthesized by using glutaraldehyde as cross-linking agent and successfully applied as biosorbents for the removal of the beta blocker metoprolol from aqueous solutions [24]. To move a step forward, in this study, two new carrageenan polymers modified by toluene-2,4-diisocyante were prepared and tested as sorbent materials for D-SPE determination of two of the most frequently detected pharmaceutical compounds. The strong novelty of this work is that this is the first report in which the carrageenan polymers have been used as extraction sorbents for determination of pharmaceuticals in aqueous media. The combination of the polymers synthesis (characterizations, etc.) along with their application as extraction sorbents, especially in the case of a very sensitive class of environmental pollutants as pharmaceuticals, provides an extra novelty to this work.

2. Materials and Methods

2.1. Materials

iCAR (Gelcarin GP-379NF) and kCAR (Gelcarin GP-812NF) were kindly supplied by FMC BioPolymer (FMC BioPolymer, Vlijmen, The Netherlands). Toluene-2,4-diisocyante (assay 95%) was purchased from Sigma-Aldrich (Sigma-Aldrich, St. Louis, MO, USA). Carbamazepine (CBZ; purity ≥ 99.0%) was purchased from Acros Organics (Acros Organics, Morris Plains, NJ, USA). Diclofenac sodium (DCF; purity ≥ 99.0%) was purchased from Sigma-Aldrich (Table 1). All other and solvents used were of analytical grade, and those used for HPLC analysis were of HPLC grade.

Table 1. General physicochemical properties of the analytes.

	Carbamazepine	Diclofenac
Molecular Formula	$C_{15}H_{12}N_2O$	$C_{14}H_{11}Cl_2NO_2$
Structure		
MW (g/mol)	236.27	296.15
Log(K_{ow})	2.45	4.51
pKa	2.3; 13.92	4.15
Aqueous solubility (mg/L)	112	2.37

2.2. Synthesis of Isocyanate-Functionalized Carrageenans (CAR-TDI)

iCAR or kCAR (15 g), respectively, was inserted in round-bottomed flask containing 50 mL of dimethylsulfoxide and stirred at room temperature until suspension was formed. Then, 100 mL of tolylene diisocyanate (TDI) was inserted into the flask, and vigorous stirring was performed at 50 °C for 4 h. The formed gel was cut into pieces and soaked in 100 mL of dichloromethane under magnetic stirring for 2 h in order to remove dimethylsulfoxide and unreacted TDI. Further purification was also done by inserting the material in distilled water at 80 °C for 3 h (for the removal of unreacted carrageenan) and finally washing with ethanol. The synthesized material was left to dry under vacuum. The prepared isocyanate-functionalized iota- (iCAR-TDI) and kappa- (kCAR-TDI) carrageenans were further chopped using cutter mill and particle size of about 100 μm was taken and stored in a desiccator till further use.

2.3. Characterization Techniques

Fourier transform infrared spectroscopy (FTIR) was used to characterize the functionalized carrageenans and possible bonds formed during drug adsorption experiments. The spectra were collected using a Perkin-Elmer FTIR spectrometer (model FTIR-2000, Perkin Elmer, Dresden, Germany). In brief, 5 mg of each sample was mixed with 180 mg of KBr in an agate mortal. The mixture was pressed under 5 tons for 2 min and pellet was formed. The pellet was then placed into an attachment in the optical compartment and FTIR spectra were obtained using. Infrared (IR) absorbance spectra were obtained between 450 and 4000 cm^{-1} at a resolution of 4 cm^{-1} using 20 co-added scans. All spectra submitted to baseline correction and normalization to 1.

X-ray diffraction (XRD) patterns were taken using a Rigaku Mini Flex diffractometer with Bragg-Brentano geometry (θ, 2θ) and a Ni-filtered CuKα radiation. Analysis was performed on net carrageenans, functionalized carrageenans with TDI and drug adsorbed samples. The samples were scanned over the internal range of 5–60°, step 0.02°, speed 2.0°/min.

Thermal Gravimetric Analysis (TGA) analysis was carried out with a SETARAM SETSYS TG-DTA 16/18. Samples (6.0 ± 0.2 mg) were placed in alumina crucibles. An empty alumina crucible was used as reference. Net and functionalized carrageenans were heated from ambient temperature to 800 °C in a 50 mL/min flow of N_2 at heating rates of 20 °C/min. Continuous recordings of sample temperature, sample weight, its first derivative and heat flow were performed.

2.4. Chromatographic Analysis

The Liquid Chromatographic (LC) system consisted of a SIL 20A autosampler with the volume injection set to 20 µL and LC-20AB pump both from Shimadzu (Kyoto, Japan). Chromatographic separation was achieved using a C_{18} (Athena) analytical column 250 × 4.6 mm with 5 µm particle size. Detection was performed using a SPD 20A DAD detector coupled in series with the LC-MS 2010EV mass selective detector, equipped with an atmospheric pressure electrospray ionization (ESI) source. The samples were analyzed using the ESI interface in positive ionization (PI) mode for CBZ and in negative mode (NI) for DCF. The mobile phase consisted of water with 0.1% formic acid (A) and methanol (B) in isocratic elution program (10% A:90% B). Column temperature was set at 40 °C and the flow rate was 0.4 mL/min. The drying gas was operated at flow 10 L/min at 200 °C. The nebulizing pressure was 100 psi, capillary voltage was 4500 V for positive ionization and −3500 V for negative ionization and the fragmentation voltage was set at 5 V. For each compound, the precursor molecular ion in the selected-ion monitoring (SIM) mode was acquired ([M + H] 237 m/z for CBZ and [M − H] 294 m/z for DCF).

2.5. Adsorption Experimental Procedure

The adsorption evaluation was done running batch experiments (all experiments were run in triplicate). However, some slight differences in experimental conditions for the adsorption of CBZ and DCF changed the whole experimental adsorption design. The latter was done due to the differences of chemical structures of model pollutants and as a result differences in pKa, solubility, decomposition, etc.

In the case of CBZ adsorption experiments, 1 g of adsorbent material was used per 1 L of aqueous solution (m = 0.030 g of adsorbent's mass were added to V = 30 mL of deionized water in a conical flask). In particular, samples were taken at predetermined time intervals and filtered using 45 µm pore size filtration membrane (Whatman, purchased by Sigma-Aldrich). However, some slight differences in experimental conditions for the adsorption of CBZ and DCF occurred. For the pH-effect experiments (C_0 = 30 mg/L), the solution pH was initially adjusted with aqueous solutions of acid or base (0.01 mol/L of HCl and/or 0.01 mol/L NaOH) to reach the appropriate pH values (2–10 for CBZ and 6–12 for DCF). The agitation rate was fixed at N = 150 rpm for all adsorption-desorption tests using shaking incubator (model Grant Instruments OLS Aqua Pro, Cambridge, UK) under a controlled temperature. Isotherms were taken running the adsorption experiments with various initial drug concentrations (pH = 2, 10; C_0 = 0–100 mg/L) at T = 20, 30, 40 °C for 24 h (contact time) (Table 2).

Table 2. L–F equilibrium fitting parameters for the adsorption of CBZ and DCF onto modified carrageenans at 20, 30, and 40 °C.

Adsorbent	Drug	pH	T (°C)	Q_{max} (mg/g)	K_{LF} ((L/mg)$^{1/b}$)	b	R^2
iCAR-TDI	CBZ	2	20	7.59	0.064	0.888	0.999
			30	7.72	0.010	0.585	0.995
			40	8.39	0.015	0.883	0.991
		10	20	7.44	0.062	0.862	0.995
			30	7.89	0.019	0.775	0.994
			40	8.51	0.005	0.640	0.994
kCAR-TDI	CBZ	2	20	9.87	0.054	0.898	0.995
			30	10.99	0.026	0.755	0.998
			40	11.12	0.011	0.631	0.999
		10	20	7.07	0.045	1.061	0.998
			30	9.27	0.022	0.826	0.998
			40	13.78	0.006	0.593	0.998
iCAR-TDI	DCF	6	20	23.41	0.095	1.485	0.998
			30	27.97	0.111	0.835	0.991
			40	37.58	0.115	0.908	0.986
kCAR-TDI		6	20	22.66	0.083	0.628	0.996
			30	35.81	0.128	0.928	0.994
			40	49.29	0.266	1.153	0.994

In the case of DCF adsorption experiments, 0.5 g of adsorbent material was used per 1 L of aqueous solution (m = 0.015 g of adsorbent's mass were added to V = 30 mL of deionized water in a conical flask). All other conditions were kept the same: C_0 = 30 mg/L for pH-effect experiments; N = 150 rpm. Isotherms were carried out (pH = 6) at T = 20, 30, 40 °C for 24 h (contact time) for varying initial DCF concentrations (C_0 = 0–70 mg/L). The value of pH (pH = 6) for carrying out isothermal experiments was selected as it was found to be the optimum value according to pH-effect tests (Table 2).

2.6. Dispersive Solid Phase Extraction Procedure (D-SPE)

The D-SPE procedure was commenced by adding 15 mg of the sorbent to 15 mL of pH adjusted sample (CBZ: pH = 10; DCF: pH = 6). A suspension was then mechanically shaken at 1000 rpm for 15 min to allow sorption of the analytes onto the sorbent. After that the suspension was centrifuged at 5000 rpm for 10 min, and the supernatant was discarded. Then, a desorption procedure was carried out as follows: 1 mL methanol was added into the vial and the analytes were desorbed via ultrasonic treatment for 10 min. Then, the supernatant was collected and evaporated to dryness under a gentle stream of nitrogen gas. Finally, the extract was re-dissolved in 50 μL of the mobile phase and injected into the LC-MS system.

2.7. Validation Study, Quality Assurance/Quality Control

Validation of the method was performed according to Document No. SANTE/11945/2015 [25]. The parameters: linearity, limits of detection (LOD), limits of quantification (LOQ), recovery, precision and uncertainty were evaluated. Selectivity of the method was estimated considering the absence of interfering peaks at the retention time of each compound. The linearity of the method was determined under optimized experimental conditions with pure solvents and matrix-matched standards (n = 5). Calibration curves were fitted by least-square regression and linearity was assumed when correlation coefficient (r^2) was higher than 0.990 with residuals lower than 20%.Responses in solvent and in matrix were compared to evaluate the matrix effect. The LOQs were defined as the minimum concentration of the analyte that can be quantified with acceptable recovery (in range 70–120%) and precision (RSD $\leq$ 20%). LOD were calculated using signal-to-noise ratio (S/N) criteria, in all cases; LOD = 3 S/N.

Accuracy and precision of the method were tested by means of recovery experiments, performed with six replicates of blank samples spiked with the target pharmaceuticals at 1, 10 and 100 μg/L. The recovery was determined by means of the measured concentration versus the spiked concentration. The mean recoveries and corresponding relative standard deviation (RSD) were calculated for accuracy and precision evaluation at three different concentration levels, corresponding to the LOQ level, to a medium and to a high concentration level of the calibration curves. Intra-day assays were performed by spiking blanks using six replicates for each concentration level in one day. Six replicates were performed for each level on six consecutive days under within-laboratory reproducibility conditions. Acceptable mean recoveries are those within the range 70–120%, with an associated repeatability RSD $\leq$ 20%, for all analytes within the scope of a method [10]. Blank samples also undergo the same procedure as with the real sample analysis.

Finally, a quality assurance and quality control were also performed. Procedural blanks were injected to monitor for background contamination. These blanks were processed in the same way as the samples and injected into the LC-MS system. Target compounds were not determined above LOQs. On the other hand, in order to validate both the calibration and the method stability, a quality control standard at an intermediate concentration was extracted and analyzed in each set of analysis.

Uncertainties were estimated on the basis of in-house validation data according to EURACHEM/ CITAC [26–28], Eurolab [29,30] and Expression of Uncertainty in Measurement (GUM) guidelines [31], at three spiking levels, as it was described in our previous works [13,15]. Uncertainties were assessed for different water matrices spiked at three concentration levels. The expanded uncertainty (U) was calculated by using the coverage factor k = 2, at the confidence level of 95%.

2.8. Central Composite Design

Response surface methodology (RSM), as an effective statistical model, has been widely used for optimization of sample preparation methods [15]. In this study, a three-level factor, Central Composite Design (CCD) (*a* = 1.681, rotatable) was applied to optimize extraction conditions for both target analytes. The design comprises 17 trials including 3 at central point, 6 at axial and 8 at factorial point from three independent variables (extraction time (A), sorbent mass (B), and eluent volume (C)) at three levels of the system. The dependent and independent variables, with their low, medium and high levels were selected based on the results from preliminary experiments and according to the literature reports and instrumental aspects. All experiments were performed in a random manner to avoid any systematic bias in the outcomes. Two replicates were performed and the averages of results were taken as a response. The main factors, their symbols, levels and design matrix as well as the response of each run are shown in Table 3.

Table 3. Experimental data for CBZ from CCD.

Standard Order	Run	Factor 1 A: Extraction Time	Factor 2 B: Sorbent Mass	Factor 3 C: Eluent Volume	Recovery R_1 DCF	R_2 CBZ
		(min)	(mg)	(mL)	(%)	(%)
12	1	10	27	1.0	89	79
8	2	15	22	1.4	99	90
1	3	5	8	0.6	40	40
4	4	15	22	0.6	85	65
2	5	15	8	0.6	53	32
11	6	10	3	1.0	50	38
9	7	1.6	15	1.0	35	27
16	8	10	15	1.0	98	92
7	9	5	22	1.4	54	36
15	10	10	15	1.0	101	95
5	11	5	8	1.4	46	38
13	12	10	15	0.3	58	47
14	13	10	15	1.7	85	80
3	14	5	22	0.6	44	27
6	15	15	8	1.4	71	55
10	16	18	15	1.0	92	85
17	17	10	15	1.0	100	90

Factor	Name	Units	Low Actual (Coded)	High Actual (Coded)	Mean	Standard Deviation
A	Extraction Time	min	5	15	10	4.481
B	Sorbent mass	mg	8	22	15	6.274
C	Eluent volume	ml	0.6	1.40	1	0.359

A program Design Expert (Trial version 7.0.0 Stat-Ease, Inc., Minneapolis, MN, USA) was applied for regression analysis of the data obtained and to estimate the response function. A quadratic model was built to describe a relationship between the DCF and CBZ extraction percent and operating parameters as defined by Equation (1):

$$Y = \beta_0 + \sum_{j=1}^{k} \beta_j X_j + \sum_{j=1}^{k} \beta_{jj} X_j^2 + \sum_{i}^{k} \sum_{<j=2} \beta_{ij} X_i X_j + e_i \tag{1}$$

where Y is the response variable, X_i and X_j are the independent variables, and k is the number of tested variables (k = 3). Regression coefficient is defined as β_0 for intercept, β_j for linear, β_{jj} for quadratic, β_{ij} for cross product term and e_i represents the residual term.

The method of signals and the analysis of variance (ANOVA) (95% confidence level) were used to estimate the significant main effects and interactions of factors (Tables 4 and 5).

Table 4. ANOVA test for CBZ by CCD.

Source	Sum of Squares	df [a]	Mean Square	F Value [b]	p-Value Prob > F [c]	Significant
Model	10,049.37	9	1116.60	15.34	0.0008	yes
A-Extraction Time	2886.44	1	2886.44	39.64	0.0004	
B-Sorbent mass	1089.03	1	1089.03	14.96	0.0062	
C-Eluent volume	894.06	1	894.06	12.28	0.0099	
AB	861.13	1	861.13	11.83	0.0109	
AC	210.13	1	210.13	2.89	0.1332	
BC	21.12	1	21.12	0.29	0.6068	
A^2	2518.01	1	2518.01	34.58	0.0006	
B^2	2228.98	1	2228.98	30.61	0.0009	
C^2	1703.76	1	1703.76	23.40	0.0019	
Residual	509.68	7	72.81			
Lack of Fit	497.02	5	99.40	15.70	0.0610	no
Pure Error	12.67	2	6.33			
Cor Total	10,559.06	16				

[a] Degree of freedom; [b] Test for comparing model variance with residual (error) variance; [c] Probability of seeing the observed F value if the null hypothesis is true.

Table 5. ANOVA test for DCF by CCD.

Source	Sum of Squares	df [a]	Mean Square	F Value [b]	p-Value Prob > F [c]	Significant
Model	9114.50	9	1012.72	55.55	<0.0001	yes
A-Extraction Time	3539.57	1	3539.57	194.15	<0.0001	
B-Sorbent mass	1386.19	1	1386.19	76.03	<0.0001	
C-Eluent volume	638.88	1	638.88	35.04	0.0006	
AB	288.00	1	288.00	15.80	0.0054	
AC	32.00	1	32.00	1.76	0.2268	
BC	0.00	1	0.00	0.00	1.0000	
A^2	2160.21	1	2160.21	118.49	<0.0001	
B^2	1548.86	1	1548.86	84.95	<0.0001	
C^2	1367.62	1	1367.62	75.01	<0.0001	
Residual	127.62	7	18.23			
Lack of Fit	122.95	5	24.59	10.54	0.0889	no
Pure Error	4.67	2	2.33			
Cor Total	9242.12	16				

[a] Degree of freedom; [b] Test for comparing model variance with residual (error) variance; [c] Probability of seeing the observed F value if the null hypothesis is true.

The significance of each coefficient and the interaction between each independent variable were evaluated according to the *p*-value at the 5% significance level. The adequacy of the model was verified using the determination coefficient R^2, the adjusted determination coefficient R^2 and the lack of fit test.

The fitted polynomial equation is expressed as surface and contour plots in order to visualize the relationship between the response and experimental levels of each factor and to evaluate the optimum conditions. Desirability function was used for simultaneous optimization of all affecting parameters in order to achieve the highest extraction efficiency (%). The adequacy of the model equation for predicting the optimum response values was validated with experimental results constructed to evaluate the optimum conditions for the response variables.

3. Results

3.1. Characterization of Materials

The morphology of the prepared materials was examined by SEM images. It was found that kCAR and iCAR have smooth surfaces without any specific shape (Figure 1a,b). The modification with TDI caused more rigid and hard surface (Figure 1c,d), while after chopping in cutting mill, the particles retained an irregular shape with sharp edges. This is an indication that the reaction had taken place.

Polymers **2017**, *9*, 595

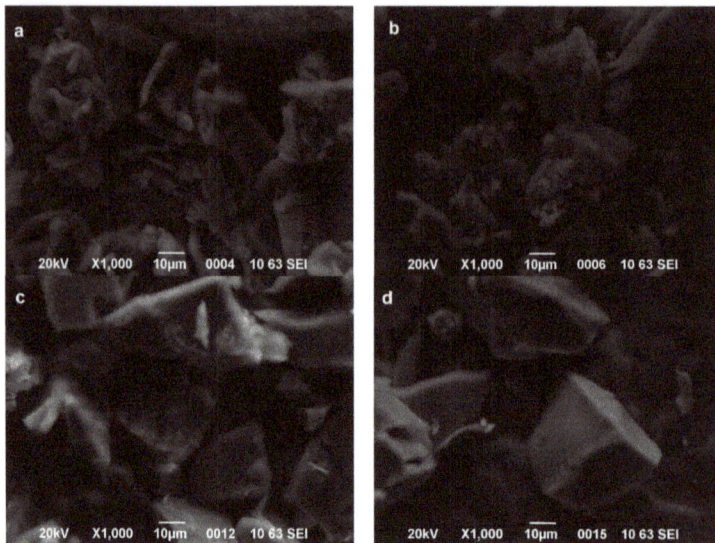

Figure 1. Scanning Electron Microscopy (SEM) images of: (**a**) kCAR; (**b**) iCAR; (**c**) kCAR-TDI; and (**d**) iCAR-TDI.

Figure 2a,b shows the FTIR spectra of neat and modified carrageenans before adsorption of CBZ, DCF. All the characteristic peaks of net carrageenans are present in the spectra: 3600–3200 cm^{-1} (broad peak owing to O–H interactions), 1645–1755 cm^{-1} (carbonyl group stretching), 1267 cm^{-1} (O=S=O asymmetric stretching), 1156 cm^{-1} (C–O–C asymmetric stretching), 1067–1074 cm^{-1} (S–O symmetric stretching), 1027–1038 cm^{-1} (C–O stretching), 928–935 cm^{-1} (C–O–C stretching in 3,6-anhydrogalactose) and 852 cm^{-1} (C–O–S) stretching in a (1-3)-D-galactose [32]. FTIR spectra of modified carrageenans are also present in Figure 2. It was observed that kCAR-TDI (Figure 2a) formed a new peak at 3248 cm^{-1} owing to >N–H stretching vibrations. In advance, two new peaks were also recorded at 1666 and 1714 cm^{-1} owing to >C=O stretching vibration. In addition, the absence of peak at 2275 cm^{-1} (–N=C=O group) confirmed the absence of unreacted TDI. Similar findings were also presented for iCAR-TDI with >N–H group to be recorded at 3260 cm^{-1} (Figure 2b).

Before analyzing the loaded-adsorbents (after adsorption of CBZ or DCF), it is necessary to present the FTIR of each drug separately. CBZ spectrum corresponds to those previously reported for the polymorph form III. Characteristic peaks were observed at 3461 (−NH valence vibration), 1676 (−CO−R vibration), 1594 and 1602 cm^{-1} (in the range of −C=C− and −C=O vibrations; −NH deformation). On the other hand, DCF spectrum exhibited distinctive peaks at 3381 cm^{-1} due to N−H stretching of the secondary amine, 1571 cm^{-1} owing to –C=O stretching of the carboxyl ion and at 748 cm^{-1} owing to C−Cl stretching.

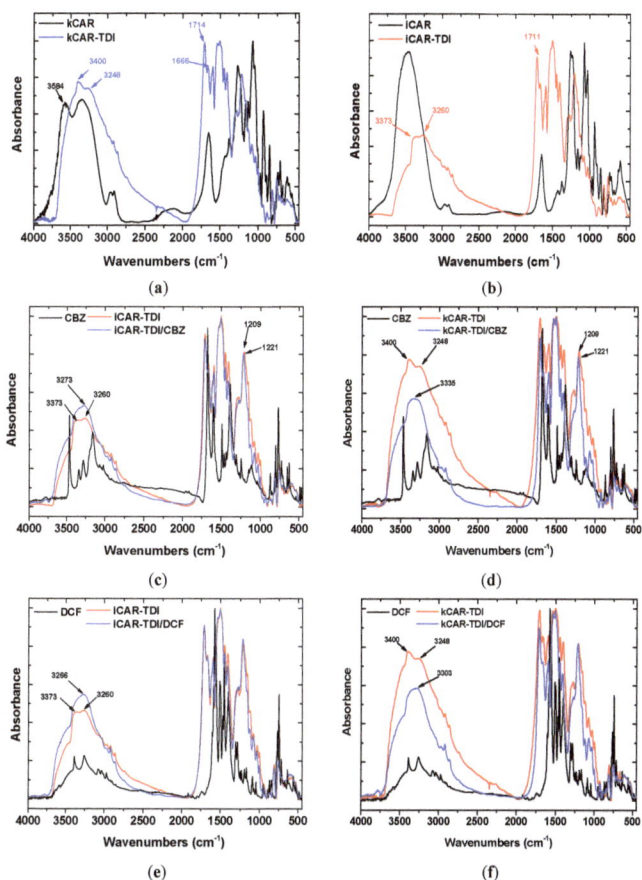

Figure 2. Fourier transform infrared spectroscopy (FTIR) spectra of iCAR and kCAR before and after drugs adsorption. (**a**) black line: kCAR; blue line: kCAR-TDI; (**b**) black line: iCAR; red line: iCAR-TDI; (**c**) black line: CBZ; red line: iCAR-TDI; blue line: iCAR-TDI after CBZ adsorption; (**d**) black line: CBZ; red line: kCAR-TDI; blue line: kCAR-TDI after CBZ adsorption; (**e**) black line: DCF; red line: iCAR-TDI; blue line: iCAR-TDI after DCF adsorption; (**f**) black line: DCF; red line: kCAR-TDI; blue line: kCAR-TDI after DCF adsorption.

After adsorption, there were shifts of wavenumbers. For CBZ adsorption, in the case of iCAR-TDI (Figure 2c), a shift from 3260 to 3273 cm^{-1} was due to the hydrogen bond formation between CBZ molecule and material. Another shift was observed from 1209 to 1221 cm^{-1} owing to interactions between sulfur and amino groups or oxygen atom of CBZ molecule. Similar findings were observed for kCAR-TDI (Figure 2d).

For DCF adsorption, in the case of iCAR-TDI (Figure 2e), a shift from 3260 to 3266 cm^{-1} was due to hydrogen bond formation between CBZ molecule and material. Analogous were the observations for kappa-modified carrageenan and recorded shifts are presenting in corresponding figures.

XRD was used to examine any changes to the physical state of carrageenans (Figure 3). As was found in a previous study [33], carrageenans show a wide broad peak indicating their amorphous state. The addition of –NCO– groups did not affect the amorphous state and the resulting cross-linked carrageenans also had a wide broad peak.

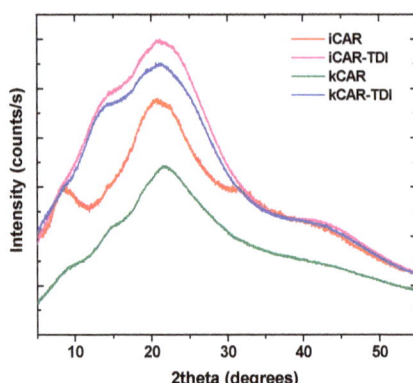

Figure 3. X-ray diffraction (XRD) patterns of iCAR and kCAR before and after TDI-modification.

To assess the thermal stability of the modified carrageenans, TGA was carried out (Figure 4). As can be seen, thermal degradation profile of –NCO– functionalized (TDI) carrageenan turned out to be quite different compared to that of neat carrageenans. The modified also showed reduced thermal degradation stability due to –NCO– groups. Similar observations were made in previous studies concerning cross-linked chitin and chitosan with HDMI [33,34].

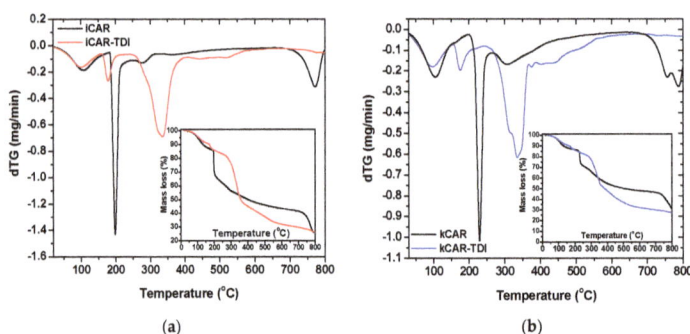

(a)   (b)

Figure 4. Differential Thermal Analysis (DTG) and Thermal Gravimetric Analysis (TGA) curved for: (a) iCAR; and (b) kCAR.

Moreover, kCAR showed to be more resistant to thermal stability than iCAR. Both carrageenans showed three decomposition stages (black lines in insets of Figure 4a,b): (i) an initial stage attributed to water evaporation and lasted at 100 °C; (ii) a main decomposition stage due to degradation of the saccharide structure of the molecule, to the dehydration of saccharide rings and decomposition of deacetylated carrageenan units; and (iii) a third stage owing to the acetylated part of the molecule. As was mentioned previously, all samples (modified and non-modified) showed an initial mass loss until 100 °C, which can be reasonably explained by the evaporation of water. DTG curves showed that decomposition for net carrageenans exhibited three stages.

On the other hand, four decomposition stages were observed for the modified carrageenans (red and blue lines in insets of Figure 4a,b): (i) the initial stage of water evaporation; (ii) a second stage owing to –NCO– segments loss; (iii) a third stage of the degradation of the saccharide structure of molecule, the dehydration of saccharide rings and decomposition of deacetylated carrageenan units; and (iv) a fourth stage owed to the acetylated part of the molecule. As was previously reported, modified carrageenans showed reduced thermal stability with iota-modified carrageenan showing the lowest one. A closer observation at the third stage of DTG curves showed that two different decomposition processes might occur during decomposition.

3.2. Adsorption Evaluation

3.2.1. pH-Effect

Solution pH is one of the important parameters affecting the properties of analytes and adsorbents, and therefore the mechanism of aqueous phase adsorption. In other words, the state (cationic/neutral/anionic) of the analyte and the functional groups present/created on the adsorbent are directly related to the corresponding pKa values and solution pH. To study the effect of solution pH on the CBZ and DCF adsorption, the initial pH values of drug solutions were varied in the range of 2–10 and 6–10, respectively. In the case of CBZ, the adsorption efficiency does not change in a marked way in the tested solutions (Figure 5).

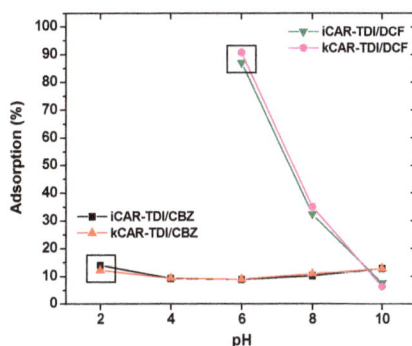

Figure 5. Effect of pH on adsorption of CBZ and DCF onto modified carrageenans (iCAR-TDI and kCAR-TDI).

This tendency of adsorption may be explained by the fact that, in this pH range, the dominant species of CBZ was neutral molecules (pKa 2.3 and 13.9) without charges and thus electrostatic interaction could be ignored in this study. The binding of CBZ onto modified carrageenans is probably controlled by hydrogen bond interactions between hydrogen bonding donor groups (i.e., $-NH_2$) and O-donor groups (i.e., $-OH$, $-OSO_3^-$). At acidic conditions (pH = 2) hydrogen bonding between phenolic OH in carrageenans and carbonyl group ($-C=O$) in CBZ may be likely dominated. At alkaline conditions, the NH_2 functional group in CBZ can interact with oxygen-containing functional groups of carrageenans, such as OH and $-OSO_3^-$ functional groups, through hydrogen bonding. H-bond binding mechanisms have been proposed for adsorption of CBZ by other authors using different polymer and porous materials [35,36].

In the case of DCF adsorption, the pH range 6–10 was evaluated in order to avoid its precipitation at acidic conditions. DCF is very water soluble in neutral-alkaline medium (50 g/L), but has low solubility (23.7 mg/L) at pH below pKa value [37]. The results demonstrated that the adsorption decreased with increasing pH values, due to the possible electrostatic repulsion between the negative surface of modified carrageenans and the anionic DCF at pH > 6 (pKa of DCF is 4.15). Similarly to CBZ, hydrogen bonding was likely the predominant adsorption mechanisms of DCF on the surface of modified carrageenans. The DCF molecule has one H-bond donor and one acceptor sites (originating from the $-COOH$ group). On the other hand, modified carrageenans have several acidic (O-/H-containing) functional groups. Therefore, DCF may interact with the carrageenan surface through H-bond formation. The carrageenans can be considered as a H-donor due to the phenol on its surface. It can be suggested that the most probable ways of H-bond formation between the phenolic OH of carrageenans and DCF are those shown in Figure 6. The effects of sulfonate groups on the adsorption of DCF have not debated and were considered steady in the studied pH range, because of the lower pKa of –OSO_3^- functional groups and their anionic forms in the studied pH range (pH = 6–10) [38]. This statement was further supported by the FTIR spectra (Figure 2). Furthermore, the H-bond interaction between the phenolic H-atom of carrageenans and the O-atoms of DCF can overcome the possible electrostatic repulsion between the negative surface charges (–COO^- of DCF and negative surface of carrageenans) at pH > 6.0 [39].

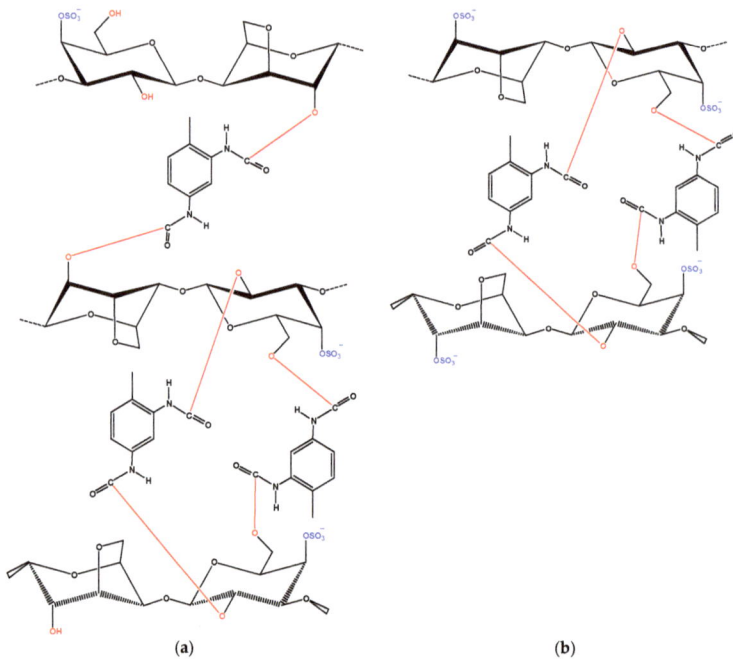

(a)   (b)

Figure 6. Structures of: (**a**) kCAR-TDI; and (**b**) iCAR-TDI.

Based on the aforementioned comments, the pH = 2 and pH = 10 for CBZ and pH = 6 for DCF, were selected as the desirable values for further examination in kinetic and isothermal studies.

3.2.2. Isotherms

One of the most important things regarding adsorption evaluation is the determination of the maximum theoretical adsorption capacity (Q_{max}). For this reason, Langmuir–Freundlich (L–F) (Equation (2)) isotherm equation [40] was applied to the experimental equilibrium points to fit them.

$$Q_e = \frac{Q_{max} K_{LF}(C_e)^{1/b}}{1 + K_{LF}(C_e)^{1/b}} \tag{2}$$

where Q_e (mg/g) is the equilibrium drug concentration in the solid phase; Q_{max} (mg/g) is the maximum amount of adsorption; K_{LF} (L/mg)$^{1/b}$ is the L–F constant; and b (dimensionless) is the L–F heterogeneity constant.

The selection of L–F equation and not Langmuir [41] or Freundlich [42] models was based on the fact that, recently, L–F is considered to be the most widely most used isotherm model presenting the most successful/accurate fitting.

The equilibrium amount in the solid phase (Q_e, mg/g) was calculated according to the following equation (where C_0 and C_e (mg/L) are the initial and equilibrium concentrations of drugs, respectively; V (L) is the volume of aqueous solution; and m (g) is the mass of carrageenans used):

$$Q_e = \frac{(C_0 - C_e)V}{m} \tag{3}$$

In addition, the equilibrium and temperature effect was also studied, as presented in Table 2 and Figure 7. All adsorbents indicate the same behavior; increasing the temperature from 20 to 30 °C, an increase of the adsorption capacity is observed, but a more drastic for the increase from 30 to 40 °C. The same behavior is observed for both CBZ and DCF. At first glance, the increase of temperature influenced more the adsorption of DCF than CBZ. In particular, iCAR-TDI enhances its Q_{max} for CBZ removal (at pH = 2) from 7.59 mg/g at 20 °C to 7.72 mg/g at 30 °C, and finally 8.39 mg/g at 40 °C. Similar adsorption behavior is revealed for the same combination of carrageenan and CBZ at pH = 10 (from 7.44 mg/g at 20 °C to 7.89 mg/g at 30 °C, and finally 8.51 mg/g at 40 °C). In the case of kCAR-TDI, the temperature effect was also slight (pH = 2: from 9.87 mg/g at 20 °C to 11.12 mg/g at 40 °C; pH = 10: from 7.07 mg/g at 20 °C to 13.78 mg/g at 40 °C). The latter confirms that these adsorbents cannot be substantially influenced by low temperatures (20 and 30 °C), but, at higher temperature (40 °C), the phenomenon becomes more intense. All of the above are listed in Table 1.

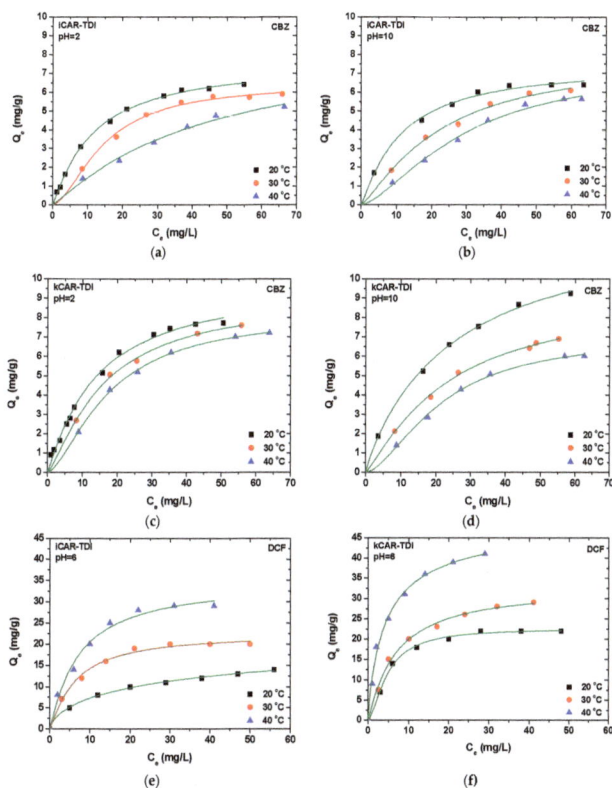

Figure 7. Isotherms for the adsorption of CBZ and DCF onto modified carrageenans. (**a**) Removal of CBZ at pH = 2 by iCAR-TDI; (**b**) Removal of CBZ at pH = 10 by iCAR-TDI; (**c**) Removal of CBZ at pH = 2 by kCAR-TDI; (**d**) Removal of CBZ at pH = 10 by kCAR-TDI; (**e**) Removal of DCF at pH = 6 by iCAR-TDI; (**f**) Removal of DCF at pH = 6 by kCAR-TDI.

3.3. Optimization of D-SPE Methodology by Central Composite Design

In extraction procedures, the optimization step is very important to increase the extraction efficiency. In the present study, the pH and the temperature of the solution have been well studied for the adsorption experiments. Desorption solvents were selected primarily using one variable at a time, while, for the other variables, a response surface methodology (RSM) using a central composite design (CCD) was applied for searching the optimal experimental conditions for both analytes. RSM is a multivariate optimization procedure which helps us to find out the optimized condition with the least number of experiments. Based on the higher adsorption capacity for both compounds, kCAR-TDI material was used for the development and optimization of the D-SPE analytical methodology.

3.3.1. Desorption Solvents

To obtain reliable and reproducible analytical results and a high enrichment factor, the eluent for the D-SPE procedure must have high affinity towards the target analytes than the sorbent. In this study, three eluents including methanol, acetonitrile and acetone were tested for elution of DCF and CBZ from the sorbents, according to the principles of green chemistry, to avoid halogenated solvents. When 1 mL of these desorption solvent was evaluated, it was observed that methanol and acetonitrile resulted in higher recoveries than acetone. Ultimately, methanol was chosen as the eluting solvents due to its better extraction efficiency (increase of 5–10% compared to ACN), lower toxicity and cost.

3.3.2. Central Composite Design

The approach of studying one factor at a time cannot describe adequately the importance of certain factors on the extraction process, because interactions between factors are not considered. In this light, to evaluate the combined effects and interactions of extraction time (A), sorbent mass (B) and eluent volume (C), the D-SPE process was further assessed by experimental design and RSM. The second order response surface, which is modeled on the resultsobtained from CCD experiments,can be expressed as:

$$R_1 = +99.97 + (16.10 \times A) + (10.07 \times B) + (6.84 \times C) + (6.00 \times A \times B) + (2.00 \times A \times C) + (0.00 \times B \times C) - 13.84A^2 - 11.72B^2 - 11.01C^2$$

$$R_2 = +92.94 + (14.54 \times A) + (8.93 \times B) + (8.09 \times C) + (10.37 \times A \times B) + (5.13 \times A \times C) + (1.62 \times B \times C) - 14.95A^2 - 14.06B^2 - 12.29C^2$$

Following to fitting the second order polynomial equation with the actual data of both responses of interest (i.e., DCF (R_1) and CBZ (R_2)), two multiple regression analyses were separately proposed in terms of all the 17 possible combinations of three independent criteria. To test the significance and adequacy of the model, the analysis of variance (ANOVA) was performed for each response and the obtained results are given in Table 4. The statistical significance of the model equations was evaluated by the F-test and p-value. According to the results, the high F-values (55.55 for DCF and 15.34 for CBZ) and small p-values (<0.0001 for DCF and 0.0008 for CBZ value, both p-values < 0.01) suggested that the regression models obtained are highly significant. The goodness of fit of regression model was carried out by determination coefficient (R^2) and adjusted determination coefficient (R^2_{adj}). R^2 values computed as 0.9862 for DCF and 0.9517 for CBZ, indicating the goodness-of-fit of the proposed models. Similarly, the values of Adj R^2 for both R_1 and R_2 were reasonably close to 1 (0.9684 for DCF and 0.8897 for CBZ), corroborating a high degree of correlation between the experimental and predicted values. Moreover, the high values of Adequate precision (DCF 21.578 and CBZ 11.323), which measures the signal to noise of the model, indicated a very high degree of precision and a good deal of reliability of the experimental values. This means that the model can be used to navigate the design space. The correlation between observed and predicted values is given in Figure 8. The points are placed very closely to the diagonal line, indicating low discrepancies between them.

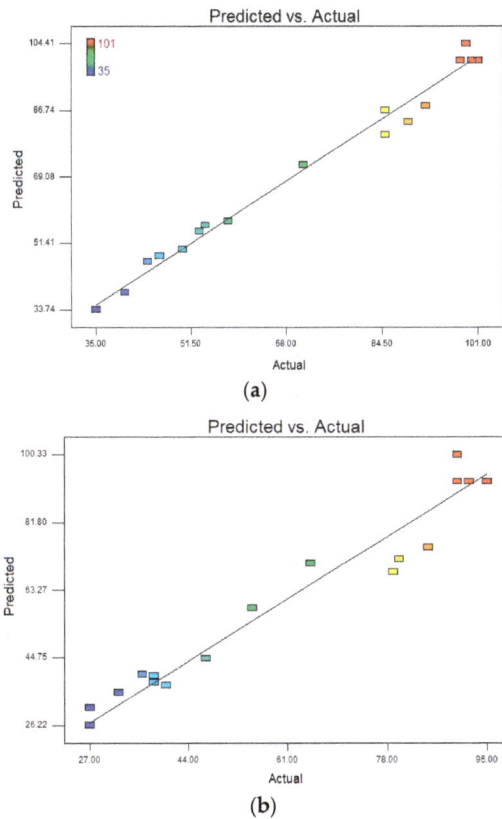

Figure 8. (a) The observed responses vs. the predicted responses for DCF; and (b) the observed responses vs. the predicted responses for CBZ.

Results from the CCD experiments, as shown in Tables 4 and 5, indicate that the extraction of both analytes depends significantly on all the studied variables, having either negative or positive effects. In particular, the linear terms of extraction time (A), sorbent mass (B) and eluent volume (C), and their quadratic terms A^2, B^2 and C^2, have significant effects on the extraction efficiency of both analytes at 5% significance level (95% confidence interval) as p-value for all these terms is less than 0.05. On the other hand, only one considerable interaction effect was established between the extraction time and the sorbent mass for both analytes. The remaining two interaction terms are non-significant as p-values for them are more than 0.05%. However, the non-significant terms were not eliminated from the model, since we are interested for the overall effect of the coefficients on the response surface.

After identifying the most significant parameters, RSM was used to find optimum condition for the best extraction efficiency. Three-dimensional surface plots (Figure 9) show the interactive effects of extraction time, sorbent mass and eluent volume on the extraction efficiency of both compounds. These plots depict the influence of any two independent variables on the response and the maximum value for each variable can be obtained. According to the results, when eluent volume is fixed at level 0, the extraction efficiency of both compounds is increased by increasing the sorbent amount, reaching a peak of ~100% for DCF and 95% for CBZ at the point of ~15 mg of the sorbent. Such increment in extraction efficiency appears to be because of the accessibility of higher surface areas together with abundant sorption sites. Meanwhile, sorbent addition beyond an adequate quantity to entirely sorb the accessible molecules of the analytes had no significant impact on further improvement in extraction efficiency, demonstrating unsaturated surface active sites of the polymer sorbent. In addition, an increase of extraction efficiency was observed as the extraction time increased up to 10–12 min and then reached a plateau. On the other hand, the extraction efficiency was nearly associated with an enhancement of elution volume from 0.3 to 1 mL and then remained almost constant.

Hence, the optimum working conditions to obtain the best response were as: 1 mL for volume of eluting solvent; 15 mg for sorbent amount; and 15 min extraction time. To confirm the model adequacy for predicting maximum extraction efficiency, three replicate experiments were performed at optimal conditions. A mean value of 97 ± 5.0% for DCF and 92 ± 8.0% for CBZ, obtained from real experiments, demonstrate the suitability of the fitted response surface model.

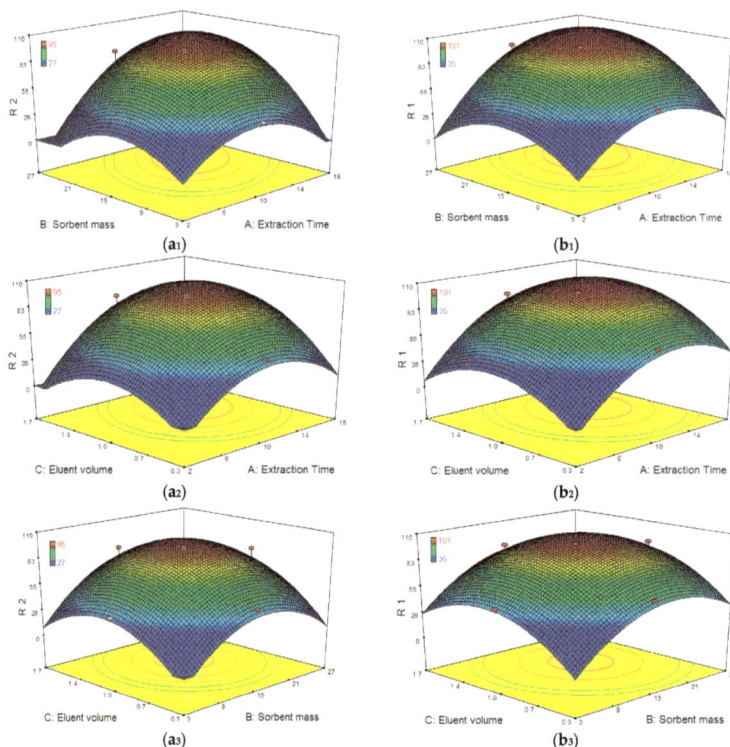

Figure 9. (a_1–a_3) Response surface plots (3D) for the effect of independent variables of D-SPE on to CBZ recovery; and (b_1–b_3) response surface plots (3D) for the effect of independent variables of D-SPE on to DCF recovery.

3.3.3. Performance Characteristics Measured in D-SPE Method Validation

The analytical performance of the developed D-SPE procedure using the novel isocyanate modified carrageenan polymer sorbent was evaluated for the pre-concentration of DCF and CBZ. Analytical figures of merit of the method are summarized in Tables 6 and 7.

Table 6. Analytical performance of D-SPE for CBZ in different matrices.

	Spiking Level (µg/L)	DW	LW	SW	RW	WWI	WWE
LOD (µg/L)	-	0.042	0.046	0.043	0.047	0.062	0.047
LOQ (µg/L)	-	0.137	0.151	0.142	0.154	0.205	0.154
RSD$_r$ (%)	1	7.2	11.2	13.1	9.6	12.1	11.2
(n = 5)	10	5.1	9.1	10.9	7.7	11.2	9.8
	100	3.7	8.8	9.2	7.5	10.7	9.0
RSD$_R$ (%)	1	10.2	15.3	16.1	14.1	17.5	17.0
(n = 5)	10	9.8	13.7	14.8	12.8	15.2	14.9
	100	9.1	10.2	14.1	11.4	14.0	12.3
U (%)	1	33.92	33.2	33.41	34.86	39.91	37.19
(k = 2, confidence level 95%)	10	15.96	16.68	17.24	17.03	19.88	18.56
	100	9.91	10.84	12.17	11.28	13.62	12.22

The calculated calibration curves gave a high level of linearity, yielding coefficients (r^2) > 0.991 for both compounds. LOD and LOQ data of analytes for all water samples are in the ranges of 0.042–0.090 µg/L and 0.137–0.298 µg/L, respectively. The precision of the method was accessed by determining relative standard deviations (RSDs) of intra-day and inter-day at three different spiked levels. The results show that the RSDs of intra-day precision are 2.8–13.5%, while that of inter-day precision are 8.5–17.5%.

Uncertainty of the analytical method was also estimated based on in-house validation data according to EURACHEM/CITAC and GUM guide for both compounds at two spiking levels, as was explained in previous works. The relative expanded uncertainty was lower than 40% for both compounds in all matrices.

Table 7. Analytical performance of D-SPE for DCF in different matrices.

	Spiking Level (µg/L)	DW	LW	SW	RW	WWI	WWE
LOD (µg/L)	-	0.060	0.067	0.063	0.068	0.090	0.068
LOQ (µg/L)	-	0.199	0.220	0.207	0.225	0.298	0.225
RSD$_r$ (%)	1	5.1	11.5	12.8	8.0	9.7	10.5
(n = 5)	10	3.9	8.9	10.5	7.1	10.2	9.3
	100	2.8	8.1	10.0	6.8	9.5	8.8
RSD$_R$ (%)	1	9.6	14.6	15.9	13.4	15.6	14.1
(n = 5)	10	8.5	12.4	13.9	11.8	13.6	12.9
	100	9.5	11.4	12.7	10.2	12.9	11.8
U (%)	1	24.51	27.49	23.05	25.28	36.57	28.42
(k = 2, confidence level 95%)	10	11.67	14.04	13.08	12.94	18.09	14.68
	100	8.06	10.0	10.25	8.99	12.44	10.51

To demonstrate the reliability and versatility of the proposed methodology coupled to LC-MS system for the analysis of real samples, five categories of aqueous samples, including distilled water (DW), river water (RW), sea water (SW), lake water (LW), and influent and effluent wastewaters (WWI and WWE) samples, were analyzed. Recoveries ranged between 70% and 108% for all the matrices, demonstrating the suitability of the proposed method (Figure 10).

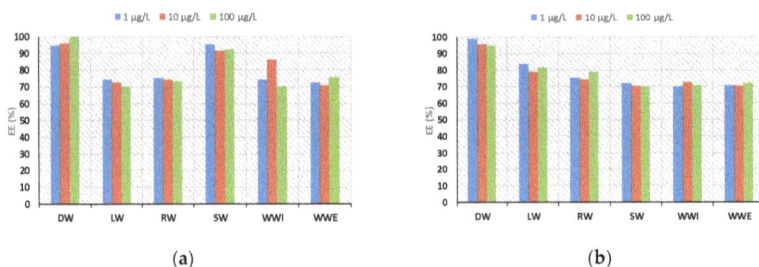

(a) (b)

Figure 10. Extraction efficiency (EE, %) of: CBZ (**a**); and DCF (**b**) by D-SPE method in different matrices.

4. Conclusions

In the present, the application of D-SPE methodology for the pre-concentration of DCF and CBZ has been demonstrated using novel synthesized isocyanate modified carrageenan polymer materials. A central composite design was applied to obtain optimal conditions. Satisfactory precision and accuracy were obtained with the proposed analytical methodology using a small amount of sample. Based on the obtained results, it is anticipated that the proposed method has a great analytical potential for accurate determination of both pharmaceutical compounds in environmental water samples.

Acknowledgments: This research has been co-financed by the European Union (European Social Fund—ESF) and Greek national funds through the Operational Program "Education and Lifelong Learning" of the National Strategic Reference Framework (NSRF)—Research Funding Program "Excellence II (Aristeia II)", Research Grant, No. 4199, which is gratefully acknowledged.

Author Contributions: Dimitra Lambropoulou, Dimitrios Bikiaris and Stavroula Nanaki conceived and designed the experiments; Stavroula Nanaki, Myrsini Papageorgiou and Christina Koulouktsi performed the experiments; Dimitra Lambropoulou, Dimitrios Bikiaris, Stavroula Nanaki, George Kyzas and Myrsini Papageorgiou analyzed the data; Stavroula Nanaki, Myrsini Papageorgiou and Christina Koulouktsi contributed reagents/materials/analysis tools; Dimitra Lambropoulou, Dimitrios Bikiaris, George Kyzas and Stavroula Nanaki wrote the paper.

Conflicts of Interest: The authors declare no conflict of interest.

References

1. Lambropoulou, D.A.; Nollet, L.M.L. *Transformation Products of Emerging Contaminants in the Environment: Analysis, Processes, Occurrence, Effects and Risks*; John Wiley and Sons Ltd.: London, UK, 2014.
2. Luo, Y.; Guo, W.; Ngo, H.H.; Nghiem, L.D.; Hai, F.I.; Zhang, J.; Liang, S.; Wang, X.C. A review on the occurrence of micropollutants in the aquatic environment and their fate and removal during wastewater treatment. *Sci. Total Environ.* **2014**, *473–474*, 619–641. [CrossRef] [PubMed]
3. Verlicchi, P.; Al Aukidy, M.; Zambello, E. Occurrence of pharmaceutical compounds in urban wastewater: Removal, mass load and environmental risk after a secondary treatment—A review. *Sci. Total Environ.* **2012**, *429*, 123–155. [CrossRef] [PubMed]
4. Evgenidou, E.N.; Konstantinou, I.K.; Lambropoulou, D.A. Occurrence and removal of transformation products of PPCPs and illicit drugs in wastewaters: A review. *Sci. Total Environ.* **2015**, *505*, 905–926. [CrossRef] [PubMed]
5. Papageorgiou, M.; Kosma, C.; Lambropoulou, D. Seasonal occurrence, removal, mass loading and environmental risk assessment of 55 pharmaceuticals and personal care products in a municipal wastewater treatment plant in central greece. *Sci. Total Environ.* **2016**, *543*, 547–569. [CrossRef] [PubMed]
6. Vieno, N.; Sillanpää, M. Fate of diclofenac in municipal wastewater treatment plant—A review. *Environ. Int.* **2014**, *69*, 28–39. [CrossRef] [PubMed]
7. Petrie, B.; Barden, R.; Kasprzyk-Hordern, B. A review on emerging contaminants in wastewaters and the environment: Current knowledge, understudied areas and recommendations for future monitoring. *Water Res.* **2014**, *72*, 3–27. [CrossRef] [PubMed]

8. Ferrari, B.; Paxéus, N.; Giudice, R.L.; Pollio, A.; Garric, J. Ecotoxicological impact of pharmaceuticals found in treated wastewaters: Study of carbamazepine, clofibric acid, and diclofenac. *Ecotoxicol. Environ. Saf.* **2003**, *55*, 359–370. [CrossRef]

9. Zhang, Y.; Geißen, S.U.; Gal, C. Carbamazepine and diclofenac: Removal in wastewater treatment plants and occurrence in water bodies. *Chemosphere* **2008**, *73*, 1151–1161. [CrossRef] [PubMed]

10. European Commission. *Implementing Decision 2015/495 of 20 March 2015 Establishing a Watch List of Substances for Union-Wide Monitoring in the Field of Water Policy Pursuant to Directive 2008/105/ec and Amending Directive 2000/60/ec*; European Commission: Brussels, Belgium, 2015.

11. Lonappan, L.; Brar, S.K.; Das, R.K.; Verma, M.; Surampalli, R.Y. Diclofenac and its transformation products: Environmental occurrence and toxicity—A review. *Environ. Int.* **2016**, *96*, 127–138. [CrossRef] [PubMed]

12. Herrero-Latorre, C.; Barciela-García, J.; García-Martín, S.; Peña-Crecente, R.M.; Otárola-Jiménez, J. Magnetic solid-phase extraction using carbon nanotubes as sorbents: A review. *Anal. Chim. Acta* **2015**, *892*, 10–26. [CrossRef] [PubMed]

13. Kyzas, G.Z.; Nanaki, S.G.; Koltsakidou, A.; Papageorgiou, M.; Kechagia, M.; Bikiaris, D.N.; Lambropoulou, D.A. Effectively designed molecularly imprinted polymers for selective isolation of the antidiabetic drug metformin and its transformation product guanylurea from aqueous media. *Anal. Chim. Acta* **2015**, *866*, 27–40. [CrossRef] [PubMed]

14. Płotka-Wasylka, J.; Szczepańska, N.; de la Guardia, M.; Namieśnik, J. Modern trends in solid phase extraction: New sorbent media. *TrAC Trends Anal. Chem.* **2016**, *77*, 23–43. [CrossRef]

15. Terzopoulou, Z.; Papageorgiou, M.; Kyzas, G.Z.; Bikiaris, D.N.; Lambropoulou, D.A. Preparation of molecularly imprinted solid-phase microextraction fiber for the selective removal and extraction of the antiviral drug abacavir in environmental and biological matrices. *Anal. Chim. Acta* **2016**, *913*, 63–75. [CrossRef] [PubMed]

16. Aghaie, A.B.G.; Hadjmohammadi, M.R. Fe$_3$O$_4$@p-Naphtholbenzein as a novel nano-sorbent for highly effective removal and recovery of berberine: Response surface methodology for optimization of ultrasound assisted dispersive magnetic solid phase extraction. *Talanta* **2016**, *156–157*, 18–28. [CrossRef] [PubMed]

17. Corazza, G.; Merib, J.; Magosso, H.A.; Bittencourt, O.R.; Carasek, E. A hybrid material as a sorbent phase for the disposable pipette extraction technique enhances efficiency in the determination of phenolic endocrine-disrupting compounds. *J. Chromatogr. A* **2017**, *1513*, 42–50. [CrossRef] [PubMed]

18. Pan, S.-D.; Zhou, L.-X.; Zhao, Y.-G.; Chen, X.-H.; Shen, H.-Y.; Cai, M.-Q.; Jin, M.-C. Amine-functional magnetic polymer modified graphene oxide as magnetic solid-phase extraction materials combined with liquid chromatography-tandem mass spectrometry for chlorophenols analysis in environmental water. *J. Chromatogr. A* **2014**, *1362*, 34–42. [CrossRef] [PubMed]

19. Wang, L.; Zhang, Z.; Zhang, J.; Zhang, L. Magnetic solid-phase extraction using nanoporous three dimensional graphene hybrid materials for high-capacity enrichment and simultaneous detection of nine bisphenol analogs from water sample. *J. Chromatogr. A* **2016**, *1463*, 1–10. [CrossRef] [PubMed]

20. Zhang, J.; Gan, N.; Chen, S.; Pan, M.; Wu, D.; Cao, Y. B-cyclodextrin functionalized meso-/macroporous magnetic titanium dioxide adsorbent as extraction material combined with gas chromatography-mass spectrometry for the detection of chlorobenzenes in soil samples. *J. Chromatogr. A* **2015**, *1401*, 24–32. [CrossRef] [PubMed]

21. Naing, N.N.; Li, S.F.Y.; Lee, H.K. Graphene oxide-based dispersive solid-phase extraction combined with in situ derivatization and gas chromatography-mass spectrometry for the determination of acidic pharmaceuticals in water. *J. Chromatogr. A* **2015**, *1426*, 69–76. [CrossRef] [PubMed]

22. Serrano, M.; Chatzimitakos, T.; Gallego, M.; Stalikas, C.D. 1-Butyl-3-aminopropyl imidazolium-functionalized graphene oxide as a nanoadsorbent for the simultaneous extraction of steroids and β-blockers via dispersive solid-phase microextraction. *J. Chromatogr. A* **2016**, *1436*, 9–18. [CrossRef] [PubMed]

23. Zia, K.M.; Tabasum, S.; Nasif, M.; Sultan, N.; Aslam, N.; Noreen, A.; Zuber, M. A review on synthesis, properties and applications of natural polymer based carrageenan blends and composites. *Int. J. Biol. Macromol.* **2017**, *96*, 282–301. [CrossRef] [PubMed]

24. Nanaki, S.G.; Kyzas, G.Z.; Tzereme, A.; Papageorgiou, M.; Kostoglou, M.; Bikiaris, D.N.; Lambropoulou, D.A. Synthesis and characterization of modified carrageenan microparticles for the removal of pharmaceuticals from aqueous solutions. *Colloids Surf. B Biointerfaces* **2015**, *127*, 256–265. [CrossRef] [PubMed]

25. European Commission. *Document Sante/11945/2015, Analytical Quality Control and Method Validation Procedures for Pesticide Residues Analysis in Food and Feed*; European Commission: Brussels, Belgium, 2015.
26. EURACHEM. *Eurachem/Citac Guide: Measurement Uncertainty Arising from Sampling: A Guide to Methods and Approaches*; EURACHEM: Olomouc, Czech, 2007.
27. EURACHEM. *Eurachem/Citac Guide: Use of Uncertainty Information in Compliance Assessment*; EURACHEM: Olomouc, Czech, 2007.
28. EURACHEM. *Eurachem/Citac Guide: Quantifying Uncertainty in Analytical Measurement*; EURACHEM: Olomouc, Czech, 2012.
29. Eurolab. *Eurolab Technical Report 1/2002, Measurement Uncertainty in Testing*; Eurolab: Brussels, Belgium, 2002.
30. Eurolab. *Eurolab Technical Report 1/2007, Measurement Uncertainty Revisited: Alternative Approaches to Uncertainty Evaluation*; Eurolab: Brussels, Belgium, 2007.
31. International Organization for Standardization. *Iso Guide 98-3, Guide to the Expression of Uncertainty in Measurement*; International Organization for Standardization: Geneva, Switzerland, 2008.
32. Duman, O.; Tunç, S.; Polat, T.G.; Bozoğlan, B.K. Synthesis of magnetic oxidized multiwalled carbon nanotube-κ-carrageenan-Fe_3O_4 nanocomposite adsorbent and its application in cationic methylene blue dye adsorption. *Carbohydr. Polym.* **2016**, *147*, 79–88. [CrossRef] [PubMed]
33. Nanaki, S.; Karavas, E.; Kalantzi, L.; Bikiaris, D. Miscibility study of carrageenan blends and evaluation of their effectiveness as sustained release carriers. *Carbohydr. Polym.* **2010**, *79*, 1157–1167. [CrossRef]
34. Gallego, R.; Arteaga, J.F.; Valencia, C.; Franco, J.M. Isocyanate-functionalized chitin and chitosan as gelling agents of castor oil. *Molecules* **2013**, *18*, 6532–6549. [CrossRef] [PubMed]
35. Suriyanon, N.; Punyapalakul, P.; Ngamcharussrivichai, C. Mechanistic study of diclofenac and carbamazepine adsorption on functionalized silica-based porous materials. *Chem. Eng. J.* **2013**, *214*, 208–218. [CrossRef]
36. Zhang, Y.L.; Zhang, J.; Dai, C.M.; Zhou, X.F.; Liu, S.G. Sorption of carbamazepine from water by magnetic molecularly imprinted polymers based on chitosan-Fe_3O_4. *Carbohydr. Polym.* **2013**, *97*, 809–816. [CrossRef] [PubMed]
37. Palomo, M.E.; Ballesteros, M.P.; Frutos, P. Analysis of diclofenac sodium and derivatives. *J. Pharm. Biomed. Anal.* **1999**, *21*, 83–94. [CrossRef]
38. Gholami, M.; Vardini, M.T.; Mahdavinia, G.R. Investigation of the effect of magnetic particles on the crystal violet adsorption onto a novel nanocomposite based on κ-carrageenan-*g*-poly(methacrylic acid). *Carbohydr. Polym.* **2016**, *136*, 772–781. [CrossRef] [PubMed]
39. Bhadra, B.N.; Ahmed, I.; Kim, S.; Jhung, S.H. Adsorptive removal of ibuprofen and diclofenac from water using metal-organic framework-derived porous carbon. *Chem. Eng. J.* **2017**, *314*, 50–58. [CrossRef]
40. Tien, C. *Adsorption Calculations and Modeling*; Butterworth-Heinemann: Boston, MA, USA, 1994.
41. Langmuir, I. The adsorption of gases on plane surfaces of glass, mica and platinum. *J. Am. Chem. Soc.* **1918**, *40*, 1361–1403. [CrossRef]
42. Freundlich, H. Over the adsorption in solution. *Z. Phys. Chem.* **1906**, *57*, 385–470.

© 2017 by the authors. Licensee MDPI, Basel, Switzerland. This article is an open access article distributed under the terms and conditions of the Creative Commons Attribution (CC BY) license (http://creativecommons.org/licenses/by/4.0/).

polymers

MDPI

Article

Hyaluronic Acid Promotes the Osteogenesis of BMP-2 in an Absorbable Collagen Sponge

Hairong Huang [1,†], **Jianying Feng** [2,†], **Daniel Wismeijer** [1], **Gang Wu** [1,*] and **Ernst B. Hunziker** [3]

1. Department of Oral Implantology and Prosthetic Dentistry, Academic Centre for Dentistry Amsterdam (ACTA), Universiteit van Amsterdam and Vrije Universiteit Amsterdam, Gustav Mahlerlaan 3004, 1081LA Amsterdam, Nord-Holland, The Netherlands; hhrstudy@126.com (H.H.); d.wismeijer@acta.nl (D.W.)
2. School of Stomatology, Zhejiang Chinese Medical University, Hangzhou 310053, China; twohorsejy@163.com
3. Departments of Osteoporosis and Orthopaedic Surgery, Inselspital (DKF), University of Bern, Murtenstrasse 35, 3008 Bern, Switzerland; ernst.hunziker@dkf.unibe.ch
* Correspondence: g.wu@acta.nl.; Tel.: +31-20-598-0866
† The authors contributed equally.

Received: 10 July 2017; Accepted: 31 July 2017; Published: 4 August 2017

Abstract: (1) Background: We tested the hypothesis that hyaluronic acid (HA) can significantly promote the osteogenic potential of BMP-2/ACS (absorbable collagen sponge), an efficacious product to heal large oral bone defects, thereby allowing its use at lower dosages and, thus, reducing its side-effects due to the unphysiologically-high doses of BMP-2; (2) Methods: In a subcutaneous bone induction model in rats, we first sorted out the optimal HA-polymer size and concentration with micro CT. Thereafter, we histomorphometrically quantified the effect of HA on new bone formation, total construct volume, and densities of blood vessels and macrophages in ACS with 5, 10, and 20 μg of BMP-2; (3) Results: The screening experiments revealed that the 100 μg/mL HA polymer of 48 kDa molecular weight could yield the highest new bone formation. Eighteen days post-surgery, HA could significantly enhance the total volume of newly-formed bone by approximately 100%, and also the total construct volume in the 10 μg BMP-2 group. HA could also significantly enhance the numerical area density of blood vessels in 5 μg BMP-2 and 10 μg BMP-2 groups. HA did not influence the numerical density of macrophages; and (4) Conclusions: An optimal combined administration of HA could significantly promote osteogenic and angiogenic activity of BMP-2/ACS, thus potentially minimizing its potential side-effects.

Keywords: hyaluronic acid; bone morphogenetic protein-2; absorbable collagen sponge; bone regeneration; angiogenesis

---

1. Introduction

Recombinant human bone morphogenetic protein-2 (BMP-2) has been in clinical use mainly for the generation of spinal fusions for more than a decade [1,2]. In recent years, BMP-2 has also been proven to be an efficacious way to promote bone regeneration in the field of dentistry and maxillofacial surgery, such as ridge augmentation [3], sinus lift [4], and periodontal and peri-implant [5] bone regeneration. It is able to accelerate bony healing processes, and substitute autologous bone transplantation [6,7]. Overall, its clinical use is quite successful; however, the use of BMP-2 is, unfortunately, associated with a number of severe undesired side effects that are able to seriously impair the health of patients and the musculoskeletal functions of the treated patients [7,8]. Such side-effects include, among others, ectopic bone formation, paralysis, and neurological disturbances [9,10]; but malignant pathologies are not involved [11,12].

BMP-2 is clinically applied topically in a free form together with an absorbable collagenous sponge (ACS) [13]. The recommended dose is exceedingly high (12 mg/ACS unit; i.e., approximately 37.3 mg of BMP-2 per gram of ACS sponge); and in this high dosage scheme probably lies the reason for many of the untoward side effects [6,9]. It is, however, not only the dosage that is able to influence the response of the targeted populations of progenitor cells and their differentiation pathways, but also the mode of application and the manner in which the agent is locally presented to the targeted cell populations. On the other hand, the microenvironment (niche conditions) in which the desired bone formation activity is aimed to take place also has a significant influence on the degree and speed of the process, as well as the type of ossification process (enchondral or desmal); for example, the local biomechanical niche conditions are able to influence this process [14], but less so with respect to the density of blood vessels present [15], even though the high numbers of blood vessels establish the presence of large numbers of perivascular adult stem cells [16] as a source of precursor cells for osteogenesis [17]. For this reason some researchers described previously [18] that a sequential release of an angiogenic factor (initial release) with the osteogenic factor (BMP-2; delayed release) is able to accelerate bone formation activities.

Respecting the methods of enhancement of BMP-2 bioactivity, glycosaminoglycans (GAGs) have been described previously to have such a potential, in particular relating to the desired osteogenesis effects [19]. Hyaluronic acid (HA) belongs, chemically, to the large groups of GAGs [20]; they are a group of large linear polysaccharides constructed of repeating disaccharide units, containing amino sugars and uronic acid, and are one of the most frequently-used tools to improve the microenvironment for BMP-induced osteogenesis. It has been found that the active components in GAGs for this desired osteogenic enhancement effects are able to bind, stabilize and present growth factors to cells for improved receptor interaction [21]. Furthermore, they can direct the immediate signaling activities of BMP2 through enhancing the subsequent recruitment of type II receptor subunits to BMP-type I receptor complexes [22]. As one of the main GAG components, HA can be a promising drug to promote the osteogenic potential of BMP-2. HA is able to stimulate osteoinduction activities in bone wound healing processes [19]. In particular, high-molecular weight HA (≈1900 KDa) was found in animal experiments to be able to promote this effect. Huang et al. [23] found that low molecular weight HA (60 kDa) and high-weight HA (900 and 2300 kDa) were able to significantly stimulate cell growth and to increase osteocalcin mRNA expression levels. In addition, it was revealed in previous research that HA is involved in several biological processes [24], such as cell differentiation [25], angiogenesis [26], morphogenesis [27], and wound healing [28]; furthermore, HA was described to be able to inhibit osteoclast differentiation [29] in addition to its downregulation potential of BMP-2 antagonists [30].

In this study we hypothesize that a combination use of BMP-2 with HA is able to promote the osteogenesis activity in a subcutaneous bone induction model at lower dosage levels of BMP-2 in ACS.

2. Materials and Methods

2.1. Experimental Design

We proceeded in two steps: initially, we performed screening experiments in a subcutaneous bone induction model to determine the optimal HA polymer size and concentration to be used for the main experiment. In the main experiment we elucidated the optimal dosage of BMP-2 to be used together with ACS and HA within a time period of 18 days.

2.2. Animals, Anesthesia, and Surgery

The animal experiment was approved by Ethical Committee of School of Stomatology, Zhejiang Chinese Medical University. All animal experiments were carried out according to the ethics laws and regulations of China and the guidelines of animal care established by Zhejiang Chinese Medical University. (Sprague-Dawly) SD rats (mean weight: 230 g, range from 190 to 250 g) were used in this study for all experiments. The animal experiments, such as anesthesia, sample randomization, and surgery were performed as we previous described [15].

2.3. Screening Experiments

The HA screening experiments were performed using five different HA polymer lengths to be tested, and each one of them was tested at 6 different concentrations of the polymer, and at three different dosages of BMP-2 (see Table 1).

Each of the HA polymer test was performed in the presence of ACS (Inductos®, Medtronic, Minneapolis, MN, USA) (identical circular ACS samples were prepared of 8 mm diameter), and with 5, 10, or 20 μg of BMP-2 (Inductos®, Medtronic, Minneapolis, MN, USA). BMP-2 portions were added to ACS sponges from syringes; thereafter, the HA-solution was added (20 μL portions per sample), just before implantation. The choice of three different dosages of BMP-2 was determined according to previous publications [31,32]. In these screening experiments one test sample was implanted in 35 SD rats on the left and right back side per animal. The evaluations of the degrees of osteoinduction obtained were performed using micro CT scans (Skyscan1176, Bruker, Kontich, Belgium) and the results were assessed by two independent observers for maximum subcutaneous bone signal intensity.

Table 1. Screening parameters.

HA-Moleculer Weights (kDa)	HA-Concentrations (μg/mL)	BMP-2-Dosages (μg)
<8	50	0
48	100	5
660	500	10
1610	1000	20
3100		

2.4. Main Experiment

Twenty-four eight-week-old male SD rats were used for the main experiment; and in each animal two 8 mm diameter BMP-2/ACS implants were placed. Eight experimental groups ($n = 6$ samples and six animals per group) were set up as following:

G1: no BMP-2, ACS alone;
G2: BMP-2/ACS, 5 μg BMP-2;
G3: BMP-2/ACS, 10 μg BMP-2;
G4: BMP-2/ACS, 20 μg BMP-2;
G5: no BMP-2, ACS alone + 2μg HA;
G6: BMP-2/ACS (5 μg BMP-2) + 2 μg HA;
G7: BMP-2/ACS (10 μg BMP-2) + 2 μg HA;
G8: BMP-2/ACS (20 μg BMP-2) + 2 μg HA.

A preimplantation control group of ACS sponges was also included in the study in order to determine the basic carrier volume before implantation as a time 0 reference volume.

In the groups containing HA, this compound was used at a concentration of 100 μg HA/mL, and the amount of 20 μL solution was added per sample. Samples were then stored overnight under aseptic conditions in a sterile hood for induction of sample drying before implantation.

2.5. Tissue Processing

Eighteen days post-operation the implanted samples were retrieved together with the surrounding tissues and chemically fixed, dehydrated, and embedded in methylmethacrylate; sections of 600 μm in thickness were produced and taken with a 1000 μm-interval between two adjacent sections. The sections were thereafter glued to Plexiglas boards, polished down (sand paper) to 100 μm thickness, and then stained with McNeal's tetrachrome, toluidine blue O, and basic fuchsin, as described previously [15].

2.6. Histomorphometry and Stereology

The histological sections were photographed at a final magnification of 200× under an Eclipse 50i light microscope (Nikon, Tokyo, Japan), and photographic subsampling was performed according to a systematic random-sampling protocol [33]. Using the photographic prints, the areas of the implants and the areas of newly-formed bone tissue were measured histomorophometrically using point counting methods [33]. Mineralized bone tissue (stained pink) and unmineralized bone tissue (light blue) (see Light micrographs of BMP-2/ACS constructs) were defined as newly-formed bone tissue; areas of collagen carrier material were measured the same way [34].

2.7. Stereological Estimators

Volume Estimators. The preimplantation reference volumes of the collagenous carrier materials ($n = 6$) were estimated using the principle of Cavalieri [35], as well as the final remaining total tissue volumes [33] at the end of the implantation time period (18 days). The degree of carrier degradation was computed by dividing the reference volume of carrier material at time point zero divided by the carrier material volume present at the end of the experiment. The areas of newly-formed bone tissue and remaining carrier materials were estimated at final magnifications of 200×, and were subsampled according to a systematic random protocol [33,35].

Numerical Estimators. Blood vessel area density and blood vessel numerical area density (number of blood vessel cross-sections per unit tissue area) (at 200× magnification) as well as macrophage numerical area densities (at 400× magnification) were estimated as previously described [33].

2.8. Statistical Analysis

All data are presented as mean values together with the standard error (SE) of the mean. Differences between the experimental groups were analyzed using the one-way ANOVA-test. Statistical significance was defined as $p < 0.05$. Correlation coefficients were determined using the Pearson product-moment correlation coefficient. Significance of correlation was defined if p-values < 0.05 were obtained. All statistical analyses were performed with SPSS® 21.0 software (SPSS, Chicago, IL, USA). The Bonferroni post-hoc test was implemented for data comparison purposes.

3. Results

The screening experiments revealed that the HA polymer of 48 kDa molecular weight was able to yield the highest osteogenesis activity, when applied at a concentration of 100 μg/mL (dosage volume: 20 μL) of HA (Figure 1), and with an added BMP-2 amount of 10 μg (BMP-2 concentration in the solution: 1 μg/μL; BMP-solution-volume added: 10 μL/sample).

5, 10 and 20 μg BMP-2 resulted in a similar total volume of newly formed bone tissue, while no bone was detected with or without HA in the absence of BMP-2 (Figure 2). The combined administration of HA significantly increased the volume of neoformed bone in the 10 μg BMP-2 group ($p = 0.024$) by approximately 100%. HA also increased new bone formation in the 20 μg BMP-2 group, which was, however, insignificant ($p = 0.3$). In the 5 μg BMP-2 group no such enhancement effect was observed.

The total construct volumes did not significantly differ among the groups without HA (Figure 3). However, among the groups with HA, the total construct volume of the 10 μg BMP-2 group in the presence of HA showed a significantly higher volume than the 5 μg BMP-2 group ($p = 0.03$) and 0 μg BMP-2 group ($p = 0.007$), respectively, but not the 20 μg BMP-2 group. Only the 10 μg BMP-2 group with HA resulted in a significantly higher total construct volume when compared to the time 0 (control group).

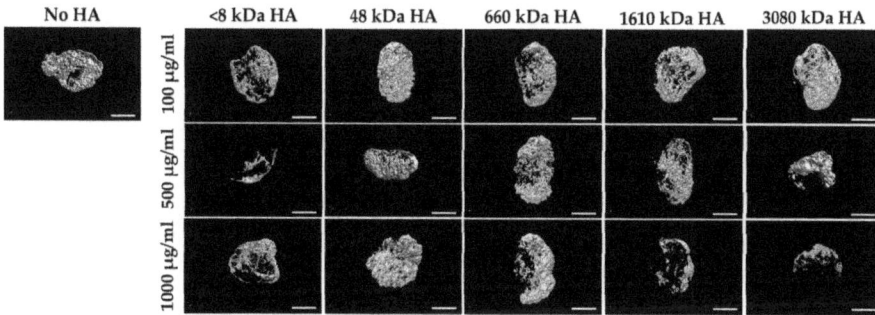

Figure 1. Micro CT images of BMP-2/ACS constructs (10 µg BMP-2 per sample) in the presence or absence of 100, 500, or 1000 µg/mL HA with different polymer sizes (<8, 48, 660, 1610, 3080 (kDa)) at 18 days after implant placement (Bar = 2.5 mm).

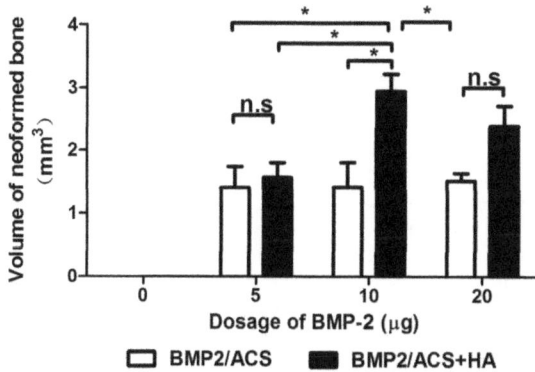

Figure 2. Mean volumes of newly formed bone tissue in the BMP-2/ACS constructs, in the presence or absence of 100 µg/mL HA (48 kDa) implanted at different BMP-2 dosages, 18 days after implant placement. Values represent means ± SEM; $n = 6$ per experimental group. The asterisks denote the level of statistical significance, i.e., * $p < 0.05$. The compared groups are indicated by brackets.

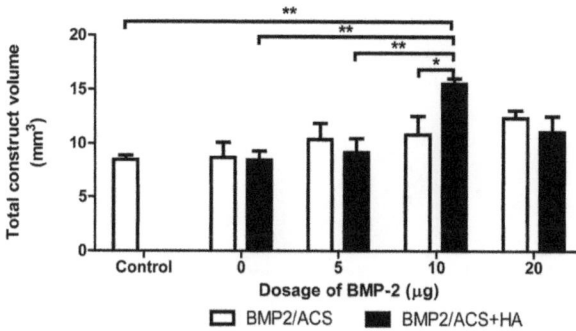

Figure 3. Mean volumes of total construct volumes of the BMP-2/ACS constructs, in the presence or absence of 100 µg/mL HA (48 kDa) implanted at different BMP-2 dosages; 18 days after implant placement. Values represent means ± SEM; $n = 6$ per experimental group. The asterisks denote the level of statistical significance, i.e., * $p < 0.05$, ** $p < 0.01$, *** $p < 0.001$). The compared groups are indicated by brackets.

The volumes of remaining ACS showed a decreasing trend from the 0 µg BMP-2 group to the 10 µg BMP-2 group; the trend then reversed to the 20 µg BMP-2 group (Figure 4). Computation of the coefficient of correlation between the first three dosages (0, 5, and 10 µg BMP-2) in the absence of HA revealed a value for $r = -0.62$ ($p = 0.006$), i.e., a significantly correlated trend was present; in the presence of HA and the same BMP-dosage groups, the correlation coefficient was $r = -0.459$ ($p = 0.075$). The combined administration of HA did not significantly influence remaining ACS volumes for each dosage group. The coefficients of variations (CV) and coefficients of errors (CE) varied between CV = 69% (CE = 35%) for the 0 µg BMP group with HA, and CV = 27.8% (CE = 13.9%) for the 10 µg BMP group without HA.

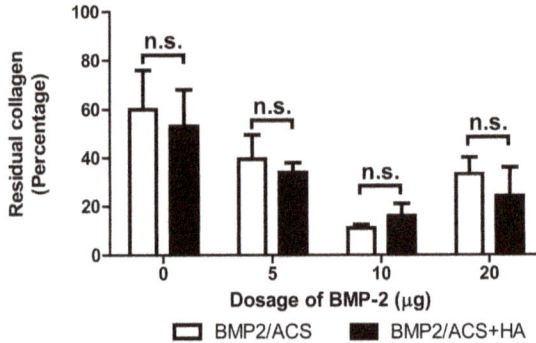

Figure 4. Mean volumes of residual collagen carrier material of the BMP-2/ACS constructs, in the presence or absence of 100 µg/mL HA (48 kDa) implanted at different BMP-2 dosages, 18 days after implant placement. Values represent means ± SEM; $n = 6$ per experimental group. n.s.: denotes the absence of significant differences ($p > 0.05$). The compared groups are indicated by brackets.

No significant differences in numerical area density of macrophages were present among these groups (Figure 5, Figure 6G). The 10 µg BMP-2 group value also was found to be significantly higher than the number of cross-sectioned blood vessels per unit tissue area in the 20 µg BMP-2 exerimental group ($p = 0.02$); but it did not significantly differ compared to the group of 5 µg BMP-2+HA (Figure 7). The combined administration of HA significantly promoted the the number of blood vessel in the 5 µg ($p = 0.017$) and 10 µg BMP-2 dosage groups ($p = 0.0001$), but not in the 20 µg BMP-2 group.

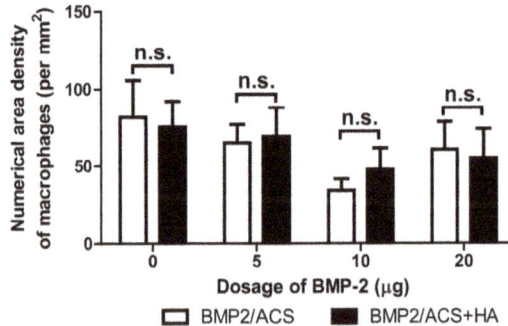

Figure 5. Mean values of the numerical area densities of macrophage cell profiles in the BMP-2/ACS constructs, in the presence or absence of 100 µg/mL HA (48 kDa) implanted at different BMP-2 dosages, 18 days after implant placement. Values represent means ± SEM; $n = 6$ per experimental group. n.s.: denotes absence of significant differences ($p > 0.05$). The compared groups are indicated by brackets.

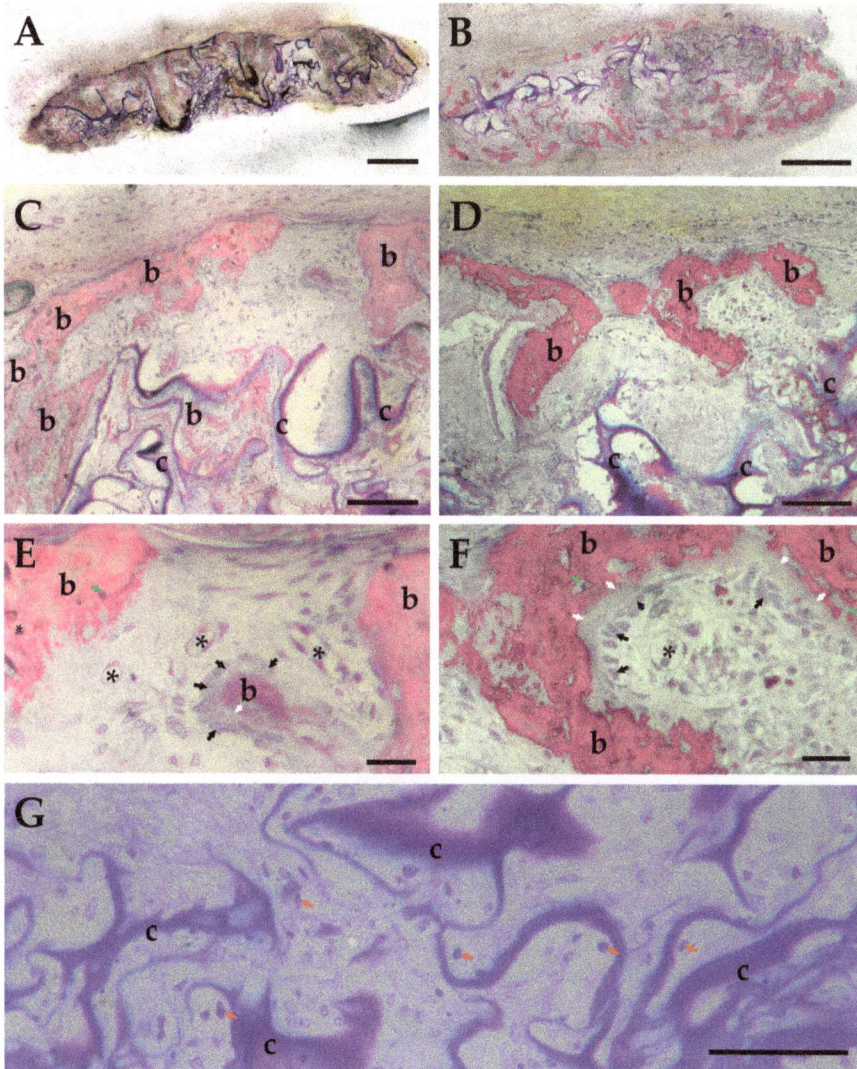

Figure 6. Light micrographs of BMP-2/ACS constructs, in the presence or absence of 100 µg/mL HA (48 kDa) at time of retrieval (18 days) at low (**A,B**) and high (**C–G**) magnifications: (**A,C,E,G**): BMP-2 10 µg+HA; (**B,D,F**): BMP-2 10 µg in the absence of HA. (**A**) illustrates homogenous bone forming activities throughout the construct, whereas in B formation of new bone tissue occurs preferentially at the interface of the construct with the native tissue. (**C,D**) illustrate the newly-formed bone tissue (b) in these two groups at higher magnifications and remaining collagen carrier material (c). (**E,F**) illustrate the blood vessels (*) and unmineralized bone areas (white arrow); osteoblasts (black arrow). In (**E,F**), the newly-formed woven bone shows a typical irregular pattern of osteocyte distribution (green arrow) within the mineralized bone matrix (pink-red stained areas). In (**E**) larger numbers of blood vessels (*) are present compared to (**D**); (**G**) illustrates the macrophages (red arrow) within BMP-2/ACS constructs. Magnification bars in (**A,C**): 500 µm; in (**C,D,G**): 100 µm; in (**E,F**): 25 µm.

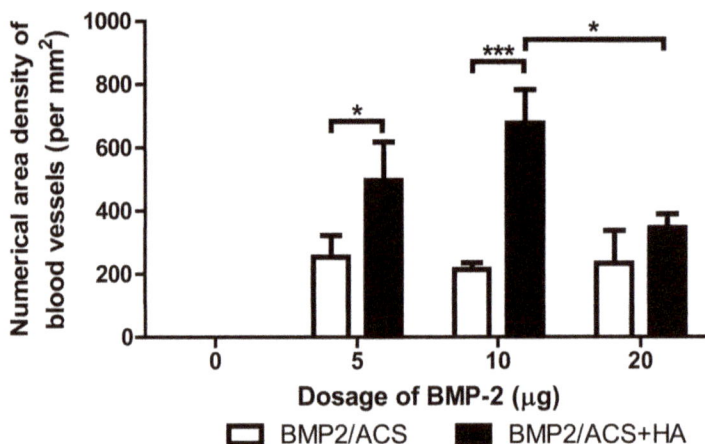

Figure 7. Mean values of the numerical area densities of blood vessel profiles in the BMP-2/ACS constructs, in the presence or absence of 100 µg/mL HA (48 kDa) implanted at different BMP-2 dosages, 18 days after implant placement. Values represent means ± SEM; n = 6 per experimental group. The asterisks denote the level of statistical significance, i.e., (* $p < 0.05$, ** $p < 0.01$, *** $p < 0.001$). The compared groups are indicated by brackets.

In the 10 µg BMP-2+HA group (Figure 6A,C), significantly less ACS and larger volumes of new bone were present when compared to the 10 µg BMP-2 group (Figure 6B,D). The number of cross-sectioned blood verssels was higher in Figure 6E than in Figure 6F, and that in Figure 6E the cross-section areas of the blood vessels are generally smaller. The computation of the average blood vessel cross-sectioned area, obtained by dividing the mean blood vessel areal density by the mean number of blood vessel cross-sections per area, revealed that the mean area per vessel for the 10 µg BMP-2 +HA group is 0.7×10^{-4} mm^2, and the mean area per blood vessel for the 10 µg BMP-2 group without HA is 2×10^{-4} mm^2; thus, the mean cross-sectioned blood vessel area is about three times larger in the experimental group in the absence of HA than in the same BMP dosage group in the presence of HA. In addition, histological observation revealed that in the 10 µg BMP-2 group without HA, the typically observed patterns of carrier degradation and new bone formation differed: whereas bone formation activities generally occured throµghout the ACS carrier materials (see Figure 6A), in the 10 µg BMP group in the absence of HA the new bone formation activities occured preferentially in the peripheral areas of the carrier materials (Figure 6B). However, the qualitiy of newly-formed bone tissue was found upon morphological examination to be the same in all experimental groups; in particular, the numerical density of osteoclasts appeared to be the same in all groups in which bone tissue had been generated, and no decline or change of the osteoclast numerical density was observed in any experimental group, in particular not in the 10 µg BMP group+HA group.

4. Discussion

HA is one of the major physiological components of the extracellular matrix (ECM), in all the connective tissues of the body. It is involved in a number of major biological processes, such as tissue organization, wound healing, angiogenesis, and remodeling of skeletal tissues [36–38]. In addition, HA is polyanionic in nature and, therefore, capable of forming ionic bonds with cationic growth factors, such as BMPs, which seems to be of significance for clinical applications [38]. In this study, we found that the combined administration of HA could significantly enhance the osteogenic potential of BMP-2/ACS, allowing a minimized unwanted side-effects [7].

Our extensive preliminary screening experiments revealed that an HA polymer length of about 48 kDa was of the optimal size range for the desired effect when used at a concentration of approximately 100 µg/mL. This might be because HA established, at these conditions, the optimal form of a gel, in which BMP-2 was most efficiently entrapped to optimally retain its bioactivity [39]. As a meshwork, HA might also reduce the free diffusion capabilites of BMP-2 and its flow, thus acting as a slow release system with an enhanced osteogenic activity potential [40].

In the present study, HA, at the optimal specifications, clearly promoted the BMP-dependent osteogenesis activity (Figure 2). In addition, the total carrier volume (Figure 3) and the number of blood vessel cross-sections per unit area of tissue, were also the highest in the 10 µg BMP group+HA group (Figure 6). Such effects were indeed absent in all other experimental groups without HA where the generated new bone mass did not even vary as a function of different BMP-2 dosage levels (Figure 2). The promoting effect of HA on new bone formation was only seen at dosages higher than the 10 µg BMP group (Figure 2), which suggested that this group might thus lie in the range of a minimal BMP dosage needed for the desired effect of higher bone volume generation in the present conditions.

The inflammatory response to BMP-2/ACS, was found to be the same in all experimental groups (Figure 5). The HA-dependent promoting effect for bone formation was unlikely attributed to a modification effect of HA on the inflammatory response. Instead, the HA-dependent facilitating effect on bone formation might be more likely associated with the degree of formation of new blood vessels, i.e., with the angiogenetic activity associated with the osteogenetic response. On one hand, we found that the number of cross-sectioned blood vessels was clearly the highest in the 10 µg BMP-2 group; on the other hand, this effect is clearly associated with the presence of a higher total surface area of blood vessel walls and, thus, of a larger blood vessel-wall associated perivascular tissue space, than when only fewer and thicker blood vessels are present; and it is indeed the peri-vascular tissue area that is the niche space carrying the pericytes and, thus, harbors the population of blood vessel associated adult stem cells of the mesenchymal type [41]; these have been previously found and identified to be able to differentiate into bone forming osteoblasts [42].

HA polymers showed an angiogenetic effect at specific polymer lengths [43], and BMP-2 itself was also shown to have itself some angiogenetic activity [44]. In addition, HA could also facilitate the migration of the perivascular stem cells [45] from their original niche to distant sites within the newly-forming tissues. HA is well-known to stimulate signal transduction pathways [46,47] that in turn facilitate cell locomotion [47]. Moreover, our data were also consistent with a recent study of Jungju Kim [48]: he found that BMP-2 activity was accompagnied only with the highest expression of osteocalcin and with a mature form of bone tissue with positive vascular markers (such as CD31 and vascular endothelial growth factors) when applied in the presence of HA, illustrating again that acitive angiogenesis was one of the key factors accounting for successful new bone formation [49].

It should always be kept in mind that BMP-activity is also associated with the recruitment, formation, and activation of osteoclasts, leading to immediate bone resorption activities. In this study no significant variation of osteoclast density in the newly formed bone tissue compartments among the groups. Thus, it appears unlikely that a lower degree of bone resorption activity would be a significant factor in supporting the formation of higher bone volumes in the 10 µg BMP group. It was, indeed, the careful dosage that was needed for BMP-2 in order to work out the required balanced-dosage of minimizing the osteoclastogeneic effects of BMP-2 and maximizing the osteogenetic effects of this pleiomorphic growth factor as we recently illustrated in sheep [40].

The clearly higher degree of blood vessel numbers and, thus, blood vessel wall surface area in the 10 µg BMP-2 group highly suggested that the HA-dependent osteogenic promotion effect of BMP-2 was related to a concomitantly associated increased angiogenetic activity. The fact that the total construct volume was also the largest one for the 10 µg BMP-2 group among all the experimental groups, supported this view since this large total construct volume was mainly due to the increased presence of bone tissue, and not to an increased volume of inflammatory area or swelling effect; moreover the volume of the residual ACS was indeed the smallest one in this group, both in relative

(Figure 4) and absolute terms (data not shown). The high degree of scatter of the mean values of the residual collagen in the experimental groups, represented by the coefficients of variations of these groups, was, however, fairly large, and again it was the smallest for the 10 µg BMP-2 groups (Figure 4); the CE of the 10 µg BMP-2 group in the absence of HA was 13.9%, and in the presence of HA was 30.6%. We were, thus, unable to put forward a clear explanation for our finding, but we are inclined to assume that this result is associated with a more rapid and efficient degradation of the collagen carrier materials deposited. However, since the degree of inflammatory response was quite similar in all groups (Figure 5), and no significant differences were encountered, it could be speculated that this phenomenon might be associated with a higher degree of osteolytic activity in this group; i.e., with a more rapid bone resorption activity in this group with the highest bone mass. There were, however, no indications found for the presence of higher numbers of osteoclasts in this group, and indeed the detailed morphological examination did not reveal any differences between groups in this respect. However, another possible (and more likely) explanation may be related to the more extensive angiogenetic activity encountered in this group: rapidly ingrowth and forming new blood vessels may be associated with the more efficient degradation of the collagen carrier materials, and indeed angiogenesis associated with tissue engineering approaches was previously described to be associated with such degradative activities [50]. Another indicator for favoring this hypothesis was the specific morphological pattern of new bone formation observed in this group: whereas, in all the other experimental groups, new bone tissue had formed mainly at the periphery of the constructs where probably most blood vessels were present, i.e., at the interface of the vascularized native tissue with the avascular construct (and bone tissue indeed does not form in the absence of a blood vasculature [51]). This pattern of bone formation relating to an osteogenic construct using ACS as carrier was observed by us also in a recent study [15]. However, the 10 µg BMP-2 group is the only one in which bone formation activities occurred by a different pattern, namely throughout the carrier construct with blood vessels being present all the way through the construct at high numerical densities (Figure 7). It appeared more probable that the more efficient degradation activities for the ACS (Figure 4) were associated with this more aggressive angiogenetic activity.

Acknowledgments: This study was supported by funds of China Scholarship Council, National Natural Science Foundation of China (81400475, 81470724, and 81600844), Zhejiang Provincial Natural Science Foundation of China (LY14H140002 and Y17H140023) and Science Technology Department of Zhejiang Province (2017C33168).

Author Contributions: Hairong Huang, Daniel Wismeijer, Ernst B. Hunziker and Gang Wu conceived and designed the experiments; Hairong Huang, Jianying Feng, Ernst B. Hunziker, and Gang Wu performed the experiments; Hairong Huang, Jianying Feng, Ernst B. Hunziker, and Gang Wu analyzed the data; Hairong Huang, Jianying Feng, Gang Wu contributed reagents/materials/analysis tools; and Hairong Huang, Jianying Feng, Daniel Wismeijer, Ernst B. Hunziker, and Gang Wu wrote the paper.

Conflicts of Interest: The authors declare no conflict of interest.

References

1. Wozney, J.M.; Rosen, V.; Celeste, A.J.; Mitsock, L.M.; Whitters, M.J.; Kriz, R.W.; Hewick, R.M.; Wang, E.A. Novel regulators of bone formation: Molecular clones and activities. *Science* **1988**, *242*, 1528–1534. [CrossRef] [PubMed]
2. Bessa, P.C.; Casal, M.; Reis, R.L. Bone morphogenetic proteins in tissue engineering: The road from the laboratory to the clinic, part i (basic concepts). *J. Tissue Eng. Regen. Med.* **2008**, *2*, 1–13. [CrossRef] [PubMed]
3. De Freitas, R.M.; Susin, C.; Tamashiro, W.M.; Chaves de Souza, J.A.; Marcantonio, C.; Wikesjo, U.M.; Pereira, L.A.; Marcantonio, E., Jr. Histological analysis and gene expression profile following augmentation of the anterior maxilla using rhbmp-2/acs versus autogenous bone graft. *J. Clin. Periodontol.* **2016**, *43*, 1200–1207. [CrossRef] [PubMed]
4. Freitas, R.M.; Spin-Neto, R.; Marcantonio Junior, E.; Pereira, L.A.; Wikesjo, U.M.; Susin, C. Alveolar ridge and maxillary sinus augmentation using rhbmp-2: A systematic review. *Clin. Implant Dent. Relat. Res.* **2015**, *17* (Suppl. 1), e192–e201. [CrossRef] [PubMed]

5. Hirata, A.; Ueno, T.; Moy, P.K. Newly formed bone induced by recombinant human bone morphogenetic protein-2: A histological observation. *Implant Dent.* **2017**, *26*, 173–177. [CrossRef] [PubMed]

6. Benglis, D.; Wang, M.Y.; Levi, A.D. A comprehensive review of the safety profile of bone morphogenetic protein in spine surgery. *Neurosurgery* **2008**, *62*, 423–431. [CrossRef] [PubMed]

7. James, A.W.; LaChaud, G.; Shen, J.; Asatrian, G.; Nguyen, V.; Zhang, X.; Ting, K.; Soo, C. A review of the clinical side effects of bone morphogenetic protein-2. *Tissue Eng. Part B Rev.* **2016**, *22*, 284–297. [CrossRef] [PubMed]

8. Faundez, A.; Tournier, C.; Garcia, M.; Aunoble, S.; Le Huec, J.C. Bone morphogenetic protein use in spine surgery-complications and outcomes: A systematic review. *Int. Orthop.* **2016**, *40*, 1309–1319. [CrossRef] [PubMed]

9. Hofstetter, C.P.; Hofer, A.S.; Levi, A.D. Exploratory meta-analysis on dose-related efficacy and morbidity of bone morphogenetic protein in spinal arthrodesis surgery. *J. Neurosurg. Spine* **2016**, *24*, 457–475. [CrossRef] [PubMed]

10. Vavken, J.; Mameghani, A.; Vavken, P.; Schaeren, S. Complications and cancer rates in spine fusion with recombinant human bone morphogenetic protein-2 (rhbmp-2). *Eur. Spine J.* **2016**, *25*, 3979–3989. [CrossRef] [PubMed]

11. Cahill, K.S.; McCormick, P.C.; Levi, A.D. A comprehensive assessment of the risk of bone morphogenetic protein use in spinal fusion surgery and postoperative cancer diagnosis. *J. Neurosurg. Spine* **2015**, *23*, 86–93. [CrossRef] [PubMed]

12. Malham, G.M.; Giles, G.G.; Milne, R.L.; Blecher, C.M.; Brazenor, G.A. Bone morphogenetic proteins in spinal surgery: What is the fusion rate and do they cause cancer? *Spine* **2015**, *40*, 1737–1742. [CrossRef] [PubMed]

13. Burkus, J.K.; Heim, S.E.; Gornet, M.F.; Zdeblick, T.A. Is infuse bone graft superior to autograft bone? An integrated analysis of clinical trials using the lt-cage lumbar tapered fusion device. *J. Spinal Disord. Tech.* **2003**, *16*, 113–122. [CrossRef] [PubMed]

14. Hagi, T.T.; Wu, G.; Liu, Y.; Hunziker, E.B. Cell-mediated bmp-2 liberation promotes bone formation in a mechanically unstable implant environment. *Bone* **2010**, *46*, 1322–1327. [CrossRef] [PubMed]

15. Huang, H.R.; Wismeijer, D.; Hunziker, E.B.; Wu, G. The acute inflammatory response to absorbed collagen sponge is not enhanced by BMP-2. *Int. J. Mol. Sci.* **2017**, *18*, 498. [CrossRef] [PubMed]

16. Murray, I.R.; Peault, B. Q&A: Mesenchymal stem cells—Where do they come from and is it important? *BMC Biol.* **2015**, *13*, 99.

17. Villanueva, J.E.; Nimni, M.E. Promotion of calvarial cell osteogenesis by endothelial cells. *J. Bone Miner. Res.* **1990**, *5*, 733–739. [CrossRef] [PubMed]

18. Bayer, E.A.; Fedorchak, M.V.; Little, S.R. The influence of platelet-derived growth factor and bone morphogenetic protein presentation on tubule organization by human umbilical vascular endothelial cells and human mesenchymal stem cells in coculture. *Tissue Eng. Part A* **2016**, *22*, 1296–1304. [CrossRef] [PubMed]

19. Sasaki, T.; Watanabe, C. Stimulation of osteoinduction in bone wound healing by high-molecular hyaluronic acid. *Bone* **1995**, *16*, 9–15. [CrossRef]

20. Mero, A.; Hyaluronic, M.C. Acid bioconjugates for the delivery of bioactive molecules. *Polymers* **2014**, *6*, 346–369. [CrossRef]

21. Rider, C.C.; Mulloy, B. Heparin, heparan sulphate and the TGF-beta cytokine superfamily. *Molecules* **2017**, *22*, 713. [CrossRef] [PubMed]

22. Kuo, W.J.; Digman, M.A.; Lander, A.D. Heparan sulfate acts as a bone morphogenetic protein coreceptor by facilitating ligand-induced receptor hetero-oligomerization. *Mol. Biol. Cell* **2010**, *21*, 4028–4041. [CrossRef] [PubMed]

23. Huang, L.; Cheng, Y.Y.; Koo, P.L.; Lee, K.M.; Qin, L.; Cheng, J.C.; Kumta, S.M. The effect of hyaluronan on osteoblast proliferation and differentiation in rat calvarial-derived cell cultures. *J. Biomed. Mater. Res. A* **2003**, *66*, 880–884. [CrossRef] [PubMed]

24. Knudson, C.B.; Knudson, W. Cartilage proteoglycans. *Semin. Cell Dev. Biol.* **2001**, *12*, 69–78. [CrossRef] [PubMed]

25. Takahashi, Y.; Li, L.; Kamiryo, M.; Asteriou, T.; Moustakas, A.; Yamashita, H.; Heldin, P. Hyaluronan fragments induce endothelial cell differentiation in a CD44- and CXCL1/GRO1-dependent manner. *J. Biol. Chem.* **2005**, *280*, 24195–24204. [CrossRef] [PubMed]

26. Goldberg, R.L.; Toole, B.P. Hyaluronate inhibition of cell proliferation. *Arthritis Rheum.* **1987**, *30*, 769–778. [CrossRef] [PubMed]

27. Vabres, P. Hyaluronan, embryogenesis and morphogenesis. *Ann. Dermatol. Venereol.* **2010**, *137*, 9–14. [CrossRef]

28. Chen, W.Y.; Abatangelo, G. Functions of hyaluronan in wound repair. *Wound Repair. Regen.* **1999**, *7*, 79–89. [CrossRef] [PubMed]

29. Chang, E.J.; Kim, H.J.; Ha, J.; Kim, H.J.; Ryu, J.Y.; Park, K.H.; Kim, U.H.; Lee, H.Z.; Kim, H.M.; Fisher, D.E.; et al. Hyaluronan inhibits osteoclast differentiation via Toll-like receptor 4. *J. Cell Sci.* **2007**, *120*, 166–176. [CrossRef] [PubMed]

30. Kawano, M.; Ariyishi, W.; Iwanaga, K.; Okinaga, T.; Habu, M.; Yoshioka, I.; Tominaga, K.; Nishihara, T. Mechanism involved in enhancement of osteoblast differentiation by hyaluronic acid. *Biochem. Biophys. Res. Commun.* **2011**, *405*, 575–580. [CrossRef] [PubMed]

31. Zhang, Y.; Yang, S.; Zhou, W.; Fu, H.; Qian, L.; Miron, R.J. Addition of a synthetically fabricated osteoinductive biphasic calcium phosphate bone graft to BMP2 improves new bone formation. *Clin. Implant Dent. Relat. Res.* **2016**, *18*, 1238–1247. [CrossRef] [PubMed]

32. Lee, K.B.; Taghavi, C.E.; Song, K.J.; Yoo, J.H.; Keorochana, G.; Tzeng, S.T.; Fei, Z.Q.; Liao, J.C.; Wang, J.C. Inflammatory characteristics of rhBMP-2 in vitro and in an in vivo rodent model. *Spine* **2011**, *36*, 149–154. [CrossRef] [PubMed]

33. Gundersen, H.J.; Bendtsen, T.F.; Korbo, L.; Marcussen, N.; Moller, A.; Nielsen, K.; Nyengaard, J.R.; Pakkenberg, B.; Sorensen, F.B.; Vesterby, A.; et al. Some new, simple and efficient stereological methods and their use in pathological research and diagnosis. *APMIS* **1988**, *96*, 379–394. [CrossRef] [PubMed]

34. Cruz-Orive, L.M.; Weibel, E.R. Recent stereological methods for cell biology: A brief survey. *Am. J. Physiol.* **1990**, *258*, 148–156.

35. Gundersen, H.J.; Jensen, E.B. The efficiency of systematic sampling in stereology and its prediction. *J. Microsc.* **1987**, *147*, 229–263. [CrossRef] [PubMed]

36. Karvinen, S.; Pasonen-Seppanen, S.; Hyttinen, J.M.; Pienimaki, J.P.; Torronen, K.; Jokela, T.A.; Tammi, M.I.; Tammi, R. Keratinocyte growth factor stimulates migration and hyaluronan synthesis in the epidermis by activation of keratinocyte hyaluronan synthases 2 and 3. *J. Biol. Chem.* **2003**, *278*, 49495–49504. [CrossRef] [PubMed]

37. Itano, N.; Atsumi, F.; Sawai, T.; Yamada, Y.; Miyaishi, O.; Senga, T.; Hamaguchi, M.; Kimata, K. Abnormal accumulation of hyaluronan matrix diminishes contact inhibition of cell growth and promotes cell migration. *Proc. Natl. Acad. Sci. USA* **2002**, *99*, 3609–3614. [CrossRef] [PubMed]

38. Peng, L.; Bian, W.G.; Liang, F.H.; Xu, H.Z. Implanting hydroxyapatite-coated porous titanium with bone morphogenetic protein-2 and hyaluronic acid into distal femoral metaphysis of rabbits. *Chin. J. Traumatol.* **2008**, *11*, 179–185. [CrossRef]

39. Hulsart-Billstrom, G.; Yuen, P.K.; Marsell, R.; Hilborn, J.; Larsson, S.; Ossipov, D. Bisphosphonate-linked hyaluronic acid hydrogel sequesters and enzymatically releases active bone morphogenetic protein-2 for induction of osteogenic differentiation. *Biomacromolecules* **2013**, *14*, 3055–3063. [CrossRef] [PubMed]

40. Hunziker, E.B.; Jovanovic, J.; Horner, A.; Keel, M.J.; Lippuner, K.; Shintani, N. Optimisation of bmp-2 dosage for the osseointegration of porous titanium implants in an ovine model. *Eur. Cell Mater.* **2016**, *32*, 241–256. [CrossRef] [PubMed]

41. Askarinam, A.; James, A.W.; Zara, J.N.; Goyal, R.; Corselli, M.; Pan, A.; Liang, P.; Chang, L.; Rackohn, T.; Stoker, D.; et al. Human perivascular stem cells show enhanced osteogenesis and vasculogenesis with nel-like molecule i protein. *Tissue Eng. Part A* **2013**, *19*, 1386–1397. [CrossRef] [PubMed]

42. James, A.W.; Zara, J.N.; Zhang, X.; Askarinam, A.; Goyal, R.; Chiang, M.; Yuan, W.; Chang, L.; Corselli, M.; Shen, J.; et al. Perivascular stem cells: A prospectively purified mesenchymal stem cell population for bone tissue engineering. *Stem Cells Transl. Med.* **2012**, *1*, 510–519. [CrossRef] [PubMed]

43. West, D.C.; Hampson, I.N.; Arnold, F.; Kumar, S. Angiogenesis induced by degradation products of hyaluronic acid. *Science* **1985**, *228*, 1324–1326. [CrossRef] [PubMed]

44. Deckers, M.M.; van Bezooijen, R.L.; van der Horst, G.; Hoogendam, J.; van Der Bent, C.; Papapoulos, S.E.; Lowik, C.W. Bone morphogenetic proteins stimulate angiogenesis through osteoblast-derived vascular endothelial growth factor a. *Endocrinology* **2002**, *143*, 1545–1553. [CrossRef] [PubMed]

45. Lei, Y.; Gojgini, S.; Lam, J.; Segura, T. The spreading, migration and proliferation of mouse mesenchymal stem cells cultured inside hyaluronic acid hydrogels. *Biomaterials* **2011**, *32*, 39–47. [CrossRef] [PubMed]

46. Turley, E.A.; Noble, P.W.; Bourguignon, L.Y. Signaling properties of hyaluronan receptors. *J. Biol. Chem.* **2002**, *277*, 4589–4592. [CrossRef] [PubMed]

47. Entwistle, J.; Hall, C.L.; Turley, E.A. HA receptors: Regulators of signalling to the cytoskeleton. *J. Cell Biochem.* **1996**, *61*, 569–577. [CrossRef]

48. Kim, J.; Kim, I.S.; Cho, T.H.; Lee, K.B.; Hwang, S.J.; Tae, G.; Noh, I.; Lee, S.H.; Park, Y.; Sun, K. Bone regeneration using hyaluronic acid-based hydrogel with bone morphogenic protein-2 and human mesenchymal stem cells. *Biomaterials* **2007**, *28*, 1830–1837. [CrossRef] [PubMed]

49. Ryan, J.M.; Barry, F.P.; Murphy, J.M.; Mahon, B.P. Mesenchymal stem cells avoid allogeneic rejection. *J. Inflamm. (London)* **2005**, *2*, 8. [CrossRef] [PubMed]

50. Walsh, W.R.; Chapman-Sheath, P.J.; Cain, S.; Debes, J.; Bruce, W.J.; Svehla, M.J.; Gillies, R.M. A resorbable porous ceramic composite bone graft substitute in a rabbit metaphyseal defect model. *J. Orthop. Res.* **2003**, *21*, 655–661. [CrossRef]

51. Calori, G.M.; Giannoudis, P.V. Enhancement of fracture healing with the diamond concept: The role of the biological chamber. *Injury* **2011**, *42*, 1191–1193. [CrossRef] [PubMed]

© 2017 by the authors. Licensee MDPI, Basel, Switzerland. This article is an open access article distributed under the terms and conditions of the Creative Commons Attribution (CC BY) license (http://creativecommons.org/licenses/by/4.0/).

polymers

MDPI

Article

Lipase-Catalyzed Synthesis, Properties Characterization, and Application of Bio-Based Dimer Acid Cyclocarbonate

Xin He, Guiying Wu, Li Xu, Jinyong Yan and Yunjun Yan *

Key Laboratory of Molecular Biophysics of the Ministry of Education, College of Life Science and Technology, Huazhong University of Science and Technology, Wuhan 430074, China; n785888@163.com (X.H.); wuguiying@hust.edu.cn (G.W.); xuli@hust.edu.cn (L.X.); yjiny@126.com (J.Y.)
* Correspondence: yanyunjun@hust.edu.cn; Tel.: +86-27-87792213

Received: 6 February 2018; Accepted: 1 March 2018; Published: 3 March 2018

Abstract: Dimer acid cyclocarbonate (DACC) is synthesized from glycerol carbonate (GC) and *Sapium sebiferum* oil-derived dimer acid (DA, 9-[(Z)-non-3-enyl]-10-octylnonadecanedioic acid). Meanwhile, DACC can be used for synthetic materials of bio-based non-isocyanate polyurethane (bio-NIPU). In this study, DACC was synthesized by the esterification of dimer acid and glycerol carbonate using Novozym 435 (*Candida antarctica* lipase B) as the biocatalyst. Via the optimizing reaction conditions, the highest yield of 76.00% and the lowest acid value of 43.82 mg KOH/g were obtained. The product was confirmed and characterized by Fourier transform-infrared spectroscopy (FTIR) and nuclear magnetic resonance spectroscopy (NMR). Then, the synthetic DACC was further used to synthesize bio-NIPU, which was examined by FTIR, thermogravimetric analysis (TGA), and differential scanning calorimetry (DSC), indicating that it possesses very good physio-chemical properties and unique material quality with a potential prospect in applications.

Keywords: glycerol carbonate; dimer acid; esterification; lipase; cyclocarbonate; bio-based non-isocyanate polyurethane

1. Introduction

Forestry oil has become an important renewable resource in the consideration of environmental concerns, fossil fuel depletion, and food production [1,2]. The major component of forestry oil is triglycerides (esters of glycerol with three fatty acids); there are different long-chain fatty acids in the different sources of oil [3], which can provide different applications in industry. Since *Sapium sebiferum* oil, a non-edible oil, is abundant in China and its production significantly benefits the environment construction, it has been attracting wide attention and has been moderately studied [4,5]. Moreover, *S. sebiferum* can give way to arable land because it can grow in alkaline, saline, droughty, and acidic soil. Its seeds contain 45–60% oil, which is mostly unsaturated, resulting in a high iodine value of 186.8 g of I_2/100 g. *S. sebiferum* oil possesses many double bonds (Figure 1), which are the appropriate groups for the synthesis of various industrial compounds and polymers, especially for the synthesis of dimer acids [5,6].

Dimer acids (DAs), produced by a Diels–Alder reaction of unsaturated fatty acids from unsaturated oil, are highly value-added industrial products and are widely used in different fields, such as adhesives, preservatives, plastic additives, and lubricants [7–9]. On the other hand, glycerol can also be easily produced from forestry oil, or sometimes is a byproduct of forestry oil processing [10–12]. A downstream synthetic product of glycerol is glycerol carbonate (GC), which also has many different industrial applications, such as coatings, surfactants, cosmetics, lubricants, and so on. Its structure contains a 2-oxo-1,3-dioxolane group and a hydroxyl group, which have high reaction activity with anhydrides, acyl chlorides, isocyanates, and the like [13–15]. Having a bio-based origin and wide reactivity, GC has become a versatile and renewable building block for chemical synthesis. To obtain highly purified GC, different synthesis methods have been investigated, including direct synthetic routes and indirect synthetic routes [16,17]. They all aim at the lowest production cost, the least pollution, and the highest yield of GC.

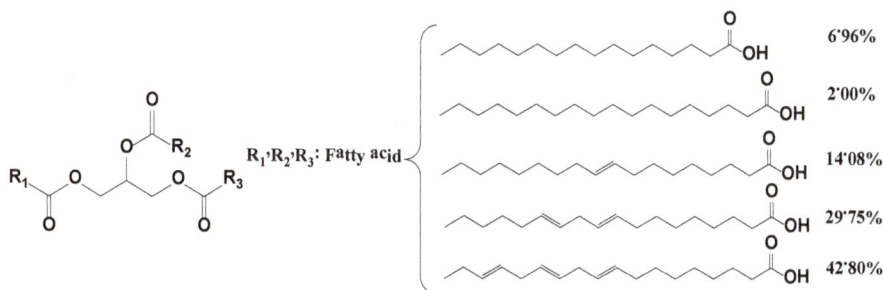

Figure 1. The structure of *Sapium sebiferum* oil and its fatty acid content.

At present, GC has been employed as a substrate to synthesize non-isocyanate polyurethane (bio-NIPU) in order to meet the increasingly serious environment regulations. The route is green and without the requirement of either toxic isocyanate monomers or phosgene [18]. However, the main synthesis pathway of cyclocarbonate is based on the catalytic synthesis of N,N'-dicyclohexylcarbodiimide (DCC) and 4-dimethylaminopyridine (DMAP), which are also poisonous to humans in some way [18–20]. Consequently, the cyclocarbonates react with amines to form urethane, or more specifically hydroxyurethane bonds [21,22]. The consumption and the recycling of the catalysts in the synthesis reaction is a huge problem, especially in large-scale industrial applications.

Dimer acid cyclocarbonate (DACC) is a monomer of non-isocyanate polyurethane. It has the following advantages: (1) simple synthetic methods; (2) biodegradability; and (3) biologic material. The route of synthesis is shown in Scheme 1. The traditional chemical synthesis of DACC requires a large number of chemical catalysts and dehydrants, resulting, in most cases, in numerous byproducts. Until now, most cyclocarbonates are chemically synthesized, which needs high pressure and more energy consumption and produces toxic waste [23,24]. On the contrary, enzymatic synthesis has garnered considerable interest with regard to its many merits, such as high regio-, chemo- and enantio-selectivity, environmental friendliness, mild reaction conditions, simple purification steps, and the enzyme being reused [16,25,26]. So far, there are considerable researches on the enzymatic synthesis of ester compounds [27]. Carlos et al. obtained the dodecyl lactate and glycolate through a lipase from *Candida antarctica* [28]. Afife et al. investigated isoamyl acetate synthesis using immobilized *Rhizomucor miehei* and *C. antarctica* lipases by esterification of acetic acid and isoamyl alcohol without organic solvent [29]. Actually, enzymatic synthesis of ester compounds is becoming a research hotspot. Novozym 435 (*Candida antarctica* lipase B physically immobilized within a macroporous resin of poly(methyl methacrylate)), a commercially available heterogeneous biocatalyst, has been used successfully for polyesters and polyamides synthesis [30].

Therefore, in this work, the enzymatic synthesis of DACC was first investigated by esterification of bio-based DA with GC. Then, experiments were designed to optimize the conditions of DACC synthesis, mainly to examine the effects of GC/DA molar ratio, reaction time, enzyme concentration, reaction temperature, molecular sieve content, agitation speed, solvent and enzyme cycling. The derivative DACC was subsequently reacted with different amines to prepare various bio-based non-isocyanate polyurethanes (bio-NIPUs), and their chemical structures and thermal properties were further characterized.

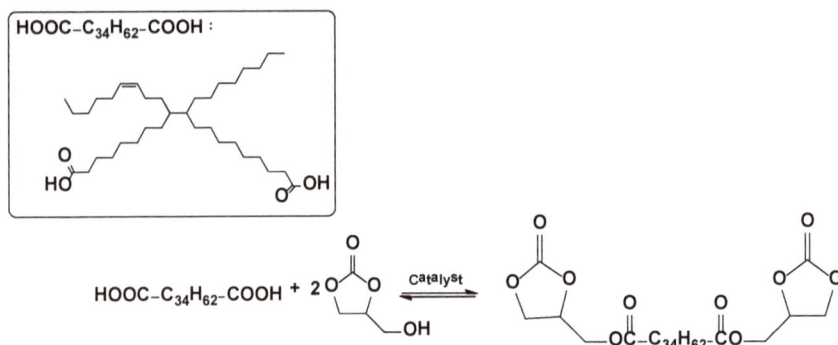

Scheme 1. Synthesis of dimer acid cyclocarbonate (DACC).

2. Materials and Methods

2.1. Materials

Dimer acid (DA, 9-[(Z)-non-3-enyl]-10-octylnonadecanedioic acid, CAS No. 61788-89-4) and glycerol carbonate (GC, CAS No. 931-40-8) were purchased from Bangcheng Chemical Ltd. (Shanghai, China) and Tokyo Chemical Industry Co., Ltd. (Tokyo, Japan), respectively. Novozym 435 (immobilized lipase B from *C. antarctica*) with a specific activity 10,000 propyl laurate units (PLUs) per gram was commercially obtained from Novozym Co. Ltd. (Zealand, Denmark). Acetonitrile (CAS No. 75-05-8), hexane (CAS No. 110-54-3), dichloromethane (CAS No. 75-09-2), molecular sieve type 4A (CAS No. 70955-01-0), ethylenediamine (EDA, CAS No. 107-15-3), diethylenetriamine (DETA, CAS No. 111-40-0), triethylenetetramine (TETA, CAS No. 112-24-3), tetraethylenepentamine (TEPA, CAS No. 112-57-2), and hexanediamine (HMDA, CAS No. 124-09-4) were analytical reagents and bought from Sinopharm Chemical Reagent Ltd. Co. (Shanghai, China).

2.2. Synthesis of Dimer Acid Cyclocarbonate (DACC)

DACC was synthesized in acetonitrile via an esterification reaction between DA and GC catalyzed by Novozym 435 [31–35]. Experiments were conducted in a conical flask placed in a thermostat shaking bed with a temperature monitor. A single factorial experiment was first designed. The effects of GC/DA molar ratio (2.00:1.00–10.00:1.00), time (4–24 h), Novozym 435 concentration (1–10 wt %, *w/w* DA), temperature (35–65 °C), molecular sieve content (30–100 wt %, *w/w* DA), agitation speed (50–300 rpm), solvent (acetonitrile, tert-butanol, tetrahydrofuran, acetone, and methylbenzene), and Novozym 435 cycle number (1–5 times) on the production of DACC were explored. The yield and acid value were chosen as the indicators.

2.3. Purification of the Esters

All the esters were purified after being prepared. These purified materials were further used in follow-up experiments to characterize their structures and calculate the conversion rates [28].

The enzyme was removed from the reaction mixture via filtration and the solvent was evaporated using a rotovap system. Then, the filtered and evaporated reaction mixtures were dissolved in *n*-hexane and the esters precipitated out, aiming to separate DA. Finally, precipitation in deionized water afforded DACC as a yellow viscous liquid; the residual water from the production was evaporated.

2.4. Synthesis of Bio-NIPU

DACC and dichloromethane were placed in the Teflon mold and stirred mechanically for 5 min. Then, different amines (EDA, DETA, TETA, TEPA, and HMDA) were added and the mixture was stirred mechanically for 5 min [36]. Thereafter, the mold was heated at 90 °C and reduced pressure for 12 h. All products were kept in a desiccator for later use.

2.5. Analysis and Characterization

The FTIR spectra of the samples were recorded in the frequency range of 4000–400 cm^{-1} with a spectral resolution of 4 cm^{-1} using a Bruker Vertex70 FTIR spectrometer equipped with a DTGS detector (Bruker Optics, Karlsruhe, Germany). Taking the KBr plate as a blank for the background, a few of the samples were dropped on it for the test.

The ^{1}H-NMR spectra of the samples were determined by a Bruker AV600 MHz NMR spectrometer (Bruker, Karlsruhe, Germany). The samples were dissolved in CDCl$_3$ (Tetramethylsilane (TMS) as internal standard) and then placed in 5 mm diameter NMR sample tubes for the determination. The measurement was performed at room temperature.

To confirm the degree of esterification, samples were characterized by acid value according to Chinese National Standard (GB/T 264-1983) [37]. The yield was calculated by the following equation. The mass of theoretical objective product was calculated according to the starting mass of DA. The actual objective product was weighed accurately after purification.

$$\text{Product yield} = \frac{\text{the weight of actual objective product}}{\text{the weight of theoretical objective product}} \times 100\%$$

Thermogravimetric analyses (TGA) were performed using a Pyris 1 TGA (Perkin-Elmer Instruments, Boston, MA, USA) at a heating rate of 10 °C /min. Approximately 5 mg of sample was subjected to temperatures from 50 to 600 °C in an N$_2$ atmosphere.

Differential scanning calorimetry (DSC) analysis was performed on a Perkin-Elmer Diamond DSC instrument (Perkin-Elmer Instruments, Boston, MA, USA) under N$_2$ atmosphere. The sample was first heated from −50 to 100 °C at 10 °C/min.

3. Results and Discussion

3.1. Single Factorial Experiments

3.1.1. Effect of GC/DA Molar Ratio

According to the chemical equation in Scheme 1, 1 mol of DACC is synthesized from 1 mol of DA and 2 mol of GC through esterification, and 2 mol water is obtained. However, according to this theoretical ratio, raw materials cannot completely react as a result of the reversible reaction. Meanwhile, DA has a long alkyl chain, resulting in steric hindrance. This repulsive hindrance lowers the electron density in the intermolecular region and disturbs the bonding interactions [38]. Therefore, with an increase in the ratio, the reaction is driven in the direction of a forward reaction. Therefore, the effect of the GC/DA molar ratio was optimized for DACC synthesis (Figure 2a). As shown in Figure 2a, the DACC yield increased with the molar ratio from 2.00 to 10.00, which was attributed to more collisions between DA and GC. When the molar ratio reached 4.00, the DACC yield remained about the same with increasing molar ratio. However, the acid value decreased, which was ascribed to the decrease in carboxylic acid (–COOH) groups. When the molar ratio reached 10.00, the acid value showed little change. Considering the economy of the raw materials and that a lower acid value led to a higher esterification degree, a GC/DA molar ratio of 8.00 was a better choice. Although the theoretical GC/DA molar ratio for synthesis of DACC was 2.00, its actual optimized GC/DA molar ratio for DACC synthesis, which resulted in a better yield and a lower acid value, was 8.00. The following tests used this optimal GC/DA molar ratio.

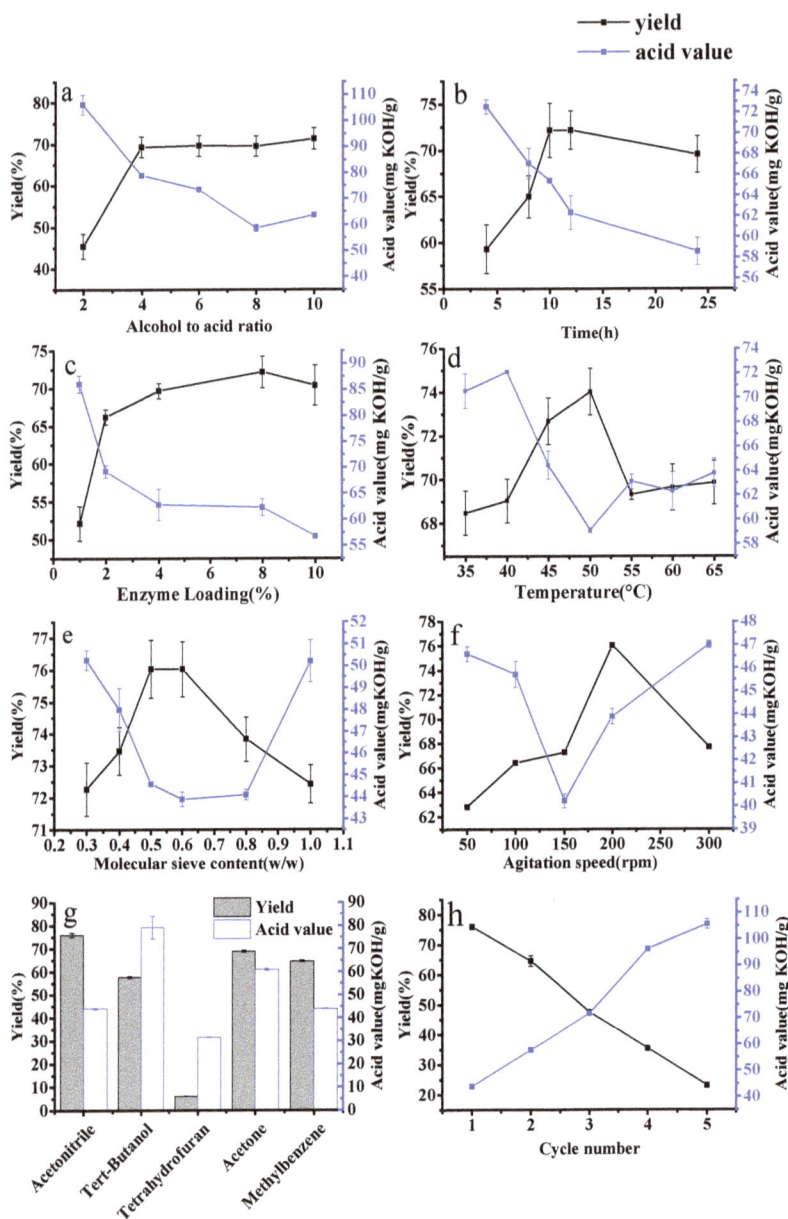

Figure 2. Effect of single factorial experiments. (**a**) GC/DA molar ratio, (**b**) time, (**c**) enzyme concentration, (**d**) temperature, (**e**) molecular sieve content, (**f**) agitation speed, (**g**) solvent, (**h**) cycle number.

3.1.2. Effect of Time

To find the proper reaction time, the reaction was performed from 4 to 24 h. As shown in Figure 2b, the reaction time had a great effect on yield. As the reaction time went on, the yield continued to increase. However, the reaction showed little change after 10 h. When the reaction time reached 24 h, the yield did not change too much because an equilibrium had been reached. Therefore, 12 h with the highest yield was selected as the optimal time. Although it had the same yield as at 10 h, the acid value was higher than that of 12 h. Given the cost of time, 12 h was better than 24 h even though the acid value was lower. Therefore, 12 h was determined as the optimal reaction time for maximizing the efficiency.

3.1.3. Effect of Enzyme Concentration

Enzymes are more efficient and environmentally friendly catalysts than chemical catalysts. Here, Novozym 435 was employed to catalyze DACC synthesis. The effect of enzyme concentration and its optimal dosage were investigated at various levels ranging from 1 to 10 wt % (Figure 2c). It can be seen from Figure 2c that the yield increased as the Novozym 435 concentration increased from 1 to 8 wt %. When the Novozym 435 concentration increased to 10 wt %, the yield decreased slightly. The adsorption of DACC on the immobilized enzyme always occurred; after reaching 10 wt %, a higher enzyme concentration led to higher adsorption. However, an excess in enzyme concentration could hamper the internal diffusion and relevant acquaintance of the substrate with the active sites [39]. Therefore, the increased concentration of immobilized enzyme was good for the synthesis of DACC, leading to a lower acid value, while the yield was affected with high enzyme dosage. Considering the better yield and an appropriate acid value, a Novozym 435 concentration of 8 wt % was chosen as the optimal enzyme concentration.

3.1.4. Effect of Temperature

The reaction temperature exerts a significant effect on enzyme activity and the economic effect of the DACC synthesis [40]. The effect of temperature on the esterification was investigated in the range from 35 to 65 °C (see Figure 2d). The maximum yield and minimum acid value were both obtained at 50 °C. A higher reaction temperature could increase the reaction rate and yield but it would lead to enzyme inactivation. At a lower reaction temperature, the substrate and intermediate cannot dissolve in sufficient quantity for synthesis [41,42]. Hence, the reaction temperature of 50 °C was selected as the optimal temperature.

3.1.5. Effect of Molecular Sieve Content

Because hydrolysis is the reverse reaction of esterification, removing water is the key to the degree of esterification. The effect of molecular sieve content was investigated over a range from 30 to 100 wt % (Figure 2e). The results show that yield reached a maximum value at 50 and 60 wt %, while the acid value reached a minimum at 60 wt %. The presence of the molecular sieves increased the yield at the initial stage because the desiccant can urge the reaction to esterification instead of hydrolysis. However, the yield decreased from 60 to 100 wt % and this could have been caused by molecular sieve adsorption. GC was adsorbed and diminished the effective concentration in the reaction. Meanwhile, the acid value increased from 60 to 100 wt %, indicating that molecular sieves adsorbed GC rather than DA. Considering the water absorption and raw materials adsorption of molecular sieves, the content at 60 wt % was defined as the optimal condition.

3.1.6. Effect of Agitation Speed

In this reaction, agitation speed is an easily overlooked single factor. Figure 2f shows that the yield reached a maximum when the agitation speed attained 200 rpm. From 50 to 200 rpm, the yield continued to increase because the mechanical agitation accelerated the collisions of the molecules. However, the yield decreased when the agitation speed approached 300 rpm; a possible reason could be that a high agitation speed would reduce the contact time and the contact area with the reaction interface. There was an interesting phenomenon that the acid value reached a minimum value when the agitation speed was 150 rpm. A reasonable explanation is that high agitation speeds could hamper the reaction between DA and the second GC, which then led to high acid values between agitation speeds of 150 and 300 rpm. Considering the yield and the acid value, 200 rpm was set as the appropriate agitation speed.

3.1.7. Effect of Solvent

In this reaction, dimer acid and glycerol carbonate do not mix. Therefore, a suitable solvent should dissolve enough raw materials for the lipase-catalyzed esterification, and the solvent should not affect lipase activity and stability. The results show that the reaction had the lowest yield in tetrahydrofuran, which might have altered the native conformation of the lipase (Figure 2g). Tert-butanal and acetone led to high acid values because GC did not dissolve entirely, which would have led to inadequate participation in the reaction. However, acetonitrile did dissolve the DA and GC at 50 °C, and had a low effect on Novozym 435. Foremost, the yield had the highest value and the acid value was low enough in acetonitrile. Thus, acetonitrile was selected for use in subsequent experiments.

3.1.8. Reuse of Novozym 435

Immobilized lipase has a reusable feature. We performed a series of experiments to examine its characteristic for the synthesis of DACC. After one cycle, the immobilized lipase was recovered by paper filtration and then dried for the next cycle. As shown in Figure 2h, the yield decreased and the acid value increased with the cycle number. The results indicate that the immobilized lipase loses its activity with cycle number; the probable reason could be that dichloromethane, which was used to wash the immobilized lipase in order to clear the remaining raw materials and DACC, could cause enzyme inactivation. Therefore, although the first two cycles can provide sufficient objective product, improving cycle efficiency will be a key research direction in the future.

3.2. Determination of DACC by FTIR

Figure 3 shows the FTIR spectra of DA and DACC. The top curve is the FTIR spectra of DA, which had a carbonyl C=O stretching vibration at 1710.26 cm^{-1} attributed to a carboxyl group. The curve below is the FTIR spectra of DACC, which had two carbonyl C=O stretching vibrations attributed to an ester group at 1742.62 cm^{-1} and carbonic ester at 1810.65 cm^{-1}. Obviously, the carboxyl group disappeared and two new carbonyl C=O stretching vibrations appeared because the carboxyl group reacted with a hydroxyl group to get an ester group. The FTIR spectra of DA and DACC clearly indicate the formation of DACC, especially the changes of the different carbonyl C=O stretching vibrations.

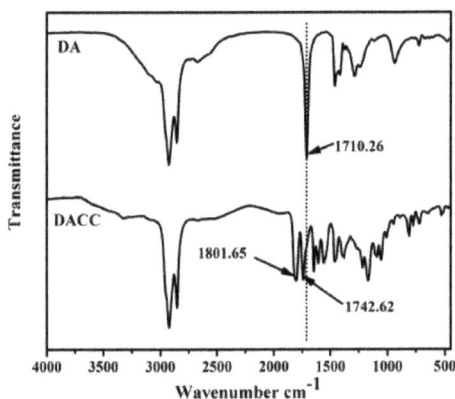

Figure 3. FTIR spectra of DA and DACC.

3.3. Determination of DACC by ^{1}H-NMR

Compared with the spectra of DA (Figure 4a), the spectra of DACC (Figure 4b) included some new peak groups attributed to methylene protons and methine protons [43], which were δ 4.21–4.28 (c, –OCH$_2$CH–, 2H), δ 4.29–4.34, 4.56–4.59 (a, –CHCH$_2$O–, 2H), δ 4.99–5.03 (b, –CH–, 1H), suggesting that the DACC had been successfully synthesized.

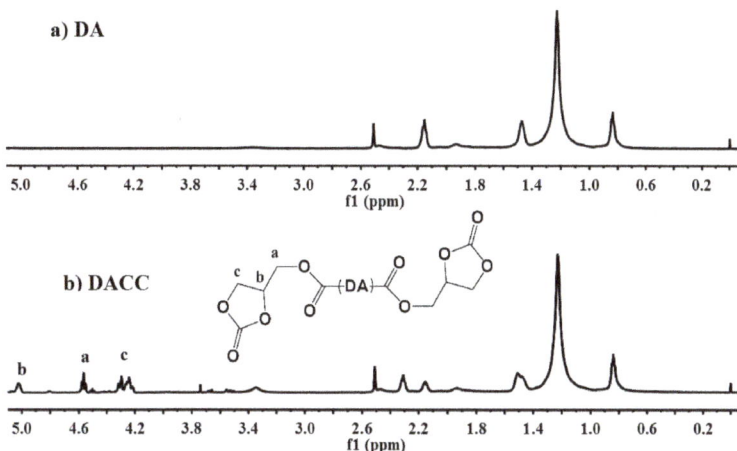

Figure 4. The ^{1}H-NMR spectra of DA (**a**) and DACC (**b**).

3.4. Characterization of Bio-NIPU

3.4.1. Characterizing Bio-NIPU by FTIR

Figure 5 shows the FTIR spectra of different bio-NIPUs and DACC. Compared with the FTIR spectra of DACC, the bio-NIPU curves had some differences. The peak at 1801.65 cm^{-1} only existed in DACC because it is the carbonyl C=O stretching vibration from carbonic ester. As can be seen from Scheme 2, the cyclocarbonates were opened by the amine groups. Two new peaks appeared at 1535.00 and 1255.50 cm^{-1}, which are the N–H stretching vibration and C–N stretching vibration, respectively, in bio-NIPU. The results confirm successful synthesis of bio-NIPU by DACC and different amines.

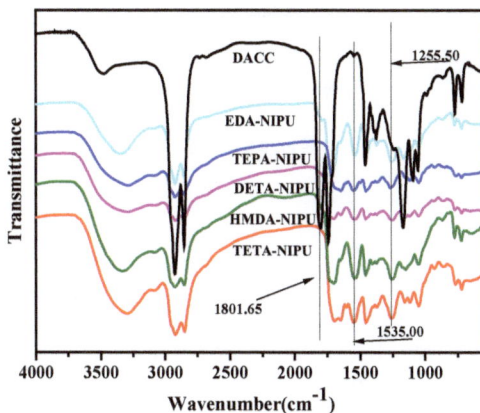

Figure 5. FTIR spectra of different NIPUs and DACC.

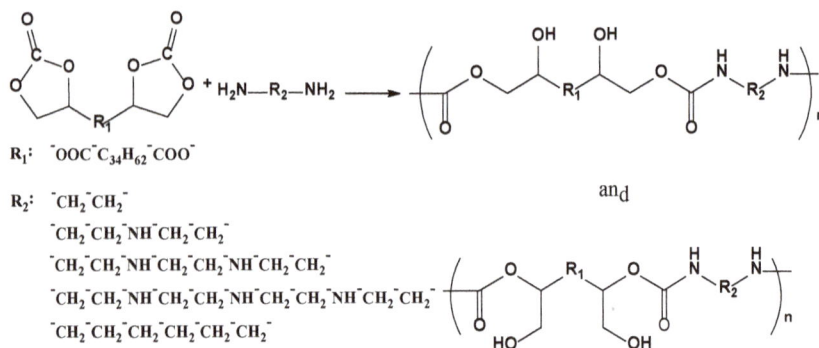

R_1: $^-OOC\,C_{34}H_{62}\,COO^-$

R_2: $^-CH_2\,CH_2^-$
$^-CH_2\,CH_2\,NH\,CH_2\,CH_2^-$
$^-CH_2\,CH_2\,NH\,CH_2\,CH_2\,NH\,CH_2\,CH_2^-$
$^-CH_2\,CH_2\,NH\,CH_2\,CH_2\,NH\,CH_2\,CH_2\,NH\,CH_2\,CH_2^-$
$^-CH_2\,CH_2\,CH_2\,CH_2\,CH_2\,CH_2^-$

and

Scheme 2. Synthesis of bio-NIPU.

3.4.2. Determination of Bio-NIPU via DSC

To study the thermal properties of different bio-NIPUs, the DSC curves are recorded in Figure 6. Glass transition temperature (T_g) values of the bio-NIPUs with secondary amines were higher than those of the bio-NIPUs with diamine. T_g value of DETA-NIPU was $-6.00\,°C$, which was higher than the other NIPUs. The results suggest that secondary amines have a positive effect on the crystallization of bio-NIPU. These bio-NIPUs have low T_g values due to the soft segment long chain structure of dimer acid. There was only one single glass transition temperature for all bio-NIPUs, indicating good miscibility.

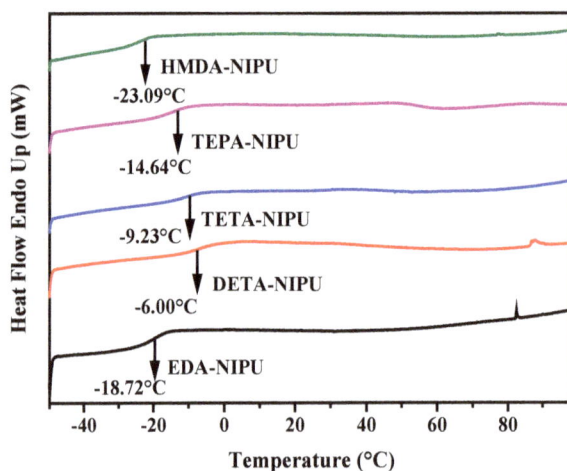

Figure 6. DSC curves of different NIPUs.

3.4.3. Determination of Bio-NIPU via TGA

The influence of the average amine functionality was investigated for the thermal stability of the synthesized polymers via thermogravimetric analysis (TGA). Figures 7 and 8 present TGA and differential thermal gravity (DTG) curves for different bio-NIPUs, respectively. The thermal stability of the EDA-NIPU, DETA-NIPU, TETA-NIPU, and HMDA-NIPU were similar with $Td_{5\%}$ (the temperature at 5% weight loss) around 210 °C. The thermal stability value of TEPA-NIPU was lower with a $Td_{5\%}$ equal to 194 °C (Table 1). The char contents at 550 °C are also shown in Table 1. These results suggest that secondary amines have no contribution to bio-NIPU synthesis. Thermal degradation of different bio-NIPUs takes place in two stages, corresponding to the hard and soft segments, because of thermodynamic incompatibility of the two segments in the bio-NIPU matrix. The temperatures for the maximum rate of degradation (T_{max}) for each of the two stages for different bio-NIPUs are given in Figure 8 and Table 1. The T_{max1} values of DETA-NIPU, TETA-NIPU, and TEPA-NIPU were about 220 °C. However, T_{max1} values of EDA-NIPU and HMDA-NIPU were higher due to no reaction between the secondary amines and cyclocarbonates. The T_{max2} value of TEPA-NIPU was 468 °C, which was the highest temperature in the different bio-NIPUs. This result is due to the high boiling point of TEPA, which may influence the maximum rate of degradation of the soft segments.

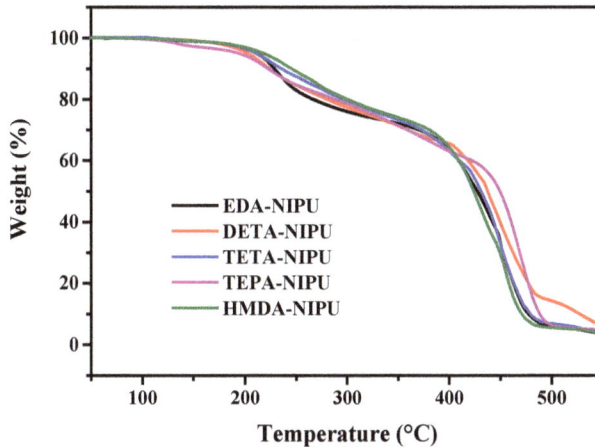

Figure 7. TGA curves of different NIPUs.

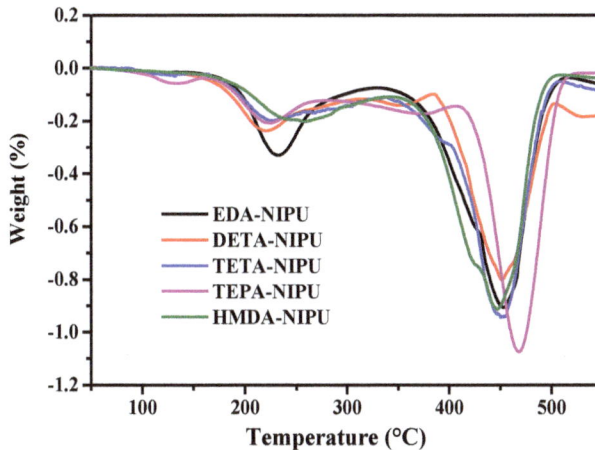

Figure 8. DTG curves of different NIPUs.

Polymers **2018**, *10*, 262

Table 1. TGA and DTG results for different bio-NIPUs.

	EDA-NIPU	DETA-NIPU	TETA-NIPU	TEPA-NIPU	HMDA-NIPU
$Td_{5\%}$ (°C)	212.31	202.90	210.85	194.09	218.10
T_{max1} (°C)	232.68	220.58	227.81	224.02	258.71
T_{max2} (°C)	453.16	451.39	450.18	468.00	447.05
Residue at 550 °C (%)	3.7	6.2	3.9	4.9	4.1

4. Conclusions

This study attempted to optimize the enzymatic catalyzed production of DACC from DA and GC derived from *S. sebiferum* oil by esterification using Novozym 435 as the catalyst. Considering the economic effect and efficiency criteria, GC/DA molar ratio, reaction time, Novozym 435 concentration, reaction temperature, molecular sieve content, agitation speed, and different solvents were optimized one by one. Meanwhile, enzyme cycling was studied. The yield was 76.00% and the lowest acid value for synthesized DACC was 43.82 mg KOH/g under optimal conditions. The product was further confirmed by FITR and NMR. Then, DACC was employed to synthesize bio-NIPU and subsequently characterized by FTIR, DSC, and TGA. These studies prove that it is feasible to synthesize DACC by lipase. Furthermore, the bio-NIPU had a low Tg and a good thermostability, which can be potentially used in the coating industry. This study demonstrates a new method to synthesize cyclocarbonate, which has a prosperous future potential for the bio-polyurethane industry.

Acknowledgments: This work was financially supported by the National Natural Science Foundation of China (Nos.: 31070089, 31170078 and J1103514), the National High Technology Research and Development Program of China (Nos.: 2011AA02A204, 2013AA065805), the National Natural Science Foundation of Hubei Province (grant No.: 2015CFA085) and the Fundamental Research Funds for Huazhong University of Science and Technology (HUST) (No.: 2014NY007). The authors would like to acknowledge the Analytical and Testing Center of HUST for their valuable assistance in FTIR, NMR, DSC, and TGA measurements.

Author Contributions: Xin He, Guiying Wu, and Yunjun Yan conceived and designed the experiments; Xin He performed the experiments; Xin He., Guiying Wu, Jinyong Yan, and Li Xu analyzed the data. Xin He wrote the paper. Xin He and Yunjun Yan contributed to the revision and proofreading of the manuscript.

Conflicts of Interest: The authors declare no conflict of interest.

References

1. Meher, L.; Vidyasagar, D.; Naik, S. Technical aspects of biodiesel production by transesterification—A review. *Renew. Sust. Energ. Rev.* **2006**, *10*, 248–268. [CrossRef]
2. Monteil-Rivera, F.; Phuong, M.; Ye, M.; Halasz, A.; Hawari, J. Isolation and characterization of herbaceous lignins for applications in biomaterials. *Ind. Crop. Prod.* **2013**, *41*, 356–364. [CrossRef]
3. Miao, S.; Wang, P.; Su, Z.; Zhang, S. Vegetable-oil-based polymers as future polymeric biomaterials. *Acta Biomater.* **2014**, *10*, 1692–1704. [CrossRef] [PubMed]
4. Wu, G.; Fan, Y.; He, X.; Yan, Y. Bio-polyurethanes from Sapium sebiferum oil reinforced with carbon nanotubes: Synthesis, characterization and properties. *RSC Adv.* **2015**, *5*, 80893–80900. [CrossRef]
5. Wu, G.; He, X.; Yan, Y. Lipase-catalyzed modification of natural Sapium sebiferum oil-based polyol for synthesis of polyurethane with improved properties. *RSC Adv.* **2017**, *7*, 1504–1512. [CrossRef]
6. Liu, Y.; Xin, H.; Yan, Y.; Xu, L. Preparation of dimeric fatty acid methyl esters and their polyamides from biodiesel. *J. Beijing Univ. Chem. Technol.* **2010**, *37*, 106–112.
7. Huang, Y.; Ye, G.; Yang, J. Synthesis and properties of UV-curable acrylate functionalized tung oil based resins via Diels-Alder reaction. *Prog. Org. Coat.* **2015**, *78*, 28–34. [CrossRef]
8. Kraack, H.; Deutsch, M.; Ocko, M.B.; Pershan, P.S. The structure of organic langmuir films on liquid metal surfaces. *Nucl. Instrum. Methods Phys. Res. Sect. B* **2003**, *200*, 363–370. [CrossRef]
9. Rix, E.; Grau, E.; Chollet, G.; Cramail, H. Synthesis of fatty acid-based non-isocyanate polyurethanes, NIPUs, in bulk and mini-emulsion. *Eur. Polym. J.* **2016**, *84*, 863–872. [CrossRef]

10. Nguyen, R.; Galy, N.; Singh, A.K.; Paulus, F.; Stoebener, D.; Schlesener, C.; Sharma, S.K.; Haag, R.; Len, C. A Simple and Efficient Process for Large Scale Glycerol Oligomerization by Microwave Irradiation. *Catalysts* **2017**, *7*, 123. [CrossRef]

11. Galy, N.; Nguyen, R.; Blach, P.; Sambou, S.; Luart, D.; Len, C. Glycerol oligomerization in continuous flow reactor. *J. Ind. Eng. Chem.* **2017**, *51*, 312–318. [CrossRef]

12. Galy, N.; Nguyen, R.; Yalgin, H.; Thiebault, N.; Luart, D.; Len, C. Glycerol in subcritical and supercritical solvents. *J. Chem. Technol. Biot.* **2017**, *92*, 14–26. [CrossRef]

13. Sonnati, M.O.; Amigoni, S.; Taffin De Givenchy, E.P.; Darmanin, T.; Choulet, O.; Guittard, F. Glycerol carbonate as a versatile building block for tomorrow: Synthesis, reactivity, properties and applications. *Green Chem.* **2013**, *15*, 236–283. [CrossRef]

14. Clements, J.H. Reactive Applications of Cyclic Alkylene Carbonates. *Ind. Eng. Chem. Res.* **2003**, *42*, 663–674. [CrossRef]

15. Kühnel, I.; Saake, B.; Lehnen, R. Oxyalkylation of lignin with propylene carbonate: Influence of reaction parameters on the ensuing bio-based polyols. *Ind. Crops Prod.* **2017**, *101*, 75–83. [CrossRef]

16. Teng, W.K.; Ngoh, G.C.; Yusoff, R.; Aroua, M.K. A review on the performance of glycerol carbonate production via catalytic transesterification: Effects of influencing parameters. *Energy Convers. Manag.* **2014**, *88*, 484–497. [CrossRef]

17. Lee, K.H.; Park, C.; Lee, E.Y. Biosynthesis of glycerol carbonate from glycerol by lipase in dimethyl carbonate as the solvent. *Bioprocess Biosyst. Eng.* **2010**, *33*, 1059–1065. [CrossRef] [PubMed]

18. Helou, M.; Carpentier, J.F.; Guillaume, S.M. Poly(carbonate-urethane): An isocyanate-free procedure from a,x-di(cyclic carbonate) telechelic poly(trimethylene carbonate)s. *Green Chem.* **2011**, *13*, 266–271. [CrossRef]

19. Lu, Y.; Shen, L.; Gong, F.; Cui, J.; Rao, J.; Chen, J.; Yang, W. Polycarbonate urethane films modified by heparin to enhance hemocompatibility and endothelialization. *Polym. Int.* **2012**, *61*, 1433–1438. [CrossRef]

20. Wang, X.; Li, M.; Wang, T.; Jin, Q.; Wang, X. An improved method for the synthesis of 2-arachidonoylglycerol. *Process Biochem.* **2014**, *49*, 1415–1421. [CrossRef]

21. Carré, C.; Bonnet, L.; Avérous, L. Original biobased nonisocyanate polyurethanes: Solvent- and catalyst-free synthesis, thermal properties and rheological behaviour. *RSC Adv.* **2014**, *4*, 54018–54025. [CrossRef]

22. Besse, V.; Camara, F.; Méchin, F.; Fleury, E.; Caillol, S.; Pascault, J.; Boutevin, B. How to explain low molar masses in PolyHydroxyUrethanes (PHUs). *Eur. Polym. J.* **2015**, *71*, 1–11. [CrossRef]

23. Bähr, M.; Bitto, A.; Mülhaupt, R. Cyclic limonene dicarbonate as a new monomer for non-isocyanate oligo- and polyurethanes (NIPU) based upon terpenes. *Green Chem.* **2012**, *14*, 1447–1454. [CrossRef]

24. Carré, C.; Bonnet, L.; Avérous, L. Solvent- and catalyst-free synthesis of fully biobased nonisocyanate polyurethanes with different macromolecular architectures. *RSC Adv.* **2015**, *5*, 100390–100400. [CrossRef]

25. Neta, N.S.; Teixeira, J.A.; Rodrigues, L.R. Sugar Ester Surfactants: Enzymatic Synthesis and Applications in Food Industry. *Crit. Rev. Food Sci.* **2015**, *55*, 595–610. [CrossRef] [PubMed]

26. Teixeira, D.A.; Da Motta, C.R.; Ribeiro, C.M.S.; de Castro, A.M. A rapid enzyme-catalyzed pretreatment of the acidic oil of macauba (*Acrocomia aculeata*) for chemoenzymatic biodiesel production. *Process Biochem.* **2017**, *53*, 188–193. [CrossRef]

27. Singh, A.K.; Remi, N.; Galy, N.; Haag, R.; Sharma, S.K.; Len, C. Chemo-Enzymatic Synthesis of Oligoglycerol Derivatives. *Molecules* **2016**, *21*, 1038. [CrossRef] [PubMed]

28. Torres, C.; Otero, C. Part I. Enzymatic synthesis of lactate and glycolate esters of fatty alcohols. *Enzyme Microb. Technol.* **1999**, *25*, 745–752. [CrossRef]

29. Güven, A.; Kapucu, N.; Mehmeto Lu, Ü. The production of isoamyl acetate using immobilized lipases in a solvent-free system. *Process Biochem.* **2002**, *38*, 379–386. [CrossRef]

30. Khan, A.; Sharma, S.K.; Kumar, A.; Watterson, A.C.; Kumar, J.; Parmar, V.S. Novozym 435-Catalyzed Syntheses of Polyesters and Polyamides of Medicinal and Industrial Relevance. *ChemSusChem* **2014**, *7*, 379–390. [CrossRef] [PubMed]

31. Yao, X.; Wu, G.; Xu, L.; Zhang, H.; Yan, Y. Enzyme-catalyzed preparation of dimeric acid polyester polyol from biodiesel and its further use in the synthesis of polyurethane. *RSC Adv.* **2014**, *4*, 31062–31070. [CrossRef]

32. Jung, H.; Lee, Y.; Kim, D.; Han, S.O.; Kim, S.W.; Lee, J.; Kim, Y.H.; Park, C. Enzymatic production of glycerol carbonate from by-product after biodiesel manufacturing process. *Enzyme Microb. Technol.* **2012**, *51*, 143–147. [CrossRef] [PubMed]

33. Su, E.; Zhang, M.; Zhang, J.; Gao, J.; Wei, D. Lipase-catalyzed irreversible transesterification of vegetable oils for fatty acid methyl esters production with dimethyl carbonate as the acyl acceptor. *Biochem. Eng. J.* **2007**, *36*, 167–173. [CrossRef]

34. Kim, S.C.; Kim, Y.H.; Lee, H.; Yoon, D.Y.; Song, B.K. Lipase-catalyzed synthesis of glycerol carbonate from renewable glycerol and dimethyl carbonate through transesterification. *Catal. B Enzym.* **2007**, *49*, 75–78. [CrossRef]

35. Ünal, M.U. A Study on the Lipase-Catalayzed Esterification in Organic Solvent. *Turk. J. Agric. For.* **1998**, *22*, 573–578.

36. Cornille, A.; Auvergne, R.; Figovsky, O.; Boutevin, B.; Caillol, S. A perspective approach to sustainable routes for non-isocyanate polyurethanes. *Eur. Polym. J.* **2017**, *87*, 535–552. [CrossRef]

37. Chen, L.; Liu, T.; Zhang, W.; Chen, X.; Wang, J. Biodiesel production from algae oil high in free fatty acids by two-step catalytic conversion. *Bioresour. Technol.* **2012**, *111*, 208–214. [CrossRef] [PubMed]

38. Liu, Y.; Lotero, E.; Goodwinjr, J. Effect of carbon chain length on esterification of carboxylic acids with methanol using acid catalysis. *J. Catal.* **2006**, *243*, 221–228. [CrossRef]

39. Bansode, S.R.; Rathod, V.K. Ultrasound assisted lipase catalysed synthesis of isoamyl butyrate. *Process Biochem.* **2014**, *49*, 1297–1303. [CrossRef]

40. Li, J.; Wang, T. Chemical equilibrium of glycerol carbonate synthesis from glycerol. *J. Chem. Thermodyn.* **2011**, *43*, 731–736. [CrossRef]

41. Seong, P.; Jeon, B.W.; Lee, M.; Cho, D.H.; Kim, D.; Jung, K.S.; Kim, S.W.; Han, S.O.; Kim, Y.H.; Park, C. Enzymatic coproduction of biodiesel and glycerol carbonate from soybean oil and dimethyl carbonate. *Enzyme Microb. Technol.* **2011**, *48*, 505–509. [CrossRef] [PubMed]

42. Nie, K.; Xie, F.; Wang, F.; Tan, T. Lipase catalyzed methanolysis to produce biodiesel: Optimization of the biodiesel production. *J. Mol. Catal. B Enzym.* **2006**, *43*, 142–147. [CrossRef]

43. Huang, Y.; Pang, L.; Wang, H.; Zhong, R.; Zeng, Z.; Yang, J. Synthesis and properties of UV-curable tung oil based resins via modification of Diels-Alder reaction, nonisocyanate polyurethane and acrylates. *Prog. Org. Coat.* **2013**, *76*, 654–661. [CrossRef]

© 2018 by the authors. Licensee MDPI, Basel, Switzerland. This article is an open access article distributed under the terms and conditions of the Creative Commons Attribution (CC BY) license (http://creativecommons.org/licenses/by/4.0/).

polymers

MDPI

Article

Effect of Glycerol Pretreatment on Levoglucosan Production from Corncobs by Fast Pyrolysis

Liqun Jiang [1,*], Nannan Wu [1], Anqing Zheng [1], Xiaobo Wang [1], Ming Liu [1], Zengli Zhao [1,*], Fang He [1], Haibin Li [1] and Xinjun Feng [2,*]

[1]   Guangdong Provincial Key Laboratory of New and Renewable Energy Research and Development, Guangzhou Institute of Energy Conversion, Chinese Academy of Sciences, Guangzhou 510640, China; wunn@ms.giec.ac.cn (N.W.); zhengaq@ms.giec.ac.cn (A.Z.); wangxb@ms.giec.ac.cn (X.W.); liuming@ms.giec.ac.cn (M.L.); hefang@ms.giec.ac.cn (F.H.); lihb@ms.giec.ac.cn (H.L.)
[2]   Key Laboratory of Bio-Based Materials, Qingdao Institute of Bioenergy and Bioprocess Technology, Chinese Academy of Sciences, Qingdao 266071, China
*   Correspondence: lqjiang@ms.giec.ac.cn (L.J.); zhaozl@ms.giec.ac.cn (Z.Z.); fengxj@qibebt.ac.cn (X.F.)

Received: 30 September 2017; Accepted: 7 November 2017; Published: 10 November 2017

Abstract: In this manuscript, glycerol was used in corncobs' pretreatment to promote levoglucosan production by fast pyrolysis first and then was further utilized as raw material for chemicals production by microbial fermentation. The effects of glycerol pretreatment temperatures (220–240 °C), time (0.5–3 h) and solid-to-liquid ratios (5–20%) were investigated. Due to the accumulation of crystalline cellulose and the removal of minerals, the levoglucosan yield was as high as 35.8% from corncobs pretreated by glycerol at 240 for 3 h with a 5% solid-to-liquid ratio, which was obviously higher than that of the control (2.2%). After glycerol pretreatment, the fermentability of the recovered glycerol remaining in the liquid stream from glycerol pretreatment was evaluated by *Klebsiella pneumoniae*. The results showed that the recovered glycerol had no inhibitory effect on the growth and metabolism of the microbe, which was a promising substrate for fermentation. The value-added applications of glycerol could reduce the cost of biomass pretreatment. Correspondingly, this manuscript offers a green, sustainable, efficient and economic strategy for an integrated biorefinery process.

Keywords: glycerol pretreatment; levoglucosan; fast pyrolysis; lignocellulose

1. Introduction

Growing global environmental concerns and the continuous depletion of fossil fuels lead to the search for sustainable alternative energy resources and technologies. Lignocellulose is the most plentiful form of biomass and the most abundant polysaccharide on Earth, which has already attracted unprecedented concern [1]. There is a widespread interest in using sugars derived from lignocellulose to produce biofuels and chemicals. Depolymerization of polysaccharides into monosaccharides is still a significant hurdle to this application [2]. Historically, research efforts on releasing sugars from lignocellulose have focused on the biochemical process and acid hydrolysis, which adopt enzymes and acids to deconstruct the lignocellulose and liberate sugars.

Fast pyrolysis is a little-explored alternative thermo-chemical depolymerization route to release sugars, which presents an excellent opportunity to establish a biorefinery [3]. Levoglucosan is the main product from cellulose fast pyrolysis, and the yield reaches as high as 59%. Levoglucosan is of great value and is mainly utilized as a chiral raw material to synthesize stereoregular polysaccharides and chiral chemicals [4,5]. Most importantly, levoglucosan can serve as feedstock for several microorganisms. Levoglucosan can be metabolized through the general glycolytic pathway after being converted to glucose 6-phosphate with the Mg-ATP-dependent levoglucosan kinase [6]. Due to the presence of levoglucosan kinase, *Aspergillus terreus, Aspergillus niger* CBX-209, oleaginous yeasts and engineered *Escherichia coli* KO11 can metabolize levoglucosan into itaconic acid, citric acid, lipid and ethanol with a comparable yield and rate as in the conversion from glucose [1,7–10]. As an anhydrosugar, levoglucosan can be hydrolyzed to glucose by using acid catalysts or solid acid catalysts, thereby providing a potentially rapid and efficient route to a biorefinery [11,12]. The fast pyrolysis process liberates a large amount of levoglucosan without any catalysts or enzymes in a short time, exhibiting a significant economic feasibility. It has been demonstrated that ethanol production from lignocellulose by following a combined thermochemical and fermentative approach is comparable to conventional processes [13].

The yield of levoglucosan from pure cellulose can reach as high as 59%. Nonetheless, fast pyrolysis of lignocellulose yields much lower levoglucosan than pure cellulose. Although cellulose-hemicellulose interaction is not significant, lignin inhibits the thermal polymerization of levoglucosan formed from cellulose and enhances the formation of the low molecular weight products form cellulose with reduced yield of char fraction [14]. Even small amounts of impurities can radically alter the rates of reaction and the products obtained. Alkali and alkaline earth metals (AAEMs) can shift the pyrolytic pathway of cellulose to promote the formation of ring scission products with little economic value at the expense of levoglucosan. Enhancing the selectivity of the thermochemical reactions toward depolymerization is critical to improve the conversion of lignocellulose into levoglucosan. Empirical pretreatments prior to fast pyrolysis have been performed to enhance levoglucosan yield from lignocellulose [15–17]. Glycerol pretreatment evidently shows an economically competitive advantage over common pretreatments, such as low boiling-point solvents, ionic liquids, hot water and dilute acid pretreatments. First, glycerol is an abundant and economic organic solvent. Furthermore, the boiling-point of glycerol is as high as 290 °C, and thus, it can be performed at atmospheric pressure. However, only a few studies have investigated glycerol pretreatment prior to fast pyrolysis. Meanwhile, a tremendous amount of glycerol presents in the liquid stream after glycerol pretreatment, resulting in a significant environmental issue and resource waste, turning into a burden for the industry. The valorization of byproduct streams of pretreatment to obtain higher value chemicals can reduce the cost of pretreatment and meet environmental requirements. Therefore, there is an urgent need to make full utilization of glycerol contained in the liquid stream from glycerol pretreatment. Till now, limited research has reported the effect of recovered glycerol from glycerol pretreatment on microbial fermentation.

Herein, the influences of glycerol pretreatment on subsequent glycerol fermentation and biomass fast pyrolysis were evaluated. Glycerol was comprehensively used for corncobs' pretreatment and microorganism fermentation (Figure 1). After glycerol pretreatment, the recovered glycerol in the liquid stream was collected and further utilized for fermentation, while the pretreated corncobs were used for levoglucosan production by fast pyrolysis. A more popular configuration like the fluidized-bed reactor was not chosen because the added complexity and variability of reactor parameters were not needed in this research. The added variability would detract the focus of investigation from the effect of glycerol pretreatment on the products distribution during the fast pyrolysis process. A pyroprobe microreactor was fit-for-purpose and was therefore selected for this study.

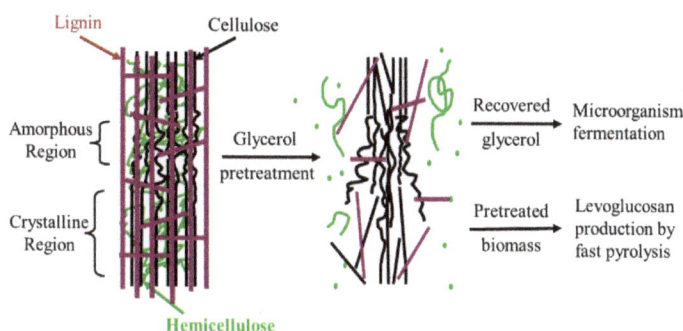

Figure 1. The scheme of this study.

2. Materials and Methods

2.1. Materials

Corncobs were bought from Baodi feed mill (Tianjin, China), air-dried until a constant weight and ground to pass through 80 mesh screen. Glycerol (analytical reagent) was bought from Fuyu Fine Chemical Co., Ltd. (Tianjin, China). The standard agents of xylose, glucose, mannose, arabinose, levoglucosan, 5-hydroxymethylfurfural (5-HMF) and furfural, and acetic acid were bought from Sigma-Aldrich (Shanghai, China).

2.2. Glycerol Pretreatment of Corncobs

For a typical pretreatment procedure, corncobs (5–20 g) were mixed with 100 g glycerol and loaded in a 250-mL round bottom three-neck flask sealed with a cork. The flask was submerged in an oil bath under vigorous stirring for 0.5–3 h with temperature control, previously heated to the desired temperature (220–240 °C). 220-0.5, 220-1, 220-2, 220-3, 240-0.5, 240-1, 240-2 and 240-3 were used to denote the pretreatment with a 5% solid-to-liquid ratio at 220 °C and 240 °C for 0.5, 1, 2 and 3 h. 240-3-5, 240-3-10, 240-3-15 and 240-3-20 were used to denote the pretreatment at 240 °C for 3 h with a 5%, 10%, 15% and 20% solid-to-liquid ratio, respectively. When the pretreatment was finished, the flask was removed from the oil bath. The glycerol pretreated corncobs were diluted with 1000 mL deionized water, then the solid and liquid fractions were separated and recovered by vacuum filtration with a 0.22-μm membrane. The solid sample was washed by deionized water to remove residual glycerol and then was dried for 24 h by a freeze dryer (Boyikang Co., Ltd., Beijing, China). The recovered glycerol contained in the liquid fraction from glycerol pretreatment was collected for fermentation.

2.3. Fermentation Experiments of the Recovered Glycerol

The strain utilized for fermentation was *Klebsiella pneumonia* (*K. pneumonia*). The strain was activated in Luria-Bertani (LB) at 37 °C. The LB medium was composed of NaCl (1 g/L), tryptone (1 g/L) and yeast extraction (0.5 g/L). For seed preparation, the strain was inoculated to a 250-mL flask containing 50 mL seed medium (same as the fermentation medium) at 180 rpm for 12 h. Then, the seed culture (5%, *v*/*v*) was then inoculated into the fermentation medium. Glycerol fermentation was performed in an orbital shaker at 200 rpm and 37 °C. The fermentation medium was composed of glycerol (20 g/L), citric acid (0.42 g/L), K_2HPO_4 (2 g/L), KH_2PO_4 (1.6 g/L), NH_4Cl (5.4 g/L), $MgSO_4 \cdot 7H_2O$ (0.2 g/L) and trace elements solution (1 mL). The trace elements solution contained $FeCl_3 \cdot 7H_2O$ (0.5 g/L), $ZnCl_2$ (0.684 g/L), $CuCl_2 \cdot 2H_2O$ (0.17 g/L), $CoCl_2 \cdot 4H_2O$ (0.476 g/L), $Na_2MoO_4 \cdot 2H_2O$ (0.005 g/L), $MnCl_2 \cdot 4H_2O$ (0.2 g/L), H_3BO_3 (0.062 g/L) and concentrated HCl (10 mL/L). After 24 h of fermentation, the broth was centrifuged and filtered for verification. The concentrations of residual glycerol and D-lactate were analyzed by an LC-10AT High Performance Liquid Chromatography (HPLC, Shimadzu, Kyoto, Japan). A Bio-Rad column

(Aminex HPX-87H, 300 × 7.8 mm, Hercules, CA, USA) was used for separation at 60 °C, and a refractive index detector (SPD-20A) was used for determination. Then, 0.005 M H_2SO_4 was used as the mobile phase with a flow rate of 0.6 mL/min. Dry cell weight (DCW, g/L) was tested by a UV visible spectroscopy system (Varian Cary 50 Bio, Palo Alto, CA, USA) at 650 nm. One unit of optical density was equal to 0.284 g DCW/L.

2.4. Elemental Analysis of Un-Treated and Glycerol Pretreated Corncobs

The organic elements were tested by a Vario EL cube analyzer (Hanau, Germany). The O content was calculated by subtracting a hundred percentage with contents of ash, C, H and N. The alkali and alkaline earth metals (AAEMs) were tested by an Optima 8000 Inductively Coupled Plasma Optical Emission Spectrometry (ICP-OES, PerkinElmer, Waltham, MA, USA). For a typical ICP-OES analysis, 0.3 g corncobs were digested by 1 mL $HClO_4$ and 3 mL HNO_3 in a test tube. The digested biomass was diluted to 10 mL using deionized water. The main analysis conditions included nebulizer flow of 1.5 L/min, flush time of 10 s, delay time of 40 s and wash time of 40 s. The normalized valencies of total AAEMs were calculated as the following equation:

$$\text{The normalized total AAEMs valencies} = \frac{(K + Na + 2Ca + 2Mg) \text{ in pretreated corncobs}}{(K + Na + 2Ca + 2Mg) \text{ in untreated corncobs}} \quad (1)$$

2.5. Compositional Analysis of Un-Pretreated and Glycerol Pretreated Corncobs

The component of corncobs was measured according to the technical report of the National Renewable Energy Laboratory [18]. After two steps of acid hydrolysis, the monosaccharides (glucose, xylose, mannose, galactose, arabinose) were analyzed by the HPLC system. An Aminex HPX-87P column (Bio-Rad, Hercules, CA, USA) was used for separation, and a refractive index detector was used for determination. Deionized water was utilized as the mobile phase, and the flow rate was set at 0.4 mL/min. The glucan content was equal to the content of cellulose, while the total content of galactan, arabinan, mannan and xylan was calculated as the content of hemicellulose. A UV spectrometer (Genesys 105, Hudson, NH, USA) was used to test acid-soluble lignin with a 320-nm wave length. Acid insoluble lignin was calculated by the difference of weight loss. The content of ash was calculated after oxidation of solid residue in a muffle furnace at 575 °C.

2.6. Thermogravimetric Analysis

The thermal analysis of un-pretreated and glycerol pretreated corncobs was conducted in a TGAQ50 thermogravimetric analyzer (TA, New Castle, DE, USA). Samples (6–8 mg) placed in alumina crucibles were heated from 50–750 °C at a rate of 20 °C/min. The analysis was conducted under nitrogen atmosphere with a flow rate of 20 mL/min.

2.7. Structural Characterization of Un-Pretreated and Glycerol Pretreated Corncobs

Fourier transform infrared spectroscopy (FTIR, Bruker TENSOR27, Optik Instruments, Brno, Czech Republic) was used to examine the main functional groups of samples. X-ray diffraction (XRD) patterns were scanned from 5°–40° with a step of 0.02° by an X'Pert PRO MPD X-ray diffractometer (PANalytical B.V., GH Eindhoven, The Netherlands). An empirical equation was used for the calculation of the crystallinity index (CrI):

$$\text{CrI (\%)} = \frac{\text{ICr} - \text{IAm}}{\text{ICr}} \times 100\% \quad (2)$$

where I_{Am} was the region of amorphous cellulose and I_{Cr} represented both the region of crystalline and amorphous cellulose [19].

2.8. Fast Pyrolysis of Un-Pretreated and Glycerol Pretreated Corncobs

The un-treated and glycerol pretreated corncobs were fast pyrolyzed in a semi-batch CDS reactor (Pyroprobe 5200, Oxford, PA, USA). The compounds of biomass fast pyrolysis were determined by 7890A gas chromatography (GC) and 5975C mass spectrometry (Agilent Technologies, Santa Clara, CA, USA). Cellulose samples were loaded into the center of quartz tubes. The temperature of the pyro-probe was set at 500 °C. The residence time and heating rate were fixed at 20 s and 10 K ms^{-1}. The temperature of the interface line was 300 °C. The split ratio was set at 50:1. Helium was used as the carrier gas. The flow rate between the pyrolyzer and GC was 20 mL/min, while the flow rate in GC was maintained at 1 mL/min. The GC column was an Agilent HP-INNO (length 30 m, internal diameter 0.25 mm, film thickness 0.25 µm). The mass spectrometer was operated in the electron impact mode (electron energy 70 eV). The oven program was 2 min at 50 °C, then 10 °C/min to 90 °C, 4 °C/min to 129 °C, 8 °C/min to 230 °C and finally held at 230 °C for 29 min. The mass was scanned from *m/z* 12–500 under the total ion current mode (TIC). The NIST Mass Spectral data library was used to identify compounds. Quantitative analysis of the main pyrolysis products was made by the external standard method. Five solutions of each standard compound were prepared by dissolving them in acetone. All experiments were tested in triplicate. The compound yield was calculated as:

$$\text{Yield of compound (wt\%)} = \frac{\text{mass of compound (g)}}{\text{mass of corncobs (g)}} \times 100\% \tag{3}$$

3. Results

3.1. The Fermentability of the Recovered Glycerol

Various temperatures, ranging from 220–240 °C, and pretreatment times, varying from 0–3 h with 5–20% solid-to-liquid ratios, were tested for glycerol pretreatment. After glycerol pretreatment, the pretreated corncobs were fast pyrolyzed for levoglucosan production, while the liquid fraction from glycerol pretreatment was collected to evaluate its fermentability. A high concentration of glycerol remained in the liquid stream from glycerol pretreatment. The common microbial inhibitors, such as furfural, 5-HMF and organic acid, were undetectable in the recovered glycerol. The glycerol was recovered and utilized as the feedstock with the initial concentration of 20 g/L. The *K. pneumoniae* could metabolize glycerol to D-lactate. As shown in Table 1, 3.4–3.7 g/L DCW and 5.2–6.1 g/L D-lactate were obtained after 24 h of fermentation, with a productivity of 1.4–1.7 g/L h. Compared to the fermentation of pure glycerol, the results indicated that the cell growth and metabolism were not inhibited in the recovered glycerol. The development of synthetic biology, metabolic engineering and systems biologic tools needs to be integrated to enhance the efficiency, including product titer, productivity, yield, consumption rate and cell growth [20,21]. Considerable additional investigation is required to boost the extensive application of the recovered glycerol.

Table 1. The fermentation of the recovered glycerol. DCW: dry cell weight.

Substrates	Cell mass (g/L)	Glycerol consumption (g/L)	D-Lactate (g/L)	Productivity (g/L DCW)
Pure glycerol	3.7	19.6	6.1	1.7
220-0.5	3.5	18.9	6.0	1.7
220-1	3.5	18.9	6.0	1.7
220-2	3.6	18.7	6.0	1.7
220-3	3.7	19.2	5.6	1.5
240-0.5	3.5	19.1	6.1	1.7
240-1	3.4	19.6	5.8	1.7
240-2	3.7	19.3	5.3	1.4
240-3	3.5	19.2	5.7	1.6
240-3-10	3.6	18.6	6.0	1.7
240-3-15	3.6	19.1	5.8	1.6
240-3-20	3.6	19.1	5.2	1.4

3.2. Main Composition and Elemental Analysis of Glycerol Pretreated Corncobs

The un-treated corncobs had a cellulose content of 32.3%; meanwhile, the hemicellulose and lignin accounted for 28.0% and 24.2%, respectively (Table 2). Glycerol pretreatment dissolved part of the lignin and hemicellulose and kept the cellulose fraction intact, leaving the solid enriched in cellulose. The glycerol showed an excellent nature and outstanding selectivity. For the pretreated corncobs (240-3), the cellulose fraction increased to 69.1%; meanwhile, the hemicellulose and lignin remained less than 10.3% and 9.2%, respectively. Further accumulation of cellulose and removal of hemicellulose and lignin occurred when longer residence time, higher temperature and a lower solid-to-liquid ratio of pretreatment were used.

Table 2. Recovery rate and chemical composition of glycerol pretreated corncobs.

Samples	Recovery rate (%)	Cellulose (wt %)	Hemicellulose (wt %)	Lignin (wt %)
Un-treated	—	32.3	28.0	24.2
220-0.5	78.9	38.2	26.8	20.6
220-1	64.7	50.8	20.8	16.3
220-2	53.4	60.4	15.6	12.5
220-3	49.6	64.4	13.1	10.9
240-0.5	70.5	45.1	23.4	18.4
240-1	57.2	56.3	16.9	14.7
240-2	50.6	63.8	13.4	11.5
240-3	47.5	69.1	10.3	9.2
240-3-10	49.1	65.6	11.7	10.6
240-3-15	52.3	60.5	13.4	12.5
240-3-20	54.0	56.9	16.1	14.1

The results of organic elements and AAEMs analysis of un-pretreated and glycerol pretreated corncobs are presented in Tables 3 and 4. The C content of corncobs was affected by glycerol pretreatment. The rank order of C content of biomass constituents is lignin > cellulose > hemicellulose. Glycerol pretreatment was able to remove part of the lignin and hemicellulose. Therefore, the contents of C in pretreated corncobs were lower than that of raw material. Similarly, the contents of H and O in pretreated corncobs were higher than those of raw material. However, the glycerol pretreatment could decrease the content of N. K^+, Na^+, Ca^{2+} and Mg^{2+} are generally the major constituents of metal ions. The total AAEMs content of raw material reached as high as 6643.5 mg/kg. After glycerol pretreatment at 220 °C for 0.5 h, the total AAEMs content was reduced to 1518.7 mg/kg; especially, the K content declined from 4809.9 down to 190.1 mg/kg. A higher pretreatment temperature, longer residence time and lower solid-to-liquid ratio of glycerol pretreatment resulted in further removal of AAEMs. After glycerol pretreatment at 240 °C for 3 h with a 5% solid-to-liquid ratio, the lowest normalized total AAEMs valence (0.1) was achieved. For the sample of 240-3, 63.7% Mg, 66.6% Ca, 77.3% Na and 98.6% K were removed from the corncobs.

Table 3. Organic elemental analysis of un-treated and glycerol pretreated corncobs.

Samples	C (wt %)	H (wt %)	N (wt %)	O (wt %)
Un-treated	44.6	5.9	0.3	38.7
220-0.5	44.2	6.3	0.1	41.3
220-1	43.7	6.3	0.1	42.5
220-2	41.8	6.0	0.1	44.9
220-3	43.1	6.2	0.1	44.6
240-0.5	42.8	6.3	0.1	43.8
240-1	42.1	6.1	0.1	45.2
240-2	42.7	6.0	0.1	44.9
240-3	43.1	6.0	0.1	45.0
240-3-10	43.3	6.1	0.1	44.0
240-3-15	43.6	6.0	0.1	42.5
240-3-20	43.8	6.1	0.1	42.1

Table 4. Alkali and alkaline earth metals (AAEMs) analysis of un-treated and glycerol pretreated corncobs.

Samples	AAEMs (mg/kg)					Normalized total AAEMs valencies
	Mg	Ca	Na	K	Total	
Un-treated	566.0	1185.2	82.4	4809.9	6643.5	—
220-0.5	294.8	993.3	40.5	190.1	1518.7	0.3
220-1	275.1	740.0	30.4	145.4	1190.9	0.3
220-2	274.7	482.0	17.7	120.2	894.6	0.2
220-3	274.9	325.6	17.7	82.9	701.1	0.2
240-0.5	306.5	802.3	39.8	182.8	1331.4	0.3
240-1	282.6	527.1	29.2	139.4	978.3	0.2
240-2	241.8	487.0	15.9	117.5	862.2	0.2
240-3	205.2	395.3	18.7	67.7	686.9	0.1
240-3-10	261.5	513.2	16.5	126.3	917.5	0.2
240-3-15	255.4	669.0	28.1	152.5	1105.0	0.2
240-3-20	282.0	823.2	24.2	138.8	1268.2	0.3

3.3. Mapping the Structural Changes of Corncobs after Glycerol Pretreatment

From the XRD analysis, both the raw and pretreated samples had the typical cellulose I diffraction angles. The CrI of the raw corncobs was 40.7%. After glycerol pretreatment, the CrI increased gradually, and the highest CrI value (72.9%) was achieved from the sample of 240-3 (Table 5). The dissolution/removal of amorphous hemicellulose and lignin should be responsible for the accumulation of crystalline cellulose and the increase of CrI.

Table 5. Characteristic parameters of un-treated and glycerol pretreated corncobs. CrI: crystallinity index.

Samples	CrI (%)	T_i (°C)	T_{max} (°C)	DTG_{max} (%/°C)	Residue (%)
Un-treated	40.7	211.2	307.6	0.9	19.8
220-0.5	45.1	246.6	330.2	1.3	16.7
220-1	56.3	266.6	330.9	1.6	14.1
220-2	66.9	260.4	328.6	1.6	12.1
220-3	69.9	283.2	327.4	1.7	11.3
240-0.5	50.7	268.7	328.0	1.6	14.8
240-1	62.9	266.3	326.9	1.8	14.4
240-2	70.6	284.4	326.5	1.9	11.5
240-3	72.9	282.7	325.0	1.7	10.3
240-3-10	71.2	281.5	327.3	1.7	10.8
240-3-15	67.1	277.2	330.6	1.5	12.6
240-3-20	64.8	281.5	330.6	1.5	12.6

Structural changes of raw and pretreated corncobs were determined by FTIR spectroscopy (Figure 2). The results illustrated that the samples before and after pretreatment had absorption at 3396–3420 and 2899–2908 cm^{-1}; corresponding to the stretching of the –OH group and C–H bonds of the alkyl group, respectively. The band around 1730 cm^{-1} corresponded to carbonyl and carboxyl stretching. The bands at 1605 and 1512 cm^{-1} were from the skeletal and stretching vibration of benzene rings. Absorptions at 1370–1205 cm^{-1} were mainly from the bending of C–H or O–H bonding in polysaccharides, including cellulose and hemicellulose. The region of 1160–1030 cm^{-1} represented the bending of C–O or C–O–C stretch and deformation bands in polysaccharides and lignin. The intensified signal in this area could be attributed to the increase of the cellulose fraction. It could be indicated that after glycerol pretreatment, the carbonyl and carboxyl stretching at 1715 cm^{-1} became weak, suggesting the removal of hemicellulose. Simultaneously, absorptions at 1370–1205 cm^{-1} and 1160–1030 cm^{-1} became strong, proving the substantial increase of polysaccharides in the pretreated biomass. With the increasing glycerol pretreatment temperature and residence time, the signals related to the functional groups in lignin and hemicellulose decreased, and the signals related to the cellulose increased. These results anticipated that glycerol pretreatment could selectively breakdown chemical bonds and functional groups, thereby modifying the microstructure of biomass, resulting in the reservation of cellulose and the removal of lignin and hemicellulose.

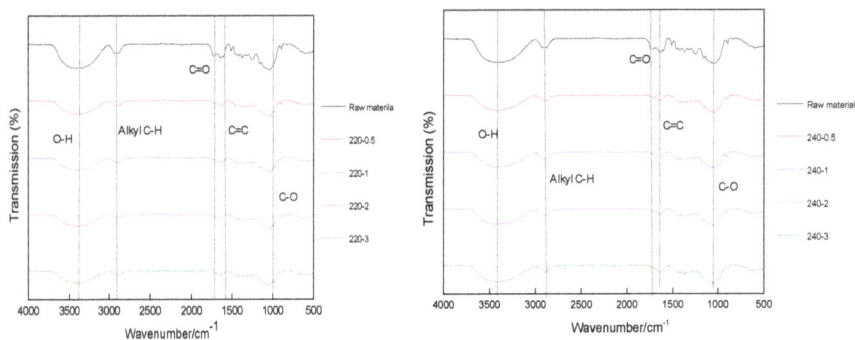

Figure 2. FTIR spectra of un-treated and glycerol pretreated corncobs.

3.4. Thermogravimetric Analysis of Corncobs after Glycerol Pretreatment

The thermogravimetry (TG) and differential TG (DTG) curves of samples are presented in Figure 3. The DTG curve of un-pretreated corncobs showed three peaks, which was attributed to the decomposition of hemicelluloses, cellulose and lignin. In glycerol pretreated corncobs, with the removal of partial hemicellulose, the first shoulder at lower temperature gradually disappeared. According to the TGA results, the raw corncobs began to degrade at 211.2 °C (Table 5). In the case of glycerol pretreated samples, the onsets of degradation were initiated at higher temperatures, approximately 246.6–284.4 °C. The maximum degradation temperature and rate of raw material were 307.6 °C and 0.9%/min, which were obviously lower than those of glycerol pretreated, approximately 330 °C and 1.6%/min. The residuals of raw material and pretreated corncobs (240-3) were different, accounting for 19.8% and 10.3%, respectively. Those phenomena could be ascribed to the partial removal of hemicellulose and lignin and effective demineralization by glycerol pretreatment, leading to a relatively low residue and a high thermal stability. Take K^+ as an example. The catalytic role of K^+ has been investigated inferring that K^+ can lower the initial decomposition temperature, decrease the maximum degradation temperature and reduce the maximum degradation rate [22]. Furthermore, biomass with higher crystallinity begins to decompose at higher temperatures, presenting sharper DTG curves and higher thermal decomposition activation energies [23,24].

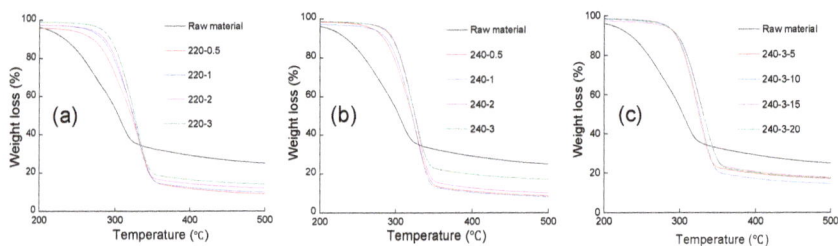

Figure 3. *Cont.*

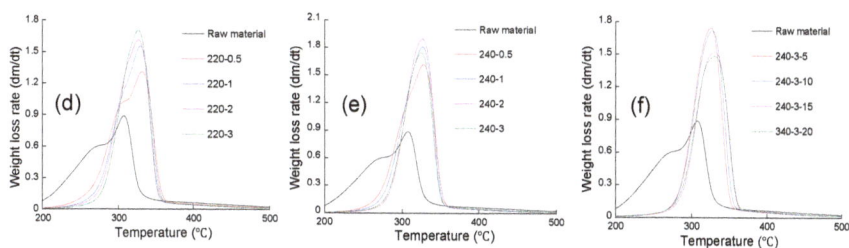

Figure 3. TG (thermogravimetry) (**a–c**) and DTG (differential TG) (**d–f**) profiles of raw material and glycerol pretreated corncobs.

3.5. Levoglucosan Production from Glycerol Pretreated Corncobs

Cellulose pyrolysis involves primary and secondary reactions. The initial breakage of glucosidic bonds and the formation of active cellulose occur in the primary reaction. In the secondary reaction, the active cellulose is further degraded to levoglucosan by dehydration of side functional groups [25]. The further decomposition of levoglucosan results in the formation of furfural, 5-HMF, hydroxyacetone, hydroxyacetaldehyde and some C_1-C_2 molecules. The yields of the main products from un-treated and glycerol pretreated corncobs' fast pyrolysis are presented in Table 6. Compared with raw material, the glycerol pretreated samples yielded more 5-HMF and simultaneously less acetic acid and furfural. Without the pretreatment step, the un-treated corncobs released only 2.2% levoglucosan during the fast pyrolysis. Levoglucosan yields from glycerol pretreated corncobs were obviously higher than that of un-treated material. Glycerol pretreatment improved the downstream levoglucosan yield in fast pyrolysis. The further increase in pretreatment temperature and residence time had a positive influence on levoglucosan formation. The maximum levoglucosan yield (35.8%) was achieved from the 240-3 sample, which was obviously higher than that from the un-treated sample.

Table 6. Yields of main compounds from un-treated and glycerol pretreated corncobs in fast pyrolysis. 5-HMF: 5-hydroxymethylfurfural.

Samples	Acetic Acid	Furfural	5-HMF	Levoglucosan
Un-treated	7.8	0.8	0.4	2.2
220-0.5	6.6	0.6	0.4	10.4
220-1	5.8	0.6	0.4	20.0
220-2	4.6	0.5	0.5	27.5
220-3	3.9	0.4	0.6	30.3
240-0.5	6.1	0.6	0.5	19.3
240-1	4.3	0.5	0.5	26.8
240-2	3.1	0.5	0.6	33.0
240-3	2.9	0.4	0.7	35.8
240-3-10	3.2	0.5	0.7	33.1
240-3-15	3.8	0.6	0.6	30.7
240-3-20	4.4	0.6	0.6	28.6

It was apparent that the glycerol pretreatment could enhance levoglucosan production from corncobs. The improvement mainly contributed to the enrichment of crystalline cellulose and the removal of AAEMs. The allomorph and relative crystallinity of cellulose can alter the slate and yield of compounds produced from biomass fast pyrolysis. The effect of biomass crystallinity structure on pyrolysis reactions has been investigated. It has been postulated that crystalline cellulose is favorable for levoglucosan production, while amorphous cellulose contributes more to gas and char production [26]. During levoglucosan formation, the crystalline structure is maintained, and the highly crystalline cellulose produced vapors dominated by levoglucosan [27]. Cellulose with a higher crystallinity tends to produce levoglucosan in a higher yield [24]. Previous researches have clearly demonstrated that the ash content significantly influences the yield of bio-oil and the product distribution of biomass fast pyrolysis. Inorganic compounds can also enhance the formation of char and gaseous species at the expense of anhydrosugars. In the presence of minerals, cellulose is mainly decomposed via depolymerization leading to low molecular weight products (especially glycolaldehyde) as the major products, whereas in the absence of minerals, levoglucosan is the main compound [28]. K^+ has a catalytic effect, promoting depolymerization/fragmentation reactions for the formation of lower molecular weight oxygenates [29]. Mg^{2+} and Ca^{2+} promote the primary production of char from cellulose and the transformation of levoglucosan into furans and light oxygenates. In terms of levoglucosan yield, various cations have been examined, and the following trends have been found in the order of strongest to mildest effect: $K^+ > Na^+ > Ca^{2+} > Mg^{2+}$ [30]. The result has demonstrated that even a trace level of AAEMs is sufficient to significantly alter the thermal degradation rate, chemical pathways and pyrolysis products' distribution. Previous research showed that a very small amount of KCl (0.004 mmol/g cellulose) resulted in steep decline from 59% down to 29% of levoglucosan yield, and the presence of as little as 0.5% switchgrass ash nearly tripled the formic acid yield and quadrupled the glycolaldehyde formation, while the yield of levoglucosan reduced to less than half of that from pure cellulose [31]. It has been postulated that the activation energy for alternative pathways is diminished as a result of inorganic addition, which can promote the formation of acetol, formic acid and glycolaldehyde directly from cellulose, thereby reducing levoglucosan yield. The use of levoglucosan in bio-oil as a fermentation feedstock depends on more than just the production of bio-oil with high sugar content, but the utilization is hindered by the presence of inhibitors for the biocatalyst [32]. Considerable efforts in decreasing toxicity and enhancing biocatalyst tolerance are vital steps for biochemical utilization of lignocellulose-derived sugars.

4. Conclusions

Herein, glycerol was utilized in corncobs' pretreatment prior to fast pyrolysis for levoglucosan production, and then, the recovered glycerol was further utilized for microbial fermentation to evaluate its fermentability. The results indicated that glycerol pretreatment could accumulate crystalline cellulose and remove most of the AAEMs. The levoglucosan yield from glycerol pretreated corncobs was obviously improved to 35.8% from 2.2%. Meanwhile, the recovered glycerol from glycerol pretreatment enriched in glycerol was an attractive and economical substrate for microorganism fermentation. The glycerol value was maximized, and the cost of biomass pretreatment was diminished to some extent. Consequently, this strategy also exhibited excellent potential to improve economic viability for the industrial production of levoglucosan and D-lactate and could be considered as a sustainable and promising route for a biorefinery.

Acknowledgments: The authors would like to thank the Natural Science Foundation of China and Guangdong Province (Grants 51606204, 51376186 and 2014A030310322), the Science and Technology Planning Project of Guangzhou City and Guangdong Province (Grant 201707010236 and 2017A020216007), the CAS Key Laboratory of Bio-based Materials (Grant KLBM2016008) and Guangdong Key Laboratory of New and Renewable Energy Research and Development (No. Y609jf1001) for their financial support.

Author Contributions: Liqun Jiang carried out the designed and completion of glycerol pretreatments, characterization and fast pyrolysis experiments and completed the manuscript drafting. Nannan Wu participated in the glycerol pretreatments. Xinjun Feng participated in the fermentation experiments. Anqing Zheng and Xiaobo Wang participated in the statistical analysis. Ming Liu made contributions to the acquisition of the data. Zengli Zhao, Fang He and Haibin Li supervised the experiments, and helped with the drafting and correction of the manuscript. All authors read and approved the final manuscript.

Conflicts of Interest: The authors declare no conflict of interest.

References

1. Liu, Y.; Via, B.K.; Pan, Y.F.; Cheng, Q.Z.; Guo, H.W.; Auad, M.L.; Taylor, S. Preparation and characterization of epoxy resin cross-linked with high wood pyrolysis bio-oil substitution by acetone pretreatment. *Polymers* **2017**, *9*, 106. [CrossRef]
2. Binder, J.B.; Raines, R.T. Fermentable sugars by chemical hydrolysis of biomass. *Proc. Natl. Acad. Sci. USA* **2010**, *107*, 4516–4521. [CrossRef] [PubMed]
3. Rover, M.R.; Johnston, P.A.; Jin, T.; Smith, R.G.; Brown, R.C.; Jarboe, L. Production of clean pyrolytic sugars for fermentation. *ChemSusChem* **2014**, *7*, 1662–1668. [CrossRef] [PubMed]
4. Bailliez, V.; Olesker, A.; Cleophax, J. Synthesis of polynitrogenated analogues of glucopyranoses from levoglucosan. *Tetrahedron* **2004**, *60*, 1079–1085. [CrossRef]
5. Cao, F.; Schwartz, T.J.; McClelland, D.J.; Krishna, S.H.; Dumesic, J.A.; Huber, G.W. Dehydration of cellulose to levoglucosenone using polar aprotic solvents. *Energy Environ. Sci.* **2015**, *6*, 1808–1815. [CrossRef]
6. Kitamura, Y.; ABE, Y.; Yasui, T. Metabolism of levoglucosan (1,2-anhydro-β-D-glucopyranose) in Microorganisms. *Agric. Biol. Chem.* **1991**, *55*, 515–521.
7. Nakagawa, M.; Sakai, Y.; Yasui, T. Itaconic acid fermentation of levoglucosan. *J. Ferment. Technol.* **1984**, *62*, 201–203.
8. Zhuang, X.L.; Zhang, H.X. Identification, characterization of levoglucosan kinase, and cloning and expression of levoglucosan kinase cDNA from *Aspergillus. niger* CBX-209 in *Escherichia coli*. *Protein Expr. Purif.* **2002**, *26*, 71–81. [CrossRef]
9. Lian, J.N.; Garcia-Perez, M.; Chen, S.L. Fermentation of levoglucosan with oleaginous yeasts for lipid production. *Bioresour. Technol.* **2013**, *133*, 183–189. [CrossRef] [PubMed]
10. Layton, D.S.; Ajjarapu, A.; Choi, D.W.; Jarboe, L.R. Engineering ethanologenic *Escherichia coli* for levoglucosan utilization. *Bioresour. Technol.* **2011**, *102*, 8318–8322. [CrossRef] [PubMed]
11. Prosen, E.M.; Radlein, D.; Piskorz, J.; Scott, D.S.; Legge, R.L. Microbial utilization of levoglucosan in wood pyrolysate as a carbon and energy source. *Biotechnol. Bioeng.* **1993**, *42*, 538–541. [CrossRef] [PubMed]
12. Bennett, N.M.; Helle, S.S.; Juff, S.J.B. Extraction and hydrolysis of levoglucosan from pyrolysis oil. *Bioresour. Technol.* **2009**, *100*, 6059–6063. [CrossRef] [PubMed]
13. Luque, L.; Oudenhoven, S.; Westerhof, R.; Rossum, G.V.; Berruti, F.; Kersten, S.; Rehmann, L. Comparison of ethanol production from corn cobs and switchgrass following a pyrolysis-based biorefinery approach. *Biotechnol. Biofuels* **2016**, *9*, 242. [CrossRef] [PubMed]
14. Hosoya, T.; Kawamoto, H.; Saka, S. Cellulose-hemicellulose and cellulose-lignin interactions in wood pyrolysis at gasification temperature. *J. Anal. Appl. Pyrolysis* **2007**, *80*, 118–125. [CrossRef]
15. Zheng, A.Q.; Zhao, Z.L.; Huang, Z.; Zhao, K.; Wei, G.Q.; Jiang, L.Q.; Wang, X.B.; He, F.; Li, H.B. Overcoming biomass recalcitrance for enhancing sugar production from fast pyrolysis of biomass by microwave pretreatment in glycerol. *Green Chem.* **2015**, *17*, 1167–1175. [CrossRef]
16. Jiang, L.Q.; Wu, N.N.; Zheng, A.Q.; Zhao, Z.L.; He, F.; Li, H.B. The integration of dilute acid hydrolysis of xylan and fast pyrolysis of glucan to obtain fermentable sugars. *Biotechnol. Biofuels* **2016**, *9*, 196. [CrossRef] [PubMed]
17. Jiang, L.Q.; Wu, N.N.; Zheng, A.Q.; Liu, A.Q.; Zhao, Z.L.; Zhang, F.; He, F.; Li, H.B. Comprehensive utilization of hemicellulose and cellulose to release fermentable sugars from corncobs via acid hydrolysis and fast pyrolysis. *ACS Susbtain. Chem. Eng.* **2017**, *5*, 5208–5213. [CrossRef]
18. Sluiter, A.; Hames, B.; Ruiz, R.; Scarlata, C.; Sluiter, B.; Templeton, D.; Crocker, D. *Determination of Structural Carbohydrates and Lignin in Biomass*; Laboratory Analytical Procedure; National Renewable Energy Laboratory: Golden, CO, USA, 2008.
19. Segal, L.; Creely, J.J.; Martin, A.E., Jr.; Conrad, C.M. An empirical method for estimation the degree of crystallinity of native cellulose using the X-ray diffractometer. *Text. Res. J.* **1959**, *29*, 786–794. [CrossRef]
20. Zambanini, T.; Kleineberg, W.; Sarikaya, E.; Buescher, J.M.; Meurer, G.; Wierckx, N.; Blank, L.M. Enhanced malic acid production from glycerol with high-cell density *Ustilago. trichophora* TZ1 cultivations. *Biotechnol. Biofuels* **2016**, *9*, 135. [CrossRef] [PubMed]
21. Zambanini, T.; Tehrani, H.H.; Geiser, E.; Merker, D.; Krabbe, J.; Buescher, J.M.; Meurer, G.; Wierckx, N.; Blank, L.M. Efficient itaconic acid production from glycerol with *Ustilago. vetiveriae* TZ1. *Biotechnol. Biofuels* **2017**, *10*, 131. [CrossRef] [PubMed]

22. Le Brech, Y.; Ghislain, T.; Leclerc, S.; Bouroukba, M.; Delmotte, L.; Brosse, N.; Snape, C.; Chaimbault, P.; Dufour, A. Effect of potassium on the mechanisms of biomass pyrolysis studied using complementary analytical techniques. *ChemSusChem* **2016**, *9*, 863–872. [CrossRef] [PubMed]

23. Wang, Z.H.; McDonald, A.G.; Westerhof, R.J.M.; Kersten, S.R.A.; Cuba-Torres, C.M.; Ha, S.; Pecha, B.; Garcia-Perez, M. Effect of cellulose crystallinity on the formation of a liquid intermediate product distribution during pyrolysis. *J. Anal. Appl. Pyrolysis* **2013**, *100*, 56–66. [CrossRef]

24. Jiang, L.Q.; Zheng, A.Q.; Zhao, Z.L.; He, F.; Li, H.B.; Wu, N.N. The comparison of obtaining fermentable sugars from cellulose by enzymatic hydrolysis and fast pyrolysis. *Bioresour. Technol.* **2016b**, *200*, 8–13. [CrossRef] [PubMed]

25. Bridgwater, A.V. Review of fast pyrolysis of biomass and product upgrading. *Biomass Bioenergy* **2012**, *38*, 68–94. [CrossRef]

26. Hosoya, T.; Sakaki, S. Levoglucosan formation from crystalline cellulose: Importance of a hydrogen bonding network in the reaction. *ChemSusChem* **2013**, *6*, 2356–2368. [CrossRef] [PubMed]

27. Mukarakate, C.; Mittal, A.; Ciesielski, P.N.; Budhi, S.; Thompson, L.; Iisa, K.; Nimlos, M.R.; Donohoe, B.S. Influence of crystal allomorph and crystallinity on products and behavior of cellulose during fast pyrolysis. *ACS Susbtain. Chem. Eng.* **2016**, *4*, 4662–4674. [CrossRef]

28. Piskorz, J.; Radlein, D.S.; Scott, D.S.; Czernik, S. Pretreatment of wood and cellulose for production of sugars by fast pyrolysis. *J. Anal. Appl. Pyrolysis* **1989**, *16*, 127–142. [CrossRef]

29. Fuentes, M.E.; Nowakowski, D.J.; Kubacki, M.L.; Cove, J.M.; Bridgeman, T.G.; Jones, J.M. Survey of influence of biomass mineral matter in thermochemical conversion of short rotation willow coppice. *J. Energy Inst.* **2008**, *81*, 234–241. [CrossRef]

30. Wang, K.G.; Zhang, J.; Shanks, B.H.; Brown, R.C. The deleterious effect of inorganic salts on hydrocarbon yields from catalytic pyrolysis of lignocellulosic biomass and its mitigation. *Appl. Energy* **2015**, *148*, 115–120. [CrossRef]

31. Patwardhan, P.R.; Satrio, J.A.; Brown, R.C.; Shanks, B.H. Influence of inorganic salts on the primary pyrolysis products of cellulose. *Bioresour. Technol.* **2010**, *101*, 4646–4655. [CrossRef] [PubMed]

32. Jarboe, L.R.; Liu, P.; Royce, L.A. Engineering inhibitor tolerance for the production of biorenewable fuels and chemicals. *Curr. Opin. Chem. Eng.* **2011**, *1*, 38–42. [CrossRef]

© 2017 by the authors. Licensee MDPI, Basel, Switzerland. This article is an open access article distributed under the terms and conditions of the Creative Commons Attribution (CC BY) license (http://creativecommons.org/licenses/by/4.0/).

MDPI

St. Alban-Anlage 66

4052 Basel

Switzerland

Tel. +41 61 683 77 34

Fax +41 61 302 89 18

www.mdpi.com

Polymers Editorial Office

E-mail: polymers@mdpi.com

www.mdpi.com/journal/polymers

www.ingramcontent.com/pod-product-compliance
Lightning Source LLC
Chambersburg PA
CBHW051700210326
41597CB00032B/5320